Quantum Communication, Computing, and Measurement 2

Quantum Communication, Computing, and Measurement 2

Edited by

P. Kumar

Northwestern University
Evanston, Illinois

G. M. D'Ariano

University of Pavia
Pavia, Italy

O. Hirota

Tamagawa University
Machida, Tokyo, Japan

SPRINGER SCIENCE+BUSINESS MEDIA, LLC

Library of Congress Cataloging-in-Publication Data

International Conference on Quantum Communication, Measurement, and Computing (4th: 1998: Northwestern University, Evanston, Ill.)
Quantum communication, computing, and measurement 2/edited by Prem Kumar, G. Mauro D'Ariano, Osamu Hirota.
p. cm.
"Proceedings of the Fourth International Conference on Quantum Communication, Measurement, and Computing, held August 22–27, 1998, in Northwestern University, Evanston, Illinois"—CIP t.p. verso.
Includes bibliographical references and index.
ISBN 978-1-4757-8675-0 ISBN 978-0-306-47097-4 (eBook)
DOI 10.1007/978-0-306-47097-4

1. Optical communications—Congresses. 2. Quantum optics—Congresses. 3. Quantum electronics—Congresses. 4. Quantum computers—Congresses. I. Kumar, Prem. II. D'Ariano, G. M. (Giacomo M.) III. Hirota, O. (Osamu), 1948- IV. Title.

TK5103.59. I55 1998
621.38—dc21

99-086155

Proceedings of the Fourth International Conference on Quantum Communication, Measurement, and Computing, held August 22–27, 1998, in Northwestern University, Evanston, Illinois

ISBN 978-1-4757-8675-0

Originally published by Kluwer Academic/Plenum Publishers, New York in 2000
Softcover reprint of the hardcover 1st edition 2000

http://www.wkap.nl/

10 9 8 7 6 5 4 3 2 1

A C.I.P. record for this book is available from the Library of Congress.

Proceedings of the
Fourth International Conference on

Quantum Communication, Measurement, and Computing

Northwestern University
Evanston, Illinois 60208, USA
August 22-27, 1998

Principal Organizers:

Prem Kumar	Northwestern University, USA
G. Mauro D'Ariano	University of Pavia, Italy
Osamu Hirota	Tamagawa University, Japan
Jeffrey H. Shapiro	Massachusetts Institute of Technology, USA
Horace P. Yuen	Northwestern University, USA

Sponsored in part by:

Tamagawa University, Japan	Northwestern University, USA
National Security Agency, USA	Office of Naval Research, USA

Local Organizing Committee:

William L. Kath	Northwestern University, USA
Anupam K. Garg	Northwestern University, USA
Jeff Z.-Y. Ou	Indiana-Purdue University, USA
Masao Osaki	Tamagawa University, Japan

1998 International Quantum Communication Award

The second Quantum Communication Award was awarded to:

H. J. KIMBLE of California Institute of Technology for *outstanding experimental advances in the areas of quantum measurements, cavity QED, and quantum logic.*

P. W. SHOR of AT&T Research for *ground-breaking contributions to the field of quantum computing.*

The awards were presented on August 26th, 1998 at the Awards Banquet of QCM'98 held in the Louis South room of Norris University Center at Northwestern University. A check in the amount of USD 4,000 was given to each award recipient.

The International Award on Quantum Communications has been established by the Research Institute of Tamagawa University to acknowledge an individual or a group of individuals for their pioneering contributions to the development of quantum communications. The award carries a plaque and a cash prize.

The first International Award on Quantum Communications was awarded in 1996 to the following four individuals:

C. H. Bennett of IBM Corporation, USA
C. W. Helstrom of University of California, San Diego, USA
A. S. Holevo of Steklov Mathematical Institute, Russia
H. P. Yuen of Northwestern University, USA

The 1996 awards were presented at the third International Conference on Quantum Communication and Measurement, September 25-30, 1996, held at Mt. Fuji-Hakone Land, Shizuoka Prefecture, Japan.

PREFACE

This volume contains the proceedings of the Fourth International Conference on Quantum Communication, Measurement, and Computing. This series of international conferences on quantum communication and measurement was established to encourage and bring together scientists working in the interdisciplinary research fields of quantum communication science and technology. The first such conference, organized by C. Benjaballah and O. Hirota under the title "Quantum Aspects of Optical Communication," assembled approximately 80 researchers in Paris in 1990. The second conference, held in Nottingham in 1994, was organized by V. P. Belavkin, R. L. Hudson, and O. Hirota and attracted about 130 participants from 22 countries. The third conference, organized by O. Hirota, A. S. Holevo, C. M. Caves, H. P. Yuen, and L. Accardi, was held in Fuji-Hakone Land, Japan, in 1996 and involved about 120 researchers from 15 countries. The present conference, organized by P. Kumar, G. M. D'Ariano, O. Hirota, J. H. Shapiro, and H. P. Yuen, was held August 22–27, 1998, in Evanston, Illinois, USA, and brought together over 140 researchers from 25 countries.

The topics at this fourth conference included quantum information theory, quantum computing, quantum communication systems, cryptography, new quantum effects and their experiments, generation and detection of nonclassical light, quantum noise, stochastic processes and filtering, and quantum measurement theory. In contrast to the previous conferences of this series, which were mainly theoretical, QCM'98 devoted more attention to the experimental aspects. In addition, quantum computation entered the conference as a main topic. The reason being that in the last few years this topic has naturally brought together the main fields of the former conferences, namely, quantum measurement theory, quantum communication and information processing, quantum noise, open quantum systems, and quantum optics. Over four days of the conference 16 oral sessions, including one plenary session, and 2 poster sessions were held in which 38 invited, 20 contributed, and 32 poster papers were presented. The plenary session featured presentations by the QCM'98 award recipients P. W. Shor and H. J. Kimble. There were no parallel sessions and the poster sessions were held in the evening after dinner. Interesting and stimulating discussions took place during coffee breaks and after sessions as well as in the evenings. The social program included a reception to welcome the participants upon arrival and a trip to Ravinia, "America's finest summer music festival," featuring an evening of Vivaldi music in the Martin Theatre at Ravinia. The conference closed with an Awards Banquet on the final evening, where the 1998 International Quantum Communication Awards were presented.

The editors of this volume thank the conference participants for stimulating presentations and thought provoking discussions at the meeting and for their contributions to this volume. We especially thank the conference secretary Kathie Heidenfelder-Cerveny and all the students of Professor Kumar for their help in organization of the conference. In the name of all the participants, we graciously thank our sponsors: Tamagawa University, Japan; Northwestern University, USA; National Security Agency, USA; and Office of Naval Research, USA. In particular, we express our sincere gratitude to Dr. P. McGrath and Dr. K. Miller of the NSA, Dr. H. Pilloff of the ONR, and I. Uslander of Northwestern University. Last but certainly not least, we thank Professor L. Villa-Komaroff, Vice President for Research at Northwestern University, Professor J. B. Cohen, Dean of the Robert R. McCormick School of Engineering and Applied Science at Northwestern University, and Professor Y. Obara, President of Tamagawa University, for their interest and generous support, both moral as well as financial, of the conference.

P. Kumar
G. M. D'Ariano
O. Hirota

Participants of the Fourth International Conference on Quantum Communication, Measurement, and Computing.

CONTENTS

PART I
QUANTUM INFORMATION AND QUANTUM COMMUNICATION

Quantum Robots Plus Environment 3
P. Benioff

Just *Two* Nonorthogonal Quantum States 11
C. A. Fuchs

On Maximum Mutual Information without Coding 17
M. Osaki, M. Ban, and O. Hirota

Aspects of Quantum Information Compression for Pure States 27
R. Jozsa

Group Covariant Signals in Quantum Information Theory 37
T. S. Usuda and I. Takumi

Optimization of a Symmetric Quantum Channel and its Optical Realization Scheme 43
M. Sasaki, S. M. Barnett, and M. Izutsu

Manipulation of Entangled States for Quantum Information Processing 49
S. Bose, S. F. Huelga, D. Jonathan, P. L. Knight, M. Murao, M. B. Plenio, and V. Vedral

Decoherence Control in Quantum Information Processing: Simple Models 59
L. Viola and S. Lloyd

Entanglement of Formation 69
W. K. Wootters

Sending Quantum Information with Gaussian States 75
A. S. Holevo

Quantum Communication with Imperfect Means: Entanglement Purification and the Quantum Repeater 83
G. Giedke, H. J. Briegel, W. Dür, J. I. Cirac, and P. Zoller

On Quantum Channel Capacity for Squeezed States 93
M. Sohma, A. S. Holevo, and O. Hirota

Asymptotic Quantum Estimation Theory for the Thermal States Family 99
M. Hayashi

The Asymptotic Efficiency of the Consistent Estimator, Berry-Uhlmann's Curvature, and Quantum Information Geometry 105
K. Matsumoto

Optimal Signal-To-Quantum Noise Ratio in Squeezed Displaced Number States 111
Y. Feng and A. I. Solomon

PART II
QUANTUM MEASUREMENT AND TOMOGRAPHY

Dual Classical and Quantum Descriptions of a Measuring Apparatus 117
O. Hay and A. Peres

Generalized Measurements, Retrodiction, and State Manipulation 125
S. M. Barnett, A. Chefles, D. T. Pegg, and L. S. Phillips

Latest Developments in Quantum Tomography 137
G. M. D'Ariano

Measuring the Quantum Polarization State of Light 147
M. G. Raymer, A. C. Funk, and D. F. McAlister

Measurement of Joint Photon-Number Distribution of a Twin-Beam State by Means of Optical Homodyne Tomography 157
M. Vasilyev, S.-K. Choi, P. Kumar, and G. M. D'Ariano

Homodyning Bell's Inequality 163
G. M. D'Ariano, L. Maccone, M. F. Sacchi, and A. Garuccio

Hydrodynamical Quantum State Reconstruction 169
L. M. Johansen

Quantum Tomography: Renormalization of Incompatible Observations ... 175
Z. Hradil, R. Myska, J. Summhammer, and H. Rauch

Accuracy in Quantum Homodyne Tomography ... 181
G. M. D'Ariano, C. Macchiavello, and N. Sterpi

Probing Quantum States of Rydberg Electrons by Half-Cycle Pulses ... 187
G. Alber and O. Zobay

Non-Markovian Quantum Trajectories for Open Systems ... 195
W. T. Strunz, L. Diósi, and N. Gisin

Maximally Robust Unravelings of Quantum Master Equations with Applications to the Atom Laser ... 201
H. M. Wiseman and J. A. Vaccaro

Quantum Noise and Dynamic Chaos in Opto-Electronic Laser Light Transformer ... 207
S. V. Larin, A. V. Masalov, A. N. Oraevsky and A. A. Putilin

Quantized Phase-Difference ... 215
J. Söderholm, A. Trifonov, T. Tsegaye, and G. Björk

Relativistic Quantum Theory with a Physical State Vector and Hypothetical Laws of Reduction ... 221
B. A. Berg

PART III
QUANTUM COMPUTING

Teleportation of General Quantum States, Squeezing and the Decoherence Problem in Quantum Computers ... 229
L. Accardi

Quantum Turing Machines: Local Transition, Preparation, Measurement, and Halting ... 241
M. Ozawa

A New Approach to Quantum Computation ... 249
H. P. Yuen

Hypersensitivity to Perturbation on an NMR Quantum Computer ... 257
R. Schack and T. A. Brun

Quantum Networks for Distributed Quantum Information Processing ... 263
H. J. Kimble

Quantum Phenomena of Trapped Ions 271
H. Walther

Progress Towards Using a Calcium Ion Trap to Perform Quantum Logic Operations 283
M. S. Gulley, J. J. Gomez, M. H. Holzscheiter, D. F. V. James, P. G. Kwiat, S. K. Lamoreaux, C. G. Peterson, V. Sandberg, M. M. Schauer, C. Simmons, D. Tupa, P. Wang, A. G. White, and R. J. Hughes

Quantum Computation with "Hot" Trapped Ions 289
D. F. V. James, S. Schneider, and G. J. Milburn

Novel Quantum Effects in Light Scattering from Cold Trapped Atoms 295
A. Orlowski, M. Gajda, P. Krekora, R. J. Glauber, and J. Mostowski

Preparation of Highly Excited Motional Energy Eigenstates with a Trapped Atom 301
S. Wallentowitz, W. Vogel, and P. E. Toschek

Optical Implementation of Grover's Algorithm: It's all Done with Mirrors 307
P. Kwiat, J. Mitchell, P. Schwindt, and A. White

Quantum State Protection in Optical Cavities 313
P. Tombesi and D. Vitali

Control of Dynamical Localization by Additional Quantum Degrees 321
K. Riedel, P. Törmä, V. Savichev, and W. P. Schleich

Two-Photon Interactions for Use in Quantum Computing 331
J. D. Franson and T. B. Pittman

Decoherence and Quantum Error Correction in Frequency Standards 337
C. Macchiavello, S. F. Huelga, J. I. Cirac, A. K. Ekert, and M. B. Plenio

Transition Phenomena in Networks of Quantum Systems 347
B. Kraus and P. Törmä

Experimental Demonstration of Coherent Quantum Feedback 353
R. J. Nelson, Y. Weinstein, S. Lloyd, and D. G. Cory

Classical Behavior of a Macroscopic Schrödinger Cat 359
C. Presilla

PART IV
QUANTUM TELEPORTATION AND CRYPTOGRAPHY

Free-Space Quantum Cryptography 367
R. J. Hughes, W. T. Buttler, P. G. Kwiat, S. K. Lamoreaux, G. L. Morgan, J. E. Nordholt, and C. G. Peterson

Parametric Schrödinger Cat with No Decoherence 375
S. Branca, F. De Martini, G. Di Giuseppe, and F. Mariani

Quantum Cryptography via Parametric Downconversion 381
G. Brassard, Tal Mor, and B. C. Sanders

Dim Coherent States as Signal States in the Bb84 Protocol: Is It Secure? 387
N. Lütkenhaus

Practical Aspects of Quantum Cryptography 393
M. Dusek, O. Haderka, and M. Hendrych

Quantum versus Classical Noise Cryptography 399
H. P. Yuen

Quantum Cryptography with Femtosecond Parametric Down-Conversion 405
A. V. Sergienko, M. Atature, B. M. Jost, J. Perina Jr., B. E. A. Saleh, and M. C. Teich

Generation of Narrow-Band Two-Photon State with Doubly Resonant Cavity in Parametric Down-Conversion 413
Z. Y. Ou

PART V
QUANTUM STATE ENGINEERING

From Entangled Photons to Entangled Atoms 423
E. S. Polzik, J. Hald, and J. L. Sørensen

Single Photon Quantum Control via High-$\chi^{(3)}$ Media 433
M. Dunstan, S. Rebic, S. Tan, S. Parkins, M. Collett, and D. Walls

Experimental Realization of Popper's Experiment: Violation of Uncertainty Principle? 443
Y.-H. Kim, R. Yu, and Y. Shih

Soliton Photon Number Squeezing: An Overview 453
S. R. Friberg and M. J. Werner

Amplitude Soliton Squeezing Using an Asymmetric Sagnac Fiber Loop ... 463
D. Krylov and K. Bergman

Time Domain Correlations and Gated Detection of Quantum Solitons ... 469
D. Levandovsky, M. Vasilyev, and P. Kumar

Observation of Squeezing in Free-Space Phase-Dependent Resonance Fluorescence ... 475
Z. H. Lu, S. Bali, and J. E. Thomas

Observation of Noiseless Image Amplification by an Optical Parametric Amplifier ... 481
S.-K. Choi, M. Vasilyev, and P. Kumar

Traveling-Wave Noiseless Amplification of Optical Images ... 487
M. I. Kolobov, I. V. Sokolov, and L. A. Lugiato

Quantum Noise Reduction in Singly Resonant Optical Devices ... 493
C. Cabrillo, J. L. Roldán, and P. García-Fernández

Influence of Dissipation Effects on Polarization-Squeezed Light Formation at Frequency Doubling ... 499
V. V. Volokhovsky and A. S. Chirkin

Degenerate Optical Parametric Amplification with Focused Gaussian Beams ... 505
K. G. Köprülü and O. Aytür

Eigen State Transmission of Birefringent Optical Fibers Using Entangled Two Photons ... 511
T. Ozeki and A. Masubuchi

Four-Photon Partition at Beamsplitter in Parametric Down-Conversion with Pulsed Pumping ... 515
Z. Y. Ou, J.-K. Rhee, and L. J. Wang

Energy Transfer in Parametric Down-Conversion ... 521
A. Bandilla, G. Drobný, and I. Jex

A Realistic Interaction-Free Resonator ... 527
M. Pavicic

Index ... 533

PART I: QUANTUM INFORMATION AND QUANTUM COMMUNICATION

QUANTUM ROBOTS PLUS ENVIRONMENTS

Paul Benioff

Physics Division, Argonne National Laboratory
Argonne, IL 60439
e-mail: pbenioff@anl.gov

A quantum robot is a mobile quantum system, including an on board quantum computer and needed ancillary systems, that interacts with an arbitrary environment of quantum systems. Quantum robots carry out tasks whose goals include carrying out measurements or physical experiments on the environment. A dynamical model of quantum robots and their interactions is discussed in which each task, as a sequence of alternating computation and action phases,is described by a unitary single time step operator $\mathcal{T} = \mathcal{T}_a + \mathcal{T}_c$ (discrete space and time are assumed). The overall system dynamics is described as a sum over paths of completed computation ($\mathcal{T}_c$) and action ($\mathcal{T}_a$) phases. A simple example of a task, measuring the distance between the quantum robot and a particle on a 1D lattice with quantum phase path dispersion present, is analyzed.

INTRODUCTION

Quantum computers are of much interest due to their increased power over classical computers in solving certain problems [1, 2]. Most studies of quantum computers consider them as stand alone systems operating in isolation from external systems as an environment. So far work on quantum computers which includes interactions with environments is limited mainly to noise effects, data base searching, and quantum oracle computing. The former considers environmental interactions as a source of noise and errors. This has stimulated the development of quantum error correcting codes to minimize this effect [3, 4] Other methods rely on the use of properties of systems with relatively long decoherence times [5, 6]. Quantum oracle computing [7] and especially data base searching [2] have been much discussed recently.

Here the emphasis is on quantum computers and their interactions with the environment in general. The interest is in quantum computers along with ancillary systems that can move in and interact with an environment of quantum systems. These are the defining characteristics of quantum robots.

Quantum robots are of interest from a foundational viewpoint [8]. If quantum mechanics is universally valid, then the systems that carry out theoretical calculations (computers) and physical experiments to test theoretical predictions (robots) must be described within quantum mechanics, i. e. as quantum computers and quantum robots. It follows that the systems that test the validity of quantum mechanics must be described by the same theory they are testing. Quantum mechanics must describe its own validity to the maximum extent possible [9].

A related reason for studying these systems is that they provide a *very small* first step towards a quantum mechanical description of systems that are aware of their environment, make decisions, are intelligent, and create theories such as quantum mechanics [10, 11, 12]. If quantum mechanics is universal, then these systems must be described within quantum mechanics to the maximum extent possible.

Another point of interest is that here there is no limitation on the types of environments included. Environments studied so far, such as oracles in quantum oracle computation[7], data bases as in Grover's algorithm [2], and quantum registers, are special types of environments. These systems share the property that their states do not change except as a result of specially designated interactions. Spontaneous changes due to interactions among the component systems and interactions with systems other those specially designated are to be minimized. For quantum registers this is emphasized by the need for quantum error correction codes and other stabilization methods to minimize this effect [3, 4].

Environments considered here are not so limited. They include interacting moving systems whose quantum states, $\Psi(t) = e^{-iHt}\Psi(0)$, are evolving with time. This is the case even if they are not interacting with a quantum robot. For these systems the expectation value for any observable O, may well have a time dependent probability $\langle\Psi(t)|O|\Psi(t)\rangle$ independent of interactions with the qauntum robot.

It is also to be noted that a quantum robot must include a quantum computer, as a system with many internal degrees of freedom, and cannot be simply a multistate head. If the head has a single internal degree of freedom that must be responsive to at least N different alternatives of environmental information, it must be possible to distinguish between N different internal states of the head. For large values of N, for example for a general purpose or universal quantum robot able to carry out many tasks in many different environments [8], this is physically unreasonable. If the head is a single spin system, it is very difficult to distinguish N different spin projection states if N is large.

In this case, and in any case where the number of alternatives that must be distinguished by the head is exponentially large, (e.g. distinguishing all bit strings of length n) the only reasonable approach is to allow the number of degrees of freedom in the head to be polynomial in $\log N$. But this is equivalent to requiring that the head is a quantum robot as it must include a quantum computer.

The next section summarizes a dynamical model of tasks carried out by quantum robots. The model generalizes that described elsewhere [8]. A simple example of a task, measuring the distance between the quantum robot and a particle, is described in following section. The example expands on an earlier description in that sums over different phase path states are included in the phase path sum. Accuracy conditions are also discussed.

The description of the example shows that it is relatively complex if specific account is taken of all the steps needed. The description in terms of elementary steps suggests the possible existence of a Church-Turing type hypothesis for physical experiments [8] similar to that [15, 16] applicable to quantum computers.

A MODEL OF QUANTUM ROBOTS WITH ENVIRONMENTS

A quantum robot consists of an on board quantum computer, a finite state output system o and a control qubit c. The dynamics of thse systems and their environmental interactions can be described by tasks whose goals include carrying out measurements and physical experiments. Each task consists of alternating computation and action phases. The goal of each computation phase is to determine the following action by generating a new state of o. Input consists of the former state of o, any stored memory, and observations of the neighborhood state of the environment. During the following action phase the action, determined by the state of o, is carried out. The states of all on board systems remain unchanged. Actions include motion of the quantum robot and neighborhood changes of the environment state. The function of the control qubit c is to turn on and off the two types of phases. The computation [action] phase is inactive when c is in state $|1\rangle[|0\rangle]$. Each phase terminates by changing the state of c.

A unitary step operator $T = T_a + T_c$ which describes the dynamical changes of the overall system during one time step is associated with each task. (To keep things simple discrete space and time are assumed.) If $\Psi(0)$ is the overall quantum robot plus environment state at time 0 the state after n time steps is given by $\Psi(n) = T^n\Psi(0)$. The action and computation phase step operators satisfy $T_a = TP_1^c$ and $T_c = TP_0^c$ where the projection operators P_i^c refer to the states of c.

Here T_c is such that it may depend on but not change the quantum robot location $\underline{x} = x, y, z$. This can be expressed by a diagonality condition $T_c = \sum_{\underline{x}} P_{\underline{x}}^{qr} T_c P_{\underline{x}}^{qr} P_0^c$. The requirement that T_a depend on but not change the states of o is given by a similar expression: $T_a = \sum_l P_l^o T_a P_l^o P_1^c$. T_a is independent of and does not change the states of the quantum computer.

Both T_c and T_a make local changes in the environment state. Those made by T_c are limited to entanglements with on board system states and other changes that are direct results of observation interactions. Changes in the environmental states made by T_a are local but are otherwise not limited. Actions of T_a also do not give entanglements with the states of the quantum computer.

The description of T given so far applies to motionless noninteracting environment systems for which the environment Hamiltonian $H_E = 0$. Extension to moving interacting systems can be done by replacing T by another step operator $\mathcal{T} = T_E^{1/2} T T_E^{1/2} = T_E^{1/2} T_a T_E^{1/2} + T_E^{1/2} T_c T_E^{1/2} = \mathcal{T}_a + \mathcal{T}_c$. Here T_a and T_c are as defined above and $T_E = e^{-iH_E\Delta}$ is a unitary step operator for the environment. This replacement of T by $\mathcal{T}$ becomes exact in the limit $\Delta \to 0$ [14].

It is useful to express the dynamical development of the system as a Feynman[13] sum over paths of computation and action phases, i.e. a phase path sum. To this end consider the matrix element $\langle w, i|\mathcal{T}^n|w_1, 0\rangle$ which gives the transition amplitude for going from state $|w_1, 0\rangle$ to state $|w, i\rangle$ in n steps. Here $|w\rangle$ denotes the state of all systems except that of the control qubit. One can use $\mathcal{T}^n = (\mathcal{T}(P_0^c + P_1^c))^n$ to obtain

$$\langle w, i|\mathcal{T}^n|w_1, 0\rangle = \sum_{t=1}^{n} \sum_{\substack{paths\ p\ of \\ length\ t+1}} \sum_{h_1,\cdots,h_t=1}^{\delta(\sum,n)} \langle p(t+1), i|(\mathcal{T}_{v_t})^{h_t}|p(t)\rangle \cdots, \langle p(3)|(\mathcal{T}_a)^{h_2}|p(2)\rangle \langle p(2)|(\mathcal{T}_c)^{h_1}|p(1), 0\rangle \quad (1)$$

Each term in this large sum give the amplitude for finding t alternating phases in the first n steps where the jth phase begins with all systems (except for c) in state $|p(j)\rangle$ and ends after h_j steps with all systems in state $|p(j+1)\rangle$. The upper limit on the h

sums shows the restriction that the sum $h_1+,\cdots,+h_t = n$. The initial and final path states $|p(1)\rangle$ and $|p(t+1)\rangle$ are $|w_1\rangle$ and $|w\rangle$.

Eq. 1 is shown for the case that the initial phase is a computation phase, or c is in the initial state $|0\rangle$. The alternation of phases is expressed by the subscripts v_j: if j is even $v_j = c$, if j is odd $v_j = a$. Also the restrictions on T_a and T_c, which also hold for $\mathcal{T}_a$ and $\mathcal{T}_c$, show that if j is even, $|p(j)\rangle$ and $|p(j+1)\rangle$ show the same robot position states. If j is odd, $|p(j)\rangle$ and $|p(j+1)\rangle$ show the same quantum computer and o system states. A similar equation holds if the initial phase is an action phase.

The equation shows clearly that for any n the overall system state is a linear sum over many phase path states of alternating computation and action phases for the task represented by $\mathcal{T}$. For each value of t and p, the equation gives the amplitude for the phase path p containing $t-1$ completed phases and one, the tth, which may or may not be complete. The h sums give the dispersion in the duration or number of time steps in each phase in p.

It follows from Eq. 1 that the overall system state $\Psi(n)$ can be expressed, for each initial state component, as an exponentially growing (with n) tree of phase paths. Each node in the tree corresponds to a state $|p(j)\rangle$. The t sum shows that some branches of the tree have very few nodes and include those having one node ($t = 1$).

A SIMPLE EXAMPLE

A simple example to illustrate the actions of a quantum robot has an environment with a single motionless particle p on a 1D lattice (i.e. $T_E = 1$ and $\mathcal{T} = T$). The task is to measure the distance between the quantum robot and p by alternating quantum robot motion with local observations for p and counting the number of nonobservations of p until p is found, (the search part). In the return part, the quantum robot goes back the same number of steps and the task ends by entering the ballast part. This part is present to preserve the unitarity of T.

The overall goal of the task, as a condition on T can be expressed as follows: Let $\phi = \sum_y c_y|y\rangle$ denote the state of p on the lattice with the quantum robot in position and internal memory state $|x, \underline{0}\rangle$. Iteration of the step operator T for this task generates to good accuracy the well known entanglement or "premeasurement" [17]

$$\phi|x,\underline{0}\rangle \longrightarrow {\sum_y}' c_y|y\rangle|x,\underline{y-x}\rangle \tag{2}$$

over a limited range of y values ($0 \leq y - x < 2^N$ -see below), denoted by the $'$ on the $\sum$. Here $|\underline{y-x}\rangle$ is the permanent memory qubit string state corresponding to the lattice distance $y-x$ (number of sites) in one direction only between p and the quantum robot. This equation is easily generalized to the case that the quantum robot is in a wave packet $\theta_{qr} = \sum_x d_x|x\rangle$ of position states to give

$$\phi_p\theta_{qr}|\underline{0}\rangle \longrightarrow {\sum_{x,y}}' c_y d_x|y\rangle|x,\underline{y-x}\rangle. \tag{3}$$

For this task the quantum computer contains two circular quantum registers, one with $N+2$ qubits and the other with $N+1$ qubits, and a head moving on the registers. Both accomodate numbers up to $2^N - 1$ with one ternary qubit in each in state $|2\rangle$ serving as an origin. The $N+2$ qubit register is a running memory for the search part counting of the number of nonobservations of p and includes a sign qubit. The other serves to hold a permanent copy of the number on the running memory when p is located. The state of o determines which task part is active. The o states $|mr1\rangle$, $|ml1\rangle$, $|dn\rangle$ correspond to the search, return, and ballast parts, and determine the action to be carried out in the following action phase. Other o states $|mr\infty\rangle$ (if p is not found in 2^N observations), and $|ml\infty\rangle$ for nonterminating action phases are also needed.

These and other details of the task can be given by a decision diagram. The diagram shown elsewhere [8] holds here if the step "$j+h=x$?" is replaced by "$x_{qr}=x_p$?". This expresses the computation phase observing if p is or is not at the quantum robot location, independent of where that is on the lattice.

In earlier work T was required to be such that just one phase path in the phase path sum $\sum_p$ of Eq. 1 contributed. Here this restriction will be retained for the computation phase step operator T_c only. The restriction is dropped for T_a in that the matrix elements $\langle x', l, i|T_a|x, l, 1\rangle$ are required to be local in the sense that their magnitude decreases as the distance $|x'-x|$ increases. The state of c is denoted by $i=0,1$. Phase path dispersion occurs because iteration of T_a generates many different action phase output states for each input state with amplitudes determined by sums of products of T_a matrix elements over all paths *within* each action phase. The dependence of the matrix elements on the different final c states $|i\rangle$ shows the action phase contributions to dispersion in the number of phases in a path (t sum) and duration of each action phase (h sum) in the phase path sum of Eq. 1.

Accuracy

It is clear from the task description that restrictions on T are needed so that the asymptotic probability of recording n on the permanent memory is given by (Eq. 2) $\sum_y' |c_y|^2\delta_{n,y-x}$. To see this let $\Psi(0)=\Theta(0)\phi_p$ represent an initial state with o and c in state $|mr1,0\rangle$ and the quantum robot and p in the lattice position state $|x\rangle\phi_p$ with $\phi_p=\sum_y c_y|y\rangle$. The probability after k steps that the search part of the task is done and $\underline{n}$ is recorded on the permanent memory is given by $P_k(n)=\langle\Psi(k)|P^{st}_{\underline{n}}(1-P^o_{mr1})|\Psi(k)\rangle=\sum_y |c_y|^2 P_k(n,y)$ where

$$P_k(n,y)=\langle y,\Theta_k(y)|P^{st}_{\underline{n}}(1-P^o_{mr1})|y,\Theta_k(y)\rangle. \tag{4}$$

In these equations $\Psi(k)=T^k\Psi(0)=\sum_y c_y\Theta_k(y)|y\rangle$ where $\Theta_k(y)$ is the state of the quantum robot after k steps corresponding to p in state $|y\rangle$. The projection operator $P^{st}_{\underline{n}}(1-P^o_{mr1})$ corresponds to completion of the search phase with n recorded in the permanent memory. The equations are based on the condition that p is motionless and its state, except for possible entanglement, remains unchanged throughout the task.

The probability $P_k(n,y)$ selects all k step phase paths in Eq. 1 containing $2n+1$ phases (n action and $n+1$ computation phases) in the completed search part of the task. The k dependence of $P_k(n)$ enters through the restriction $\delta(\sum,k)$ on the h sums of Eq. 1, which express the quantum dispersion in the durations of the different phases. It depends sensitively on the properties of T and on the distance $y-x$. For sufficiently large k and reasonable T_a (local, etc.) the amplitudes of paths, that are still in search parts of the task with $< n+1$ completed computation phases, should be quite small. In this case if ϕ is a wave packet localized around some value y_0, then for values of y around y_0, the infinite time limit $P_\infty(n,y)$ should exist and be sensibly equal to $P_k(n,y)$ for large k if $0\le y_0-x<2^N$.

Additional restrictions on T_a are needed so that for large k the distribution of $P_k(n,y)$, as a function of n has a peak value related to the distance $y-x$. This can be attained by requiring that the T_a matrix elements have the form $\langle mr1,x',i|T_a|1,x,mr1\rangle=$

$a_i e^{-\alpha(x'-x-1+i)^2}$ where a_i is an i dependent coefficient.. In this case, for large α and k, $P_k(n,y)$ should be peaked around the value $n = y - x$ with a narrow dispersion (provided $0 \leq y - x < 2^N$). In the limit $\alpha,\ k = \infty$ the distance measurement would be completely accurate with no dispersion in the result. In this case $P_\infty(n,y) = \delta_{n,y-x}$ which agrees with Eq. 2. The limit $k \to \infty$ is needed because phase duration dispersion is present in the phase path sum (the exponent is 0 if either $x' = x + 1$ and $i = 0$ or $x' = x$ and $i = 1$).

Generalization to the case in which Eq. 3 applies is straightforward. In this case $P_k(n,y)$ is replaced in the above by $\sum_{x,x'} d^*_{x'} d_x P_k(n, x', x, y)$. The sum is nondiagonal in x and diagonal in y because in this simple example, the quantum robot moves and p is motionless. For large values of α, k the values of $P_k(n,x',x,y)$ would be quite small for $x' \neq x$. In the limit $\alpha = \infty$, $P_\infty(n,x',x,y) = \delta_{x'-x}\delta_{n,y-x}$ which agrees with Eq. 3 for $0 \leq y - x < 2^N$.

Acknowledgements

This work is supported by the U.S. Department of Energy, Nuclear Physics Division, under contract W-31-109-ENG-38.

REFERENCES

1. P. Shor, Algorithms for quantum computation. Discrete log and factoring *in* "Proceedings of the 35th Annual Symposium on the Foundations of Computer Science", S. Goldwasser, ed., IEEE Computer Society, Los Alamitos, CA 1994, p. 124-134.
2. L.K.Grover, Quantum Mechanics Helps in Searching for a Needle in a Haystack, *Phys. Rev. Letters*, 79:325 (1997); G. Brassard, Searching a Quantum Phone Book, *Science*, 275:627 (1997).
3. R. Laflamme, C. Miquel, J. P. Paz, and W. H. Zurek, Perfect Quantum Error Correction Code, *Phys. Rev. Letters*, 77:198 (1996); D. P. DiVincenzo and P. W. Shor, Fault-Tolerant Error Correction with Efficient Quantum Codes *Phys. Rev Letters*, 77:3260 (1996).
4. E. Knill, R. Laflamme, and W. H. Zurek, Resilient Quantum Computation, *Science*, 279:342 (1998).
5. D. P. DiVincenzo, Quantum Computation, *Science*, 270:255 (1995).
6. N. A. Gershenfeld and I. L. Chuang, Bulk Spin-Resonance Quantum Computation *Science*, 275:350 (1997); D. G. Cory, A. F. Fahmy, and T. F. Havel, Ensemble Quantum Computing by NMR Spectroscopy, *Proc. Natl. Acad. Sci.*, 94:1634 (1997).
7. C. H. Bennett, G. Brassard, E. Bernstein, and U. Vazirani, Strengths and Weaknesses of Quantum Computing, *SIAM Jour. Computing*, 26:1510, (1997); W. Van Dam, Quantum Oracle Interrogation, Los Alamos Archives rept. No. quant-ph/9805006.
8. P. A. Benioff, Quantum Robots and Environments, *Phys Rev. A*, to Appear Aug. 1998; Some Foundational Aspects of Quantum Computers and Quantum Robots, *Superlattices and Microstructures*, 23:407 (1998).
9. A. Peres and W. Zurek, Is quantum Theory Universally Valid?, *Amer. Jour. Phys.* 50:807 (1982).
10. R. Penrose, The Emperor's New Mind, Penguin Books, New York, 1991.
11. H. P. Stapp, Mind, Matter, and Quantum Mechanics, Springer Verlag, Berlin 1993.
12. E. Squires, Conscious Mind in the Physical World IOP Publishing, Bristol England, 1990
13. R.P. Feynman and A. R. Hibbs, Quantum Mechanics and Path Integrals, McGraw-Hill Book Co. New York 1965.
14. M. Reed and B. Simon, Methods of Modern Mathematical Physics, Vol. 1, Academic Press, New York, 1972, Chap. VIII; I Montavy and G. Münster, Quantum Fields on a Lattice, Cambridge University Press, New York, 1994, Chap. 1.
15. A. Church, An Unsolvable Problem of Elementary Number Theory, *Am. Jour. Math.* 58:345 (1936); A. M. Turing, On Computable Numbers with an Application to the Entscheidungsproblem *Proc. Lond. Math. Soc.* 42:230 (1936).

16. D. Deutsch, Quantum Theory, the Church-Turing Principle and the Universal Quantum Computer, *Proc. Roy. Soc. (London) A* 400:997 (1985).
17. A. Peres, On Quantum Mechanical Automata, *Physics Letters* 101A:249 (1984).

JUST *TWO* NONORTHOGONAL QUANTUM STATES

Christopher A. Fuchs

Norman Bridge Laboratory of Physics, 12-33
California Institute of Technology
Pasadena, California 91125
email: cfuchs@cco.caltech.edu

From the perspective of quantum information theory, a system so simple as one restricted to just two nonorthogonal states can be surprisingly rich in physics. In this paper, we explore the extent of this statement through a review of three topics: (1) "nonlocality without entanglement" as exhibited in binary quantum communication channels, (2) the tradeoff between information gain and state disturbance for two prescribed states, and (3) the quantitative clonability of those states. Each topic in its own way quantifies the extent to which two states are "quantum" with respect to each other, i.e., the extent to which the two together violate some classical precept. It is suggested that even toy examples such as these hold some promise for shedding light on the foundations of quantum theory.

INTRODUCTION

The total set of states available to a quantum system corresponds to the uncountably infinite set of density operators over a given Hilbert space. With that set and a sufficiently general notion of measurement and time evolution, one can say *everything* that can be said about the system.[1] Nevertheless, as one gains experience in our field, it becomes hard to shake the feeling that much of the *essence* of quantum theory already makes itself known in the case of just two nonorthogonal states.[2] This is because of the overpowering importance of the quantum no-cloning theorem[3]: a set of two nonorthogonal states is the smallest set of states for which the theorem is active.[4] More generally, such a set forms the smallest set of states for which *no* information can be gathered without a conjugate disturbance.[5] They fulfill a role that the founding fathers tried so hard to pin on a single, solitary quantum state.[6]

In this connection, the ultimate questions we should like to ask are the following. To what extent does the newfound language of quantum information allow us to sharpen our understanding of this example and, more importantly, what can it tell us about the foundations of quantum theory itself? What hint might it give us of the tools

Quantum Communication, Computing, and Measurement 2
Edited by Kumar et al., Kluwer Academic / Plenum Publishers, New York, 2000.

required for digging even deeper in the coming century? These, of course, are difficult questions. But certainly no progress can be made in their answering without the courage of one small step. Here, we shall start in that direction by reviewing what is known about two nonorthogonal states that is already expressible in the language of quantum information. In particular, we will pay attention to how this allows us to express when two states are the most "quantum" with respect to each other. We will do this, in turn, from the perspective of (1) "nonlocality without entanglement" in binary quantum channels, (2) the tradeoff between information gain and state disturbance in quantum eavesdropping, and (3) the imperfect clonability of two states by various criteria. At the paper's conclusion, we will use these perspectives to attempt a tighter formulation of the grand questions above.

Throughout we will consider two nonorthogonal pure states $|\psi_0\rangle$ and $|\psi_1\rangle$ separated in a Hilbert space $\mathcal{H}_S$ by some angle θ. Without loss of generality for our considerations, we assume that the overlap $x = \langle\psi_0|\psi_1\rangle = \cos\theta$ is a positive real number. The variable x will be the most important parameter for our problems, expressing in one way or the other the degree of quantumness the two states hold with respect to each other. For the problems below that require the assumption of some a priori probabilities for the two states, we will assume them equal. To say that the identity of a state is "unknown" is to say that $|\psi_0\rangle$ and $|\psi_1\rangle$ are each assigned an a priori probability of 1/2. Whenever it is more convenient to think of the two quantum states as density operators, we shall denote them by $\rho_0 = |\psi_0\rangle\langle\psi_0|$ and $\rho_1 = |\psi_1\rangle\langle\psi_1|$.

NONLOCALITY w/o ENTANGLEMENT FOR BINARY CHANNELS

Consider using the alternate preparations $|\psi_0\rangle$ and $|\psi_1\rangle$ as the physical implementation of a binary alphabet in some communication scheme. Why would one ever want to do this? Well, there are various reasons based on practical considerations. For instance, the transmitter may have only low-energy coherent states available for signaling—these are necessarily nonorthogonal.[7] Also, nonorthogonal signals are sometimes able to achieve higher capacities in noisy quantum channels than orthogonal signals.[8] But let us consider this possibility purely for its own aesthetics.

With the adoption of a nonorthogonal alphabet, the signals will, of necessity, be imperfectly distinguishable by the receiver. For instance, if the criterion of distinguishability is the smallest possible error probability P_e in a decision about the signal's identity (following some measurement), then[6,9]

$$P_e = \frac{1}{2} - \frac{1}{4}\mathrm{tr}|\rho_1 - \rho_0| = \frac{1}{2}\Big(1 - \sqrt{1-x^2}\Big) . \tag{1}$$

This measure, in fact, shows just what one expects: as the overlap between the states increases, their mutual distinguishability decreases.

What is *quantum* about this lack of distinguishability in the signal states? One might be tempted to say, "Everything." If one draws an analogy between the quantum state and a point in a classical phase space, then one has that classical states can always be distinguished with perfect reliability and quantum states cannot. However, a more proper analogy is between quantum states and the Liouvillean probability distributions on phase space.[10] That is to say, overlapping quantum states are more analogous to the outputs of a noisy classical communication channel, where the receiver must distinguish between two probability distributions $p_0(y)$ and $p_1(y)$ over the output alphabet. From this point of view, the answer to the question above is, "Nothing." The distinction between quantum and classical must be seen in other ways.

One natural way crops up in a different aspect of the communication scenario: it is in the concept of *nonlocality without entanglement.*[11] As the signals in a long message start to accumulate, the receiver may be tempted to start the decoding process signal by signal. For classical channels, where the task is to accumulate information about long strings of the probability distributions $p_0(y)$ and $p_1(y)$, it turns out that such a strategy is never harmful. Signal-by-signal decoding never decreases the number of reliable bits per transmission.[12] For quantum mechanical messages composed from a nonorthogonal alphabet, however, this is not the case. A higher channel capacity can be achieved by allowing the receiver the capability to perform large collective quantum measurements on multiple transmissions.[13]

More specifically, if the receiver is restricted to perform a fixed generalized measurement signal by signal, or even an adaptive measurement signal by signal,[14] the greatest capacity achievable with a fixed alphabet is given numerically by the alphabet's *accessible information*[15] maximized over all a priori probability distributions. In our case, this number turns out to be

$$C_1 = \frac{1}{2}\left(1+\sqrt{1-x^2}\right)\log\left(1+\sqrt{1-x^2}\right) + \frac{1}{2}\left(1-\sqrt{1-x^2}\right)\log\left(1-\sqrt{1-x^2}\right) . \quad (2)$$

On the other hand, if the receiver can perform collective quantum measurements over arbitrarily large numbers of signals, then the greatest capacity is calculated by maximizing the alphabet's von Neumann entropy over all a priori probability distributions.[16] The resultant in our case is[17]

$$C_\infty = 1 - \frac{1}{2}\Big((1-x)\log(1-x) + (1+x)\log(1+x)\Big) . \quad (3)$$

The meaning of this result is that when one is speaking of correlations between nonorthogonal states—as one would be in the situation where these states are concatenated into codewords for a communication channel—the whole is greater than the sum of the parts. Extra correlation, and hence extra information, can be ferreted out of these codewords by collective measurements on the whole.* When the signals are orthogonal to each other—a situation in which one is tempted to say that they are classical—then the whole possesses nothing that the parts do not already contain.

This distinction in channel capacities suggests that the difference

$$Q = C_\infty - C_1 \quad (4)$$

defines an effective notion of "quantumness" for the two states. It signifies the extra information the two states carry with respect to each other that can be unlocked only by nonlocal means on separate transmissions.

Notice that, by this reckoning, two states are the most quantum with respect to each other when $x = 1/\sqrt{2}$, i.e., when the two states are separated by an angle $\theta = 45°$. Here $Q \approx 0.202$. In ways, this result is quite pleasing. Since $C_\infty = C_1$ when $\theta = 0°$ and $\theta = 90°$, one might well expect the states to be maximally quantal when their separation is exactly between these two extremes in the sense of Hilbert-space geometry.

*It is an open question whether these channel examples exhibit the strongest form of "nonlocality without entanglement." In the strongest version,[11,18] one is not only concerned with the discrepancy between collective and sequentially adaptive measurements, but between collective measurements and any measurements whatsoever that are purely local with respect to the separate transmissions. For instance, within the largest class of local measurements the receiver might perform weak measurements that ping-pong back and forth between the separate transmissions: first collect a little information from signal 1, then adjust the measurement and collect a little information from signal 2. Following that, adjust again and return to signal 1 to collect a little more, and so on and so on.

INFORMATION GAIN vs. QUANTUM STATE DISTURBANCE

The founding fathers of quantum mechanics were fond of saying things like this typical example of Pauli's:[19]

> The indivisibility of elementary quantum processes ... finds expression in an indeterminacy of the interaction between [the] instrument of observation ... and the system observed ..., which cannot be got rid of by determinable corrections. For every act of observation is an interference, of undeterminable extent ...

However, given the difficulty in ascribing objective properties to quantum systems independently of measurement (as indicated by the Kochen-Specker theorem and the violation of the Bell inequalities[20]), what can the terminology of "interference" or "disturbance" possibly refer to? What precisely is it, if anything, that is disturbed by measurement?

One of the more interesting things about the applied field of quantum cryptography as far as the foundations of quantum mechanics is concerned is that it provides the tools to breath some real life into this old question. To get somewhere, however, one must realize that one cannot simply speak of performing measurements on a single quantum system prepared in a single quantum state: one must, at the very least, consider two nonorthogonal states as in the B92 protocol.[21] Then the referents of the words "information gain" and "disturbance" can have precise meanings.

The scheme is the following. Alice encodes the various secret bits she wishes to share with Bob into the quantum states $|\psi_0\rangle$ and $|\psi_1\rangle$ and sends them on their way. The eavesdropper Eve interacts some probe with the systems while they are en route. This leaves her probe variously either in one of two (mixed) quantum states, ρ_0^{E} and ρ_1^{E}. In the process, Alice's states are perturbed variously into ρ_0^{A} and ρ_1^{A}.

These four states taken together provide a basis for an information-disturbance tradeoff principle. For instance, one might gauge the amount of information that Eve has received by her potential for guessing the individual key bits through measurements on her probe. Her best probability of error P will given by the leftmost expression in Eq. (1) with the density operators suitably replaced by ρ_0^{E} and ρ_1^{E}. Similarly one might gauge the disturbance D to Alice's system by Bob's probability of identifying Eve's intervention as he performs the standard maneuvers for extracting a key from Alice's signals. Holding P fixed while optimizing Eve's probe's interaction, one obtains the rather complicated tradeoff principle:[6]

$$D=\frac{1}{2}-\frac{1}{2}\sqrt{1+x^2\Big(-1-4P+4P^2+2x^2+2\sqrt{(1-x^2)(4P-4P^2-x^2)}\Big)}\;. \qquad (5)$$

At the endpoint corresponding to a maximal information gain by Eve, this tradeoff is especially interesting for defining a notion of "quantumness" . There, Eve's probability of error in identifying the bit must be given by Eq. (1); the minimal disturbance that can be imparted to Alice's states in this case is

$$D_{\mathrm{QMI}}=\frac{1}{2}\Big(1-\sqrt{1-x^2+x^4}\Big)\;. \qquad (6)$$

This number—the minimum disturbance at maximum information—expresses the two states' relative fragility when exposed to information-gathering measurements. Notice that once again the two states are most quantum with respect to each other when $x=1/\sqrt{2}$. In that case, $D\approx 0.067$ while Eve's probability of error is $P_e\approx 0.146$. The angle $\theta=45°$ starts to look quite robust as far as "most quantum" is concerned.

QUANTUM CLONING MACHINES

Lest one become complacent in accepting the "obviousness" of $\theta = 45°$ signifying when two states are the most quantum with respect to each other, let us consider one more notion of quantumness. Lately it has become a popular pastime to consider the issue of how close one can come to ideal cloning for an unknown quantum state.[22] In some ways this is closely related to the information-disturbance question; for if one could clone ideally, then one could create the potential for gathering information without disturbing. However, upon closer inspection, one finds quite a divergence between the two issues.

Consider the issue at hand. One would like to take the given system, secretly prepared in either $|\psi_0\rangle$ or $|\psi_1\rangle$, attach it to some ancillary system prepared in a standard state $|s\rangle \in \mathcal{H}_A$, and have the two together evolve to the state $|\psi_0\rangle|\psi_0\rangle$ or $|\psi_1\rangle|\psi_1\rangle$ as should be the case. Instead, the best one can hope for is some states $|\psi_0^{\mathrm{SA}}\rangle$ and $|\psi_1^{\mathrm{SA}}\rangle$ on the combined space $\mathcal{H}_S \otimes \mathcal{H}_A$ that obtain some *but not all* the characteristics of ideal clones. The only characteristic we shall require here is that the marginal states isolated to each system be identical, i.e., $\mathrm{tr}_{\mathrm{S}}|\psi_i^{\mathrm{SA}}\rangle\langle\psi_i^{\mathrm{SA}}| = \mathrm{tr}_{\mathrm{A}}|\psi_i^{\mathrm{SA}}\rangle\langle\psi_i^{\mathrm{SA}}|$ for $i = 0$ and 1.

With this as the sole criterion of a cloning device, what is a good clone if it is not ideal? There are several measures that one might imagine for gauging this, but we shall consider only two: the average global fidelity $|\langle\psi_i^{\mathrm{SA}}|(|\psi_i\rangle|\psi_i\rangle)|^2$, and the average local fidelity $\langle\psi_i|\left(\mathrm{tr}_{\mathrm{A}}|\psi_i^{\mathrm{SA}}\rangle\langle\psi_i^{\mathrm{SA}}|\right)|\psi_i\rangle$. Remarkably, these two measures are not optimized by the same cloning interaction; they give distinct notions of optimal cloning.[23] For the case at hand, the optimal values of the global and local fidelities turn out to be

$$F_g = \frac{1}{2}\left(1 + x^3 + (1 - x^2)\sqrt{1 + x^2}\right), \qquad \text{and} \tag{7}$$

$$F_l = \frac{1}{2} + \frac{\sqrt{2}}{32x}(1+x)\left(3 - 3x + \sqrt{1 - 2x + 9x^2}\right)\sqrt{-1 + 2x + 3x^2 + (1 - x)\sqrt{1 - 2x + 9x^2}}\,,$$

respectively. Each of these measures now define a notion of quantumness for our two states. With respect to F_g, two states are the most quantum with respect to each other when $x = 1/\sqrt{3}$. With respect to F_l, they are the most quantum when $x = 1/2$. In neither case do we find the coveted $x = 1/\sqrt{2}$ value.

CONCLUSION

What is the essence of quantum theory? What crucial features of the phenomena about us lead ineluctably to just this formalism? These are questions that have been asked since the earliest days of the theory. Each generation has its answer; ours no doubt will find part of it written in the language of quantum information. What is striking about the newest turn—the quantum information revolution—is that it provides a set of tools for this analysis from *within* quantum theory. The example of the tradeoff between information and disturbance in quantum eavesdropping is typical. Words about "measurements causing disturbance" have been with us since 1927, but those always in reference to outdated, illegitimate classical concepts. The time is ripe to consider turning the tables, to ask "What is quantum mechanics trying to tell us?"[24] Why is the world so constituted as to allow single-bit information transfers to be disturbed by outside information-gatherers, but never *necessarily* more so than by an amount $D_{\mathrm{OMI}} \approx 0.067$? Why is the world so constituted that binary preparations can be put together in a way that the whole is more than a sum of the parts, but never more so than by $Q \approx 0.202$ bits? The answers surely cannot be that far away.

ACKNOWLEDGMENTS

This paper is dedicated to Jeff Kimble for joking that "there just can't be so much to say about only two nonorthogonal states." I thank Michał Horodecki for useful discussions. I acknowledge a Lee A. DuBridge Fellowship and the support of DARPA through the Quantum Information and Computing Institute (QUIC).

REFERENCES

1. K. Kraus, "States, Effects, and Operations," Springer-Verlag, Berlin (1983).
2. For a very basic result in this respect, see Theorem 6 and Section V of I. Pitowsky, Infinite and finite Gleason's theorems and the logic of indeterminacy, *J. Math. Phys.* 39:218 (1998).
3. W. K. Wootters and W. H. Zurek, A single quantum cannot be cloned, *Nature* 299:802 (1982); D. Dieks, Communication by EPR devices, *Phys. Lett. A* 92:271 (1982).
4. H. P. Yuen, Amplification of quantum states and noiseless photon amplifiers, *Phys. Lett. A* 113:405 (1986); H. Barnum, C. M. Caves, C. A. Fuchs, R. Jozsa, and B. Schumacher, Noncommuting mixed states cannot be broadcast, *Phys. Rev. Lett.* 76:2818 (1996).
5. C. H. Bennett, G. Brassard, and N. D. Mermin, Quantum cryptography without Bell's theorem, *Phys. Rev. Lett.* 68:557 (1992).
6. C. A. Fuchs, Information gain vs. state disturbance in quantum theory, *Fort. der Phys.* 46:535 (1998); C. A. Fuchs and A. Peres, Quantum state disturbance vs. information gain: Uncertainty relations for quantum information, *Phys. Rev. A* 53:2038 (1996).
7. M. Sasaki, T. S. Usuda, O. Hirota, and A. S. Holevo, Applications of the Jaynes-Cummings model for the detection of nonorthogonal quantum states, *Phys. Rev. A* 53:1273 (1996).
8. C. A. Fuchs, Nonorthogonal quantum states maximize classical information capacity, *Phys. Rev. Lett.* 79:1163 (1997).
9. C. W. Helstrom, "Quantum Detection and Estimation Theory," Academic Press, NY (1976).
10. C. M. Caves and C. A. Fuchs, Quantum information: How much information in a state vector?, *in*: "The Dilemma of Einstein, Podolsky and Rosen – 60 Years Later," A. Mann and M. Revzen, eds., Israel Physical Society (1996).
11. C. H. Bennett, D. P. DiVincenzo, C. A. Fuchs, T. Mor, E. Rains, P. W. Shor, J. A. Smolin, and W. K. Wootters, Quantum nonlocality without entanglement, `quant-ph/9804053`.
12. T. M. Cover and J. A. Thomas, "Elements of Information Theory," Wiley, NY (1991), Sect. 8.9.
13. A. S. Holevo, Capacity of a quantum communication channel, *Prob. Info. Trans.* 15:247 (1979).
14. A. Fujiwara and H. Nagaoka, Operational capacity and pseudoclassicality of a quantum channel, *IEEE Trans. Inf. Theory* 44:1071 (1998).
15. L. B. Levitin, Optimal quantum measurements for two pure and mixed states, *in*: "Quantum Communications and Measurement," V. P. Belavkin, O. Hirota, and R. L. Hudson, eds., Plenum Press, NY (1995); C. A. Fuchs and C. M. Caves, Ensemble-dependent bounds for accessible information in quantum mechanics, *Phys. Rev. Lett.* 73:3047 (1994).
16. P. Hausladen, R. Jozsa, B. Schumacher, M. Westmoreland, and W. K. Wootters, Classical information capacity of a quantum channel, *Phys. Rev. A* 54:1869 (1996); A. S. Holevo, The capacity of the quantum channel with general signal states, *IEEE Trans. Inf. Theory* 44:269 (1998); B. Schumacher and M. D. Westmoreland, Sending classical information via noisy quantum channels, *Phys. Rev. A* 56:131 (1997).
17. C. A. Fuchs, "Distinguishability and Accessible Information in Quantum Theory," Ph.D. thesis, University of New Mexico, 1996. LANL archive `quant-ph/9601020`.
18. A. Peres and W. K. Wootters, Optimal detection of quantum information, *Phys. Rev. Lett.* 66:1119 (1991).
19. W. Pauli, "Writings on Philosophy and Physics," Springer-Verlag, Berlin (1995).
20. A. Peres, "Quantum Theory: Concepts and Methods," Kluwer, Dordrecht (1993).
21. C. H. Bennett, Quantum cryptography using any two nonorthogonal states, *Phys. Rev. Lett.* 68:3121 (1992).
22. For a deluge of articles on cloning, see the LANL `quant-ph` archive.
23. D. Bruß, D. P. DiVincenzo, A. Ekert, C. A. Fuchs, C. Macchiavello, and J. A. Smolin, Optimal universal and state-dependent quantum cloning, *Phys. Rev. A* 57:2368 (1998).
24. N. D. Mermin, What is quantum mechanics trying to tell us?, *Am. J. Phys.* 66:753 (1998).

ON MAXIMUM MUTUAL INFORMATION WITHOUT CODING

Masao Osaki,[1] Masashi Ban,[2] and Osamu Hirota[1]

[1]Research Center for Quantum Communications,
Tamagawa University, Japan.
[2]Advanced Research Laboratory, Hitachi, Ltd., Japan.

INTRODUCTION

We consider the communication systems sending the classical (Shannon) information through the quantum channel. The classical information source X is represented with symbols $\{x_i\}$ and their *a priori* probabilities $\{\xi_i\}$. At a transmitter, each symbol x_i is modulated to a quantum state $\hat{\rho}_i$ and detected at a receiver by the quantum mechanically generalized measurement, so-called the probability operator-valued measure (POM) $\{\hat{\Pi}_j\}$.[1, 2] Here we call $\hat{\Pi}_j$ a detection operator since it represents a detection process deciding a received signal to be 'j'. The detection operator is no-negative, $\hat{\Pi}_j \geq 0$, and resolution of identity, $\sum_j \hat{\Pi}_j = \hat{I}$, where $\hat{I}$ is the identity operator on the signal Hilbert space H_s spanned by signal quantum states. This detection process can be regarded as a quantum channel characterized by the conditional probability $P(j|i)$ defined as follows:

$$P(j|i) = \mathrm{Tr}\hat{\Pi}_j\hat{\rho}_i. \tag{1}$$

Namely, $P(j|i)$ represents the probability to decide a received signal as 'j' while the signal 'i' is transmitted truly.

The mutual information between the transmitter and the receiver is defined with the conditional probabilities $\{P(j|i)\}$ and *a priori* probabilities $\{\xi_i\}$.

$$I_1(X;Y) = \sum_i \sum_j \xi_i P(j|i) \log \left[\frac{P(j|i)}{\sum_k \xi_k P(j|k)} \right], \tag{2}$$

where the subscript '1' indicates the length of codeword, namely, the transmission without coding. The maximum value of the mutual information with respect to the detection operators and *a priori* probabilities is called "the maximum mutual information without coding" C_1. That is,

$$C_1 = \max_{\xi} \max_{\hat{\Pi}} I_1(X;Y). \tag{3}$$

In the same way, the maximum mutual information for each codeword length n can be defined and represented by C_n.

Since these maximum mutual informations exhibit the super-additivity,[3]

$$C_m + C_n \leq C_{m+n}, \tag{4}$$

the channel capacity of the quantum channel can be defined as follows:

$$C = \lim_{n\to\infty} \frac{C_n}{n}. \tag{5}$$

This quantum channel capacity is represented as follows:[4, 5, 6]

$$C = \max_{\xi} \left[S\left(\sum_i \xi_i \hat{\rho}_i\right) - \sum_i \xi_i S(\hat{\rho}_i) \right], \tag{6}$$

where $S(\hat{\rho})$ is von Neumann entropy given with $\{\lambda_i\}$ the eigen-values of a density operator $\hat{\rho}$.

$$S(\hat{\rho}) = -\mathrm{Tr}\hat{\rho}\log\hat{\rho} = -\sum_i \lambda_i \log \lambda_i. \tag{7}$$

Our resent interest is to find a coding by which the mutual information approaches to the quantum channel capacity. To go beyond C_1 and to approach C, the super-additivity should be effectively generated by the coding. However, there is little information how to generate the super-additivity. Some codings with super-additivity have been found by heuristic manner.[7] In order to investigate the mechanism of the super-additivity, finding so many examples of the coding with super-additivity is an important step. Under the such situation, C_1 can be used as a criterion to judge the existence of the super-additivity. That is, if some coding with its codeword length n and some measurement give the mutual information which satisfies $I_n(X;Y) > nC_1$, such a coding is said to exhibit the super-additivity. Hence the derivation of C_1 is one of the most essential problems in quantum information theory.

The judgment of the existence of super-additivity is sometimes very delicate problem, so that the exact value of C_1 should be obtained. Hence the analytical derivation of C_1 is much more preferable than the numerical one. As shown in the definition of C_1 in Eq. (3), however, it is obtained by means of the double maximization with respect to the detection operators and *a priori* probabilities. This simultaneous optimization makes the problem so difficult. In the following, we show some tools to investigate C_1. After the summary of the analytical derivation of C_1 in the case of binary pure state signals, some directions are discussed to derive C_1 for the cases of more than binary pure state signals.

SOME TOOLS TO SEEK THE SOLUTIONS

Accessible Information

Since the simultaneous maximization in Eq. (3) is pretty difficult, it seems better to separate the problem into the individual maximization with respect to the detection operators and *a priori* probabilities. The accessible information I_{ac} is the mutual information maximized with respect to the detection operators for some given *a priori* probabilities.

$$I_{\mathrm{ac}} = \max_{\hat{\Pi}} I_1(X;Y). \tag{8}$$

While the derivation of I_{ac} must be easier than that of C_1, there is no necessary and sufficient condition for detection operators to give the accessible information. Necessary conditions and theorem about the detection operators are given as follows:

Necessary Condition for Detection Operators to Give the Accessible Information (by Holevo)[2]: The optimum detection operators must satisfy the following equation

$$\hat{\Pi}_j\left(\hat{F}_j-\hat{F}_i\right)\hat{\Pi}_i=0, \forall i,j, \quad \hat{F}_j=\sum_l \xi_l \log\left[\frac{P(j|l)}{\sum_k \xi_k P(j|k)}\right]\hat{\rho}_l. \tag{9}$$

This condition is derived based on the condition that the first variation of the mutual information with respect to the detection operators is equal to zero. Since the mutual information is a convex function on a compact convex set of $\{\hat{\Pi}_j\}$,[9] this condition is just a necessary condition.

2nd Necessary Condition for Detection Operators to Give the Accessible Information (by Holevo)[8]: This condition is derived based on the second variation of the mutual information.

$$\begin{aligned}
\delta^2 I &= \mathrm{Tr}\left\{\left(\sqrt{\hat{\Pi}_i}\left[\hat{F}_j-\hat{F}_i\right]\sqrt{\hat{\Pi}_i}\right)\hat{A}^\dagger\hat{A}\right\}-\mathrm{Tr}\left\{\left(\sqrt{\hat{\Pi}_j}\left[\hat{F}_j-\hat{F}_i\right]\sqrt{\hat{\Pi}_j}\right)\hat{A}\hat{A}^\dagger\right\}\\
&\quad +2\sum_k \xi_k\left(\frac{1}{P(i|k)}+\frac{1}{P(j|k)}\right)\left(\mathrm{Re}\left[\mathrm{Tr}\left\{\sqrt{\hat{\Pi}_i}\hat{\rho}_k\sqrt{\hat{\Pi}_j}\hat{A}\right\}\right]\right)^2\\
&\quad -2\left(\frac{1}{\sum_m \xi_m P(i|m)}+\frac{1}{\sum_m \xi_m P(j|m)}\right)\left(\mathrm{Re}\left[\mathrm{Tr}\left\{\sqrt{\hat{\Pi}_i}\overline{\sigma}\sqrt{\hat{\Pi}_j}\hat{A}\right\}\right]\right)^2\\
&\le 0,
\end{aligned} \tag{10}$$

where $\overline{\sigma}=\sum_k \xi_k\hat{\rho}_k$ and inequality must be satisfied for any bounded operator $\hat{A}$. Even if this condition is satisfied by some set of detection operators, it may give the local (not global) maximum of the mutual information.

The Number of Detection Operators (by Davies)[9]: The maximization of the mutual information with respect to the detection operators includes the optimization of their number. Since I_{ac} is attained at extreme points of the convex set of $\{\hat{\Pi}_j\}$, all the detection operators which attains I_{ac} can be rank 1, i.e. $\hat{\Pi}_j=|\omega_j\rangle\langle\omega_j|$. The number of different $|\omega_j\rangle$ exists between d and d^2 when the signal quantum states span d-dimensional Hilbert space, namely, $\dim H_s = d$.

Signal Quantum States and Detection Operators

The following assumptions of signal quantum states and of detection operators also help the derivation not only of I_{ac} but also of C_1.

Symmetric Signal Quantum States: The preparation of the signal quantum states in symmetric, such as group covariant quantum states,[9] makes the problem much simpler. In the case of M-ary symmetric signal, we consider here, each quantum state is represented with an initial quantum state $\hat{\rho}_1$ and some unitary (or anti-unitary) operator $\hat{V}$ which satisfies $\hat{V}^M=\hat{I}$ (or $-\hat{I}$) as follows:

$$\hat{\rho}_k=\hat{V}^{k-1}\hat{\rho}_1\hat{V}^{\dagger k-1}. \tag{11}$$

M-ary phase shift keyed (M-PSK) signals employed in the practical microwave communication systems are the case of this symmetry.

Square-Root Detection (SRD): The set of detection operators which is automatically generated from the prepared signal quantum states is sometimes useful. The square-root detection (SRD), we call here, is defined for a set of pure state signals. While it may have long historical stories, the recent application to the information optimum detection was done by Hausladen and Wootters.[10] The detection operators of SRD are defined with a set of signal quantum states $\{|\phi_i\rangle\}$ as follows:

$$\hat{\Pi}_j = |\mu_j\rangle\langle\mu_j|, \quad |\mu_j\rangle = \left(\sum_{k=1}^{M} |\phi_k\rangle\langle\phi_k|\right)^{-1/2} |\phi_j\rangle. \tag{12}$$

It is reported that SRD is information optimum when the signal quantum states are nearly orthogonal.[10] Especially for the symmetric quantum states with equal *a priori* probabilities, SRD is the error probability optimum and satisfies the necessary condition to give the accessible information.[11, 12]

Group Covariant Detection (GCD): The group covariant detection (GCD) is defined for some group covariant signals.[9] It is also proved that GCD attains the maximum mutual information for group covariant signal quantum states with equal *a priori* probabilities when the representation of the group on the set of all states on H_s is irreducible.[9] For the cases of M-ary symmetric signal explained above, the detection operator of GCD can be given as follows:

$$\hat{\Pi}_j = \frac{d}{M}\hat{V}^{k-1}|\omega\rangle\langle\omega|\hat{V}^{\dagger k-1}, \tag{13}$$

where $d = \dim H_s$, M is the number of signals, and $|\omega\rangle$ is a unit vector in H_s. Hence GCD has a degree of freedom to change this unit vector. For some $|\omega\rangle$, detection operators in Eq. (13) coincides with those by the square-root detection in Eq. (12). That is, the square-root detection is a special case of the group covariant one for M-ary symmetric signal.

BINARY PURE STATE SIGNAL

The signal quantum states are represented as follows:

$$\hat{\rho}_1 = |\phi_1\rangle\langle\phi_1|, \quad \hat{\rho}_2 = |\phi_2\rangle\langle\phi_2|. \tag{14}$$

According to the Davies' theorem, the number of detection operators should be considered between 2 and 4 since $\dim H_s = 2$. However, Levitin has revealed that two detection operators are enough to attain I_{ac}.[13] His proof can be interpreted that all the possible solutions of the necessary condition in Eq. (9) are examined for their second necessary condition in Eq. (10). Consequently, he found that only one solution which consists of two detection operators satisfies Eq. (10). Hence two detection operators attain I_{ac} for each *a priori* probabilities. Fuchs also derived the information optimum detection operators for I_{ac} as projectors.[14]

Our numerical analysis about C_1 was done based on the Davies' theorem, namely using 4, 3, and 2 detection operators, and on the maximization with respect to *a priori* probabilities.[12] It revealed two points: "C_1 is obtained when *a priori* probabilities are equal," and "the mutual information attained by 4 or 3 detection operators does not exceed that by 2." More precisely, when 4 or 3 detection operators give C_1, some of the detection operators are proportional to the other.

The final proof about C_1 was done by one of authors (M. Ban). He proved that the equal *a priori* probability gives the maximum of I_{ac} for two-input and two-output quantum channel (see Appendix). The maximum mutual information without coding, C_1 can be obtained as follows:

$$C_1 = \frac{1+\sqrt{1-|\kappa|^2}}{2}\log\left[1+\sqrt{1-|\kappa|^2}\right] + \frac{1-\sqrt{1-|\kappa|^2}}{2}\log\left[1-\sqrt{1-|\kappa|^2}\right], \quad (15)$$

where $\kappa = \langle\phi_1|\phi_2\rangle$. If the binary phase shift keyed (BPSK) signal in coherent state ($|\phi_1\rangle = |\alpha\rangle$, $|\phi_2\rangle = |-\alpha\rangle$) is assumed, $\kappa = \exp[-2n_s]$, where $n_s = |\alpha|^2$ represents the signal photon numbers. The detection which give C_1 is SRD, so that it is also the error probability optimum one.

Table 1. Conditions to attain C_1 for binary pure state signal.

C_1	analytical solution
a priori probabilities	$\xi_i = 1/2$
detection	square-root
quantum channel	2×2

TERNARY SYMMETRIC PURE STATE SIGNAL

Here we consider the ternary pure states as a signal. It seems hopeless to solve or to evaluate the necessary condition in Eq. (9) for a case of arbitrary set of three quantum states. Hence we treat the set of quantum states with certain symmetry expressed in Eq. (11) so that it makes the problem easier.

Two-Dimensional Hilbert Space

When $\dim H_s = 2$, the ternary symmetric quantum states can be represented in vectors with some base.

$$|\phi_1\rangle \Rightarrow \begin{bmatrix} 1 \\ 0 \end{bmatrix}, |\phi_2\rangle \Rightarrow \begin{bmatrix} \frac{1}{2} \\ \frac{\sqrt{3}}{2} \end{bmatrix}, |\phi_3\rangle \Rightarrow \begin{bmatrix} \frac{1}{2} \\ -\frac{\sqrt{3}}{2} \end{bmatrix}. \quad (16)$$

This set of signal quantum states was used in the papers[15, 16] at which *a priori* probabilities are set to be equal. The accessible information results that $I_{ac} = 0.5850$[bits]. The detection attaining I_{ac} is not SRD but a kind of group covariant detections explained in Eq. (13).

According to the definition of C_1, on the other hand, *a priori* probabilities are the parameters to maximize the mutual information. Since $\dim H_s = 2$, the number of information optimum detection operators seems to exist between 2 and 4. However, it is a two-dimensional "real" Hilbert space, so that 2 or 3 detection operators are enough for the optimization.[8, 17] Our numerical analysis including the cases of 2 and 3 detection operators and changing *a priori* probabilities resulted that "C_1 is attained when two of three signal quantum states are sent with equal probabilities and two detection operators are used to detect them." Then we have $C_1 = 0.6454$[bits]. But it is interesting that the quantum channel capacity, C, of this signal set is obtained when all the three quantum states have the same *a priori* probabilities. That is, $C = 1$[bits].

Table 2. Conditions to attain C_1 for ternary symmetric pure state signal ($\dim H_s = 2$).

C_1	numerical solution
a priori probabilities	$\xi_i = 1/2$
detection	square-root
quantum channel	2×2

Three-Dimensional Hilbert Space

Here we consider the quantum states as follows:

$$|\phi_1\rangle = |\alpha\rangle, |\phi_2\rangle = |\alpha e^{\frac{2\pi}{3}i}\rangle, |\phi_3\rangle = |\alpha e^{-\frac{2\pi}{3}i}\rangle, \tag{17}$$

where all the quantum states are in coherent state. The signal photon number is represented by $n_s = |\alpha|^2$. This set of quantum states is called three phase shift keyed (3PSK) signal. Each state is related by the unitary operator as follows:

$$\hat{V} = \exp\left[\frac{2\pi i}{3}\hat{n}\right], \tag{18}$$

where $\hat{n}$ is the photon number operator.

Unfortunately C_1 for this set of quantum states is not obtained yet. The essential difficulty exits in the number of detection operators. That is, in accordance with Davies' theorem, we have to maximize the mutual information by 3 to 9 different detection operators. However, Fuchs and Peres have analyzed it numerically for the cases of $\dim H_s = 3, 4$ and it resulted that 3 and 4 detection operators are enough to attain I_{ac}, respectively.[19] For the situation of nearly orthogonal signal quantum states ($n_s \approx \infty$), it seems reasonable that three detection operators are enough to maximize the mutual information. This is because the error probability is almost zero by SRD under the situation, and then the mutual information is approximately equal to the entropy of the information source. That is, C_1 for the nearly orthogonal quantum states is attained by SRD with $\xi_i = 1/3, \forall i$.

When the signal photon number decrease ($n_s < \infty$), there is no candidates for C_1. If *a priori* probabilities are fixed to be equal, one of the candidates for attaining I_{ac} is SRD since it satisfies not only the necessary condition Eq. (9) but also the second necessary condition Eq. (10) with some $\hat{A}$ (not all) for any signal photon number ($0 \leq n_s \leq \infty$). Hence it can be considered that the smaller the error probability is, the larger the mutual information is.

In the region of small signal photon number ($n_s \approx 0.1$), however, it is clarified that SRD does not attain I_{ac}.[18] The detection which gives larger mutual information is found in the other variations of group covariant detection. According to the numerical analysis, the best detection attaining the largest mutual information in the group covariant detections is SRD for $n_s \geq 0.245$. For $0.08 < n_s < 0.245$, the best detection varies continuously, and then, saturates a specific group covariant detection for $n_s \leq 0.08$.

On the other hand, if *a priori* probabilities are changed and finally two of three signal quantum states are sent with equal *a priori* probabilities, the larger mutual information than that by any group covariant detection is obtained for $n_s \leq 0.13$. This can be interpreted that the dimension of signal Hilbert space is reduced and results approximately two ($\dim H_s \approx 2$) when the signal photon number becomes much small.

Hence the optimization problem may be changed from three-dimensional Hilbert space into two-dimensional one as discussed in the previous section. These properties are depicted in Fig. 1 and summarized in Table .

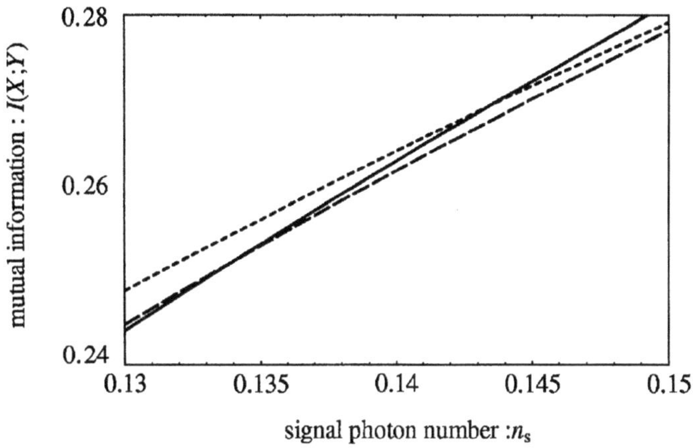

Figure 1. Comparison of each mutual information. Solid line: by SRD, Break line: by specific group covariant detection, Dotted line: by 2×2 channel.

Table 3. Summary of maximum mutual information for ternary symmetric pure state signal ($\dim H_s = 3$).

signal photon number	$n_s \ll 0.1$	$0.1 < n_s < \infty$	$n_s \approx \infty$
$\max I_1(X;Y)$	C_1	$I_{ac}(?)$	C_1
a priori probabilities	$\xi_i = 1/2$	$\xi_i = 1/3$	$\xi_i = 1/3$
detection	square-root	group covariant	square-root
quantum channel	2×2	3×3	3×3

To identify C_1 of 3PSK signal, the transition process from 3×3 quantum channel to 2×2 one must be clarified.

CONCLUSION

In this paper, we have investigated the derivations of C_1 for some simplest signal quantum states. For binary pure state signal, the analytical solution of C_1 is given. The conditions for *a priori* probabilities, detection process, and quantum channel are summarized in Table .

When the ternary symmetric pure state signals are considered in two-dimensional Hilbert space, it is confirmed numerically that C_1 is attained by sending two of three quantum states with equal *a priori* probability. This property can be seen in the case of three-dimensional Hilbert space with small signal photon number.

In the case of ternary symmetric pure state signals in three-dimensional Hilbert spaces, derivation of C_1 becomes much difficult. The most essential problem is the number of information optimum detection operators. We have to consider it from 3 to

9 except for the limiting cases in which the signal photon number is enough large or small. In the both limiting cases, large and small signal photon numbers, the candidates for C_1 can be assumed based on the error free case and the binary case, respectively. While the square-root detection satisfies not only the first necessary condition but also the second one, there is no proof such detection attains C_1. It is also found that the other group covariant detection attains the larger mutual information than that by the square-root detection at the transition process from 3×3 quantum channel to 2×2 one.

Table 4. Derivations of C_1, and I_{ac}. ○ : analytically derived. △ : numerically derived. × : not derived yet.

signal quantum states	C_1	I_{ac} ($\xi_i = 1/M$)
binary pure	○ $\xi_i = 1/2$	○
ternary symmetric pure ($\dim H_s = 2$)	△ $\xi_i = 1/2$	○
ternary symmetric pure ($\dim H_s = 3$)	×	×

When we try to derive C_1 for much more complicated M-ary signal sets, the quantum channel attaining C_1 will change from $M \times M$ to 2×2 depending on the orthogonality (signal photon number) of signal quantum states. This is because we have never found $M \times N(> M)$ quantum channel which gives C_1 while Davies' theorem indicates this possibility.

APPENDIX

What are input probabilities that attain the maximum value of the mutual information of two-input and two-output quantum channel?

When a signal detection process for two pure quantum states $|\psi_1\rangle$ and $|\psi_2\rangle$ with input probabilities p_1 and p_2 ($p_1 + p_2 = 1$, $p_1 \geq 0, p_2 \geq 0$) is described by two decision operators, the maximum value of the mutual information for the given input probabilities is obtained

$$I = -\ln 2 - p_1 \ln p_1 - p_2 \ln p_2 + \frac{1}{2}\left(1 + \sqrt{1 - 4\alpha p_1 p_2}\right) \ln\left(1 + \sqrt{1 - 4\alpha p_1 p_2}\right) + \frac{1}{2}\left(1 - \sqrt{1 - 4\alpha p_1 p_2}\right) \ln\left(1 - \sqrt{1 - 4\alpha p_1 p_2}\right), \tag{19}$$

where we set $\alpha = \kappa^2 = |\langle\psi_1|\psi_2\rangle|^2$ $(0 \leq \alpha \leq 1)$. The derivation of this equation is given in Appendix A of Phys. Rev. A **54**, 2718 (1996) [see Eq. (A14)]. Since the mutual information I is symmetric with respect to the input probabilities p_1 and p_2, we can restrict the range of the input probability p_1 to $0 \leq p_1 \leq \frac{1}{2}$. If we put $p_1 = x$ and $p_2 = 1 - x$, the mutual information I given by Eq. (19) can be expressed in the following form:

$$I(x) = H(x) - H(f(x)), \tag{20}$$

where $H(x)$ is the binary entropic function and $f(x)$ is the average probability of error,

$$H(x) = -x \ln x - (1 - x) \ln(1 - x), \tag{21}$$

$$f(x) = \frac{1}{2}\left(1 - \sqrt{1 - 4\alpha(1 - x)x}\right). \tag{22}$$

Since $0 \le \alpha \le 1$, $0 \le x \le \frac{1}{2}$ and $f(x)$ increases monotonously with the parameter α, we obtain the inequality $f(x) \le f(x)|_{\alpha=1} = x$, that is,

$$0 \le f(x) \le x \le \frac{1}{2}. \tag{23}$$

In the following, we will derive the inequality $dI(x)/dx \ge 0$ for $0 \le f(x) \le x \le \frac{1}{2}$ to show that the maximum value of the mutual information $I(x)$ is attained by the equal input probabilities $p_1 = p_2 = \frac{1}{2}$ $(x = \frac{1}{2})$.

Differentiating the mutual information $I(x)$ with respect to x, we obtain

$$\frac{dI(x)}{dx} = (1-2x)\left[\frac{1}{1-2x}\ln\left(\frac{1-x}{x}\right) - \frac{\alpha}{1-2f}\ln\left(\frac{1-f}{f}\right)\right], \tag{24}$$

where $f = f(x)$ and we have used the relations,

$$\frac{dH(x)}{dx} = \ln\left(\frac{1-x}{x}\right), \tag{25}$$

$$\frac{df(x)}{dx} = \frac{\alpha(1-2x)}{\sqrt{1-4\alpha(1-x)x}} = \alpha\frac{1-2x}{1-2f}. \tag{26}$$

Since we have $\alpha = (1-f)f/(1-x)x$ from Eq. (22), we can eliminate the parameter α from Eq. (24) and we obtain the following expression,

$$\frac{dI(x)}{dx} = \frac{1-2x}{(1-x)x}\left[G(x) - G(f)\right], \tag{27}$$

with

$$G(x) = \frac{(1-x)x}{1-2x}\ln\left(\frac{1-x}{x}\right). \tag{28}$$

Since the inequality $0 \le f \le x \le \frac{1}{2}$ holds, $G(x) \ge G(f)$ is equivalent to $dG(x)/dx \ge 0$. Thus we obtain the relation,

$$\frac{dI(x)}{dx} \ge 0 \iff \frac{dG(x)}{dx} \ge 0. \tag{29}$$

To show the inequality $dG(x)/dx \ge 0$, we calculate $dG(x)/dx$,

$$\frac{dG(x)}{dx} = \frac{1}{1-2x}\left[\frac{1-2x+2x^2}{1-2x}\ln\left(\frac{1-x}{x}\right) - 1\right] = \frac{1}{1-2x}M(z), \tag{30}$$

where we set $\sqrt{z} = (1-x)/x$ $[x = 1/(1+\sqrt{z})]$ and the function $M(z)$ is defined by

$$M(z) = \frac{1}{2}\left(\frac{z+1}{z-1}\ln z - 2\right). \tag{31}$$

The range of z is given by $z \ge 1$ because of $0 \le x \le \frac{1}{2}$. Here we further introduce a function $K(z)$ by

$$K(z) = (z+1)\ln z - 2(z-1), \tag{32}$$

which yields the relation $M(z) = \frac{1}{2}K(z)/(z-1)$. Since $z \ge 1$, the inequality $dG(x)/dx \ge 0$ is equivalent to $K(z) \ge 0$. Therefore we obtain the relation,

$$\frac{dI(x)}{dx} \ge 0 \iff \frac{dG(x)}{dx} \ge 0 \iff K(z) \ge 0. \tag{33}$$

To show the inequality $K(z) \geq 0$, we calculate $dK(z)/dz$ and $d^2K(z)/dz^2$ and we obtain the following result,

$$\frac{dK(z)}{dz} = \ln z + \frac{1}{z} - 1, \qquad \frac{d^2K(z)}{dz^2} = \frac{1}{z}\left(1 - \frac{1}{z}\right). \tag{34}$$

Since $z \geq 1$, the inequality $d^2K(z)/dz^2 \geq 0$ holds and we obtain $dK(z)/dz \geq dK(1)/dz = 0$. This indicates that the function $K(z)$ increases monotonously with z and we obtain the inequality $K(z) \geq K(1) = 0$. Thus we find from Eq. (33) that $dI(x)/dx \geq 0$ in the range $0 \leq x \leq \frac{1}{2}$ and the mutual information $I(x)$ takes the maximum value at $x = \frac{1}{2}$.

Therefore we have shown that the maximum value of the mutual information given by Eq. (19) is attained by the equal input probabilities $p_1 = p_2 = \frac{1}{2}$.

ACKNOWLEGMENTS

The authors would like to thank Professor A. S. Holevo of the Steklov Mathematical Institute for enlightening discussions. One of the authors (M.O.) was supported by Grants of Ministry of Education, Japan.

REFERENCES

1. C. W. Helstrom, *Quantum Detection and Estimation Theory* (Academic Press, NY, 1976).
2. A. S. Holevo, J. Multivar. Anal., **3**, 337, (1973).
3. A. S. Holevo, Prob. Peredachi Inf., **15**, 3, (1979), Engl. Transl., Probl. Inf. Trnsm. (USSR), **15**, 247, (1979).
4. P. Hausladen, R. Jozsa, B. Schumacher, M. Westmoreland, and W. K. Wootters, Phys. Rev. A **54**, 1869, (1996).
5. A. S. Holevo, IEEE, Trans. Inform. Theory, **IT-44**, 269, (1998).
6. B. Schumacher, and M. D. Westmoreland, Phys. Rev. A **56**, 131, (1997).
7. M. Sasaki, K. Kato, M. Izutsu, and O. Hirota, Phys. Rev. A **58**, 146, (1998).
8. A. S. Holevo, private communication.
9. E. B. Davies, IEEE Trans. Inform. Theory, **IT-24**, 596, (1978).
10. P. Hausladen, and W. K. Wootters, J. Mod. Opt., **41**, 2385, (1994).
11. M. Ban, K. Kurokawa, R. Momose, and O. Hirota, Int. J. Theor. Phys., **36**, 1269, (1997).
12. M. Osaki, M. Ban, and O. Hirota, J. Mod. Opt., **45**, 269, (1998).
13. L. B. Levitin, *Quantum Communications and Measurement*, Ed. by V. P. Belavkin *et al.*, pp. 439-448, (Plenum Press, NY, 1995).
14. C. A. Fuchs, and C. M. Caves, Phys. Rev. Lett. **73**, 3047, (1994).
15. A. S. Holevo, Prob. Peredachi Inf., **9**, 31, (1973), Engl. Transl., Probl. Inf. Trnsm. (USSR), **9**, 110, (1973).
16. A. Peres, and W. K. Wootters, Phys. Rev. Lett. **66**, 1119, (1991).
17. M. Sasaki, S. M. Barnett, and O. Hirota, In preparation, (1998).
18. M. Osaki, T. S. Usuda, and O. Hirota, Phys. Lett. A, to appear, (1998).
19. C. Fuchs, and A. Peres, Phys. Rev. A **53**, 2038, (1996).
20. M. Ban, M. Osaki, and O. Hirota, Phys. Rev. A **54**, 2718, (1996).

ASPECTS OF QUANTUM INFORMATION COMPRESSION FOR PURE STATES

Richard Jozsa*

School of Mathematics and Statistics
University of Plymouth
Plymouth, Devon PL4 8AA
England
Email: rjozsa@plymouth.ac.uk

Consider a source of pure quantum states with density matrix ρ and von Neumann entropy S. It is known that the quantum information of the source may be faithfully compressed to S qubits/signal (asymptotically for long sequences) and that this limit is optimal if we demand high fidelity for long strings of signals. The usual method for achieving this compression, based on the typical subspace of the source, requires knowledge of the density matrix. We describe a new "universal" compression scheme which does not even require knowledge of ρ. More precisely, suppose that the source is known to have von Neumann entropy less than some given bound S_0 but is otherwise completely unspecified. Our method will faithfully compress the quantum information of any such source to S_0 qubits/signal.

It is known that S is always less than or equal to the Shannon entropy H of the prior distribution of the source. According to Shannon's theorem, H provides the limit for compression of classical signals, so one may ask the question: what is the physical origin of the extra compression available in the quantum case compared to the classical case (where the prior probabilities are the same)? One may suggest that this effect is a consequence of possible *non-orthogonality* of signal states. However we demonstrate that it is possible to make each pair of signal states of a source *more parallel* (keeping the prior probabilities the same) while *increasing* the von Neumann entropy. This casts doubt on the conventional wisdom that overlap of quantum states is a measure of their (non-)distinguishability.

INTRODUCTION

The problem of quantum information compression that we will be considering is the following. Let $\mathcal{H}_d$ denote a Hilbert space of d dimensions. Alice has a source Σ which produces an (unending) sequence of d dimensional pure quantum states. Each state

*The results reported here were obtained in collaboration with M. Horodecki, P. Horodecki, R. Horodecki[6] and with J. Schlienz[9].

Quantum Communication, Computing, and Measurement 2
Edited by Kumar et al., Kluwer Academic / Plenum Publishers, New York, 2000.

is drawn independently from the distribution $|\psi_1\rangle, \ldots, |\psi_n\rangle \in \mathcal{H}_d$ with probabilities $p_1, \ldots, p_n$ respectively. She wishes to store the states as economically as possible – using the least possible number of Hilbert space dimensions per state – so that later another user (Bob) can reconstruct the state sequence from the stored version with arbitrarily high fidelity. In an equivalent problem Alice may wish to transmit the sequence of states to Bob through a noiseless quantum channel as cheaply as possible, and the cost of using the channel is measured by the number of qubits transmitted. Thus she wishes to compress the quantum information of the state sequence as much as possible.

Alice might just store the states themselves using d dimensions, i.e. $\log_2 d$ qubits, per state. However it is known that this solution is not optimal and by applying suitable block-coding schemes she may code the states more economically. The complete solution of this problem, given by Schumacher et al.[3, 4, 5] is tantamount to a quantum analogue of Shannon's source coding theorem[1, 2]. Here we will summarise the essential features for our later purposes.

Let $\rho = \sum_i p_i |\psi_i\rangle \langle\psi_i|$ be the density matrix of the source and let $S(\rho) = -tr \rho \log_2 \rho$ be its von Neumann entropy. We may distinguish two situations for Alice: she may or may not know the identities of the individual states in the sequence. If she does not know them, then she will be unable to reliably gain this (classical) information (for non-orthogonal $|\psi_i\rangle$'s) by examining the states in any way. However it turns out, remarkably, that the optimal possible compression is not improved by knowledge of the state identities – it depends only on knowledge of the density matrix of the source and it remains optimal whether or not she knows the state identities. (This feature is not expected to remain true for sources generating *mixed* states). In this paper, for definiteness we will consider the situation in which Alice knows the identities of the individual states. Hence for her coding scheme she may construct any (mixed) state $\rho_{i_1 \ldots i_K}$ whatever, of her choosing, to represent the quantum information of each possible K-sequence $|\psi_{i_1}\rangle \ldots |\psi_{i_K}\rangle$. Bob on the other hand knows only the statistical structure of the source (and also Alice's coding procedure). His decoding procedure is restricted to any possible physical evolution i.e. any CP map, on the state stored by Alice. If Bob decodes $\rho_{i_1 \ldots i_K}$ to a state $\sigma_{i_1 \ldots i_K}$ on $(\mathcal{H}_d)^K$ then the average block fidelity is defined by

$$\overline{F} = \sum_{i_1 \ldots i_K} p_{i_1} \ldots p_{i_K} \langle\psi_{i_K}| \ldots \langle\psi_{i_1}| \sigma_{i_1 \ldots i_K} |\psi_{i_1}\rangle \ldots |\psi_{i_K}\rangle$$

(noting that $p_{i_1} \ldots p_{i_K}$ is the probability of occurrence of the K-sequence $|\psi_{i_1}\rangle \ldots |\psi_{i_K}\rangle$).

The fundamental theorem of quantum information compression may then be stated as follows (c.f. Schumacher et al.[3, 4, 5] for a proof and further discussion).

Theorem 1: For any $\epsilon > 0$ and $\delta > 0$ (however small)

(a) If $S(\rho) + \delta$ qubits per state are available then for all sufficiently large K there is a coding/decoding scheme which has average block fidelity $> 1 - \epsilon$.

(b) If only $S(\rho) - \delta$ qubits per state are available then for all sufficiently large K, any coding/decoding scheme will have block fidelity $< \epsilon$.

Thus the von Neumann entropy of the source provides the limit for high fidelity compression. Furthermore we may describe an explicit coding/decoding scheme which achieves (a). It is based on the classical theory of typical sequences[2] associated with a probability distribution $\mathcal{R} = \{r_1, \ldots, r_m\}$ on m symbols $a_1, \ldots, a_m$. Consider sampling the distribution independently K times, resulting in outcome sequence $a_{i_1} \ldots a_{i_K}$ with probability $r_{i_1 \ldots i_K} = r_{i_1} \ldots r_{i_K}$. Let $SEQ(K)$ be the set of all such sequences of length K and let $H = -\sum_i r_i \log_2 r_i$. Then we have the fundamental result[2]: For any given $\epsilon > 0$ and $\delta > 0$ and for all sufficiently large K there is a subset $TYP(K) \subseteq SEQ(K)$

which has size $2^{K(H+\delta)}$ (i.e. an exponentially small fraction of $SEQ(K)$) but whose total probability exceeds $1-\epsilon$ (i.e. is as high as desired). The sequences in $TYP(K)$ are called *typical* sequences and those not in $TYP(K)$ are called *atypical* sequences. Intuitively this theorem asserts that (for all sufficiently large K) any sequence generated by the sampling may be assumed with arbitrarily high probability, to be a typical sequence. The set $TYP(K)$ may be explicitly described as follows[2]: a sequence $a_{i_1}\dots a_{i_K}$ is typical if and only if its probability satisfies

$$2^{-K(H+\epsilon)} \leq r_{i_1\dots i_K} \leq 2^{-K(H-\epsilon)}$$

i.e. each of the (approximately) 2^{KH} typical sequences is approximately equally likely with probability about $1/2^{KH}$.

Now for our quantum information compression protocol the essential insight is to apply this result *not* to the prior probabilities $\{p_1,\dots,p_n\}$ but to $\{\lambda_1,\dots,\lambda_d\}$, the probability distribution of eigenvalues of ρ. Note that the density matrix for state sequences of length K is the K-fold tensor product $\rho^K = \rho\otimes\dots\otimes\rho$ and the eigenvalues of ρ^K are just the K-fold products $\lambda_{i_1}\dots\lambda_{i_K}$. Given any $\epsilon > 0$ and $\delta > 0$ let $\Lambda(K)$ be the subspace of $(\mathcal{H}_d)^K$ given by the span of all eigenstates $|\lambda_{i_1}\dots\lambda_{i_K}\rangle$ corresponding to all *typical* sequences $i_1\dots i_K$ of eigenvalues. $\Lambda(K)$ is called the *typical subspace* (for block length K). Since the Shannon entropy of $\{\lambda_1,\dots,\lambda_d\}$ is just the von Neumann entropy $S(\rho)$ we see that $dim\,\Lambda(K) = 2^{K(S(\rho)+\delta)}$ i.e. the typical subspace occupies about $KS(\rho)$ qubits. Let Π denote projection onto $\Lambda(K)$. By considering ρ^K in its eigenbasis we see that the properties of typical sequences give

$$tr\rho^K\Pi > 1-\epsilon \tag{1}$$

Thus Alice accumulates blocks of K states and performs a measurement to see whether their joint state lies in the typical subspace or not. If the state projects to $\Lambda(K)$ then she stores the resulting $K(S(\rho)+\delta)$ qubits. If it projects to the orthogonal complement (which occurs with low probability $< \epsilon$) then she stores any chosen state of $K(S(\rho)+\delta)$ qubits. Then it follows[4] from eq. (1) that the average block fidelity between the original sequence and the stored state, will exceed $1-2\epsilon$.

UNIVERSAL COMPRESSION OF QUANTUM INFORMATION

The optimal compression scheme outlined above requires knowledge of the density matrix of the source (in order to locate the typical subspace). But suppose that we do not have complete knowledge of the density matrix of the source. For example we may know only the expectation values $tr\rho A$ of some observables A which are insufficient to reconstruct ρ. How much faithful compression of the quantum information is possible in this case? This question was first raised by Horodecki et al.[8] and solved there for the case of qubit sources. Let S_0 be the maximum possible von Neumann entropy of any density matrix consistent with the given data about ρ. Then theorem 1(b) implies that faithful compression beyond S_0 qubits per state will be impossible so that S_0 is certainly a lower bound. Horodecki et al.[8] showed that this is indeed the *greatest* lower bound for qubit sources but the case of higher dimensions appeared intractible with the methods used there, which involved explicit analysis of the constraints imposed on ρ. Shortly thereafter Jozsa et al.[6] showed that a stronger result could be obtained more easily! Suppose that all we know about ρ (in general dimension d) is that its entropy $S(\rho)$ is less than some bound S_0 (and we do not even know the constraints that may have led to this bound). Then the quantum information may always be compressed (asymptotically) to S_0 qubits per state i.e. there is an optimal compression scheme for which the particular form or structure of any constraints on ρ (apart from imposing the bound S_0) are irrelevant! Below we will give a proof of this result written in more elementary terms than that given in Jozsa et al.[6] and also outline the basic classical information theoretic ingredient – the theory of types[7, 2] of Csiszar and Körner. This will lead to a new subset of K-sequences which play a role analogous to the typical sequences above.

We begin by considering again the set $SEQ(K)$ of all sequences $\underline{a} = a_{i_1} \dots a_{i_K}$ of length K generated by K independent samplings of a probability distribution $\mathcal{R} = \{r_1, \dots, r_m\}$ on m symbols $a_1, \dots, a_m$. In each sequence $\underline{a}$, each symbol a_i occurs some number of times, denoted $n(a_i)$. Let $p_{\underline{a}}(a_i) = n(a_i)/K$. Then the probability distribution $P(\underline{a}) = \{p_{\underline{a}}(a_1), \dots, p_{\underline{a}}(a_m)\}$ is called the *type* of $\underline{a}$. We will need three basic properties of types (c.f. §12.3 in Cover and Thomas[2] for a detailed discussion).

Our first property states that the number of different possible types for sequences of length K is $\leq (K+1)^m$. Indeed each of the m symbols a_i may appear 0 or 1 or ... or K times giving at most $(K+1)^m$ possibilities for the occurrences. Note that there are extra conditions e.g. that the total number of occurrences must be K, which we have not used. Thus for fixed m, the number of types grows only polynomially with the length K of the sequence.

Our second property concerns the number of sequences that have a given type $P = \{p_1, \dots, p_m\}$. Let $H(P)$ be the Shannon entropy of the type P and let $n(P)$ be the number of sequences (of length K) with type P. Then property 2 states that

$$n(P) \leq 2^{KH(P)}$$

This may be seen for example, by applying Stirlings formula to the multinomial coefficient

$$n(P) = \begin{pmatrix} K \\ Kp_1 \ Kp_2 \ \dots \ Kp_m \end{pmatrix}$$

The details may be found in the reference above.

Let $T(P) \subseteq SEQ(K)$ denote the set of all K-sequences of type P. Our third property gives a bound for the total probability $prob(T(P))$ of $T(P)$ (i.e. the sum of the probabilities of all the sequences in $T(P)$):

$$prob(T(P)) \leq 2^{-KD(P||\mathcal{R})} \tag{2}$$

Here $D(P||\mathcal{R})$ is the relative entropy (or Kullback-Leibler distance) between the type P and the starting distribution $\mathcal{R}$:

$$D(P||\mathcal{R}) = -\sum_i p_i \log_2 \frac{p_i}{r_i}$$

We omit the proof, which may be found in the reference above.

Now let $H_0 < \log_2 m$ be any prescribed bound and consider the set $A(K) \subseteq SEQ(K)$ of all K-sequences whose type has entropy $\leq H_0$ i.e.

$$A(K) = \{\underline{a} \in SEQ(K) : H(P(\underline{a})) \leq H_0\} \tag{3}$$

Note that the definition of $A(K)$ is completely independent of the prior distribution of the symbols in the sequences (i.e. of $\mathcal{R}$ above). We will show[2, 7] that the set $A(K)$ satisfies the essential properties enjoyed by $TYP(K)$, not just for some one probability distribution with entropy H_0 but simultaneously for all distributions with entropy less than H_0!

Theorem 2: Let $\epsilon > 0$ and $\delta > 0$ be given. Then for all sufficiently large K
(a)The size of $A(K)$ is $\leq 2^{K(H_0+\delta)}$.
(b) If $\mathcal{R}$ is *any* probability distribution for the symbols $a_1, \ldots, a_m$ and $H(\mathcal{R}) < H_0$ then the total probability of $A(K)$ is $\geq 1 - \epsilon$.
Proof of Theorem 2: (a) follows from the first and second properties of types. The total number of types is $\leq (K+1)^m$ and the number of sequences of type P is $\leq 2^{KH(P)}$. Hence the size of $A(K)$ is bounded above by

$$(K+1)^m 2^{KH_0} = 2^{K(H_0 + m\frac{\log(K+1)}{K})}$$

Then choose K large enough that $m\frac{\log(K+1)}{K} < \delta$ to get (a).
(b) follows from the first and third properties of types. We estimate the total probability of the complement $B = SEQ(K) - A(K)$. By definition of $A(K)$, B consists of all those sequences whose type P has entropy exceeding H_0. Thus property 3 of types states that $prob(B)$ is less than the sum of $2^{-KD(P||\mathcal{R})}$ over all such types P, and using property 1 we get

$$prob(B) \leq (K+1)^m \max_{P:H(P)>H_0} 2^{-KD(P||\mathcal{R})} = (K+1)^m 2^{-K \min_{P:H(P)>H_0} D(P||\mathcal{R})}$$

Now since $H(\mathcal{R}) < H_0$ i.e. $H(\mathcal{R}) = H_0 - \eta$ for some fixed $\eta > 0$, and $H(P) > H_0$ it follows that $\min_{P:H(P)>H_0} D(P||\mathcal{R}) > 0$ (as the relative entropy is 0 iff the two distributions coincide). Hence for all sufficiently large K we get $prob(B) < \epsilon$ for any prescribed $\epsilon > 0$.

Theorem 2 is the basis of a classical universal compression scheme[7, 2]. Consider m symbols $a_1, \ldots, a_m$ and let $\mathcal{S}$ be any source of these symbols, with entropy $H(\mathcal{S}) < H_0$. Let $A(K)$ be defined as above. Then theorem 2 states that for all sufficiently large K we may assume (with arbitrarily high probability $1 - \epsilon$) that any K-sequence generated by any such source, is in $A(K)$. Furthermore there are only about 2^{KH_0} such sequences. Thus we have compressed the information of the source to $\frac{1}{K}\log 2^{KH_0}$ bits per symbol, i.e. H_0 bits per symbol. We emphasise again here that the definition of $A(K)$ is independent of the details of the distribution of $\mathcal{S}$. It is in this sense that the compression protocol is *universal* i.e. applying simultaneously to all sources with entropy $< H_0$.

Let us now return to *quantum* information compression. If we look again at the optimal quantum information compression scheme described previously (based on $TYP(K)$ and the typical subspace) and replace the role of $TYP(K)$ by $A(K)$ then theorem 2 immediately gives the following:
Let Σ_0 be any quantum source with density matrix ρ_0 and von Neumann entropy $S_0 = S(\rho_0)$. Then ther exists a quantum information compression scheme which will asymptotically faithfully compress to S_0 qubits per state, any quantum source Σ whose density matrix ρ *commutes with* ρ_0 and has $S(\rho) < S_0$. The method involves, as before, projecting K-blocks of states into a subspace (of size $K(S_0+\delta)$ qubits), but now defined as the span of all those K-sequences of eigenstates of ρ_0 which correspond to members of $A(K)$ (rather than $TYP(K)$).

To achieve our final result – a *fully* universal quantum information compression scheme – we now show how to remove the condition that ρ commutes with ρ_0. Recall that d denotes the dimension of the state space of our source. Consider the set $A(K)$ as defined in eq. (3) for a fixed constant $H_0 = S_0$ and for K-sequences of d symbols $a_1, \ldots, a_d$. For any orthonormal basis $B = \{|a_1\rangle, \ldots, |a_d\rangle\}$ of $\mathcal{H}_d$ let $\Upsilon(B)$ be the subspace of $(\mathcal{H}_d)^K$ spanned by all K-sequences $|a_{i_1}\rangle |a_{i_2}\rangle \ldots |a_{i_k}\rangle$ of basis states such that $a_{i_1}a_{i_2}\ldots a_{i_K} \in A(K)$. According to theorem 2(b), for all sufficiently large K, given any $\delta > 0$, $\Upsilon(B)$ has dimension less than $2^{K(S_0+\delta)}$. Now given any density matrix ρ we may associate to it $\Upsilon(B_\rho)$ where B_ρ is the basis of eigenvectors of ρ. Conversely given any orthonormal basis B there is a ρ with $B = B_\rho$.

Theorem 3: Let Υ be the smallest subspace of $(\mathcal{H}_d)^K$ which contains $\Upsilon(B)$ for all choices of orthonormal basis B. Then $dim\, \Upsilon \leq (K+1)^{d^2} 2^{K(S_0+\delta)}$.

This theorem finally provides our universal quantum information compression scheme for all quantum sources with von Neumann entropy less than S_0. We noted above that projection into $\Upsilon(B)$ will faithfully compress all sources, with von Neumann entropy less than S_0, whose density matrices are diagonal in the basis B. Thus projection into Υ will simultaneously faithfully compress all sources with von Neumann entropy less than S_0. The compression factor is given by

$$\frac{\log dim\, \Upsilon}{K} = \frac{d^2 \log(K+1) + K(S_0+\delta)}{K} \tag{4}$$

which tends to $S_0 + \delta$ qubits per state (for any $\delta > 0$).

Proof of Theorem 3: The basic idea is that although the $\Upsilon(B)$'s are rather large spaces lying in different directions inside $(\mathcal{H}_d)^K$, their total span is limited by the fact that the variety of different $\Upsilon(B)$'s is labelled by the choice of basis in $\mathcal{H}_d$ (and not $(\mathcal{H}_d)^K$!) i.e. by a d-dimensional unitary transformation and this can be parameterised by less than d^2 parameters, independent of K. For clarity we will give the argument for $d = 2$. The generalisation to $d > 2$ is immediate.

Let $B_0 = \{|0\rangle, |1\rangle\}$ be any fixed chosen basis and let $B = \{|0'\rangle, |1'\rangle\}$ be any other basis. Then we may write

$$|0'\rangle = a\,|0\rangle + b\,|1\rangle \qquad |1'\rangle = c\,|0\rangle + d\,|1\rangle$$

where a, b, c, d form a unitary matrix. For B_0 consider each possible sequence of basis states given by $A(K)$, defining the basis states for $\Upsilon(B_0)$. Suppose for example that one such sequence is

$$|0\rangle\,|1\rangle\,|0\rangle \ldots |1\rangle \tag{5}$$

Then for $\Upsilon(B)$ we will have the corresponding basis state

$$|0'\rangle\,|1'\rangle\,|0'\rangle \ldots |1'\rangle = (a\,|0\rangle + b\,|1\rangle)(c\,|0\rangle + d\,|1\rangle)(a\,|0\rangle + b\,|1\rangle) \ldots (c\,|0\rangle + d\,|1\rangle) \tag{6}$$

and indeed $\Upsilon(B)$ is spanned by all such states as eq. (5) ranges over all sequences in $A(K)$. Varying the coefficients a, b, c, d corresponds to varying the choice of B in $\Upsilon(B)$. Now consider multiplying out eq. (6) with a, b, c, d as algebraic parameters, and collect together all terms which have the same coefficient as a function of a, b, c, d. These coefficients will all have the form $a^{n_1} b^{n_2} c^{n_3} d^{n_4}$ where each of the powers is 0 or 1 or ... or K. Hence there are at most $(K+1)^4$ such possibilities. Thus we obtain an expression of the form

$$\sum a^{n_1} b^{n_2} c^{n_3} d^{n_4}\, |\psi_{n_1 n_2 n_3 n_4}\rangle \tag{7}$$

where each ket $|\psi_{n_1 n_2 n_3 n_4}\rangle$ is independent of a, b, c, d, being just the sum of various K-sequences of $|0\rangle$'s and $|1\rangle$'s. Now each of the $2^{K(S_0+\delta)}$ states in eq. (5) generates at most $(K+1)^4$ states $|\psi_{n_1 n_2 n_3 n_4}\rangle$ and as we vary over all possible B's by varying the parameters a, b, c, d we remain within the span of these $(K+1)^4 2^{K(S_0+\delta)}$ states. Hence for $d = 2$ we get $dim\, \Upsilon \leq (K+1)^4 2^{K(S_0+\delta)}$. For general $d > 2$ we have d^2 parameters $a, b, \ldots$ leading just to a factor of $(K+1)^{d^2}$ in this expression.

Note that one may obtain[6] tighter estimates on $dim\, \Upsilon$ by considering the further constraint $n_1 + n_2 + \ldots = K$ on the sum of powers but this is unnecessary for our purposes. Indeed it is sufficient here for $dim\, \Upsilon$ to be bounded by any *polynomial* multiple of $2^{K(S_0+\delta)}$ (c.f. eq. (4)).

SOME REMARKS ON THE INTERPRETATION OF QUANTUM INFORMATION COMPRESSION

Why is information compression possible at all? How can we heuristically explain its underlying mechanism?

For classical information, according to Shannon's source coding theorem[1, 2] compression is always possible if the prior probability distribution of the signals is not uniform i.e. has non-maximal entropy. The entropy of the source then gives the best possible compression factor. For example suppose that Alice's source emits 0's and 1's with respective probabilities 0.8 and 0.2. Then Shannon's theorem states that this classical information may be compressed from 1 bit per raw signal to $H(0.8, 0.2) = 0.72$ bits per signal and no further. We may say that even before Bob receives anything from Alice, he may guess that a symbol will be 0 and be correct more often than not. In this sense Bob has some prior (albeit probabilistic) information about the identity of the signal. Hence if Alice simply sends the signals themselves she is sending some redundant information which might have been "compressed out" first.

In the case of quantum information the optimal compression factor is given by $S(\rho)$ the von Neumann entropy of the source, rather than $H(p_1, \ldots, p_n)$, the Shannon entropy of the prior distribution of quantum states. Now it is well known that $S(\rho) \leq H(p_1, \ldots, p_n)$ (with equality iff the states are mutually orthogonal) so that in the quantum scenario, further compression is possible beyond the classically expected limit of $H(p_1, \ldots, p_n)$ (if we regard the quantum states $|\psi_i\rangle$ as analogous to the classical signals). For example if we have two states $|0\rangle$ and $\frac{1}{\sqrt{2}}(|0\rangle + |1\rangle)$ with equal prior probabilities of a half then $H(p_1, p_2) = 1$ while $S(\rho) = 0.601$. What is the origin of the extra quantum compression?

The fact that $S(\rho) = H(p_1, \ldots, p_n)$ if the states are mutually orthogonal suggests that *non-orthogonality* is the key to the possibility of extra quantum compression. Indeed suppose that $|\phi\rangle$ and $|\chi\rangle$ are two non-orthogonal states of a quantum system. We may decompose $|\phi\rangle$ as a superposition of components parallel and perpendicular to $|\chi\rangle$

$$|\phi\rangle = a\,|\chi\rangle + b\,\left|\chi^{\perp}\right\rangle \tag{8}$$

where $\left\langle \chi^{\perp} | \chi \right\rangle = 0$ and $a = \langle \chi | \phi \rangle$. Since any time evolution in quantum mechanics is unitary and linear the evolutions of $|\chi\rangle$ and $\left|\chi^{\perp}\right\rangle$ will remain orthogonal and the decomposition in (8) is preserved. We might interpret this as saying that the overlap $|\langle \chi | \phi \rangle|$ measures the extent to which the state $|\phi\rangle$ behaves as though it were actually *equal* to the state $|\chi\rangle$. This view is further formalised in the many worlds interpretation of quantum theory according to which (8) may be thought of as a "splitting" into two "worlds". In one of these worlds the state $|\chi\rangle$ is indeed actually precisely *equal* to $|\phi\rangle$.

Thus in the compression of quantum states we may intuitively expect to be able to compress out a common overlapping part from a collection of non-orthogonal signals, with increasing compression possible as the states become more parallel. Upon reception, to reconstitute the states, Bob "re-instates" the common quantum part which is the same for each signal. For the case of two signal states $|\psi_1\rangle$ and $|\psi_2\rangle$ with fixed prior probabilities p_1 and p_2 say, the compression factor $S(\rho)$ does indeed vary monotonically with the overlap $|\langle \psi_1 | \psi_2 \rangle|$ from its maximum $H(p_1, p_2)$ for orthogonal states to 0 for parallel states. However, somewhat enigmatically, for collections of three or more quantum states the above heuristic explanation fails completely!

It is possible for signal states in a family to become pairwise more parallel (keeping the prior probabilities fixed) yet have the von Neumann entropy *increase*! This casts some doubt on the interpretation of the overlap of quantum states given above (in the context of more than two states) and leaves open the question of the underlying mechanism of quantum information compression. As an explicit example of this enigmatic phenomenon consider the following three states taken with equal prior probabilities of one third each:

$$
\begin{aligned}
|\psi_1\rangle &= |0\rangle \\
|\psi_2\rangle &= \tfrac{1}{\sqrt{2}}(|0\rangle + |1\rangle) \\
|\psi_3\rangle &= \tfrac{\sqrt{2}-1}{\sqrt{6}}\,|0\rangle + \tfrac{\sqrt{2}+1}{\sqrt{6}}\,|1\rangle
\end{aligned}
$$

and the following three states taken with the same prior probabilities:

$$
\begin{aligned}
|\psi'_1\rangle &= |0\rangle \\
|\psi'_2\rangle &= \tfrac{1}{\sqrt{2}}(|0\rangle + |1\rangle) \\
|\psi'_3\rangle &= \tfrac{2}{3\sqrt{3}}\,|0\rangle + \tfrac{4}{3\sqrt{3}}\,|1\rangle + \tfrac{\sqrt{7}}{3\sqrt{3}}\,|2\rangle
\end{aligned}
$$

Then it may be readily verified that $|\langle\psi_i|\psi_j\rangle| \leq |\langle\psi'_i|\psi'_j\rangle|$ for all i and j i.e. the primed collection is pairwise more parallel than the corresponding states of the unprimed collection, yet $S(\rho') = 0.92$ and $S(\rho) = 0.85$ i.e. $S(\rho') > S(\rho)$, the pairwise-more-parallel collection has higher von Neumann entropy!

The overlap $|\langle\phi|\chi\rangle|$ of non-orthogonal states is also a fundamental ingredient in the quantification of the (non-)distinguishability of quantum states. Although $|\phi\rangle$ and $|\chi\rangle$ are distinct states in the mathematical formalism of quantum theory, there is no physical process that can distinguish them with certainty[10] and indeed the overlap provides a quantitative measure of the extent to which the states cannot be distinguished. This is in contrast to distinct states in classical physics which are always perfectly distinguishable in principle. The above example now appears even more paradoxical if we recall that the von Neumann entropy $S(\rho)$ also characterises the maximum classical information capacity of an ensemble of pure quantum states[11]. Thus in our example, by making the signal states *pairwise less* distinguishable (i.e. more parallel) we are able to *increase* the classical information capacity! This counterintuitive effect shows that the concept of distinguishability within a family of quantum states is a global property of the whole family and cannot be reduced to separately examining the distinguishability of each possible constituent *pair* of states! A further investigation and discussion of these issues is given in Jozsa and Schlienz[9]. In the present paper we just wish to point out that the issue of understanding the basis of quantum information compression remains quite unresolved, even in the relatively simple case of three pure states.

Acknowledgements

This research was supported in part by the European TMR Network ERB–FMRX–CT96–0087.

REFERENCES

1. C. E. Shannon, *Bell Syst. Tech. J.* 27:379 (1948).
2. T. M. Cover and J. A. Thomas, "Elements of Information Theory," John Wiley and Sons, New York (1991).
3. B. Schumacher, *Phys. Rev. A* 51:2378 (1995).
4. R. Jozsa and B. Schumacher, *J. Mod. Opt.* 41:2343 (1994).
5. H. Barnum, Ch. Fuchs, R. Jozsa and B. Schumacher, *Phys. Rev. A* 54:4707 (1996).
6. R. Jozsa, M. Horodecki, P. Horodecki and R. Horodecki, *Phys. Rev. Lett.* 81:1714 (1998).

7. I. Csiszar and J. Körner, "Coding Theory for Discrete Memoryless Systems," Academic Press, New York (1981).
8. M. Horodecki, R. Horodecki and P. Horodecki, *Acta Phys. Slovaca* 48:133 (1998).
9. R. Jozsa and J. Schlienz, "Distinguishability of States and von Neumann Entropy," preprint in preparation (1998).
10. A. Peres, *Phys. Lett. A* 128:19 (1988).
11. P. Hausladen, R. Jozsa, B. Schumacher, M. Westmoreland and W. Wootters, *Phys. Rev. A* 54:1869 (1996).

GROUP COVARIANT SIGNALS IN QUANTUM INFORMATION THEORY

Tsuyoshi Sasaki Usuda[1] and Ichi Takumi[1]

[1]Dept. A. I. and Computer Science
Nagoya Institute of Technology
Gokiso-cho, Showa-ku, Nagoya, 466-8555, JAPAN

Necessary and sufficient condition for group covariant signals is given. This condition shows what kind of signals correspond to "group covariant signals." We show by applying this condition that many signals, such as arbitrary binary signals, PPM, FSK, and PSK coherent-state signals, equidistant codes, and binary linear codes, are all group covariant.

INTRODUCTION

Group covariant signals[1] was defined by Davies when he consider the maximization problem of the mutual information of the quantum channel for a classical information. If signals in a quantum communication system are group covariant, we have advantages[1,2] in computing or calculating several quantities in quantum information theory, such as the minimum error probability and the maximum mutual information. Davies showed in Ref. 1 that one can easily compute the maximum mutual information when signals are group covariant and have an irreducible representation. Ban *et al.* showed in Ref. 2 that the minimum error probability is achieved by so-called square-root measurement[3,4] when signals are group covariant with respect to a cyclic group. However, for given signals, it is not trivial whether they are group covariant or not. To consider what kind of signals correspond to "group covariant signals" is helpful, together with studying what advantages we have gotten when signals are group covariant.[1,2]

In this paper, we will give a necessary and sufficient condition of group covariant signals in order to check group covariancy of given signals and show some examples.

GROUP COVARIANT SIGNALS

In this section, we present a condition that group covariant signals must be satisfied. Here we assume that the quantum state signals are pure and use the following definition of group covariant signals.

Definition 1 (Group covariant signals) : Let G be a group with an operation (or a multiplication) $\circ$ and be a set of parameters characterizing signals $\{|\psi_i\rangle | i \in G\}$. The

Quantum Communication, Computing, and Measurement 2
Edited by Kumar et al., Kluwer Academic / Plenum Publishers, New York, 2000.

set of signals is called *group covariant* if there exist unitary operators $\hat{U}_k (k \in G)$ such that

$$\hat{U}_k |\psi_i\rangle = |\psi_{k \circ i}\rangle, \ \forall i, k \in G. \tag{1}$$

The M-ary signals called 'symmetric states' in Ref. 2 are examples of group covariant signals in which the operation is defined as addition modulo M.

We give the following proposition in order that one can easily tell whether given signals are group covariant or not.

Proposition 1 : *M-ary signals $\{|\psi_i\rangle | i = 0, 1, \cdots, M-1\}$ are group covariant with respect to a group $(G; \circ)$ of order M if and only if*

$$\forall i, j, k \in G, \ \ \langle \psi_{k \circ i} | \psi_{k \circ j} \rangle = \langle \psi_i | \psi_j \rangle, \tag{2}$$

where $G = \{0, 1, \cdots, M-1\}$ and $\circ$ is the operation of the group G.

Proof of Proposition 1 : First, the necessity directly follows from the definition of group covariant signals.

To show the sufficiency, let

$$\forall i, j, k \in G, \ \ \langle \psi_{k \circ i} | \psi_{k \circ j} \rangle = \langle \psi_i | \psi_j \rangle, \tag{3}$$

and define bounded linear operators $\hat{U}_k (k \in G)$ in the signal space $\mathcal{H}_S = \{|\psi\rangle = \sum_{i=0}^{M-1} c_i |\psi_i\rangle | c_i : \text{complex number}\}$ spanned by the signals as

$$\forall |\psi\rangle = \sum_{i=0}^{M-1} c_i |\psi_i\rangle \in \mathcal{H}_S, \quad \hat{U}_k |\psi\rangle \equiv \sum_{i=0}^{M-1} c_i |\psi_{k \circ i}\rangle. \tag{4}$$

In order to say the signals $\{|\psi_i\rangle\}$ to be group covariant, it is sufficient to show the above $\hat{U}_k (k \in G)$ are unitary operators on $\mathcal{H}_S$. For any $|\psi\rangle = \sum_{i=0}^{M-1} c_i |\psi_i\rangle, |\psi'\rangle = \sum_{i=0}^{M-1} c_i' |\psi_i\rangle \in \mathcal{H}_S$,

$$\begin{aligned} \langle \psi | \hat{U}_k^\dagger \hat{U}_k | \psi' \rangle &= \left(\sum_{i=0}^{M-1} c_i^* \langle \psi_i | \hat{U}_k^\dagger \right) \left(\sum_{i=0}^{M-1} c_i' \hat{U}_k |\psi_i\rangle \right) = \sum_{i=0}^{M-1} \sum_{j=0}^{M-1} c_i^* c_j' \langle \psi_{k \circ i} | \psi_{k \circ j} \rangle \\ &= \sum_{i=0}^{M-1} \sum_{j=0}^{M-1} c_i^* c_j' \langle \psi_i | \psi_j \rangle = \langle \psi | \psi' \rangle. \end{aligned} \tag{5}$$

Hence,

$$\hat{U}_k^\dagger \hat{U}_k = \hat{\mathrm{I}}, \tag{6}$$

where $\hat{\mathrm{I}}$ is the identity operator on $\mathcal{H}_S$. Similarly, $\hat{U}_k \hat{U}_k^\dagger = \hat{\mathrm{I}}$. Thus, $\hat{U}_k (k \in G)$ are unitary and the signals $\{|\psi_i\rangle\}$ are group covariant. □

Applying Proposition 1, we can tell group covariancy of given signals by merely checking the relation of inner products as expressed in Eq. (2). In the next section, we will show that there are many group covariant signals in classically modulated or coded signals.

EXAMPLES OF GROUP COVARIANT SIGNALS

Characteristics of the Gram Matrices

For M-ary signals $\{|\psi_i\rangle | i = 0, 1, \cdots, M-1\}$, the Gram matrix is defined as $\Gamma_M = [\langle \psi_i | \psi_j \rangle]$. If the Gram matrix of signals is given, we can imagine corresponding signals and can define so-called square-root measurement[3,4] which is known as asymptotically optimal or pretty good measurement. Here we consider the characteristics of the Gram matrices for group covariant signals. First, we have the following remark for M-ary signals.

Remark 1 : *If all of the non-diagonal components of the Gram matrix of signals are equal, that is,*

$$\Gamma_M = \begin{bmatrix} 1 & a & \dots & a \\ a & 1 & \dots & a \\ \vdots & \vdots & \ddots & \vdots \\ a & a & \dots & 1 \end{bmatrix}, \tag{7}$$

the signals are group covariant.

Remark 1 is trivial from Proposition 1. As signals having the Gram matrix in Eq. (7), there are M-ary (classically) orthogonal signals,[5] e.g., M-ary pulse-position-modulated (PPM) and frequency-shift-keyed (FSK) coherent-state signals, and equidistant codes, e.g., orthogonal and simplex codes.

For binary and 3-ary signals, we have the following remarks.

Remark 2 : *The Gram matrix of binary group covariant signals has the form*

$$\Gamma_2 = \begin{bmatrix} 1 & a \\ a & 1 \end{bmatrix}. \tag{8}$$

Conversely, binary signals with the Gram matrix in Eq. (8) are group covariant.

Remark 3 : *The Gram matrix of 3-ary group covariant signals has the form*

$$\Gamma_3 = \begin{bmatrix} 1 & a & a^* \\ a^* & 1 & a \\ a & a^* & 1 \end{bmatrix}. \tag{9}$$

Conversely, 3-ary signals with the Gram matrix in Eq. (9) are group covariant.

For M-ary signals ($M \geq 4$), the Gram matrices are not simply characterized as Eqs. (8) and (9) since operations defined on the groups are not unique. In the following, we will consider typical two operations.

Group Covariant Signals with respect to a Group with Addition Modulo M

Definition 2 : For $n, m \in G = \{0, 1, \cdots, M-1\}$,

$$n \circ m \equiv n + m (\mathrm{mod} M), \tag{10}$$

where $+$ in right side of Eq. (10) is ordinary addition.

The operation in Definition 2 is addition modulo M. We have the following remark for the operation.

Remark 4 : *The Gram matrix for group covariant signals with respect to a group with the operation in Definition 2 is a circulant and Hermitian.*

As example of such signals, there is M-ary PSK coherent-state signals. This fact was also mentioned in Refs. 2 and 6.

Group Covariant Signals with respect to a Group with Exclusive-OR

Definition 3 : For $x, y \in G = \{0, 1, \cdots, 2^n - 1\}$,

$$x \oplus y \equiv x + y(\mathrm{mod}2) = (x_{n-1} \oplus y_{n-1})(x_{n-2} \oplus y_{n-2})...(x_0 \oplus y_0), \tag{11}$$

where n is a natural number, $x_{n-1}x_{n-2}...x_0$ and $y_{n-1}y_{n-2}...y_0$ are the binary notations of x and y, respectively, and $\oplus$ in right side of Eq. (11) is ordinary Exclusive-OR.

We call the operation in Definition 3 "Exclusive-OR" in this paper. We find by applying Proposition 1 that the Gram matrix for the group covariant signals with respect to a group with the above operation has the following characteristic.[7]

Remark 5 : *The $2^n \times 2^n$ Gram matrix Γ_{2^n} for group covariant signals with respect to a group $(G; \oplus)$ is partitioned into a block-matrix as*

$$\Gamma_{2^n} = \begin{bmatrix} A & B \\ B & A \end{bmatrix}, \tag{12}$$

where A and B are $2^{n-1} \times 2^{n-1}$ submatrices and have the "same structure" as $\Gamma_{2^{n-1}}$, and therefore, are partitioned into block-matrices as for $\Gamma_{2^{n-1}}$. Here the statement that a matrix A has the "same structure" as $\Gamma_{2^{n-1}}$ means the following relation holds.

$$\forall i, j, k \in \{0, 1, \cdots, 2^{n-1} - 1\}, \quad A_{i,j} = A_{k\oplus i, k\oplus j}. \tag{13}$$

As examples of such signals, there are sets of codewords of length n. In the following, we would like to show some results.

Remark 6 : *The binary code of all codewords of length n is group covariant with respect to a group $(G; \oplus), G = \{0, 1, \cdots, 2^n - 1\}$.*

Proof of Remark 6 : Let 0 and 1 be classical letters and $\{00...0, \cdots, 11...1\}$ be a set of all classical codewords of length n. Corresponding set of quantum codewords is $\{|\psi_0\rangle = |0\rangle \otimes |0\rangle \otimes ... \otimes |0\rangle, \cdots, |\psi_{2^n-1}\rangle = |1\rangle \otimes |1\rangle \otimes ... \otimes |1\rangle\}$, where $|0\rangle$ and $|1\rangle$ are the letter states.

Assume that each codeword corresponds to the binary notation of each element in the group $G = \{0, 1, \cdots, 2^n - 1\}$.

The inner product between two codeword states are

$$\langle\psi_i|\psi_j\rangle = \kappa^{d_{\mathrm{H}}}, \tag{14}$$

where $\kappa = \langle 0|1\rangle$ is the inner product between the letter states and is assumed to be real and

$$d_{\mathrm{H}} = \sum_{m=0}^{n-1} i_m \oplus j_m, \tag{15}$$

is Hamming distance between the codewords, here $i_{n-1}i_{n-2}...i_0$ and $j_{n-1}j_{n-2}...j_0$ are the binary notations of $i, j \in G$, respectively. For $k \in G$, consider

$$\langle\psi_{k\oplus i}|\psi_{k\oplus j}\rangle = \kappa^{d'_{\mathrm{H}}}. \tag{16}$$

Let $k_{n-1}k_{n-2}...k_0$ be the binary notation of $k \in G$, then

$$d'_{\mathrm{H}} = \sum_{m=0}^{n-1} (k_m \oplus i_m) \oplus (k_m \oplus j_m) = \sum_{m=0}^{n-1} i_m \oplus j_m = d_{\mathrm{H}}. \tag{17}$$

From Eqs. (14), (16), and (17), the set of all codewords of length n satisfies the condition in Proposition 1 and is group covariant. □

Remark 7 : *We pick up some codeword states from 2^n possible sequences of length n and construct a code which is the set of selected codeword states. If the subset H of the group $(G;\oplus), G = \{0, 1, \cdots, 2^n - 1\}$, that corresponds to the code, is a subset of G, then the code is group covariant with respect to $(H;\oplus)$.*

Remark 7 is trivial from Remark 6. From Remark 6 and Remark 7, we have the following proposition.

Proposition 2 : *Arbitraly binary linear codes are all group covariant.*

Proof of Proposition 2 : For a binary linear code $\boldsymbol{C}$, we have

$$\forall x, y \in \boldsymbol{C}, \quad x + y \in \boldsymbol{C}, \tag{18}$$

where $+$ is addition of codewords. From Eq. (18), we obtain

$$x + x = 0 \in \boldsymbol{C}, \tag{19}$$

where 0 is the codeword whose digits are all zero. Equation (19) implies that the code $\boldsymbol{C}$ includes the identity element 0 and the inverse element $x^{-1} = x$ for each element x, so that $(\boldsymbol{C};+)$ is a group.

It is well known that addition $+$ of codewords is the same as Exclusive OR $\oplus$. Therefore, $(\boldsymbol{C};+)$ is a subset of the group $(G;\oplus)$ and group covariant by Remark 7. □

Non-Covariant Signals

Here we consider examples of signals which are not group covariant. The following proposition is directly lead from Proposition 1.

Proposition 3 : *If M-ary signals $\{|\psi_i\rangle | i = 0, 1, \cdots, M-1\}$ are group covariant, the i-th row (or column) vector of the Gram matrix is a permutation of the 0-th row (or column) vector of the matrix. As a result, the elements of the Gram matrix are constructed by $(\Gamma_M)_{0,j} (j = 0, 1, \cdots, M-1)$.*

This is a necessary condition for group covariant signals. If M-ary signals $\{|\psi_i\rangle\}$ are group covariant, the distances (inner products) between a signal $|\psi_i\rangle$ and the other $M-1$ signals $\{|\psi_j\rangle | j \neq i\}$ are the same as the distances (inner products) between another signal $|\psi_{i'}\rangle$ and the other $M-1$ signals $\{|\psi_j\rangle | j \neq i'\}$, so that there is no "special" signal and all the signals are equal in the sense of the signal distances.

Clearly, M-ary amplitude-shift-keyed (ASK) signals ($M \geq 3$) and quadrature-amplitude-modulated (QAM) coherent-state signals are not covariant. Because, considering the locations of the signals in the phase plane (the classical signal space), a signal near the origin and a signal far from the origin have different inner products to the other signals. For example, the Gram matrix for 3-ary ASK coherent-state signals $\{|0\rangle, |\alpha\rangle, |-\alpha\rangle\}$ is

$$\Gamma_3 = \begin{bmatrix} 1 & \kappa & \kappa \\ \kappa & 1 & \kappa^4 \\ \kappa & \kappa^4 & 1 \end{bmatrix}, \tag{20}$$

where $\kappa = \exp[-\frac{1}{2}|\alpha|^2]$. In Eq. (20), κ^4 appear at the first and the second rows but does not appear at the zeroth row. The result of "check" of group covariancy of several signals is summarized in Table 1.

Table 1. Examples of group covariant and non-covariant signals.

Covariant	Non-covariant
Binary signals M-ary orthogonal signals (PPM and FSK signals, equidistant codes, etc.) M-ary PSK signals Binary linear codes (Parity check, Cyclic, and BCH codes, etc.)	M-ary ASK signals ($M \geq 3$) L^2-ary QAM signals ($L \geq 3$)

CONCLUSION

We have shown what kind of signals correspond to "group covariant signals." We gave necessary and sufficient condition for group covariant signals in order to check group covariancy of given signals. We see from this condition that many signals, such as arbitrary binary signals, PPM, FSK, and PSK coherent-state signals, equidistant codes, and binary linear codes, are all group covariant. This has significance if we link it with studies of advantages when signals are group covariant, which has been considered by Davies,[1] Ban,[2] etc. For example, the minimum error probability is achieved by the square-root measurement when signals are group covariant with respect to a group not only with addition modulo M [2] but also with Exclusive-OR. The last fact can be shown by applying Theorem 1 in Ref. 8 and will be presented in Ref. 7.

Acknowledgments

The authors would like to thank Prof. O. Hirota, Dr. M. Osaki, and Miss R. Momose of Tamagawa University and Dr. M. Sasaki of Communication Reserach Laboratory of Post and Telecommunications for their helpful discussions.

This research is supported in part by the Hori Information Science Promotion Foundation and Grants-in Aid for Encouragement of Young Scientists (No.09750410) from the Ministry of Education, Science, Sports and Culture of Japan.

REFERENCES

1. E. B. Davies, Information and quantum measurement, *IEEE Trans. Inf. Theory* IT-24:596 (1978).
2. M. Ban, K. Kurokawa, R. Momose, and O. Hirota, Optimum measurements for discrimination among symmetric quantum states and parameter estimation, *Inter. J. Theor. Phys.* 36:1269 (1997).
3. P. Hausladen and W. K. Wootters, A 'pretty good' measurement for distinguishing quantum states, *J. Mod. Opt.* 41:2385 (1994).
4. P. Hausladen, R. Jozsa, B. Schumacher, M. Westmoreland, and W. K. Wootters, Classical information capacity of a quantum channel, *Phys. Rev. A* 54:1869 (1996).
5. C. W. Helstrom, "Quantum Detection and Estimation Theory," Academic Press, NY (1976).
6. M. Osaki, M. Ban, and O. Hirota, The maximum mutual information without coding for binary quantum-state signals, *J. Mod. Opt.* 45:269 (1998).
7. T. S. Usuda, I. Takumi, and O. Hirota, manuscript in preparation.
8. M. Sasaki, K. Kato, M. Izutsu, and O. Hirota, Quantum channels showing superadditivity in classical capacity, *Phys. Rev. A* 58:146 (1998).

OPTIMIZATION OF A SYMMETRIC QUANTUM CHANNEL AND ITS OPTICAL REALIZATION SCHEME

Masahide Sasaki,[1] Stephen M. Barnett,[2] and M. Izutsu[1]

[1]Communications Research Laboratory
Ministry of Posts & Telecommunications,
Koganei, Tokyo 184-8795, Japan
[2]Department of Physics and Applied Physics, University of Strathclyde,
Glasgow G4 0NG, Scotland

INTRODUCTION

One of the basic problems of transmission of classical information on a quantum channel is the maximization problem of the Shannon mutual information $I(X:Y)$ between input letters $\{x_i \in X\}$ and output letters $\{y_j \in Y\}$. Given an ensemble of input letters of density operators $\{\hat{\rho}_i\}$, it is a complicated nonlinear problem to find the optimum detection strategy, i.e., the POVM $\{\hat{\pi}_j\}$. In this paper, we consider the case where the input states are M-ary symmetric states $\{\hat{\rho}_0, \hat{\rho}_1, \cdots, \hat{\rho}_{M-1}\}$ made of a two-state system. We prepare a source state as a thermally mixed state

$$\hat{\rho}_0 \equiv (1-n_{\rm th})|\uparrow\rangle\langle\uparrow| + n_{\rm th}|\downarrow\rangle\langle\downarrow| \tag{1}$$

where $|\uparrow\rangle$ and $|\downarrow\rangle$ represent the basis vectors of the system, and $n_{\rm th}$ is the thermal noise fraction. The symmetric inputs are made by rotating this source state by the angle $2k\pi/M$ ($k=0, 1, \cdots, M-1$) around the y-axis as

$$\hat{\rho}_k = \hat{V}^k \hat{\rho}_0 \hat{V}^{\dagger k}, \quad \hat{V} \equiv \exp(-i\frac{\pi}{M}\hat{\sigma}_y), \quad \hat{V}^k = \begin{pmatrix} \cos(\frac{k\pi}{M}) & -\sin(\frac{k\pi}{M}) \\ \sin(\frac{k\pi}{M}) & \cos(\frac{k\pi}{M}) \end{pmatrix}. \tag{2}$$

Prior probabilities are all equal to $\frac{1}{M}$.

These inputs are group covariant with respect to a finite cyclic group G. The maximization of $I(X:Y)$ for group covariant inputs was already studied by Davies[1] and the following theorem is already known: If the representation of G on $\mathcal{H}_s$ is *irreducible* in the sense that the G-invariant point on $\mathcal{H}_s$ is only the multiples of identity operator,

$$\hat{U}_g^\dagger \hat{A} \hat{U}_g = \hat{A} \quad \forall g \in G \quad \Longrightarrow \quad \hat{A} \propto \hat{I}, \tag{3}$$

then their exists a unit state vector $|\psi\rangle$ on $\mathcal{H}_s$ such that the mutual information $I(X:Y)$ is maximized by the covariant output

$$\hat{\Pi}_g = \frac{d}{M}\hat{U}_g|\psi\rangle\langle\psi|\hat{U}_g^\dagger, \quad \sum_{g\in G}\hat{\Pi}_g = \hat{I} \tag{4}$$

Quantum Communication, Computing, and Measurement 2
Edited by Kumar et al., Kluwer Academic / Plenum Publishers, New York, 2000.

where d is the dimension of $\mathcal{H}_s$, and M is the order of G, i.e., the number of inputs. This theorem will be very useful to reduce possible forms of the optimum POVM.

Unfortunately, however, one cannot directly apply this theorem to the case of Eq. (2). In fact, the unitary representation $\{\hat{I}, \hat{V}, \cdots, \hat{V}^{M-1}\}$ is not irreducible in the sense of Eq. (3). Thus in exploring the optimum strategy, we have to rely on other arguments.

OPTIMUM DETECTION OPERATORS

In maximizing the Shannon mutual information

$$I(X:Y) = \sum_{i=0}^{M-1} \xi_i \sum_{j=0}^{N-1} P(j|i) \ln \frac{P(j|i)}{\sum_{k=1}^{M-1} \xi_k P(j|k)}, \tag{5}$$

where $P(j|i) = \mathrm{Tr}(\hat{\pi}_j \hat{\rho}_i)$, the number of outputs N is free to vary in general. However we shall begin with the restricted maximization of $I(X:Y)$ within the class of $\hat{V}$-symmetric POVM

$$\hat{\pi}_j = |\pi_j\rangle\langle\pi_j|, \quad |\pi_j\rangle = \hat{V}^j|\pi_0\rangle, \quad \sum_{j=0}^{M-1} \hat{\pi}_j = \hat{I}. \tag{6}$$

The state vector $|\pi_0\rangle$ should have a form of $|\pi_0\rangle \equiv \sqrt{2/M}\ {}^t(\cos\theta, \sin\theta)$ to satisfy the resolution of the identity. The channel matrix $P(j|i)$ can be fully specified by the following M components of the first column,

$$p(\theta - \frac{k\pi}{M}) \equiv P(0|k) \equiv \frac{1}{M} f(2\theta - \frac{2k\pi}{M}), \quad f(x) \equiv\ 1 + (1 - 2n_{\text{th}})\cos x. \tag{7}$$

Then Eq. (5) can be reduced to

$$I(\theta) = \frac{1}{M} \sum_{k=0}^{M-1} f(2\theta - \frac{2k\pi}{M}) \ln f(2\theta - \frac{2k\pi}{M}), \tag{8}$$

which is a periodic fuction of θ with a period $\frac{\pi}{M}$. Fig. 1 shows numerical examples of $M = 2, 3, 4$, and 5 taking $n_{\text{th}} = 0.05$. It is clear that $I(\theta)$ is always maximized at $\theta = \frac{\pi}{2}$ regardless of M.

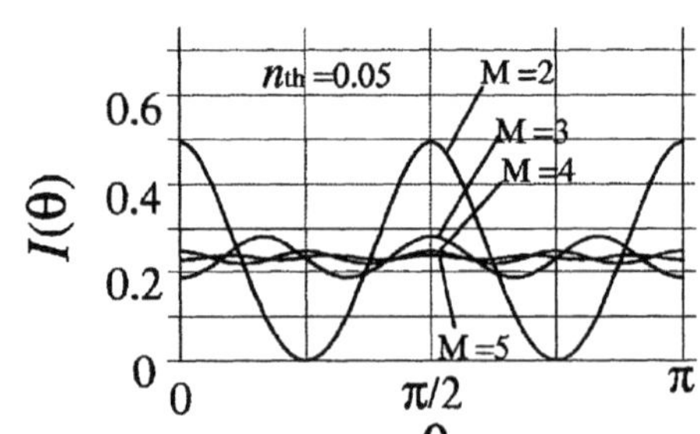

Figure 1. The Shannon mutual information versus the optimization paremeter θ for M =2, 3, 4 and 5 in the case of $n_{\text{th}} = 0.05$.

Lemma 1
$I(\theta)$ has the global maximum at $\theta = \frac{\pi}{2}$.

Proof

The detail of the proof will be presented elsewhere. Here we only summarize the main points. $I(\theta)$ can be expanded in terms of $|(1-2n_{\rm th})\cos(2\theta-\frac{2k\pi}{M})|$ (<1). By using the following property

$$\sum_{k=0}^{M-1} \exp\left[-iL\frac{2k\pi}{M}\right] = \begin{cases} M & \text{for } L/M = q \text{ (integer)} \\ 0 & \text{for } L/M \neq \text{ integer} \end{cases}, \tag{9}$$

it is seen that $I(\theta)$ can be expressed in the form of

$$I(\theta) = C_1 + \sum_{n=0}^{\infty} C_2(nM)(-1)^{nM}\cos(2\theta nM), \tag{10}$$

where the coefficients C_1 and $C_2(nM)$ are positive definite. Then $I(\theta)$ is maximized when $(-1)^{nM}\cos(2\theta nM) = 1$, that is, $\theta = \frac{\pi}{2}$ for all M. ∎

This lemma by itself cannot ensure the optimality. There still remains a possibility that the true maximum of the Shannon mutual information may be larger than the global maximum of $I(\theta)$:

$$\max I(X:Y) \geq I(\theta = \frac{\pi}{2}).$$

Actually, the true maximum must be attained at least by the measurement having N outputs bounded by $2 \leq N \leq 4$ which is not covered by Eq. (6).[1] What is it like?

The following lemma might be useful.[2]

Lemma 2

If the signal states are made on the d dimensional real subspace, the Shannon mutual information can be maximized by the detection with N outputs bounded by $d \leq N \leq d(d+1)/2$.

Proof

The proof can be made parallel to the original one in ref. 1 with a slight replacement which is the number of the supporting points of the POVM defined by

$$\sum_j \mu_j \bar{\pi}_j = \hat{I}/d, \quad \sum_j \mu_j = 1, \tag{11}$$

where $\mathrm{Tr}\bar{\pi}_j = 1$. Denoting by D the number of real parameters needed to specify each $\bar{\pi}_j$, the number of the supporting points is $D+1$. Since $\bar{\pi}_j = \bar{\pi}_j^\dagger$ and $\mathrm{Tr}\bar{\pi}_j = 1$, $D = d(d+1)/2 - 1$. ∎

In this case the bound can be titer such as $2 \leq N \leq 3$. So the POVM we are seeking has 3 outputs at most and is of rank 1. Therefore it can be covered by the following set of state vectors,

$$\begin{aligned} |\omega_0(\theta)\rangle &= \sqrt{2-a^2-b^2}\begin{pmatrix}\cos\theta \\ \sin\theta\end{pmatrix}, & 0 \leq a^2+b^2 \leq 2 \\ |\omega_1(\theta+\varphi_a)\rangle &= a\begin{pmatrix}\cos(\theta+\varphi_a) \\ \sin(\theta+\varphi_a)\end{pmatrix}, & a^2 = \frac{\cos\varphi_b}{\sin\varphi_a \sin(\varphi_a-\varphi_b)} \\ |\omega_2(\theta+\varphi_b)\rangle &= b\begin{pmatrix}\cos(\theta+\varphi_b) \\ \sin(\theta+\varphi_b)\end{pmatrix}, & b^2 = \frac{\cos\varphi_a}{\sin\varphi_b \sin(\varphi_b-\varphi_a)}. \end{aligned} \tag{12}$$

Note that the lengths of these three vecors are not free to vary. They are determined by the two angle parameters φ_a and φ_b by imposing $\sum_j |\omega_j\rangle\langle\omega_j| = \hat{I}$. The channel matrix is given by the following elements;

$$\begin{aligned} \langle\omega_0(\theta)|\hat{\rho}_k|\omega_0(\theta)\rangle &= \frac{2-a^2-b^2}{2} f(2\theta - \frac{2k\pi}{M}), \\ \langle\omega_1(\theta+\varphi_a)|\hat{\rho}_k|\omega_1(\theta+\varphi_a)\rangle &= \frac{a^2}{2} f(2\theta + 2\varphi_a - \frac{2k\pi}{M}), \\ \langle\omega_2(\theta+\varphi_b)|\hat{\rho}_k|\omega_2(\theta+\varphi_b)\rangle &= \frac{b^2}{2} f(2\theta + 2\varphi_b - \frac{2k\pi}{M}). \end{aligned} \tag{13}$$

Then the mutual information for M-N (=2, 3) channel can be expressed as the convex combination of the mutual information $I(\theta)$ for the M-M symmetric channel (Eq. (8)):

$$I(X:Y) = (1-\frac{a^2}{2}-\frac{b^2}{2})I(\theta)+\frac{a^2}{2}I(\theta+\varphi_a)+\frac{b^2}{2}I(\theta+\varphi_b), \\ \leq \max_{\theta} I(\theta). \tag{14}$$

Therefore by lemma 1,

$$\max I(X:Y) = I(\theta=\frac{\pi}{2}). \tag{15}$$

The equality in Eq. (14) holds when

$$I(\frac{\pi}{2}) = I(\frac{\pi}{2}+\varphi_a) = I(\frac{\pi}{2}+\varphi_b). \tag{16}$$

This could happen when φ_a and φ_b are $\frac{\pi}{M}$ multiplied by integers, because of the periodicity of $I(\theta)$. This result is summarized by the following theorem.

Theorem

For the M-ary symmetric inputs $\{\hat{\rho}_k\}$ defined by Eq. (2), $I(X:Y)$ is maximized by the M-M symetric channel made of the POVM

$$\hat{\pi}_j^* = |\pi_j^*\rangle\langle\pi_j^*|, \quad |\pi_j^*\rangle = \sqrt{\frac{2}{M}}\begin{pmatrix} -\sin(\frac{j\pi}{M}) \\ \cos(\frac{j\pi}{M}) \end{pmatrix}, \tag{17}$$

and the maximum value is $I(\frac{\pi}{2})$.

$I(X:Y)$ is also maximized by the M-N (N=2, 3) channel made of the POVM $\hat{\omega}_j^* = |\omega_j^*\rangle\langle\omega_j^*|$ such as

$$\begin{aligned} |\omega_0^*\rangle &= \begin{pmatrix} 0 \\ \sqrt{2-a^2-b^2} \end{pmatrix}, & 0 \leq a^2+b^2 \leq 2, \\ |\omega_1^*\rangle &= a\begin{pmatrix} -\sin(\frac{m\pi}{M}) \\ \cos(\frac{m\pi}{M}) \end{pmatrix}, & a^2 = \frac{\cos(\frac{n\pi}{M})}{\sin(\frac{m\pi}{M})\sin(\frac{(m+n)\pi}{M})} \geq 0, \\ |\omega_2^*\rangle &= b\begin{pmatrix} \sin(\frac{n\pi}{M}) \\ \cos(\frac{n\pi}{M}) \end{pmatrix}, & b^2 = \frac{\cos(\frac{m\pi}{M})}{\sin(\frac{n\pi}{M})\sin(\frac{(m+n)\pi}{M})} \geq 0, \end{aligned} \tag{18}$$

where m and n are positive integers.

By using this theorem, the following is straightforward: if M is even, the optimum strategy becomes the standard von Neumann measurement constructed by the binary orthogonal state vectors, and if M is odd, at least 3 outputs are required. It is worth demonstrating the cases of $M=5$ and $M=7$ (Fig. 2). The measurememt state vectors are represented by the solid lines. The dashed lines represent the input state vectors taking $n_{\text{th}}=0$ (pure states). There can be several configurations of the measurement state vectors according to choices of parameters (m,n) in the theorem. But by simply changing the sign of either one of the state vectors, which do not change the channel matrix, they can be classified into minimal class(es) of topology. When $M=5$, there is a single class of measurement of $N=3$. When $M=7$, there are two classes of the mesurememt of $N=3$. One corresponds to $a^2=b^2=1/(2\sin^2\frac{2\pi}{7})$, and the other corresponds to $a^2=b^2=1/(2\sin^2\frac{3\pi}{7})$.

It might be worth mentioning another aspect of optimization, that is, the minimization of the average error probability

$$P_e(\theta) \equiv 1-\sum_{k=0}^{M-1} P(k|k) = 1-p(\theta). \tag{19}$$

This has a period of π with respect to θ being independent of M. The average error probability is minimized by the covariant detection having M outputs $\{|\pi_k(\theta{=}0)\rangle\}$ $(k=0,\cdots,M-1)$. The proof will be presented elsewhere.

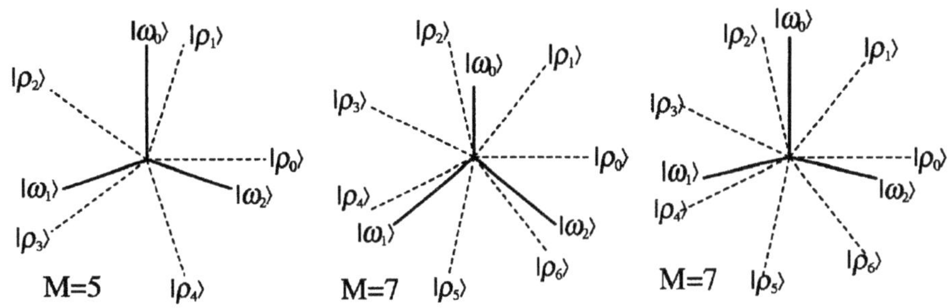

Figure 2. The optimum measurement state vectors (thick solid lines) in the case of $M = 5$ and $M = 7$. The input state vectors (pure states) are indicated by the dashed lines.

IMPLEMENTATION

When the signal system is a single mode photon polarization state, the POVM derived in the preceding section can be implemeted by a current technology. Fig. 3 shows the optimum decoding circuit for the case of $M = N = 3$. It has the two input ports "a" and "b". The signal photon is guided into the port "a" while the other port "b" is initialized as the vacuum state at the left end.

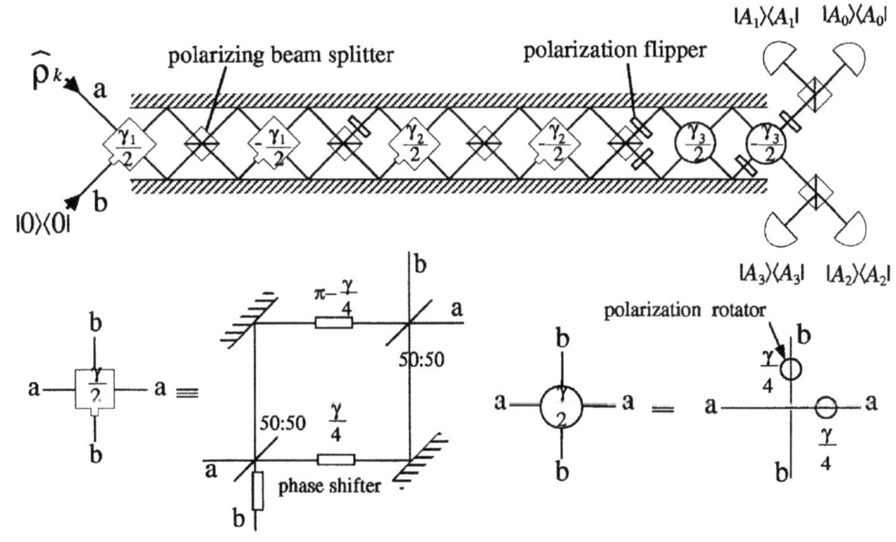

Figure 3. The decoding circuit for $N = 3$. They consists of the unitary transformation $\hat{U}$ and the measurement by $|A_j\rangle$.

This setup makes the 4 dimensional Hilbert space $\mathcal{H}_{\mathrm{ex}}$ spanned by the basis $\{|\uparrow\rangle_a|0\rangle_b, |\downarrow\rangle_a|0\rangle_b, |0\rangle_a|\uparrow\rangle_b, |0\rangle_a|\downarrow\rangle_b\}$ which we denote as $\{|A_0\rangle, |A_1\rangle, |A_2\rangle, |A_3\rangle\}$. The signal photon passes through the circuit consisting of the Mach-Zehnder type interferometer, the polarizing beam splitter, the polarization flipper, and the polarization rotator, and is then detected at the right end being observed *which path* and *which polarization* simultaneously which corresponds to $\{|A_j\rangle\langle A_j|\}$.

The circuit design is based on the Naimark extension $\{|\Pi_j\rangle\}$ of the nonorthogonal measurement $\{|\pi_j\rangle\}$ given by Eq. (17):

$$\begin{aligned} |\Pi_0\rangle &\equiv |\pi_0\rangle_a|0\rangle_b + \tfrac{1}{\sqrt{3}}|0\rangle_a|\uparrow\rangle_b, \\ |\Pi_1\rangle &\equiv |\pi_1\rangle_a|0\rangle_b - \tfrac{1}{\sqrt{3}}|0\rangle_a|\uparrow\rangle_b, \\ |\Pi_2\rangle &\equiv |\pi_2\rangle_a|0\rangle_b + \tfrac{1}{\sqrt{3}}|0\rangle_a|\uparrow\rangle_b, \\ |\Pi_3\rangle &\equiv |0\rangle_a|\downarrow\rangle_b. \end{aligned} \tag{20}$$

The orthogonal basis $\{|\Pi_j\rangle\}$ can be transformed to $\{|A_j\rangle\}$ by $|\Pi_j\rangle = \hat{U}^\dagger|A_j\rangle$ where

$$\hat{U} \equiv \begin{pmatrix} 0 & \sqrt{\frac{2}{3}} & \frac{1}{\sqrt{3}} & 0 \\ -\frac{1}{\sqrt{2}} & \frac{1}{\sqrt{6}} & -\frac{1}{\sqrt{3}} & 0 \\ -\frac{1}{\sqrt{2}} & -\frac{1}{\sqrt{6}} & \frac{1}{\sqrt{3}} & 0 \\ 0 & 0 & 0 & 1 \end{pmatrix}. \tag{21}$$

The circuit in Fig. 3 is just an implementation of this unitary transformation.[3, 4, 5] The three parameters in the circuit are γ_1 =arctan$(1/\sqrt{2})$ $(0 < \gamma_1 < \pi/2)$, $\gamma_2 = \pi/2$, and $\gamma_3 = \pi$. When $M > 3$, the different unitary transformation is required for the 3-output optimum measurement but it can be effected by the similar 2-input circuit with slight modification.

CONCLUDING REMARKS

We considered the optimum detection strategy for the case of M-ary symmetric quantum states made of a two-state system. In maximizing the Shannon mutual information, this case is out of application of the Davies theorem Eq. (4) as explained in the introduction. However, the optimum strategy of the M-M symmetric channel obtained in this paper is quite similar to the one specified by the Davies theorem. This may be due to the dimentionality $d = 2$. For a higher dimensional system, things seem to be more complicated. Maximization of the Shannon mutual information of such cases is a remaining open question.

From the view point of putting quantum detection theory to the test, the detection circuit in Fig. 3 may be of great importance. So far only the two kinds of criteria, the minimum *average* error probability[6] and the maximum rate of retrodicting quantum state[7] have been confirmed. The proposed circuit is concerned with the criterion of the maximum mutual information which has never been confirmed by experiment.

The authors would like to thank Dr. Osaki, Prof. Hirota, Prof. A. S. Holevo, Dr. Usuda, and Dr. A. Chefles for giving crucial comments on this work. SMB thanks the UK Engineering and Physical Science Research Council for financial support.

REFERENCES

1. E. B. Davies, IEEE Trans. Inf. Theory **IT-24**, 596 (1978).
2. The authors are indebted to a private communication from A. S. Holevo for Lemma 2.
3. N. J. Cerf, C. Adami, and P. G. Kwiat, Phys. Rev. **A57**, R1477 (1998).
4. M. Reck, A. Zeilinger, H. J. Bernstein, and P. Bertani, Phys. Rev. Lett. **73**, 58 (1994).
5. A. Barenco, C. H. Bennett, R. Cleve, D. P. M. DiVincenzo, N. Margolus, P. Shor, T. Sleator, J. A. Smolin, and H. Weinfurter, Phys. Rev. **A52**, 3457, (1995).
6. S. M. Barnett and E. Riis, J. Mod. Opt. **44**, 1061 (1997).
7. B. Huttner, A. Muller, J. D. Gautier, H. Zbinden, and N. Gisin, Phys. Rev. **A54**, 3783 (1996).

MANIPULATION OF ENTANGLED STATES FOR QUANTUM INFORMATION PROCESSING

S. Bose[1], S.F. Huelga[1], D. Jonathan[1], P.L. Knight[1], M. Murao[1], M.B. Plenio[1] and V. Vedral[1,2]

[1]Optics Section, The Blackett Laboratory, Imperial College, London, SW7 2BZ, United Kingdom
[2]Centre for Quantum Computing, University of Oxford, Parks Road OX1 3PU, Oxford, United Kingdom

Entanglement manipulation, and especially Entanglement Swapping is at the heart of current work on quantum information processing, purification and quantum teleportation. We will discuss how it may be generalized to multiparticle systems and how this enables multi-user quantum cryptographic protocols to be developed. Our scheme allows us to establish multiparticle entanglement between particles which belong to distant users in a communication network through a prior distribution of Bell state singlets followed by local measurements. We compare our method for generating entanglement with existing schemes using simple quantum networks, and highlight the advantages and applications in cryptographic conferencing and in reading messages from more than one source through a single quantum measurement. We also discuss how entanglement leads to the idea of 'telecloning', in which a teleportation-like protocol can be found which reproduces the output of an optimal quantum cloning machine

INTRODUCTION

Quantum teleportation, proposed five years ago by Bennett and co-workers[1], has become one of the key tools of quantum information theory. From its original conception as a means of transmitting quantum information, teleportation has now found applications in various areas such as quantum channel capacities[3] and distributed quantum computing[4]. The recent successful experiments of Bouwmeester *et al* and Boschi *et al*[2] have demonstrated practical realizations of this idea.

In this paper we present two new applications of teleportation-like schemes. The first, a generalization of entanglement swapping[5], uses teleportation to manipulate and transfer entanglement between the constituent subsystems of multiparticle entangled states[7]. This offers the possibility to construct a 'quantum telephone exchange', allowing secure multiuser conferencing over remote channels. The second application, 'telecloning'[10], exploits a generalization of the teleportation protocol, where the quantum information contained in the input qubit is transferred not to one new qubit, but to a subspace of many. We show that, by an appropriate choice of this subspace, a teleportation-like protocol can be found which reproduces the output of an optimal quantum cloning machine[12, 13]. The net result is that optimal copies of an input qubit are simultaneously created at spatially separated locations.

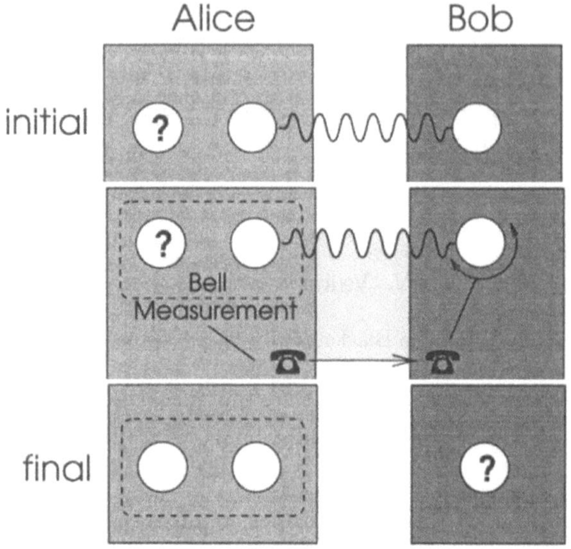

Figure 1. Standard teleportation protocol as proposed by Bennett *et al.* Alice and Bob initially share a maximally entangled pair of qubits. Bob reconstructs Alice's unknown state by applying the appropriate rotation after receiving the results of her Bell measurement.

Before proceeding, it is convenient to recall the key ideas of the original teleportation protocol[1]. This allows an unknown state $|\phi\rangle_X$ of a quantum system X to be faithfully transmitted between two spatially separated parties (a sender, Alice, and a receiver, Bob). The essential steps of this procedure (say in the simplest case where X is a 1-qubit system) are as follows: first and foremost, Alice and Bob must share a maximally entangled state of two qubits S (sender) and R (receiver), such as $|\Phi^+\rangle = \frac{1}{\sqrt{2}}(|00\rangle_{SR} + |11\rangle_{SR})$. Next, Alice performs a joint measurement of the 2-qubit system $X \otimes S$ in the Bell basis:

$$\left|\Phi^{\pm}\right\rangle = \frac{1}{\sqrt{2}}(|00\rangle \pm |11\rangle), \tag{1}$$

$$\left|\Psi^{\pm}\right\rangle = \frac{1}{\sqrt{2}}(|01\rangle \pm |10\rangle). \tag{2}$$

Finally, Alice sends a two-bit message to Bob informing him of her measurement result. Bob then rotates his qubit using one of the unitary operators $\mathbf{1}$, σ_z, σ_x or σ_y, according to whether Alice's result was respectively $|\Phi^+\rangle$, $|\Phi^-\rangle$, $|\Psi^+\rangle$ or $|\Psi^-\rangle$. The final state of Bob's qubit is then equal to the original state $|\phi\rangle_X$, regardless of the measurement result. This insensitivity to measurement results is the crucial property of the teleportation protocol, and one which we shall also require for our entanglement swapping and telecloning schemes.

STANDARD ENTANGLEMENT SWAPPING

We shall begin by briefly reviewing standard entanglement swapping as proposed by Zukowski *et al*[5] and experimentally demonstrated by Pan*et al*[6], which in itself can be viewed as a simple generalization of teleportation.

Consider a modification of teleportation in which two changes are made: the state to be teleported is replaced by an entangled Bell state (an algebraic change) and the pair of particles forming this Bell state are kept spatially separated (a configurational change). Let particles 1 and 2 be mutually entangled (in a Bell state), and particles 3 and 4 be mutually entangled (also in a Bell state). If one conducts a measurement of the Bell operator on particles 2 and 3 (which projects particles 2 and 3 to a Bell state), then the particles 1 and 4 are instantaneously projected to one of the Bell states as well. Whereas prior to the measurement, the Bell pairs were (1,2) and (3,4), after the measurement the Bell pairs are (2,3) and (1,4). A pictorial way of representing the above process is given in Fig. 2. It is clear that the most interesting aspect of this scheme is that particles 1 and 4, which do not share any common past, are entangled after the swapping.

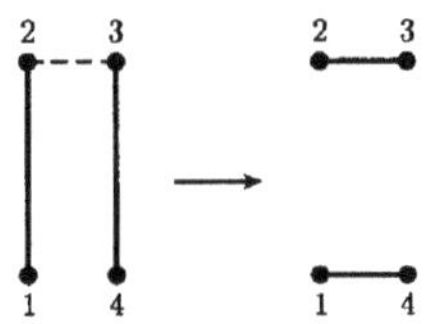

Figure 2. The swapping of entanglement between pairs of particles due to a Bell state measurement on two of them is shown. The bold lines connect particles in Bell states and the dashed lines connect particles on which the Bell state measurement is made.

GENERALIZED ENTANGLEMENT SWAPPING

We now proceed to an even further generalization of teleportation in which we start with entangled states of a greater number of particles and do joint measurements on a greater number of particles[7]. But before that we turn to some necessary notation and terminology. In terms of a binary variable $u_i \in \{0,1\}$ and its complement u_i^c (defined as $1 - u_i$), one can write down any Bell state (not normalised) of two particles i and j as

$$|\Psi(i,j)\rangle_\pm = |u_i, u_j\rangle \pm |u_i^c, u_j^c\rangle \ . \tag{3}$$

In the above it is understood that $|u_i\rangle$ and $|u_i^c\rangle$ are two orthogonal states of a two state system. Then the N-particle generalisation of Bell states will be states of the type

$$|\psi\rangle = \prod_{i=1}^{N} |u_i\rangle \pm \prod_{i=1}^{N} |u_i^c\rangle. \tag{4}$$

For N=2 they reduce to the Bell states and for N = 3 and 4 they represent the GHZ states. For a general N we shall call them cat states. We shall show that the original entanglement swapping scheme can be generalised to the case of starting with cat states involving any number of particles, doing local measurements by selecting any number of particles from the different cat states and also ending up with cat states involving any number of particles. To see that consider an initial state in which there are N different sets of entangled particles in cat states. Let each of these sets be labelled by m (where

$m = 1, 2, .., N$), the ith particle of the mth set be labelled by $i(m)$ and the total number of particles in the mth set be n_m. Then the initial state can be represented by

$$|\Psi\rangle = \prod_{m=1}^{N} |\Psi\rangle_m, \tag{5}$$

in which each of the cat states $|\Psi\rangle_m$ is given by

$$|\Psi\rangle_m = \prod_{i=1}^{n_m} |u_{i(m)}\rangle \pm \prod_{i=1}^{n_m} |u^c_{i(m)}\rangle \tag{6}$$

where the symbols $u_{i(m)}$ stand for binary variables $\in \{0, 1\}$ with $u^c_{i(m)} = 1 - u_{i(m)}$. Now imagine that the first p_m particles from all the entangled sets are brought together (i.e a total of $p = \sum_{m=1}^{N} p_m$ particles) and a joint measurement is performed on all of them. Note that the set of all cat states of p particles forms a complete orthonormal basis. Let the nature of the measurement on the selected particles be such that it projects them to this basis. Such a basis will be composed of states of the type

$$|\Psi(p)\rangle = \prod_{m=1}^{N} \prod_{i=1}^{p_m} |u_{i(m)}\rangle \pm \prod_{m=1}^{N} \prod_{i=1}^{p_m} |u^c_{i(m)}\rangle. \tag{7}$$

By simply operating with $|\Psi(p)\rangle\langle\Psi(p)|$ on $|\Psi\rangle$ of Eq.(5), we find that the rest of the particles (i.e those not being measured) are projected to states of the type

$$|\Psi(\sum_{m=1}^{N} n_m - p)\rangle = \prod_{m=1}^{N} \prod_{i=p_m+1}^{n_m} |u_{i(m)}\rangle \pm \prod_{m=1}^{N} \prod_{i=p_m+1}^{n_m} |u^c_{i(m)}\rangle, \tag{8}$$

which represents a cat state of the rest of the particles. In a schematic way the above process can be represented as

$$\prod_{m=1}^{N} |E(n_m)\rangle \rightarrow |E(p)\rangle \otimes |E(\sum_{m=1}^{N} n_m - p)\rangle \tag{9}$$

where $|E(n)\rangle$ denotes a n particle cat state. As a specific example, in Fig. 3, we have shown the conversion of a collection of two Bell states and a 3 particle GHZ state to a 3 particle GHZ state and a 4 particle GHZ state due to a projection of 3 of these particles to a 3 particle GHZ state.

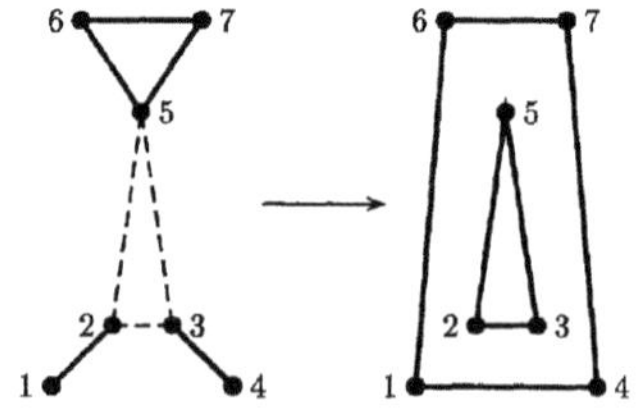

Figure 3. The conversion of two Bell states and a 3 particle GHZ state to a 3 particle GHZ state and a 4 particle GHZ state due to a GHZ state projection on three particles (one taken from each of the initially entangled sets) is shown. The bold lines connect mutually entangled particles and the dashed lines connect particles on which the GHZ state projection is made.

As must be evident from Fig. 3, there is a general rule of thumb for finding out the result when our method of entanglement manipulation is applied to a certain collection of cat states of particles. One just has to connect the particles being measured to frame a polygon and those not being measured to frame a complementary polygon. These two polygons represent the two multiparticle cat states obtained after the manipulation.

Note that this scheme, being a significant generalization of teleportation, allows new applications. The configurational changes (i.e spreading out the constituent particles of the state to be teleported) result in applications such as constructing a *quantum telephone exchange* and speeding up the distribution of entangled particles between two parties[7]. The algebraic change (i.e the fact that more particles are involved), facilitates the preparation of entangled states of higher number of particles from entangled states of lower number of particles[7]. We describe the first application in some detail below.

QUANTUM TELEPHONE EXCHANGE

Suppose there are N users in a communication network. Suppose each of them share an entangled pair of particles (in a Bell state) with a common central exchange. Consider Fig. 4 : A, B, C and D are users who share the Bell pairs (1,2), (3,4), (5,6) and (7,8) respectively with a central exchange O. Now suppose that A, B and C wish to share a GHZ triplet. Then a measurement which projects particles 2, 3 and 5 to GHZ states will have to be performed at O. Immediately, particles 1, 4 and 6 belonging to A, B and C respectively will be reduced to a GHZ state. In a similar manner one can entangle particles belonging to any N users of the network and create a N particle cat state.

The main advantages of using this technique for establishing entanglement over the simple generation of N particle entangled states at a source and their subsequent distribution are as follows.

(A) Firstly, each user can at first purify a large number of partially decohered Bell pairs shared with the central exchange to obtain a smaller number of pure shared Bell pairs. These can then be used as the starting point for the generation of any types of multiparticle cat states of the particles possessed by the users. The problems of decoherence during propagation of the particles can thus be avoided in principle. Also the necessity of having to purify N-particle cat states can be totally avoided. Purification of singlets followed by our scheme will generate N-particle cats in their purest form.

(B) Secondly, our method allows a certain degree of freedom to entangle particles belonging to any set of users only if the necessity arises. It may not be known in advance exactly which set of users will need to share a N particle cat state. To arrange for all possibilities in an a priori fashion would require selecting all possible combinations of users and distributing particles in multiparticle entangled states among them. That is very uneconomical. On the other hand, generating entangled N-tuplets at the time of need and supplying them to the users who wish to communicate is definitely time consuming.

It is pertinent to compare our scheme with the Biham-Huttner-Mor cryptographic network with exchanges[8]. There are two main differences: Firstly, they used a time reversed EPR scheme for setting up the connections and had quantum memories to protect their states. We use a multiparticle generalisation of entanglement swapping. Secondly, their prime focus was to connect any pair of users of a N-user network on request, while our main focus is to establish multiparticle entangled states of the particles possessed by the users. Any number of users can then engage in activities like

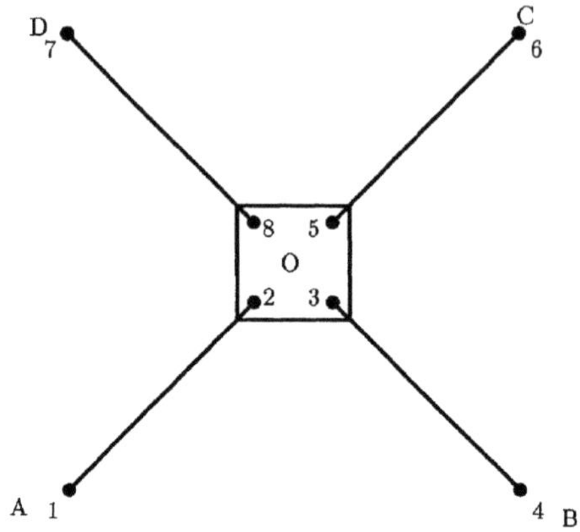

Figure 4. The configuration used for the distribution of entanglement. Initially users A,B,C and D share Bell pairs with the central exchange O. Subsequently, a local measurement at O is sufficient to entangle particles belonging to any subset of users chosen from A, B, C and D.

multiparty cryptographic conferencing and multiparty version of superdense coding[9].

TELECLONING

We now present a new quantum information processing scheme, a quantum telecloning process[10], which generalizes teleportation by combining it with optimal quantum cloning. This allows the optimal broadcasting of quantum information from one sender (Alice) to M spatially separated recipients, requiring only a single measurement by Alice followed by classical communication and local 1-qubit rotations.

Before we detail our scheme, let us briefly review some facts about optimal quantum cloning. While teleportation aims to *transmit* quantum information faithfully, optimal cloning seeks to *spread* it among several parties in the most efficient way possible. The 'no-cloning' theorem[11] prevents this spreading from being perfect; nevertheless, it is still reasonable to ask *how accurately* such copies can be made[12]. If the quality of the copies (measured, for instance, by their fidelity with respect to the original state $|\phi\rangle_X$) is chosen to be independent of $|\phi\rangle_X$, then the answer is given by the so-called $N \to M$ Universal Quantum Copying Machines (UQCMs)[13].

These 'machines' are unitary transformations that transform N input systems, identically prepared in state $|\phi\rangle_X$, onto M output systems ($M \geq N$), each of which ends up in a mixed state described by the reduced density operator

$$\rho_{out} = \gamma|\phi\rangle_X\langle\phi| + (1-\gamma)\,|\phi^\perp\rangle_X\langle\phi^\perp| \tag{10}$$

(where $|\phi^\perp\rangle_X$ is a state orthogonal to $|\phi\rangle_X$)[13]. The fidelity factor γ of these imperfect copies has a definite upper limit imposed by quantum mechanics. In the case where each input system consists of one qubit, this optimal value is given by[13]

$$\gamma = \frac{M\,(N+1)+N}{M\,(N+2)}. \tag{11}$$

Unitary transformations which realize this bound have also been explicitly constructed[13]. In general, they involve the N 'original' qubits, $M-N$ 'blank paper' qubits B (initially prepared in some fixed state $|0\cdots 0\rangle_B$), and an ancilla system A containing at least $M-N+1$ levels[13] (also initially in some fixed state $|0\cdots0\rangle_A$). In this paper, we shall be mainly interested in the situation where only one original qubit X is available, that is, $N=1$. In this case, the cloning transformation U_{1M} is defined as follows: for an initial state $|\phi\rangle_X = a|0\rangle_X + b|1\rangle_X$, we have

$$\begin{aligned} &U_{1M}\left(|\phi\rangle_X \otimes |0\cdots0\rangle_A|0\cdots0\rangle_B\right) \\ &\qquad = a|\phi_0\rangle_{AC} + b|\phi_1\rangle_{AC}, \end{aligned} \tag{12}$$

where

$$\begin{aligned} |\phi_0\rangle_{AC} &= U_{1M}|0\rangle_X|0\cdots0\rangle_A|0\cdots0\rangle_B \\ &= \sum_{j=0}^{M-1} \alpha_j |A_j\rangle_A \otimes |\{0, M-j\}, \{1, j\}\rangle_C, \end{aligned} \tag{13}$$

$$\begin{aligned} |\phi_1\rangle_{AC} &= U_{1M}|1\rangle_X|0\cdots0\rangle_A|0\cdots0\rangle_B \\ &= \sum_{j=0}^{M-1} \alpha_j |A_{M-1-j}\rangle_A \otimes |\{0, j\}, \{1, M-j\}\rangle_C, \end{aligned} \tag{14}$$

$$\alpha_j = \sqrt{\frac{2(M-j)}{M(M+1)}} \tag{15}$$

and where C denotes the M qubits holding the copies (originally the X and B qubits). Here, $|A_j\rangle_A$ are M orthogonal normalized states of the ancilla and $|\{0, M-j\}, \{1, j\}\rangle$ denotes the symmetric and normalized state of M qubits where $(M-j)$ of them are in state $|0\rangle$ and j are in the orthogonal state $|1\rangle$. For example, for $M=3,\ j=1$:

$$|\{0,2\}, \{1,1\}\rangle = \frac{1}{\sqrt{3}}\left(|001\rangle + |010\rangle + |100\rangle\right). \tag{16}$$

We note that, even though the minimum number of ancilla qubits required to support the M levels $|A_j\rangle_A$ is of the order of $\log_2 M$, these can be more conveniently represented as the symmetrized states of $(M-1)$ qubits[13]:

$$|A_j\rangle_A \equiv |\{0, M-1-j\}, \{1, j\}\rangle_A. \tag{17}$$

In this form, states $|\phi_0\rangle$ and $|\phi_1\rangle$ above become $(2M-1)$-qubit states, obeying the following simple symmetries:

$$\sigma_z \otimes \cdots \otimes \sigma_z |\phi_0\rangle = |\phi_0\rangle, \tag{18}$$

$$\sigma_z \otimes \cdots \otimes \sigma_z |\phi_1\rangle = -|\phi_1\rangle, \tag{19}$$

$$\sigma_x \otimes \cdots \otimes \sigma_x |\phi_{0(1)}\rangle = |\phi_{1(0)}\rangle. \tag{20}$$

In other words, the states $|\phi_i\rangle$ transform under simultaneous action of the Pauli operators on all $(2M-1)$ qubits just as a single qubit transforms under the corresponding single Pauli operator. We also note that these operations are strictly local, that is, factorized into a product of independent rotations on each qubit.

We now show how the existence of these local symmetries allows $1 \to M$ quantum cloning to be realized remotely, via a 'telecloning' scheme analogous to the teleportation protocol. The scheme works by exploiting the multiparticle entanglement structure of particular joint states of $2M$ particles ('telecloning states'), of the form:

$$|\psi_{TC}\rangle = \left(|0\rangle_P \otimes |\phi_0\rangle_{AC} + |1\rangle_P \otimes |\phi_1\rangle_{AC}\right)/\sqrt{2}, \tag{21}$$

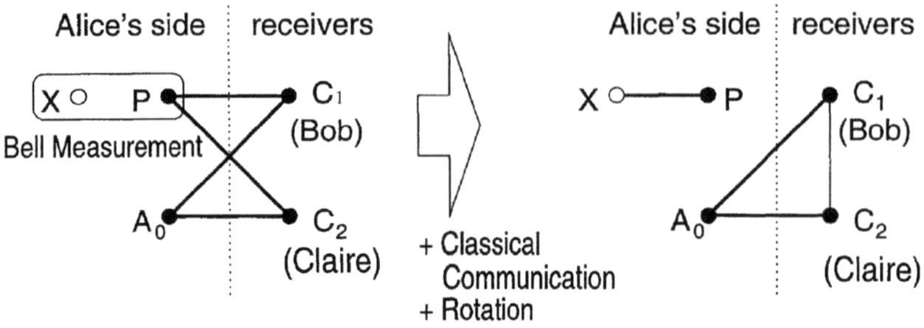

Figure 5. Quantum telecloning $M = 2$ copies of an unknown 1-qubit state denoted X. Alice and her associates Bob and Claire (the 'receivers') initially share a multiparticle entangled state (telecloning state) consisting of the qubits P (the 'port'), A_0 (the ancilla), C_1 and C_2 (outputs, or 'copy' qubits). The solid lines indicate the existence of entanglement between pairs of qubits when the remaining ones are traced out. Alice performs a Bell measurement of the port along with the 'input' qubit X; subsequently, the receivers perform appropriate rotations on the output qubits, obtaining two optimal quantum clones. Since these rotations are independent, each clone can be at a different location.

where $|\phi_0\rangle_{AC}$ and $|\phi_1\rangle_{AC}$ are the optimal cloning states given by Eqs. (13) and (14). Here, C denotes the M qubits which shall hold the copies, each of which is held by one of Alice's associates. For convenience, we shall refer to them collectively as 'the receivers' (though it should be kept in mind that they may all be far away from each other). P represents a single qubit held by Alice, which we shall refer to as the 'port' qubit. Finally, A denotes an $M - 1$ qubit ancilla, which for convenience we will also assume to be on Alice's side (even though, once again, each qubit may in reality be at a different location).

The telecloning of $|\phi\rangle_X$ can now be accomplished by the following generalization of the standard teleportation procedure:

1. Alice performs a Bell measurement of qubits X and P, obtaining one of the four results $|\Psi^{\pm}\rangle_{XP}$, $|\Phi^{\pm}\rangle_{XP}$. If the result is $|\Phi^{+}\rangle_{XP}$, then subsystem AC is projected precisely into the optimal cloning state given in Eq. (12). In this case, our task is accomplished.

2. In case one of the other Bell states is obtained, we can still recover the correct state of AC by exploiting the symmetries of states $|\phi_0\rangle_{AC}$ and $|\phi_1\rangle_{AC}$ under the Pauli matrix operations (Eqs. (18)-(20)). Specifically, if $|\Phi^{-}\rangle_{XP}$ is obtained, we must perform σ_z on each of the $2M - 1$ qubits in AC; similarly, if $|\Psi^{+}\rangle_{XP}$ or $|\Psi^{-}\rangle_{XP}$ are obtained, they must all be rotated by σ_x and $\sigma_x\sigma_z = i\sigma_y$, respectively.

We can see that this protocol is formally identical to the standard teleportation. The difference in the present case is that the quantum information contained in the input state $|\phi\rangle_X$ is transferred onto the 'effective qubit' represented by the orthogonal states $|\phi_0\rangle_{AC}$ and $|\phi_1\rangle_{AC}$, and thus becomes spread over several separate 'physical' qubits. This procedure is illustrated in Fig. 5 for the case of $M = 2$ copies. We stress that, apart from Alice's Bell measurement, only local 1-qubit operations are required in the telecloning procedure. In this way, all of the qubits except the input X and the port P can be spatially separated from each other. It is also worthwhile to add that rotating the ancilla qubits in step (2) above is not strictly necessary. The correct copy states of each output (given by Eq. (10)) are obtained at the output regardless of these operations, since local rotations on one qubit cannot affect another qubit's reduced density operator.

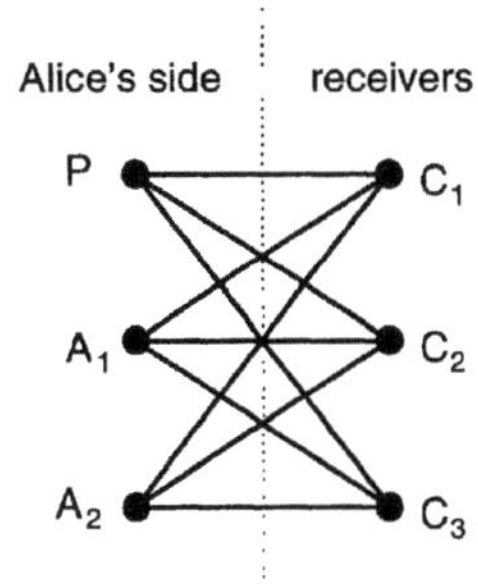

Figure 6. The telecloning state for $M = 3$ clones, consisting of one 'port' qubit, P two ancilla qubits (A_1 and A_2) and three output qubits (C_{1-3}). Solid lines indicate the existence of two-qubit entanglement. Due to the symmetries of the state, the roles of the port and ancilla qubits may be interchanged, as well as those of the transmitting and receiving sides.

Further insight into the telecloning procedure can be gained by examining the detailed entanglement structure of the telecloning state itself. First of all, we note that by rewriting this state so that the qubits on Alice's and the receivers' sides are explicitly separated, we obtain

$$|\psi_{TC}\rangle = \frac{1}{\sqrt{M+1}} \sum_{j=0}^{M} |\{0, M-j\}, \{1,j\}\rangle_{PA} \otimes |\{0, M-j\}, \{1,j\}\rangle_{C} . \tag{22}$$

This form highlights the high degree of symmetry of the telecloning state: it is completely symmetric under the permutation of any two particles on the same side, and also under the exchange of both sides. This implies that, in fact, any of the $2M$ qubits can be used as the telecloning port, with the clones being created on the opposite side. We can also see that the telecloning state contains only $\log_2(M+1)$ e-bits of entanglement between the two sides, representing a much more efficient use of entanglement than the more straightforward approach where Alice first clones her particle M times and then uses M singlets to transmit these states to the different receivers.

While entanglement between the two sides gives a measure of the resources necessary to accomplish telecloning, the entanglement between each pair of particles helps track how information from Alice's unknown state is conveyed to the clones. Due to the symmetries of the telecloning state, there are only two different classes of pairs: those where both qubits are on opposite sides (Alice's and the receivers') and those where they are on the same side. Using the Peres-Horodecki theorem[14] we find[10] that the first kind of pair are entangled, while the second are disentangled.

These calculations allow us to view the telecloning state as a 'network' of entangled qubits, each of which is only connected to the M qubits on the opposite side (so the total number of 'links' is M^2; see Fig. 6). Essentially, we may think of these 2-qubit connections as 'communication channels' through which quantum information may travell). In this sense, the multiparticle entanglement structure functions as a *multiuser channel*, allowing quantum information from Alice's input state to be conveyed to all the output clones.

CONCLUSIONS

We have studied various applications of multiparticle entanglement and teleportation in this article. Given the rapid experimental progress in this area, we would anticipate at least some of the ideas above being implemented in the forseeable future. They offer interesting possibilities for secure communication between many users and the optimal broadcasting of quantum information.

Acknowledgements

This work was supported by European TMR Research Networks, the UK Engineering and Physical Sciences Research Council, the Brazilian agency Conselho Nacional de Desenvolvimento Científico e Tecnológico (CNPq), the ORS Award Scheme, the Inlaks Foundation, the Japan Society for Promotion of Science, and the Elsag-Bailey Company.

REFERENCES

1. C.H. Bennett, G. Brassard, C. Crépeau, R. Josza, A. Peres and W.K. Wootters, Teleporting an unknown quantum state via dual classic and Einstein-Podolsky-Rosen channels, Phys. Rev. Lett. 70:1895 (1993).
2. D. Bouwmeester, J.-W. Pan, K. Mattle, M. Eibl, H. Weinfurter and A. Zeilinger, Experimental quantum teleportation, Nature 390:575 (1997); D. Boschi, S. Branca, F. DeMartini, L. Hardy and S. Popescu Experimental realization of teleporting an unknown pure quantum state via dual classical and Einstein-Podolsky-Rosen channels, Phys. Rev. Lett. 80:21 (1998).
3. C. H. Bennett, G. Brassard, S. Popescu, B. Schumacher, J. A. Smolin, W. K. Wootters, Purification of noisy entanglement and faithful teleportation via noisy channels, Phys. Rev. Lett. 76:722 (1996); C. H. Bennett, D. P. DiVincenzo, J. A. Smolin, W. K. Wootters, Mixed State Entanglement and Quantum Error Correction, Phys. Rev. A 54:3824 (1996).
4. L. Grover, Quantum telecomputation, e-print quant-ph/9704012; A. Ekert, S. F. Huelga, C. Macchiavello, J. I. Cirac, Distributed Quantum Computation over Noisy Channels, e-print quant-ph/9803017.
5. M. Zukowski, A. Zeilinger, M. A. Horne, and A. K. Ekert, Event-ready-detections in a Bell experiment via entanglement swapping, Phys. Rev. Lett 71:4287 (1993).
6. J. Pan, D. Bouwmeester, H. Weinfurter and A. Zeilinger, Experimental Entanglement Swapping: Entangling Photons That Never Interacted, Phys. Rev. Lett. 80:3891 (1998).
7. S. Bose, V. Vedral, and P. L. Knight, Multiparticle generalization of entanglement swapping, Phys. Rev. A 57:822 (1998).
8. E. Biham, B. Huttner and T. Mor, Quantum cryptographic network based on quantum memories, Phys. Rev. A 54:2651 (1996).
9. C.H. Bennett and S. J. Wiesner, Communication via one-particle and 2-particle operations on Einstein-Podolsky-Rosen states, Phys. Rev. Lett. 69:2881 (1992).
10. M. Murao, D. Jonathan, M. B. Plenio and V. Vedral, Quantum telecloning and multiparticle entanglement, e-print quant-ph/9806082
11. W.K. Wootters and W.H. Zurek, A single quantum cannot be cloned, Nature 299:802 (1982).
12. V. Bužek and M. Hillery, Quantum copying - beyond the no-cloning theorem, Phys. Rev. A 54:1844 (1996).
13. N. Gisin and S. Massar, Optimal Quantum Cloning Machines, Phys. Rev. Lett. 79:2153 (1997); D. Bruß, D.P. DiVincenzo, A. Ekert, C.A. Fuchs, C. Macchiavello and J.A. Smolin, Optimal Universal and State-Dependent Quantum Cloning, Phys. Rev. A. 57:368 (1998); D. Bruß, A. Ekert and C. Macchiavello, Optimal universal quantum cloning and state estimation, e-print quant-ph/9712019; V. Bužek, M. Hillery and P. L. Knight, Flocks of quantum clones: multiple copying of qubits, Fort. Phys. 46:521 (1998); V. Bužek and M. Hillery, Universal optimal cloning of qubits and quantum registers, e-print quant-ph/9801009; R. Werner, Optimal Cloning of Pure States, e-print quant-ph/9804001; P. Zanardi, A Note on Quantum Cloning in *d* dimensions, e-print quant-ph/9804011.
14. A. Peres, Separability criterion for density matrices, Phys. Rev. Lett. 77:1413 (1996); M. Horodecki, P. Horodecki and R. Horodecki, Information-theoretic aspects of inseparability of mixed states, Phys. Lett. A 223:1 (1996).

DECOHERENCE CONTROL IN QUANTUM INFORMATION PROCESSING: SIMPLE MODELS

Lorenza Viola and Seth Lloyd

d'Arbeloff Laboratory for Information Systems and Technology,
Department of Mechanical Engineering,
Massachusetts Institute of Technology,
Cambridge, Massachusetts 02139

We explore a strategy for protecting the evolution of a qubit against the effects of environmental noise based on the application of controlled time-dependent perturbations. In the case of a purely decohering coupling, an explicit sequence of control operations is designed, able to average out the decoherence of the qubit with high efficiency. We argue that, in principle, the effects of arbitrary qubit-environment interactions can be removed through suitable decoupling perturbations acting on the system dynamics over time scales comparable to the correlation time of the environment.

INTRODUCTION

Decoherence remains one of the most serious obstacles to the exploitation of the speed-up promised by quantum computation [1]. Broadly speaking, two different philosophies are being investigated to overcome the decoherence problem. On one hand, *passive* error-prevention schemes have been proposed, based on the idea of encoding logical quantum bits (qubits) within subspaces which do not decohere owing to symmetry properties [2, 3]. On the other hand, *active* error-correction approaches have been formalized within a sophisticated theory of quantum error-correcting codes (QECC), where a logical qubit is encoded in the larger Hilbert space of several physical qubits and suitable feedback operations are conditionally carried out[4, 5].

Although a purposeful manipulation is implied in the latter case, quantum error-correcting codes can be properly interpreted in terms of a clever redundancy in the software architecture rather than a physical way to operate on decoherence. In this work, we explore the possibility of using control techniques to modify and eliminate decoherence. Unlike recent proposals for feedback (or closed-loop) control schemes of decoherence in quantum optical systems[6], we apply control in the simpler open-loop configuration to general models of quantum information processing systems[7]. The underlying idea is suggested by high-resolution pulsed Nuclear Magnetic Resonance (NMR), where astonishingly versatile refocusing and decoupling techniques are nowadays available to remove the effects of interactions among the spins that are considered

Quantum Communication, Computing, and Measurement 2
Edited by Kumar et al., Kluwer Academic / Plenum Publishers, New York, 2000.

unwanted or uninteresting[8]. In our analysis, we outline the conditions under which analogous procedures can be extended from eliminating interactions *internal* to the system to suppressing interactions of the system with an *external* quantized environment. In particular, the role of the environment correlation time will be pointed out as a further parameter to be engineered in the struggle for preserving quantum coherence.

QUANTUM BANG-BANG CONTROL OF QUBIT DECOHERENCE

We start by investigating a prototype situation that conveys the basic idea in the simplest form. We will focus on the dynamics of a single memory cell of quantum information (qubit) undergoing decoherence due to the coupling to a thermal reservoir. The physical qubit can be associated either to a fictitious or to a real spin-1/2 system, the latter case allowing for a direct reference to the language of Nuclear Magnetic Resonance and NMR quantum computation[8, 9]. We assume that the fastest relaxation process originated by the interaction with the quantized environment is a *purely dephasing* process i.e., in NMR terminology, no dissipative T_1-type of decay takes place. A general model for the dynamics of the overall qubit + bath system is then provided by the purely decohering spin-boson Hamiltonian ($\hbar = 1$):

$$H_0 = H_S + H_B + H_{SB} = \frac{\omega_0}{2}\sigma_z + \sum_k \omega_k b_k^\dagger b_k + \sigma_z \sum_k (g_k b_k^\dagger + g_k^* b_k) \,. \tag{1}$$

Here, σ_z is the standard diagonal Pauli matrix, with qubit basis states denoted as $|i\rangle$, $i = 0, 1$, while $b_k^\dagger, b_k, g_k$ are bosonic operators and coupling parameters for the k-th field mode respectively. Hamiltonian (1) is widely used in the quantum computation literature to investigate the effect of phase errors, representing the most nonclassical and dangerous source of errors for quantum qubits[10, 11, 12]. Since $[\sigma_z, H_0] = 0$, spin populations are not affected by time evolution and decoherence dynamics is characterized completely by the qubit coherence with respect to the computational basis:

$$\rho_{01}(t) \equiv \langle 0|\, \rho_S(t)\, |1\rangle = \langle 0|\, \mathrm{Tr}_B\{\rho_{tot}(t)\}\, |1\rangle = \mathrm{Tr}_{B,S}\{\rho_{tot}(t_0)\sigma_+(t)\} \equiv \langle \sigma_+(t)\rangle \,, \tag{2}$$

where, starting from the left and using standard notations, the relevant reduced density matrix element $\rho_{01}(t) = \rho_{10}^*(t)$ in the Schrödinger picture is linked to the expectation value of the time-evolved ladder operator in the Heisenberg representation

$$\sigma_+(t) = \frac{1}{2}(\sigma_x(t) + i\,\sigma_y(t)) \,. \tag{3}$$

The evolution of the qubit coherence (2) can be calculated exactly under the customary assumptions about the initial state of the overall system, i.e. qubit and environment are initially uncorrelated and the environment is in thermal equilibrium at a temperature T. In the Heisenberg representation, one may write formally

$$\langle \sigma_+(t)\rangle = \langle G_{tot}(t_0, t)\, \sigma_+(t_0)\rangle \,, \tag{4}$$

with the propagator $G_{tot}(t_0, t)$ determined by the solution of the Heisenberg equations for the coupled spin + bath motion:

$$\left\{ \begin{array}{rcl} \dot\sigma_z(t) &=& 0 \,, \\ \dot\sigma_+(t) &=& i\omega_0\sigma_+(t) + 2i\sum_k (g_k b_k^\dagger(t) + g_k^* b_k(t))\,\sigma_+(t) \,, \\ \dot b_k(t) &=& -i\omega_k b_k(t) - i g_k \sigma_z(t) \,, \\ \dot b_k^\dagger(t) &=& +i\omega_k b_k^\dagger(t) + i g_k^* \sigma_z(t) \,. \end{array} \right. \tag{5}$$

The result can be written in the following form[7]:

$$\rho_{01}(t) = \mathrm{e}^{i\omega_0(t-t_0)-\Gamma_0(t-t_0)}\, \rho_{01}(t_0) \,, \tag{6}$$

the loss of phase information being characterized by the damping function

$$\Gamma_0(t-t_0) \equiv \sum_k \Gamma_0(k; t-t_0) = \sum_k 4|g_k|^2 \coth\left(\frac{\omega_k}{2T}\right) \frac{1-\cos\omega_k(t-t_0)}{\omega_k^2} \,, \tag{7}$$

in units where the Boltzmann constant $k_B = 1$. In the limit of a truly macroscopic environment, a description in terms of a continuum of modes is appropriate and the dependence of the decoherence function (7) on reservoir properties can be cast in a compact form after introducing the spectral density function $I(\omega)$,

$$\Gamma_0(t-t_0) \equiv \int_0^\infty d\omega\, \Gamma_0(\omega; t-t_0) = \int_0^\infty d\omega\, I(\omega)\, 4\coth\left(\frac{\omega}{2T}\right) \frac{1-\cos\omega(t-t_0)}{\omega^2} \,. \tag{8}$$

Depending on the temperature T and the spectral density $I(\omega)$, qualitatively different open-system evolutions arise in general, with a different interplay between quantum fluctuation and dissipation phenomena. Regardless the details of the spectral density function, however, the existence of a certain ultraviolet cut-off frequency ω_c is always demanded on physical grounds, leading to

$$I(\omega) \to 0 \qquad\qquad \text{for } \omega > \omega_c \,. \tag{9}$$

Although the specific value of ω_c depends on a natural cut-off frequency varying from system to system, ω_c can be generally associated to a characteristic time $\tau_c \sim \omega_c^{-1}$ setting the fastest (finite !) time scale of the irreversible dynamics. τ_c is known as the *correlation time* of the environment. The dynamics of the decoherence process arising from (8) for various choices of $I(\omega)$ has been investigated in detail elsewhere[7, 11]. A pictorial representation for the important class of Ohmic reservoirs, $I(\omega) \propto \omega\, \mathrm{e}^{-\omega/\omega_c}$, is shown in Fig. 1.

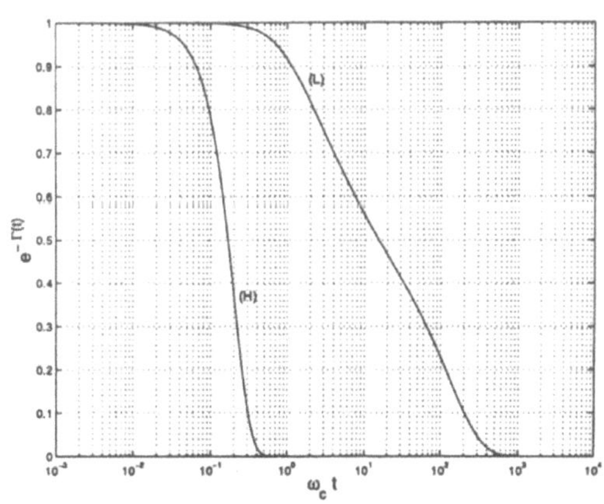

Figure 1. Qubit decoherence as a function of time for an Ohmic environment. Time is in units of T^{-1} and $\omega_c = 100$. High- and low-temperature behaviors are depicted, (H) $\omega_c/T = 10^{-2}$ and (L) $\omega_c/T = 10^2$ respectively.

We introduce now a procedure aimed at improving the coherence properties of the qubit by adding a controllable time-dependent interaction to the original Hamiltonian:

$$H(t) = H_0 + H_1(t) = H_S + H_B + H_{SB} + H_1(t) \,. \tag{10}$$

In the same spirit underlying multiple-pulse techniques in the manipulation of nuclear spin Hamiltonians[8], we try to average out the unwanted effects of the qubit-reservoir coupling H_{SB} by applying a sequence of coherent π-pulses that repetitively flip the state of the system. Under the assumptions that the duration of the pulses is short enough compared to the typical decoherence time and the strength of the control field is sufficient to override the H_{SB} coupling, the pulsed-mode operation allows one to separate the actions of the bath and the external field, by neglecting H_{SB} while $H_1(t)$ is on. Specifically, $H_1(t)$ is assumed to schematize a train of n_P identical π-pulses along the $\hat{x}$-axis applied on resonance at instants $t = t_P^{(n)}$, $n = 1, \ldots, n_P$, with pulse separation $t_P^{(n+1)} - t_P^{(n)} = \Delta t$. By invoking, as usual, the rotating-wave approximation, we have

$$H_1(t) = \Xi(t) \left[\cos(\omega_0 t)\sigma_x + \sin(\omega_0 t)\sigma_y \right] , \tag{11}$$

with envelope function

$$\Xi(t) = \sum_{n=1}^{n_P} V \left[\vartheta(t - t_P^{(n)}) - \vartheta(t - t_P^{(n)} - \tau_P) \right] , \quad t_P^{(n)} = t_0 + n\Delta t, \; n = 1, \ldots, n_P \, . \tag{12}$$

In Eq. (12), $\vartheta(\cdot)$ denotes the Heaviside step function, and the height V and the width τ_P of each pulse satisfy $2V\tau_P = \pi$. To simplify things, we work henceforth in the limit of infinitely narrow pulses $\tau_P \to 0$, assuming the kicks of radiofrequency control field large enough to produce instantaneous spin rotations. By analogy with the classical technique of bang-bang (or on-off) controls, whereby piecewise controls with extremal values are exploited[13], one may look at this strategy as an implementation of *quantum bang-bang control.*

In order to depict the evolution associated to a given pulse sequence, it is convenient to think the latter as formed by repeated elementary cycles of spin-flips, a complete cycle being able to return the spin back to the starting configuration. For definiteness, let us analyze the first cycle, made of the following steps: evolution under H_0 during $t_0 \leq t \leq t_P^{(1)}$; π-pulse P_1 at time $t_P^{(1)}$; evolution under H_0 during $t_P^{(1)} \leq t \leq t_P^{(2)}$; π-pulse P_2 at time $t_P^{(2)}$. After a total time $t_1 = t_0 + 2\Delta t$, the first cycle is complete. The description of π-pulses turns out to be extremely simple in the Heisenberg representation. Nothing happens to the bath operators $b_k, b_k^\dagger$ in the limit of instantaneous pulses, while, by denoting with $t_P^{-(+)}$ the times immediately before (after) a pulse respectively, spin operators are transformed as follows:

$$\begin{cases} \sigma_z(t_P^+) &= -\sigma_z(t_P^-) \, , \\ \sigma_+(t_P^+) &= [\sigma_+(t_P^-)]^\dagger \, . \end{cases} \tag{13}$$

In terms of the free propagator $G_{tot}(t_i, t_j)$ introduced in (4) to evolve coherence from t_i to t_j, the time development during the cycle can be represented as

$$\langle \sigma_+(t_0 + 2\Delta t) \rangle = \langle G_{tot}(t_0, t_0 + \Delta t) \, \sigma_+(t_0) \, G_{tot}^\dagger(t_0 + \Delta t, t_0 + 2\Delta t) \rangle \, , \tag{14}$$

to be compared with

$$\langle \sigma_+(t_0 + 2\Delta t) \rangle = \langle G_{tot}(t_0 + \Delta t, t_0 + 2\Delta t) \, G_{tot}(t_0, t_0 + \Delta t) \, \sigma_+(t_0) \rangle \tag{15}$$

in the absence of pulses. Since instantaneous rotations introduce discontinuous changes in operators (13), care must be taken in evaluating the two propagators $G_{tot}(t_0, t_0 + \Delta t)$, $G_{tot}(t_0 + \Delta t, t_0 + 2\Delta t)$ separately, by solving Heisenberg equations of motion (5) with initial conditions at $t = t_0$, $t = t_P^+ = t_0 + \Delta t$ respectively. Only at the end of the calculation everything can be expressed with respect to the initial time of the cycle. Omitting the details, the result for the coherence evolution over the first complete cycle is[7]

$$\rho_{01}(t_0 + 2\Delta t) = \mathrm{e}^{-\Gamma_P(N=1,\Delta t)}\, \rho_{01}(t_0) \,, \tag{16}$$

where a new decoherence function for $N = 1$ spin cycles has been introduced:

$$\Gamma_P(N = 1, \Delta t) = \sum_k \Gamma_0(k; 2\Delta t) \left|1 - 2\frac{1 - \mathrm{e}^{i\omega_k \Delta t}}{1 - \mathrm{e}^{2i\omega_k \Delta t}}\right|^2 . \tag{17}$$

Since, for each mode, the additional factor arising from the pulses is of order $O(\omega_k^2 \Delta t^2) \ll 1$ for small Δt, we may guess that something interesting is happening in a regime where the state of the qubit is tipped very rapidly. This is made clear by generalizing the description to an arbitrary number N of spin-flip cycles, involving a total number of $n_P = 2N$ π-pulses. After straight forward calculations along the same line outlined above, the expression for the qubit coherence at the final time $t_N = t_0 + 2N\Delta t$ is the following:

$$\rho_{01}(t_0 + 2N\Delta t) = \mathrm{e}^{-\Gamma_P(N,\Delta t)}\, \rho_{01}(t_0) \,, \tag{18}$$

with

$$\Gamma_P(N, \Delta t) = \sum_k \Gamma_0(k; 2N\Delta t) \left|1 - f_k(N, \Delta t)\right|^2 , \tag{19}$$

$$f_k(N, \Delta t) = 2\,\frac{1 - \mathrm{e}^{i\omega_k \Delta t}}{1 - \mathrm{e}^{2i\omega_k \Delta t}} \sum_{n=1}^{N} \mathrm{e}^{2i(n-1)\omega_k \Delta t} . \tag{20}$$

From inspection of Eq. (19), the contribution due to the pulse sequence turns out to manifest in the typical form of an interference factor. The implications for the decoherence properties are easily stated by considering the mathematical limit where $\Delta t \to 0$, $N \to \infty$, subjected to the constraint $2N\Delta t = t_N - t_0$. Under these conditions, one can prove that[7]

$$\lim_{\Delta t \to 0} f_k(N, \Delta t) = 1 \qquad \forall k \qquad \Rightarrow \qquad \lim_{\Delta t \to 0} \Gamma_P(N, \Delta t) = 0 \,. \tag{21}$$

Thus, in the limit of continuous flipping, decoherence is completely washed out for any temperature and any spectral density function.

Obviously, a continuous limit of this kind is scarcely meaningful from a physical point of view. However, this ideal situation should be approached if Δt is made small compared to the fastest characteristic time operating within the environmental noise, i.e. the reservoir correlation time τ_c. Hence, we expect that a sufficient condition in order to meet (21) and suppress decoherence is

$$\Delta t \ll \tau_c \,. \tag{22}$$

More explicitly, this result implies that, given an arbitrary time t, one can always recover the initial state and the coherence of the qubit by making t the end time of a pulse sequence and by adjusting the parameters to satisfy $t = t_N = 2N\Delta t$ and $\Delta t \ll \tau_c$. Then, at time t, a *coherence echo* is formed. Alternatively, by keeping the qubit flipped and restricting the observation to cycle times t_N, $N = 1, 2, \ldots$, the system is found to evolve ideally as it would do in the absence of the coupling H_{SB} responsible for decoherence. A typical behavior originated by the pulsing procedure for the prototype high-temperature Ohmic environment of Fig. 1 is displayed in Fig. 2.

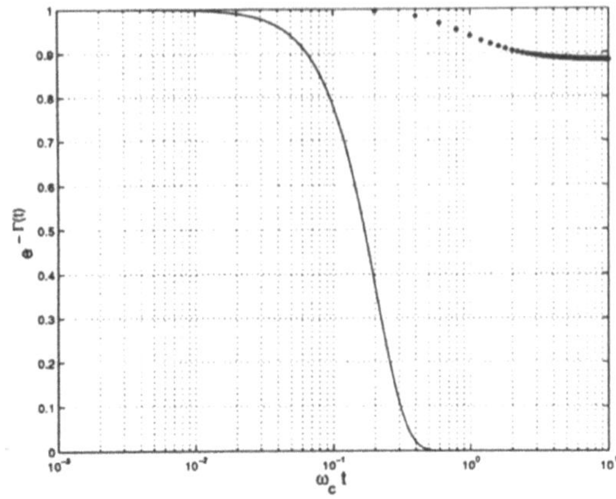

Figure 2. Qubit pulsed decoherence as a function of time for the Ohmic high-temperature environment of Fig. 1. A pulse separation $\Delta t = \tau_c/10$ has been used and coherence stroboscopically evaluated using Eqs. (18)-(19). Long-time deviations from unit value arise from cumulation of errors in the presence of a small but finite Δt.

So far, the suppression of decoherence has been derived in a rather formal way. Actually, a simple physical explanation can be provided as well. Similarly to the original well-known spin-echo phenomenon[14], and to the more sophisticated solid-echoes or magic echoes experiments[8], the basic argument here is a time-reversal argument. The examination of a single elementary spin-flip cycle suffices to capture the underlying mechanism. Roughly speaking, and looking back at the representation (4), it is the presence of the transformed propagator $G^{-1}_{tot}(t_0+\Delta t, t_0+2\Delta t)$, generated by the couple of π-pulses, that *simulates* the effect of a time-reversal. Would the evolution during the second half of the cycle be identical to the one in the first Δt interval, then it would be

$$G^{-1}_{tot}(t_0+\Delta t, t_0+2\Delta t) = G^{-1}_{tot}(t_0, t_0+\Delta t) \ , \tag{23}$$

and, therefore, $\langle \sigma_+(t_0+2\Delta t)\rangle = \langle \sigma_+(t_0)\rangle$ as a consequence of the cyclic property in the trace. Instead, this reversal is only approximate in general since the two propagations differ by a dephasing factor $e^{i\omega_k \Delta t}$ in the evolution of each reservoir mode[7]. However, if the condition (22) is met, then the cycle is effectively equivalent to an exact time-reversal and, by iteration on every cycle, the elimination of decoherence (21) is achieved.

DYNAMICAL DECOUPLING OF QUBIT-ENVIRONMENT INTERACTIONS

In this section, we rederive the result established above in a form that opens up the way to further generalization. The first step is to formally reinterpret the method of reducing environment-induced decoherence by successive application of π-pulses within the general framework of decoupling techniques based on *controlled averaging*. In NMR, sophisticated decoupling schemes are routinely used to simplify complex spectra by manipulating the underlying spin Hamiltonian to an extent allowing for a successful analysis[8, 15]. In particular, a relevant class of decoupling procedures, including spin-decoupling and multiple-pulse experiments, involves selective averaging in the internal spin space. The idea is to introduce controlled motions into the system, with the time-dependence designed in such a way that undesired terms in the Hamiltonian are averaged out. In extending similar techniques to the decoupling of interactions between a system and its environment, the major difference stems from the fact that the decoupling action can be easily exerted only on the system variables, the bath degrees of freedom being generally uncontrollable.

Looking back at the Hamiltonian (1), we start by seeking a perturbation $H_1(t)$ to be added as a suitable decoupling interaction in order to remove H_{SB}, Eq. (10). We restrict to a situation where $H_1(t)$ is *cyclic*, i.e. satisfying the following conditions ($t_0 = 0$ henceforth):

$$(i) \quad H_1(t) = H_1(t + \Delta t) \qquad \text{for some } \Delta t \; ; \tag{24}$$

$$(ii) \quad U_1(t) \equiv T \exp\Big\{ - i \int_0^t ds\, H_1(s) \Big\} = U_1(t + T_c) \qquad \text{for some } T_c \, . \tag{25}$$

From (ii), $U_1(T_c) = \mathbb{I}$ and T_c is called the *cycle time*. An elegant description of the dynamics arising in the presence of a cyclic perturbation is provided by the so-called average Hamiltonian theory[8, 15]. We only recall here the basic ingredients in a language that is appropriate to quantum information processing. In particular, we schematize the perturbation as a sequence on n_P ideal pulses (or logic gates), represented by unitary transformations $P_1, \ldots, P_{n_P}$, and separated by free evolution periods under H_0. The operators $\{P_i\}$ fulfill $U_1(T_c) = P_{n_P} \ldots P_2 P_1 = \mathbb{I}$ by cyclicity. Average Hamiltonian theory is based on moving to the time-dependent interaction representation with respect to $H_1(t)$ defined, as usual, by

$$\rho_{tot}(t) = U_1(t) \tilde{\rho}_{tot}(t) U_1^\dagger(t) \, , \tag{26}$$

with

$$\frac{d}{dt}\tilde{\rho}_{tot}(t) = -i\big[\tilde{H}(t), \tilde{\rho}_{tot}(t)\big] \, , \qquad\qquad \tilde{H}(t) = U_1^\dagger(t) H_0 U_1(t) \, . \tag{27}$$

In (27), H_0 is given by (1) and, according to the standard NMR literature, the transformed Hamiltonian $\tilde{H}(t)$ is also known as the *toggling frame Hamiltonian*. If $U_{tot}(t)$ and $\tilde{U}_{tot}(t)$ denote the time evolution operators in the Schrödinger and interaction picture respectively, due to (25) one gets

$$U_{tot}(nT_c) = U_1(nT_c)\tilde{U}_{tot}(nT_c) = \tilde{U}_{tot}(nT_c) = \big[\tilde{U}_{tot}(T_c)\big]^n \equiv e^{-i\overline{H}t_n} \, , \tag{28}$$

i.e. provided the observation of the dynamics is restricted to *stroboscopic* and *synchronized* sampling, $t_n = nT_c$, it is sufficient to know the evolution over a single cycle as described by the transformed Hamiltonian $\tilde{H}(t)$. In (28), the last equality defines the average Hamiltonian and contains the main result of average Hamiltonian theory: for suitable repeated observations, the motion of the system under the influence of the time-dependent field $H_1(t)$ can be represented by a *constant* average Hamiltonian $\overline{H}$. The calculation of $\overline{H}$ is usually performed on the basis of a standard Magnus expansion of the time-ordered exponential defining $\tilde{U}_{tot}(T_c)$[15], i.e.

$$\tilde{U}_{tot}(T_c) = T \exp\Big\{ - i \int_0^{T_c} ds\, \tilde{H}(s) \Big\} \equiv e^{-i\overline{H}T_c} = e^{-i[\overline{H}^{(0)} + \overline{H}^{(1)} + \ldots]T_c} \, . \tag{29}$$

In our case, the evaluation of (29) is simplified since the toggling frame Hamiltonian $\tilde{H}(t)$ is piecewise constant during the intervals Δt_k separating consecutive rotations and one can introduce stepwise transformed Hamiltonians:

$$\tilde{U}_{tot}(T_c) = e^{-i\tilde{H}_{n_P}\Delta t_{n_P}} \ldots e^{-i\tilde{H}_0 \Delta t_0} \, , \qquad\qquad \tilde{H}_k = (P_k \ldots P_1)^{-1} H_0 \, (P_k \ldots P_1) \, , \tag{30}$$

where, obviously, $\sum_k \Delta t_k = T_c$. Thus, the lowest-order approximation $\overline{H}^{(0)}$ to the average Hamiltonian in (29) has a particularly simple form:

$$\overline{H}^{(0)} = \frac{1}{T_c}\Big\{\tilde{H}_0\Delta t_0 + \ldots + \tilde{H}_{n_P}\Delta t_{n_P}\Big\} = \frac{1}{T_c}\sum_{k=0}^{n_P} \Delta t_k\, P_1^{-1}\ldots P_k^{-1}\, H_0\, P_k \ldots P_1 \,. \qquad (31)$$

In addition, explicit expressions for the corrections arising from higher-order terms are systematically available from the Magnus expansion[15]. Loosely speaking, the goal of decoupling is to devise a tranformation to a toggling frame (26), where the unwanted coupling H_{SB} no longer appears up to a certain order $\overline{H}^{(r)}$ and higher-order contributions $\overline{H}^{(r+1)}, \ldots$ are made neglegible.

We are now in a position to apply this formalism to the evolution of the decohering qubit, whose Hamiltonian we rewrite for convenience as follows:

$$H_0 = H_S + H_B + H_{SB} = \frac{\omega_0}{2}\sigma_z + \sum_k \omega_k b_k^\dagger b_k + \sigma_z \mathcal{B}_z \,, \qquad \mathcal{B}_z = \sum_k (g_k b_k^\dagger + g_k^* b_k) \,. \quad (32)$$

Using (27), the transformation to the toggling frame associated to $H_1(t)$ leads to

$$\tilde{H}(t) = H_B + U_1^\dagger(t)\, H_S\, U_1(t) + [U_1^\dagger(t)\, \sigma_z\, U_1(t)]\, \mathcal{B}_z \,, \qquad (33)$$

and to a zero-th order average Hamiltonian given by (31):

$$\overline{H}^{(0)} = H_B + \frac{1}{T_c}\sum_{k=0}^{n_P} \Delta t_k \mathcal{P}_k^{-1} H_S \mathcal{P}_k + \Big[\frac{1}{T_c}\sum_{k=0}^{n_P} \Delta t_k \mathcal{P}_k^{-1}\sigma_z \mathcal{P}_k\Big]\mathcal{B}_z \,, \qquad (34)$$

where, in (33) and (34), the fact that reservoir operators are unaffected by the control field has been evidenced and the short notation $\mathcal{P}_k = P_k \ldots P_1$ has been introduced. The second and third terms in (34) correspond, in general, to a transformed qubit Hamiltonian and a transformed qubit-bath interaction. It is immediate to realize that the effect of the π-pulse sequence of Sec. II viewed in this frame is to cause the latter interaction term to vanish. To make the identification explicit, we rearrange the elementary spin-flip cycle as follows:

$$\underbrace{-\,\frac{\Delta t}{2}\,-\,P_x^{180}\,-\,\frac{\Delta t}{2}\,-\,\frac{\Delta t}{2}\,-\,P_x^{180}\,-\,\frac{\Delta t}{2}}_{cycle}\,-\,\frac{\Delta t}{2}\,\cdots \qquad (35)$$

where the rotation axis of the pulses has been indicated and, to compare with (31), $\Delta t_0 = \Delta t_2 = \Delta t/2$, $\Delta t_1 = \Delta t$, $T_c = 2\Delta t$. Written in the form (35), the decoupling sequence for (32) is nothing but a variant of the famous *Carr-Purcell* (CP) sequence, that is ordinarily exploited to get rid of static applied-field inhomogeneities[8, 15]. At variance with this standard usage of the CP-sequence, however, where the size of Δt is of no importance, the averaging of H_{SB} at zeroth-order does not guarantee by itself the elimination of decoherence. Obviously, higher-order terms in (29) have to be quenched. The question under which circumstances the Magnus series can be truncated after the leading term or the few lowest-order corrections is nontrivial. It is possible to show[16] that, since the r-th order contribution $\overline{H}^{(r)} = O(\Delta t^r)$, a sufficient condition is $\Delta t\,\omega_c \ll 1$, which is identical to (22). Then, from (28), one ideally gets the stroboscopic equality

$$\rho_S(nT_c) = \mathrm{Tr}_B\{\, e^{-i\overline{H}^{(0)} nT_c}\, \rho_S(0)\rho_B(0)\, e^{i\overline{H}^{(0)} nT_c}\,\} = \rho_S(0) \,. \qquad (36)$$

The generalization of the above scheme to decouple arbitrary qubit-environment interactions is, in principle, straight forward since the most general bilinear coupling can be expressed as a mixed sum of *error-generators*[2, 17]:

$$H_{SB} = \sigma_x \mathcal{B}_x + \sigma_y \mathcal{B}_y + \sigma_z \mathcal{B}_z \ , \tag{37}$$

for suitable reservoir operators. Then, provided there is no constraint on the rate of control so that condition (22) can be assumed, one has to ensure the existence of a gate sequence generating the required temporal average:

$$\sum_{k=0}^{n_P} \Delta t_k \, \mathcal{P}_k \sigma_\alpha \mathcal{P}_k = 0 \ , \qquad \alpha \in \{x, y, z\} \ . \tag{38}$$

The purely decohering coupling corresponds to $\mathcal{B}_x = \mathcal{B}_y = 0$. Actually, binary sequences of the Carr-Purcell type also suffice to decouple any form of interaction (37) involving at most two error generators. A specially relevant case is a Jaynes-Cummings-like dissipative coupling with $\mathcal{B}_z = 0$ and[2]

$$H_{SB} = \sum_k (g_k b_k^\dagger \sigma_- + h.c.) \ \Rightarrow \ \mathcal{B}_x = \sum_k (g_k b_k^\dagger + h.c.) \, , \ \mathcal{B}_y = \sum_k (-i g_k b_k^\dagger + h.c.) \ , \tag{39}$$

which can be eliminated, in principle, by a sequence of π-pulses along the $\hat{z}$-axis. A slightly more elaborated sequence is necessary to decouple the qubit from the simultaneous action of the three error generators. It turns out that it can be derived on the basis of a simple group-theoretic argument. A more detailed and general formulation of the method is presented elsewhere[16].

CONCLUSIONS

Our work demonstrates the possibility to modify the evolution of a quantum open system by applying external controllable interactions. From the perspective of quantum information, the analysis suggests a different promising direction compared to conventional quantum error-correction techniques. The practical usefulness of the proposed approach strongly depends, in its present status, on the time scale of the motional processes causing relaxation. The effectiveness of analogous schemes under less idealized assumptions and in the presence of a finite bound on the control rate deserves further investigation, together with the possibility of examining decoherence properties within a fully quantum mechanical control configuration as recently proposed in [18].

Acknowledgments

The authors are indebted to E. Knill for many insightful comments. This work was supported by ONR, by AFOSR, and by DARPA/ARO under the Quantum Information and Computation initiative (QUIC) and the NMR Quantum Computing initiative (NMRQC).

REFERENCES

1. D. P. DiVincenzo, Quantum computing, *Science* 270:255 (1995).
2. P. Zanardi and M. Rasetti, Error avoiding quantum codes, *Mod. Phys. Lett. B* 11:1085 (1997); Noiseless quantum codes, *Phys. Rev. Lett.* 79:3306 (1997); P. Zanardi, Dissipation and decoherence in a quantum register, *Phys. Rev. A* 57:3276 (1998).

3. D. A. Lidar, I. L. Chuang, and K. B. Whaley, Decoherence free subspaces for quantum computation, LANL e-print quant-ph/9807004, submitted to *Phys. Rev. Lett.*
4. P. W. Shor, Scheme for reducing decoherence in quantum computer memory, *Phys. Rev. A* 52:R2493 (1995); E. Knill and R. Laflamme, Theory of quantum error-correcting codes, *Phys. Rev. A* 55:900 (1997), and references therein.
5. D. G. Cory *et al.*, Experimental quantum error correction, LANL e-print quant-ph/9802018, submitted to *Phys. Rev. Lett.*
6. D. Vitali, P. Tombesi, and G. J. Milburn, Quantum-state protection in cavities, *Phys. Rev. A* 57:4930 (1998).
7. L. Viola and S. Lloyd, Dynamical suppression of decoherence in two-state quantum systems, LANL e-print quant-ph/9803057, to be published in *Phys. Rev. A* 58, (1998, September issue).
8. R. R. Ernst, G. Bodenhausen, and A. Wokaun. "Principles of Nuclear Magnetic Resonance in One and Two Dimensions", Clarendon Press, Oxford (1987).
9. N. Gershenfeld and I. L. Chuang, Bulk spin-resonance in quantum computation, *Science* 275:350 (1997); Quantum computing with molecules, *Sci. Am.* 278:66 (1998).
10. W. G. Unruh, Maintaining coherence in quantum computers, *Phys. Rev. A* 51:992 (1995).
11. G. M. Palma, K.-A. Suominen, and A. K. Ekert, Quantum computers and dissipation, *Proc. R. Soc. London A* 452:567 (1996).
12. M. Ban, Photon-echo technique for reducing the decoherence of a quantum bit, to be published in *J. Mod. Opt.* (1998).
13. Y. Takahashi, M. J. Rabins, and D. M. Auslander. "Control and Dynamic Systems", Addison-Wesley, Reading, MA (1970); J. Macki and A. Strauss. "Introduction to Optimal Control Theory", Springer-Verlag, New York (1982).
14. E. L. Hahn, Spin echoes, *Phys. Rev.* 80:580 (1950).
15. U. Haeberlen. "High Resolution NMR in Solids", New York, Academic Press (1976).
16. L. Viola, E. Knill, and S. Lloyd, Dynamical decoupling of quantum open systems interactions, in preparation.
17. E. Knill, Group representation, error bases and quantum codes, LANL report LAUR-96-2807, http://www.c3.lanl.gov/ knill.
18. S. Lloyd, Controllability and observability of quantum systems, LANL e-print quant-ph/9703042, submitted to *Phys. Rev. A*.

ENTANGLEMENT OF FORMATION

William K. Wootters

Department of Physics
Williams College
Williamstown, MA 01267

At least three distinct information-theoretic measures of quantum entanglement have been proposed in recent years. This paper highlights one of them, the entanglement of formation, as well as a related quantity called the tangle. The tangle can be used to express a simple law of entanglement, limiting the extent to which a single qubit may be simultaneously entangled with each of two other qubits.

Quantum entanglement is one of the central concepts in quantum information theory and is perhaps the feature that distinguishes it most dramatically from its classical counterpart. Ultimately one would like to develop a full quantitative theory of entanglement, but even to begin to develop such a theory requires a measure of entanglement. This paper highlights one such measure, the entanglement of formation, and a related quantity called the "tangle," and presents one example of a quantitative law of entanglement.

A pure state of a bipartite system is called entangled if neither part of the system has a pure state of its own, that is, if the state is not the tensor product of states of the two subsystems. A classic example is the singlet state of a pair of spin-$\frac{1}{2}$ particles, $\frac{1}{\sqrt{2}}(|\uparrow\downarrow\rangle - |\downarrow\uparrow\rangle)$, in which the two particles exhibit perfect anticorrelation along any spin axis but have no definite spin directions of their own. A *mixed* state is called entangled if it cannot be expressed as a probabilistic mixture of unentangled pure states.[1, 2]

For a *pure* state $|\psi\rangle$ of a bipartite system, there is a unique natural information-theoretic measure of entanglement, namely, the von Neumann entropy of either of the two parts:[3, 4]

$$E(\psi) = S(\rho_A) = -\mathrm{Tr}\rho_A \log_2 \rho_A = S(\rho_B). \tag{1}$$

Here A and B are the two subsystems, ρ_A is the partial trace of $|\psi\rangle\langle\psi|$ over subsystem B, and ρ_B is defined similarly. Intuitively this measure makes sense because the only reason either part of the system has any entropy is that it is entangled with the other part, the system as a whole having no entropy. A more rigorous justification is the following:[3] E as defined above measures the asymptotic conversion ratio between the state $|\psi\rangle$ and a standard, completely entangled two-qubit state such as the singlet state. That is, if two separated participants Alice and Bob share n singlets, they may

Quantum Communication, Computing, and Measurement 2
Edited by Kumar et al., Kluwer Academic / Plenum Publishers, New York, 2000.

use these singlets to create, in an asymptotically reversible way, roughly $\frac{n}{E(\psi)}$ pairs in a state approximating $|\psi\rangle$, using only local operations and classical communication.

For a mixed state ρ of a bipartite system the situation is more complicated. The entropy of each part no longer measures entanglement, because the parts can have entropy even if they are not entangled. At least three distinct information-theoretic measures of entanglement have been proposed for mixed states: (i) the entanglement of formation,[5] which is intended to quantify the resources needed to create a given state, (ii) the relative entropy of entanglement,[6, 7] which measures the distinguishability of the given state from the most similar unentangled state, and (iii) the distillable entanglement,[8] which measures the amount of pure-state entanglement that can be distilled from the given state.

To define the entanglement of formation, we note first that a mixed state, specified by a density matrix ρ, can be expressed in many ways as a probabilistic mixture of pure states:

$$\rho = \sum_i p_i |\psi_i\rangle\langle\psi_i|, \tag{2}$$

where $\sum_i p_i = 1$. For example, the completely mixed state of two qubits, whose density matrix is $\rho = \frac{1}{4}I$, can be written either as an equal mixture of the four standard basis states $|00\rangle, |01\rangle, |10\rangle, |11\rangle$ (using "0" and "1" now in place of "$\uparrow$" and "$\downarrow$") or as an equal mixture of the four completely entangled states $\frac{1}{\sqrt{2}}(|00\rangle \pm |11\rangle)$ and $\frac{1}{\sqrt{2}}(|01\rangle \pm |10\rangle)$, or in any of infinitely many other ways. Clearly not all pure-state decompositions of a given density matrix have the same average entanglement: in the example just given, the two decompositions have average entanglements equal to zero and one, respectively. We define the entanglement of formation $E(\rho)$ to be the minimum average entanglement over all pure-state decompositions of ρ:[5]

$$E(\rho) = \inf \sum_i p_i E(\psi_i). \tag{3}$$

For the completely mixed state, for example, the entanglement of formation is zero.

Entanglement of formation derives its name from the following argument. Alice and Bob can create a large number of pairs of particles in the state ρ by first preparing pairs in the pure states that define a particular decomposition of ρ, and then collecting these pairs into a large set without keeping a record of the identity of each pair. The entanglement of formation measures the asymptotic number of singlets required, per pair, to create many pairs in the state ρ by this method. (Again we assume that Alice and Bob can perform only local operations and can communicate only classically.)

There is, however, a weakness in this argument, revealing a sense in which entanglement of formation could conceivably not live up to its name.[9] One can imagine other ways of creating many pairs in the state ρ, and the method just outlined may not be the best. For example, Alice and Bob could choose to create the pairs two at a time by building the mixed state $\rho \otimes \rho$ out of pure states. This method allows more possibilities than the first method, so that one can imagine it being more efficient. In fact no one has yet determined whether or not it is more efficient. In other words, no one knows whether $E(\rho \otimes \rho)$ is equal to $2E(\rho)$ or is less than $2E(\rho)$. If the latter is the case, then presumably one would do even better by building the pairs n at a time, so that the true measure of the number of singlets required to create the state ρ would be

$$E_\infty(\rho) = \lim_{n\to\infty} E(\rho \otimes \cdots \otimes \rho)/n, \tag{4}$$

where the tensor product contains n factors. If it turns out that $E(\rho)$ is not always equal to $E_\infty(\rho)$, then E_∞ would be the more meaningful measure of entanglement. For the rest of this paper we will deal only with E itself, despite this uncertainty about its interpretation, since very little is known about E_∞.

The definition of entanglement of formation calls for a minimization over all decompositions of the given density matrix. For a general bipartite system no analytic expression for the entanglement of formation is known, and the minimization simply has to be done numerically. Fortunately, however, for the special case of two qubits, we do have an analytic formula for E.[10, 11] This formula is conveniently expressed in terms of a quantity called the "tangle" of the density matrix,[12] which I now introduce.

For a general pure state of two qubits A and B,

$$|\psi\rangle = a|00\rangle + b|01\rangle + c|10\rangle + d|11\rangle, \tag{5}$$

the entanglement, that is, the entropy of either qubit, can be written as

$$E = h\Big(\frac{1}{2}(1 + \sqrt{1-\tau})\Big), \tag{6}$$

where $h(x) = -x\log_2 x - (1-x)\log_2(1-x)$ and the tangle τ is

$$\tau = 4|ad - bc|^2 = 4\det \rho_A = 4\det \rho_B. \tag{7}$$

The tangle runs from 0 (no entanglement) to 1 (maximal entanglement), and E is a monotonically increasing function of τ, so that τ may itself be taken as a measure of entanglement.

The formula for entanglement of formation of a *mixed* state of two qubits is the same as Eq. (6), but with a more general, and rather more complicated, definition of τ. (The definition given above has no meaning for a mixed state.) Let ρ be a general mixed state of a pair of qubits; thus ρ is a 4x4 density matrix. Let $\tilde{\rho}$ be the "spin-flipped" image of ρ, defined by

$$\tilde{\rho} = (\sigma_y \otimes \sigma_y)\,\rho^*(\sigma_y \otimes \sigma_y), \tag{8}$$

where ρ^* is the complex conjugate of ρ in the standard basis, and σ_y expressed in that same basis is $\left(\begin{smallmatrix} 0 & -i \\ i & 0 \end{smallmatrix}\right)$. The matrix $\rho\rho^*$, which is not necessarily hermitian, has real non-negative eigenvalues which we label $\lambda_1^2, \ldots, \lambda_4^2$ in order of decreasing size. Finally, we define the tangle of the mixed state ρ to be[10, 11]

$$\tau = (\max\{\lambda_1 - \lambda_2 - \lambda_3 - \lambda_4, 0\})^2. \tag{9}$$

It is this τ that one inserts into Eq. (6) to get the entanglement of formation. Again, since the tangle is monotonically related to entanglement of formation, it can be used as a measure of entanglement in its own right.

The formula given by Eqs. (6) and (9) makes it easy to evaluate the entanglement of formation of any state of two qubits. Unfortunately, neither this formula nor the tangle concept has been generalized to pairs of qutrits or larger systems. Moreover, even for a two-qubit system, of the three information-theoretic measures of entanglement mentioned above, only the entanglement of formation has so far been expressible in analytic form. The others in general require numerical extremization, and in the case of distillable entanglement even such extremization is very difficult, because we have no simple mathematical characterization of the local operations that Alice and Bob are allowed to perform.

We can, however, use Eq. (9) to gain insight into entanglement. Consider a triple of qubits labeled A, B, and C, the whole system being in a pure state. By tracing out qubit C, we obtain the density matrix of A and B and can use Eq. (9) to find the tangle τ_{AB} between them. It also makes sense to define, via Eq. (7), the tangle $\tau_{A(BC)}$ between qubit A and the pair BC: $\tau_{A(BC)} = 4 \det \rho_A$. One can then prove the following inequality relating $\tau_{A(BC)}$, τ_{AB}, and τ_{AC}.[13]

$$\tau_{AB} + \tau_{AC} \leq \tau_{A(BC)}. \tag{10}$$

In words, this inequality makes the following two assertions: (i) the entanglement between A and B (or between A and C) is limited by the extent to which A is entangled with the pair BC; (ii) there is a trade-off between A's entanglement with B and its entanglement with C.

Interestingly, given any values of the tangles τ_{AB}, τ_{AC}, and $\tau_{A(BC)}$ such that $\tau_{AB} + \tau_{AC} = \tau_{A(BC)}$, that is, such that the inequality (10) becomes an equality (and such that each tangle is between zero and one), there exists a quantum state having these values. Specifically, the quantum state

$$|\psi\rangle = a|100\rangle + b|010\rangle + c|001\rangle, \tag{11}$$

in which the three entries in each $|\ldots\rangle$ refer to the qubits A, B, and C and the numbers a, b, and c are real, has the following values for the tangles:

$$\begin{aligned} \tau_{AB} &= 4a^2b^2 \\ \tau_{AC} &= 4a^2c^2 \\ \tau_{A(BC)} &= 4a^2(b^2 + c^2), \end{aligned} \tag{12}$$

so that the equality holds. In this sense the inequality (10) is as strong as it could be. One could not, for example, replace the tangle in Eq. (10) with a non-trivial function of the tangle without either weakening the inequality or rendering it false.

One can take this inequality a step further by looking at the difference between the two sides. That is, for any given state one can check to see how far the sum $\tau_{AB} + \tau_{AC}$ falls short of the tangle between A and the pair BC taken as a whole. This difference could be interpreted as the part of the entanglement between A and BC that is not accounted for by the entanglement of A with B and C individually. Remarkably, one finds that for any pure state of three qubits, this difference is invariant under permutations of the labels A, B, and C.[12] It therefore expresses a kind of three-way entanglement among the objects without favoring any one of them. It would interesting to study this quantity further, to see whether one can flesh out its interpretation.

The trade-off between entanglements as expressed in Eq. (10) seems likely to apply in some sense to all quantum systems, not just to qubits. One has the intuition that every quantum object has a certain capacity for entanglement, determined by the dimension of its state space, and that the more it is entangled with one system the less it can be entangled with others. However, to give this intuition an exact mathematical formulation, we will need to learn more about the quantification of entanglement. In particular, we will need to find out whether there is a natural generalization of the tangle to larger systems. It would also be interesting to know whether the tangle has a direct physical interpretation, or whether it derives an interpretation only through its relationship with the entanglement of formation.

REFERENCES

1. A. Peres, *Phys. Rev. Lett.* **77**, 1413 (1996).
2. M. Horodecki, P. Horodecki, and R. Horodecki, *Phys. Lett. A* **223**, 1 (1996).
3. C. H. Bennett, H. J. Bernstein, S. Popescu, and B. Schumacher, *Phys. Rev. A* **53**, 2046 (1996).

4. S. Popescu and D. Rohrlich, *Phys. Rev. A* **56**, R3319 (1997).
5. C. H. Bennett, D. P. DiVincenzo, J. Smolin, and W. K. Wootters, *Phys. Rev. A* **54**, 3824 (1996).
6. V. Vedral, M. B. Plenio, M. A. Rippin, and P. L. Knight, *Phys. Rev. Lett.* **78**, 2275 (1997).
7. V. Vedral, M. B. Plenio, K. Jacobs, and P. L. Knight, *Phys. Rev. A* **56**, 4452 (1997).
8. C. H. Bennett, G. Brassard, S. Popescu, B. Schumacher, J. Smolin, and W. K. Wootters, *Phys. Rev. Lett.* **76**, 722 (1996).
9. S. Popescu, personal communication
10. S. Hill and W. K. Wootters, *Phys. Rev. Lett.* **78**, 5022 (1997).
11. W. K. Wootters *Phys. Rev. Lett.* **80**, 2245 (1998).
12. W. K. Wootters *Phil. Trans. R. Soc. Lond. A* **356**, 1717 (1998).
13. V. Coffman and J. Kundu, unpublished.

SENDING QUANTUM INFORMATION WITH GAUSSIAN STATES

Alexander S. Holevo,[1]

[1]Steklov Mathematical Institute,
Russian Academy of Sciences, Moscow

Quantum information characteristics, such as quantum mutual information, loss, noise and coherent information are explicitly calculated for Bosonic attenuation/amplification channel with input Gaussian state. The coherent information is shown to be negative for the values of the attenuation coefficient $k < 1/\sqrt{2}$.

INFORMATION CHARACTERISTICS OF A QUANTUM CHANNEL

During the last couple of years an impressive progress has been achieved in the theory of transmission of classical information through quantum communication channels (see Ref. 1 for a comprehensive survey). The problem of sending quantum information is much less understood; we refer in particular to the papers in Refs. 2, 3, 4, initiating the study of this problem, where the reader can find further references. In this paper we make a contribution to this study by considering rather concrete situation: sending Gaussian (quasifree) states through linear Bosonic channels.

Consider quantum system in a Hilbert space $\mathcal{H}$, with a fixed density operator ρ. A *channel* is a transformation of quantum states as presented by density operators, given by the relation

$$T[\rho] = \sum_j A_j \rho A_j^*,$$

where A_j are bounded operators in $\mathcal{H}$ satisfying $\sum_j A_j^* A_j = I$. Let us denote $H(\rho) = -\mathrm{Tr}\rho \log \rho$ the von Neumann entropy of a density operator ρ. We call ρ the input state, and $T[\rho]$ the output state of the channel. There are three important entropy quantities related to the couple (ρ, T) :

1) The entropy of the input state $H(\rho)$;
2) The entropy of the output state $H(T[\rho])$;
3) The entropy exchange $H(\rho, T)$.

While the definition and the meaning of the first two entropies is obvious, the third quantity is somewhat more sophisticated. To define it one introduces the *reference system*, described by the Hilbert space $\mathcal{H}_R$, isomorphic to the Hilbert space $\mathcal{H}_Q = \mathcal{H}$ of the initial system. Then according to Refs. 2, 3 there exists *purification* of the state ρ, i.e. a unit vector $|\psi\rangle \in \mathcal{H}_Q \otimes \mathcal{H}_R$ such that

Quantum Communication, Computing, and Measurement 2
Edited by Kumar et al., Kluwer Academic / Plenum Publishers, New York, 2000.

$$\rho = \mathrm{Tr}_R|\psi\rangle\langle\psi|.$$

The *entropy exchange* is then defined as

$$H(\rho, T) = H((T \otimes Id)[|\psi\rangle\langle\psi|]),$$

that is as the entropy of the output state of the dilated channel $(T \otimes Id)$ applied to the input which is purification of the input state ρ. One can then show that $H(\rho, T)$ is equal to the entropy increase in the channel environment E provided the channel is represented by a unitary interaction with the environment system being initially in a pure state.[2, 3]

From these three entropies one can construct three other quantities, bearing some analogy with the classical mutual information. In general, if ρ_{12} is a density operator in a tensor product Hilbert space $\mathcal{H}_1 \otimes \mathcal{H}_2$, and ρ_1, ρ_2 are the partial states of ρ_{12},resp., in $\mathcal{H}_1, \mathcal{H}_2$,then one can introduce the quantity

$$C_{12} = H(\rho_1) + H(\rho_2) - H(\rho_{12}).$$

This quantity is nonnegative by subadditivity of quantum entropy, and to certain extent reflects information correlation between the two systems,[2] although its operational meaning is still not completely clarified. Then from the above three entropies one can construct three quantum information quantities:[2, 4]

1) *Quantum mutual information*

$$I = H(\rho) + H(T[\rho]) - H(\rho, T),$$

reflecting quantum information transfer from the reference system R to the output of the initial system Q'. The important component of it is the *coherent information* $H(T[\rho]) - H(\rho, T)$, supremum of which with respect to input states ρ was conjectured as the quantum capacity of the channel T.[3]

2) *Loss*

$$L = H(\rho) + H(\rho, T) - H(T[\rho]),$$

which reflects quantum information transfer from the reference system R to the output of the environment E'.

3) *Noise*

$$N = H(T[\rho]) + H(\rho, T) - H(\rho),$$

which reflects quantum information transfer from the output of the environment E' to the output of the initial system Q'. In Ref. 4 quantum Wenn's diagrams were introduced to visualize the relations between the entropy and the information quantities. However, in contrast to classical case, some areas in these diagrams representing conditional entropies may have negative measure. We use another graphic representation via the *information triangle.* In this representation the entropies $H(\rho), H(T[\rho]), H(\rho, T)$ are associated with the sides of the triangle, and the information quantities I, L, N are attached to its vertices. The deficiency of this picture is that the representation of the information quantities is only qualitative: roughly, the bigger is the quantity - the bigger is distance from the corresponding vertex to the opposite side of the triangle, and vice versa.

Although the entropy and information quantities described above were studied in some detail from the general point of view, they are far from being completely understood, and concrete examples in which they can be explicitly evaluated are certainly welcome. In quantum statistics there is one large class of states for which many explicit calculations are possible – the so called quasifree states of canonical commutation relations, in many respect analogous to the classical Gaussian probability distributions. They are the states of the maximal entropy among all states with fixed second moments, for example, mean energy for a quadratic Hamiltonian. The aim of the present paper is the study the behavior of the information triangle for Gaussian input state and the most common attenuation/amplification channel.

QUANTUM GAUSSIAN STATES

In this Section we repeat some results of Refs. 5, 6, 7, and give a new variant of the expression for the entropy of a general quantum Gaussian state. Let q_j, p_j be the canonical observables satisfying the Heisenberg CCR

$$[q_j, p_k] = i\delta_{jk}\hbar I, \quad [q_j, q_k] = 0, \quad [p_j, p_k] = 0.$$

Let us introduce the column vector

$$R = [q_1, \ldots, q_s; p_1, \ldots, p_s]^T.$$

We also introduce real column $2s$-vector $z = [x_1, \ldots, x_s; y_1, \ldots, y_s]^T$, and the unitary operators in $\mathcal{H}$

$$V(z) = \exp\, i \sum_{j=1}^{s} (x_j q_j + y_j p_j) = \exp\, i\, R^T z.$$

The operators $V(z)$ satisfy the Weyl-Segal CCR

$$V(z)V(z') = \exp[i/2\Delta(z, z')]V(z + z'), \tag{1}$$

where

$$\Delta(z, z') = \hbar \sum_{j=1}^{s} (x'_j y_j - x_j y'_j)$$

is the canonical symplectic form. The Weyl -Segal CCR is the rigorous counterpart of the Heisenberg CCR, involving only bounded operators. We denote by

$$\Delta = \left[\begin{array}{cccc|cccc} 0 & & & & \hbar & & & 0 \\ & 0 & & & & h & & \\ & & \ddots & & & & \ddots & \\ & & & 0 & 0 & & & \hbar \\ \hline -\hbar & & & 0 & 0 & & & \\ & -\hbar & & & & 0 & & \\ & & \ddots & & & & \ddots & \\ 0 & & & -\hbar & & & & 0 \end{array}\right] \tag{2}$$

the $(2s) \times (2s)$-skew-symmetric *commutation matrix* of components of the vector R. Most of the results below are valid for the case where the commutation matrix is arbitrary skew-symmetric matrix, not necessarily of the canonical form (2).

The density operator ρ is called *Gaussian*, if its quantum characteristic function has the form

$$\mathrm{Tr}\rho V(z) = \exp(i\, m^T z - \frac{1}{2} z^T \alpha z),$$

where m is column $(2s)$-vector and α is real symmetric $(2s) \times (2s)$-matrix.

One can show that

$$m = \mathrm{Tr}\rho R\,;\ \alpha - \frac{i}{2}\Delta = \mathrm{Tr}R\rho R^T$$

(cf {aa5,aa6}). The *mean* m can be arbitrary vector; in what follows we will be interested in the case $m = 0$. The necessary and sufficient condition on the *correlation matrix* α is the matrix uncertainty relation

$$\alpha - \frac{i}{2}\Delta \geq 0. \tag{3}$$

This condition is equivalent to its transpose $\alpha + \frac{i}{2}\Delta \geq 0$,and to the following matrix generalization of the Heisenberg uncertainty relation

$$\Delta^{-1}\alpha\Delta^{-1} + \frac{1}{4}\alpha^{-1} \geq 0, \tag{4}$$

which is obtained by combining together (3) and its transpose. The state ρ is pure if and only if the equality holds in this equation, or

$$(\Delta^{-1}\alpha)^2 = -\frac{1}{4}I. \tag{5}$$

Let us introduce the function

$$g(x) = (x+1)\log(x+1) - x\log x, \qquad x > 0.$$

We shall also use the matrix function abs$(\cdot)$, which is defined as follows: for a diagonalizable matrix $M = T\mathrm{diag}(m_j)T^{-1}$, we put abs$M = T\mathrm{diag}(|m_j|)T^{-1}$. In Ref. 7 it was shown that the entropy of the Gaussian state is equal to

$$H(\rho) = \frac{1}{2}\mathrm{Sp}G(-(\Delta^{-1}\alpha)^2),$$

where

$$G(a^2) = (a+\frac{1}{2})\log(a+\frac{1}{2}) - (a-\frac{1}{2})\log(a-\frac{1}{2}),$$

and Sp denotes trace of a matrix, as distinct from trace of operator. The matrix $\Delta^{-1}\alpha$ has purely imaginary eigenvalues $\pm ia_j$ and is diagonalizable. Since $G(a^2) = g(|a| - \frac{1}{2})$, we obtain another expression

$$H(\rho) = \frac{1}{2}\mathrm{Sp}g(\mathrm{abs}(\Delta^{-1}\alpha) - \frac{I}{2}), \tag{6}$$

which will be used in the sequel.

PURIFICATION OF GAUSSIAN STATES

Let us denote $\mathcal{H}_Q = \mathcal{H}_1$ the Hilbert space of irreducible representation $z \to V_1(z)$ of the CCR (1), $\mathcal{H}_R = \mathcal{H}_2$ the Hilbert space of irreducible representation $z \to V_2(z)$ of the CCR

$$V_2(z)V_2(z') = \exp[-i/2\Delta(z,z')]V_2(z+z').$$

For example, $V_2(z) = \exp i\sum_{j=1}^{s}(x_j p_j^{(2)} + y_j q_j^{(2)})$, where $q_j^{(2)}, p_j^{(2)}$ satisfy the Heisenberg CCR in $\mathcal{H}_2$. In $\mathcal{H}_1 \otimes \mathcal{H}_2$ the operators $V(z_1, z_2) = V_1(z_1) \otimes V_2(z_2)$ satisfy the CCR

$$V(z_1, z_2)V(z_1', z_2') = \exp[i/2\Delta(z_1, z_2; z_1', z_2')]V(z_1 + z'_1, z_2 + z'_2),$$

where

$$\Delta(z_1, z_2; z_1', z_2') = \Delta(z_1, z_1') - \Delta(z_2, z_2').$$

Following Ref. 8 we introduce Gaussian state ρ_{12} in $\mathcal{H}_1 \otimes \mathcal{H}_2$ with the correlation matrix

$$\alpha_{12} = \begin{bmatrix} \alpha & \Delta\sqrt{-(\Delta^{-1}\alpha)^2 - I/4} \\ -\Delta\sqrt{-(\Delta^{-1}\alpha)^2 - I/4} & \alpha \end{bmatrix},$$

that is

$$\begin{aligned} &\mathrm{Tr}\rho_{12}V(z_1, z_2) \\ &= \exp[-\frac{1}{2}(z_1^T\alpha z_1 + z_2^T\alpha z_2 + z_1^T\Delta\sqrt{-(\Delta^{-1}\alpha)^2 - I/4}z_2 - z_2^T\Delta\sqrt{-(\Delta^{-1}\alpha)^2 - I/4}z_1)]. \end{aligned}$$

Obviously, $\rho_1 = \mathrm{Tr}_2\rho_{12}$.

Let us show that ρ_{12} is pure. With

$$\Delta_{12} = \begin{bmatrix} \Delta & 0 \\ 0 & -\Delta \end{bmatrix},$$

we have

$$\Delta_{12}^{-1}\alpha_{12} = \begin{bmatrix} \Delta^{-1}\alpha & \sqrt{-(\Delta^{-1}\alpha)^2 - I/4} \\ \sqrt{-(\Delta^{-1}\alpha)^2 - I/4} & -\Delta^{-1}\alpha \end{bmatrix},$$

and it is easy to check that $(\Delta_{12}^{-1}\alpha_{12})^2 = -I/4$. By the criterium (5) ρ_{12} is pure.

We shall be interested in the particular subclass of Gaussian states most familiar in quantum optics, which we call gauge-invariant. These are the states having the P-representation

$$\rho = \pi^{-s}|\det N|^{-1}\int \exp(-\zeta^\dagger N^{-1}\zeta)|\zeta\rangle\langle\zeta|d^{2s}\zeta \tag{7}$$

(see, e.g., Ref. 9, Sec. V, 5. II). Here $\zeta \in \mathbf{C}^s$, $|\zeta\rangle$are the coherent vectors in $\mathcal{H}$, $a|\zeta\rangle = \zeta|\zeta\rangle$, N is positive Hermitian matrix such that

$$N = \mathrm{Tr} a\,\rho\, a^\dagger$$

(we use here vector notations, where $a = [a_1, \ldots, a_s]^T$ is a column vector and $a^\dagger = [a_1^\dagger, \ldots, a_s^\dagger]$ is a row vector) and $a_j = \frac{1}{2\hbar}(q_j + ip_j)$. As shown in Ref. 7, the correlation matrix of such states is

$$\alpha = \hbar\begin{bmatrix} \mathrm{Re}N + I/2 & -\mathrm{Im}N \\ \mathrm{Im}N & \mathrm{Re}N + I/2 \end{bmatrix},$$

The real $2s \times 2s-$ matrices of such form can be rewritten as complex $s \times s-$ matrices, by using the correspondence

$$\begin{bmatrix} A & -B \\ B & A \end{bmatrix} \leftrightarrow A + iB,$$

which is in fact algebraic isomorphism, provided $A^T = A, B^T = -B$. Apparently,

$$\frac{1}{2}\mathrm{Sp}\begin{bmatrix} A & -B \\ B & A \end{bmatrix} = \mathrm{Sp}(A + iB).$$

By using this correspondence, we have

$$\alpha \leftrightarrow \hbar(N + I/2), \qquad \Delta \leftrightarrow -i\hbar I,$$

and

$$\Delta^{-1}\alpha \leftrightarrow i(N + I/2).$$

In particular, the formula (6) becomes

$$H(\rho) = \mathrm{Sp}g(N),$$

which is well known (see, e.g., Ref. 7) and confirms (6).

For future use we also need the correspondence

$$\Delta_{12}^{-1}\alpha_{12} \leftrightarrow \begin{bmatrix} i(N + I/2) & \sqrt{N^2 + N} \\ \sqrt{N^2 + N} & -i(N + I/2) \end{bmatrix}. \tag{8}$$

For the case of one degree of freedom we shall be interested in the following Section, N is just nonnegative number and ρ is *elementary* Gaussian state with the characteristic function

$$\exp\left[-\frac{\hbar}{2}\left(N + \frac{1}{2}\right)|z|^2\right], \tag{9}$$

where we put $|z|^2 = (x^2 + y^2)$.

ATTENUATION/AMPLIFICATION CHANNEL

Let us consider CCR with one degree of freedom described by one mode annihilation operator $a = \frac{1}{2\hbar}(q + ip)$, and let a_0 be another mode in the Hilbert space $\mathcal{H}_0 = \mathcal{H}_E$ describing environment. Let the environment be initially in the vacuum state, which is described by the characteristic function (9) with $N = 0$ i.e. $\exp[-\frac{\hbar}{4}|z|^2]$,

The linear attenuator with coefficient $k < 1$ is described by the transformation

$$a \to ka + \sqrt{1 - k^2}a_0$$

in the Heisenberg picture. Similarly, the linear amplifier with coefficient $k > 1$ is described by the transformation

$$a \to ka + \sqrt{k^2 - 1}a_0.$$

It follows that the corresponding transformations $T_k[\rho]$ of states in the Schroedinger picture have, correspondingly, the characteristic functions

$$\mathrm{Tr}T_k[\rho]V(z) = \mathrm{Tr}\rho V(kz)\exp[-\frac{\hbar}{4}(1 - k^2)|z|^2], \qquad k < 1, \tag{10}$$

$$\mathrm{Tr}T_k[\rho]V(z) = \mathrm{Tr}\rho V(kz)\exp[-\frac{\hbar}{4}(k^2 - 1)|z|^2], \qquad k > 1, \tag{11}$$

see Ref. 10. Let the input state ρ of the system have the characteristic function (9), i.e. $\exp[-\frac{\hbar}{4}(N+\frac{1}{2})|z|^2]$. The entropy of ρ is

$$H(\rho) = g(N).$$

From (10), (11) we find that the output state $T_k[\rho]$ is again elementary Gaussian with N replaced by

$$\widetilde{N} = k^2 N, \qquad k < 1; \qquad \widetilde{N} = k^2 N + (k^2 - 1), \qquad k > 1.$$

Thus

$$H(T_k[\rho]) = g(k^2 N), \qquad k < 1; \qquad H(T_k[\rho]) = g(k^2 N + (k^2 - 1)), \qquad k > 1.$$

Now we calculate the entropy exchange $H(\rho, T_k)$. The (pure) input state ρ_{12} of the extended system $\mathcal{H}_1 \otimes \mathcal{H}_2$ is characterized by the 2×2–matrix (8). The action of the extended channel $(T \otimes Id)$ transforms this matrix into

$$\Delta_{12}^{-1}\widetilde{\alpha}_{12} \leftrightarrow \begin{bmatrix} i(\widetilde{N}+\frac{1}{2}) & k\sqrt{N^2+N} \\ k\sqrt{N^2+N} & -i(N+\frac{1}{2}) \end{bmatrix}.$$

From formula (6) we deduce $H(\rho, T_k) = g(|\lambda_1| - \frac{1}{2}) + g(|\lambda_2| - \frac{1}{2})$, where λ_1, λ_2 are the eigenvalues of the complex matrix in the right-hand side. The eigenvalues are: $\lambda_1 = \frac{i}{2}$,

$$\lambda_2 = -i[(1-k^2)N + \frac{1}{2}], \qquad k < 1; \qquad i[(k^2-1)(N+1) + \frac{1}{2}], \qquad k > 1.$$

Therefore we obtain

$$H(\rho, T_k) = g((1-k^2)N) \qquad k < 1; \qquad g((k^2-1)(N+1)), \qquad k > 1.$$

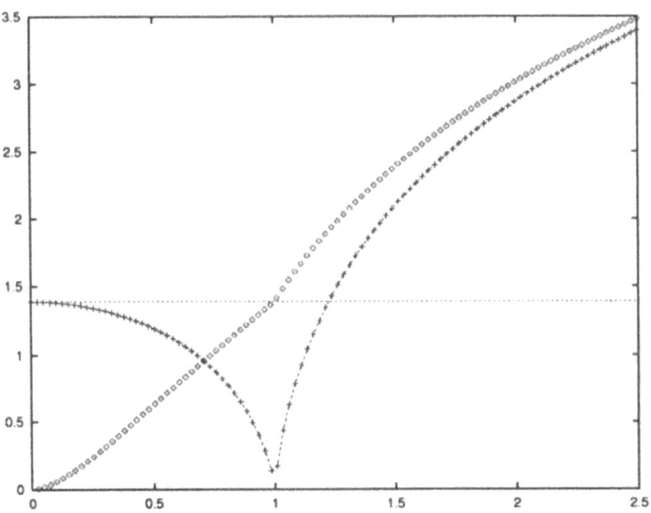

The behavior of the entropies $H(T_k[\rho]), H(\rho, T_k)$ as functions of k is clear from Fig. 1. In particular, for all N the coherent information $H(T_k[\rho]) - H(\rho, T_k)$ turns out to be positive for $k > 1/\sqrt{2}$ and negative otherwise. It tends to $-H(\rho)$ for $k \to 0$, is equal to $H(\rho)$ for $k = 1$,and quickly tends to zero as $k \to \infty$ (Fig. 2; on both

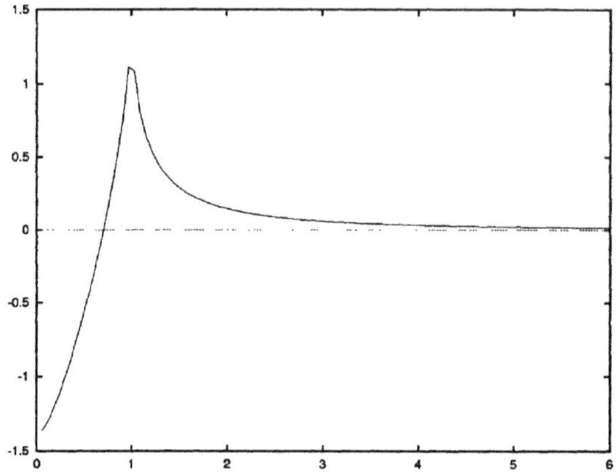

Figure 2. The coherent information as function of k.

plots $N = 1$). The behavior of the information triangle shows that loss dominates for $k \to 0$, mutual information for $k \div 1$, while the noise – as $k \to \infty$. This agrees with what one should expect on physical grounds from quantities presenting quantum mutual information, loss and noise and gives further support for their use in quantum information theory. However, negativity of the coherent information for $k < 1/\sqrt{2}$ looks somewhat mysterious and waits for a physical explanation.

Acknowledgments

This work was initiated when the author was visiting Physical Department of the University of Milan under the contract with Italian Ministry of Foreign Affairs provided by A. Volta Center for Scientific Culture. The author is grateful to Prof. L. Lanz for his hospitality and stimulating discussions. Part of this work was completed during the 1998 Elsag-Bailey – I.S.I. Foundation research meeting on quantum computation.

REFERENCES

1. A. S. Holevo, Coding theorems for Quantum Channels, *Tamagawa University Research Review*, 4, (1998).
2. G. Lindblad, Quantum entropy and quantum measurements, *in*: "Quantum Aspects of Optical Communication," C. Benjaballah, O. Hirota, S. Reynaud, ed., Springer-Verlag, Berlin (1991).
3. H. Barnum, M. A. Nielsen, B. Schumacher, Information transmission through noisy quantum channels, LANL Report No. quant-ph/9702049, (1997).
4. C. Adami and N. J. Cerf, Capacity of noisy quantum channels, *Phys. Rev. A*, A56:3470 (1997).
5. A. S. Holevo, Some statistical problems for quantum Gaussian states, *IEEE Transactions on Information Theory*, IT-21:533 (1975).
6. A. S. Holevo, "Probabilistic and statistical aspects of quantum theory," Chapter 5, North-Holland, Amsterdam, (1982).
7. A. S. Holevo, M. Sohma and O. Hirota, The capacity of quantum Gaussian channels, Preprint, Tamagawa University (1998).
8. A. S. Holevo, Generalized free states of C*-algebra of the CCR., *Theor. Math. Phys.*, 6:3 (1971).
9. C. W. Helstrom, "Quantum Detection and Estimation Theory," Chapter 5, Academic Press, New York (1976).
10. A. S. Holevo, Towards the mathematical theory of quantum communication channels, *Problems of Information Transm.*, 8:63 (1972).

QUANTUM COMMUNICATION WITH IMPERFECT MEANS: Entanglement Purification And The Quantum Repeater

G. Giedke,[1] H. J. Briegel,[1,2,3] W. Dür[1], J. I. Cirac[1], and P. Zoller[1]

[1] Institut für Theoretische Physik, Universität Innsbruck, Technikerstr. 25, A–6020 Innsbruck, AUSTRIA
[2] Departamento de Fisica, Universidad de Castilla–La Mancha, 13071 Ciudad Real, SPAIN
[3] Sektion Physik, Ludwig-Maximilians-Universität, Theresienstr. 37, D–80333 München, GERMANY

Recent results forentanglement purification with imperfect local operations and measurements are revisited and applied to the quantum repeater problem.

1. INTRODUCTION

Quantum communication offers a number of advantages over its classical counterpart such as speed-up for certain communication tasks [1], the ability to construct provably secure cryptographic keys [2], or dense coding schemes [3]. To exploit these advantages it is necessary to faithfully transmit qubits between locations A and B, which can be reduced to the task of distributing *maximally entangled (EPR) pairs* of qubits between A and B. Quantum information may subsequently be transmitted by *teleportation* [4] using these pairs. The creation of EPR pairs between distant locations is therefore of central importance for quantum communication.

Sending qubits via some channel introduces noise and reduces the quality of the pairs. As a result, A and B share pairs in some mixed state ρ instead of the desired maximally entangled state $|\phi\rangle$. We use the *fidelity* $F(\rho) = \langle\phi|\rho|\phi\rangle$ of ρ with respect to the desired state ϕ to measure the quality of the pairs.

If the channel is not too noisy (such that the fidelity of the pairs is $F \geq F_{\min} > 0.5$) then entanglement purification protocols (EPP) like those invented by Bennett *et al.* [5] or Deutsch *et al.* [6] allow to produce high fidelity pairs from a larger supply of lower fidelity pairs employing local operations at A and B and classical communication between them. Otherwise, if $F < F_{\min}$ the distance between A and B cannot be bridged in one leap. Especially for long-distance communication this will inevitably be the case, since both absorption losses and depolarization errors scale exponentially with the length of the channel. These problems can be solved using a quantum repeater protocol as in Ref. [7].

The central idea of this quantum repeater is the use of both entanglement purification and entanglement swapping to fight the accumulation of noise across a long

Quantum Communication, Computing, and Measurement 2
Edited by Kumar et al., Kluwer Academic / Plenum Publishers, New York, 2000.

channel. This allows one to operate with a much smaller number of auxiliary particles, placed at certain points within the (compound) channel, compared to approaches that are based on quantum error correction and concatenated coding [8].

As the local operations and measurements that are used in entanglement purification protocols are generally *imperfect*, they themselves represent *sources of noise* in the compound channel. An essential ingredient of the repeater is therefore an analysis of entanglement purification schemes that operate with imperfect local operations.

In Section 2, we will first analyze this problem – which is interesting on its own, regardless of its role in the repeater problem – in the the context of a quite general characterization of *imperfect realizations of purification protocols*. This allows us to derive lower bounds on the fidelity of the pairs that can be achieved with the aid of these protocols, and thresholds for their applicability. In Section 3, we relate this analysis to a generic error model that has been used in the treatment of the quantum repeater problem. Section 4 gives a brief review of the quantum repeater based on entanglement purification.

2. PURIFICATION WITH IMPERFECT MEANS

Entanglement purification [5, 9, 10] is one of the most important tools in the theory of quantum information and, in particular, in quantum communication. It allows, in principle, to create maximally entangled states of particles at different locations, even if the channel that connects those locations is noisy [11].

After reviewing the EPP [5] and introducing a parameter δ to characterize the errors in local operations and measurement, we will study the EPP in the presence of general errors and derive an error threshold below which purification is possible and a lower bound for the asymptotically attainable fidelity.

Entanglement Purification Protocols

The basic idea in entanglement purification is to "distill" a few N' pairs of qubits in highly entangled states ρ' out of $N \geq N'$ pairs in a state ρ with lower fidelity using local operations and measurements, and classical communication. If the initial pairs are in a state that is non–separable (cannot be written as a convex combination of factorized density operators [12]), then one can obtain asymptotically (in the limit $N \to \infty$) maximally entangled states [13]. In particular, if $F(\rho) > 1/2$ one can reach this goal by using the *purification procedure* devised by Bennett *et al.* [5] and improved by Deutsch *et al.* [6]. It consists of a concatenation of *purification steps* involving two pairs of qubits, which give (if successful) rise to a single pair with higher fidelity. In all these procedures, local operations and measurements are assumed to be error free [9, 14]. In practice, however, there will be errors due to the coupling to the environment, imprecise apparatus, etc. Although small, they will limit the maximum attainable fidelity and will dictate whether purification is possible or not. The purpose of this section is to analyze this problem for the purification protocol introduced in Ref. [5].

The EPP [5] consists of a sequence of steps in which local operations are applied to two pairs of qubits, followed by a measurement of one of the pairs which is then discarded. Depending on the outcome of the measurement, the other pair is discarded or not. In the latter case the fidelity F_1 of the remaining pair is larger than that of the original ones. This step is applied to the N pairs obtaining $N_1 \leq N/2$ pairs of fidelity F_1. Then it is applied to the resulting N_1 pairs obtaining N_2 pairs of fidelity $F_2 > F_1$. Continuing in this vein, one can reach asymptotically $F_n \to 1$ as $n \to \infty$.

Let us consider a single purification step. It starts out with two pairs 1 and 2 in the state $\rho_{12} = \rho \otimes \rho$, applies the local operations described by the superoperator $\mathcal{U}$ and then measures each of the qubits of the second pair in the basis $\{|0\rangle, |1\rangle\}$. We denote by x the outcome of the measurement: $x = 0$ if the qubits are found in the state $|0\rangle_2 \equiv |00\rangle_2$; $x = 1$ if they are in $|1\rangle_2 \equiv |11\rangle_2$; $x = 2$ if they are in $|2\rangle_2 \equiv |01\rangle_2$; and $x = 3$ if they are in $|3\rangle \equiv |10\rangle_2$ (the subscript 2 denotes the second pair). We denote by $\mathcal{P}_x$ ($x = 0, \ldots, 3$) the map defined as follows

$$\mathcal{P}_x(\rho_{12}) \equiv {}_2\langle x|\mathcal{U}(\rho_{12})|x\rangle_2. \tag{1}$$

This map is linear and completely positive. The probability of obtaining the outcome x is $p_x(\rho_{12}) = \mathrm{tr}\,[\mathcal{P}_x(\rho_{12})]$. If the outcome is $x = 2, 3$, then the first pair is discarded and otherwise it is kept. In the latter case, the state of the first pair will be

$$\rho_1' = \frac{\mathcal{P}_0(\rho_{12}) + \mathcal{P}_1(\rho_{12})}{p_0(\rho_{12}) + p_1(\rho_{12})}. \tag{2}$$

Thus, each (successful) step of the purification protocol is completely characterized by the maps $\mathcal{P}_{0,1}$.

On the other hand, if one is only interested in the fidelity at each step, one can use a simpler characterization of each purification step in terms of a single real number. The local operations characterized by $\mathcal{U}$ consist of a bilateral XOR gate and specific single qubit rotations. The diagonal elements of the density operator ρ' in the Bell basis depend (on average) only on the diagonal elements of the density operator ρ, and therefore each purification step can be characterized by a non–linear map between these four diagonal matrix elements. We denote the state after the nth iteration by ρ_n and its diagonal elements in the Bell basis $|\phi^i\rangle$, $(i = 0, 1, 2, 3)$

$$|\phi^{0,3}\rangle = \frac{1}{\sqrt{2}}\left(|00\rangle \pm |11\rangle\right), \quad |\phi^{1,2}\rangle = \frac{1}{\sqrt{2}}\left(|01\rangle \pm |10\rangle\right)$$

by $A_n^i = \langle\phi^i|\rho_n|\phi^i\rangle$. Clearly, we have

$$A_{n+1}^i = \frac{\langle\phi^i|\mathcal{P}_0(\rho_n \otimes \rho_n) + \mathcal{P}_1(\rho_n \otimes \rho_n)|\phi^i\rangle}{\mathrm{tr}\left[\mathcal{P}_0(\rho_n \otimes \rho_n) + \mathcal{P}_1(\rho_n \otimes \rho_n)\right]}. \tag{3}$$

At the end of each step the resulting state is brought into Werner form, that is the three diagonal elements A^1, A^2, A^3 are made equal to $(1 - A^0)/3$. Therefore one can concentrate on the first diagonal element, the fidelity $F \equiv A^0$, only. The fidelity after the nth purification step is then given by

$$A_{n+1}^0 = \frac{f\left(A_n^0\right)}{g\left(A_n^0\right)}, \tag{4}$$

where

$$f(x) = x^2 + (1-x)^2/9 \text{ and } g(x) = \left[(1+2x)^2 + (2-2x)^2\right]/9. \tag{5}$$

This map has an attractive fixed point at $A^0 = 1$, and all $A_0^0 > 1/2$ are attracted to it. This expresses the fact that in the absence of errors, one can use this purification protocol to purify states with $F > 1/2$ and reach a fidelity as close to one as we please.

Characterization of Errors

In practice, while performing the purification protocols, errors will occur, both in the local operation and in the measurements. The imperfections in the local operations can be accounted for by substituting the action of the superoperator $\mathcal{U}$ in Eq. (1) by the action of some other completely positive, trace preserving linear map. The errors in the measurements will be related to the following fact: in practice, the outcomes $x = 0, 1$ will be ultimately attributed to the presence/absence of clicks in some kind of detectors. Due to imperfections, the projection operators (or, more generally, POVMs) corresponding to those clicks are not exactly the same as the ideal ones [see Eq. (1)]. Consequently, the probabilities of the outcomes x as well as the state remaining after the measurement will differ from the ideal case. In general, we can describe both these erroneous operations and measurements in terms of a single completely positive linear map $\tilde{\mathcal{P}}_x$ which does not necessarily preserve the trace. That is, if the two pairs are initially in the state $\rho_{12} = \rho \otimes \rho$, a purification step yields the outcome x with a probability $\tilde{p}_x(\rho_{12}) = \mathrm{tr}[\tilde{\mathcal{P}}_x(\rho_{12})]$. The state of the pair after the measurement is

$$\tilde{\rho}_1' = \frac{\tilde{\mathcal{P}}_0(\rho_{12}) + \tilde{\mathcal{P}}_1(\rho_{12})}{\tilde{p}_0(\rho_{12}) + \tilde{p}_1(\rho_{12})}. \tag{6}$$

Thus, as before, the maps $\tilde{\mathcal{P}}_{0,1}$ completely characterize each purification step. We characterize the errors by a single parameter

$$\delta := \max_{x=0,1} d(\mathcal{P}_x, \tilde{\mathcal{P}}_x), \tag{7}$$

where $d(\mathcal{P}, \tilde{\mathcal{P}})$ denotes a distance between $\mathcal{P}$ and $\tilde{\mathcal{P}}$, and is defined as

$$d(\mathcal{P}, \tilde{\mathcal{P}}) = \max_{\rho \in C(H)} ||\mathcal{P}(\rho) - \tilde{\mathcal{P}}(\rho)||. \tag{8}$$

(Here $||\cdot||$ denotes the trace norm $||A|| = \mathrm{tr}(|A^\dagger A|^{1/2})$ on the convex space $C(H)$ of positive linear operators on the Hilbert space H with $\mathrm{tr}(\rho) \leq 1$.)

The motivation for this definition with respect to other possible definitions is that it easily gives lower bounds even for physical processes where there are measurements and post selection as it is in the case of entanglement purification, i.e. when the map describing the physical process is not trace preserving. Some useful properties of d are collected in [15]. In the following we will make use of the fact that for all $\rho \in C(H)$ and $\phi \in H$ (normalized state) we have

$$\langle\phi|\mathcal{P}(\rho)|\phi\rangle - d(\mathcal{P}, \tilde{\mathcal{P}}) \leq \langle\phi|\tilde{\mathcal{P}}(\rho)|\phi\rangle \leq \langle\phi|\mathcal{P}(\rho)|\phi\rangle + d(\mathcal{P}, \tilde{\mathcal{P}}) \tag{9}$$

and

$$\mathrm{tr}\left[\mathcal{P}(\rho)\right] - d(\mathcal{P}, \tilde{\mathcal{P}}) \leq \mathrm{tr}\left[\tilde{\mathcal{P}}(\rho)\right] \leq \mathrm{tr}\left[\mathcal{P}(\rho)\right] + d(\mathcal{P}, \tilde{\mathcal{P}}). \tag{10}$$

From these inequalities we can derive upper and lower bounds for the diagonal matrix elements in the Bell basis of the density operator ρ_n after the nth purification step, and in particular for the fidelity $\tilde{A}_n^0 = \langle\phi^0|\rho_n|\phi^0\rangle$. Analogous to (3), we have

$$\tilde{A}_{n+1}^0 = \frac{\langle\phi^0|\tilde{\mathcal{P}}_0(\tilde{\rho}_n \otimes \tilde{\rho}_n) + \tilde{\mathcal{P}}_1(\tilde{\rho}_n \otimes \tilde{\rho}_n)|\phi^0\rangle}{\mathrm{tr}\left[\tilde{\mathcal{P}}_0(\tilde{\rho}_n \otimes \tilde{\rho}_n) + \tilde{\mathcal{P}}_1(\tilde{\rho}_n \otimes \tilde{\rho}_n)\right]}. \tag{11}$$

Using (9) and Eq. (4) it follows that

$$\frac{f\left(\tilde{A}_n^0\right) - 2\delta}{g\left(\tilde{A}_n^0\right) + 2\delta} \leq \tilde{A}_{n+1}^0 \leq \frac{f\left(\tilde{A}_n^0\right) + 2\delta}{g\left(\tilde{A}_n^0\right) - 2\delta}, \tag{12}$$

where f, g are defined in (5). In the following subsection we will use these inequalities to derive expressions for the attainable fidelity and the error threshold.

Results: Error Threshold, Lower Bound for Attainable Fidelity

Using (12) and (4) we get

$$\tilde{A}_{n+1}^{0} \geq \frac{(\tilde{A}_n^0)^2 + \left(1 - \tilde{A}_n^0\right)^2/9 - 2\delta}{\left(1 + 2\tilde{A}_n^0\right)^2/9 + \left(2 - 2\tilde{A}_n^0\right)^2/9 + 2\delta}. \tag{13}$$

The rhs of (13) is monotonically increasing with $\tilde{A}_n^0$ for all $\tilde{A}_n^0 \geq 1/8$. Therefore replacing $\tilde{A}_n^0$ by $\frac{1}{8} \leq a_n \leq \tilde{A}_n^0$ in (13) yields a lower bound for $\tilde{A}_{n+1}^0$. Since the interval $[1/8, 1]$ is mapped into itself by the lhs of (13) this replacement is allowed for all n, and we arrive at the dynamical system defined by $a_0 = A_0^0$ and

$$a_{n+1} = \frac{a_n^2 + (1 - a_n)^2/9 - 2\delta}{(1 + 2a_n)^2/9 + (2 - 2a_n)^2/9 + 2\delta}. \tag{14}$$

For all n, the value of a_n is a lower bound of the fidelity after n purification steps.

In the case $\delta = 0$ the original map of Bennett *et al.* is recovered. The three fixed points of that map at $a_l(\delta) \approx 0.25, a_i(\delta) \approx 0.5$, and $a_u(\delta) \approx 1$ survive for nonzero δ and are given by the roots of a cubic polynomial. They are plotted as a function of δ in Fig. 1 (broken line). For $\delta \geq 0.008$ only the lowest fixpoint survives. This marks the threshold, above which the above considerations can no longer guarantee purification to be possible for all types of errors.

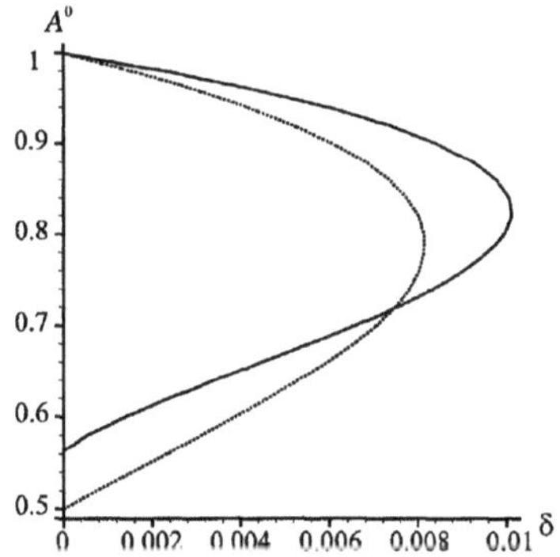

Figure 1. Fixed points of the nonlinear map: the intersections of a vertical line at δ with the plotted curve give the a-coordinates of two of the fixed points for scheme [5] (broken) and scheme [6] (solid).

A similar analysis has been carried out in [15] for the EPP of Deutsch *et al.* [6]. The derivation is somewhat more involved, since all four diagonal elements A^i come into play. For $\delta \leq \delta_{th} \approx 0.01$, one finds a non–linear map with three fixed points with fidelities F_l, F_i, F_u (Fig. 1, solid line). For all states that initially fulfill $F_u > A^0 > F_{min}$ and $A^1 < 0.5$ the fidelity is increased by repeated successful purification steps, and asymptotically reaches some value F_∞, bounded from below by F_u.

These results show that even in the presence of arbitrary errors – restricted only by δ and the requirement that the dynamics be positive definite on the 4-qubit Hilbert space – and without any kind of error correction the purification schemes [5, 6] still allow to distill a few high fidelity pairs from a large lower fidelity supply for errors up to half a percent.

This approach includes systematic and random errors and takes into account the possibility that errors build up in the course of the iteration of the purification step. In particular, for the derivation of the bounds on the diagonal elements, we assume that in each step *that* error occurs which is worst for future purification (while consistent with δ). These remarks make clear, that additional knowledge of the erroneous dynamics may significantly ease the requirements for successful purification. In the next section we study the purification protocol for a simple generic error model.

3. A GENERIC ERROR MODEL

The Error Model

A simple model for imperfect gate operations can be derived from a Jaynes-type principle [17]: we look for the operation that maximizes the entropy of the resulting state while keeping its trace and average fidelity constant. This leads to imperfect operations given by a mixture of the ideal operation and the completely depolarizing operation $\rho \mapsto I/\mathrm{tr}(I)$. We define *imperfect operations* on states of one or more qubits by the following maps

$$\tilde{\mathcal{O}}_1\rho = p_1\mathcal{O}_1\rho + \frac{1-p_1}{2}\mathrm{tr}_1\{\rho\} \otimes I_1, \tag{15}$$

$$\tilde{\mathcal{O}}_{12}\rho = p_2\mathcal{O}_{12}\rho + \frac{1-p_2}{4}\mathrm{tr}_{12}\{\rho\} \otimes I_{12}\,, \tag{16}$$

the first of which describes an imperfect one-qubit operation on particle 1, and the second an imperfect two-qubit operation on particles 1 and 2. In these expressions, $\mathcal{O}_i$ is the ideal (perfect) operation, and I_1 and I_{12} denote identity operators on the subspace where the ideal operation acts. The quantities p_1 and p_2 measure the *reliability* of the operations. The expressions (15) and (16) describe a situation where we have no knowledge about the result of an error occuring during some operation ("white noise"), except that it happens with a certain probability $(1-p_j)$.

We finally describe an *imperfect measurement* on a single qubit (in the computational basis) by a POVM corresponding to the 'imperfect projectors'

$$\begin{aligned} P_0^\eta &= \eta|0\rangle\langle 0| + (1-\eta)|1\rangle\langle 1| \\ P_1^\eta &= \eta|1\rangle\langle 1| + (1-\eta)|0\rangle\langle 0|\,. \end{aligned} \tag{17}$$

The parameter η is a measure for the quality of the projection onto the basis states. Assume the qubit to be measured is in the state $\rho = |0\rangle\langle 0|$ and we are trying to measure its state with the aid of an measurement apparatus described by (17). The expectation values $\langle P_0^\eta\rangle = \eta$ and $\langle P_1^\eta\rangle = 1-\eta$ simply mean that the apparatus will give us the wrong result ("1") with probability $1-\eta \geq 0$. That is, the result is not completely reliable. An ideal measurement, in contrast, is described by $\eta = 1$. It is clear that the effect of the measurement on the measured qubit is not fully specified by the POVM (17). In the present context, however, this description is sufficient as all measured particles are removed from the system i.e. we trace over their degrees of freedom after the measurement.

The operations (15) and (16) together with single-qubit measurements (17) are sufficient to describe all operations occurring in the context of teleportation and entanglement purification. For example, a Bell measurement (the measurement of a projector in the Bell basis) on two particles, say 1 and 2, can be realized by a two-qubit operation $\mathcal{O}_{12} \equiv \mathrm{CNOT}_{1\to 2}^{\mathrm{imperf}}$ (controlled NOT) followed by a Hadamard rotation of particle 1 and two single-qubit measurements $P^1_{0(1)}$, $P^2_{0(1)}$ on particles* 1 and 2. Instead of performing the Hadamard transformation one can also measure particle 1 in a rotated basis. In summary, an imperfect Bell measurement is described by an imperfect two-qubit operation followed by two imperfect single-qubit measurements, effecting e.g. $|0\rangle|0\rangle \pm |1\rangle|1\rangle \rightarrow (|0\rangle \pm |1\rangle)|0\rangle$. Before turning to the results obtained with this model, we want to relate it to the more general description of Section 2.

For the purpose of this argument, the state after measuring $x = 0$ with the imperfect apparatus is taken to be $\eta P_0 \rho P_0 + (1-\eta) P_1 \rho P_1$. With this expression for the imperfect measurement $\tilde{\mathcal{M}}$ (one-qubit measurement on an arbitrarily large Hilbert space) and with the definition (8) of d we get

$$d(\mathcal{M}, \tilde{\mathcal{M}}) = 1 - \eta. \tag{18}$$

For an imperfect two-qubit operation as in (16) we calculate that

$$\frac{3}{2}(1-p_2) \le d(\mathcal{O}_{12}, \tilde{\mathcal{O}}_{12}) \le 2(1-p_2). \tag{19}$$

The complete purification step $\mathcal{P}$ is a concatenation of two two-qubit gates (combining all of A's and B's operations, resp.) each with reliability p_2 and two one-qubit measurements (η). For typical values of $\eta, p_2 > 0.9$ and $\eta = p_2$ one calculates that δ satisfies the bounds $1.8(1-p_2) \le \delta \le 4.5(1-p_2)$.

We will see that within the generic error model, purification is possible even for parameters η, p_2 that would put δ on the order of several percent – well above the threshold δ_{th} for general (i.e. worst case) errors.

Purification with the Generic Error Model

Consider again the purification protocol of [5]. With the adopted error model we can give a closed expression for the fidelity after one successful purification step:

$$F' = \frac{[F^2 + (\frac{1-F}{3})^2][\eta^2 + (1-\eta)^2] + [F(\frac{1-F}{3}) + (\frac{1-F}{3})^2][2\eta(1-\eta)] + (\frac{1-p_2^2}{8p_2^2})}{[F^2 + \frac{2}{3}F(1-F) + \frac{5}{9}(1-F)^2][\eta^2 + (1-\eta)^2] + [F(\frac{1-F}{3}) + (\frac{1-F}{3})^2][8\eta(1-\eta)] + 4(\frac{1-p_2^2}{8p_2^2})}. \tag{20}$$

The curve $F'(F)$ is plotted in Figure 4 (solid line). It has two intersections with the diagonal, which are the fixpoints of the map (20). The upper point, $F_{\max} < 1$ is an attractor and gives the maximum value of the fidelity beyond which no pair can be purified and represents the asymptotically attainable fidelity. The repelling fixed point at $F_{\min} > 1/2$ gives the lower endpoint of the interval within which purification is possible. Figure 2 shows this interval as a function of the error parameter $\eta = p_2$. One sees, for example, that even for $1 - p_2 = 2\%$ (well outside the region covered by the results of Section 2) there is still a large number of purifiable states. The threshold is at $1 - p_2 \approx 3.3\%$.

The purification protocol [6], converges faster than scheme [5] in the ideal case. In the case of imperfect operations, it does also converge to a higher attainable fidelity for a larger number of initial states (see [7]), as demonstrated in Fig. 2 (dotted line). Having prepared the necessary tools, we now turn to a brief discussion of the quantum repeater.

*This procedure is not necessary, but to have a consistent model for both connection and purification, we want to describe all operations in terms of (15–17).

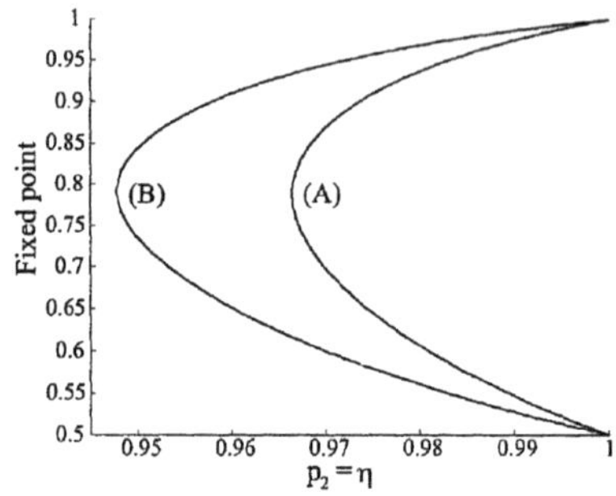

Figure 2. Fixpoint ($F_{\min}$ and $F_{\max}$) of scheme [5] (A) and scheme [6] (B) plotted against error parameters $p_2 = \eta$.

4. THE QUANTUM REPEATER

The quantum repeater as introduced in [7] is designed to distribute highly entangled pairs of qubits over long distances. Since absorption losses and depolarization errors scale exponentially with the length of the, e.g., optical fiber connecting the distant nodes, one cannot hope to cover any long distance between A and B in one leap. Rather, the distance between the nodes may not be much larger than the absorption length of the fiber (roughly 10km for standard optical fibers [18]).

Our solution of this problem consists of a method for the creation of entanglement between particles at distant nodes, which uses auxiliary particles at intermediate "connection points" and a *nested purification protocol.* We first divide the channel into N segments with connection points (i.e. auxiliary nodes) in between. We then create N elementary EPR pairs of fidelity F_1 between the nodes A & C_1, C_1 & C_2, $\ldots C_{N-1}$ & B, as in Fig. 3. The number N is chosen such that $F_{\min} < F_1 \approx F_{\max}$. Subsequently, we *connect* these pairs by making Bell measurements at the nodes C_i and classically communicating the results between the nodes as in the schemes for teleportation [4] and entanglement swapping [4, 16].

Figure 3. Connection of a sequence of N EPR pairs with $N-1$ connection points C_i.

Unfortunately, with every connection the fidelity F' of the resulting pair will decrease: on the one hand, the connection process involves imperfect operations which introduce noise; on the other hand, even for perfect connections, the fidelity decreases. As a consequence of both effects the fidelity F_N of the final pair shared between A and B decreases exponentially with N.

It is possible to solve this problem of exponential decay in fidelity, by using a *nested purification protocol* as described in detail in Ref. [7]. The idea is to construct an alternating sequence of purification and connection procedures, which allows to distribute entanglement across larger and larger distances while keeping its fidelity constant. The protocol is based upon two basic elements, namely pair connections and purification. The purification part has already been discussed in Sec. 3.2. The connection of L neighboring pairs via entanglement swapping, as explained above, results in a pair of length L (in short an "L-pair" – in units of the elementary pairs we start with) and fidelity[†]

$$F_L = \frac{1}{4} + \frac{3}{4}\left(\frac{p_1^2 p_2(4\eta^2-1)}{3}\right)^{L-1}\left(\frac{4F-1}{3}\right)^L . \tag{21}$$

This formula describes in fact an exponential decrease of the resulting fidelity, unless both the elementary pairs and all the operations involved in the connection process are perfect. One can therefore not connect all EPR pairs in one stroke, but has to repurify these L-pairs before their fidelity drops below $F_{\min}$.

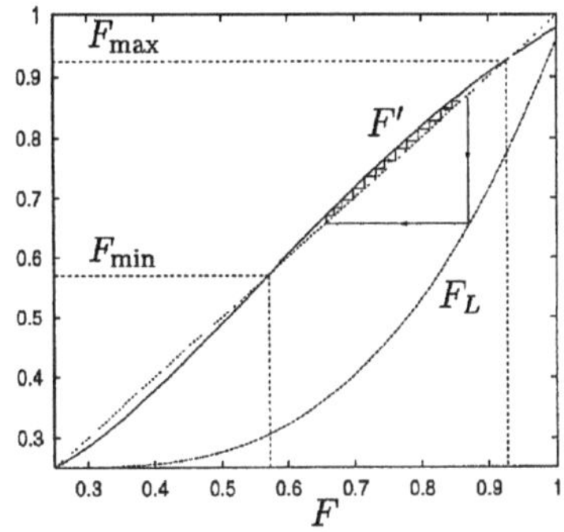

Figure 4. 'Purification loop' for connecting and purifying EPR pairs. Parameters are $L = 3$, $\eta = p_1 = 1$, and $p_2 = 0.97$.

The two-step process connection–purification can be visualized as in Fig. 4. Starting from F, the fidelity F_L after connecting L pairs can be read off from the curve below the diagonal. Reflecting this value back to the diagonal line, as indicated by the arrows in Fig. 4, sets the starting value for the purification curve. If F_L lies within the purification interval, then iterated application of (20) leads back to the initial value F (staircase). Once the initial value F is reobtained, we have a smaller number of pairs of length L. We then proceed by connecting again groups of L such pairs, obtaining a yet smaller number of pairs of length L^2, and so on.

In summary, each level in the nested protocol corresponds to one cycle in Fig. 4. As long as it is possible to construct a closed cycle in Fig. 4, it is possible to distribute the entanglement over arbitrary long distances, even though the operations that are used in the distribution process are themselves imperfect (i.e. noisy).

One can show that the overhead in time needed to distribute the entanglement grows polynomially, while the number of particles that has to be stored at each of the locations A, C_1, C_2, etc. in Fig. 3 needs only to grow logarithmically with the distance.

The formula for the endpoints $F_{\max}, F_{\min}$ (derived from Eq. 20) of the purification interval (depicted in Fig. 2) allows to derive an error threshold for the quantum repeater: if one connection operation alone would decrease even the highest attainable fidelity $F_{max}(\eta, p_2)$ below the minimal value $F_{min}(\eta, p_2)$ required for purification the repeater protocol breaks down. The equation $F_2(F_{\max}) = F_{\min}$ allows to calculate the corresponding error parameters. Putting (rather pessimistically) $\eta = p_1 = p_2$ we find that equality holds for $p_2 \approx 0.971$. Hence the repeater can tolerate errors of nearly 3%.

[†]The parameter p_2 for two-qubit operations occurs since we decompose the Bell measurement in a two-qubit gate followed by two one-qubit measurements.

5. SUMMARY

In summary, we have studied the effect of errors on certain quantum communication protocols. We have introduced an error parameter δ that allows to characterize a general class of errors for quantum operations and measurements. We have shown that below some error threshold the entanglement purification scheme [5] does allow to increase the fidelity significantly and we have given a lower bound for the attainable fidelity as a function of δ. We have related this analysis to a generic error model that corresponds to gates that are subject to a certain amount of white noise during their operation. The quantum repeater has been studied for this error model. A (meta-) protocol consisting of a nested sequence of entanglement purification and entanglement swapping operations allows to distribute highly entangled pairs of qubits between distant locations. The length of the channel connecting these locations may thereby be much greater than its coherence length and/or absorption length. The scheme tolerates errors in the percent region and requires an overhead in resources (time) that grows only polynomially with the distance between A and B.

Both results demonstrate that essential quantum communication protocols can tolerate errors of more than 0.5% and may be realized with a moderate number of qubits. Thus they represent, hopefully, encouraging results for future experimental implementations.

One of us (HJB) enjoyed discussions with G. M. D'Ariano, C. H. Bennett, M. Horodecki, E. Knill, and A. M. Steane.

REFERENCES

1. H. Buhrman *et al.*, quant–ph/9802040, to appear in Proceedings of the 30th Annual ACM Symposium on Theory of Computing (ACM Press) (1998).
2. A. Ekert, Phys. Rev. Lett., **67**, 661 (1991).
3. C. H. Bennett and S. Wiesner, Phys. Rev. Lett., **69**, 2881 (1992).
4. C. H. Bennett *et al.*, Phys. Rev. Lett., **70**, 1895 (1993).
5. C. H. Bennett *et al.*, Phys. Rev. Lett., **76**, 722 (1996).
6. D. Deutsch *et al.*, Phys. Rev. Lett., **77**, 2818 (1996).
7. H. J. Briegel *et al.*, quant–ph/9803056 (1998); W. Dür *et al.* quant–ph/9808065 (1998).
8. E. Knill and R. Laflamme, quant–ph/9608012 (1996); E. Knill *et al.*, quant–ph/9702058 (1997). See also D. Aharonov and M. Ben-Or, quant-ph/9611025; C. Zalka, quant-ph/ 9612028.
9. C.H. Bennett *et al.*, Phys. Rev. A, **54**, 3824 (1996).
10. N. Gisin, Phys. Lett. **A210**, 151 (1996).
11. B. Schumacher, Phys. Rev. A, **54**, 2614 (1996); B. Schumacher *et al.*, *ibid.*, **56**, 131 (1997).
12. A. Peres, Phys. Rev. Lett., **77**, 1413 (1996); M. Horodecki *et al.*, Phys. Lett. **A223**, 1 (1996).
13. P. Horodecki and R. Horodecki, Phys. Rev. Lett., **76**, 2196 (1996).
14. V. Vedral and M. Plenio, Phys. Rev. A, **57**, 1619 (1998).
15. G. Giedke *et al.* quant–ph/9809017 (under review with Phys. Rev. A) (1998).
16. M. Zukowski *et al.*, Phys. Rev. Lett., **71** 4287 (1993).
17. H. J. Briegel *et al.*, unpublished
18. W. Tittel *et al.*, quant-ph/9707042.

ON QUANTUM CHANNEL CAPACITY FOR SQUEEZED STATES

M. Sohma,[1] A. S. Holevo,[2] and O. Hirota[3]

[1]Matsushita Research Institute Inc., Tokyo, Japan.
[2]Steklov Mathematical Institute, Moscow.
[3]Research Center for Quantum Communications
Tamagawa University, Japan.

This paper presents the classical capacities of quantum Gaussian channels using squeezed states. The calculation is based on a general formula for the entropy of quantum Gaussian state

INTRODUCTION

The explicit relation between the quantum channel capacity and the von Neumann entropy is very intersting topics and it has been discussed by many scientists in the fields of physics and information science. In 1996, Hausladen *et al.*[2] showed that the capacity of the quantum channel with discrete alphabets is equal to Holevo bound (von Neumann entropy of the letter ensemble). Then in the case of continuous alphabet, it was proved that the channel capacity is also equal to the Holevo bound.[7] When the signal state is of coherent state, the capacity becomes the entropy of Bose-Einstein distribution(Yuen-Ozawa bound[12]) predicted by Gordon.[1] Thus we are concerned with the maximization of von Neumann entropy for various signal states. However, one has no general method of calculating the maximum von Neumann entropy of both cases for discrete and continuous. Recently the formula for the von Neumann entropy of the general Gaussian state has been clarified.[8] In this paper, we shall show the capacity of the channel with continuous alphabets of squeezed states by using Holevo's formula, and discuss on the capacity of squeezed state channel in the quantum formulation and the case of semi classical in which the measurement is assumed as standard measurement process.

THE ENTROPY OF GENERAL GAUSSIAN STATE

Here we give a brief discussion on the formula for calculation method of von Neumann entropy clarified by one of authors. Let $\mathcal{H}$ be a Hilbert space and let $\rho(0)$ be a density operator in $\mathcal{H}$. Consider the family of density operators

$$\rho(\mu) = \mathcal{D}(\mu)\rho(0)\mathcal{D}(\mu)^{\dagger}; \mu \in \mathbf{C}, \tag{1}$$

Quantum Communication, Computing, and Measurement 2
Edited by Kumar et al., Kluwer Academic / Plenum Publishers, New York, 2000.

where **C** is the complex plane, and

$$\mathcal{D}(\mu) = \exp(\mu a^\dagger - \overline{\mu} a), \tag{2}$$

is the unitary *displacement operator* in $\mathcal{H}$.

In general, $\rho(0)$ describes background noise, comprising quantum noise, and μ is the classical signal. We impose the power constraint on the signal and the noise:

$$\hbar\omega \int |\mu|^2 P(d\mu) + \mathrm{Tr}\rho(0)\mathbf{H} \leq \hbar\omega(N_t + \frac{1}{2}). \tag{3}$$

where **H** is the Hamiltonian

$$\mathbf{H} = \hbar\omega(a^\dagger a + \frac{1}{2}). \tag{4}$$

According to Holevo's theorem,[7] the capacity of such a channel is equal to

$$C = \sup_{P \in \mathcal{P}_1} \{H(\rho_P) - \int H(\rho(\mu))P(d\mu)\}, \tag{5}$$

where $H = -\mathrm{Tr}\rho \log \rho$ is the von Neumann entropy, $\rho_P = \int \rho(\mu)P(d\mu)$, and $\mathcal{P}_1$ is a convex subset of probability distributions $P(d\mu)$ on **C**, satisfying the inequality (3). In the case (1) the operators $\rho(\mu)$ and $\rho(0)$ are unitarily equivalent and we have important simplification $H(\rho(\mu)) \equiv H(\rho(0))$, resulting in

$$C = \sup_{P \in \mathcal{P}_1} H(\rho_P) - H(\rho(0)). \tag{6}$$

In this paper, we calculate the capacity in the case where $\rho(0)$ is the Gaussian state defined as follows, in particular the squeezed state.

Definition. *The density operator ρ is called Gaussian, if its quantum characteristic function has the form*

$$\mathrm{Tr}\rho \, \exp iR^T z = \exp(i\, m^T z - \frac{1}{2} z^T \alpha z), \tag{7}$$

where $z = [x, y]^T$, $R = [q, p]^T$, m is column 2 -vector and α is real symmetric 2×2-matrix:

$$\alpha = \begin{bmatrix} \alpha^{qq} & \alpha^{qp} \\ \alpha^{pq} & \alpha^{pp} \end{bmatrix} \tag{8}$$

One can show that[3,4]

$$m = \mathrm{Tr}\rho R \; ; \; \alpha - \frac{i}{2}\Delta = \mathrm{Tr} R \rho R^T. \tag{9}$$

The *mean* m can be arbitrary vector; the necessary and sufficient condition on the *correlation matrix* α is

$$\alpha^{qq}\alpha^{pp} - (\alpha^{qp})^2 \geq \frac{1}{4}. \tag{10}$$

The state ρ is pure if and only if the equality holds in this equation.

Note that the relation between the real notation and the usual physical parametrization of squeezed states is given as follows. Consider the squeezed state given by the vector $S(\zeta)|0>$, where $\zeta = \gamma e^{i\theta}$ and $S(\zeta) = \exp((\zeta^* a^2 - \zeta(a^\dagger)^2)/2)$. Then α is represented by squeezing parameters γ, θ as follows

$$\begin{aligned} \alpha^{qq} &= \frac{\hbar}{2\omega}\{\cosh 2\gamma - \sinh 2\gamma \cos\theta \}, \\ \alpha^{pp} &= \frac{\hbar\omega}{2}\{\cosh 2\gamma + \sinh 2\gamma \cos\theta \}, \\ \alpha^{qp} &= \frac{\hbar}{2}\sinh 2\gamma \, \sin\theta. \end{aligned} \tag{11}$$

It was shown that[8] the von Neumann entropy of the Gaussian state is given by

$$H(\rho) = G\left(\frac{\alpha^{qq}\alpha^{pp} - (\alpha^{qp})^2}{\hbar^2}\right), \tag{12}$$

where G is the function defined by the following relation for $d \geq \frac{1}{2}$.

$$G(d^2) = (d+\frac{1}{2})\log(d+\frac{1}{2}) - (d-\frac{1}{2})\log(d-\frac{1}{2}) \tag{13}$$

$$= \frac{1}{2}\log d^2 + \log\left[1 - \sum_{k=1}^{\infty}\frac{1}{2k(2k+1)}(\frac{1}{4d^2})^k\right]. \tag{14}$$

Using this formula, we shall calculate the capacity in the next section.

CALCULATING THE CAPACITY

Let $\rho(0)$ be Gaussian density operator with $m = 0$ and the correlation matrix α. Let P be Gaussian probability distribution with the correlation matrix β. Without loss of generality we may assume it has zero mean. The inequality (3) then takes the form (in terms of real variables):

$$\frac{1}{2}(\omega^2(\alpha^{qq} + \beta^{qq}) + \beta^{pp} + \alpha^{pp}) \leq \hbar\omega(N_t + \frac{1}{2}). \tag{15}$$

The mixture ρ_P will be again Gaussian density operator with zero mean and the correlation matrix $(\alpha + \beta)$. And in the case of Gaussian $\rho(0)$, we can restrict optimization in (6) to Gaussian probability distribution P. Thus, the capacity of the general Gaussian channel is equal to

$$C = \max_{\beta\in B_1} G\left(\frac{(\alpha^{qq}+\beta^{qq})(\alpha^{pp}+\beta^{pp}) - (\alpha^{qp}+\beta^{qp})^2}{\hbar^2}\right)$$
$$-G\left(\frac{\alpha^{qq}\alpha^{pp} - (\alpha^{qp})^2}{\hbar^2}\right), \tag{16}$$

where B_1 is the convex set of real positive matrices β, satisfying (15). Because $G(d^2)$ is monotonously increasing function of d^2, the capacity is given by

$$C = G(d^2_{max}) - G\left(\frac{\alpha^{qq}\alpha^{pp} - (\alpha^{qp})^2}{\hbar^2}\right), \tag{17}$$

where

$$d^2_{max} = \frac{1}{\hbar^2}\max_{\beta\in B_1}\left[(\alpha^{qq}+\beta^{qq})(\alpha^{pp}+\beta^{pp}) - (\alpha^{qp}+\beta^{qp})^2\right]. \tag{18}$$

We can easily calculate d^2_{max}, and obtain the capacity as follows.
If the inequality

$$(\frac{1}{2}(\omega\alpha^{qq} - \frac{\alpha^{pp}}{\omega}))^2 + (\alpha^{qp})^2 \leq (\hbar(N_t + \frac{1}{2}) - \frac{1}{2}(\omega\alpha^{qq} + \frac{1}{\omega}\alpha^{pp}))^2. \tag{19}$$

holds, then the capacity is given by

$$C = G\left((N_t + \frac{1}{2})^2\right) - G\left(\frac{\alpha^{qq}\alpha^{pp} - (\alpha^{qp})^2}{\hbar^2}\right). \tag{20}$$

On the other hand, if (19) does not hold, the capacity is given by

$$C = G\left((N_t + \frac{1}{2})^2 - D_{min}\right) - G\left(\frac{\alpha^{qq}\alpha^{pp} - (\alpha^{qp})^2}{\hbar^2}\right), \tag{21}$$

where

$$D_{min} = \frac{1}{\hbar^2}\left(\sqrt{(\frac{1}{2}(\omega\alpha^{qq} - \frac{\alpha^{pp}}{\omega}))^2 + (\alpha^{qp})^2} - (\hbar(N_t + \frac{1}{2}) - \frac{1}{2}(\omega\alpha^{qq} + \frac{\alpha^{pp}}{\omega}))\right)^2 \tag{22}$$

In particular, when $\rho(0)$ is a squeezed state with the vector $S(\zeta)|0>$, where $\zeta = \gamma e^{i\theta}$, the capacity is given as follows.
Now we shall consider the case where $\theta = 0$. Then from (11) we have

$$\alpha^{qq} = \frac{\hbar}{2\omega}e^{-2\gamma}, \quad \alpha^{pp} = \frac{\hbar\omega}{2}e^{2\gamma}, \quad \alpha^{qp} = 0. \tag{23}$$

Substituting (23) to (19), (20), we can get the following results.
If the inequality

$$\log\sqrt{\frac{1}{2N_t + 1}} \leq \gamma \leq \log\sqrt{2N_t + 1} \tag{24}$$

holds, then the capacity is given by

$$C = G\left((N_t + \frac{1}{2})^2\right) = (N_t + 1)\log(N_t + 1) - N_t \log N_t, \tag{25}$$

which is equal to the Yuen-Ozawa bound (von Neumann entropy of the thermal state). And the optimum Gaussian distribution β_{opt} is given by

$$\beta^{qq}_{opt} = \frac{\hbar}{2\omega}(-e^{-2\gamma} + (2N_t + 1)), \tag{26}$$

$$\beta^{pp}_{opt} = \frac{\omega\hbar}{2}(-e^{2\gamma} + (2N_t + 1)), \tag{27}$$

$$\beta^{qp}_{opt} = 0. \tag{28}$$

In this case use of squeezed states cannot improve capacity.[12] It is interesting, that this does not make worse, provided the condition (24) holds. Figure 1 illustrates capacity C as function of squeezing parameter in the case of $N_t = 1$.

Next we shall consider the question whether there is a squeezing parameter such that the capacity becomes equal to the Yuen-Ozawa bound, when the coherent amplitude of squeezing state is restricted to a real number, i.e.,

$$\beta^{pp} = 0, \quad \beta^{qp} = 0. \tag{29}$$

The squeezing parameter satisfying $\beta^{qq}_{opt} = 0$ and the inequality (24), is

$$\gamma_0 = \log\sqrt{1 + 2N_t}. \tag{30}$$

So we can find the capacity is equal to the Yuen-Ozawa bound if and only if (30) holds. And then we have

$$\beta^{qq}_{opt} = \frac{2\hbar}{\omega}\frac{N_t(N_t + 1)}{2N_t + 1}. \tag{31}$$

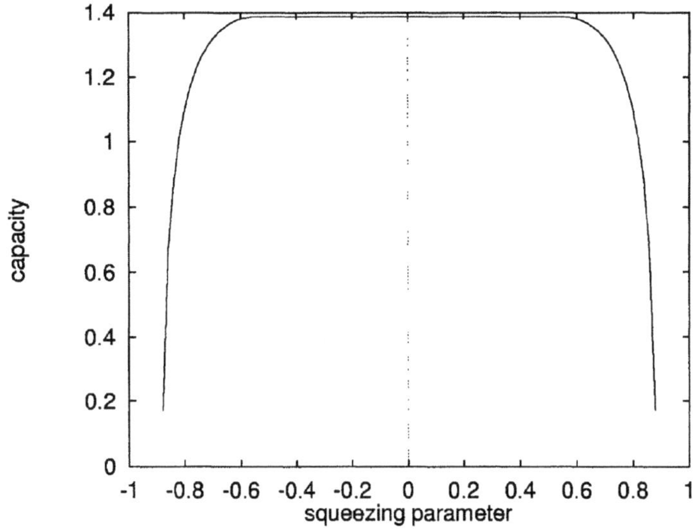

Figure 1. Dependence of the capacity of the complex amplitude squeezed state channel with constrained output power, $N_t = 1$, on the squeezing parameter γ.

DISCUSSIONS

When the signal state is coherent state and the receiver is Heterodyne, the maximum mutual information is

$$I_{max} = \log(1 + N_t). \tag{32}$$

On the other hand, when the signal state is squeezed state with optimum parameter and the receiver is Homodyne, the maximum mutual information is

$$I_{max} = \log(1 + 2N_t). \tag{33}$$

Thus use of squeezed states can improve the maximum mutual information under the output power constraint. However, we have shown that the channel capacity of squeezed state is the same as that of coherent state. The capacity of the coherent state can be realized by photon counting system as shown by Gordon. Such a result looks like a very reasonable one, because the photon number distribution of the Gaussian mixed state of coherent state is Bose-Einstein distribution, and the maximum von Neumann entropy is the entropy of Bose-Einstein distribution. However, the photon number distribution of the Gaussian mixed state of squeezed state is not Bose-Einstein distribution.[9] Hence in the case of squeezed state, the conventional photon counting system is not optimum. In subsequent paper, we will discuss the problem for the realization of the capacity of squeezed states.

REFERENCES

1. J. P. Gordon, "Noise at optical frequencies; information theory," in Quantum Electronics and Coherent Light, Proc. Int. School Phys. "Enrico Fermi", Course XXXI, ed. P. A. Miles, New York: Academic Press 1964, pp.156–181.
2. P. Hausladen, R. Jozsa, B. Schumacher, M. Westmoreland, W. Wootters, "Classical information capacity of a quantum channel," Phys. Rev. A, vol. 54, no. 3, pp. 1869–1876 1996.

3. A. S. Holevo, "Some statistical problems for quantum Gaussian states", *IEEE Transactions on Information Theory*, vol. **IT-21**, no.5, pp. 533–543, 1975.
4. A. S. Holevo, *Probabilistic and statistical aspects of quantum theory*, chapter 5, North-Holland, 1982.
5. A. S. Holevo, "The capacity of quantum communication channel with general signal states," LANL Report no. quant-ph/9611023, Nov. 1996. (Appeared in IEEE Trans. Inform. Theory, vol. 44, no. 1, pp. 296–272 1998).
6. A. S. Holevo, "On quantum communcation channel with constrained inputs," quant-ph/9705054.
7. A. S. Holevo, "Coding theorems for Quantum Channels," *Tamagawa University Research Review*, no. 4, 1998.
8. A. S. Holevo, M. Sohma, O. Hirota, "The capacity of quantum Gaussian channels," in preparation.
9. O. Hirota and S. Ikehara, "The photon distribution of the spontaneous two-photon emission," Phys. Lett. A, vol. 57 A, no. 4, pp. 317–319, 1976.
10. D. S. Lebedev, L. B. Levitin, "The maximal amount of information transmissible by an electromagnetic field," Information and Control, vol. 9, pp. 1–22, 1966.
11. B. Schumacher, M. D. Westmoreland, "Sending classical information via noisy quantum channel," Phys. Rev. A., vol. 56, no. 1, pp. 131–138, 1997.
12. H. P. Yuen and M. Ozawa, "Ultimate information carrying limit of quantum systems," Phys. Rev. Lett., vol. 70, no. 4, pp. 363–366, 1993.
13. H. P. Yuen, "States that give the maximum signal-to-quantum noise ratio for a fixed energy," Phys. Lett., vol. 56A, no. 2, pp. 105–106, 1976.

ASYMPTOTIC QUANTUM ESTIMATION THEORY FOR THE THERMAL STATES FAMILY

Masahito Hayashi[1]

[1]Department of Mathematics, Kyoto University, Kyoto, 606-8502, Japan
e-mail address: masahito@kusm.kyoto-u.ac.jp

Concerning state estimation, we will compare two cases. In one case we cannot use the quantum correlations between samples. In the other case, we can use them. In addition, under the later case, we will propose a method which simultaneously measures the complex amplitude and the expected photon number for the thermal states.

INTRODUCTION

Quantum estimation is essentially different from classical estimation regarding the following two points. The first point is that we cannot simultaneously construct the optimal estimators corresponding to respective parameters because of non-commutativity between them. It has been a serious problem since the beginning of the quantum estimation [1, 2, 3]. The second point is that we can reduce the estimation error under the assumption that we can prepare independent and identical samples of the unknown quantum state. It was pointed by Nagaoka [4, 5] concerning the large deviation theory in one-parameter estimation. The purpose of this paper is to clear the second point concerning the mean square error (MSE).

Our situation is divided into the following two cases. In the first case, we estimate the unknown state by independently measuring every sample. In this case, we may decide the n-th POVM from $n-1$ data which have been already given. In the second case, we estimate the unknown state by regarding n-sample system as a single composite system. In this case, we may use POVMs which are indivisible into every sample system. In order to construct these POVMs, we need to use quantum correlations between every sample. The former is called the non-quantum correlation case and the later the quantum correlation case. When the unknown state is a pure state, the errors of both are asymptotically equivalent in the first order [6]. Concerning the spin 1/2 system, see Hayashi [7].

In this paper, we formulate a general theory for the asymptotic quantum estimation. It is applied to the simultaneous estimation of the expected photon number and the complex amplitude for the quantum thermal state.

Quantum Communication, Computing, and Measurement 2
Edited by Kumar et al., Kluwer Academic / Plenum Publishers, New York, 2000.

ASYMPTOTIC ESTIMATION THEORY

In this paper, we use a quantum state family $\mathcal{S}$ parameterized by finite parameters $\theta^1, \cdots, \theta^d$:

$$\mathcal{S} := \{\rho_\theta \in \mathcal{S}(\mathcal{H}) | \theta = (\theta^1, \ldots, \theta^d) \in \Theta \subset \mathbf{R}^d\},$$

where the set $\mathcal{S}(\mathcal{H})$ denotes the set of densities on $\mathcal{H}$. For simplicity, we assume that ρ_θ is nondegenerate.

Non-quantum correlation case

The non-quantum correlation case is formulated as follows. A pair $\mathcal{E}_n = (\{M_k\}_{k=1}^n, \hat{\theta}_n)$ is called a *recursive estimator* where $\hat{\theta}_n$ is a function estimating the unknown parameter from n data, and $\{M_k\}_{k=1}^n$ is a sequence of POVMs $M_1, M_2(\omega_1), \ldots, M_n(\omega_1, \ldots, \omega_{n-1})$ as follows: the n-th POVM $M_k(\omega_1, \ldots, \omega_{k-1})$ is determined by $k-1$ data which have been already given. A sequence $\{\mathcal{E}_n\}_{n=1}^\infty$ of recursive estimators is called a *recursive MSE consistent estimator* if

$$\lim_{n\to\infty} \underbrace{\int\int \cdots \int}_{n} \left\| \hat{\theta}_n(\omega_1, \omega_2, \ldots, \omega_n) - \theta \right\|^2 P_{\rho_\theta}^{\mathcal{E}_n}(d\omega_1, d\omega_2, \ldots, d\omega_n) = 0, \ \forall \theta \in \Theta,$$

where

$$P_\rho^{\mathcal{E}_n}(d\omega_1, d\omega_2, \ldots, d\omega_n) = \operatorname{tr} \rho M_1(d\omega_1) \operatorname{tr} \rho M_2(\omega_1)(d\omega_2) \ldots \operatorname{tr} \rho M_n(\omega_1, \ldots, \omega_{n-1})(d\omega_n).$$

We define the *non-quantum-correlational Cramér-Rao type bound* $C_\theta^{NQC}(G)$ for a weighted matrix G(G is a $d \times d$ real positive symmetric matrix.) as:

$$C_\theta^{NQC}(G) := \inf \left\{ \liminf_{n\to\infty} \operatorname{Tr} nGV_\theta(\mathcal{E}_n) \middle| \{\mathcal{E}_n\}_{n=1}^\infty \text{ is a recursive MSE consistent estimator} \right\},$$

where the MSE matrix $V_\theta(\mathcal{E}_n)$ is given by:

$$V_\theta^{i,j}(\mathcal{E}_n) = \underbrace{\int \cdots \int}_{n} \left(\hat{\theta}_n^{\,i}(\omega_1, \ldots, \omega_n) - \theta^i\right) \left(\hat{\theta}_n^{\,j}(\omega_1, \ldots, \omega_n) - \theta^j\right) P_{\rho_\theta}^{\mathcal{E}_n}(d\omega_1, \ldots, d\omega_n).$$

We have the following equation:

$$C_\theta^{NQC}(G) = \inf \left\{ \operatorname{Tr} G \left(J_\theta^M\right)^{-1} \middle| M \text{ is a POVM on } \mathcal{H} \right\},$$

where J_θ^M denotes the Fisher information matrix at θ of $\{\operatorname{tr} \rho_\theta M(d\omega) | \theta \in \Theta\}$. It is derived by Jensen's inequality [8]. Under some regular condition, we show that there exists a recursive MSE consistent estimator $\{\mathcal{E}_n\}_{n=1}^\infty$ such that [8]:

$$n \operatorname{Tr} GV_\theta(\mathcal{E}_n) \to C_\theta^{NQC}(G) \text{ as } n \to \infty, \ \forall \theta \in \Theta.$$

According Holevo[3], we have $J_\theta^M \le \tilde{J}_\theta$, where $\tilde{J}_\theta$ is the *RLD Fisher information matrix* defined as: $\tilde{J}_{\theta;i,j} := \operatorname{tr}(\tilde{L}_{\theta;i})^* \rho_\theta \tilde{L}_{\theta;j}$, $\tilde{L}_{\theta;i} := (\rho_\theta)^{-1} \frac{\partial \rho_\theta}{\partial \theta^i}$. Therefore we have the following inequality

$$C_\theta^{NQC}(G) \ge C_\theta^R(G),$$

where

$$\begin{aligned} C_\theta^R(G) &:= \inf \left\{ \operatorname{tr} GV | V \text{ is a } d \times d \text{ real symmetric matrix } V \ge \tilde{J}_\theta^{-1} \right\} \\ &= \operatorname{Tr} G \operatorname{Re} \tilde{J}_\theta^{-1} + \operatorname{Tr} \left| \sqrt{G} \operatorname{Im} \tilde{J}_\theta^{-1} \sqrt{G} \right|. \end{aligned}$$

Quantum correlation case

Next, we formulate the quantum correlation case. For this purpose, we consider a quantum counterpart of independent and identically distributed condition. If $\mathcal{H}_1, \ldots, \mathcal{H}_n$ are n Hilbert spaces which correspond to the physical systems, then their composite system is represented by the tensor Hilbert space.

$$\mathcal{H}^{(n)} := \mathcal{H}_1 \otimes \cdots \otimes \mathcal{H}_n.$$

Thus, a state on the composite system is denoted by a density operator $\rho^{(n)}$ on $\mathcal{H}^{(n)}$. In particular if n element systems $(\mathcal{H}_1, \ldots, \mathcal{H}_n)$ of the composite system $\mathcal{H}^{(n)}$ are independent of each other, there exists a density ρ_k on $\mathcal{H}_k$ such that

$$\rho^{(n)} = \rho_1 \otimes \cdots \otimes \rho_n, \text{ on } \mathcal{H}^{(n)}.$$

The condition:

$$\mathcal{H}_1 = \cdots = \mathcal{H}_n = \mathcal{H}, \ \rho_1 = \cdots = \rho_n = \rho$$

corresponds to the independent and identically distributed condition in the classical case. Therefore, we consider the parameter estimation problem for the family $\{\rho_\theta^{(n)} \text{ on } \mathcal{H} | \theta \in \Theta\}$ which is called the *n-i.i.d. extended family*.

In this case, we use a sequence $\{M^n\}_{n=1}^\infty$ of POVMs where M^n is a POVM on $\mathcal{H}$ whose measurable set is $\mathbf{R}^d$ as an estimator. A sequence $\{M^n\}_{n=1}^\infty$ is called an *MSE consistent estimator* if $\{M^n\}_{n=1}^\infty$ satisfies

$$\lim_{n\to\infty} \|\hat{\theta} - \theta\|^2 \operatorname{tr} \rho_\theta^{(n)} M^n(\, d\hat{\theta}) = 0, \ \forall \theta \in \Theta.$$

A recursive MSE consistent estimator can be regarded as an MSE consistent estimator because a recursive estimator $\mathcal{E}_n = (\{M_k\}_{k=1}^n, \hat{\theta}_n)$ is regarded as a POVM $M(\mathcal{E}_n)$ as follows:

$$M(\mathcal{E}_n)(B) := \int_{\hat{\theta}_n^{-1}(B)} \bigotimes_{k=1}^n M_k(\omega_1, \ldots, \omega_{k-1})(\, d\omega_k), \ \forall B \subset \mathbf{R}^d \text{ on } \mathcal{H}^{(n)}.$$

We define the *quantum-correlational Cramér-Rao type bound* $C_\theta^{QC}(G)$ for a weighted matrix G as:

$$C_\theta^{QC}(G) := \inf \left\{ \liminf_{n\to\infty} n \operatorname{Tr} G V_\theta(M^n) \middle| \{M^n\}_{n=1}^\infty \text{ is an MSE consistent estimator } \right\}$$

where the MSE matrix $V_\theta(M^n)$ is given by:

$$V_\theta^{i,j}(M^n) = \int_{\mathbf{R}^d} (\hat{\theta}^i - \theta^i)(\hat{\theta}^j - \theta^j) \operatorname{tr} \rho_\theta^{(n)} M^n(\, d\hat{\theta}).$$

We have the following equation

$$C_\theta^{QC}(G) = \liminf_{n\to\infty} n C_\theta^n(G),$$

where $C_\theta^n(G)$ denotes the non-quantum-correlational Cramér-Rao type bound for the n-i.i.d. extended family [8]. From the definition of the n-i.i.d. extended family, we have $C_\theta^{NQC}(G) \geq n C_\theta^n(G)$. Therefore, we have the first inequality of (1).

$$C_\theta^{NQC}(G) \geq C_\theta^{QC}(G) \geq C_\theta^R(G). \tag{1}$$

It shows the second inequality of (1) that $J_\theta^{M^n} \leq n \tilde{J}_\theta$ for any POVM M^n on $\mathcal{H}^{(n)}$. Therefore the difference between $C_\theta^{NQC}(G)$ and $C_\theta^{QC}(G)$ means the difference of the quantum correlation case from the non-quantum correlation case. Under some regular condition, we can show that there exists an MSE consistent estimator $\{M^n\}_{n=1}^\infty$ such that [8]:

$$n \operatorname{Tr} G V_\theta(M^n) \to C_\theta^{QC}(G) \text{ as } n \to \infty, \ \forall \theta \in \Theta.$$

Now we consider the estimation for the the complex amplitude ζ and expected photon number N for the quantum thermal states family defined as:

$$\mathcal{S} := \left\{ \rho_{\zeta,N} := \frac{1}{\pi N} \int_{\mathbf{C}} \exp\left(-\frac{|\zeta-\alpha|^2}{N}\right) |\alpha\rangle\langle\alpha|\, d^2\alpha \,\middle|\, \zeta \in \mathbf{C}, N > 0 \right\}.$$

Estimation of complex amplitude ζ

In the case of that photon number N is known, we estimate the tow unknown parameters $\zeta = (\theta^1 + i\theta^2)/\sqrt{2}$. This estimation problem is investigated by Yuen, Lax and Holevo [2, 3]. In this case they calculated the inverse $\tilde{J}_\theta^{-1}$ of the RLD Fisher information matrix as:

$$\tilde{J}_\theta^{-1} = \begin{pmatrix} N+\frac{1}{2} & \frac{i}{2} \\ -\frac{i}{2} & N+\frac{1}{2} \end{pmatrix}.$$

They calculated the non-quantum-correlational Cramér-Rao type bound $C_\theta^{NQC}(G)$ as follows:

$$C_\theta^{NQC}(G) = C_\theta^R(G) = 2\left(N+\frac{1}{2}\right) g_1 + \sqrt{g_1^2 - g_2^2 - g_3^2}, \tag{2}$$

where the weighted matrix G is parameterized as:

$$G = \begin{pmatrix} g_1+g_2 & g_3 \\ g_3 & g_1-g_2 \end{pmatrix}.$$

From (1) and (2), we have the following equations.

$$C_\theta^{NQC}(G) = C_\theta^{QC}(G) = C_\theta^R(G). \tag{3}$$

In this case, the optimal estimator is the squeezed heterodyne.

Simultaneous estimation of complex amplitude ζ and expected photon number N

Next we consider the case of that both of the expected photon number N and the complex amplitude ζ are unknown. In this case, we estimate three unknown parameters θ^1, θ^2 and $\theta^3 = N$. The first equation of (3) isn't held. Therefore, the squeezed heterodyne isn't optimal. The inverse $\tilde{J}_\theta^{-1}$ of the RLD Fisher information matrix is calculates as:

$$\tilde{J}_\theta^{-1} = \begin{pmatrix} N+\frac{1}{2} & \frac{i}{2} & 0 \\ -\frac{i}{2} & N+\frac{1}{2} & 0 \\ 0 & 0 & N(N+1) \end{pmatrix}.$$

Therefore we can calculated $C_\theta^R(G)$ as:

$$C_\theta^R(G) = g_0 N(N+1) + 2\left(N+\frac{1}{2}\right) g_1 + \sqrt{g_1^2 - g_2^2 - g_3^2},$$

if the weighted matrix G can be parameterized as:

$$G = \begin{pmatrix} g_1+g_2 & g_3 & 0 \\ g_3 & g_1-g_2 & 0 \\ 0 & 0 & g_0 \end{pmatrix}. \tag{4}$$

If the weighted matrix G can be parameterized as (4), we obtain the following equations:

$$C_\theta^{NQC}(G) > C_\theta^{QC}(G) = C_\theta^R(G). \tag{5}$$

A proof for $C_\theta^{NQC}(G) > C_\theta^{QC}(G)$ is omitted. The inequality $C_\theta^{NQC}(G) > C_\theta^{QC}(G)$ means that we cannot the simultaneous measurement of the photon number counting and heterodyne for a single sample.

Construction of an MSE consistent estimator $\{M^n\}_{n=1}^\infty$ attaining $C_\theta^R(I)$

Now, for a weighted matrix I, we construct an MSE consistent estimator $\{M^n\}_{n=1}^\infty$ such that

$$\lim_{n\to\infty} n \operatorname{tr} V_\theta(M^n) = C_\theta^R(I), \ \forall\theta \in \Theta.$$

It is sufficient for $C_\theta^{QC}(I) = C_\theta^R(I)$ to construct such an MSE consistent estimator.

Every POVM M^n is constructed in the following step:

(1) Evolve the unknown state $\underbrace{\rho_{\zeta,N} \otimes \cdots \otimes \rho_{\zeta,N}}_{n}$ as:

$$\underbrace{\rho_{\zeta,N} \otimes \cdots \otimes \rho_{\zeta,N}}_{n} \to U_n \underbrace{\rho_{\zeta,N} \otimes \cdots \otimes \rho_{\zeta,N}}_{n} U_n^* = \rho_{\sqrt{n}\zeta,N} \otimes \underbrace{\rho_{0,N} \otimes \cdots \otimes \rho_{0,N}}_{n-1} \text{ on } \mathcal{H}^{(n)},$$

where

$$U_n = \exp\phi_{n-1}(a_n^* a_1 - a_1^* a_n) \cdots \exp\phi_2(a_3^* a_1 - a_1^* a_3) \exp\phi_1(a_2^* a_1 - a_1^* a_2) \text{ on } \mathcal{H}^{(n)}$$

$$\phi_i = \arctan\frac{1}{\sqrt{i}}, \quad i = 1, 2, \ldots, n-1.$$

a_i denotes the annihilation operator on $\mathcal{H}_i$.

(2) Measure the first sample $\rho_{\sqrt{n}\zeta,N}$ by the heterodyne, then we get the estimate $\hat{\zeta}$ of the complex amplitude.

(3) Measure the others by the photon counting, then we obtain $n-1$ data which obey the probability distribution $P^N(k)$:

$$P^N(k) = \frac{1}{N+1}\left(\frac{N}{N+1}\right)^k, \quad k = 0, 1 \ldots.$$

(4) We obtain the estimate $\hat{N}$ of the expected photon number N by the maximum likelihood estimator of the probability distribution $P^N(k_1), \ldots, P^N(k_{n-1})$.

CONCLUSION

We formulate an asymptotic quantum estimation theory. This theory is applied to the simultaneous measurement of the photon number counting and the heterodyne for thermal states. It is a future study to realize the MSE consistent estimator proposed in this paper in an actual physical system.

Acknowledgments

This work was supported by the Research Fellowships of the Japan Society for the Promotion of Science for Young Scientists No. 9404. The author would like to thank Dr. K. Matsumoto for useful discussions of these topics.

REFERENCES

1. C. W. Helstrom. "Quantum Detection and Estimation Theory," Academic Press, New York (1976).
2. H. P. Yuen and M. Lax, Multiple-parameter quantum estimation and measurement of nonselfadjoint observables, *IEEE trans. Inform. Theory,* **IT 19**, 740 (1973).
3. A. S. Holevo. "Probabilistic and Statistical Aspects of Quantum Theory," North_Holland, Amsterdam (1982).
4. H. Nagaoka, On the relation Kullback divergence and Fisher information -from classical systems to quantum systems-, *in:* "Proc. Society Information Theory and its Applications in Japan" (1992)(in Japanese).
5. H. Nagaoka, Two quantum analogues of the large deviation Cramŕr-Rao inequality, *in:* "Proc. 1994 IEEE Int. Symp. on Information Theory" p.118 (1994).
6. M. Hayashi, Asymptotic estimation theory for a finite-dimensional pure state model, *J. Phys. A: Math. Gen.* **31** 4633 (1998).
7. M. Hayashi, Asymptotic Quantum Parameter Estimation in Spin 1/2 System, LANL e-print quant-ph/9710040 (1997).
8. M. Hayashi and K. Matsumoto, Quantum mechanics as a statistical model which allows a free choice of measurements *in:* "Large Deviation and Statistical Inference" RIMS koukyuuroku No. 1055 RIMS, Kyoto (1998) (in Japanese).

THE ASYMPTOTIC EFFICIENCY OF THE CONSISTENT ESTIMATOR, BERRY-UHLMANN'S CURVATURE, AND QUANTUM INFORMATION GEOMETRY

Keiji Matsumoto,[1]

[1]University of Tokyo, Tokyo 113, Japan

It is pointed out that Berry-Uhlmann's parallelism plays key role in the first order asymptotic theory of quantum statistical estimation, and that the parallelism is closely related to Nagaoka's quantum information geometry, which is already successfully applied to quantum estimation theory.

INTRODUCTION

In the classical statistical estimation theory, Amari, Nagaoka, and others have shown (Amari, 1985) that 'dual connection' plays key role in the the higher order asymptotic theory. The quantum version of the theory is constructed by Nagaoka (Nagaoka, 1987; Nagaoka, 1989; Nagaoka, 1994) and is successfully applied to the characterization of quantum statistical models which have efficient estimators.

On the other hand, in the paper, it is shown that Berry-Uhlmann's curvature (Berry, 1984; Simon, 1983; Aharonov and J. Anandan, 1987; Shapere, Wilczek, 1989; Uhlmann, 1986; Uhlmann, 1993) is a good measure of 'umcertainty', or the difficulty of simultaneous estimation of the several components of multi dimensional unknown parameter, in the framework of the first order asymptotic theory. Moreover, we shed light on the intrinsic relation between Berry-Uhlmann's parallelism and Nagaoka's information geometry.

The paper is organized as follows. The first section is brief summary of the first order asymptotic theory of statistical estimation. In the second section, Nagaoka's quantum information geometry and Berry-Uhlmann's curvature are introduced, in such a manner that the intrinsic relation between them is apparent. Estimation theoretical meanings of Berry-Uhlmann's parallelism is explained in the third section.

In the paper, the dimension d of the Hilbert space $\mathcal{H}$ is assumed to be finite, for mathematical simplicity. The author conjectures, however, that the essence of the theory is unchanged by this assumption. In fact, some of the results are proved also in the infinite dimensional case. To keep the limit on the length, proofs are omitted. For the complete discussion, see Matsumoto(1997a-d; 1998a-b).

Quantum Communication, Computing, and Measurement 2
Edited by Kumar et al., Kluwer Academic / Plenum Publishers, New York, 2000.

ASYMPTOTIC THEORY OF QUANTUM STATISTICAL ESTIMATION

Framework of the Theory

In the quantum estimation theory, we try to estimate the density matrix of the given system from the data $\xi \in \Xi$ produced from a measuring apparatus. For simplicity, it is assumed that the system belongs to a *model* $\mathcal{M} = \{\rho(\theta) | \theta \in \Theta \subset \mathbf{R}^m\}$, and that the true value of the parameter θ is to be estimated. For example, $\mathcal{M}$ is a set of spin states with given wave function part and unknown spin part.

Let us assume N samples of the unknown state $\rho(\theta)$ are given, and denote by $M_{(k)}$ the positive operator valued measure (POM) (Holevo, 1982) which corresponds to the measurement applied to kth sample. Outcoming data of kth measurement $M_{(k)}$ is denoted by $\xi_{(k)} \in \Xi$. Here, the kth measurement $M_{(k)}$ is chosen dependent on $\xi_{(1)}, \xi_{(2)}, \ldots, \xi_{(k-1)}$. After measuring all of N samples, we obtain estimate $\hat{\theta}_{(N)}(\xi_{(1)}, \xi_{(2)}, \ldots, \xi_{(N)})$ of θ. A pair $(\{M_{(k)}\}_{k=1}^{N}, \hat{\theta}_N)$ is called *estimator*. An estimator $(\{M_{(k)}\}_{k=1}^{N}, \hat{\theta}_N)$ is said to be *consistent* if $\mathrm{Prob}_\theta\{\|\hat{\theta}_{(N)} - \theta\| > \epsilon\} \to 0$, as $N \to \infty$. In the paper, we focus on consistent estimators.

Our purpose is to modify $(\{M_{(k)}\}_{k=1}^{N}, \hat{\theta}_N)$ to minimize the 'probability of error',

$$\mathrm{Prob}_\theta\{d(\theta, \hat{\theta}_{(N)}) > \epsilon\}, \tag{1}$$

where $d(\cdot,\cdot)$ is a distance defined on $\mathbf{R}^m$. As N tends to infinity, the error probability (1) decreases exponentially. In the first order asymptotic theory, we calculate the rate of decrease and and modify estimator to minimize the rate. In this paper, we do not use quantum correlation in measuring N samples. Many of results, however, holds true even if the usage of quantum correlation is allowed.

Basic Theorems

We define *Bahadur type bound* $B_\theta(G)$ by

$$B_\theta(G) := \inf\{\text{the largest eigenvalue of } GJ_\theta^{M-1} | M \text{ is a POM}\},$$

where, G is related to the distance $d(\cdot,\cdot)$ by,

$$d(\theta, \theta + d\theta) = \sum_{i,j} [G]_{i,j} d\theta^i d\theta^j + o(d\theta)^2,$$

and J_θ^{M-1} is the Fisher information matrix of the family $\{\mathrm{tr}\rho(\theta)M(d\xi) | \theta \in \Theta \subset \mathbf{R}^m\}$ of probability distributions. Under some regularity conditions, it is proved that the inverse of Bahadur type bound gives attainable lower bound of the error (1) in the sence of the first order asymptotic theory (Nagaoka, 1992; Matsumoto, 1998b).

Theorem 1

$$\inf_{consistent\ est.} \varliminf_{\epsilon\downarrow 0} \varliminf_{N\to\infty} \frac{1}{N\epsilon^2} \log \mathrm{Prob}_\theta\{d(\theta, \hat{\theta}_N) > \epsilon\} = -(B_\theta(G))^{-1}. \tag{2}$$

We define *SLD Fisher information matrix* $J^S(\theta)$ by the following theorem;

Theorem 2 *For $\theta \in \Theta$, there is a $m \times m$ real symmetric matrix $J^S(\theta)$ such that*

$$\forall \mathbf{v} \in \mathbf{R}^m \quad B_\theta(\mathbf{v}\mathbf{v}^T) = \mathbf{v}^T J^{S-1}(\theta)\mathbf{v}. \tag{3}$$

Here, note that $J^S(\theta)$ is independent of $\mathbf{v} \in \mathbf{R}^m$. When $G = \mathbf{v}\mathbf{v}^T$, the distance $d(\cdot,\cdot)$

degenerate except for the direction of $\mathbf{v}$, and the measure (1) of the error is concerned only with the error in the direction of $\mathbf{v}$.

A deference between classical estimation theory and quantum estimation theory is existence of 'uncertainty', or the difficulty to estimate several components of the unknown parameter simultaneously. A possible measure of 'uncertainty' is the difference between both sides of the inequality,

$$\begin{aligned} &\inf_{\text{consistent est.}} \max_{\|\mathbf{e}\|=1} \varliminf_{\epsilon\downarrow 0} \varliminf_{N\to\infty} \frac{1}{N\epsilon^2} \log \mathrm{Prob}_\theta\{d_{\mathbf{e}}(\theta, \hat{\theta}_{(N)}) > \epsilon\} \\ &\geq \max_{\|\mathbf{e}\|=1} \inf_{\text{consistent est.}} \varliminf_{\epsilon\downarrow 0} \varliminf_{N\to\infty} \frac{1}{N\epsilon^2} \log \mathrm{Prob}_\theta\{d_{\mathbf{e}}(\theta, \hat{\theta}_{(N)}) > \epsilon\}, \end{aligned} \tag{4}$$

where

$$d_{\mathbf{e}}(\theta, \theta + d\theta) = \sum_{i,j} [\sqrt{G}\mathbf{e}\mathbf{e}^T\sqrt{G}]_{i,j} d\theta^i d\theta^j + o(d\theta)^2.$$

The equality

$$\varliminf_{\epsilon\downarrow 0} \varliminf_{N\to\infty} \frac{1}{N\epsilon^2} \log \mathrm{Prob}_\theta\{d(\theta, \hat{\theta}_{(N)}) > \epsilon\} = \max_{\|\mathbf{e}\|=1} \varliminf_{\epsilon\downarrow 0} \varliminf_{N\to\infty} \frac{1}{N\epsilon^2} \log \mathrm{Prob}_\theta\{d_{\mathbf{e}}(\theta, \hat{\theta}_{(N)}) > \epsilon\},$$

the inequality (4), and the definition SLD Fisher information matrix lead to, for any nonnegative matrix G,

$$-(B_\theta(G))^{-1} \geq -(\text{the largest eigenvalue of } GJ^{S-1})^{-1} \tag{5}$$

We use the deference of both sides of the inequality (5) as a measure of the uncertainty. When the parameter θ is one dimensional the equality always holds.

GEOMETRY OF QUANTUM STATISTICAL MODELS

Tangent Bundle $\mathcal{T}(\mathcal{W}_r)$

Let $\mathcal{W}_r$ be the space of $d\times r$ complex and full-rank matrix W such that $\mathrm{tr}WW^* = 1$, $\mathcal{P}_r$ the space of density operators whose rank is r, and π the map from $\mathcal{W}_r$ to $\mathcal{P}_r$ such that $\rho = \pi(W) := WW^*$. One possible physical interpretation of W can be also seen as a state vector $|W\rangle$ in a bigger Hilbert space $\mathcal{H}\otimes\mathcal{H}'$, such that $\mathrm{tr}_{\mathcal{H}'}|W\rangle\langle W| = \rho$. Here, the dimension of $\mathcal{H}'$ is r and the operation $\pi(\cdot)$ corresponds to the partial trace of $|W\rangle\langle W|$ over $\mathcal{H}'$.

In this subsection, geometrical structures are introduced into the tangent bundle $\mathcal{T}(\mathcal{W}_r)$ over $\mathcal{W}_r$, which is a real manifold, with the coordinate system $\zeta = (\zeta^1, ..., \zeta^{2rd-1})^T$. The *matrix representation* $\mathbf{M}(\partial/\partial\zeta^i)$ of the tangent vector $\partial/\partial\zeta^i$ (throughout the thesis, the tangent vector is understood as the differential operator) is a $d\times r$ complex matrix such that

$$\mathbf{M}(\partial/\partial\zeta^i) := 2\partial W/\partial\zeta^i \tag{6}$$

A natural inner product $\langle\langle\cdot,\cdot\rangle\rangle_W$ is introduced to $\mathcal{T}(\mathcal{W}_r)$,

$$\langle\langle\hat{X},\hat{Y}\rangle\rangle_W := \sum_{i,j} (\mathrm{Re}(\mathbf{M}\hat{X})_{ij}\mathrm{Re}(\mathbf{M}\hat{Y})_{ij} + \mathrm{Im}(\mathbf{M}\hat{X})_{ij}\mathrm{Im}(\mathbf{M}\hat{X})_{ij}). \tag{7}$$

The logarithmic derivative $\mathbf{L}(\hat{X})$ of $X \in \mathcal{T}_W(\mathcal{W}_d)$ is a $d \times d$ complex matrix which satisfies the equation,

$$\mathbf{L}(\hat{X})W = \mathbf{M}(\hat{X}). \tag{8}$$

We introduce an affine connection by the following parallel transport:

$$\mathbf{L}\left(\prod_{W_0 \to W_1}^{(w)} \hat{X}\right) := \mathbf{L}\hat{X} - \operatorname{Re}\operatorname{tr}(\pi(W_1)\mathbf{L}\hat{X}). \tag{9}$$

Notice that the definition of logarithmic derivative (, and the connection $\prod^{(w)}$ also), has arbitrariness caused by the kernel of W^*. Therefore, when r is smaller than d, the equation (9) defines an equivalent class of affine connections.

As is easily seen from the definition, $\mathcal{W}_r$ is curvature free in terms of $\prod^{(w)}$. The torsion $T^{(w)}(\hat{X}, \hat{Y})$, however, does not vanish.

Nagaoka's Quantum Information Geometry

In this subsection, we introduce geometrical structure (Nagaoka's quantum information geometry) to the tangent bandle $\mathcal{T}(\mathcal{P}_r)$. Letting π_* be a differential map of π, the *horizontal lift* $h_W(X)$ of $X \in \mathcal{T}_\rho(\mathcal{P}_r)$ is a member of $(\ker \pi_*)^\perp$ which satisfies $\pi_*(h_W(X)) = X$. The inner product $\langle \cdot, \cdot \rangle$ in $\mathcal{T}(\mathcal{P}_r)$ is introduced by

$$\langle X, Y \rangle_{\pi(W)} = \langle\langle h_W(X), h_W(Y) \rangle\rangle_W .$$

As for an affine connection, *e-parallel transport* $\prod^{(e)}$ is defined as follows:

$$\prod_{\rho \to \sigma}^{(e)} X := \pi_* \left(\prod_{W \to V}^{(w)} h_W X\right), \quad \pi(W) = \rho,\, \pi(V) = \sigma. \tag{10}$$

These definition of inner product and affine connection is self-consistent because

$$\begin{aligned} \langle\langle h_W X, h_W Y \rangle\rangle &= \langle\langle h_{WU} X, h_{WU} Y \rangle\rangle , \\ \pi_* \left(\prod_{W \to V}^{(w)} h_W(X)\right) &= \pi_* \left(\prod_{WU \to VU'}^{(w)} h_{WU}(X)\right), \quad (U,\, U' \in U(r)) \end{aligned}$$

holds true. When the density matrix ρ is not faithful, the definition of $\prod^{(e)}$ has arbitrariness caused by arbitrariness of $\prod^{(w)}$.

As easily seen from the definition, $\mathcal{P}_r$ is curvature free in terms of e-connection. The torsion of e-connection $T^{(e)}$ is obtained by the projection of $T^{(w)}$,

$$T^{(e)}(X, Y) = \pi_* \left(T^{(w)}\left(h_W X, h_W Y\right)\right). \tag{11}$$

Nagaoka has shown that e-connection and its dual connection nicely characterizes the estimation theoretical property of quantum statistical model (Nagaoka,1987; Nagaoka, 1989; Nagaoka, 1994).

Berry-Uhlmann's Parallelism

Because $\pi(WU)$ $(U \in U(r))$ is identical to $\pi(W)$ it is natural to see the space $\mathcal{W}_r$ as the total space of the *principal fiber bundle* with the *base space* $\mathcal{P}_r$ and the *structure group* $U(r)$ (Kobayashi and Nomizu, 1963).

Let $\eta = (\eta^1, \eta^2, \dots, \eta^{2d^2-1}) \in \mathbf{R}^{2d^2-1}$ be a coordinate of $\mathcal{P}_r$. For a closed loop $\eta \to \eta + d\eta^i \to \eta + d\eta^i + d\eta^j \to \eta + d\eta^j \to \eta$, let us consider a curve $C_h = \{W(t)|0 \leq t \leq 1\}$ in $\mathcal{W}_r$ such that $\pi(W(t))$ goes roud the inifinitesimal loop and

$$dW(t)/dt = \mathbf{M}h_{W(t)}(d/dt) \tag{12}$$

is satisfied. Then, the curve C_h is not closed generally, and $W(1)$ writes,

$$\begin{aligned} W(1) &= W(0) + \frac{1}{2}W(0)A_{ij}d\eta^i d\eta^j + o(d\eta)^2, \\ W(0)A_{ij} &= \mathbf{M}\left(T^{(w)}(h_W\,\partial/\partial\eta^i,\, h_W\,\partial/\partial\eta^j) - h_W T^{(e)}(\partial/\partial\eta^i,\, \partial/\partial\eta^j)\right). \end{aligned} \tag{13}$$

Here, A_{ij} is a $r \times r$ skew Hermitian matrix and is a kind of representation of Berry-Uhlmann's curvature. In (13), each term in the right hand side of the second equation has arbitrariness, but their deference, A_{ij}, is uniquely defined.

GEOMETRICAL THEORY OF STATISTICAL ESTIMATION

SLD Fisher Information Matrix as a Natural Metric

First, it should be pointed out that SLD Fisher information matrix is the metric tensor of the inner product $\langle\cdot,\cdot\rangle$, $J^S(\theta) = [\langle\partial/\partial\theta^i,\, \partial/\partial\theta^j\rangle_\theta]$. Notice also that, by virtue of the theorem 2, the equality in (5) is achievable if the parameter of the model is one dimensional. This fact can be interpreted as: the more distant $\rho(\theta)$ and $\rho(\theta + d\theta)$ are, the easier it is to distinguish them.

Berry-Uhlmann's Curvature

Let us call the model the faithful state model (the pure state model) if all the members of the model is faithful (pure).

Theorem 3*If the model is the faithful state model or the pure state model, then the necessary and sufficient condition for the equality in* (5) *to be achievable is* $A_{ij}(\theta) = 0$ *at any* $\theta \in \Theta$.

In other words, if and only if the model is free of Berry-Uhlmann's curvature, there is no 'non commutativity' in the model. Is the deference between both sides of (5) the increasing function of the magnitude of Berry-Uhlmann's curvature? In the pure state model, we have theorems which affirmatively answer this inquiry.

When pure state model is considered, $\mathcal{W}_1$ is the set of d dimensional complex vectors, and $A_{ij}(\theta)$ is pure imaginary scalar. Let us denote by $[A_{ij}]$ the $m \times m$ matrix whose (i,j)th component is A_{ij}, and by $\pm\alpha_1, \pm\alpha_2, \dots$ $(1 \geq \alpha_1 \geq \alpha_2 \geq \dots \geq 0)$ the eigenvalues of the matrix $J^{S-1}[A_{ij}]$. Especially, if $m = 2$, $J^{S-1}[A_{ij}]$ has only two eigenvalues $\pm\alpha_1$ which is equal to $|A_{12}|/\sqrt{det J^S}$, or the Berry-Uhlmann's curvature per unit area.

Theorem 4*Consider the pure state model* $\mathcal{M}$ *with two dimensional parameter. If* $\alpha_1(\theta) \geq \alpha_1(\theta')$ *and* $J^S(\theta) = J^S(\theta')$, *then* $B_\theta(G) \geq B_{\theta'}(G)$ *for any nonnegative real matrix* G.

If $\alpha_1(\theta) = \alpha_1(\theta')$ *and* $J^S(\theta) \leq J^S(\theta')$, *then* $B_\theta(G) \geq B_{\theta'}(G)$ *for any nonnegative real matrix* G.

Because SLD Fisher information matrix $J^S(\theta)$ is estimation theoretically natural metric, it is reasonable to take $G = J^S(\theta)$. The following theorem is valid for pure state models with arbitrary dimension of parameter.

Theorem 5 *Consider pure state models $\mathcal{M}$. Then, $B_\theta(J^S(\theta)) = 2\left(1+\sqrt{1-\alpha_1^2}\right)^{-1}$, which is a decreasing function of α_1*

REFERENCES

Aharonov, Y., and Anandan. J., 1987, Phase change during a cyclic quantum evolution, Phys. Rev. Lett. **58**, 1593-1596.

Amari.S, 1985, "Differential-Geometrical Methods in Statistics," Lecture Notes in Statistics, Vol. 28 Springer, Berlin.

Berry, M. V., 1984, Quantal phase factors accompanying adiabatic changes, Proc. Roy. Soc. London **A392**, 45-57.

Fujiwara, A., and Nagaoka, H., 1995, Quantum Fisher metric and estimation for pure state models, Phys. Lett. **201A**,119-124.

Holevo, A. S., 1982, "Probabilistic and Statistical Aspects of Quantum Theory," North-Holland, Amsterdam.

Kobayashi, S., and Nomizu, K., 1963 "Foundations of Differential Geometry, I, II," John Wiley, New York.

Matsumoto, K., 1997a, A new approach to the Cramer-Rao type bound of the pure state model, quant-ph/9711008.

Matsumoto, K., 1997b, Uhlmann's parallelism in quantum estimation theory, LANL e-print quant-ph/9711027.

Matsumoto, K., 1997c, Berry's phase in the quantum estimation theory, and its intrinsic relation with the complex structure, METR **97-10**, University of Tokyo.

Matsumoto, K., 1997d, Uhlmann's parallelism and Nagaoka's quantum information geometry, METR **97-09**, University of Tokyo.

Matsumoto, K., 1998a, "A Geometrical Approach to Quantum Estimation Theory," doctoral thesis, Graduate School of Mathematical Sciences, University of Tokyo.

Matsumoto, K., 1998b, A study of a first order asymptotic theory of statistical estimation when the experimental scheme is controlled adaptively, IBIS'98, 67-70 (in Japanese).

Nagaoka, H., 1987, On Fisher information of quantum statistical models, SITA'87, 19-21 (in Japanese).

Nagaoka, H., 1989, On the parameter estimation problem for quantum statistical models, SITA'89, 577-582.

Nagaoka, H., 1992, On the relation between Kullback divergence and Fisher information - from classical systems to quantum systems -, Proc. Workshop on open problems in information theory, 63-72, (in Japanese)

Nagaoka, H., 1994, Differential geometrical aspects of quantum state estimation and Relative Entropy," METR 94-14.

Shapere, A., and Wilczek, F., 1989, "GEOMETRIC PHASES IN PHYSICS," Advanced Series in Mathematical Physics,vol. 5, World Scientific.

Simon, B., 1983, Holonomy, the quantum adiabatic theorem, and Berry's phase, Phys. Rev. Lett. **51**, 2167-2170.

Uhlmann, A., 1986, Parallel transport and 'quantum holonomy' along density operators, Rep. Math. Phys., vol.24, 229-240.

Uhlmann, A., 1993, Density operators as an arena for differential geometry, Rep. Math. Phys., vol.33, 253-263.

OPTIMAL SIGNAL-TO-QUANTUM NOISE RATIO IN SQUEEZED DISPLACED NUMBER STATES

Yinqi Feng[1] and Allan I. Solomon[2]

Faculty of Mathematics and Computing
The Open University
Milton Keynes, MK7 6AA, United Kingdom
[1]E-mail:y.feng@open.ac.uk
[2]E-mail:a.i.solomon@open.ac.uk

For an arbitary quantum state of radiation with frequency ω, the optimum signal-to-quantum noise ratio ρ for fixed energy (or power per unit frequency) $\hbar\omega N_s$ has the value $4N_s(N_s+1)$, attainable by the squeezed coherent vacuum. The question arises as to what ρ is attainable when the vacuum is imperfect, and there are n photons present. We show that in such a situation the attainable signal-to-quantum noise ratio ρ is reduced by a factor $(2n+1)^2$.

INTRODUCTION

In his classic paper of 1976, Yuen[1] proved the following important results:

- for an arbitrary quantum state of radiation with frequency ω, the optimum signal-to-quantum noise ratio ρ for fixed energy (or power per unit frequency) $\hbar\omega N_s$ has the value $4N_s(N_s+1)$.
- this optimal value is attainable by the squeezed coherent vacuum.

This fundamental result has recently been extended to the case of deformed photons[2]. The question naturally arises as to what extent the optimal value must be modified for other states. We derive here the optimal signal-to-quantum noise ratio for squeezed photon number states, more accurately, squeezed displaced number states (SDN states). These may be taken as a paradigm for the displacing and squeezing of an imperfect vacuum. We show that for such states with photon number n, the ratio ρ is reduced by a factor $(2n+1)^2$ from Yuen's result.

SQUEEZED DISPLACED NUMBER STATES

In the following we summarize the definitions and some properties of photon states necessary in calculating the signal-to-quantum noise ratio. Our squeezed displaced

Quantum Communication, Computing, and Measurement 2
Edited by Kumar et al., Kluwer Academic / Plenum Publishers, New York, 2000.

number states (SDN states) are defined by

$$|z,\theta,\alpha,n>= D(\alpha)U(z,\theta)|n> \tag{1}$$

where $D(\alpha)$ is the displacement operator given by

$$D(\alpha) = \exp(\alpha a^\dagger - \alpha^* a) \tag{2}$$

and

$$U(z,\theta) = S(z)P(\theta) \tag{3}$$

is the general squeezing operator, which is a three-parameter element of the group SU(1,1). $S(z)$ is the more usual (restricted, 2-parameter) squeezing operator, given by

$$S(z) = \exp[\frac{1}{2}(za^{\dagger 2} - z^* a^2)], \quad z = r\exp(i\phi) \tag{4}$$

and $P(\theta)$ is phase transformation operator, given by

$$P(\theta) = \exp[i\theta(N + \frac{1}{2})], \quad N = a^\dagger a \tag{5}$$

For $\theta = 0$, the states $|z,\theta,\alpha,n>$ reduce to $D(\alpha)S(z)|n>$, called squeezed number states. For $\lambda = 1, \mu = 0$, the states $|z,\theta,\alpha,n>$ reduce to $D(\alpha)|n>$, called displaced number states. And for $\alpha = 0$ and $\lambda = 1, \mu = 0$, the states $|z,\theta,\alpha,n>$ reduce to number states $|n>$. For $n = 0$, the squeezed number states $|z,\alpha,n>$ and the displaced number states $|\alpha,n>$ reduce to the more familiar coherent squeezed states $D(\alpha)S(z)|0>$ and coherent states $D(\alpha)|0>$.

The displacement operator produces the transformations[3]:

$$D^\dagger(\alpha)aD(\alpha) = a + \alpha \quad D^\dagger(\alpha)a^\dagger D(\alpha) = a^\dagger + \alpha^* \tag{6}$$

and the unitary transformation of the operators a and $a^\dagger$ by $U(z,\theta)$ and $U^\dagger(z,\theta)$ is given by:

$$\begin{aligned} U^\dagger(z,\theta)aU(z,\theta) &= \lambda^* a + \mu a^\dagger \\ U^\dagger(z,\theta)a^\dagger U(z,\theta) &= \lambda a^\dagger + \mu^* a \end{aligned} \tag{7}$$

where we have put

$$\lambda = \exp(-i\theta)\cosh r, \quad \mu = \exp[i(\phi - \theta)]\sinh r$$

and $|\lambda|^2 - |\mu|^2 = 1$. Quadrature operators are defined as usual by

$$X = \frac{1}{\sqrt{2}}(a + a^\dagger) \quad P = \frac{1}{i\sqrt{2}}(a - a^\dagger) \tag{8}$$

From Equations (1) -(8) we obtain the following mean values in the SDN state $|z,\theta,\alpha,n>$:

$$\begin{aligned} \langle N\rangle &= |\lambda|^2 n + |\mu|^2(n+1) + |\alpha|^2 \\ \langle X\rangle &= \frac{1}{\sqrt{2}}(\alpha + \alpha^*) \\ \langle P\rangle &= \frac{1}{i\sqrt{2}}(\alpha - \alpha^*) \\ (\Delta X)_n^2 &= \frac{1}{2}|\lambda + \mu|^2(2n+1) = |\lambda + \mu|^2(\Delta X)_0^2 \\ (\Delta P)_n^2 &= \frac{1}{2}|\lambda - \mu|^2(2n+1) = |\lambda - \mu|^2(\Delta P)_0^2 \end{aligned} \tag{9}$$

where $(\Delta X)^2 = \langle X^2\rangle - \langle X\rangle^2$, $\quad (\Delta P)^2 = \langle P^2\rangle - \langle P\rangle^2$. And $(\Delta X)_0^2, (\Delta P)_0^2$ are the values for the number state $|n>$.

THE SIGNAL-TO-QUANTUM NOISE RATIO ρ

The signal-to-quantum noise ratio in the state $\rho_{|>}$ is defined by

$$\rho_{|>} = \frac{\langle X\rangle^2}{(\Delta X)^2} \tag{10}$$

So for the SDN state $|z, \theta, \alpha, n>$, we have

$$\rho_{|z,\theta,\alpha,n>} = \frac{|\alpha + \alpha^*|^2}{|\lambda + \mu|^2(2n+1)}. \tag{11}$$

Note that $X^2 + P^2 = 2N + 1$, so that $\langle X^2\rangle + \langle P^2\rangle = \langle 2N+1\rangle$, whence

$$\langle X\rangle^2 + (\Delta X)^2 + \langle P\rangle^2 + (\Delta P)^2 = \langle 2N+1\rangle. \tag{12}$$

Then expression Eq. (10) for ρ becomes

$$\rho_{|>} = \frac{[(2\langle N\rangle + 1) - \langle P\rangle^2] - (\Delta P)^2}{(\Delta X)^2} - 1. \tag{13}$$

Under the energy (or power per unit frequency) constraint refnotel

$$\hbar\omega\langle N\rangle \leq \hbar\omega N_s$$

the signal-to-quantum noise ratio can be maximised by using all the available energy and allocating no energy to $\langle P\rangle$; that is

$$\langle N\rangle = N_s, \quad \langle P\rangle = 0 \tag{14}$$

Using Eqs. (9) and (14), and noting $|\lambda^2 - \mu^2|^2 \geq 1$, we can write Eq. (13) in terms of $|\lambda + \mu|^2$:

$$\rho_{|z,\theta,\alpha,n>} \leq \frac{(\Delta P)_0^2}{(\Delta X)_0^2}\left[\frac{2N_s+1}{|\lambda+\mu|^2(\Delta P)_0^2} - \frac{1}{|\lambda+\mu|^4}\right] - 1. \tag{15}$$

We thus find that the maximum value is given by

$$\rho_{|\alpha,n>max} = \frac{2N_s+1}{(\Delta X)_0^2} - 1 = \frac{4(N_s - n)}{2n+1} \tag{16}$$

for the displaced number state $|\alpha, n>$ with

$$\begin{aligned} \alpha &= \sqrt{N_s - n} \\ \lambda &= 1 \\ \mu &= 0 \end{aligned} \tag{17}$$

and

$$\rho_{|z,\alpha,n>max} = \frac{(2N_s+1)^2}{4(\Delta X)_0^2(\Delta P)_0^2} - 1 = \frac{4(N_s-n)(N_s+1+n)}{(2n+1)^2} \tag{18}$$

obtained with the squeezed number state $|z, \alpha, n>$ with

$$\begin{aligned} \alpha &= \sqrt{\frac{(N_s-n)(N_s+1+n)}{(2N_s+1)}} \\ \lambda &= \frac{N_s+1+n}{\sqrt{(2N_s+1)(2n+1)}} \\ \mu &= \frac{n - N_s}{\sqrt{(2N_s+1)(2n+1)}} \end{aligned} \tag{19}$$

For the case $n = 0$ in Eqs. (16) and (18) , one finds the same results as Yuen[1].

DISCUSSION

From Eq. (11), we obtain

$$\rho_{|\alpha>} > \rho_{|\alpha,n>} > |n>$$
$$\rho_{|z,\alpha>} > \rho_{|z,\alpha,n>} > |n>$$

for given complex values of α and z. From Equation (18) we have

$$\frac{\rho_{|z,\alpha,n>max}}{\rho_{|z,\alpha>max}} = \frac{(N_s - n)(N_s + 1 + n)}{N_s(N_s + 1)(2n + 1)^2} = \frac{1}{(2n + 1)^2}[1 - \frac{n(n + 1)}{N_s(N_s + 1)}] \tag{20}$$

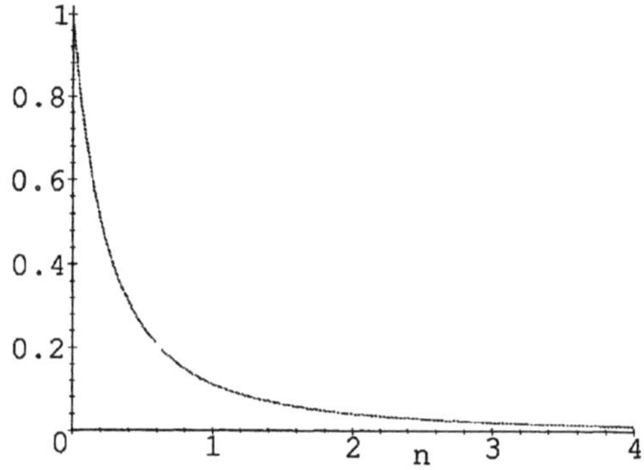

Figure 1. Ratio of $\frac{\rho_n}{\rho}$

This ratio is plotted in Fig.1. We can see when photo number n is increased, the ratio decreases as $\frac{1}{(2n+1)^2}$ (for any $N_s >> n$), whence a slight deviation from a squeezed vacuum results in large diminution of ρ.

Considering the physical interpretation, suppose an apparatus for performs the operations of phase-shifting ($P(\theta)$), squeezing ($S(z)$) and displacing ($D(\alpha)$). Then, more generally, the best signal-to-quantum noise ratio for final states $|f>$ which can be obtained by applying this sequence of operations to an initial state $|i>$, with "position" and "momentum" spreads $(\Delta X)_i(\Delta P)_i$, is the ratio in Equation (18):

$$\rho_{|f>max} = \frac{(2N_s + 1)^2}{4(\Delta X)_i^2(\Delta P)_0^i} - 1 = (\frac{N_s + 1/2}{(\Delta X)_i(\Delta P)_i})^2 - 1 \tag{21}$$

That is, the ratio decreases quadratically with the phase-space "area" $\sim (\Delta X)_i(\Delta P)_i$ of the initial state $|i>$. Equation (21) is realised for any initial state $|i>$ (such as thermal states) for which $(\Delta X)_i^2$ and $(\Delta P)_i^2$ have the following relationships with the final state $|f>= D(\alpha)U(z,\theta)|i>$:

$$(\Delta X)_f^2 = |\lambda + \mu|^2(\Delta X)_i^2$$
$$(\Delta P)_f^2 = |\lambda - \mu|^2(\Delta P)_i^2$$

REFERENCES

1. H.P.Yuen, States that give the maximum signal-to-quantum noise ratio for a fixed energy, *Phys. Lett. A*, 56:4281 (1976)
2. Allan I. Solomon, Optimal signal-to-quantum noise ratio for deformed photons, *Phys. Lett. A*, 188:215 (1994)
3. Mandel and Wolf, " Optical Coherent and Quantum Optics ", Cambridge University Press (1995)

PART II: QUANTUM MEASUREMENT AND TOMOGRAPHY

DUAL CLASSICAL AND QUANTUM DESCRIPTIONS OF A MEASURING APPARATUS

Ori Hay and Asher Peres

Department of Physics
Technion—Israel Institute of Technology
32 000 Haifa, Israel

The measuring process is the interface of classical and quantum phenomena. A measuring apparatus must be described by quantum mechanics while it interacts with the quantum system under observation. Then, it has to be be given a classical description so that the result of the measurement appears as objective reality. Alternatively, that apparatus may be treated at all times by quantum mechanics. In that case, it never achieves the "objective" status, but it can be measured by a second apparatus which has such a dual description. The question is whether these two different descriptions of the same apparatus are mutually consistent. A detailed analysis leads to the conclusion that not every physical system is a legitimate measuring apparatus. The quantum measurement problem has a satisfactory solution only if the measuring apparatus is properly designed.

FORMULATION OF THE PROBLEM

Quantum phenomena do not occur in a Hilbert space. They occur in a laboratory. Quantum theory provides statistical predictions for results of measurements performed on physical systems that have been prepared in a specified way. The mathematical tools that are used for this purpose involve complex matrices, functional analysis, C^*-algebras, etc. On the other hand, the preparation and measurement are performed by macroscopic devices, and these are described in classical terms. The necessity of using a classical terminology was emphasized by Bohr since the very early days of quantum mechanics.[1] Bohr's insistence on a classical description was very strict. He wrote:[2]

> ... it is decisive to recognize that, *however far the phenomena transcend the scope of classical physical explanation, the account of all evidence must be expressed in classical terms.* The argument is simply that by the word "experiment" we refer to a situation where we can tell others what we have done and what we have learned and that, therefore, the account of the experimental arrangement and of the results of the observations must be expressed in unambiguous language, with suitable application of the terminology of classical physics.

Quantum Communication, Computing, and Measurement 2
Edited by Kumar et al., Kluwer Academic / Plenum Publishers, New York, 2000.

Bohr never described the measuring process as a dynamical interaction between an apparatus and the system under observation. He was of course fully aware that measuring apparatuses are made of the same kind of matter as everything else, and they obey the same physical laws. It is therefore tempting to use quantum theory in order to investigate their behavior during a measurement. However, if this is done, the quantized apparatus loses its status of a measuring instrument. It becomes a mere intermediate system in the measuring process, and there must still be a *final* instrument which has a purely classical description:[3]

> In the system to which the quantum mechanical formalism is applied, it is of course possible to include any intermediate auxiliary agency employed in the measuring process. Since, however, all those properties of such agencies which, according to the aim of the measurement, have to be compared with the corresponding properties of the object, must be described on classical lines, their quantum mechanical treatment will for this purpose be essentially equivalent with a classical description. The question of eventually including such agencies within the system under investigation is thus purely a matter a matter of practical convenience ... The only significant point is that in each case some ultimate measuring instruments, like the scales and clocks which determine the frame of space-time coordination — on which, in the last resort, even the definitions of momentum and energy quantities rest – must always be described entirely on classical lines, and consequently kept outside the system subject to quantum mechanical treatment.

Measurement was understood by Bohr as a primitive notion. He could thereby elude questions which caused considerable controversy among other authors.[4,5] Note that Bohr never claimed that there were classical systems and quantum systems. For him, there were physical systems that could be given, under appropriate circumstances, a classical description, or a quantum description.

A quantum dynamical description of the measuring process was first attempted by von Neumann, in his treatise on the mathematical foundations of quantum theory.[6] In the last section of that book, as in an afterthought, von Neumann represented the apparatus by a single degree of freedom, whose value was correlated to that of the dynamical variable being measured. Such an apparatus is not, in general, left in a definite pure state, and it does not admit a classical description. Therefore, von Neumann introduced a second apparatus which observes the first one, and possibly a third apparatus, and so on, until there is a final measurement, which is *not* described by quantum dynamics and has a definite result (for which quantum mechanics can only give statistical predictions). The essential point that was suggested, but not proved by von Neumann, is that the introduction of this sequence of apparatuses is irrelevant: the final result is the same, irrespective of the location of the "cut" between classical and quantum physics.

At this point, von Neumann also speculated that the final step would involve the consciousness of the observer—a rather bizarre statement in a mathematically rigorous monograph:

> First, it is inherently entirely correct that the measurement or the related process of the subjective perception is a new entity relative to the physical environment and is not reducible to the latter. Indeed, subjective perception leads us into the intellectual inner life of the individual, which is extra-observational by its very nature ... It is a fundamental requirement of the scientific viewpoint — the

so-called principle of the psycho-physical parallelism — that it must be possible so to describe the extra-physical process of the subjective perception as if it were in reality in the physical world.

In the present article, we introduce a dual description for the measuring apparatus. It obeys quantum mechanics while it interacts with the system under observation, and then it is "dequantized" and is described by a classical Liouville density, which provides the probability distribution for the results of the measurement. Alternatively, the apparatus may always be treated by quantum mechanics, and be measured by a second apparatus which has such a dual description. The question is whether these two different methods of calculation give the same result.[7]

It turns out that a sufficient condition for agreement between the two methods is that the dynamical variable used as a "pointer" by the first apparatus be represented by an operator of the Weyl-Wigner type.[8] These "quasi-classical" operators are defined as follows: Let a classical dynamical variable, $A(q,p)$, be expressed as a Fourier transform,

$$A(q,p) = \iint d\sigma\, d\tau\, e^{i(\sigma q+\tau p)}\, \alpha(\sigma,\tau). \tag{1}$$

The correponding Weyl-Wigner operator is obtained by replacing, in the above expression, the classical variables q and p by the corresponding quantum operators $\hat{q}$ and $\hat{p}$. It can then be shown that the expectation value of $\hat{A}(\hat{q},\hat{p})$ for any quantum state, pure or mixed, is equal to the classical expression

$$\langle A\rangle = \iint W(q,p)\, A(q,p)\, dq\, dp, \tag{2}$$

where $W(q,p)$ is Wigner's quasi-probability distribution.[8,9] If the latter is nowhere negative, it can be interpreted as a classical Liouville distribution. The same symbols, q and p, will henceforth be used both for classical variables and for operators, since the meaning of the symbol is always clear from the context and there is no risk of confusion.

Two examples will be examined. In both cases, the quantum system that is observed is a particle of spin j. We want to measure the J_z component. In the first example, we couple J_z to the linear position, q, of a pointer. The latter is then measured by a second pointer, whose linear position is Q. The problem is to find the probability distribution of Q, for a given initial state of the quantum system. Both q and Q are operators of the Weyl-Wigner type. As shown explicitly, it makes no difference if we dequantize Q at the end of the process, or we dequantize q after the first measurement, and we always treat Q classically.

In the second example, we couple J_z to the phase, θ, of a harmonic oscillator. The second apparatus, again a linear pointer with position Q, measures $\cos\theta$ (not θ itself because the phase is not a well behaved self-adjoint operator in quantum mechanics). The usual quantum operator for $\cos\theta$ is well behaved. However, it is not the Weyl-Wigner operator corresponding to the classical variable $\cos\theta$. In this case, we find that the expectation value $\langle Q\rangle$ is not the same when the first apparatus is treated quantum mechanically, or it is treated classically, while it is measured by the second one. That is, when we perform the required calculations for such a measuring process, the result depends on the location chosen for the von Neumann cut. It is likely that the validity of these conclusions is not restricted to the particular examples for which we provide detailed calculations.

To avoid any misunderstanding, we emphasize that the classical description of a pointer is not by means of a point in phase space, but by a *Liouville density*. Quantum theory makes only statistical predictions, and any semiclassical treatment that simulates it must also be statistical. Our approach involves only strictly orthodox quantum mechanics. We never speculate about modifications of the conventional theory, as have been proposed by some authors. While the present work may not satisfy the desiderata of some physicists (those who want to be "realists"), it does prove the viability of those of Bohr and von Neumann, provided that the physical system that is employed as the measuring instrument is indeed suitable for filling that role.

LINEAR POINTER

Let the system under observation be a spin j particle. We want to measure the spin component J_z which satisfies, in natural units ($\hbar = 1$),

$$J_z\,|m\rangle = m\,|m\rangle, \qquad m = j, j-1, \dots, -j. \tag{3}$$

The initial state of the system is $\sum a_m|m\rangle$. In elementary discussions of quantum measurements, there is no explicit decription of the apparatus. The typical textbook just says that the result of the measurement is m, with probability $|a_m|^2$. One may imagine a pointer, jumping from $q = 0$ to $q = m$ (in suitable units), with probability $|a_m|^2$, as a result of the measuring process. (In the language of statistical mechanics, the Liouville function of the pointer has peaks of size $|a_m|^2$ near $q = m$.) It is then possible to imagine a second apparatus which measures the first one, and has its pointer moving from $Q = 0$ to $Q = q$. The readings of the two apparatuses of course agree with each other.

Let us now provide a quantum dynamical description for the apparatuses. The initial state of the first pointer is specified by a wave function $\phi(q)$. The position q and its conjugate momentum $p = -i\partial/\partial q$ are linear operators in Hilbert space. Likewise, the second apparatus is a linear pointer with position operator Q, momentum operator $P = -i\partial/\partial Q$, and initial state $\Phi(Q)$.

The joint state of the complete setup is, initially,

$$\psi_0 = \sum_m a_m\,|m\rangle \otimes \phi(q) \otimes \Phi(Q). \tag{4}$$

The interaction between the system and the first apparatus is represented by the unitary operator

$$U_1 = e^{-iJ_z p} = e^{-J_z(\partial/\partial q)}, \tag{5}$$

leading to a new state

$$\psi_1 = U_1\,\psi_0 = \sum_m a_m\,|m\rangle \otimes \phi(q-m) \otimes \Phi(Q). \tag{6}$$

For any given $|m\rangle$, the pointer q has moved by m length units. Likewise, the second pointer senses the value of q and moves by q units. The interaction of the two pointers is generated by

$$U_2 = e^{-iqP} = e^{-q(\partial/\partial Q)}, \tag{7}$$

so that

$$\psi_2 = U_2\,\psi_1 = \sum_m a_m\,|m\rangle \otimes \phi(q-m) \otimes \Phi(Q-q). \tag{8}$$

The probability distribution of Q (while we ignore the other variables) is

$$\int_{-\infty}^{\infty} dq\, \psi_2^\dagger \psi_2 = \int_{-\infty}^{\infty} dq \sum_m |a_m|^2 \, |\phi(q-m)|^2 \, |\Phi(Q-q)|^2. \tag{9}$$

It is easily seen that the same result is obtained if a "von Neumann cut" is placed after the first apparatus. That is, the probability density for q, resulting from Eq. (6), will be considered as a classical probability distribution for the position of the first pointer. The initial distribution for the second one is $|\Phi(Q)|^2$. The two pointers interact classically by a canonical transformation generated by a Hamiltonian qP, as in Eq. (7). The final result for the probability distribution of Q is the same as in the preceding quantum mechanical calculation.

ENCODING A MEASUREMENT IN A PHASE

We shall now measure the same quantum system with a different apparatus. Instead of a linear pointer, we use the phase of a harmonic oscillator, whose Hamiltonian is $H_{\rm osc} = \frac{1}{2}(p^2 + q^2)$. In classical mechanics, the phase is given by $\theta = \arctan(p/q)$. In quantum mechanics, the issue is more complicated, as we shall see.

First, let us give, as in the preceding section, an elementary classical description of the quantum measurement (it will later be needed for comparison with the semiclassical and the purely quantum treatments). The final phase of the classical oscillator, which plays the role of a pointer, is given by

$$\theta = \theta_0 - m\chi, \tag{10}$$

with probability $|a_m|^2$. Here, χ is any constant (we shall take $\chi < \pi/2j$, so that there is no overlap in the final values of θ). It will be convenient to take $\theta_0 = \pi/2$.

The second apparatus is, as before, a linear pointer. It is coupled to $\cos\theta$ (not to θ itself, because the phase is not a well defined quantum operator, but its cosine can be properly defined, so as to be self-ajoint and therefore measurable, in principle). The final position of the second pointer, treated classically, thus is

$$Q = Q_0 + \cos\theta = Q_0 + \sin m\chi. \tag{11}$$

This elementary classical result, for which no dynamical justification was given, will now be compared with the one obtained by treating both apparatuses as quantum systems.

Two quantum apparatuses

The first apparatus is a harmonic oscillator (e.g., one of the modes of an electromagnetic field in a cavity), initially prepared in a coherent state $|\alpha\rangle$. The choice of a coherent state is convenient, because these states are as close as possible to classical physics. In particular, their Wigner function is everywhere positive, and therefore it can readily be reinterpreted as a classical Liouville density.

The second apparatus is, as before, a linear pointer prepared in a state $\Phi(Q)$. The joint state of the complete setup is, initially

$$\psi_0 = |m\rangle \otimes |\alpha\rangle \otimes \Phi(Q). \tag{12}$$

Here, we have assumed for simplicity that the quantum system is in one of the eigenstates $|m\rangle$ of J_z (the goal of the measurement is to determine m). It is obvious that any linear combination $\sum a_m|m\rangle$ would give, after the quantum system is traced out, a statistical mixture with weights $|a_m|^2$.

The calculations then proceed as in the preceding case, but become much more complicated.[10] The final result for the expected nature of Q is

$$\langle Q\rangle - \langle Q\rangle_0 = \sin m\chi\, e^{-r^2} \sum_{n=0}^{\infty} r^{2n+1} \Big/ \sqrt{n!(n+1)!}\,. \tag{13}$$

The coefficient $\sin m\chi$ is the classical result (11) for the displacement of Q. The quantum nature of the first apparatus reduces the average value of this displacement by a factor

$$S(r) = e^{-r^2} \sum_{n=0}^{\infty} r^{2n+1} \Big/ \sqrt{n!(n+1)!}\,, \tag{14}$$

where $r \equiv |\alpha|$ is the amplitude of the coherent state in which the oscillator was prepared. For small r, we have $S(r) \to r$. For large r, the sum in (14) can be written as an asymptotic series[11]

$$S(r) = 1 - \frac{1}{8r^2} - \frac{7}{128r^4} - \frac{75}{1024r^6} - \cdots. \tag{15}$$

Thus, when r is large, $S(r) \to 1$. This is the expected result, since a harmonic oscillator in a coherent state with large r is almost classical.

Semiclassical description

The above results will now be compared with a semiclassical treatment similar to the one in the preceding section. The second apparatus is always described by classical statistical mechanics. It is prepared in a given Liouville distribution, and it interacts with the first apparatus, for which we also assume a Liouville distribution, which is initially identical to the Wigner function $W(q,p)$ that results from the first stage of the measurement. It is indeed possible to equate these two distributions, because the first apparatus is in a coherent state, so that its Wigner function is everywhere positive. If we had chosen another state, pure or mixed, whose Wigner function had negative regions, it would have been inconsistent to dequantize the first apparatus. This would simply mean that, in such a state, it is not a legitimate measuring apparatus.

Detailed calculations[10] lead to the final result

$$\langle Q\rangle - \langle Q\rangle_0 = \sin m\chi\, \sqrt{\pi/2}\, r e^{-r^2}\, [I_0(r^2) + I_1(r^2)]. \tag{16}$$

The expression that multiplies $\sin m\chi$ (which was the classical result) tends to $r\sqrt{\pi/2}$ when r is small, and to 1 when r is large.

Why is this result different from the preceding one, in Eq. (13)? The reason is that the two classically equivalent expressions for the variable that is measured by the second apparatus, namely

$$\cos\theta = q \Big/ \sqrt{p^2+q^2}. \tag{17}$$

are not equivalent when these expressions become operators in quantum mechanics. The semiclassical result (16) was obtained by using the Wigner function $W(q,p)$ of the first apparatus (the harmonic oscillator) if it were a classical probability density. However, the operators corresponding to the classical expressions in Eq. (17), whose expectation value we seek, are not of the Weyl-Wigner form. It is therefore not surprising that in this case the semiclassical approximation gives a final result which is different from the quantum prediction.

SUMMARY AND OUTLOOK

If you expected to find in this article a solution of the so-called "quantum measurement problem," you may be disappointed. Indeed, that problem is ill defined, and it is understood in different ways by various authors.[4,5] Our way of formulating it, for which we can indicate a solution, simply is to say: quantum theorists describe the physical world by means of a complex Hilbert space (vectors and operators) that defies any realistic interpretation, while experimenters find plain numbers—they count "clicks." The experimenters use measuring instruments made of ordinary matter, for which quantum theory is assumed valid, but the ultimate outcome of the measuring process is essentially classical.[1–3] Therefore, at some stage, a transition has to be made from the quantum formalism to a classical language.

Here, the role of decoherence should be clarified. Some authors claim that decoherence provides the solution of the measurement problem (with the particular meaning that they attribute to that problem), but many others dispute this point of view. Yet, decoherence has an essential role in the preceding discussion. We have repeatedly taken the liberty of discarding one link in the von Neumann chain, after moving over to the next link. For example, in Eq. (9), we obtained the probability distribution for the second apparatus by averaging out the dynamical variables of the spin and the first apparatus. This was not a deliberate omission on our part. In the real world, with many interacting degrees of freeedom, correlations between consecutive links are exceedingly difficult to follow. Furthermore, the replacement of Wigner's function $W(q,p)$ by a Liouville function is consistent only if $W(q,p) \geq 0$. Owing to decoherence, this condition is likely to be fulfilled by any macroscopic apparatus which is not in a pure state, but rather in a mixed state with $\Delta q\, \Delta p \gg \hbar$ (this inequality is the hallmark of being "macroscopic"). All the negative parts of W are completely washed away by the coarseness of the apparatus.

Yet, technological advances occasionally permit us to proceed one more step along the von Neumann chain, and it is important to examine whether this makes any difference in the final result. We have shown that if the measuring apparatus is suitably chosen, the transition from quantum mechanics to classical *statistical* mechanics can proceed in a consistent way. However, a "bad" choice of apparatus is incompatible with a classical description (more precisely, the semiclassical results do not coincide with those predicted by quantum theory, though they may asymptotically agree for quasi-classical quantum states).

This brings us to the unavoidable fundamental question: what are the properties that are necessary for a physical system to be a legitimate measuring apparatus? Our results indicate that if an apparatus uses as its "pointer" a dynamical variable represented by an operator of the Weyl-Wigner form, it is legitimate to dequantize it and to proceed as if its Wigner function were a classical probability density. For other types of operators, the transition from quantum to classical mechanics usually is only an approximation (which may be excellent if the quantum state of the apparatus is quasi-classical).

In summary, there is nothing mysterious in the transition from the quantum world to the classical one. There is no need of invoking anthropomorphic concepts, such as consciousness. Plain orthodox quantum mechanics and classical statistical mechanics correctly reproduce all statistical predictions that can be verified in experiments.

Acknowledgments

OH was supported by a grant from the Technion Graduate School. Work by AP was supported by the Gerard Swope Fund, and the Fund for Encouragement of Research.

REFERENCES

1. N. Bohr, *in* "Atti del Congresso Internazionale dei Fisici," Como (1927); reprinted in *Nature* 121:78 and 121:580 (1928).
2. N. Bohr, *in* "Albert Einstein, Philosopher-Scientist," P. A. Schilpp, ed., (Library of Living Philosophers, Evanston (1949) pp. 201–241.
3. N. Bohr, *in* "New Theories in Physics," International Institute of International Cooperation, Paris (1939) pp. 11–45.
4. J. A. Wheeler and W. H. Zurek, eds., "Quantum Theory and Measurement," Princeton University Press, Princeton (1983).
5. J. Bub, "Interpreting the Quantum World," Cambridge University Press, Cambridge (1997).
6. J. von Neumann, "Mathematische Grundlagen der Quantenmechanik," Springer, Berlin (1932); transl. by R. T. Beyer, "Mathematical Foundations of Quantum Mechanics," Princeton University Press, Princeton (1955).
7. A. Peres, "Quantum Theory: Concepts and Methods," Kluwer, Dordrecht (1993) p. 376.
8. M. Hillery, R. F. O'Connell, M. O. Scully, and E. P. Wigner, *Phys. Reports* 106:121 (1984).
9. E. Wigner, *Phys. Rev.* 40:749 (1932).
10. O. Hay and A. Peres, *Phys. Rev. A* 58:116 (1998).
11. B. C. Berndt, "Ramanujan's Notebooks," Springer, New York (1985) Vol. I, p. 58.

GENERALISED MEASUREMENTS, RETRODICTION AND STATE MANIPULATION

Stephen M. Barnett[1], Anthony Chefles[1] , David T. Pegg[2] and Lee S. Phillips[1]

[1]Department of Physics and Applied Physics University of Strathclyde, Glasgow G4 0NG, Scotland
[2]Faculty of Science, Griffith University, Nathan, Brisbane Q 4111, Australia

We develop and discuss novel types of generalised quantum measurement. The associated non-unitary time evolution enables the manipulation of quantum states in ways that are not achievable by unitary operations. Particular emphasis is given to quantum optical realisations.

INTRODUCTION

For the purposes of this article, a conventional quantum measurement is one which may be described in terms of von Neumann's formulation[1]. An observable O is represented by the Hermitian operator $\hat{O}$ having (for simplicity) a non-degenerate set of eigenvalues O_l with corresponding eigenstates $|O_l\rangle$. The probability that the measurement will lead to the result O_l is

$$P(O_l) = \langle O_l|\hat{\rho}|O_l\rangle = \mathrm{Tr}(|O_l\rangle\langle O_l|\hat{\rho}) \tag{1}$$

where $\hat{\rho}$ is the density matrix for the initial state. A measurement of O giving the result O_l leads us to infer that, immediately after the measurement, the system is in the state with density matrix

$$\hat{\rho}' = \frac{|O_l\rangle\langle O_l|\hat{\rho}|O_l\rangle\langle O_l|}{\langle O_l|\hat{\rho}|O_l\rangle} \tag{2}$$

that is, in the eigenstate and that a subsequent measurement of O will again give the value O_l. Most real quantum measurements, such as photodetetection, are more destructive than this in that they destroy the quantum state. They are, nevertheless, examples of von Neumann measurements is that they respect the probabilities given in (1).

Quantum Communication, Computing, and Measurement 2
Edited by Kumar et al., Kluwer Academic / Plenum Publishers, New York, 2000.

A more general approach to measurement results if we supplement the von Neumann measurement by a unitary evolution, typically involving coupling to other quantum systems[2, 3]. Such considerations lead to a generalization of the von Neumann measurement[4, 5], which is most conveniently described in terms of Probability Operator Measures (or POMs). In this formulation, the probability associated with a measurement result n is given by

$$P(n) = \mathrm{Tr}(\hat{\Pi}_n \hat{\rho}). \tag{3}$$

In order to ensure that this is a true probability, $\hat{\Pi}_n$ must be non-negative definite and self adjoint. The normalisation of the probability distribution for all possible states implies that the POM elements, $\hat{\Pi}_n$, obey the following resolution of the identity:

$$\sum_n \hat{\Pi}_n = \hat{1}. \tag{4}$$

The POM elements are not always orthogonal even though they correspond to distinct experimental outcomes[5]. It is sometimes convenient to factorise the POM elements in the form

$$\hat{A}_n = \hat{U}_n \hat{\Pi}_n^{1/2} \tag{5}$$

where $\hat{U}_n$ is some unitary operator so that $\hat{\Pi}_n = \hat{A}_n^\dagger \hat{A}_n$. The POM generalization of the von Neumann post-measurement density matrix (2) then becomes

$$\hat{\rho}' = \frac{\hat{A}_n \hat{\rho} \hat{A}_n^\dagger}{\mathrm{Tr}(\hat{\rho} \hat{A}_n^\dagger \hat{A}_n)}. \tag{6}$$

In this paper, we shall examine some aspects and applications of the theory of generalised measurements with particular reference to quantum optical systems. The general theme is the exploitation of generalised measurements to manipulate states in ways which are impossible with conventional unitary evolution.

PREDICTION AND RETRODICTION

Conventionally we think of a physical system (such as a field mode) being prepared in a known (pure) state $|i\rangle$ at time t_0, evolving to the state $\hat{U}(t_1, t_0)|i\rangle$ at time t_1, when a measurement is made. The (von Neumann) measuring device then projects the system onto the state $|f\rangle$ with a corresponding measurement result. The probability for the system to be in state $|i\rangle$ at t_0 and in state $|f\rangle$ at t_1 is

$$P(f,i) = P(f|i)P(i) = |\langle f|\hat{U}(t_1,t_0)|i\rangle|^2 \tag{7}$$

because $P(i) = 1$, that is we know the prepared state. At time t, such that $t_1 > t \geq t_0$, we would assign the system the state $\hat{U}(t, t_0)|i\rangle$, which changes smoothly until $t = t_1$, when it is suddenly collapsed or projected onto $|f\rangle$. We call this assignment of the state the *predictive* formalism. Given the initial state at t_0 we predict the probability of a collapse onto $|f\rangle$ at t_1.

We can also use a *retrodictive* formalism[6] in which we know the outcome of the measurement, that is we know $|f\rangle$ and assign the system the state $\hat{U}(t, t_1)|f\rangle = \hat{U}^\dagger(t_1, t)|f\rangle$ for $t_1 \geq t > t_0$. This evolves smoothly back in time until it is projected onto the state $|i\rangle$ at t_0. With this formalism the probability for the system to be in the state $|i\rangle$ at t_0 and in state $|f\rangle$ at t_1 is

$$P(f,i) = P(i|f)P(f) = |\langle i|\hat{U}^\dagger(t_1,t_0)|f\rangle|^2 \tag{8}$$

because $P(f) = 1$, that is we know the outcome of the measurement. The equivalence of these two formalisms, which involve very different assignations of the state and quite distinct times when wavefunction collapse occurs, shows the tenuous nature of these descriptions of what really happens. In many cases in practice it is actually the measurement result which is known and so some form of retrodiction is useful. In the next section we show how retrodiction can usefully be employed in the analysis of generalized optical measurements. In section IV we examine the effects of imperfect measurements on the retrodicted state and show how retrodiction can be applied for state preparation.

OPTICAL PROJECTION SYNTHESIS

An important element in developing generalized measurements is the unitary interaction between the system to be measured and a second or ancilla quantum system. One particularly simple optical realization of this is provided by the action of a beam splitter which coherently superposes two input modes to form a pair of output modes. Consider the arrangement depicted in figure 1, in which two modes, a and b, are superposed using a 50:50 beam splitter and then the number of photons in each output mode is measured by photocounting. If the detectors are ideal then we can associate the appearance of n_a and n_b counts in the output modes a and b with the retrodicted state of the output modes

$$|\psi_{OUT}\rangle = |n_a\rangle_a |n_b\rangle_b. \tag{9}$$

This retrodicted state unitarily evolves back through the beam splitter to become an entangled state of the input modes

$$\begin{aligned} |\psi_{IN}\rangle &= \hat{R}^\dagger |n_a\rangle_a |n_b\rangle_b \\ &= \frac{2^{-(n_a+n_b)/2}}{\sqrt{n_a! n_b!}} (\hat{a}^\dagger - i\hat{b}^\dagger)^{n_a} (\hat{b}^\dagger - i\hat{a}^\dagger)^{n_b} |0\rangle_a |0\rangle_b. \end{aligned} \tag{10}$$

Here

$$\hat{R} = \exp\left[i\frac{\pi}{4}(\hat{b}^\dagger \hat{a} + \hat{a}^\dagger \hat{b})\right] \tag{11}$$

is the unitary operator describing the coherent interaction between the modes[7]. (Note that we are employing an interaction picture in which the free propagation of the light does not appear.) If mode b was prepared in the known state $|B\rangle_b$ then the detection of n_a and n_b counts in the output modes a and b is associated with the POM element

$$\hat{\Pi}(n_a, n_b) = {}_b\langle B|\psi_{IN}\rangle\langle\psi_{IN}|B\rangle_b \tag{12}$$

which is proportional to the projector onto the pure state ${}_b\langle B|\psi_{IN}\rangle / |{}_b\langle B|\psi_{IN}\rangle|$ of mode a. In the retrodictive formalism, the state of mode a is collapsed onto this state before reaching the beam splitter. Our ability to retrodictively synthesise these projectors is now limited by our ability to fabricate suitable input states $|B\rangle_b$[8]. For example, the states required in order to measure the phase probability distribution by this method have photon number probabilities given by the reciprocal of the binomial distribution[9, 10].

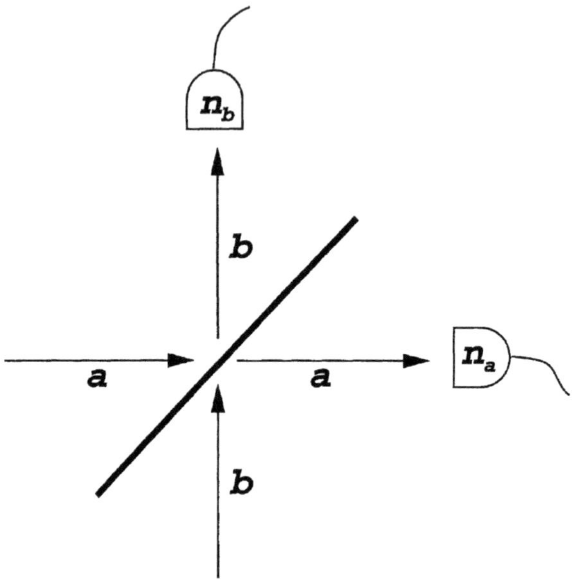

Figure 1. Experimental arrangement for optical projection synthesis. Mode a is mixed at a 50:50 beam splitter with mode b in the state $|B\rangle_b$. Photocount measurements on both output modes induce the generalised measurement on mode a which is described by (12).

IMPERFECT DETECTION AND RETRODICTION

Suppose an imperfect detector records an outcome n when a perfect detector would have given the result m. The retrodicted state will be a statistical mixture

$$\hat{\rho}_n = \sum_m p(m|n)|m\rangle\langle m| \tag{13}$$

where $|m\rangle$ is the state corresponding to the result m and $p(m|n)$ is the conditional probability that a perfect detector would have yielded a result m if the imperfect detector registers n. From Bayes' theorem

$$P(n|m)p(m) = p(m|n)P(n) \tag{14}$$

where $P(n|m)$ is the probability of a result n if a perfect detector would have yielded m, $P(n)$ is the probability of a result n and $p(m)$ is the probability of a result m for a perfect detector. Because we are retrodicting the past state, we cannot assign values to $p(m)$ on the basis of any prior information about the past state, so we must treat $p(m)$ as equal for all values of m. Then from (13) and (14) we have at the detection time t_1[11]

$$\hat{\rho}_n = \frac{\sum_m P(n|m)|m\rangle\langle m|}{\sum_{m'} P(n|m')} \tag{15}$$

which is a function only of the recorded result n and of the imperfections of the detector. That is, as a purely retrodicted state it contains no information about prior preparation of the system. The retrodicted state at earlier times is $\hat{\rho}_n(t) = \hat{U}^\dagger(t_1,t)\hat{\rho}_n\hat{U}(t_1,t)$.

As a specific example, $P(n|m)$ for a photodetector with an efficiency η is given by[7, 11]

$$P(n|m) = \eta^n(1-\eta)^{m-n}C_n^m \tag{16}$$

where C_n^m is the binomial coefficient. Here outcome n represents the recording of n photocounts. The probability that the prepared field contained m photons based only on the fact that the subsequent measurement gave n photocounts is

$$\langle m|\hat{\rho}_n|m\rangle = \frac{P(n|m)}{\sum_{m'} P(n|m')} \tag{17}$$

that is, the normalised conditional probability that n photocounts are detected given that m photons were prepared.

APPLICATION TO STATE PREPARATION

The collapse of the retrodicted state of the field at t_0 can be used to prepare states. If $\hat{\rho}$ is an entangled state involving two modes a and c with the detector in mode a, then the projection of the retrodicted state $\hat{\rho}_n^a$ at time t_0 onto this entangled state collapses the state of mode c into

$$\hat{\rho}_n^c = \frac{\mathrm{Tr}_a(\hat{\rho}_n^a \hat{\rho})}{\mathrm{Tr}_{a,c}(\hat{\rho}_n^a \hat{\rho})}. \tag{18}$$

We note that in this formalism, this state comes into existence at the time t_0 which is *before* the detection event at time t_1 which 'causes' it. We now apply this idea to the specific example of quantum state preparation using quantum scissors[12]. Consider the projection synthesis device depicted in figure 1. In order to introduce an entangled state into mode a we add another 50:50 beam splitter BS1 as shown in figure 2, with input states as shown. The entangled state of the two modes output from BS1 is

$$|\psi\rangle_{a,c} = \frac{1}{\sqrt{2}}(|1\rangle_c|0\rangle_a + i|0\rangle_c|1\rangle_a). \tag{19}$$

The input mode b is prepared in the state $|f\rangle_b = \sum c_n|n\rangle_b$.

Firstly let the two photodetectors have unit efficiencies and let one photon be detected in output mode a and zero photons be detected in output mode b. The retrodicted field just prior to the beam splitter is $\hat{R}^\dagger|1\rangle_a|0\rangle_b$, where $\hat{R}$ is given by (11). Projection onto the known field in state $|f\rangle_b$ collapses mode a into the state

$$\kappa'\ {}_b\langle f|\hat{R}^\dagger|1\rangle_a|0\rangle_b = \kappa(c_1^*|0\rangle_a + ic_0^*|1\rangle_a). \tag{20}$$

where κ and κ' are normalization constants. This can be regarded as the retrodicted field state from the measuring apparatus comprising the two detectors, the beam splitter BS2 and the known input field in mode b. At the beam splitter BS1 this retrodicted field state is projected onto $|\psi\rangle_{a,c}$ and the output state of mode c collapses to the (unnormalized) state $c_0|0\rangle_c + c_1|1\rangle_c$. This result is interesting for a number of reasons. We see that it is simply a truncation after the one-photon component of the original input state $|f\rangle_b$. Thus the qubit of information contained in the first two components of $|f\rangle_b$ has been transmitted to the light leaving the beam splitter in mode c. No light from mode b reaches mode c. Further, in the retrodictive formalism, the collapse of the output mode c into this state takes place *before* the light in state $|f\rangle_b$ reaches the beam splitter BS2. Thus the qubit of information has been transported backwards in time. Because this effect depends on uncontrollable detection events at the photodetectors, however, the device cannot be used to send messages into the past. Essentially, an observer at beam splitter BS1 would need to wait for information about the detection events to reach him by normal means before he could know whether or not he had received the appropriate qubit.

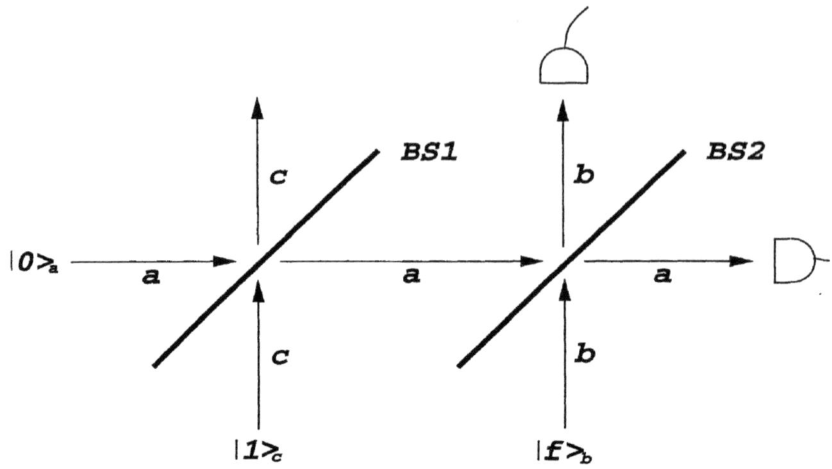

Figure 2. Extension of the projection synthesis setup which realises a quantum scissors device. When the photodetectors register one count in output mode a and no counts in output mode b, the state of mode c collapses into a superposition of the first two terms in the number state expansion of $|f\rangle$, rescaled for normalisation.

For the device to find practical use in truncating easily preparable states, the fidelity[13] of the output with the desired state will need to be reasonable for attainable photodetector efficiencies. When both detectors have non-unit efficiency η, the retrodicted state will be a statistical mixture as discussed earlier and thus its projection onto the output of beam splitter BS1 will also result in a mixture. When the state $|f\rangle_b$ is the coherent state $|\alpha\rangle_b$ we can show that this mixed state has a density operator with the form

$$\hat{\rho}_c = \frac{1}{1+|\alpha|^2(2-\eta)}\left[(|0\rangle + \alpha|1\rangle)(\langle 0| + \alpha^*\langle 1|) + |\alpha|^2(1-\eta)|0\rangle\langle 0|\right]. \quad (21)$$

This is a *prepared* state and thus its experimental measurement can confirm our approach. The probability that a measurement will find this mixed state to be in the desired state $|\chi\rangle = (1+|\alpha|^2)^{-1/2}(|0\rangle + \alpha|1\rangle)$ is given by the fidelity

$$F = \langle\chi|\hat{\rho}_c|\chi\rangle = 1 - \frac{|\alpha|^4(1-\eta)}{(1+|\alpha|^2)\left[1+|\alpha|^2(2-\eta)\right]}. \quad (22)$$

This is a surprisingly good result. If we wish to prepare an equally weighted superposition of the vacuum and the one photon state we require $|\alpha|^2 = 1$ and the fidelity becomes $(5-\eta)/(6-2\eta)$ which takes values between unity at $\eta = 1$ and $5/6$ as η tends to zero. For example a detection efficiency of 0.5 yields $F = 0.9$. The fidelity with the equally weighted superposition can be slightly improved by setting $|\alpha|^2 = (2-\eta)^{-1}$. We should mention that the above results can also be calculated using the usual predictive formalism but here the collapse of the output state into a truncated coherent state does not occur until the time of the actual detection events. This is well after the output light has left the beam splitter BS1 and is a sizeable distance away. In the retrodictive approach, this collapse takes place at the beam splitter itself.

DISCRIMINATING BETWEEN COHERENT STATES

A simple example of optical projection synthesis is the error-free discrimination between the two coherent states $|\alpha\rangle_a$ and $|-\alpha\rangle_a$[14, 15, 8]. If mode b is prepared in the coherent state $|i\alpha\rangle_b$ then the beam splitter will transform the input state into

$$\hat{R}|\pm\alpha\rangle_a \otimes |i\alpha\rangle_b = \left|\frac{1}{\sqrt{2}}(\pm\alpha-\alpha)\right\rangle_a \left|\frac{i}{\sqrt{2}}(\alpha\pm\alpha)\right\rangle_b . \tag{23}$$

One of the output modes is in its vacuum state, while the other is in a coherent state with mean photon number $2|\alpha|^2$. Ideal photocounting will unambiguously reveal the unknown coherent state unless there are no photons at either detector. The probability for this to occur is $\exp(-2|\alpha|^2) = \langle\alpha|-\alpha\rangle$ and corresponds with the Ivanovic-Dieks-Peres bound for error-free state discrimination[16, 17, 18].

We can explicitly construct the POM elements associated with the possible outcomes of this state discrimination experiment by substituting the coherent state $|i\alpha\rangle_b$ for $|B\rangle_b$ in equation (12)[8]. The POM element corresponding to the inconclusive result, $\hat{\Pi}(?)$, follows from setting n_a and n_b equal to zero

$$\hat{\Pi}(?) = \hat{\Pi}(0,0) =: \exp(-\hat{a}^\dagger\hat{a}) : \exp(-|\alpha|^2) \tag{24}$$

where the colons denote normal ordering. The POM elements corresponding to the conclusive inference that the state was $|\alpha\rangle$ is

$$\begin{aligned} \hat{\Pi}(\alpha) &= \hat{\Pi}(0, any) - \hat{\Pi}(0,0) \\ &= : \exp\left[-\frac{1}{2}(\hat{a}^\dagger - \alpha^*)(\hat{a}-\alpha)\right] : - : \exp\left[-(\hat{a}^\dagger\hat{a} + |\alpha|^2)\right] : \end{aligned} \tag{25}$$

and the corresponding element for the state $|-\alpha\rangle$ can be found from this by replacing α with $-\alpha$. The resolution of the identity requires the POM element corresponding to detection events in *both* counters, but the probability for this to happen, given the chosen input states, is zero. We can apply the ideas of retrodiction and imperfect photodetection to analyse a possible experiment with finite efficiency detectors. Under these circumstances the measurement will still be error free but the probability of an inconclusive result is increased to $\exp(-2\eta|\alpha|^2)$[8].

STATE SEPARATION

It is natural that retrodiction should be related to state discrimination. After all, the problem of state discrimination is to determine in which of a known set of possibly non-orthogonal states a system was prepared, which obviously refers to the past. In this and the following sections, we shall look more closely at state discrimination, in particular the properties of the associated measurement transformations. We shall also examine exact cloning, which is related to state discrimination and, as we shall see, operates by a similar principle.

In the more conventional, predictive formalism, unambiguous state discrimination maps each member of a set of known, non-orthogonal, quantum states onto a corresponding member of an orthogonal set[19]. When the operation fails, we obtain an inconclusive result. It is interesting to ask, what other useful quantum operations are there which, with some probability, reduce the square overlap between states? We shall say that such operations perform *state separation*[20].

In this section, we discuss state separating operations in general for two non-orthogonal states. In particular, we give the maximum value of the probability, P_S, that two non-orthogonal states can be separated by an arbitrary degree. In the next section, we discuss the relationship between state separation, unambiguous discrimination and exact cloning. In particular, we show that the Ivanovic-Dieks-Peres bound[16, 17, 18] on the probability of state discrimination is a special case of the general bound on the separation probability, and discuss some properties of the optimum operation. We also make use of state separation to interpolate between unambiguous and optimal state discrimination measurements. We then show that the demonstration by Duan and Guo[21] that two non-orthogonal states can, with some probability, be exactly cloned, is another example of an operation which carries out state separation. The Duan-Guo bound[21] on the probability of exactly cloning a pair of non-orthogonal states is then also a particular case of the separation bound. The latter also readily gives the more general least upper bound on the probability of producing N copies of the state given M initial ones, where $N \geq M$. Finally, we show that the state discrimination and N from M cloning bounds imply each other.

Consider a quantum system prepared in one of the two states $|\psi_\pm^1\rangle$. We are not told which of the states the system is in, although we do know that both states are equally probable. We aim to transform the state into $|\psi_\pm^2\rangle$, where

$$|\langle\psi_+^2|\psi_-^2\rangle|^2 \leq |\langle\psi_+^1|\psi_-^1\rangle|^2, \tag{26}$$

that is, the operation decreases the square-overlap between the two possible states of the system. This makes them more distinct. This operation cannot be successful all of the time, and there must be a probability of failure. We wish to maximise the probability P_S of this separation of the states being successfully carried out. To analyse this problem, it is again convenient to employ the Kraus representation of quantum operations. Here, each of the possible, distinguishable outcomes of an operation, which are labelled by the index n, is associated with a linear transformation operator $\hat{A}_n$. These form a resolution of the identity

$$\sum_n \hat{A}_n^\dagger \hat{A}_n = \hat{1}. \tag{27}$$

Note that this expression is equivalent to (4). If the system is prepared with the initial density operator $\hat{\rho}$, then the probability P_n of the nth outcome is $\mathrm{Tr}\hat{\rho}\hat{A}_n^\dagger\hat{A}_n$, upon which the density operator is transformed according to $\hat{\rho} \rightarrow \hat{A}_n\hat{\rho}\hat{A}_n^\dagger/P_n$.

The state separation operation can have two possible outcomes, success or failure[22]. The corresponding transformation operators, $\hat{A}_S$ and $\hat{A}_F$ change the states $|\psi_\pm\rangle$ in the following way:

$$\hat{A}_S|\psi_\pm^1\rangle = \mu_\pm|\psi_\pm^2\rangle, \tag{28}$$
$$\hat{A}_F|\psi_\pm^1\rangle = \nu_\pm|\phi_\pm\rangle, \tag{29}$$

where the $\mu_\pm$ and $\nu_\pm$ are complex coefficients and the failure operator $\hat{A}_F$ transforms $|\psi_\pm^1\rangle$ into some other normalised states $|\phi_\pm\rangle$. The resolution of the identity Eq. (27) becomes

$$\hat{A}_S^\dagger\hat{A}_S + \hat{A}_F^\dagger\hat{A}_F = \hat{1}. \tag{30}$$

Consequently, we have $|\mu_\pm|^2 + |\nu_\pm|^2 = 1$. We denote by $P_{S\pm} = |\mu_\pm|^2$ the conditional probability that the desired transformation takes place given the initial state $|\psi_\pm^1\rangle$, and see that the total success probability is

$$P_S = \frac{P_{S+} + P_{S-}}{2} = \frac{1}{2}(|\mu_+|^2 + |\mu_-|^2). \tag{31}$$

This probability is bounded by the positivity of the operator $\hat{A}_F^\dagger \hat{A}_F$. As is shown in [20], the least upper bound on P_S is given by

$$P_S \leq \frac{1 - |\langle\psi_+^1|\psi_-^1\rangle|}{1 - |\langle\psi_+^2|\psi_-^2\rangle|}. \tag{32}$$

For this bound to be attained, both states must have the conditional separation probabilities ($P_{S\pm} = P_S$).

The maximum value of the separation probability has a natural composition property. Consider three pairs of quantum states, $|\psi_\pm^i\rangle, i = 1, 2, 3$, such that

$$|\langle\psi_+^3|\psi_-^3\rangle|^2 \leq |\langle\psi_+^2|\psi_-^2\rangle|^2 \leq |\langle\psi_+^1|\psi_-^1\rangle|^2. \tag{33}$$

Let P_S^{12}, P_S^{23} and P_S^{13} be the maximum probabilities of carrying out the transformations $|\psi_\pm^1\rangle \rightarrow |\psi_\pm^2\rangle$, $|\psi_\pm^2\rangle \rightarrow |\psi_\pm^3\rangle$ and $|\psi_\pm^1\rangle \rightarrow |\psi_\pm^3\rangle$ respectively. Evidently, $P_S^{13} = P_S^{12} P_S^{23}$, so it is possible to attain the maximum separation probability by mapping the initial states directly onto the final ones, or through one or more sets of intermediate states.

It is important to examine what happens when the separation does not succeed, in particular whether or not one can make a further attempt. If the attempt fails, then the initial density operator is transformed into

$$\hat{\rho}_{F\pm} = \frac{\hat{A}_F|\psi_\pm^1\rangle\langle\psi_\pm^1|\hat{A}_F^\dagger}{\langle\psi_\pm^1|\hat{A}_F^\dagger \hat{A}_F|\psi_\pm^1\rangle}. \tag{34}$$

If the operation is optimised so that P_S takes its maximum value, corresponding to saturation of the inequality (32), then the states $\hat{\rho}_{F\pm}$ are identical. We can see this by noting that if the equality in (32) is satisfied, the operator $\hat{A}_S^\dagger \hat{A}_S$ has an eigenvector lying in the subspace spanned by $|\psi_\pm^1\rangle$ with eigenvalue 1. Eq. (30) implies that this eigenvector lies in the nullspace of $\hat{A}_F^\dagger \hat{A}_F$. We may expand this eigenvector as $|\psi\rangle = c_+|\psi_+^1\rangle + c_-|\psi_-^1\rangle$ where the coefficients $c_\pm$ are easily scaled to obtain normalisation. The positivity of $\hat{A}_F^\dagger \hat{A}_F$ implies that $\hat{A}_F|\psi\rangle = 0$, or

$$c_+\hat{A}_F|\psi_+^1\rangle = -c_-\hat{A}_F|\psi_-^1\rangle. \tag{35}$$

This expression, when applied to Eq. (34), gives $\hat{\rho}_{F+} = \hat{\rho}_{F-}$. Thus, if an optimum state separating operation fails, it is impossible for a further attempt to succeed, since a failure erases the bit of information describing the initial preparation.

A separation operator which attains the limit in (32) is

$$\hat{A}_S = \left(\frac{1 - |\langle\psi_+^1|\psi_-^1\rangle|}{1 - |\langle\psi_+^2|\psi_-^2\rangle|}\right)^{1/2} \sum_{r=\pm} \frac{|\psi_r^2\rangle\langle\psi_r^{1\perp}|}{\langle\psi_r^{1\perp}|\psi_r^1\rangle}. \tag{36}$$

Here, we have introduced the *reciprocal states* $|\psi_+^{1\perp}\rangle$ and $|\psi_-^{1\perp}\rangle$[23]. These are the states in the subspace spanned by the $|\psi_\pm^1\rangle$ which are orthogonal to $|\psi_-^1\rangle$ and $|\psi_+^1\rangle$ respectively. The corresponding failure operator may be taken to have the form $\hat{A}_F = \hat{U}(\hat{1} - \hat{A}_S^\dagger \hat{A}_S)^{1/2}$, where $\hat{U}$ may be any unitary operator.

The separation bound (32) simplifies when $|\psi^2_\pm\rangle$ are orthogonal. A von Neumann measurement would be able to distinguish perfectly between these states, so we would know with certainty the initial state of the system. Given two non-orthogonal states $|\psi^1_\pm\rangle$ and setting $\langle\psi^2_+|\psi^2_-\rangle = 0$, the bound (32) on the separation probability is the limit on the probability of such an unambiguous discrimination attempt succeeding. The maximum probability so obtained is equal to that derived by Ivanovic, Dieks and Peres[16, 17, 18]:

$$P_{IDP} = 1 - |\langle\psi^1_+|\psi^1_-\rangle|. \tag{37}$$

Failure of the discrimination attempt gives an inconclusive result. The IDP limit is not the absolute maximum of the discrimination probability, but is rather the maximum subject to the constraint that the measurement never gives incorrect results. The absolute maximum probability of discriminating between two states $|\psi_\pm\rangle$ is instead given by the well-known Helstrom limit[4]

$$P_H = \frac{1}{2}\left(1 + \sqrt{1 - |\langle\psi_+|\psi_-\rangle|^2}\right). \tag{38}$$

The Helstrom measurement always gives a definite answer to the question of the initial state of the system, but this will not necessarily be correct. The probability of an incorrect result is $1-P_H$. Polarisation discrimination to the Helstrom limit has recently been demonstrated for optical pulses with a mean photon number of 0.1[25].

Let us consider transforming the states $|\psi^1_\pm\rangle$ into $|\psi^2_\pm\rangle$, where $|\psi^2_\pm\rangle$ are not orthogonal but are more distinct than the initial states. The output state is then sent to an optimal detector. Such a detector will correctly identify the state with probability P_H in Eq. (38), where $|\psi_\pm\rangle = |\psi^2_\pm\rangle$. The probability of an incorrect result, given that the separation takes place, is $1 - P_H$. Thus, the probability of correctly determining the state of the system is $P_D = P_S P_H$. We see that the probability of obtaining an incorrect result $P_E = P_S(1 - P_H)$, and that the probability of an inconclusive result is $P_I = 1 - P_S$. These are an exhaustive set of possible outcomes, so

$$P_S + P_E + P_I = 1. \tag{39}$$

We can use the state separation and Helstrom bounds to find the minimum error probability P_E given a fixed value of the separation probability P_S, or equivalently, the probability P_I of an inconclusive result. Rearrangin the separation bound in (32) gives an inequality for $|\langle\psi^2_+|\psi^2_-\rangle|$ in terms of P_S and $|\langle\psi^1_+|\psi^1_-\rangle|$. This can be written in terms of P_{IDP}, and we obtain

$$P_E \geq \frac{1}{2}\left(P_S - \sqrt{P_S^2 - (P_S - P_{IDP})^2}\right). \tag{40}$$

When $P_S = 1$, corresponding to no state separation, the minimum value of P_E is $1 - P_H$, which corresponds to the Helstrom measurement. For $P_S = P_{IDP}$, we find that P_E can be zero, which gives the IDP measurement. This inequality is equivalent to that derived in [24], and corresponds to a family of measurements which optimally interpolates between the Helstrom and IDP limits.

A further practical use of state separation is the production of exact copies of a quantum state. The quantum formalism implies that it is impossible to devise a means of detministically cloning a quantum system, unless we know an orthogonal basis to which the state belongs[26]. However, Duan and Guo[21] showed that two non-orthogonal states can, with some probability, be cloned by the same device. Here, a quantum system is prepared either of the states $|\psi_\pm\rangle$ with equal *a priori* probabilities. An ancillary system is also prepared in some 'blank' state $|\chi\rangle$. A successful cloning operation will transform the product states $|\psi_\pm\rangle|\chi\rangle$ into $|\psi_\pm\rangle|\psi_\pm\rangle$. The square-overlap of the final states is the square of that of the initial states, and so is reduced by the operation. Exact cloning is then another application of state separation, and its success probability is bounded accordingly by (32). Taking $|\psi^1_\pm\rangle = |\psi_\pm\rangle|\chi\rangle$ and $|\psi^2_\pm\rangle = |\psi_\pm\rangle|\psi_\pm\rangle$, we find that the separation bound in (32), viewed as a bound on the probability of producing an exact copy of the initial state, gives the Duan-Guo [21] limit

$$P_{DG} = \frac{1}{1 + |\langle\psi_+|\psi_-\rangle|}. \tag{41}$$

The state separation bound yields the more general least upper bound on producing N copies of $|\psi_\pm\rangle$ from M initial copies where $N \geq M$, with equal ease. Here, the initial states $|\psi^1_\pm\rangle$ are the products $|\psi_\pm\rangle_1 \ldots |\psi_\pm\rangle_M |\chi\rangle$, where the blank state $|\chi\rangle$ is composed of $N - M$ subsystems. The final states $|\psi^2_\pm\rangle$ are $|\psi_\pm\rangle_1 \ldots |\psi_\pm\rangle_N$, the N copies of $|\psi_\pm\rangle$. It immediately follows from (32) that the maximum probability of producing N copies of the state given the smaller number M of initial copies $|\psi_\pm\rangle$ is

$$P_{MN} = \frac{1 - |\langle\psi_+|\psi_-\rangle|^M}{1 - |\langle\psi_+|\psi_-\rangle|^N}. \tag{42}$$

The general cloning bound P_{MN} and the Ivanovic-Dieks-Peres bound P_{IDP} have an interesting relationship, which we now examine. We will see in particular how they may be derived from each other. To derive the latter from the former, suppose that a system is prepared in one of the states $|\psi_\pm\rangle$. The maximum probability that we can make N copies of the state is then given by P_{1N}. It follows from Eq. (42) that as $N \to \infty$, $P_{1N} \to P_{IDP}$ from above. In this limit, the state could be determined through the statistics of appropriate measurements on the copies, so Eq. (42) implies that the states can be discriminated unambiguously with probability P_{IDP}. Consistency with the cloning bound implies that we cannot exceed P_{IDP}. If state discrimination could be accomplished with higher probability than P_{IDP}, then with the same probability, we could manufacture an arbitrarily large number of copies of the state given one initial realisation. This probability would exceed P_{1N} for sufficiently large N. The discrimination limit P_{IDP} and the infinite cloning bound $P_{1\infty}$ are then equal due to the possibility of performing unambiguous state discrimination given an arbitrarily large number of copies.

It is also the case that the general cloning bound P_{MN} follows from the discrimination limit P_{IDP}. If we have M quantum systems all prepared in one of the states $|\psi_\pm\rangle$, and P_{IDP} is the maximum probability that we can discriminate between two quantum states, then it is impossible to discriminate between the M-particle products with probability greater than $P_{M\infty}$. We then ought to be unable to improve upon this bound by first attempting an N from M cloning operation, with $N \geq M$, followed by an attempt to discriminate between the N-particle products. The latter cannot be accomplished with probability greater than $P_{N\infty}$. Writing the cloning probability as P_C, this must be constrained by the fact the compound operation cannot be accomplished with probability greater than $P_{M\infty}$. Thus, $P_{M\infty} \geq P_C P_{N\infty}$, so P_C cannot be greater than P_{MN} in (42).

In conclusion, we have examined multiple aspects and applications of generalised measurements, including projection synthesis, imperfect detection, state preparation, state discrimination, state separation and state cloning. Some of the techniques we have used may seem counterintuitive at first sight, but fit in naturally with the concept of a generalised measurement. Advances in experimental technique, especially within the field of quantum optics, mean that these applications of generalised measurement may soon be realised.

Acknowledgements

This work was supported by the UK Engineering and Physical Sciences Research Council and the Australian Research Council.

REFERENCES

1. J. von Neumann, Mathematical Foundations of Quantum Mechanics (Princeton University Press, Princeton, 1955).
2. E. Arthurs and J. L. Kelly Jr., *Bell Syst. Tech. J.* **44**, 725 (1965).
3. P. A. Benioff, *J. Math. Phys.* **13**, 231 (1972).
4. C. W. Helstrom, *Quantum Detection and Estimation Theory*, (Academic Press, New York, 1976).
5. K. Kraus, *States, Effects and Operations: Fundamental Notions of Quantum Theory*, Springer-Verlag, Berlin Heidelberg, 1983.
6. Compare with Y. Aharonov, P. G. Bergmann and J. L. Lebowitz, *Phys. Rev.* **134B**, 1410 (1964).
7. S. M. Barnett and P. M. Radmore, Methods in Theoretical Quantum Optics (Oxford University Press, Oxford, 1997).
8. L. S. Phillips, S. M. Barnett and D. T. Pegg, *Phys. Rev. A* (in press).
9. S. M. Barnett and D. T. Pegg, *Phys. Rev. Lett.*, **76** 4148 (1996).
10. D. T. Pegg, S. M. Barnett and L. S. Phillips, *J. Mod. Opt.* **44**, 2135 (1997).
11. S. M. Barnett, L. S. Phillips and D. T. Pegg, (submitted to Opt. Commun.).
12. D. T. Pegg, L. S. Phillips and S. M. Barnett, To appear in Physical Review Letters.
13. R. Josza, *J. Mod. Opt.* **41**, 2315-2323 (1994).
14. B. Huttner, N. Imoto, N. Gisin and T. Mor, *Phys. Rev. A* **51**, 1863 (1995).
15. S. M. Barnett, *Phil. Trans. R. Soc. Lond. A* **355**, 2279 (1997).
16. I. D. Ivanovic, *Phys. Lett. A* **123** 257 (1987).
17. D. Dieks, *Phys. Lett. A* **126** 303 (1988).
18. A. Peres, *Phys. Lett. A* **128** 19 (1988).
19. A. Chefles, *Phys. Lett. A* **239** 339 (1998).
20. A. Chefles and S. M. Barnett, LANL Report No. quant-ph/9808018, submitted to Physical Review Letters.
21. L-M. Duan and G-C. Guo, LANL Report No. quant-ph/9704020.
22. As is shown in [20], it is unnecessary to consider operations which have multiple outcomes corresponding to success and failure.
23. A. Chefles and S. M. Barnett, LANL Report No. quant-ph/9807023, submitted to Physics Letters A.
24. A. Chefles and S. M. Barnett, *J. Mod. Opt.* **45** 1295 (1998).
25. S. M. Barnett and E. Riis, *J. Mod. Opt.* **44** 1061 (1997).
26. W. K. Wootters and W. H. Zurek, *Nature* **299** 802 (1982).

LATEST DEVELOPMENTS IN QUANTUM TOMOGRAPHY

Giacomo Mauro D'Ariano

Department of Electrical and Computer Engineering,
Northwestern University 2145 North Sheridan Road,
Evanston, IL 60208-3118

Istituto Nazionale di Fisica della Materia,
Unità di Pavia, via Bassi 6, 27100 Pavia, Italy

Latest developments in quantum tomography are presented. The method for measuring the state of radiation is derived in a simple group-theoretical framework that allows generalization to arbitrary quantum systems. Some recently developed topics are synthetically reviewed, including tomography of many radiation modes using only one local oscillator, generalization to N-level systems, and new "adaptive" techniques for noise reduction. A set of newly proposed experiments is presented, based on a *conditioned* tomographic technique.

INTRODUCTION

Optical homodyne tomography is now a well assessed method to measure the quantum state of radiation (for a review see Ref. 1). The density operator ρ can be measured in a given representation by averaging a set of special functions over homodyne data. The method has been extended to estimate the expectation value $\langle O \rangle$ of an arbitrary operator O, making homodyne tomography the first universal detector for radiation.[2]

In this paper I will present some recent progress in quantum tomography, with the method extended to any quantum system, as any number of radiation modes or N-level systems. The method is based on the possibility of measuring a set of observables—socalled *quorum*—which are irreducible for the unitary representation of the dynamical group of the system. For a set of radiation modes the quorum is given by all linear combinations of the creation and annihilation operators of the modes. For a set of N-level systems it is the set of all linear combinations of the angular momentum operators of each system. The symmetries of the quorum can be exploited to reduce the statistical errors of the method, and this is the basis of recently discovered "adaptive" techniques.

In the first section, devoted to homodyne tomography, I will present a synthetic derivation of the method in a simple group-theoretical framework that allows generalization to arbitrary quantum system. In the same section I will also illustrate: *i*) the emergence of bounds for quantum efficiency (on the basis of a simple example); *ii*) the

Quantum Communication, Computing, and Measurement 2
Edited by Kumar et al., Kluwer Academic / Plenum Publishers, New York, 2000.

equivalence classes of estimators, and the generating function of all s-ordered monomials in the field operator; *iii)* the new adaptive method for reducing statistical errors; *iv)* the generalization to many modes of radiation using only one local oscillator. In the subsequent section I will outline the generalization to other dynamical systems, treating spin tomography as an example. Finally, in the last section I will shortly outline a set of new experiments which are now feasible with the homodyne tomographic method, including the possibility of detecting small Schrődinger cats of radiation, checking Bell inequalities, observing the Greenberger-Horne-Zeilinger state, and finally checking the quantum state-reduction rule.

HOMODYNE TOMOGRAPHY

The method allows to estimate the ensemble average $\langle O\rangle$ of a given (generally complex and unbounded) operator O of the radiation field in a unknown state ρ. In this section I focus attention on a single mode described by boson operators a and $a^\dagger$, with $[a,a^\dagger]=1$; the generalization to many modes will be given in the following section.

In homodyne tomography the problem is to estimate the ensemble average $\langle O\rangle$ from homodyne measurement of quadrature operators $X_\phi=\frac{1}{2}(a^\dagger e^{i\phi}+ae^{-i\phi})$ at different phases ϕ with respect to the local oscillator (LO). I assume that the homodyne detector is properly used in the strong LO limit: this is the only assumption of the method, which works also for nonunit quantum efficiency η of the homodyne detector (the overall quantum efficiency η including also the effect of any source of Gaussian noise[3]). The problem is to estimate the expectation $\langle O\rangle$ of a given operator O for an unknown state ρ by averaging a suitable function over homodyne data. The method provides a rule that to every operator O assigns an unbiased estimator $\mathcal{E}_\eta[O](x;\phi)$, such that for arbitrary state one has

$$\langle O\rangle=\int_0^\pi\frac{d\phi}{\pi}\int_{-\infty}^{+\infty}dx\,p_\eta(x;\phi)\mathcal{E}_\eta[O](x;\phi)\,, \tag{1}$$

$p_\eta(x;\phi)$ denoting the probability distribution of the outcomes x for the quadrature X_ϕ detected with quantum efficiency η. Notice that, due to the symmetry $X_{\phi+\pi}=-X_\phi$, the phase averaging can be restricted to the window $\phi\in[0,\pi]$. According to the central-limit theorem, the mathematical expectation (1) can be estimated by averaging the estimator $\mathcal{E}_\eta[O](x;\phi)$ over experimental data only if $\mathcal{E}_\eta[O](x;\phi)$ has moments bounded up to the third order. As in the strong LO approximation the probability $p_\eta(x;\phi)$ must decay as a Gaussian for large x, it follows that the integral in Eq. (1) can be experimentally sampled for any *a priori* unknown probability distribution $p_\eta(x;\phi)$ only if $\mathcal{E}_\eta[O](x;\phi)$ increases slower than $\exp(kx^2)$ for large x and is bounded for $|x|<+\infty$. In this case one is guaranteed that the integral in Eq. (1) can be statistically sampled over a sufficiently large set of data, and the mean values for different experiments will be Gaussian distributed around the mathematical expectation (1), allowing estimation of confidence intervals. On the contrary, if the kernel $\mathcal{E}_\eta[O](x;\phi)$ is unbounded at some $(x;\phi)$, then the ensemble average $\langle O\rangle$ cannot be measured using homodyne tomography.

The analytic form for the estimator $\mathcal{E}_\eta[O](x;\phi)$ is given by[1]

$$\mathcal{E}_\eta[O](x;\phi)=\frac{1}{4}\int_0^\infty dt\,e^{\frac{1-\eta}{8\eta}t}\,\mathrm{Tr}\left[O\cos\sqrt{t}(X_\phi-x)\right]\,. \tag{2}$$

Eq. (2) can be easily derived from the identity

$$O=\int\frac{d^2\alpha}{\pi}\mathrm{Tr}[OD(\alpha)]D^\dagger(\alpha)\,, \tag{3}$$

where $D(\alpha) = \exp(\alpha a^\dagger - \alpha^* a)$ denotes the displacement operator. Changing to polar variables $\alpha = \frac{i}{2} k e^{i\phi}$, and using the relation between moment generating functions

$$\langle \exp(-ikX_\phi) \rangle = e^{\frac{1-\eta}{8\eta} k^2} \int_{-\infty}^{+\infty} \mathrm{d}x\, p_\eta(x;\phi) e^{-ikx} , \tag{4}$$

one obtains

$$\langle O \rangle = \int_0^{2\pi} \frac{\mathrm{d}\phi}{\pi} \int_0^{+\infty} \frac{\mathrm{d}k\, k}{4}\, e^{\frac{1-\eta}{8\eta} k^2}\, \mathrm{Tr}[O e^{ikX_\phi}] \int_{-\infty}^{+\infty} \mathrm{d}x\, p_\eta(x;\phi) e^{-ikx} . \tag{5}$$

Eq. (5) can be rewritten as the expectation of an estimator if the integrals over k and x can be exchanged, namely

$$\langle O \rangle = \int_0^{2\pi} \frac{\mathrm{d}\phi}{\pi} \int_{-\infty}^{+\infty} \mathrm{d}x\, p_\eta(x;\phi) \int_0^{+\infty} \frac{\mathrm{d}k\, k}{4}\, e^{\frac{1-\eta}{8\eta} k^2}\, \mathrm{Tr}[O e^{ik(X_\phi - x)}] , \tag{6}$$

and using the symmetry $p(x;\phi+\pi) = p(-x;\phi)$ one obtains Eqs. (1) and (2).

The route for generalization to arbitrary quantum systems

In this subsection I briefly analyze the derivation of the starting identity (3), which is the core of the method, and which will be exploited later for generalization to arbitrary quantum system other than the bosonic field. The basic definition of the quantum tomographic method is to measure a *quorum* of observables from which the ensemble average of any operator can be obtained for any state of the system. In the case of homodyne tomography the quantum system is a harmonic oscillator, and the quorum is given by the set of all linear combinations of creation and annihilation operators $a^\dagger$ and a; such set, apart from a trivial multiplication constant, corresponds to the set of all quadratures X_ϕ for $\phi \in [0,\pi]$. In order to obtain a trace identity of the form (3) one can exploit a unitary irreducible representation of a Lie group generated by the quorum.[4] In the present case the group is given by the central extension of the abelian group of displacements over the complex plane, with unitary irreducible representation given by the displacement operators $D(\alpha)$, and with composition law $D(\alpha)D(\beta) = D(\alpha+\beta)\exp[2\mathrm{Im}(\alpha\beta^*)]$. Since the representation is irreducible (there is no Hilbert subspace which is left invariant by the group) according to the Shur's lemma only the identity commutes with the whole group representation. Now it is easy to show that the integral $\int \mathrm{d}^2\alpha\, D(\alpha) O D^\dagger(\alpha)$ commutes with all displacement operators (by multiplying the integral by a fixed $D(z)$, using the composition law and shifting the integration variable), which means that the integral is a multiple of the identity. The value of the integral can be obtained by taking its vacuum expectation, and using completeness of coherent states $|\alpha\rangle = D(\alpha)|0\rangle$. In this way one is lead to the identity

$$\mathrm{Tr}(O) = \int \frac{\mathrm{d}^2\alpha}{\pi} D(\alpha) O D^\dagger(\alpha) . \tag{7}$$

Using identity (7) to evaluate the trace under the integral in Eq. (3) one obtains

$$\begin{aligned} \int \frac{\mathrm{d}^2\alpha}{\pi} \mathrm{Tr}[O D(\alpha)] D^\dagger(\alpha) &= \int \frac{\mathrm{d}^2\alpha}{\pi} \int \frac{\mathrm{d}^2\beta}{\pi} D(\beta) D(\alpha) O D^\dagger(\beta) D^\dagger(\alpha) \\ &= \int \mathrm{d}^2\beta D(\beta) O D^\dagger(\beta) \int \frac{\mathrm{d}^2\alpha}{\pi^2} e^{\beta\alpha^* - \beta^*\alpha} = O , \end{aligned} \tag{8}$$

which proves identity (3).

An example of unbiased estimator, and the emergence of bounds for η

Eq. (2) can be used for estimating the matrix element $\rho_{n+l,n}$ of the density operator of the unknown state. Here the operator O is given by $O = |n\rangle\langle n+l|$. By taking the normal ordering in Eq. (2) one obtains

$$\mathcal{E}_\eta[|n\rangle\langle n+l|](x;\phi) = \frac{1}{4}\int_0^\infty dt\, e^{-\frac{2\eta-1}{8\eta}t}\langle n+l|{:}\cos(\sqrt{t}(X_\phi - x){:}|n\rangle =$$

$$= \frac{1}{4}e^{il\phi}\sqrt{\frac{n!}{(n+l)!}}\mathrm{Re}\left[\left(\frac{i}{2}\right)^l \int_0^\infty dt\, e^{-\frac{2\eta-1}{8\eta}t - i\sqrt{t}x}\, t^{l/2} L_n^l(t/4)\right], \quad (9)$$

$L_n^l(x)$ denoting the customary generalized Laguerre polynomials. From Eq. (9) we see that $\mathcal{E}_\eta[|n\rangle\langle m|](x;\phi)$ is bounded only if $\eta > \frac{1}{2}$, which means that the matrix element can be experimentally estimated only for quantum efficiency η above the bound $\eta_b = \frac{1}{2}$. This example shows a typical feature of the tomographic method when applied to infinite dimensional Hilbert space, namely that for every operator O generally there is a bound $\eta_b[O] > 0$ below which the ensemble average $\langle O\rangle$ cannot be estimated. This is not an artifact of the method, but is due to the perfect unbiasedness of the estimator, which works without any *a priori* knowledge of the state. For η approaching the bound $\eta_b[O]$ the statistical error in the estimation of $\langle O\rangle$ becomes unbounded (in Ref. 5 an asymptotic evaluation of errors for the diagonal matrix elements $\rho_{n,n}$ is derived for large n). Finally, it is worth noticing that for estimating the matrix elements $\rho_{n+l,n}$ an algorithm numerically more efficient than Eq. (9) is used, based on a factorization formula for the estimator that holds for $\eta = 1$, and exploiting the inversion of the Bernoulli convolution for $\frac{1}{2} < \eta < 1$.[1]

O	$\mathcal{E}_\eta[O](x;\phi)$	η_b
${:}a^{\dagger n}a^m{:}_s$	Eq. (13)	0
${:}D(\alpha){:}_s$	Eq. (12)	0
$W_s \doteq \frac{2}{\pi(1-s)}\left(\frac{s+1}{s-1}\right)^{a^\dagger a}$	$-\frac{2}{\pi s_\eta}\Phi\left(1,\frac{1}{2};\frac{2x^2}{s_\eta}\right)$	$(1-s)^{-1}$
$\|n\rangle\langle n+l\|$	Eq. (9)	$\frac{1}{2}$
$\|\alpha\rangle\langle\beta\|$	$\kappa\langle\alpha\|\beta\rangle\Phi\left(1,\frac{1}{2};-\kappa(x-w_\phi)^2\right)$	$\frac{1}{2}$
$\|x\rangle\langle x'\|$	Ref. 6	1
$\|\alpha;\zeta\rangle\langle\beta;\zeta\|$	Ref. 6	$(1+e^{-\|\zeta\|^2})^{-1}$

Table 1. Estimators for various field operators and relative bounds for quantum efficiency $\eta > \eta_b$. Legenda: $\Phi(\alpha;\beta;z)$ confluent hypergeometric function, $s_\eta = s - 1 + \eta^{-1}$, $\kappa = 2\eta/(2\eta-1)$, $w_\phi = \frac{1}{2}(\beta e^{i\phi} + \overline{\alpha}e^{-i\phi})$.

Equivalence classes of unbiased estimators

A remarkable consequence of the inversion symmetry for the quadrature operator $p_\eta(x;\phi+\pi) = p_\eta(-x;\phi)$ is that there are "null estimators", which have zero expectation for arbitrary probability $p_\eta(x;\phi)$, i. e. for arbitrary state ρ. Null estimators are obtained as linear combinations of the following functions[7]

$$\mathcal{N}_{k,n}(x;\phi) = x^k e^{\pm i(k+2+2n)\phi} \qquad k,n \geq 0\,. \quad (10)$$

The functions $\mathcal{N}_{k,n}(x;\phi)$ have zero expectation for arbitrary probability as a consequence of the Wilcox formula

$$\langle X_\phi^k\rangle = \frac{k!}{2^k}\sum_{p=0}^{[[k/2]]}\sum_{s=0}^{k-2p}\frac{\langle a^{\dagger s}a^{k-2p-s}\rangle}{2^p p! s!(k-2p-s)!}e^{i(2p+2s-k)\phi} \tag{11}$$

along with the identity $\int_0^\pi d\phi\, e^{iq\phi}=0$ for $q\neq 0$ even, $[[x]]$ denoting the integer part of x. Hence, for every operator O one actually has an equivalence class of infinitely many unbiased estimators, which differ by a linear combination of functions $\mathcal{N}_{k,n}$. Here I denote the equivalence relation by the symbol $\simeq$, i. e. $\mathcal{N}_{k,n}\simeq 0$ for $k,n\geq 0$. Non trivial examples of equivalence are:[8] *i)* $e^{\pm 2(n+1)i\phi}f(xe^{\pm i\phi})\simeq 0$, with $f(z)$ analytic function of z; the relations involving the Dirac comb over $[0,\pi]$ *ii)* $\delta_\pi(\phi)\simeq\frac{1}{\pi}$, and *iii)* $f(x^2)\delta_\pi(\phi)\simeq\frac{2}{\pi}\mathrm{Re}\left[f\left(x^2e^{2i\phi}\right)/\left(1-e^{-2i\phi}\right)-\frac{1}{2}f(x^2)\right]$ with $f(z)$ analytic.

Using the above equivalences it is straightforward to evaluate the generating function of all s-ordered monomials. One has[8]

$$\mathcal{E}_\eta[:D(\alpha):_s](x;\phi) = \exp\left(\frac{1}{2}s_\eta|\alpha|^2\right)\frac{\alpha e^{-i\phi}e^{2x\alpha e^{-i\phi}}+\alpha^* e^{i\phi}e^{-2x\alpha^* e^{i\phi}}}{\alpha e^{-i\phi}+\alpha^* e^{i\phi}}\,, \tag{12}$$

and

$$\mathcal{E}_\eta[:a^{\dagger n}a^m:_s](x;\phi) = \frac{e^{i(m-n)\phi}H_{m+n}^{\min(n,m)}\left(\sqrt{\frac{2}{s_\eta}}x\right)}{\binom{n+m}{n}\left(\sqrt{\frac{2}{s_\eta}}x\right)^{n+m}}\,, \tag{13}$$

where $s_\eta\doteq s-1+\eta^{-1}$ and $H_n^h(x)=\sum_{l=0}^h\frac{(-)^l n!}{l!(n-2l)!}(2x)^{n-2l}$ denotes the "truncated" Hermite polynomial. Notice the equivalence $e^{i(m-n)\phi}H_{m+n}^{\min(n,m)}(\kappa x)\simeq e^{i(m-n)\phi}H_{m+n}(\kappa x)$, between truncated and customary Hermite polynomials, hence the agreement with the previous result by Richter[9] for the normal ordered case $s=1$.

A list of estimators with their quantum efficiency bounds η_b is given in Table 1. For polynomials in the field operators there is no bound. For the Wigner function $W_s(\alpha,\alpha^*)=\langle D^\dagger(\alpha)W_sD(\alpha)\rangle$ (the operator W_s is defined in table 1) the bound depends on the ordering parameter s. For coherent-state and number matrix elements one has $\eta_b=\frac{1}{2}$; for squeezed representations $\eta_b>\frac{1}{2}$; finally, for the quadrature representation $\eta_b=1$. Essentially, the bound is related to the "fuzziness" of the state representation[1], and at present no analytical state representation is known having bound $\eta_b<\frac{1}{2}$ (see also Ref. 5).

Adaptive method

The estimator can be chosen within the equivalence class in order to minimize the statistical error. This is the basis of the recently discovered "adaptive tomography" of Refs. 7, 10. For arbitrary operator O a particular representative $\mathcal{E}_\eta[O](x;\phi)$ of the equivalence class of estimators is given in Eq. (2). However, one can add a linear combination of null estimators $\mathcal{N}_{k,n}(x;\phi)$ given in Eq. (10), leading to the new estimator

$$\mathcal{J}_\eta[O](x;\phi)=\mathcal{E}_\eta[O](x;\phi)+\mu\cdot\mathcal{F}(x;\phi)+\nu\cdot\mathcal{F}^*(x;\phi)\simeq\mathcal{E}_\eta[O](x;\phi)\,, \tag{14}$$

where $\mathcal{F}(x;\phi)$ and $\mathcal{F}^*(x;\phi)$ denote the vectors of null estimators $\mathcal{F}(x;\phi)\doteq\{\mathcal{N}_{k,n}(x;\phi)\}$, and μ and ν are vectors of coefficients to be determined. Minimization of the variance $\overline{\Delta\mathcal{J}_\eta^2}$ leads to the linear set of equations for μ and ν

$$\mathbf{A}\mu=\mathbf{b}\,,\qquad \mathbf{A}\nu=\mathbf{c}\,, \tag{15}$$

where

$$\mathbf{A} = \overline{\mathcal{F}\mathcal{F}^*}\ , \qquad \mathbf{b} = -\overline{\mathcal{E}_\eta[O]\mathcal{F}^*}\ , \qquad \mathbf{c} = -\overline{\mathcal{E}_\eta[O]\mathcal{F}}\ , \tag{16}$$

the overbar denoting the experimental average $\overline{\mathcal{F}} \doteq \lim_{N\to\infty} N^{-1}\sum_{n=1}^{N}\mathcal{F}(x_n;\phi_n)$ over homodyne outcomes x_n at (random) phases ϕ_n. Solving the linear system (15) one can show that the variance is reduced by the amount

$$\overline{\Delta\mathcal{J}_\eta^2} - \overline{\Delta\mathcal{E}_\eta^2} = -\mathbf{b}\cdot\mathbf{A}^{-1}\mathbf{b}^* - \mathbf{c}\cdot\mathbf{A}^{-1}\mathbf{c}^*\ . \tag{17}$$

The method works as follows. One first obtains the matrix **A** and vectors **b** and **c** by averaging over homodyne outcomes according to Eqs. (16). Then the linear system (15) is solved, and the optimized estimator in Eq. (14) is obtained. Finally, the ensemble average $\langle O\rangle$ is recovered by averaging the optimized estimator. In this way, the estimator is "adapted" to data and the method becomes nonlinear. For simple operators the optimized estimator can be derived analytically for some classes of states. For example, for the number operator $O \equiv a^\dagger a$ one can prove[7] that for coherent states, squeezed vacuum, and Schrődinger-cat states only the null function $\mathcal{F} = \exp(2i\phi)$ contributes to the optimization, and the optimized estimator is given by[7, 10]

$$\mathcal{J}_\eta[a^\dagger a](x;\phi) = 2x^2 - \frac{1}{2} + 2\mathrm{Re}[\mu\exp(2i\phi)]\ , \qquad \mu = -\overline{2x^2\exp(2i\phi)} \equiv -\frac{1}{2}\langle a^{\dagger 2}\rangle\ , \tag{18}$$

with variance

$$\overline{\Delta\mathcal{J}_\eta^2[a^\dagger a]} \equiv \langle\Delta(a^\dagger a)^2\rangle + \frac{1}{2}[\langle a^{\dagger 2}a^2\rangle - \langle a^{\dagger 2}\rangle\langle a^2\rangle + 2\langle a^\dagger a\rangle + 1]\ . \tag{19}$$

Notice that the leading noise term $\langle a^{\dagger 2}a^2\rangle$ is canceled by the nonlinear term $\langle a^{\dagger 2}\rangle\langle a^2\rangle$. Analogous cancellations are found for other operators, as for the quadrature X_ϕ and the field operator a,[7, 10], where, remarkably, the ideal heterodyne noise is achieved for coherent states. Another noticeable feature is that, differently from the representative (2), the optimized estimator is peaked and symmetrically distributed around the mean value.[10] In Ref. 10 the method has been numerically implemented for estimating the matrix elements $\rho_{n,m}$ in the number representation, proving that a noise reduction up to 60% can be achieved, especially for low n and m. However, it is likely that a much better noise reduction can be achieved by a suitable choice of the basis of null functions.

Multimode homodyne tomography with one local oscillator

For $M+1$ radiation modes the method is easily generalized by using estimators for tensor product operators which are just the products of their relative estimators, i. e. $\mathcal{E}_\eta[\otimes_{n=0}^{M}O_n](\{x_n\};\{\phi_n\}) = \prod_{n=0}^{M}\mathcal{E}_\eta[O_n](x_n;\phi_n)$. The case of a general operator is then obtained by linearity. However, this method needs a separate measurement—whence a separate LO—for each mode. In Ref. 11 it is shown that it is possible to estimate the expectation value of any multimode observable using a single LO, scanning all possible linear combinations of modes on it. Here I don't give the derivation of the method, but just present the final results. The estimator is given by

$$\mathcal{E}_\eta[O](x;\theta,\psi) = \frac{\kappa^{M+1}}{M!}\int_0^\infty \mathrm{d}t\, e^{-t+2i\sqrt{\kappa t}x}\, t^M\, \mathrm{Tr}\{O\!:\exp[-2i\sqrt{\kappa t}X(\theta,\psi)]\!:\}\ , \tag{20}$$

where $:\ldots:$ denotes normal ordering, $\kappa = \frac{2\eta}{2\eta-1}$, and the quadrature operator $X(\theta,\psi)$ is the following linear combination of single-mode quadratures

$$X(\theta,\psi) = \frac{1}{2}\left[A^\dagger(\theta,\psi) + A(\theta,\psi)\right]\ , \qquad A(\theta,\psi) = \sum_{l=0}^{M} e^{-i\psi_l}u_l(\theta)a_l\ , \tag{21}$$

a_l and $a_l^\dagger$ $(l = 0,\ldots,M)$ being the annihilation and creation operators of the $M+1$ independent modes with $[a_l, a_{l'}^\dagger] = \delta_{ll'}$, $\theta = (\theta_0,\ldots,\theta_M)$ and $\psi = (\psi_0,\ldots,\psi_M)$ denoting hyper-polar angles with ranges $\psi_l \in [0, 2\pi]$ and $\theta_l \in [0, \pi/2]$, whereas $u_l(\theta)$ are hyperspherical coordinates, such that $\sum_{l=0}^M u_l^2(\theta) = 1$, with $u_0(\theta) = \cos\theta_1$, $u_1(\theta) = \sin\theta_1\cos\theta_2$, $u_2(\theta) = \sin\theta_1\sin\theta_2\cos\theta_3$, $\ldots$, $u_{M-1}(\theta) = \sin\theta_1\sin\theta_2\ldots\sin\theta_{M-1}\cos\theta_M$, $u_M(\theta) = \sin\theta_1\sin\theta_2\ldots\sin\theta_{M-1}\sin\theta_M$. The ensemble average $\langle O\rangle$ is obtained by averaging the estimator (20) as follows

$$\langle O\rangle = \int d\mu[\psi] \int d\mu[\theta]\, p(x,\theta,\psi)\, \mathcal{E}_\eta[O](x;\theta,\psi)\,, \tag{22}$$

where $\int d\mu[\psi] \doteq \prod_{l=0}^M \int_0^{2\pi} \frac{d\psi_l}{2\pi}$, and $\int d\mu[\theta] \doteq 2^M M! \prod_{l=1}^M \int_0^{\pi/2} d\theta_l\, \sin^{2(M-l)+1}\theta_l \cos\theta_l$.

In particular, one can estimate the matrix element $\langle\{n_l\}|R|\{m_l\}\rangle$ of the joint density matrix of modes. This will be obtained by averaging the following estimator

$$\begin{aligned}\mathcal{E}_\eta[|\{m_l\}\rangle\langle\{n_l\}|](x;\theta,\psi) &= e^{-i\sum_{l=0}^M (n_l-m_l)\psi_l}\, \frac{\kappa^{M+1}}{M!} \prod_{l=0}^M \left\{ [-i\sqrt{\kappa}u_l(\theta)]^{\mu_l-\nu_l} \sqrt{\frac{\nu_l!}{\mu_l!}} \right\} \\ &\times \int_0^\infty dt\, e^{-t+2i\sqrt{\kappa t}x}\, t^{M+\frac{1}{2}\sum_{l=0}^M(\mu_l-\nu_l)} \prod_{l=0}^M L_{\nu_l}^{\mu_l-\nu_l}[\kappa u_l^2(\theta) t]\end{aligned}\ ,\tag{23}$$

where $\mu_l = \max(m_l, n_l)$, and $\nu_l = \min(m_l, n_l)$. Using simple identities for Laguerre polynomials one can easily derive the estimator for the probability distribution of the total number of photons $N = \sum_{l=0}^M a_l^\dagger a_l$

$$\mathcal{E}_\eta[|n\rangle\langle n|](x;\theta,\psi) = \frac{\kappa^{M+1}}{M!} \int_0^\infty dt\, e^{-t+2i\sqrt{\kappa t}x}\, t^M L_n^M[\kappa t]\,, \tag{24}$$

where $|n\rangle$ denotes the eigenvector of N for eigenvalue n. Notice that the estimator in Eq. (24) does not depend on phases ϕ_l and angles θ_l, and thus their knowledge is not needed in this measurement. Other examples of two-mode estimators are

$$\mathcal{E}_\eta[z^{a^\dagger a+b^\dagger b}](x;\theta,\psi) = \frac{1}{(z+\frac{1-z}{\kappa})^2}\Phi\left(2, \frac{1}{2} - \frac{1-z}{z+\frac{1-z}{\kappa}}x^2\right)\,, \tag{25}$$

where a and b denote the annihilator operators of the two modes, and $\Phi(\alpha;\beta;z)$ denotes the customary confluent hypergeometric function. For the first two moments one obtains the simple expressions

$$\mathcal{E}_\eta[a^\dagger a + b^\dagger b](x;\theta,\psi) = 4x^2 + \frac{2}{\kappa} - 2\,, \tag{26}$$

$$\mathcal{E}_\eta[(a^\dagger a + b^\dagger b)^2](x;\theta,\psi) = 8x^4 + \left(\frac{24}{\kappa} - 20\right)x^2 + \frac{6}{\kappa^2} - \frac{10}{\kappa} + 4\,. \tag{27}$$

It is worth noticing that analogous estimators for the difference of photon numbers are singular, and in order to recover the correlation between modes a cutoff procedure is needed, analogous to the one used in Ref. 12.

GENERALIZATION TO SPINS AND OTHER DYNAMICAL SYSTEMS

The tomographic method can be easily generalized to other dynamical systems, looking for a quorum of observables that allow the estimation of any ensemble average.

The starting point is to generalize the identity (3) exploiting a unitary irreducible representation of a Lie group generated by the quorum.[4] Here, for the sake of simplicity, I consider only the case of a semisimple compact Lie group $\mathbb{G} = \{g\}$, with irreducible unitary representation R of dimension $d < \infty$. However, with some technicalities, the method can be extended to the cases of infinite dimensional representations, noncompact groups, and reducible groups with central extension.

The equivalent of Eq. (7) can be obtained by noticing that the integral over $\mathbb{G}$ $\int \mathrm{d}\mu(g) R^\dagger(g) O R(g)$ commutes with all unitary operators $R(h)$, $h \in \mathbb{G}$ (this can be checked by group composition $R(h)R(g) = R(hg)$, and by shifting the integration variable g using invariance of the Haar's measure $\mathrm{d}\mu(g)$). As the representation is irreducible, according to the Shur's lemma the integral is a multiple of the identity: this constant can then be obtained by evaluating the trace of the integral, using the trace invariance under cyclic permutation, and normalization of the Haar's measure over $\mathbb{G}$. In this way one obtains the identity

$$\int \mathrm{d}\mu(g) R^\dagger(g) O R(g) = \frac{1}{d}\mathrm{Tr}(O) , \tag{28}$$

Similarly, using orthogonality of characters, one can show that $\int \mathrm{d}\mu(g)\, \mathrm{Tr}[R(g)] R^\dagger(g) = d^{-1}$, and through the following steps

$$\begin{aligned} &\int \mathrm{d}\mu(g)\, \mathrm{Tr}[O R^\dagger(g)] R(g) = d \int \mathrm{d}\mu(g) \int \mathrm{d}\mu(h)\, R^\dagger(h) O R^\dagger(g) R(h) R(g) \\ &= \int \mathrm{d}\mu(h)\, R^\dagger(h) O\, \mathrm{Tr}[R(h)] = \frac{1}{d} O , \end{aligned} \tag{29}$$

one proves the general tomographic identity

$$\int \mathrm{d}\mu(g)\, \mathrm{Tr}[O R^\dagger(g)] R(g) = \frac{1}{d} O , \tag{30}$$

which is the equivalent of Eq. (3). Identity (30) is used to obtain the estimator for $\langle O \rangle$, by taking the ensemble average of both sides of the identity. As an example, here I report the estimator for the measurement of the matrix element of the density operator of a spin J. Upon denoting by $|m\rangle$ the eigenvector of J_z with eigenvalue $m \in [-J, J]$, one has[13]

$$\mathcal{E}[|l\rangle\langle i|](m; \vec{n}) = (2J+1) \left(\lambda_{i,m} \lambda^*_{l,m} - \frac{\lambda_{i,m+1}\lambda^*_{l,m+1} + \lambda_{i,m-1}\lambda^*_{l,m-1}}{2} \right) , \tag{31}$$

where

$$\begin{aligned} \lambda_{l,m} &= e^{i\varphi(m-l)} \sqrt{(J+m)!(J-m)!(J+l)!(J-l)!} \\ &\times \sum_\nu \frac{(-1)^\nu (\cos\frac{\vartheta}{2})^{2J+m-l-2\nu} (-\sin\frac{\vartheta}{2})^{l-m+2\nu}}{(J-l-\nu)!(J+m-\nu)!(\nu+l-m)!\nu!} , \end{aligned} \tag{32}$$

$\vec{n}$ being a unit vector with polar angles θ and ϕ, and the average being performed over all possible measurement results m and over all possible spin rotations around $\vec{n}$ according to the formula

$$\langle O \rangle = \sum_{m=-J}^{J} \int \frac{d\vec{n}}{4\pi}\, p(m; \vec{n})\, \mathcal{E}[O](m; \vec{n}) , \tag{33}$$

$p(m; \vec{n})$ denoting the probability of outcome m for the measurement of $\vec{J} \cdot \vec{n}$. One can easily recognize the correspondence with conventional optical quantum tomography: $x \leftrightarrow m$, $\phi \leftrightarrow \vec{n}$. On the basis of Eq. (33) an experimental setup can be devised for tomography of spin observables, which is a simple modification of the Stern-Gerlach experiment.[13]

NEW EXPERIMENTS

The quantum tomographic technique opens the possibility of a new type of experiments and tests of quantum mechanics. In Ref. 14 a test of Bell's inequality is proposed that is based on two-mode homodyne tomography, with the possibility of achieving very good detection quantum efficiencies. Using three-mode homodyne tomography, in principle it is now possible to make a complete test of the preparation of a Greenberger-Horne-Zeilinger state,[15] which cannot be checked by simple coincidence measurements. Finally, it is now possible a direct test of nonclassicality on various one-mode and two-modes states, by tomographically measuring some special observables of the field.[16]

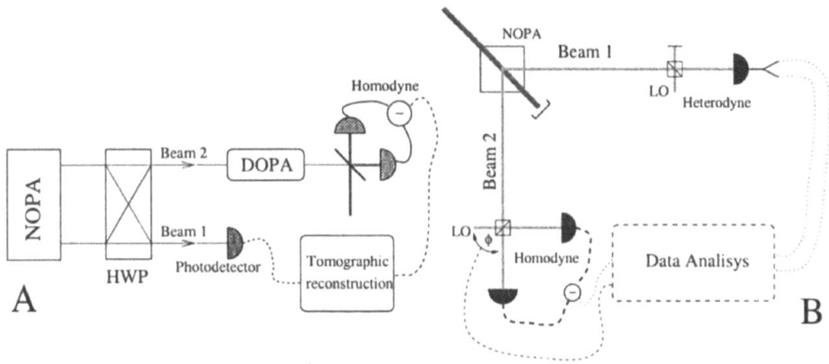

Figure 1. Two examples of *conditioned tomography.* The general scheme involves a nondegenerate optical parametric amplifier (NOPA) that produces a couple of correlated twin beams. A quantum measurement is performed on one beam, and a tomographic reconstruction is made on the second beam, conditioned on the result of the first measurement. Figure A: Experimental scheme for generation and tomographic detection of Schrödinger cats (see text).[18] Scheme for testing the state reduction for heterodyne detection.[19]. After heterodyning the beam 1 of the twin couple, the reduced state of beam 2 is tomographically reconstructed conditioned by the heterodyne outcome. In place of the heterodyne detector one can put any other kind of detector for testing the state reduction on different observables.

Using parametric downconversion, a new set of experiment is now possible, which we can categorize as *conditioned tomography.* The general scheme is the following. A nondegenerate optical parametric amplifier (NOPA) produces a couple of correlated twin beams 1 and 2 from vacuum downconversion. A quantum measurement is performed on beam 1, and a tomographic reconstruction is made on beam 2, conditioned on the result λ of the first measurement, namely using an estimator $\mathcal{E}_\eta[O](x;\phi;\lambda)$ which depends on the outcome λ of the measurement on beam 1.

An example of conditioned tomographic measurement scheme is depicted in Fig. 1A, which represents a tomographic improvement of a scheme proposed in Ref. 17 in order to generate and detect *Schrödinger-cat* states. The experiment consists in feeding the twin beams—here two orthogonally polarized modes of radiation, the *signal* and the *readout*—into a half-wave plate (HWP) which rotates the polarization direction. The rotation angle and the gain of the NOPA are related by the back-action-evading condition. When a number of photons is detected at the readout mode, a Schrödinger-cat appears on the signal mode (an additional squeezer (DOPA) is inserted on the signal mode to "stretch" the cat). In Ref. 18 it is shown that the tomographic technique tolerates very realistic values for quantum efficiency at the readout photodetector, and a precise reconstruction of the cat at the signal mode is possible, recovering the visibility of the homodyne probability oscillation, which otherwise would have been completely washed out by the low quantum efficiency at the readout detector.

Another new experiment based on the conditioned tomographic scheme is the proposal of Ref. 19 for a test of state reduction. The scheme of the experiment is depicted in Fig. 1B. Again a couple of twin beams is generated by a NOPA. After heterodyning beam 1, the reduced state of beam 2 is tomographically reconstructed conditioned by the heterodyne outcome. In place of the heterodyne detector one can put any other kind of detector for testing the state reduction on different observables: for heterodyne detection the reduced state is a coherent state, whereas, for example, for photodetection it is a number eigenstate. The state reduction can be tested by a direct measurement of the fidelity between the theoretically expected reduced state and the experimental state, using a suitable conditioned estimator that takes into account also state distortion due to finite NOPA gain and nonunit quantum efficiencies at detectors. Monte Carlo simulated experiments[19] show that a decisive test can be performed even with only a few thousand measurements, with low gains at the NOPA and low quantum efficiencies at the readout photodetector.

Acknowledgments

This work is supported by Istituto Nazionale di Fisica della Materia and cosponsored by Ministero dell'Università e della Ricerca Scientifica e Tecnologica under the project *Amplificazione e rivelazione di radiazione quantistica.*

REFERENCES

1. G. M. D'Ariano, in Quantum Optics and Spectroscopy of Solids, ed. by T. Hakioglu and A. S. Shumovsky, (Kluwer Academic Publisher, Amsterdam 1997), p. 175.
2. G. M. D'Ariano, in Quantum Communication,Computing, and Measurement, ed. by O. Hirota, A. S. Holevo, and C. M. Caves, (Plenum Publishing, New York 1997), p. 253.
3. G. M. D'Ariano, and N. Sterpi, *J. Mod. Opt.* 44:2227 (1997).
4. M. Paini, Thesis, University of Pavia (1999)
5. G. M. D'Ariano, C. Macchiavello, and N. Sterpi, in this volume.
6. G.M. D'Ariano, U. Leonhardt, and H. Paul, *Phys. Rev. A* 52:R1801 (1995).
7. G. M. D'Ariano and M. G. A. Paris, *Acta Physica Slovaca* 48:191 (1998)
8. G. M. D'Ariano, unpublished.
9. Th. Richter, *Phys. Lett.* A221:327 (1996).
10. G. M. D'Ariano and M. G. A. Paris, unpublished.
11. G. M. D'Ariano, M. F. Sacchi, M. Vasilyev, and P. Kumar, unpublished.
12. G. M. D'Ariano, M. Vasilyev, and P. Kumar, *Phys. Rev. A* 58:636 (1998).
13. G. M. D'Ariano, L. Maccone, and M. Paini, unpublished
14. G. M. D'Ariano, L. Maccone, M. F. Sacchi, and A. Garuccio, in this volume.
15. G. M. D'Ariano, M. F. Sacchi, and Y. Shih, unpublished
16. G. M. D'Ariano, M. F. Sacchi, and P. Kumar, *Phys. Rev. A* (1999), in press.
17. S. Song, C.M. Caves and B. Yurke, *Phys. Rev. A* 41:5261 (1990).
18. G.M. D'Ariano, C. Macchiavello and L. Maccone, unpublished.
19. G. M. D'Ariano, P. Kumar, C. Macchiavello, and N. Sterpi, unpublished

MEASURING THE QUANTUM POLARIZATION STATE OF LIGHT

M. G. Raymer, A. C. Funk, and D. F. McAlister

Oregon Center for Optics and Department of Physics
University of Oregon
Eugene, OR 97403 USA

Quantum-state tomography is proposed as a means to achieve a complete measurement of the quantum polarization state of a light wave. A set of measurements using dual-polarization balanced homodyne detection is shown to be *tomographically complete* with respect to the statistics of the SU(2) Stokes operators on the Poincaré sphere. Complete reconstruction of the polarization sector of the density matrix of a partially polarized optical field can be achieved while randomizing the overall phase of the dual-polarization local oscillator.

QUANTUM STATE MEASUREMENT

The conventional interpretation of a quantum state is that it is an *underlying structure of information* about a physical system. While it cannot be measured directly in a single trial, its knowledge is necessary and sufficient for making predictions about the probability distributions for any and all measurable quantities pertaining to the system. This can be viewed as the *forward* statement of the measurement hypothesis. An *inverse problem* can be stated: is it possible to infer the true quantum state defined by a general, unknown preparation procedure by making separate measurements of a certain set of different quantities over a large ensemble of individual systems, all prepared by the same unknown procedure? The answer is *yes*, and a set of quantities whose statistical characterization allows for full state reconstruction has been termed *tomographically complete.*[1] Thus, if the probability distributions (or densities) are precisely known for a tomographically complete set of observables then a deterministic, linear mathematical operation can be used to reconstruct precisely the quantum state of the measured ensemble (which is necessarily infinite in this limiting case). For discrete degrees of freedom, such as angular momentum[2] a finite number of discrete variables make up a tomographically complete set. For systems with a continuous degree of freedom any tomographically complete set must contain in general an uncountable set of variables. If certain prior information is known about the state then a finite number of variables may again be sufficient. For example, it has been shown that if an optical field is known a priori to contain fewer than a given number of photons then the state can be reconstructed exactly using a finite set of distributions.[3, 4]

Quantum Communication, Computing, and Measurement 2
Edited by Kumar et al., Kluwer Academic / Plenum Publishers, New York, 2000.

Since the early quantum-optical state reconstruction experiments,[5] a great advance has occurred in our understanding of the theoretical and experimental issues involving reconstruction of systems described by a *single degree of freedom*. Note that here a quantum-optical state refers to a state of the electromagnetic field (which is a continuous degree of freedom), not the state of photons treated as fundamental particle-like entities. For further background see the Special Issue of J. Modern Optics on "Quantum State Preparation and Measurement"[6] and two reviews.[7, 8]

The quantum mechanics of *two degrees of freedom* is much richer than that of a single one because of the possibility for quantum correlations, leading in their strongest form to Bell-type inequalities. Because of the nonclassical nature of these correlations it might be thought that "even a statistical description including correlations between subsystems cannot fully encompass quantum correlations."[9] There is a sense in which this claim is incorrect. It is possible to find a tomographically complete set of observables for the combined system whose statistical characterization will yield the combined quantum state. One need only measure joint distributions of certain pairs of commuting variables to determine completely the state. (There might be a very large number of distributions requiring measurement in practice.) Even if the measurements on each subsystem are made independently and are local to each subsystem, still the combined state can be determined by later comparing the results. In fact, the needed measurements are just those needed to determine the state of each subsystem separately. This point was made clearly by Wootters for the case of spin systems with discrete variables.[10] The analogous result for two systems having continuous variables has also been proven.[11] An example with continuous variables is the two electric-field amplitudes of a pair of optical modes.[4, 12, 13]

Here we will discuss a particularly interesting sub-case of two-mode optical state reconstruction – the measurement of the quantum state of polarization of an optical beam. Rather than requiring a full reconstruction of the two-mode state (i.e., the density matrix) the measurement of the quantum state of polarization requires reconstruction only of a subset of the density matrix. We call this subset of the density matrix the "polarization sector," since its knowledge allows a complete characterization of the state of polarization on the Poincaré sphere (actually on the set of Poincaré spheres). This representation corresponds to the SU(2) group, and its use to describe a pair of harmonic oscillators is associated with Schwinger. We will outline the basics of the proposed measurement technique, in which we introduce a new type of optical detection – dual-polarization balanced homodyne detection. Various forms of "nonclassical" states of polarization have been proposed.[14, 15] Experiments using two-mode squeezed light have been carried out in the context of polarization interferometry,[16] but no previous efforts at complete characterization of a nonclassical state have been reported to our knowledge.

In fact, no clear definition for the concept of quantum polarization state of a light wave seems to have been developed (in contrast to the polarization state of a single photon, which seems to be unambiguous[17]). We propose a definition applicable to light waves.

The ideas discussed here may have use outside of polarization characterization. For example, optical interferometry for measuring small phase shifts is known to be described by SU(2).[18] Also, two degenerate molecular vibrational modes can display so-called pseudo-rotation, which is described by SU(2).[19] Further, squeezed states of collective atomic angular momentum can be defined within SU(2).[20]

THE POLARIZATION OF LIGHT

A quasi-monochromatic, well collimated light beam with frequency ω and wave number k is characterized by the amplitudes of the two transverse, orthogonal field components. The field (positive-frequency part) can be written

$$\vec{E}^{(+)}(z,t) = a_1\vec{u}_1 \exp\left[-i(\omega t - kz)\right] + a_2\vec{u}_2 \exp\left[-i(\omega t - kz)\right], \tag{1}$$

where $\vec{u_1}$ and $\vec{u_2}$ are orthogonal, unit polarization vectors (e.g. linear). (The form of transverse beam profile is suppressed and is assumed to be smooth – for example a TEM_{00} gaussian mode.) The phenomenon of partial polarization of light occurs when the field amplitudes (photon annihilation operators) a_1, a_2 are partially correlated statistically.[21]

The Schwinger SU(2) representation for two harmonic oscillators is given in terms of the pseudo-angular-momentum vector operator[18] $\vec{J} = (J_1, J_2, J_3)$, where the Stokes operators[14, 15, 17] are given by

$$J_1 = \frac{1}{2}(a_1^\dagger a_1 - a_2^\dagger a_2), \qquad J_2 = \frac{1}{2}(a_1^\dagger a_2 + a_2^\dagger a_1), \qquad J_3 = \frac{1}{2i}(a_1^\dagger a_2 - a_2^\dagger a_1), \tag{2}$$

which obey the familiar commutation relations $[J_1, J_2] = iJ_3$, et cycl. The total photon number operator is $N = a_1^\dagger a_1 + a_2^\dagger a_2$. This operator commutes with each J_i and is related to $\vec{J}$ by $J^2 = \frac{1}{2}N(N+1)$.

Note J_1 is one-half the difference number, and J_2, J_3 contain relative-phase information. To illustrate this consider the two-mode optical coherent state[22] $|A_1 \exp(-i\phi_1)\rangle_1 |A_2 \exp(-i\phi_2)\rangle_2$, for which

$$\langle J_1\rangle = \frac{1}{2}\left(A_1^2 - A_2^2\right), \qquad \langle J_2\rangle = A_1A_2\cos(\Delta\phi), \qquad \langle J_3\rangle = A_1A_2\sin(\Delta\phi), \tag{3}$$

where $\Delta\phi = \phi_1 - \phi_2$ is the relative phase, and brackets mean expectation value. This leads to the relation (for coherent states only) $|\langle\vec{J}\rangle|^2 = \langle J_1\rangle^2 + \langle J_2\rangle^2 + \langle J_3\rangle^2 = \langle\frac{1}{2}N\rangle^2$, that is, the vector $\langle\vec{J}\rangle$ points to the surface of a sphere of radius $\langle\frac{1}{2}N\rangle$. In classical optics this is the Poincaré sphere, and $\vec{S} = \langle\vec{J}\rangle$ is called the Stokes vector.[23] Each point on this sphere corresponds to a different classical state of elliptical polarization, as illustrated in Fig.1. The angles specifying the direction of the Stokes vector in the abstract space are $\Delta\phi$, which is the relative phase between the modes, and γ, which we term the "relative-amplitude parameter," since it satisfies $\tan(\gamma/2) = A_2/A_1$. By varying γ and $\Delta\phi$ one can fully sweep $\vec{S}$ over the sphere. An important physical point is that the overall phase of the pair of modes (which will be taken to be ϕ_1) does not physically affect the polarization of the light. This is reflected in the fact that the Stokes operators do not depend on the overall phase.

The usual "Stokes parameters" are the components of $\vec{S}$. For completely polarized light (an optical coherent state) they satisfy $S_1^2+S_2^2+S_3^2 = \langle\frac{1}{2}N\rangle^2$. Classically, partially polarized light can be characterized by a distribution of points on the Poincaré sphere, and then[24] $S_1^2+S_2^2+S_3^2 < \langle\frac{1}{2}N\rangle^2$. The degree of polarization is defined generally to be

$$D = \frac{S_1^2+S_2^2+S_3^2}{\langle\frac{1}{2}N\rangle^2} = \frac{\langle J_1\rangle^2 + \langle J_2\rangle^2 + \langle J_3\rangle^2}{\langle\frac{1}{2}N\rangle^2} \le 1. \tag{4}$$

For completely unpolarized light, $D = 0$.

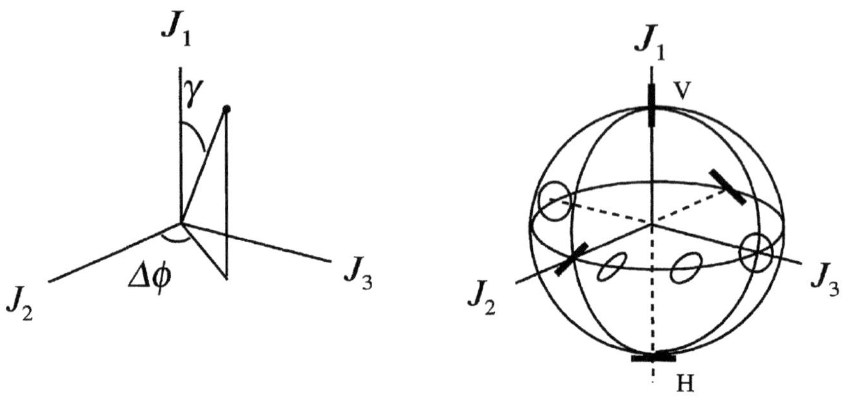

Figure 1. Coordinates and angles-$\Delta\phi$; γ; Polarization ellipses on Poincaré sphere.

THE QUANTUM POLARIZATION OF LIGHT

We wish to go beyond this characterization which uses only mean values, and consider fluctuations of the J_i, the components of $\vec{J}$. This leads to our proposal for a general definition of the quantum polarization of light. The variances of the J_i can easily be shown to be given by

$$\mathrm{var}(J_i) = \frac{1}{4}\langle N\rangle + \langle: (\Delta J_i)^2 :\rangle \qquad (i = 1, 2, 3), \tag{5}$$

where $\Delta J_i \equiv J_i - \langle J_i\rangle$ and : : indicates normal operator ordering of the a's and $a^\dagger$'s. As in the generalized Einstein relation,[21] the first term on the rhs of Eq.(5) represents particle (shot) noise while the second term is called "wave noise."

For a two-mode optical coherent state the wave noise is zero, leaving equal (shot-noise) variances for each component of $\vec{J}$. This state can be pictured in the Poincaré J_1, J_2, J_3 space as a gaussian ball with equal radii in each direction. The fact that the length of $\vec{J}$ fluctuates in a coherent state means that the state occupies at least several Poincaré spheres. A squeezed-polarization state has been defined to be one for which $\mathrm{var}(J_i) < \frac{1}{4}\langle N\rangle$ for at least one component.[15] These can be produced using second- or third-order optical nonlinearities.[14, 15]

Consider all of the moments of the components of $\vec{J}$ (e.g., $\langle (J_i)^k\rangle$), or most generally $\langle: (J_1)^p(J_2)^r(J_3)^s :_{arb}\rangle$, where here the dots indicate an arbitrary operator ordering. To know these moments it is sufficient to know all moments of the form $\langle (a_1^\dagger)^a(a_1)^b(a_2^\dagger)^c(a_2)^d\rangle$, where $a + c = b + d$, that is, there are equal numbers of raising and lowering operators, as seen from Eq.(2). It is simple to show that to know these moments it is sufficient to know all matrix elements (in Fock basis) of the density matrix of the type

$${}_1\langle n_1|{}_2\langle n_2|\rho|n_1'\rangle_1|n_2'\rangle_2 \qquad (\text{restricted to } n_1 + n_2 = n_1' + n_2'), \tag{6}$$

where the subscripts label the two modes. We call this subset of the full density matrix the "polarization sector," since it fully specifies the fluctuations of the components of $\vec{J}$, as well as their cross correlations $\langle: (J_1)^p(J_2)^r(J_3)^s :_{arb}\rangle$.

We can better understand the meaning of the polarization sector by representing the Fock basis in terms of the simultaneous eigenstates $|j, m\rangle$ of J^2, J_1. These satisfy

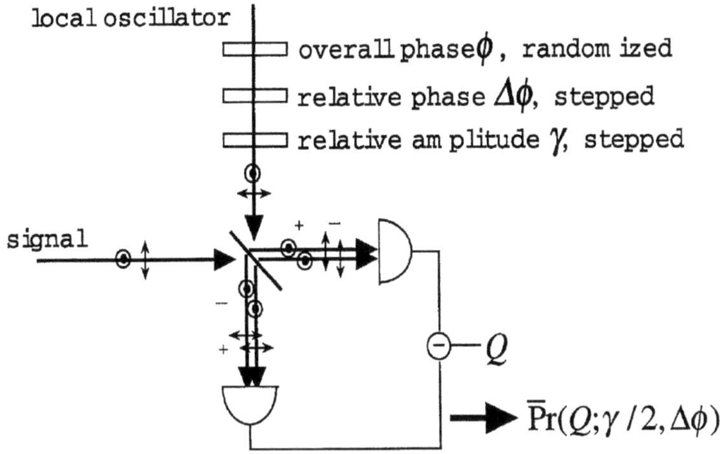

Figure 2. Schematic of dual-polarization balanced homodyne apparatus.

$J^2|j,m\rangle = j(j+1)|j,m\rangle$ and $J_1|j,m\rangle = m|j,m\rangle$, where[18] $j = \frac{1}{2}(n_1 + n_2)$ and $m = \frac{1}{2}(n_1 - n_2)$. Then the polarization sector [Eq.(6)] can be expressed as

$$\langle j,m|\rho|j,m'\rangle \qquad (\text{all } j,m,m'). \tag{7}$$

The polarization sector of the density matrix describes coherence between m and m' states on each individual j-sphere, but not between different j-spheres.

This means that calculating averages of any function of $\vec{J}$ involves two kinds of sums: quantum expectations involving sums over m, m' nondiagonal elements within a j-block; and an average over different values of j, representing total energy.[25] Because it allows one to calculate any average involving the Stokes operators, we refer to the block-diagonal polarization sector $\langle j,m|\rho|j,m'\rangle$ as specifying the "quantum polarization state of light."

MEASURING THE QUANTUM POLARIZATION STATE OF LIGHT

We wish to measure all of the moments of the J_i, in order to characterize polarization at the quantum level. For example, both thermal light and two-mode squeezed vacuum light have a degree of polarization [Eq.(4)] equal to zero. But higher-order moments will distinguish them.[14, 26]

We show here that the moments of the J_i, as well as the full polarization sector $\langle j,m|\rho|j,m'\rangle$ of the density matrix $\langle j,m|\rho|j',m'\rangle$, can be measured without measuring the full density matrix. We show that this can be accomplished using dual-mode balanced homodyne detection with a randomized relative phase between the local oscillator (LO) field and the signal field. This method is especially useful in cases that the signal field is from a source which has no external reference phase, such as a typical laser.

The apparatus is sketched in Fig.2. The signal beam, comprised of two linear-polarization components, enters one port of a (nonpolarizing) beamsplitter, which is 50% reflecting and transmitting for both s and p polarizations. The LO, also comprised of two polarizations, enters the other input port. Two phase shifters alter independently the phases of the two LO modes in such a way that the overall phase ϕ is randomized, while the relative phase $\Delta\phi$ is systematically stepped by known increments. The LO field is given by

$$\vec{E}_{LO}^{(+)}(z,t) = A_{LO}\left\{\cos(\gamma/2)\vec{u}_1\exp[-i\phi] + \sin(\gamma/2)\vec{u}_2\exp[-i(\phi-\Delta\phi)]\right\}\exp\left[-i(\omega t - kz)\right], \quad (8)$$

where A_{LO} and γ together set the amplitudes of the two LO modes. The photodetectors integrate the photon numbers incident during a chosen time interval. After subtraction and scaling, the quantity that is measured on a single trial is the "combined quadrature amplitude," whose operator representation is[13, 27]

$$Q(\gamma/2,\phi,\Delta\phi) = \cos(\gamma/2)q_1(\phi) + \sin(\gamma/2)q_2(\phi-\Delta\phi), \quad (9)$$

where the familiar single-mode quadrature operators are $q_i(\theta) = a_i\exp(-i\theta) + a_i^\dagger\exp(i\theta)$, for $i = 1, 2$.

The measurement procedure is: (i) Set the polarization state of the LO, via γ and $\Delta\phi$; (ii) While randomizing ϕ, measure $Q(\gamma/2,\phi,\Delta\phi)$ repeatedly to build up a histogram $\overline{P}(Q;\gamma/2,\Delta\phi)$ of outcomes Q; (iii) Process the histograms to yield the moments of the J_i or the polarization sector of the density matrix.

The reconstruction of the polarization sector is a sub-case of the reconstruction of the entire density matrix, which is achieved by[27]

$$_1\langle n_1|_2\langle n_2|\rho|n_1'\rangle_1|n_2'\rangle_2 = \frac{1}{2}\int_{-\infty}^{\infty}dQ\int_0^{\pi}d\gamma\int_0^{\pi}d\Delta\phi\int_0^{2\pi}d\phi P(Q;\gamma/2,\Delta\phi,\phi)R_{n_1n_1'}^{n_2n_2'}(Q;\gamma/2,\Delta\phi,\phi), \quad (10)$$

where $P(Q;\gamma/2,\Delta\phi,\phi)$ is the ϕ-dependent distribution (histogram) that would be measured if one did *not* randomize overall phase ϕ. The sampling function is[27]

$$R_{n_1n_1'}^{n_2n_2'}(Q;\gamma/2,\Delta\phi,\phi) = r_{n_1n_1'}^{n_2n_2'}(Q;\gamma/2)\exp\left[i(n_1+n_2-n_1'-n_2')\phi - i(n_2-n_2')\Delta\phi\right], \quad (11)$$

where $r_{n_1n_1'}^{n_2n_2'}(Q;\gamma/2)$ is a known function. In the polarization sector we have $n_1+n_2-n_1'-n_2' = 0$, so $R_{n_1n_1'}^{n_2n_2'}$ does not depend on ϕ. This allows Eq.(10) to be simplified, using the $|j,m\rangle$ representation, to

$$\langle j,m|\rho|j,m'\rangle = \pi\int_{-\infty}^{\infty}dQ\int_0^{\pi}d\gamma\int_0^{\pi}d\Delta\phi\overline{P}(Q;\gamma/2,\Delta\phi)r_{mm'}^{j}(Q;\gamma/2)\exp\left[i(m-m')\Delta\phi\right], \quad (12)$$

where $\overline{P}(Q;\gamma/2,\Delta\phi) = (1/2\pi)\int_0^{2\pi}d\phi P(Q;\gamma/2,\Delta\phi,\phi)$ is the ϕ-averaged distribution. Note that the γ and $\Delta\phi$ integrals in Eq.(12) represent sweeping around the full Poincaré sphere for a given j; this allows a tomographic reconstruction of the polarization sector.

To reconstruct all moments of the J_i we need to reconstruct all of the moments $\langle(a_1^\dagger)^a(a_1)^b(a_2^\dagger)^c(a_2)^d\rangle$ with $a+c = b+d$. These are obtained from the measured histograms using[4, 13]

$$\langle(a_1^\dagger)^a(a_1)^b(a_2^\dagger)^c(a_2)^d\rangle = \sum_{l=0}^{l_{max}}\sum_{i=0}^{l}\tilde{C}_{l,i}\langle Q^l\exp\left[i(c-d)\Delta\phi\right]\rangle\left.\right|_{\gamma_{l,i}}, \quad (13)$$

where the $\tilde{C}_{l,i}$ are known coefficients which depend on a, b, c, and d.[4] The bracket symbol on the rhs means an average over the Q data as well as a sum over $\Delta\phi$, for a fixed value of γ, that is,

$$\langle F(Q,\Delta\phi)\rangle \mid_\gamma \equiv (1/2\pi)\int_{-\infty}^{\infty} dQ \int_0^{2\pi} d\Delta\phi \overline{P}(Q;\gamma/2,\Delta\phi)F(Q,\Delta\phi), \tag{14}$$

for any function $F(Q,\Delta\phi)$. Useful examples of this sampling formula are –
Single-mode photon number means and fluctuations:[12]

$$\langle a_1^\dagger a_1\rangle = \langle Q^2\rangle \mid_{\gamma=0} - 1/2; \qquad \langle a_2^\dagger a_2\rangle = \langle Q^2\rangle \mid_{\gamma=\pi} - 1/2, \tag{15}$$

$$\langle (a_1^\dagger a_1)^2\rangle = \frac{2}{3}\langle Q^4\rangle \mid_{\gamma=0} - \langle Q^2\rangle \mid_{\gamma=0}; \quad \langle (a_2^\dagger a_2)^2\rangle = (\text{same with } \gamma=\pi), \tag{16}$$

Number cross correlations:[12]

$$\begin{aligned}\langle (a_1^\dagger a_1)(a_2^\dagger a_2)\rangle = & \frac{2}{3}\langle Q^4\rangle \mid_{\gamma=\pi/2} - \frac{1}{6}\langle Q^4\rangle \mid_{\gamma=0} - \frac{1}{6}\langle Q^4\rangle \mid_{\gamma=\pi} \\ & -\frac{1}{2}\langle Q^2\rangle \mid_{\gamma=0} - \frac{1}{2}\langle Q^2\rangle \mid_{\gamma=\pi} + \frac{1}{4},\end{aligned} \tag{17}$$

Field cross correlations:

$$\langle a_1^\dagger a_2\rangle = 2\langle Q^2 \exp[-i\Delta\phi]\rangle \mid_{\gamma=\pi/2}, \tag{18}$$

$$\langle (a_1^\dagger)^2 (a_2)^2\rangle = \frac{8}{3}\langle Q^4 \exp[-i2\Delta\phi]\rangle \mid_{\gamma=\pi/2}. \tag{19}$$

By combining these moments we can construct the means $\langle J_i\rangle$ and variances $\mathrm{var}(J_i)$ of each component of $\vec{J}$. Similar calculation also yields the cross correlations $\langle : (J_1)^p (J_2)^r (J_3)^s :_{arb}\rangle$. Note that only three values of relative amplitudes γ are needed to reconstruct $\langle J_i\rangle$ and $\mathrm{var}(J_i)$, while a continuous set of γ values is needed to reconstruct $\langle j,m|\rho|j,m'\rangle$.

EXPERIMENTS

We are presently working to implement dual-polarization balanced homodyne detection. Preliminary results for a two-mode coherent state in the ps regime are encouraging, but further reductions of technical noise are needed to reach the shot-noise limit.

Acknowledgments.

This work is supported by the National Science Foundation.

REFERENCES

1. For an elementary review see M. G. Raymer, Measuring the Quantum Mechanical Wave Function, *Contemp. Physics* 38:343 (1997).
2. J. Ashburn, R. Cline, P. van der Burgt, W. Westerveld, and J. Risley, Experimentally determined density matrices for H(n=3) formed in H+-He collisions from 20 to 100 keV, *Phys. Rev. A* 41:2407 (1990).
3. U. Leonhardt and M. Munroe, Number of phases required to determine a quantum state in optical homodyne tomography, *Phys. Rev. A* 54:3682 (1996).
4. D. F. McAlister and M. G. Raymer, Correlation and joint density matrix of two spatial-temporal modes from balanced-homodyne sampling, *J. Mod. Opt.* 44:2359 (1997).
5. D. T. Smithey, M. Beck, M. G. Raymer, and A Faridani, Measurement of the Wigner distribution and the density matrix of a light mode using optical homodyne tomography: application to squeezed states and the vacuum, *Phys. Rev. Lett.* 70:1244 (1993).
6. Special Issue of the J. Mod. Opt. on "Quantum State Preparation and Measurement," eds. W. Schleich and M. G. Raymer (vol. 44, 1997).
7. U. Leonhardt, "Measuring the Quantum State of Light," Cambridge Univ. Press, Cambridge (1997).
8. D. -G. Welsch, W. Vogel, and T. Opatrny, Homodyne detection and quantum state reconstruction, (to appear in *Progress in Optics*).
9. E. Joos, Decoherence through interaction with the environment, *in:* "Decoherence And The Appearance Of A Classical World In Quantum Theory," D. Giulini et al., eds., Springer, Berlin (1996), pg. 35.
10. W. Wootters, Local accessibility of quantum states, *in:* "Complexity, Entropy, And The Physics Of Information," Wojciech H. Zurek, ed., Addison-Wesley, Redwood City (1990).
11. M. G. Raymer, D. T. Smithey, M. Beck, M. Anderson, and D. F. McAlister, Measurement of the Wigner function in quantum optics, *Proceedings of the Third Annual International Wigner Symposium* Sep. (1993).
12. D. F. McAlister and M. G. Raymer, Ultrafast photon-number correlations from dual pulse, phase-averaged homodyne detection, *Phys. Rev. A* 55:R1609 (1997).
13. T. Opatrny, D. -G. Welsch, and W. Vogel, Multi-mode density matrices of light via amplitude and phase control, *Opt. Commun.* 134:112 (1997); Homodyne detection for measuring internal quantum correlations of optical pulses, *Phys. Rev. A* 55:1416 (1996).
14. V. P. Karasev and A. V. Masalov, Unpolarized light states in quantum optics, *Opt. Spectrosc.* 74:551 (1993).
15. V. N. Beskrovnyi and A. S. Chirkin, Polarization-squeezed light generation in a second order nonlinear medium, *in:* "Quantum Communication, Computing, and Measurement," O. Hirota, A. S. Holevo, and C. M. Caves, eds., Plenum, New York (1997), pg. 483; Light source with nonclassical polarization state based on an optical frequency doubler, *Quantum Semiclass. Opt.* 10:263 (1998).
16. P. Grangier, R. E. Slusher, B. Yurke, and A. LaPorta, Squeezed light enhanced polarization interferometer, *Phys. Rev. Lett.* 59:2153 (1987).
17. J. M. Jauch and F. Rohrlich, "The Theory of Photons and Electrons," Addison-Wesley, Reading (1955).
18. B. Yurke, S. L. McCall, and J. R. Klauder, SU(2) and SU(1,1) interferometers, *Phys. Rev. A* 33:4033 (1986).
19. J. A. Cina, Phase-controlled optical pulses and the adiabatic electronic sign change, *Phys. Rev. Lett.* 66:1146 (1991).
20. K. Wodkiewicz and J. H. Eberly, Coherent states, squeezed fluctuations, and the SU(2) and SU(1,1) groups in quantum optics applications, *J. Opt. Soc. Am.* B 2:458 (1985).
21. For a review see L. Mandel and E. Wolf, "Optical Coherence and Quantum Optics," Cambridge University Press, Cambridge (1995), Chap. 6.
22. Not to be confused with a coherent state of angular momentumm, as discussed in F. T. Arrechi, E. Courtens, R. Gilmore, and H. Thomas, Atomic coherent states in quantum optics, *Phys. Rev. A* 6:2211 (1972).
23. For a review see G. B. Malykin, Use of the Poincaré sphere in polarization optics and classical and quantum mechanics. Review, *Radiophysics and Quantum Electronics* 40:175 (1997).
24. R. Azzam and N. Bashara, "Ellipsometry and Polarized Light," Oxford, north-Holland, Amsterdam (1977).
25. We wish to thank S. Barnett, D. Pegg, and W. Shleich for helpful discussions on this point.

26. J. Lehner, U. Leonhardt, and H. Paul, Unpolarized light: classical and quantum states, *Phys. Rev. A* 53:2727 (1996).
27. M. G. Raymer, D. F. McAlister and U. Leonhardt, Two-mode quantum-optical state measurement: Sampling the joint density matrix, *Phys. Rev. A* 54:2397 (1996).

MEASUREMENT OF JOINT PHOTON-NUMBER DISTRIBUTION OF A TWIN-BEAM STATE BY MEANS OF OPTICAL HOMODYNE TOMOGRAPHY

Michael Vasilyev, Sang-Kyung Choi, Prem Kumar, and G. M. D'Ariano

Department of Electrical and Computer Engineering,
Northwestern University, Evanston, Illinois 60208-3118

We report the first measurement of the joint photon-number probability distribution for a two-mode quantum state created by a nondegenerate optical parametric amplifier. The measured distribution exhibits inherently quantum correlations between the signal and idler photon numbers, whereas the marginal distributions are thermal as expected for parametric fluorescence.

INTRODUCTION

The method of optical homodyne tomography (OHT), developed in the last several years, has become an important tool in studying the quantum properties of light.[1-3] Its reconstruction algorithm has evolved from a computationally unstable inverse-Radon transform to a powerful direct sampling approach.[4] The latter allows one to directly determine the density-matrix elements ρ_{nm} of the quantum state in the Fock representation by averaging the so-called pattern functions

$$F_{nm}(X_\phi, \phi) = f_{nm}(X_\phi) \exp[i(n-m)\phi] \tag{1}$$

over the experimental quadrature outcomes X_ϕ of the balanced-homodyne detector and over the local-oscillator (LO) phases ϕ. This procedure is greatly simplified if one is interested only in the diagonal elements of the density matrix, which give the photon-number distribution of the quantum state.[2] Since the pattern functions for the diagonal elements ρ_{nn} are independent of ϕ, they can be averaged over quadrature outcomes that are taken at random LO phases, thus simplifying the experiment by eliminating the need for phase locking.

For the measurement of photon-number distributions, the direct sampling approach represents a very powerful alternative to the direct detection method. It allows one to use fast, high-quantum-efficiency *p-i-n* photodiodes instead of the slower and less efficient avalanche photodetectors. The issue of high quantum efficiency becomes extremely important in studies of the states of light possessing inherently quantum features. These features, for example, the recently-observed oscillations in the photon-number distribution of squeezed vacuum,[3] wash out very rapidly with degradation of the quantum efficiency from unity.

Quantum Communication, Computing, and Measurement 2
Edited by Kumar et al., Kluwer Academic / Plenum Publishers, New York, 2000.

In this paper, we report the first measurement of the joint photon-number distribution of a two-mode state—the twin-beam state emerging from a nondegenerate optical parametric amplifier (NOPA)—with inherently quantum features. This state is of great interest to researchers owing to the quantum correlation that is imposed onto the two modes by the parametric interaction. In a NOPA, one pump photon simultaneously produces a pair of parametrically down-converted photons that belong to two different (signal and idler) modes. This corresponds to the creation of the following two-mode state:[5]

$$|\Psi\rangle = \frac{1}{\sqrt{\bar{n}+1}} \sum_{n=0}^{\infty} \left(\frac{\bar{n}}{\bar{n}+1}\right)^{n/2} |n,n\rangle \,, \tag{2}$$

where $\bar{n} = g - 1$ is the average number of photons in each mode with g being the gain of the parametric amplifier. The total photon number in the two modes is always even because of the pairwise nature of the photon creation process. Hence, its probability distribution exhibits even-odd oscillations similar to those for a squeezed-vacuum state.[3] Although the photons in the signal and idler modes are perfectly correlated, their statistics in each mode alone is thermal, yielding the following joint probability distribution:

$$P(n,m) \equiv |\langle n,m|\Psi\rangle|^2 = \frac{\delta_{nm}}{\bar{n}+1} \left(\frac{\bar{n}}{\bar{n}+1}\right)^{n} \,. \tag{3}$$

The distribution (3) has zero probabilities everywhere except along the main diagonal, as shown in Fig. 1(left).

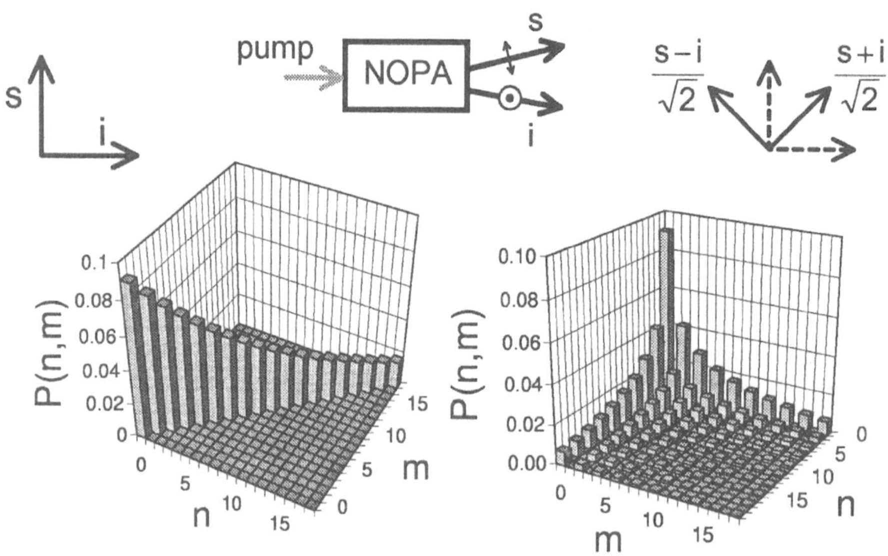

Figure 1. Theoretical plots of the joint photon-number distributions for the signal and idler modes (left), and the +45° and −45°-polarized modes (right). The average number of photons in each mode is 10.

In a polarization non-degenerate parametric amplifier, the signal and idler modes are orthogonally polarized. Such an amplifier can be shown to be equivalent to two independent degenerate parametric amplifiers for modes that are polarized at ±45° with respect to the signal polarization. Each of these modes is in a squeezed-vacuum state, and Figure 1(right) shows their joint distribution

$$P(n,m) = \begin{cases} 0 & \text{for } n = 2k+1 \text{ or } m = 2l+1 \ , \\ \frac{(2k-1)!!\,(2l-1)!!}{2^{k+l}\,k!\,l!}\,\frac{1}{\bar{n}+1}\left(\frac{\bar{n}}{\bar{n}+1}\right)^{k+l} & \text{for } n = 2k,\ m = 2l \ , \end{cases} \qquad (4)$$

where $\bar{n}$ is the same as that in Eq. (2). It is easy to see that the distribution (4) factorizes into a product of marginal distributions.

EXPERIMENTAL SETUP

The schematic of our experimental setup is shown in Fig. 2. The NOPA, consisting of a 5-mm-long KTP crystal, is pumped by the second harmonic of a Q-switched and mode-locked Nd:YAG laser. The laser produces a 100-MHz train of 120-ps pulses at 1064 nm (85 ps for the second harmonic at 532 nm) with a 205-ns Q-switch envelope (145 ns for 532 nm) having a 1 kHz repetition rate. The 1064-nm orthogonally-polarized

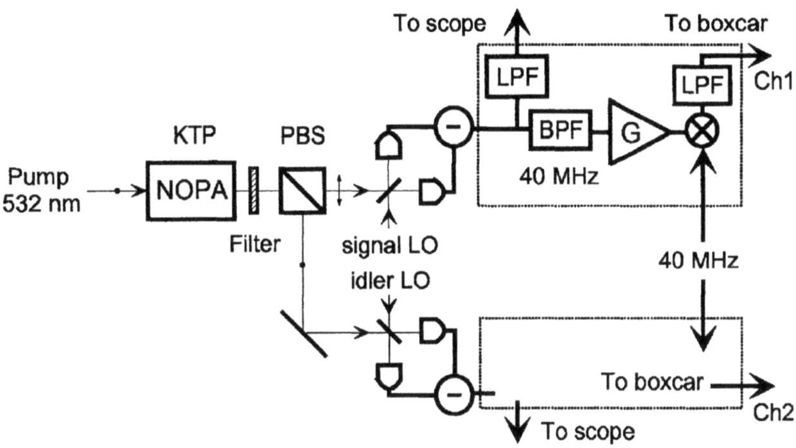

Figure 2. A schematic of the experimental setup.

twin beams emitted by the NOPA are detected separately by two balanced-homodyne detection setups using two independent LO's.[6] Low- and high-frequency parts of the photocurrents are separated. The peak amplitudes of the 5-MHz low-pass-filtered photocurrents in the signal and the idler arms are monitored by the oscilloscope. A 10-MHz-wide band of radio frequencies near $\Omega/2\pi = 40$ MHz is selected in each arm by means of a bandpass filter and amplified with a low-noise amplifier. The amplified noise photocurrent is then down-converted to the near-DC region by use of an RF mixer and sampled by the boxcar integrator (signal arm—by channel 1, idler arm—by channel 2). The outputs of the boxcar channels are a measure of the quadrature amplitudes X_ϕ^S and X_ψ^I, where ϕ and ψ are the phases of the signal and idler LO's, respectively.[7] The joint photon-number probability distribution $P(n,m)$ of the twin beams is then obtained by averaging the two-mode pattern function $f_{nn}(X_\phi^S)f_{mm}(X_\psi^I)$ over the quadrature samples X_ϕ^S and X_ψ^I and over the independently randomly-varying LO phases ϕ and ψ.[5]

The overall quantum efficiencies for the twin beams, including propagation losses, homodyne efficiencies, and detector efficiencies, are estimated to be $\eta_S = 0.38$ and $\eta_I = 0.35$, respectively. The most significant factor in deviation of our quantum efficiency from unity is the efficiency of homodyne overlap, which is approximately 0.45

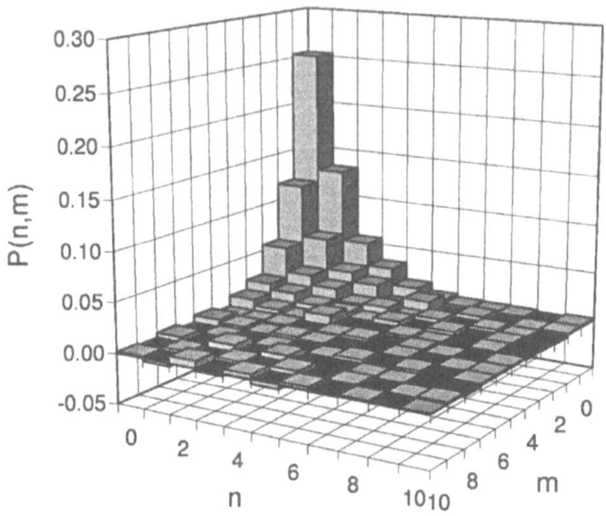

Figure 3. Joint photon-number distribution of the twin-beam state reconstructed from 320,000 homodyne data samples. The mean photon numbers are: $\bar{n}$=1.1 for the signal beam and $\bar{m}$=1.0 for the idler beam.

in our present setup. This is because the traveling-wave pulsed amplification process causes a change in the spatio-temporal profile of the amplified field, resulting in its mismatch with the LO. We have shown previously that the photon statistics of each of the parametric beams alone can be measured very efficiently using a self-generated matched LO (self-homodyne tomography).[8] This approach, however, can not be used for a joint measurement of the twin beams because it renders the NOPA phase-sensitive, which would distort the self-generated LO, making its matching with the mode of the quantum state of interest inefficient. The observation of the even-odd oscillations of the total photon number is extremely sensitive to the quantum efficiency and requires the overall efficiency to be at least 0.8.[5] The correlation between the signal and the idler beams, on the other hand, is less sensitive to η_S and η_I, and could be observed with our current setup.

MEASUREMENT RESULTS

The measured joint photon-number distribution is shown in Fig. 3. The detected mean photon numbers are $\bar{n} = 1.1$ for the signal beam and $\bar{m} = 1.0$ for the idler beam. The non-unity quantum efficiency results in spreading of the distribution around the main diagonal $n = m$, where a delta-like correlation is expected for $\eta_S = \eta_I = 1$. The marginal distributions for the signal or the idler beam alone are shown in Fig. 4. They indicate good agreement with the theoretically-predicted thermal distributions for the same mean photon numbers.

To show the quantum character of the measured distribution, we used it to find

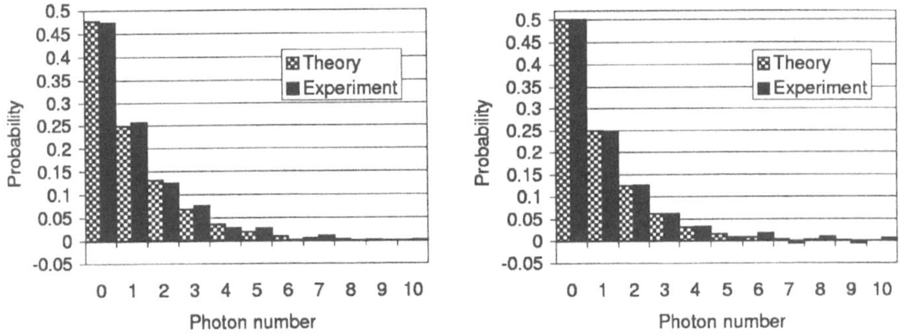

Figure 4. Marginal distributions for the signal (left) and the idler (right) beams reconstructed from the same data as the distribution in Fig. 3. Theoretical distributions for the same mean photon numbers $\bar{n}$=1.1 (left) and $\bar{m}$=1.0 (right) are also shown.

the photon-number correlation $d(n)$ between the two modes:

$$d(n) = \sum_{k=\max(-n,0)}^{N} P(k, k+n) \ , \tag{5}$$

which is shown in Fig. 5. The number $N = 18$ is determined by the size of our reconstructed distribution. In the limit of $N \to \infty$, $d(n)$ is the probability of finding the difference between the signal and the idler photon numbers to be n. In the case of ideal homodyne detection, $d(n)$ is expected to be the Kronecker δ_{n0}. For a non-unity quantum efficiency, however, the correlation $d(n)$ is no longer a delta-function; it spreads around $n = 0$. Nevertheless, it remains narrower than the correlation function for two independent coherent-state beams having the same mean photon numbers as the twin beams, which represents the standard quantum limit of correlation between classical states. We compare the photon-number correlation observed in our measurement with the standard quantum limit. The twin-beam correlation function shown in Fig. 5 is narrower than the coherent-state correlation function, which indicates the inherently quantum character of the twin-beam state. For substantial deviations $|n|$ from the main diagonal, $d(n)$ becomes randomly oscillating due to the increasingly large contribution of the statistical errors in the measurement of the joint distribution.

CONCLUSIONS

We have measured, for the first time, the joint photon-number distribution of a twin-beam state. While the marginal distributions of the signal or the idler beam alone are thermal, the photon-number correlation between them shows an inherently quantum character. An improvement of the homodyne efficiency will allow us to also observe the oscillations of the total photon number. A detailed study is underway to overcome the mode-matching limitations of the current setup.

The work was supported in part by the U.S. Office of Naval Research.

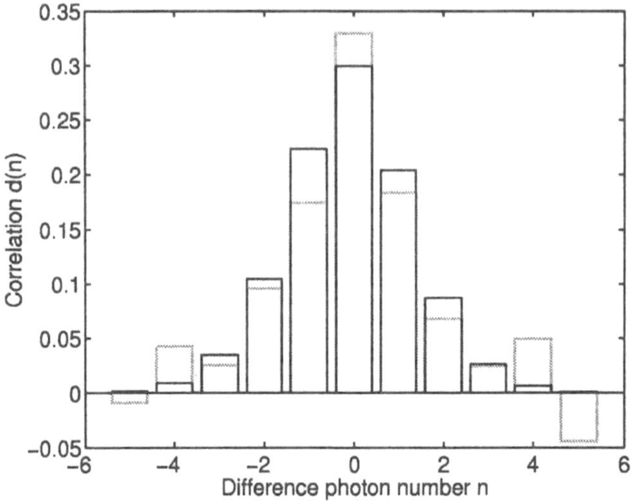

Figure 5. Correlation function $d(n)$ of the signal and idler photon numbers (solid gray line), obtained from the joint distribution of Fig. 3. Here $N = 18$. Solid black line — correlation of two independent coherent states with mean photon numbers $\bar{n}$=1.1 and $\bar{m}$=1.0.

REFERENCES

1. D. T. Smithey, M. Beck, M. G. Raymer, and A. Faridani, "Measurement of the Wigner distribution and the density matrix of a light mode using optical homodyne tomography: application to squeezed states and the vacuum," Phys. Rev. Lett. **70**, 1244 (1993); G. Breitenbach, T. Müller, S. F. Pereira, J.-Ph. Poizat, S. Schiller, and J. Mlynek, "Squeezed vacuum from a monolithic optical parametric oscillator," J. Opt. Soc. Am. B **12**, 2304 (1995).
2. M. Munroe, D. Boggavarapu, M. E. Anderson, and M. G. Raymer, "Photon-number statistics from the phase-averaged quadrature-field distribution: theory and ultrafast measurement," Phys. Rev. A **52**, R924 (1995).
3. S. Schiller, G. Breitenbach, S. F. Pereira, T. Müller, and J. Mlynek, "Quantum statistics of the squeezed vacuum by measurement of the density matrix in the number state representation," Phys. Rev. Lett. **77** 2933 (1996).
4. G. M. D'Ariano, C. Macchiavello, and M. G. A. Paris, "Detection of the density matrix through optical homodyne tomography without filtered back projection," Phys. Rev. A **50**, 4298 (1994); G. M. D'Ariano, U. Leonhardt, and H. Paul, "Tomographic detection of the density matrix," Phys. Rev. A **52**, R1801 (1995); U. Leonhardt, M. Munroe, T. Kiss, Th. Richter, and M. G. Raymer, "Sampling of photon statistics and density matrix using homodyne detection," Opt. Comm. **127**, 144 (1996).
5. G. M. D'Ariano, M. Vasilyev, and P. Kumar, "Self-homodyne tomography of a twin-beam state," Phys. Rev. A **58**, 636 (1998).
6. Z. Y. Ou, S. F. Pereira, H. J. Kimble, and K. C. Peng, "Realization of the Einstein-Podolsky-Rosen paradox for continuous variables," Phys. Rev. Lett. **68**, 3663 (1992).
7. O. Aytür and P. Kumar, "Pulsed squeezed-light measurement: a new technique," Optics Letters **15**, 390 (1990).
8. M. V. Vasilyev, S.-K. Choi, P. Kumar, and G. M. D'Ariano, "Investigation of the photon statistics of parametric fluorescence in a traveling-wave parametric amplifier by means of self-homodyne tomography," Opt. Lett. **23**, 1393 (1998).

HOMODYNING BELL'S INEQUALITY

G. M. D'Ariano[1,2], L. Maccone[2], M. F. Sacchi[2] and A. Garuccio[3]

[1]Department of Electrical and Computer Engineering,
Northwestern University, Evanston, IL 60208

[2]Theoretical Quantum Optics Group
Università degli Studi di Pavia and INFM — Unità di Pavia
via A. Bassi 6, I-27100 PAVIA, Italy

[3]Dipartimento Interateneo di Fisica
Università degli Studi di Bari and
INFN, Sezione D, I-70126 BARI, Italy

We propose to test Bell's inequality through homodyne tomography. The experimental apparatus is mainly composed of a nondegenerate optical parametric amplifier and four photodiodes. The experimental data are gathered through a self–homodyne scheme and are processed by quantum tomography. Our procedure does not introduce supplementary hypotheses (fair sampling assumption), since we do not need to collect auxiliary data to normalize probabilities. Moreover in the proposed experiment, the minimum value of detector quantum efficiency η is virtually reduced down to $\eta = .5$ (in practice to $\eta \sim 65 \div 70\%$). The feasibility of the experiment is shown by some numerical results based on Monte–Carlo simulations.

INTRODUCTION

In a typical experiment to test Bell's inequality that involve optical radiation the source emits a pair of correlated photons and two detectors separately check the presence of the two photons after polarizing filters at angles α and β, respectively. Alternatively, one can use four photodetectors with polarizing beam splitters in front, with the advantage of checking through coincidence counts that photons come in pairs. Let $p_{\alpha,\beta}$ be the joint probability of finding one photon at each detector with polarization angle α and β, respectively. In terms of the correlation function

$$C(\alpha,\beta) = p_{\alpha,\beta} + p_{\bar{\alpha},\bar{\beta}} - p_{\bar{\alpha},\beta} - p_{\alpha,\bar{\beta}} \,, \tag{1}$$

Bell's inequality [1] writes as follows

$$B(\alpha,\beta,\alpha',\beta') \doteq |C(\alpha,\beta) - C(\alpha,\beta')| + |C(\alpha',\beta') + C(\alpha',\beta)| \leq 2 \,, \tag{2}$$

Quantum Communication, Computing, and Measurement 2
Edited by Kumar et al., Kluwer Academic / Plenum Publishers, New York, 2000.

$\bar{\alpha}$ and $\bar{\beta}$ denoting the polarization angles orthogonal to α and β respectively. In this paper we propose a new kind of test for Bell's inequality based on quantum homodyne tomography [2,3] (for a review see Ref. 4). In our setup the photodetectors are replaced by homodyne detectors, which allows measuring the joint probabilities $p_{\alpha,\beta}$ through the tomographic technique. The main advantage of the tomographic test is that linear photodiodes with available quantum efficiency η higher than 90% [5] can be used. On the other hand, the method works effectively even with η as low as 70%, without the need of a "fair sampling" assumption,[6,7] since all data are collected in a single experimental run. In the present case the customary homodyne technique would need many additional beam splitters and local oscillators (LO) that are coherent each other. As we will show, the setup is greatly simplified by using the recent self-homodyne technique.[8] Another advantage of self-homodyning is the more efficient signal-LO mode-matching, with improved overall quantum efficiency.

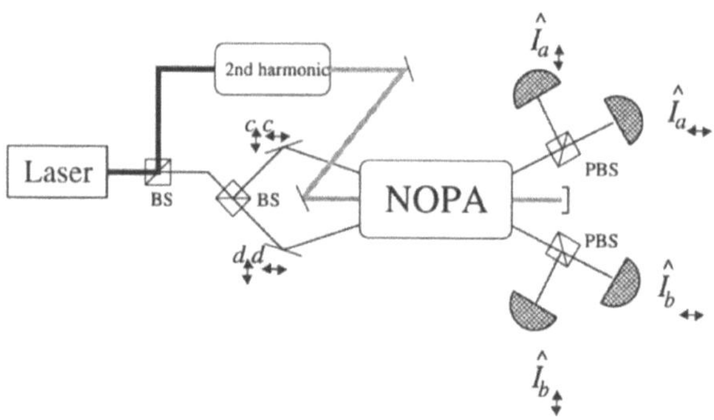

Figure 1. Self-homodyne scheme for the tomographic test of Bell's inequality. PBS and BS denote 'polarizing beam splitter' and 'conventional beam splitter' respectively. Input radiation modes $a_{\updownarrow}$, $b_{\leftrightarrow}$, $a_{\leftrightarrow}$ and $b_{\updownarrow}$ are in the vacuum state, while modes $c_{\updownarrow}$, $c_{\leftrightarrow}$, $d_{\updownarrow}$, $d_{\leftrightarrow}$ (at laser frequency ω_0) are in a coherent state. At the output of the nondegenerate parametric amplifier (NOPA) the four photocurrents $\hat{I}$ at radiofrequency Ω are detected, yielding the value of the quadratures of the field modes $a_{\updownarrow}$, $b_{\leftrightarrow}$, $a_{\leftrightarrow}$ and $b_{\updownarrow}$. The outcome quadratures are then used to reconstruct the probabilities of interest through quantum tomography.

THE EXPERIMENTAL SET-UP

The apparatus for generating the correlated beams is a nonlinear crystal, cut for Type-II phase–matching, acting as a nondegenerate optical parametric amplifier (NOPA). The NOPA is injected with excited coherent states (see Fig. 1) in modes $c_{\leftrightarrow}, c_{\updownarrow}, d_{\leftrightarrow}, d_{\updownarrow}$ all with equal intensities and at the same frequency ω_0, c and d denoting mode operators for the two different wave-vector directions, and $\updownarrow$ and $\leftrightarrow$ representing vertical and horizontal polarization, respectively. The NOPA is pumped at the second harmonic $2\omega_0$. At the output of the amplifier four photodetectors separately measure the intensities $\hat{I}_{a_\updownarrow}$, $\hat{I}_{b_\leftrightarrow}$, $\hat{I}_{a_\leftrightarrow}$, $\hat{I}_{b_\updownarrow}$ of the mutual orthogonal polarization components of the fields propagating at different wave vectors. The output photocurrent is narrow-band filtered at radiofrequency $\Omega \ll \omega_0$. In the process of direct detection, the central modes $c_{\updownarrow,\leftrightarrow}$ and $d_{\updownarrow,\leftrightarrow}$ beat with $\omega_0 \pm \Omega$ sidebands, thus playing the role of the LO in homodyne detectors. The four photocurrents $\hat{I}_{a_\updownarrow}$, $\hat{I}_{b_\leftrightarrow}$, $\hat{I}_{a_\leftrightarrow}$, $\hat{I}_{b_\updownarrow}$ yield the value of the quadratures of the four modes [8]

$$s_\pi = \frac{1}{\sqrt{2}}\Big(s_\pi(+) + s_\pi(-)\Big)\,, \qquad s = \{a, b\} \qquad \pi = \{\leftrightarrow, \updownarrow\}\,, \tag{3}$$

where $a_\pi(\pm)$ and $b_\pi(\pm)$ denote the sideband modes at frequency $\omega_0 \pm \Omega$, which are in the vacuum state at the input of the NOPA. The quadrature is defined by the operator $\hat{x}_\phi \doteq \frac{1}{2}(ae^{-i\phi} + a^\dagger e^{i\phi})$, where ϕ is the relative phase between the signal and the LO. The value of the quadratures is used as input data for the tomographic measurement of the correlation function $C(\alpha, \beta)$. The direction of polarizers (α, β) in the experimental setup does not need to be varied over different trials, because, as we will show in the following, such direction can be changed tomographically.

Under condition of phase-matching and in the parametric approximation, the spontaneous down-conversion at the NOPA is described in terms of the field modes in Eq. (3) by the unitary evolution operator

$$\hat{U}(\xi) = \exp\left[\xi\left(a^\dagger_\updownarrow b^\dagger_\leftrightarrow + e^{i\varphi} a^\dagger_\leftrightarrow b^\dagger_\updownarrow\right) - \text{h. c.}\right]\,, \tag{4}$$

where $\xi = \chi\gamma L/c$ is a rescaled interaction time evaluated by the nonlinear susceptibility χ of the medium, the crystal length L, the pump amplitude γ and the speed c of light in the medium, whereas φ represents the relative phase between the orthogonal polarization components of the pump field. Correspondingly, the state at the output of the NOPA writes as follows

$$|\psi\rangle = (1 - |\Lambda|^2) \sum_{n=0}^{\infty} \sum_{m=0}^{\infty} \Lambda^{n+m} e^{i\varphi m} |n, n, m, m\rangle \equiv |\psi_{1,2}\rangle \otimes |\psi_{3,4}\rangle\,, \tag{5}$$

where $\Lambda = \xi \tanh|\xi|/|\xi|$ and $|i, l, m, n\rangle$ denotes the common eigenvector of the number operators of modes $a_\updownarrow$, $b_\leftrightarrow$, $a_\leftrightarrow$, $b_\updownarrow$, with eigenvalues i, l, m and n, respectively. The average photon number *per* mode is given by $N = |\Lambda|^2/(1 - |\Lambda|^2)$. The four-mode state vector in Eq. (5) factorizes into a couple of twin beams $|\psi_{1,2}\rangle$ and $|\psi_{3,4}\rangle$, each one entangling a couple of spatially divergent modes ($a_\updownarrow$, $b_\leftrightarrow$ and $a_\leftrightarrow$, $b_\updownarrow$, respectively).

HOMODYNING BELL'S INEQUALITY

The conventional experiments that involve a two-photon polarization-entangled state generated by spontaneous down-conversion, consider the four-mode entangled state which can be obtained by keeping only the first-order terms of the sums in Eq. (5), and by ignoring the vacuum component, as usually only intensity correlations are measured. Here, on the contrary, we measure the joint probabilities on the state (5) to test Bell's inequality through homodyne tomography, which yields the value of $B(\alpha, \beta, \alpha', \beta')$ in Eq. (2). The tomographic technique is a kind of universal detector, which can measure any observable $\hat{O}$ of the field, by averaging a suitable "pattern" function $\mathcal{R}[\hat{O}](x, \phi)$ over homodyne data x at random phase ϕ. The "pattern" function is obtained through the trace relation [9]

$$\mathcal{R}[\hat{O}](x, \phi) = \text{Tr}\left[\hat{O} K_\eta(x - \hat{x}_\phi)\right]\,, \tag{6}$$

where $K_\eta(x)$ is a distribution given in Ref. 10. For factorized many-mode operators $\hat{O} = \hat{O}_1 \otimes \hat{O}_2 \otimes \ldots \otimes \hat{O}_n$ the pattern function is just the product of those corresponding to each single-mode operator $\hat{O}_1, \ldots, \hat{O}_n$ labelled by variables $(x_1, \phi_1), \ldots, (x_n, \phi_n)$. By linearity the pattern function is extended to generic many-mode operators.

Now we consider the observables that are involved in testing Bell's inequality (2). Let us denote by $p_{\alpha,\beta}(i,l,m,n)$ the probability of having i,l,m,n photons in modes $a_\updownarrow, b_\leftrightarrow, a_\leftrightarrow, b_\updownarrow$ for the "rotated" state

$$|\psi\rangle_{\alpha,\beta} \equiv \hat{U}_{1,3}(\alpha)\hat{U}_{2,4}(\beta)|\psi\rangle\ , \tag{7}$$

$\hat{U}_{1,3}(\alpha)$ and $\hat{U}_{2,4}(\beta)$ being the unitary operators

$$\hat{U}_{1,3}(\alpha) = \exp\left[\alpha\left(a_\updownarrow^\dagger a_\leftrightarrow - a_\updownarrow a_\leftrightarrow^\dagger\right)\right]\ , \tag{8}$$

$$\hat{U}_{2,4}(\beta) = \exp\left[\beta\left(b_\updownarrow^\dagger b_\leftrightarrow - b_\updownarrow b_\leftrightarrow^\dagger\right)\right]\ . \tag{9}$$

The probabilities in Eq. (1) can be written as $p_{\alpha,\beta}=p_{\alpha,\beta}(1,1)$, $p_{\bar{\alpha},\bar{\beta}}=p_{\alpha,\beta}(0,0)$, $p_{\bar{\alpha},\beta}=p_{\alpha,\beta}(0,1)$, and $p_{\alpha,\bar{\beta}}=p_{\alpha,\beta}(1,0)$, with

$$p_{\alpha,\beta}(n,m) = \frac{p_{\alpha,\beta}(n,1-m,1-n,m)}{P(1,1)}\ , \qquad \{n,m=0,1\}\ . \tag{10}$$

The denominator $P(1,1)$ in Eq. (10) represents the absolute probability of having at the output of the NOPA one photon in modes $a_\updownarrow, a_\leftrightarrow$ and one photon in modes $b_\updownarrow, b_\leftrightarrow$, independently on the polarization, namely

$$P(1,1) = \sum_{n=0,1}\sum_{m=0,1} p_{\alpha,\beta}(n,1-m,1-n,m)\ . \tag{11}$$

In our procedure both the numerator and the denominator of Eq. (10) are measured in only one run, hence we do not need a fair sampling assumption, namely we do not have to collect auxiliary data to normalize probabilities. Moreover, since we can exploit quantum efficiencies as high as $\eta = 90\%$ and the tomographic pattern functions in Eq. (6) already take into account η, we do not need supplementary hypothesis for it.[7]

The observables that correspond to probabilities $p_{\alpha,\beta}(i,l,m,n)$ in Eqs. (10) and (11) are the projectors

$$|\,i,l,m,n\rangle_{\alpha,\beta}\ {}_{\alpha,\beta}\langle\,i,l,m,n| = \hat{U}_{1,3}^\dagger(\alpha)\,\hat{U}_{2,4}^\dagger(\beta)\,|\,i,l,m,n\rangle\langle\,i,l,m,n|\,\hat{U}_{2,4}(\beta)\,\hat{U}_{1,3}(\alpha)\ . \tag{12}$$

¿From Eqs. (6) and (10,11,12), one finds that $P(1,1)$ is measured through the following average $\mathcal{AV}$ of homodyne data

$$P(1,1) = \mathcal{AV}\left\{\left(K_1^1\,K_0^3 + K_0^1\,K_1^3\right)\left(K_1^2\,K_0^4 + K_0^2\,K_1^4\right)\right\}\ , \tag{13}$$

where K_n^j denotes the diagonal tomographic kernel function for mode j, namely

$$K_n^j \equiv \langle n|K_\eta(x-\hat{x}_{\phi_j})|n\rangle\ . \tag{14}$$

The kernel function for the numerator of Eq. (10) involves both the diagonal terms (14) and the following off-diagonal terms

$$K_+^j \equiv \langle 0|K_\eta(x-\hat{x}_{\phi_j})|1\rangle\ , \qquad K_-^j = (K_+^j)^*\ . \tag{15}$$

Finally, the expression for $C(\alpha,\beta)$ in Eq. (1) is given by

$$\begin{aligned} C(\alpha,\beta) = \mathcal{AV}\Big\{&[\,\cos\,(2\alpha)\left(K_1^1\,K_0^3 - K_0^1\,K_1^3\right) + \sin(2\alpha)\left(K_+^1\,K_-^3 + K_-^1\,K_+^3\right)] \\ &\times[\,\cos\,(2\beta)\left(K_0^2\,K_1^4 - K_1^2\,K_0^4\right) + \sin(2\beta)\left(K_+^2\,K_-^4 + K_-^2\,K_+^4\right)]\Big\}/P(1,1)\ . \end{aligned} \tag{16}$$

The statistical error for $B(\alpha,\beta,\alpha',\beta')$ in Eq. (2) can be obtained from the variance calculated upon dividing the set of data into large statistical blocks. However, $C(\alpha,\beta)$— and thus $B(\alpha,\beta,\alpha',\beta')$— are non linear averages (they are ratios of averages). Hence, since the nonlinearity of B introduces a systematical error which is vanishingly small for increasingly larger sets of data, the estimated mean value of B is obtained from the full set of data, instead of averaging the mean value of the blocks.

SOME NUMERICAL RESULTS

We now present some numerical results obtained from Monte–Carlo simulations of the proposed experiment. For the simulation we use the theoretical homodyne probability pertaining to the state $|\psi\rangle$ in Eq. (5).

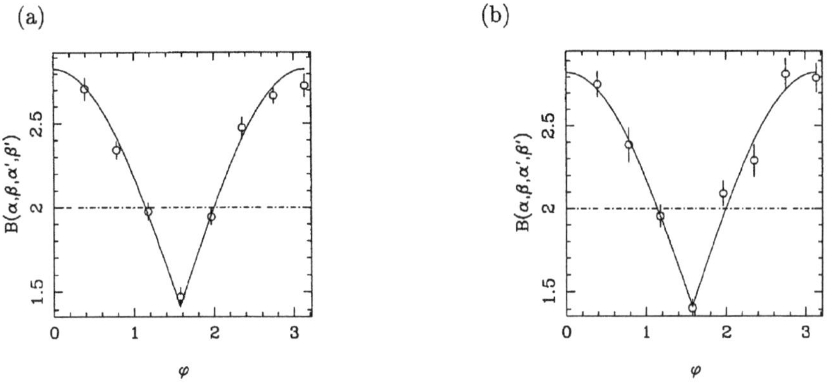

Figure 2. Plot of $B(\alpha,\beta,\alpha',\beta')$ *vs* the phase φ in the state of Eq. (5) with $N = 0.5$ average photon number *per* mode. The dot-dashed line and the full line represent the classical bound and the quantum theoretical value for B, respectively. The parameters of the simulation are: $\alpha = 0$; $\beta = \frac{3}{8}\pi$; $\alpha' = \frac{\pi}{4}$; $\beta' = \frac{\pi}{8}$. For (a) The quantum efficiency and the number of simulated homodyne data are $\eta = 85\%$ and $n = 10^7$ in (a), and $\eta = 75\%$ and $n = 10^8$ in (b).

In Fig. 2 we present the simulation results for B in Eq. (2) *vs* the phase φ in the state of Eq. (5). The full line represents the value of B evaluated through the quantum theoretical value $C(\alpha,\beta)$ given by

$$C(\alpha,\beta) = \cos\varphi \sin 2\alpha \sin 2\beta - \cos 2\alpha \cos 2\beta \,. \tag{17}$$

Quantum efficiency $\eta = 85\%$ has been used in Fig. 2(a), nonetheless notice that for $\varphi = \pi$ (corresponding to a maximum violation with respect to the classical bound 2), the obtained value is over 10 σ distant from the bound. By increasing the number of homodyne data, it is possible to obtain good results also for lower quantum efficiency [see Fig. 2(b), wherein $\eta = 75\%$]. By increasing the number of data to 8×10^8, the value of $B(0, \frac{3}{8}\pi, \frac{\pi}{4}, \frac{\pi}{8}) = 2.834 \pm 0.268$ has been obtained for $\varphi = \pi$, $N = 0.5$ average photon number *per* mode and η as low as 65%. This result is to be compared with the quantum theoretical value of $2\sqrt{2}$. Fig. 3 shows the results of B *vs* different values of the mean photon number *per* mode N for the state in Eq. (5). Notice that the statistical errors are nearly independent on N.

For an order of magnitude of the data acquisition rate in a real experiment, one can consider that in a typical setup with a NOPA pumped by a 2nd harmonic of a Q-switched mode-locked Nd:YAG the pulse repetition rate is 80 MHz, with a 7 ps pulse duration, the effective number of data depending on the speed of the boxcar integrator.

In conclusion we have proposed a test of Bell's inequality, based on self–homodyne tomography. The rather simple experimental apparatus is mainly composed of a NOPA crystal and four photodiodes. The experimental data are collected through a self–homodyne scheme and processed by quantum tomography.

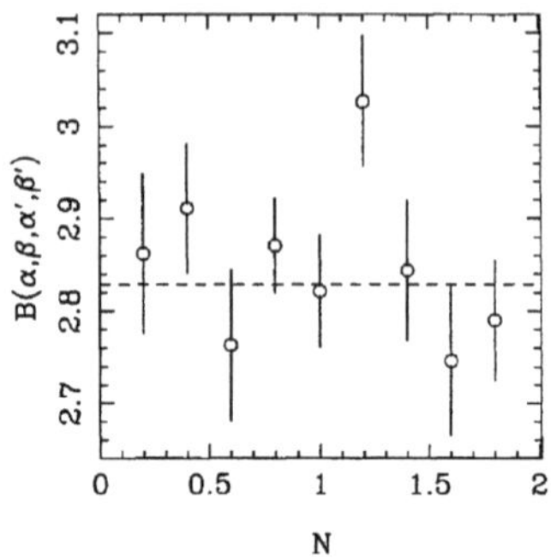

Figure 3. Plot of $B(\alpha,\beta,\alpha',\beta')$ with $\alpha = 0$; $\beta = \frac{3}{8}\pi$; $\alpha' = \frac{\pi}{4}$; $\beta' = \frac{\pi}{8}$ *vs* different values of the mean photon number *per* mode N in the state of Eq. (5). For each simulated experiment, the number of homodyne data is $n = 10^7$ and the quantum efficiency is $\eta = 85\%$. The dotted line represents the quantum theoretical value $2\sqrt{2}$.

No supplementary hypotheses are introduced, a quantum efficiency η as high as 90% is currently available, and, anyway, η as low as 70% is tolerated for $10^7 \div 10^8$ experimental data. We have presented some numerical results based on Monte–Carlo simulations that confirm the feasibility of the experiment, showing violations of Bell's inequality for over 10 σ with detector quantum efficiency $\eta = 85\%$.

Acknowledgments

The Quantum Optics Group of Pavia acknowledges the INFM for financial support (PRA-CAT97).

REFERENCES

1. J. S. Bell, Physics **1**, 195 (1965).
2. D. T. Smithey, M. Beck, M. G. Raymer and A. Faridani, Phys. Rev. Lett. **70**, 1244 (1993).
3. G. Breitenbach, S. Schiller and J. Mlynek, Nature **387**, 471 (1997).
4. G. M. D'Ariano, *Measuring quantum states*, in *Quantum Optics and the Spectroscopy of Solids*, ed. by T. Hakioğlu and A.S. Shumovsky, Kluwer Academic Publishers (1997), p. 175.
5. C. Kim and P. Kumar, Phys. Rev. Lett. **73**, 1605 (1994).
6. J. F. Clauser and M. A. Horne, Phys. Rev. D **10**, 256 (1974).
7. L. De Caro and A. Garuccio, Phys. Rev. A **54**, 174 (1996).
8. G. M. D'Ariano, M. Vasilyev and P. Kumar, Phys. Rev. A **58**, 636 (1998).
9. G. M. D'Ariano, in "Quantum Communication, Computing, and Measurement", ed. by O. Hirota, A. S. Holevo and C. M. Caves, Plenum Publishing (New York and London 1997), p. 253.
10. G. M. D'Ariano, U. Leonhardt and H. Paul, Phys. Rev. A **52**, R1801 (1995).

HYDRODYNAMICAL QUANTUM STATE RECONSTRUCTION

Lars M. Johansen

Faculty of Engineering
Buskerud College
N-3601 Kongsberg
Norway

A new and general quantum state reconstruction method is proposed. It can be applied to reconstruct the state of a particle in an arbitrary, time-dependent potential. The state is reconstructed by measuring the position probability distribution at $n+1$ different time values. This yields an n-th order Taylor polynomial expansion of the density matrix in the off-diagonal variable.

INTRODUCTION

In recent years a rapid development has taken place in the field of quantum state reconstruction. Various new methods have been developed, and some of them have been applied in real experiments. The first method of practical utility, optical homodyne tomography[1, 2], has been used to reconstruct the quantum state of coherent and squeezed states in quantum optics[3, 4]. The direct sampling method[5, 6] has been used to reconstruct the motional state of ions in traps, measuring for the first time a Wigner function with negative values[7]. A number of other state reconstruction methods have also been developed[8]. The method of homodyne tomography has been generalized to the reconstruction of material particles in arbitrary, time independent potentials[9, 10, 11]. Recently, the method has also been generalized to a restricted class of time-dependent potentials[12].

In homodyne tomography, the quantum state is in effect retrieved by rotating the Wigner distribution so that the marginal distributions can be obtained for various complementary angles. When the distributions have been measured for a complete set of angles, the data can be inverted to yield the complete Wigner distribution (or the density matrix). In the direct sampling method, a single point in phase space is retrieved for a single setting of the experimental parameters.

Recently, a new state reconstruction method was found[13]. It can be used to retrieve the state of a material particle in an arbitrary, time dependent potential. By measuring the position probability distribution for $n + 1$ different time values, an n-th order Taylor polynomial of the density matrix in the off-diagonal variable is obtained. Thus, this method allows reconstruction of parts of the density matrix. By increasing the number of time values used, increasing portions of the density matrix are retrieved.

Quantum Communication, Computing, and Measurement 2
Edited by Kumar et al., Kluwer Academic / Plenum Publishers, New York, 2000.

The method can also be applied to quantum optical states. Interestingly, essentially the same method can be applied both for quantum systems and for classical statistical mechanical systems. Thus, one might employ this method to retrieve, e.g., the state of a particle in a pure state as well as a classical, non-equilibrium gas of non-interacting particles. The basic difference between the two cases is that the reconstruction algorithm for quantum systems involves some quantum correction terms if the potential is of order higher than 2 in position.

THE HYDRODYNAMICAL MOMENTS

The density matrix can be expressed in terms of various kinds of moments[14, 15, 16, 17]. In this paper, we shall employ an expansion in terms of local moments, i.e., moments depending on position. Assuming that the density matrix is an analytic function, we may expand the density matrix in the Taylor series[18, 19, 20]

$$\langle x+y|\,\hat{\rho}\,|x-y\rangle = \sum_{n=0}^{\infty} \frac{f_n(x,t)}{n!} \left(\frac{2iy}{\hbar}\right)^n . \tag{1}$$

We have here defined the moments[18]

$$f_n(x,t) = \int_{-\infty}^{\infty} dp\, p^n\, W(x,p,t). \tag{2}$$

We recognize the zeroth order moment f_0 as the position probability density, the first order moment f_1 can be shown to be equal to j/m, where j is the probability current density, the second order moment f_2 can be interpreted as being proportional to a kinetic energy density, and so on. Such local moments are frequently called hydrodynamical, although this name should not be taken too literally. Firstly, the moments f_n relate to an ensemble of single particles, whereas the common use of such terms relate to macroscopic assemblies of particles. Secondly, the moments f_n may exhibit nonclassical properties. Thus, for instance, the moment f_2 may take on negative values[21].

Madelung showed that the Schrödinger equation can be reformulated as two coupled equations for the probability density and the probability current density[22]. One of the equations was the conservation equation for probability, the other was a nonlinear equation resembling classical equations for fluid motion. Thus, for pure states the two first moments f_n can describe the state completely. Furthermore, it is known that a state in local thermal equilibrium is described by the first three hydrodynamical moments[23]. But for an arbitrary state which is neither pure nor in thermal equilibrium, an infinite number of moments may be required to describe the state. Fortunately, for practical state reconstruction purposes it will be shown that a finite number of moments is sufficient.

A RECURSIVE RECONSTRUCTION ALGORITHM

In order to reconstruct the quantum state, we must find a way of observing the moments f_n. Just as in classical statistical mechanics, the moments f_n are connected through an infinite set of coupled equations[19, 20]

$$\frac{\partial f_n}{\partial t} = -\frac{1}{\mu}\frac{\partial f_{n+1}}{\partial x} - \sum_{k=0}^{[(n-1)/2]} \left(\frac{\hbar}{2i}\right)^{2k} \binom{n}{2k+1} \frac{\partial^{2k+1} V}{\partial x^{2k+1}} f_{n-2k-1} \tag{3}$$

where μ is the particle mass, and where the potential V may be an arbitrary function of both position and time. By setting $n = 0$ here, we obtain the well known conservation equation for the probability density.

The reconstruction algorithm is found by performing a simple integral of this set of equations. We then find that

$$f_{n+1}(x,t) = -\mu\frac{\partial}{\partial t}\int_{-\infty}^{x} dx'\, f_n(x',t)$$
$$- \mu \sum_{k=0}^{[(n-1)/2]} \left(\frac{\hbar}{2i}\right)^{2k} \binom{n}{2k+1} \int_{-\infty}^{x} dx'\, \frac{\partial^{2k+1} V(x',t)}{\partial x'^{2k+1}} f_{n-2k-1}(x',t). \quad (4)$$

Note that the moment f_{n+1} is expressed in terms of lower order moments only. The lower order moments are integrated over position. Also, we need to know the time derivative of the moment f_n . Therefore, we must measure this moment for two time values. A simple argument shows[13] that the n-th order moment can be retrieved recursively by measuring the zeroth order moment for $n + 1$ different time values.

A NONRECURSIVE ALGORITHM

The recursive algorithm found in the previous section requires extensive numerical processing of the experimental data. In this section, we consider a very simple, nonrecursive algorithm for a free particle. It can be used to retrieve the moments without performing a numerical integration. To this end, consider the generalized function[24]

$$u_n(x) = \frac{1}{n!}x^n u(x), \quad n \geq 0, \quad (5)$$

where $u(x)$ is Heaviside's step-function. Assuming that a generalized function is always to be integrated over a proper test-function, we find that[24]

$$\frac{d^{n+1}u_n(x)}{dx^{n+1}} = \delta(x), \quad (6)$$

where $\delta(x)$ is Dirac's δ-function. We now define the moments

$$F_n(x,t) = \int dx'\, u_n(x-x')\, f_0(x',t). \quad (7)$$

We recognize F_0 as the usual cumulative distribution. An arbitrary moment F_n is observed as the average of the moment u_n, and it can thus be observed without performing any numerical integration. By performing some simple manipulations, we now find that[24]

$$f_n(x,t) = \mu^n \frac{\partial^n F_{n-1}(x,t)}{\partial t^n}, \quad n \geq 1. \quad (8)$$

We see that the hydrodynamical moment f_n simply is the n-th order derivative with respect to time of the average F_{n-1}. Thus, the moment f_n can be retrieved from an observation of the average value of u_{n-1} for $n + 1$ time values.

This nonrecursive algorithm can be generalized to harmonic oscillator potentials[24], and consequently also to quantum optics.

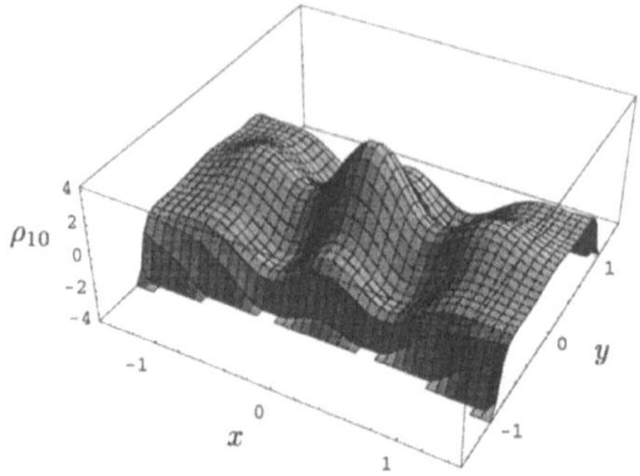

Figure 1. The 10-th order Taylor-polynomial, obtained from a measurement of f_0 at 11 time-values.

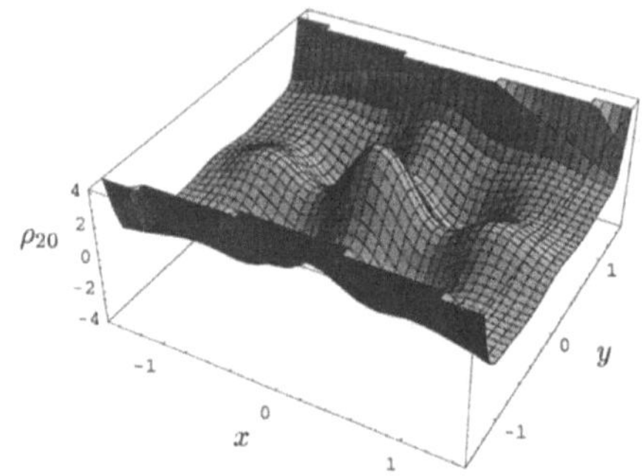

Figure 2. The 20-th order Taylor-polynomial, obtained from a measurement of f_0 at 21 time-values.

AN EXAMPLE STATE

We have just seen how the moments f_n can be retrieved experimentally. We now introduce the N-th order Taylor polynomial

$$\rho_N(x, y, t) = \sum_{n=0}^{N} \frac{f_n(x,t)}{n!} \left(\frac{2iy}{\hbar}\right)^n . \tag{9}$$

For a given N, this will be a good approximation of the density matrix up to a certain value of the off-diagonal variable y. This is illustrated in figures 1-3, where Taylor polynomials of different order of the density matrix corresponding to the pure superposition state

$$\psi(x) = e^{-[x/(2\sigma)]^2 + ik_0 x} + e^{-[x/(2\sigma)]^2 - ik_0 x} \tag{10}$$

have been plotted. As we see, the density matrix is retrieved for increasing portions of the off-diagonal variable by increasing the number of time values for which the probability density f_0 is observed.

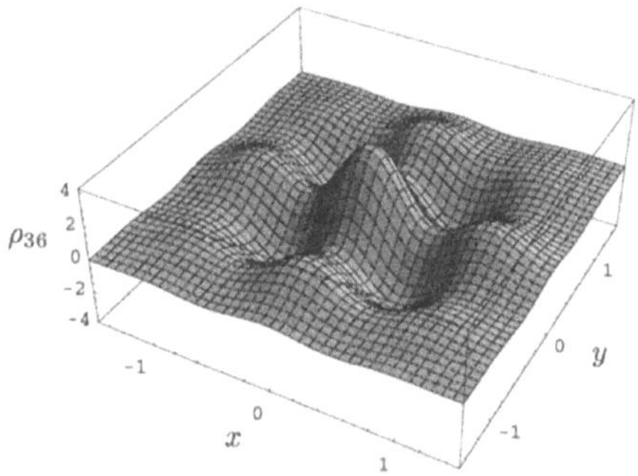

Figure 3. The 36-th order Taylor-polynomial, obtained from a measurement of f_0 at 37 time-values. It is basically identical to the exact density matrix within the chosen plotting region.

CONCLUSION

The method of hydrodynamical quantum state reconstruction can be performed on a particle in an arbitrary, time-dependent potential in one dimension. The reconstruction can be performed directly on the experimental data, and the same procedure can in principle be employed for an ensemble of quantum mechanical particles as well as for a macroscopic assembly of classical particles such as a gas. The general algorithm given here was recursive. Simple, nonrecursive algorithms have been found for a free particle and a harmonic oscillator.

ACKNOWLEDGEMENTS

I am grateful for useful discussions with Ulf Leonhardt and Stefan Weigert. Also, I would like to thank Leonhardt, Kiss and Bardroff for making available a preprint of the paper[12].

REFERENCES

1. J Bertrand and P. Bertrand, A tomographic approach to Wigner's function, *Found. Phys.* 17:397 (1987).
2. K. Vogel and H. Risken, Determination of quasiprobability distributions in terms of probability distributions for the rotated quadrature phase, *Phys. Rev. A* 40:2487(1989).
3. D. Smithey, M. Beck, M. Raymer, and A. Faridani, Measurement of the Wigner distribution and the density matrix of a light mode using optical homodyne tomography: Application to squeezed states and the vacuum, *Phys. Rev. Lett.* 70:1244 (1993).
4. G. Breitenbach, S. Schiller, and J. Mlynek, Measurement of the quantum states of squeezed light, *Nature* 387:471 (1997).
5. K. Banaszek and K. Wódkiewicz, Direct probing of quantum phase space by photon counting, *Phys. Rev. Lett.* 76:4344 (1996).
6. S. Wallentowitz and W. Vogel, Unbalanced homodyning for quantum state measurements, *Phys. Rev. A* 53:4528 (1996).
7. D. Leibfried, D. M. Meekhof, B. E. King, C. Monroe, W. M. Itano, and D. J. Wineland, Experimental determination of the motional quantum state of a trapped atom, *Phys. Rev. Lett.* 77:4281 (1996).
8. Special issue on "Quantum State Preparation and Measurement", *J. Mod. Opt.* 44:2021, No. 11/12 (1997).

9. M. G. Raymer, M. Beck, and D. F. McAlister, Complex wave-field reconstruction using phase-space tomography, *Phys. Rev. Lett.* 72:1137 (1994).
10. U. Leonhardt and M. G. Raymer, Observation of moving wave packets reveals their quantum state, *Phys. Rev. Lett.* 76:1985 (1996).
11. D. S. Krähmer and U. Leonhardt, State reconstruction of one-dimensional wave packets, *Appl. Phys. B* 65:725 (1997).
12. U. Leonhardt, T. Kiss, and P. J. Bardroff, State reconstruction of wave packets moving in time-dependent potentials and the existence of Wronskian pairs, submitted to *J. Phys. A.*
13. L. M. Johansen, Hydrodynamical quantum state reconstruction, *Phys. Rev. Lett.* 80:5461 (1998).
14. H. M. Nussenzweig, "Introduction to Quantum Optics". Gordon & Breach Science Publishers, London (1973).
15. W. Band and J. L. Park, Quantum state determination: Quorum for a particle in one dimension, *Am. J. Phys.* 47:188 (1979).
16. A. Wünsche, Reconstruction of operators from their normally ordered moments for a single boson mode, *Quantum Opt.* 2:453 (1990).
17. C. T. Lee, Moment problem for a density matrix and a biorthogonal set of operator bases, *Phys. Rev. A* 46:6097 (1992).
18. J. E. Moyal, Quantum mechanics as a statistical theory, *Proc. Cambridge Philos. Soc.* 45:99 (1949).
19. J. V. Lill, M. I. Haftel, and G. H. Herling, Semiclassical limits in quantum-transport theory, *Phys. Rev. A* 39:5832 (1989).
20. J. V. Lill, M. I. Haftel, and G. H. Herling, Mixed state quantum mechanics in hydrodynamical form, *J. Chem. Phys.* 90:4940 (1989).
21. L. M. Johansen, Nonclassical evolution of a free particle, *in* "Fifth International Conference on Squeezed States and Uncertainty Relations", D. S. Han, J. Janszky, Y. S. Kim, and V. I. Man'ko, eds., NASA Goddard Space Flight Center, Greenbelt, Maryland (1998).
22. E. Madelung, Quantentheorie in hydrodynamischer form *Z. Phys.* 40:322 (1926).
23. D. Hilbert, Begründung der kinetischen gastheorie, *Matematische Annalen*, 72:562 (1912).
24. L. M. Johansen, Nonrecursive hydrodynamical quantum state reconstruction, submitted to *Phys. Rev. A.*

QUANTUM TOMOGRAPHY: RENORMALIZATION OF INCOMPATIBLE OBSERVATIONS

Z. Hradil,[1,2] R. Myška,[3] J. Summhammer,[2] and H. Rauch[2]

[1]Department of Optics, Palacky University,
17. listopadu 50, 772 07 Olomouc, Czech Republic
[2]Atominstitut der Österreichischen Universitäten,
Stadionallee 2, A-1020, Austria
[3]Joint Laboratory of Optics of Palacký University
and Phys. Inst. Czech Acad. Sci.,
17. listopadu 50, 772 07 Olomouc, Czech Republic

Standard deterministic techniques for quantum state reconstruction, as for example optical homodyne tomography, photon chopping, unbalanced homodyning etc., are based on the deterministic inversion of measured data. Since the frequencies obtained in realistic experiments always differ from probabilities predicted by quantum theory due to fluctuations, imperfections and realistic restrictions, the algorithm of inversion cannot guarantee the positive definiteness of the reconstructed density matrix. Hence the estimation of the noises may appear as doubtful.

Quantum states may be successfully reconstructed within quantum and information theories using the maximum likelihood estimation. The question of deterministic schemes: "What quantum state is determined by that measurement?" is replaced by the formulation consistent with quantum theory: "What quantum state(s) seems to be most likely for that measurement?" Nonlinear equation for reconstructed state is formulated. An exact solution may be approached by subsequent iterations. Reconstruction is formulated as a problem of proper normalization of incompatible (nonorthogonal) measurements. The results obtained by this novel method may differ significantly from the standard predictions. Data are fitted better keeping the constraint of positive definiteness of reconstructed density matrix. However, this interpretation may enlarge uncertainty in prediction of quantum state in comparison with deterministic schemes, since, in general, there is a whole family of states which fit the measured data equally well. The novel technique is nonlinear and the reinterpretation of existing reconstruction schemes represents an advanced program.

INTRODUCTION

Quantum theory brings information about observable events on the most fundamental level currently available. The statistical nature of almost all quantum phe-

Quantum Communication, Computing, and Measurement 2
Edited by Kumar et al., Kluwer Academic / Plenum Publishers, New York, 2000.

nomena seems to be its characteristic feature. This intrinsic uncertainty cannot be considered in accordance with the "classical" experience as a lack of knowledge about the internal structure of the system, since this does not exhaust the richness of the quantum world. The intrinsic uncertainty is hidden in the wave function, the origin of which remains unknown and unexplained by quantum theory. The pragmatic interpretation of quantum theory concentrates on the observable aspects, which may be successfully addressed within existing techniques. Determination of quantum state on the basis of the performed measurement may be acknowledged as one of those topical problems.

Although the history of state reconstruction may be traced back to the early days of quantum mechanics to Pauli problem,[1] only quantum optics opened the new era. Theoretical prediction of Vogel and Risken[2] was closely followed by the experimental realization of the suggested algorithm by Smithey *et al.*[3] Since that time many improvements and new techniques have been proposed[4,5] and similar techniques are currently being used also in atomic physics as quantum endoscopy.[6] The potential gain of this treatment is tremendous. Provided that certain quantum measurement enables us to determine the wave function of the system, the statistics of any further possible measurement may be forecast. The quantum state reconstruction plays therefore the role of a universal measurement. and is considered nowadays as a standard technique.

Nevertheless, there are some potential problems associated with standard treatment. Available realistic measurements are always limited as far as the amount and accuracy of data is concerned. Consequently, any scheme for posterior estimation is affected by these imperfections. Particularly, the standard techniques based on inversion of quantum prediction do not preserve the semipositive definiteness of reconstructed density matrix, a necessary condition of quantum state definition. This may be accomplished using the statistical approach. There are several proposals based on various statistical concepts, including Bayesian approach,[8] Maximum Entropy[9] principle or the Maximum Likelihood estimation (MaxLik). They all release the relation between the observed data and the quantum state generating them. The MaxLik estimation will be focused in this contribution. Its formulation follows closely the motivation of standard treatments.

The technique of MaxLik estimation is widely used in various branches of technology and science.[10] It is usually used as a tool for fitting a few parameters which maximizes the likelihood under the given constraints. In quantum theory, MaxLik has already been used for quantum phase estimation[11,12] and recently, it has been adopted for estimation of more parameters as well.[13,14] Nevertheless, in all these applications additional constraints were crucial for successful application of this method. For example, this is done by restrictions put on the dimension of the problem,[13] or by additional assumptions about the nature of the noise.[14] Without these conditions the problem of MaxLik estimation is considered as intractable due to the multidimensional nonlinear optimization. Nevertheless, quantum theory of MaxLik estimation without any other additional assumptions is addressed in this contribution. As a result, this complex optimization can be interpreted as quantum prediction for renormalized projectors. Hence, the procedure of mathematical statistics may be interpreted purely in the language of quantum theory. This feature indicates another close and fundamental relation between geometry of Hilbert space and concepts of mathematical statistics.[15,16] Various issues of this novel treatment are explained in this contribution in the form of a dozen frequently asked questions concerning quantum tomography.

A DOZEN FREQUENTLY ASKED QUESTIONS

1. WHAT IS THE QUANTUM TOMOGRAPHY?

   Tomography is routinely used in medicine. A picture of body is obtained by a synthesis of many various X-ray projections of the object. The registration of rotated quadrature operator $\hat{x}_\theta = (\hat{a}e^{-i\theta} + \hat{a}^\dagger e^{i\theta})/2$ using homodyne detection represents an analogy to the absorption of X-rays in medicine. Complete probability $P(x_\theta, \theta)$ determines, for example, the Wigner function of quantum state by the Radon transformation[2]

$$W(\alpha_r, \alpha_i) = \frac{1}{4\pi^2} \int_{-\infty}^{\infty} \int_{-\infty}^{\infty} dx|\eta|d\eta \int_0^\pi d\theta P(x,\theta) e^{i\eta x - i\alpha_r \cos\theta - i\alpha_i \sin\theta}.$$

   Similarly, there are many other experimental techniques in quantum theory. That enables us to observe "various faces" of a quantum system: homodyne detection,[3,17,18] photon chopping,[19] quantum state measurement via heterodyne detection,[20] unbalanced homodyning,[21] direct probing by photon counting,[22] quantum endoscopy,[6] etc.

2. WHAT IS THE PRAGMATIC INTERPRETATION OF QUANTUM TOMOGRAPHY?

   Quantum tomography is nothing else than a theory of conditional measurements. Already performed experiments condition the future outcomes of different forthcoming measurements done on the same system.

3. WHAT IS THE COMMON MATHEMATICAL FORMULATION OF STANDARD APPROACHES?

   Standard approaches are based on the formal inversion of quantum prediction

$$\langle y_i|\hat{\rho}|y_i\rangle = f_i,$$

   where $|y_i\rangle$ represent the registered projectors and f_i are their counted frequencies. Particularly, in the case of tomography based on homodyne detection, the probability $P(x,\theta)$ is simply replaced by discretized frequencies $f(x_i, \theta_j)$.

4. ARE THERE ANY APPARENT FLAWS IN STANDARD APPROACHES?

   Yes. The reconstructed object is not a density matrix $\hat{\rho} \geq 0$ in general! Hence, some probabilities are predicted as negative. Standard approaches do not describe the statistical nature of quantum observations properly and do not distinguish sufficiently what can and what cannot be predicted from the given data.[7]

5. WHAT IS THE REMEDY AGAINST THE DIFFICULTIES OF STANDARD APPROACHES?

   An approach based on statistical interpretation. The detected data could be generated by nearly any quantum state, nevertheless, with various likelihoods. Likelihood is given by the product of probabilities for all independently counted outcomes (enumerated by index i here)

$$\mathcal{L} = \prod_i p_i^{n_i}$$

   provided that data enumerated by index i are registered n_i times, $\sum n_i = n$.

6. COULD THE NOVEL APPROACH TO QUANTUM TOMOGRAPHY BE DEMONSTRATED ON A SIMPLE STATISTICAL MODEL?

Yes, on the model of estimation of prior probabilities. Assume that data n_i are generated with some prior probabilities $p_i, \sum_i p_i = 1$, which should be found from the data. The likelihood for associating the data $\{n_i\}$ with some probabilities $\{p_i\}$ is given by the multinomial distribution

$$\mathcal{P}(\{p_i\}|\{n_i\}) = n! \prod_i \frac{1}{n_i!} p_i^{n_i}.$$

The most probable guess is given by frequencies $f_i = n_i/n$.

7. IS THE STATISTICAL APPROACH SOMEHOW RELATED TO ENTROPY?

Yes, to the relative entropy (Kullback–Leibler divergence). The logarithm of the ratio of likelihood functions
$\mathcal{P}(\{f_i\})/\mathcal{P}(\{p_i\})$ yields the Kullback–Leibler divergence

$$K(f_i/p_i) = n \sum_i f_i \log \frac{f_i}{p_i}.$$

This equals up to the sign to the relative entropy.[23]

8. WHAT IS THE MAXIMUM LIKELIHOOD ESTIMATION?

MaxLik estimation searches for the state which provides the largest likelihood for the given data. Equivalently, such states minimize the relative entropy for the given data. In the above mentioned example the MaxLik estimation of probabilities is given by frequencies f_i. However, there are some significant differences in quantum theory:

- The estimated probabilities depend on the same quantum state and are therefore not independent.
- The probabilities are not normalized to one since projectors need not be complete.

9. WHAT IS THE MATHEMATICAL FORMULATION OF MAXLIK ESTIMATION?

Probabilities are given in quantum theory as

$$p_i = \langle y_i|\hat{\rho}|y_i\rangle.$$

Maximization of likelihood on the class of possible states (density matrices) corresponds to the nonlinear operator equation

$$\hat{R}(\hat{\rho})\hat{\rho} = \hat{\rho}.$$

Unknown density matrix is assumed in its diagonal form

$$\hat{\rho} = \sum_k r_k |\phi_k\rangle\langle\phi_k|,$$

and operator $\hat{R}$ is given as

$$\hat{R} = \sum_i \frac{f_i}{\rho_{ii}} |y_i\rangle\langle y_i|,$$

$$\rho_{ii} = \sum_k r_k |\langle \phi_k | y_i \rangle|^2.$$

Reconstruction is done in the subspace where operator $\hat{R}$ equals to the identity operator.[24]

10. WHAT IS THE PHYSICAL INTERPRETATION OF MAXLIK RECONSTRUCTION?

MaxLik estimation may be interpreted as renormalization of incompatible (non-commuting) observables for which the synthesis of various projections is done. Assume the rescaling of the projectors as

$$|y_i\rangle\langle y_i| \rightarrow |y_i'\rangle\langle y_i'| = \frac{f_i}{\rho_{ii}} |y_i\rangle\langle y_i|.$$

The rescaled projectors satisfy the quantum prediction

$$\langle y_i' | \hat{\rho} | y_i' \rangle = f_i.$$

The operator $\sum_i |y_i'\rangle\langle y_i'|$ characterizes the overlapping of rays $|y_i'\rangle$. Hence, the MaxLik reconstruction reproduces the quantum prediction for suitable renormalized projectors. This is achieved on a subspace–field of view–enclosed by condition $\hat{R} = \hat{1}$.

11. WHAT ARE THE PHYSICAL CONSEQUENCES OF MAXLIK ESTIMATION?

MaxLik approach generalizes the standard treatment. Whenever the latter one has a solution on the manifold of density matrices, then this is the solution of the former one as well. Besides this, MaxLik estimation provides a whole family of extremum states not distinguishable by the given measurement. Averaging over this family enhances the uncertainty of state prediction. MaxLik reconstruction of diagonal elements of density matrix has been already done for homodyne detection with fluctuating phases.[25,26]

12. DOES THIS FIELD OF VIEW HAVE ANY SIMPLE INTERPRETATION?

Realistic data can never provide complete information about quantum state in infinite dimensional Hilbert space. Any prediction should be restricted to certain field of view. Assume a simple example of reconstruction of diagonal elements of density matrix via photon counting with ideal detector. Suppose n times repeated counting, always with zero registered photoelectrons. The "standard" prediction of quantum state reads $\hat{\rho} = |0\rangle\langle 0|$, where $|0\rangle$ denotes the vacuum state. Nevertheless this interpretation is not the only one. Assume an additional classical noise represented by a projector into the strong coherent state $\hat{\mathcal{N}}_\epsilon = |\frac{\alpha}{\sqrt{\epsilon}}\rangle_{coh}\langle\frac{\alpha}{\sqrt{\epsilon}}|$ appearing with the negligible probability ϵ. The state

$$\hat{\rho}_\epsilon = (1 - \epsilon)\hat{\rho} + \epsilon \hat{\mathcal{N}}_\epsilon$$

cannot be distinguished from the standard one for sufficiently small $\epsilon < 1/n$. This may appear as crucial for some observations. For example, the average numbers of particles differ significantly for both the states. Field of view is specified as the subspace where operator $\hat{R}$ reproduces the identity operator. This is evident in the case of orthogonal measurements but rather nontrivial in the case of nonorthogonal measurements.

Acknowledgments

This work was supported by the TMR–Network ERB–FMRX–CT96–0057, PECNO "Perfect Crystal Neutron Optics," East–West program of the Austrian Academy of Sciences, and by the grant of Czech Ministry of Education VS 96028.

REFERENCES

1. S. Weigert, *Phys. Rev. A* 45: 7688 (1992).
2. K. Vogel, H. Risken, *Phys. Rev. A* 40: 2847 (1989).
3. D. T. Smithey, M. Beck, M. G. Raymer, A. Faridani, *Phys. Rev. Lett.* 70: 1244 (1993).
4. U. Leonhardt, "Measuring of the Quantum State of Light," Cambridge Press, Cambridge, (1997).
5. D.–G. Welsch, W. Vogel, T. Opatrný, Homodyne detection and quantum state reconstruction, *to appear in* "Progress in Optics," E. Wolf, ed., Kluwer Publishing, Amsterdam (1999).
6. S. Wallentowitz, W. Vogel, *Phys. Rev. Lett.* 75: 2932 (1996); P. J. Bardoff, C. Leichtle, G. Schrade, W. P. Schleich, *Phys. Rev. Lett.* 77: 2198 (1996); D. Leibfried, D. M. Meekhof, B. E. King, C. Monroe, W. M. Itano, D.J. Wineland, *Phys. Rev. Lett.* 77: 4281 (1996).
7. Z. Hradil, *Phys. Rev. A* 55: R1561 (1997).
8. K. R. W. Jones, *Phys. Rev. A* 50: 3682 (1994).
9. V. Bužek, G. Adam, G. Drobný, *Phys. Rev. A* 54: 804 (1996).
10. R. Barlow, "Statistics," J. Wiley, Chichester (1989).
11. J. H. Shapiro, S. R. Shepard, N. C. Wong, *Phys. Rev. Lett.* 62: 2377 (1989).
12. A. S. Lane, S. L. Braunstein, and C. M. Caves, *Phys. Rev. A* 47: 1667 (1993).
13. D. Mogilevtsev, Z. Hradil, J. Peřina, *Rev. Mod. Opt.* 44: 2261 (1997).
    D. Mogilevtsev, Z. Hradil, J. Peřina, *Quantum Semiclass. Opt.* 10: 345 (1998).
14. S. M. Tan, *Rev. Mod. Opt.* 44: 2233 (1997).
15. W. K. Wooters, *Phys. Rev. D* 23: 357 (1981).
16. S. L. Braunstein, C. M. Caves, *Phys. Rev. Lett.* 72: 3439 (1994).
17. G. M. D'Ariano, U. Leonhardt, H. Paul, *Phys. Rev. A* 52: R1801 (1995).
18. S. Schiller, G. Breitenbach, S. F. Pereira, T. Müller, J. Mlynek, *Phys. Rev. Lett.* 77: 2933 (1996).
19. H. Paul, P. Törma, T. Kiss, I. Jex, *Phys. Rev. Lett.* 76: 2464 (1996).
20. M. G. A. Paris, *Phys. Rev. A* 53: 2658 (1996).
21. S. Wallentowitz, W. Vogel, *Phys. Rev. A* 53:4528 (1996).
22. K. Banaszek, K. Wódkiewicz, *Phys. Rev. Lett.* 76: 4344 (1996).
23. C. M. Caves, P. D. Drummond, *Rev. Mod. Phys.* 66: 481 (1994).
24. Z. Hradil, J. Summhammer, H. Rauch, not published.
25. K. Banaszek, *Phys. Rev. A* 57: 5013 (1998); K. Banaszek, *acta phys. slov.* 48: 195 (1998).
26. Z. Hradil, R. Myška, *Acta Phys. Slov.* 48: 199 (1998).

ACCURACY IN QUANTUM HOMODYNE TOMOGRAPHY

G.M. D'Ariano,[1,2] C. Macchiavello[2] and N. Sterpi[2]

[1] Department of Electrical and Computer Engineering,
Northwestern University, Evanston, IL 60208

[2]Dipartimento di Fisica "A. Volta"
and Istituto Nazionale di Fisica della Materia
via Bassi 6, 27100 Pavia, Italy

We study the statistical errors in homodyne tomography of radiation density matrix in the photon number representation. We give an asymptotic estimate for large matrix indexes at different values of the quantum efficiency η of homodyne detectors. We show that for fixed η the errors increase exponentially as functions of the matrix index.

INTRODUCTION

Optical homodyne tomography is by now a well assessed method to measure the quantum state of radiation. The density operator $\hat{\rho}$ is measured in some representation by averaging the so-called kernel functions (or pattern functions) over homodyne data; in particular, the matrix element between generic states $|\varphi\rangle$ and $|\psi\rangle$ is evaluated under the following integral [1, 2]

$$\langle\varphi|\hat{\rho}|\psi\rangle = \int_0^\pi \frac{d\phi}{\pi} \int_{-\infty}^{+\infty} dx\, p_\eta(x,\phi)\, \langle\varphi|K_\eta(x-\hat{x}_\phi)|\psi\rangle\,. \tag{1}$$

In Equation (1) $p_\eta(x,\phi)$ is the probability distribution of the field quadrature $\hat{x}_\phi = (a^\dagger e^{i\phi} + a e^{-i\phi})/2$ for overall efficiency η of the homodyne detector (ϕ is the phase of the field mode with respect to the local oscillator, $a^\dagger$ and a are the creation and annihilation operators of the mode). The kernel function $\langle\varphi|K_\eta(x-\hat{x}_\phi)|\psi\rangle$ is determined by the operator

$$K_\eta(x-\hat{x}_\phi) = \frac{1}{2}\mathrm{Re}\int_0^{+\infty} dk\, k\, e^{\frac{1-\eta}{8\eta}k^2}\, e^{ik(x-\hat{x}_\phi)}\,. \tag{2}$$

The behavior of the kernel function depends both on the particular chosen representation (i.e. $|\varphi\rangle$ and $|\psi\rangle$) and on the value of η. The boundedness of the averaged kernel in Equation (1) sets the validity limits of the tomographic reconstruction. In a previous work[1] the most used representations were considered, corresponding to number, coherent, squeezed, and quadrature states. The bounds for η were established, below which

Quantum Communication, Computing, and Measurement 2
Edited by Kumar et al., Kluwer Academic / Plenum Publishers, New York, 2000.

the matrix elements cannot be measured: the coherent and the Fock representations turned out to be the best choice, with bound $\eta = 0.5$.

It is not clear if the bound $\eta = 0.5$ is the minimum for any representation, and "exotic" representations having some lower bound would be very interesting for tomography. Here we just mention that also unusual representations, such as the multi-photon representation,[3] have the same bound $\eta = 0.5$, whereas for the eigenvectors of the squeeze operator[4] $\Lambda = -i/2(a^2 - a^{\dagger 2})$ the bound is $\eta = 1$, as for the eigenvectors of the quadrature. On the other hand, representations based on eigenvectors of polynomials in a and $a^\dagger$ with degree larger than 2 are not analytical on the Fock space.[5]

The aim of this paper is to analyse the statistical errors in the measurement of the density operator in the Fock representation. We give an *a priori* asymptotic estimate of the error $\sigma(n,m)$ for large indexes n and m of the matrix element $\langle n|\hat{\rho}|m\rangle$. We show that for $0.5 < \eta < 1$ the error increases exponentially as functions of the matrix index. For $\eta = 0.5$ the errors diverge. Although this estimate is obtained after neglecting the correlations between different matrix elements, the asymptotic formula compares favorably with exact numerical evaluations. In the end, we give an analytic estimate of such correlations.

ACCURACY OF THE MEASUREMENT

Recently the statistical errors of the tomographic measurement of the density matrix in the number representation were numerically evaluated[6] on the basis of Equation (1). In the same paper it was also shown that for unit overall efficiency η of the homodyne detector the errors on the diagonal of the matrix saturate to a fixed value, independent of the radiation state. Now, in order to calculate the errors analytically, we exploit the inversion of the Bernoulli convolution,[7] that gives

$$\langle n|\hat{\rho}|m\rangle = \sum_{j=0}^{\infty} C^{\eta}_{n,m}(j)\, \overline{f_{n+j,m+j}}\,, \tag{3}$$

where we defined

$$C^{\eta}_{n,m}(j) = \eta^{-\frac{1}{2}(n+m)} \left[\binom{n+j}{n} \binom{m+j}{m} \right]^{\frac{1}{2}} (1-\eta^{-1})^j \tag{4}$$

$$f_{n+j,m+j}(x,\phi) = \langle n+j|K_1(x-\hat{x}_\phi)|m+j\rangle\,, \tag{5}$$

and

$$\overline{f_{n+j,m+j}} = \int_0^{\pi} \frac{d\phi}{\pi} \int_{-\infty}^{+\infty} dx\, p(x,\phi;\hat{\rho}_\eta)\, f_{n+j,m+j}(x,\phi)\,. \tag{6}$$

In Equation (6) $p(x,\phi;\hat{\rho}_\eta)$ is the homodyne probability distribution corresponding to the "dressed" state $\hat{\rho}_\eta$, that is related to the "bare" state $\hat{\rho}$ by the Liouvillian transformation*

$$\hat{\rho}_\eta = \sum_{s=0}^{\infty} \frac{(\eta-1)^s}{s!} a^s \eta^{-\frac{a^\dagger a}{2}} \hat{\rho}\, \eta^{-\frac{a^\dagger a}{2}} (a^\dagger)^s\,. \tag{7}$$

An *a priori* estimation of the measurement error is obtained by calculating the variances of the real and the imaginary part of the series (3). The variance of the real

*A discussion about the measurement of dressed states by homodyne tomography for non unit quantum efficiency at detectors has been recently reported.[8]

part is

$$\sigma_R^2(n,m) = \sum_{j,l=0}^{\infty} C^{\eta}_{n,m}(j) C^{\eta}_{n,m}(l)$$
$$\times \left[\overline{\mathrm{Re}[f_{n+j,m+j}] \cdot \mathrm{Re}[f_{n+l,m+l}]} - \overline{\mathrm{Re}[f_{n+j,m+j}]} \cdot \overline{\mathrm{Re}[f_{n+l,m+l}]}\right] . \tag{8}$$

The variance of the imaginary part $\sigma_I^2(n,m)$ is defined in analogous way. For a number of measurements N, the experimental error on the density matrix element $\langle n|\hat{\rho}|m\rangle$ is given by $\sigma(n,m)/\sqrt{N}$, with

$$\sigma(n,m) = \sqrt{\sigma_R^2(n,m) + \sigma_I^2(n,m)} . \tag{9}$$

Upon neglecting the statistical correlations for $j \neq l$ in Eq. (8), one obtains

$$\sigma^2(n,m) \simeq \sum_{j=0}^{\infty} [C^{\eta}_{n,m}(j)]^2 \, \Delta^2_{n+j,m+j} , \tag{10}$$

where

$$\Delta^2_{n+j,m+j} = \overline{|f_{n+j,m+j}|^2} - |\overline{f_{n+j,m+j}}|^2 . \tag{11}$$

For large indexes n,m the quantity $\Delta_{n+j,m+j}$ saturates to the fixed value $\sqrt{2}$ for any j. In this case the series in Equation (10) can be summed and the error behaves as follows:

$$\sigma^2(n,m) \simeq 2\,\eta^{-(n+m)} \Phi[n+1, m+1; 1; (1-\eta^{-1})^2] , \tag{12}$$

where $\Phi(\alpha,\beta;\gamma;z)$ is the customary hypergeometric function. The estimate (12) is independent of the radiation state for sufficiently large n and m. Notice that the convergence radius of the hypergeometric series in (12) is $(1-\eta^{-1})^2 < 1$, that means $\eta > 0.5$. Therefore, the errors diverge for $\eta = 0.5$, as also previously shown.[9]

For the diagonal errors, we can use a very good asymptotic approximation of the hypergeometric function $\Phi[n+1, n+1; 1; (1-\eta^{-1})^2]$ leading, for $n \gg (2\eta-1)/(1-\eta)$, to the following asymptotic expression

$$\sigma^2(n,n) \simeq \frac{\eta^{3/2}}{\sqrt{\pi(1-\eta)}} e^{\frac{1}{4n}\frac{(2\eta-1)^2}{\eta(1-\eta)}} \left(\frac{1}{2\eta-1}\right)^{2n+1} \frac{1}{\sqrt{n}} . \tag{13}$$

The estimate (13) shows the exponential growth of the error $\sigma(n,n)$ versus n, with rate $-\ln(2\eta-1)$.

In Fig. 1 we report the asymptotic expressions for the diagonal errors obtained from Eqs. (12) and (13) for tomography of a coherent state, in comparison to the exact numerical evaluation obtained from Eq. (1). The growth rate versus the matrix index is quite correctly reproduced. The difference between the asymptotic errors and the actual ones is due to the statistical correlations between different matrix elements.

Now we estimate the correlations that contribute to the diagonal errors. Since for large index n we have $\overline{f_{n+j,n+j}} = 0$, the contribution of the correlations is

$$\lambda(n,n) = \sum_{\substack{j,l=0 \\ j\neq l}}^{\infty} C^{\eta}_{n,n}(j) C^{\eta}_{n,n}(l) \overline{f_{n+j,n+j} f_{n+l,n+l}} \tag{14}$$

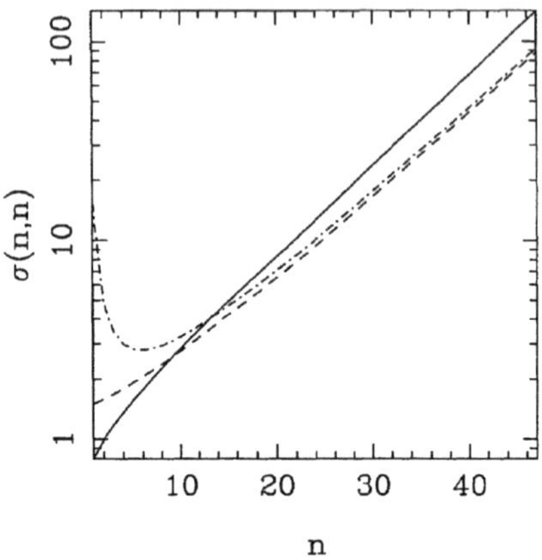

Figure 1. Diagonal statistical errors $\sigma(n,n)$ for a coherent state with 10 photons and $\eta = 0.95$ (log-linear scale). Solid line: actual errors; dashed line: estimate given by (12); dot–dashed line: estimate given by (13).

If we consider the asymptotic approximation for the kernel functions †

$$f_{n+j,n+j}(x) \simeq (-1)^{n+j} 2\cos(k_{n+j}x) \tag{15}$$

(with $k_{n+j} \propto \sqrt{n+j}$), Eq. (14) becomes

$$\lambda(n,n) = 2\sum_{j\neq l}(-1)^{j+l}C^{\eta}_{n,n}(j)C^{\eta}_{n,n}(l)\int_0^{\pi}\frac{\mathrm{d}\phi}{\pi}\int_{-\infty}^{+\infty}\mathrm{d}x\, p(x,\phi;\hat{\rho}_\eta)\,\cos\left[\delta k_n(j,l)x\right], \tag{16}$$

where we neglected the fast oscillating part in $f_{n+j,n+j}(x)f_{n+l,n+l}(x)$ and we defined $\delta k_n(j,l) \doteq k_{n+j} - k_{n+l}$. Thus, the integration is easily performed and we have

$$\lambda(n,n) = 2\sum_{j\neq l}(-1)^{j+l}C^{\eta}_{n,n}(j)C^{\eta}_{n,n}(l)\, e^{-\frac{1}{8}\delta k_n(j,l)^2}\sum_{s=0}^{\infty}\langle s|\hat{\rho}_\eta|s\rangle\, L_s\left(\frac{\delta k_n(j,l)^2}{4}\right), \tag{17}$$

where $L_s(z)$ denote the Laguerre polynomials.

Let us give few examples. For highly excited states, if the photon probability distribution is considered approximately constant, the sum over s is zero. This is the only case where the correlations give a null contribution. For coherent states with average number of photons $|\alpha|^2$ we have

$$\lambda(n,n) = 2\sum_{j\neq l}(-1)^{j+l}C^{\eta}_{n,n}(j)C^{\eta}_{n,n}(l)\, e^{-\frac{1}{8}\delta k^2_n(j,l)}J_0\left(\sqrt{\eta\,|\alpha|^2\,\delta k_n(j,l)^2}\right), \tag{18}$$

where $J_0(z)$ denote the zeroth–order Bessel function. For thermal states with average number of photons $\bar{n}$ the correlations are given by

$$\lambda(n,n) = 2\sum_{j\neq l}(-1)^{j+l}C^{\eta}_{n,n}(j)C^{\eta}_{n,n}(l)\, e^{-\frac{1+2\eta\bar{n}}{8}\delta k^2_n(j,l)}. \tag{19}$$

From the definition (4) we see that in this case the correlations always give a positive contribution. Thus, for thermal states the expression in Eq. (12) is not only an asymptotic estimate, but also an asymptotic lower bound for the actual errors expected in homodyne measurement of the density matrix.

†We remind that the diagonal kernel functions are independent of ϕ.

CONCLUSIONS

We obtained an asymptotic analytical expression for the *a priori* estimated experimental errors in homodyne tomography of the density operator in the Fock representation. For fixed overall efficiency at homodyne detectors η, the diagonal errors are exponentially growing as functions of the matrix index, with rate $-\ln(2\eta - 1)$. In the end, we estimated the contribution to the errors of the statistical correlations between different matrix elements.

Acknowledgments

This work is supported by Istituto Nazionale di Fisica della Materia (PRA-CAT97).

REFERENCES

1. G.M. D'Ariano, U. Leonhardt, and H. Paul, Homodyne detection of the density matrix of the radiation field, *Phys. Rev. A* 52:R1801 (1995).
2. G.M. D'Ariano, Latest developments in quantum tomography, in this Volume.
3. G.M. D'Ariano, M. Rasetti, and M. Vadacchino, *Phys. Rev. D* 32:1034 (1985).
4. C.G. Bollini, and L.E. Oxman, Shannon entropy and the eigenstates of the single mode squeeze operator, *Phys. Rev. A* 47:2339 (1993).
5. R.A. Fisher, M.M. Nieto, and V.D. Sandberg, *Phys. Rev. D* 29:110 (1984).
6. G.M. D'Ariano, C. Macchiavello, and N. Sterpi, Systematic and statistical errors in homodyne measurements of the density matrix, *Quant. Semicl. Opt.* 9:929 (1997).
7. T. Kiss, U. Herzog, and U. Leonhardt, Compensation of losses in photodetection and in quantum state measurements, *Phys. Rev. A* 52:2433 (1995).
8. G.M. D'Ariano, M. Vasilyev, and P. Kumar, Self-homodyne tomography of a twin beam state, *Phys. Rev. A* 58:636 (1998).
9. G.M. D'Ariano and C. Macchiavello, Loss–error compensation in quantum state measurement, *Phys. Rev. A* 57:3131 (1998).

PROBING QUANTUM STATES OF RYDBERG ELECTRONS BY HALF-CYCLE PULSES

G. Alber[1] and O. Zobay [2]

[1]Abteilung für Quantenphysik, Universität Ulm, D-89069 Ulm, Germany
[2] Optical Sciences Center, University of Arizona, Tucson, AZ 85721

Recently performed experiments[1] have demonstrated that half-cycle pulses (HCPs), i.e. unimodular electromagnetic pulses, are a useful new spectroscopic tool which is particularly well suited for investigating the dynamics of weakly bound Rydberg electrons. Typically their pulse durations range from the subpicosecond to the nanosecond regime and these pulses have already been produced with electric field strengths up to 10^5V/cm. So far work in this context has concentrated mainly on studies of total ionization or survival probabilities and on energy-resolved ionization spectra of Rydberg electrons [1,2,3]. Thus it has been shown with the help of a classical picture of the ionization process that energy-resolved ionization spectra yield direct information about the initial momentum distribution of a Rydberg electron. However, this way any phase information about the initial quantum state is lost.

In the following we address the question whether this phase information can be obtained from energy- and angle-resolved ionization probabilities. For this purpose a (multidimensional) semiclassical description of the ionization process of Rydberg electrons by half-cycle pulses is presented. In this theoretical approach it is particularly apparent how phase information about the initial quantum state of the Rydberg electron manifests itself in the angle- and energy-resolved ionization spectra. Furthermore, this way a detailed understanding of the ionization dynamics is obtained which is based on the underlying classical dynamics. In order to emphasize the essential physical aspects our subsequent discussion focuses on the sudden-ionization approximation [2] in which the ionizing HCP can be approximated by a delta-function in time. However, it should be mentioned that besides numerical advantages as far as the treatment of the Coulomb problem is concerned the presented semiclassical approach is also well suited for describing all effects which might arise from finite pulse durations or from spatial variations of realistic HCPs.

QUANTUM MECHANICAL IONIZATION SPECTRA IN THE SUDDEN-IONIZATION APPROXIMATION

Before we present the main ideas of the general multidimensional semiclassical description let us first of all summarize the quantum mechanical description of ionization

Quantum Communication, Computing, and Measurement 2
Edited by Kumar et al., Kluwer Academic / Plenum Publishers, New York, 2000.

by an HCP in the sudden-ionization approximation [2]. In the subsequent treatment we shall use Hartree atomic units with $e = \hbar = m_e = 1$ (e and m_e denote the electronic charge and mass).

Let us consider an atom initially prepared in a highly excited energy eigenstate $\mid nl\rangle$. Thereby its principal quantum number n is assumed to be large and its angular momentum quantum number l is small. This Rydberg electron is ionized at time $t = 0$ by a HCP. In the sudden-ionization approximation the HCP changes the momentum of the Rydberg electron abruptly by the amount

$$\vec{p}_0 = -\int_{-\infty}^{\infty} dt\vec{E}(t). \tag{1}$$

Thereby $\vec{E}(t)$ denotes the electric field strength of the HCP. The sudden ionization approximation is valid provided the Kepler period $T_{orb} = 2\pi(n-\alpha)^3$ of the Rydberg electron is much larger than the pulse duration of the HCP (α denotes the quantum defect of the Rydberg electron [4]). The electronic state $\mid \psi_0\rangle$ immediately after the application of the HCP is given by

$$\mid \psi_0\rangle = e^{i\vec{p}_0\cdot\hat{\vec{x}}} \mid nl\rangle. \tag{2}$$

Its subsequent time evolution is determined by the atomic Hamiltonian (in the absence of the HCP). Thus the angle- and energy resolved ionization probability is given by

$$\frac{d^3P_{ion}}{d\epsilon\wedge d\Omega} = \sqrt{2\epsilon} \mid \langle\vec{p}(\epsilon,\Omega) \mid \psi(t\to\infty)\rangle \mid^2 \tag{3}$$

with $\epsilon = \vec{p}^2/2$ and Ω denoting the final energy and solid angle of the emitted electron. In the case of a hydrogen atom, for example, which is prepared in an s-state initially and which is ionized by a linearly polarized HCP, i.e. $\vec{p}_0 = p_0\vec{e}_z$, a partial wave analysis yields

$$\langle\vec{p}(\epsilon,\Theta,\phi) \mid \psi(t\to\infty)\rangle = \tag{4}$$
$$\frac{1}{\sqrt{2\pi}|\vec{p}|}\sum_{l'=0}^{\infty} e^{i\sigma_{l'}}P_{l'}(\cos\Theta)\int_0^{\infty} drS_{n,l=0}(r)j_{l'}(|\vec{p}_0|r)F_{l'}(\epsilon;r).$$

Eq.(4) describes the probability amplitude of detecting an ionized electron with asymptotic momentum $\vec{p}$. The spherical angles of the emitted electron are denoted (Θ,ϕ) with $d\Omega = \sin\Theta d\Theta\wedge d\phi$. $P_{l'}(x)$ is the Legendre polynomial of order l', $S_{n,l=0}(r)$ is the radial wave function of the initial state $\mid nl = m = 0\rangle$ and $F_{l'}(\epsilon;r)$ is the regular Coulomb wave function with energy $\epsilon > 0$ and angular momentum quantum number l'. The Coulomb scattering phase is denoted $\sigma_{l'}$ and $j_{l'}(x)$ is a spherical Bessel function of order l'. Eqs.(3)and (4) determine the quantum mechanical angle- and energy resolved ionization probability in the sudden-ionization approximation.

SEMICLASSICAL DESCRIPTION OF IONIZATION BY A HCP

For highly excited Rydberg states, i.e. $n \gg 1$, and for sufficiently weak field strengths of the HCP, i.e. $\mid \vec{p}_0 \mid^2 /2 \ll 1$, the ionization process can be described with a high degree of accuracy by semiclassical methods. This way one obtains insight into the underlying classical dynamics of the ionization process. Furthermore, in this theoretical approach it becomes particularly apparent how phase information about the initial quantum state of the Rydberg electron manifests itself in the angle- and energy-resolved ionization spectra.

Let us first of all consider briefly the main theoretical ingredients of such a semiclassical description for the case of hydrogenic Rydberg states.

Semiclassically the energy eigenstate of a Rydberg electron with energy $\epsilon_n = -1/2n^2$ which is prepared in an s-state is approximately given by [5]

$$\langle \vec{x} \mid n, l = m = 0 \rangle = \frac{1}{\sqrt{8\pi} n^{3/2} r \sqrt{p(r)}} [\exp(i \int_0^r drp(r) - i\pi/2 - i\pi/4) + c.c.]$$
$$= A_{cl}(\vec{x}) \{\exp[iW_+(\vec{x})] + \exp[iW_-(\vec{x})]\}. \quad (5)$$

Thereby $p(r) = \sqrt{2(\epsilon_n - 1/r)}$ and $r = \mid \vec{x} \mid$ denote the electronic radial momentum and the distance from the nucleus. Eq.(5) is valid for positions of the Rydberg electron which are located well inside the classically allowed region. The state immediately after the application of a HCP is given by Eq.(2). Within the framework of a semiclassical description [6] this quantum state corresponds to two Lagrangian manifolds $L_\pm = \{(\vec{x}, \vec{p}_\pm(\vec{x})) | \vec{p}_\pm(\vec{x}) = \nabla W_\pm(\vec{x}) + \vec{p}_0 \equiv \pm p(r)\vec{e}_r + \vec{p}_0\}$ in the classical phase space of the electron.

After the application of the HCP the time evolution of this semiclassical quantum state is characterized by the bi-valued family of classical trajectories $\vec{x}(t; \vec{x}^{(0)}, \vec{p}_\pm(\vec{x}))$ which start at all possible points $\vec{x}^{(0)}$ in the support of $A_{cl}(\vec{x})$ with initial momenta $\vec{p}_\pm(\vec{x})$ and which evolve according to the classical Coulomb Hamiltonian $H = \vec{p}^2/2 - 1/r$. The probability amplitude $\langle \vec{p} \mid \psi(t \to \infty) \rangle$ of detecting an ionized Rydberg electron with momentum $\vec{p}$ is determined by all those classical trajectories with assume the asymptotic, final momentum $\vec{p}$. With each of these trajectories one can associate [6]

- a classical action (eikonal)

$$S = \int \vec{x} \cdot d\vec{p} + \vec{x}^{(0)} \cdot \vec{p}^{(0)}, \quad (6)$$

- a classical probability density

$$P^{(cl)} = \left| \frac{dp_x^{(f)} \wedge dp_y^{(f)} \wedge dp_z^{(f)}}{dx^{(0)} \wedge dy^{(0)} \wedge dz^{(0)}} \right|^{-1}, \quad (7)$$

- and a Morse index μ which is equal to the number of sign changes of $\frac{dp_x^{(f)} \wedge dp_y^{(f)} \wedge dp_z^{(f)}}{dx^{(0)} \wedge dy^{(0)} \wedge dz^{(0)}}$ times their multiplicities.

Thereby $\vec{p}^{(f)}$ denotes the final momentum of the ionized Rydberg electron. In terms of these classical quantities the probability amplitude of detecting an ionized Rydberg electron with asymptotic momentum $\vec{p}$ is given by

$$\langle \vec{p} \mid \psi(t \to \infty) \rangle = \sum_j \sqrt{P_j^{(cl)}} e^{-i(S_j + \frac{\pi}{2}\mu_j)} A_{cl}(\vec{x}_j^{(0)}) e^{i[W_j(\vec{x}_j^{(0)}) + \vec{p}_0 \cdot \vec{x}_j^{(0)}]}. \quad (8)$$

Depending on whether the initial momentum of the classical trajectory j is positive or negative one has to choose for W_j either the outgoing or incoming component $W_\pm$.

CHARACTERISTIC PROPERTIES OF ANGLE- AND ENERGY-RESOLVED IONIZATION SPECTRA

In order to exhibit characteristic properties of the ionization spectra and their relation to the underlying classical dynamics of the Rydberg electron let us consider a linearly polarized HCP with $\vec{p}_0 = p_0\vec{e}_z$ and an initial hydrogenic Rydberg state $\mid nl = 0\rangle$.

Due to axial symmetry $d^3P_{ion}/d\epsilon \wedge d\Omega$ is independent of ϕ. The initial positions $(r^{(0)}, \Theta^{(0)})$ of the classical trajectories with final asymptotic energy $\epsilon > 0$ are determined by energy conservation, i.e.

$$\frac{1}{2}[\vec{p}_\pm(r^{(0)}, \Theta^{(0)}) + \vec{p}_0]^2 - 1/r^{(0)} = \epsilon \tag{9}$$

or

$$1/r^{(0)} = \pm[\epsilon_n - \frac{(\epsilon - \epsilon_n - \vec{p}_0^2/2)^2}{2\vec{p}_0^2\cos^2(\Theta^{(0)})}]. \tag{10}$$

Thus there is one value of $r^{(0)}$ for each value of $\Theta^{(0)}$. The initial positions of those classical trajectories which yield a particular final energy ϵ and emission angle Θ are a dynamical property of the Coulomb problem and can be determined in a straightforward way.

In particular, one can distinguish two characteristic dynamical regimes:

1. $0 < \epsilon < \epsilon_n + \vec{p}_0^2/2$
   In this case trajectories with $0 < \Theta^{(0)} < \pi/2$ ($\pi/2 < \Theta^{(0)} < \pi$) start with negative (positive) radial momenta.

2. $\epsilon_n + \vec{p}_0^2/2 < \epsilon$
   In this regime of high asymptotic energies trajectories with $0 < \Theta^{(0)} < \pi/2$ ($\pi/2 < \Theta^{(0)} < \pi$) start with positive (negative) radial momenta.

Characteristic examples for the dependences of the final observation angle Θ on the initial emission angle $\Theta^{(0)}$ are depicted in Figs. 1 and 2. The deflection function shown

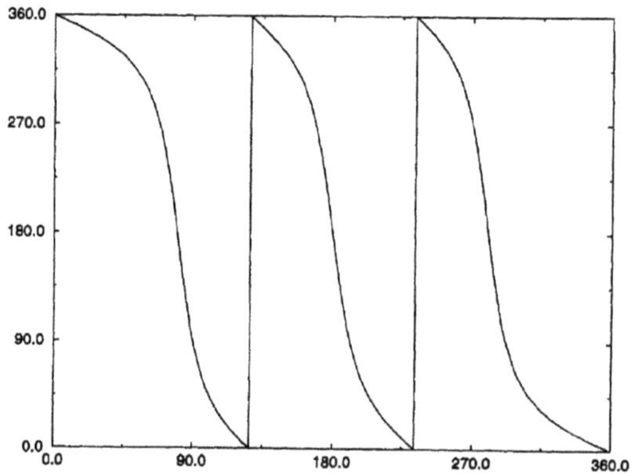

Figure 1. Deflection function, i.e. final angle Θ as a function of the initial emission angle $\Theta^{(0)}$ (in degrees); parameters are $n = 50$, $\mid \vec{p}_0 \mid= 0.05$a.u., $\epsilon = 10$meV.

in Fig. 1 corresponds to a case of small electron energies. It is apparent that there are three different initial emission angles $\Theta^{(0)}$ for each final angle Θ. Thus on the basis of Eq.(8) it is expected that the resulting angular- and energy resolved ionization probability displays a complex interference structure arising from quantum mechanical interferences between the probability amplitudes of these three contributing classical trajectories. Furthermore, there are classical trajectories with initial emission angles $\Theta^{(0)} \neq 0, \pi$ which are ionized along the direction of polarization of the HCP, i.e. with $\Theta = 0, \pi$. Due to the axial symmetry of the problem these trajectories give rise to the semiclassical glory phenomenon [7] which manifests itself in a divergence of the primitive semiclassical ionization amplitude of Eq.(8). However, this divergence can be removed in a straightforward way with the help of semiclassical uniformization methods [7].

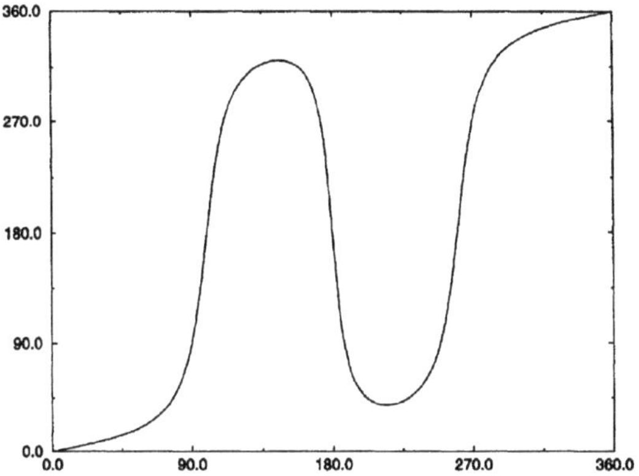

Figure 2. Deflection function, i.e. final angle Θ as a function of the initial emission angle $\Theta^{(0)}$ (in degrees); parameters are $n = 50$, $|\vec{p}_0| = 0.05$a.u., $\epsilon = 40$meV.

The characteristic behaviour of the deflection function for high energies of the ionized electron, i.e. $\epsilon_n + \vec{p}_0^2/2 < \epsilon$, is shown in Fig. 2. At these energies electrons are emitted in the polarization direction of the HCP with initial angle $\Theta^{(0)} = 0$ only so that there is no longer a semiclassical glory phenomenon at $\Theta = 0$. However, in the opposite direction, i.e. for $\Theta = \pi$, a semiclassical glory phenomenon is still observable as there are classical trajectories leaving the atom in this direction with initial angles $\Theta^{(0)} \neq 0, \pi$. In addition, also an extremum of the deflection function is noticable around an emission angle of approximately $\Theta^{(0)} \approx 3\pi/4$ which gives rise to a semiclassical rainbow phenomenon [7]. At this emission angle the contributions of two classical trajectories coalesce and the primitive semiclassical ionization amplitude of Eq.(8) diverges. Again this divergence can be removed in a straightforward with the help of semiclassical uniformization techniques [7].

As an example, the resulting energy-resolved angular distribution is depicted in Fig. 3. The overall good agreement between the semiclassical result obtained on the basis of Eq.(8) and the exact quantum mechanical result of Eqs.(3) and (4) is apparent. The primitive semiclassical result of Eq.(8) deviates from the quantum mechanical one only in a small region around the rainbow and backward glory angles. According to Eq.(8) the oscillations of the ionization probability can be explained in a straightforward way in terms of quantum mechanical interferences between the probability amplitudes of the contributing classical trajectories. Thus the angle- and energy resolved ionization spectrum can be related directly to the underlying classical dynamics of the Rydberg electron under the influence of the ionizing HCP.

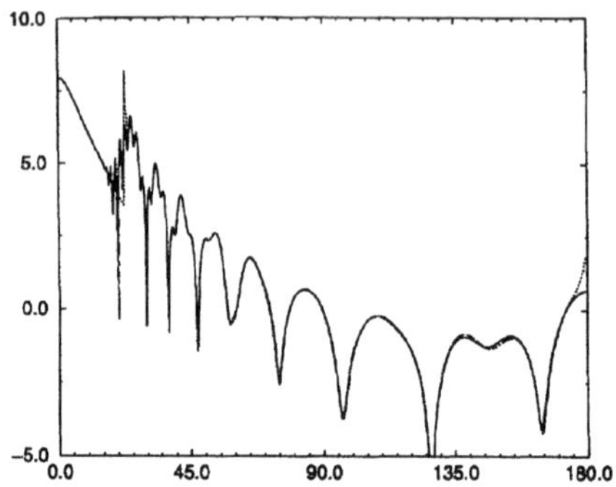

Figure 3. Energy-resolved angular distribution $\ln[d^3P_{ion}/d\epsilon \wedge d\Omega]$ in the sudden-ionization approximation: exact quantum mechanical result of Eq.(4)(full), primitive semiclassical result of Eq.(8) (dotted) and uniform semiclassical result (dashed); parameters are $n = 50$, $|\vec{p}_0| = 0.05$a.u., $\epsilon = 40$meV.

OUTLOOK AND CONCLUSION

The presented multidimensional semiclassical approach offers an adequate description of ionization of Rydberg electrons by HCPs. Thereby the angle- and energy-resolved ionization spectra can be related directly to the underlying classical dynamics of the Rydberg electron. Within this semiclassical treatment the phase information contained in the quantum state is taken into account properly. Thus on the basis of this semiclassical treatment it is expected that also all phase properties of the initial quantum state of the Rydberg electron might be reconstructed from observed angle- and energy resolved ionization spectra. Work in this direction is currently in progress. This semiclassical approach offers also significant numerical advantages as it avoids the usual problems connected with the large extensions of Rydberg states and with the Coulomb singularity. It can be extended in a straightforward way to cases where the pulse durations are finite and where spatial variations of the ionizing HCP are no longer negligible.

Acknowledgments

This work is supported by the Deutsche Forschungsgemeinschaft within the Schwerpunktprogramm 'Zeitabhängige Phänomene und Methoden' and by the U.S. Office of Naval Research Contract No. 14-91-J1205, and by the U.S. Army Research Office.

REFERENCES

1. R. R. Jones, D. You, and P. H. Bucksbaum, Phys. Rev. Lett. **70**, 1236 (1993); G. M. Lankhuijzen and L. D. Noordam, Phys. Rev. Lett. **74**, 35 (1994); N. E. Tielking and R. R. Jones, Phys. Rev. A. **52**, 1371 (1995); R. B. Vrijen, G. M. Lankhuijzen, and L. D. Noordam, Phys. Rev. Lett. **79**, 617 (1997).
2. C. O. Reinhold, H. Shaw, and J. Burgdörfer, J. Phys. B. **27**, L469 (1994); K. J. La Gattuta and P. B. Lerner, Phys. Rev. A. **49**, R1547 (1994); A. Bugacov, B. Piraux, M. Pont, and R. Shakeshaft, Phys. Rev. A. **51**, 1490 (1995); C. O. Reinhold, J. Burgdörfer, M. T. Frey, and F. B. Dunning, Phys. Rev. A. **54**, R33 (1996); R. Gebarowski, J. Phys. B. **30**, 2143 (1997).
3. C. D. Schwieters and J. B. Delos, Phys. Rev. A **51**, 1023 (1995); Phys. Rev. A **51**, 1030 (1995)

4. M. J. Seaton, Rep. Prog. Phys. **46**, 167 (1983).
5. H. A. Bethe and E. Salpeter, *Quantum Mechanics of one- and two- electron atoms* (Plenum, N.Y., 1977).
6. V. P. Maslov and M. V. Fedoriuk, *Semiclassical Approximation in Quantum Mechanics* (Reidel, Boston, 1981).
7. M. V. Berry and K. E. Mount, Rep. Prog. Phys. **35**, 315 (1972).

NON-MARKOVIAN QUANTUM TRAJECTORIES FOR OPEN SYSTEMS

Walter T. Strunz[1], Lajos Diósi[2], and Nicolas Gisin[3]

[1] Fachbereich Physik, Universität GH Essen, 45117 Essen, Germany
[2] Research Institute for Particle and Nuclear Physics, 1525 Budapest 114, POB 49, Hungary
[3] Group of Applied Physics, University of Geneva, 1211 Geneva 4, Switzerland

We present a theory of non-Markovian quantum trajectories that describe the dynamics of a quantum system coupled to an environment of harmonic oscillators. When averaged over the noise, they reproduce the standard reduced density matrix without any approximation, in particular without Markov approximation. The time evolution of the environment can also be determined from these non-Markovian quantum trajectories. We present a simple example to demonstrate the power of our approach. As a remarkable feature of non-Markovian quantum trajectories, we show that for strong coupling to a dissipative environment with memory, the asymptotic state can be reached in a finite time. Further examples are described in the text.

INTRODUCTION

The dynamics of open quantum systems is a very timely problem, both to address fundamental questions (quantum decoherence, measurement problem) as well as to tackle the more practical problems of engineering the quantum devices necessary for the emerging fields of nanotechnology and quantum computing. So far, the true dynamics of open systems has almost always been simplified by the Markov approximation: environmental correlation times are assumed negligibly short compared to the system's characteristic time scale.

The time evolution of the reduced density operator ρ_t of such Markovian open systems is described by a master equation of Lindblad form

$$\frac{d}{dt}\rho_t = -i[H_s, \rho_t] + \frac{1}{2}\sum_m \left([L_m\rho_t, L_m^\dagger] + [L_m, \rho_t L_m^\dagger]\right), \tag{1}$$

where H_s is the Hamiltonian of the 'system' and the operators L_m describe the effect of the environment in the Markov approximation.

In recent years, a breakthrough in the solution and understanding of Markovian master equations was achieved through the discovery of *stochastic unravellings*[1, 2].

Quantum Communication, Computing, and Measurement 2
Edited by Kumar et al., Kluwer Academic / Plenum Publishers, New York, 2000.

These are stochastic Schrödinger equations for 'system' states $\psi_t(z)$ (*quantum trajectories*), driven by a certain classical noise z_t. Crucially, the ensemble mean $M[\ldots]$ over the noise recovers the 'system' density operator,

$$\rho_t = M\Big[|\psi_t(z)\rangle\langle\psi_t(z)|\Big]. \tag{2}$$

Hence, the solution of eq.(1) is reduced from a problem in the matrix space of ρ_t to a much simpler Monte Carlo simulation of quantum trajectories $\psi_t(z)$ in the state space.

For the Markovian master eq.(1), several such unravellings are known. Some involve jumps at random times[1], others have continuous, diffusive solutions[2]. They have been used extensively over recent years as they provide useful insight into the dynamics of continuously monitored (individual) quantum processes[3], or into the mechanism of decoherence[4]. In addition, they provide an efficient tool for the numerical solution of the master eq.(1). It is thus desirable to extend the powerful concept of stochastic unravellings to the more general case of non-Markovian evolution. In this paper we present a non-Markovian quantum trajectory approach to open systems which we have developed recently[5]. In the Markov limit, it reduces to the diffusive 'quantum state diffusion' (QSD) stochastic Schrödinger equation[2], which is why we refer to our result as 'non-Markovian quantum state diffusion' (NMQSD)[5].

Most commonly, the master equation (1) is derived by solving the Schrödinger equation

$$\frac{d}{dt}|\Psi_{tot}\rangle = -iH_{tot}|\Psi_{tot}\rangle \tag{3}$$

of some total system including the environment and then tracing over the environmental degrees of freedom. Assuming that environmental correlation times are much shorter than relevant system time scales, one arrives at (1). Then, *starting* from the master equation (1), the Markov stochastic unravellings are constructed in such a way as to ensure the correct ensemble mean (2).

Our approach to a non-Markovian unravelling is quite different. We find quantum trajectories for the 'system' by solving the total Schrödinger equation (3) for both 'system' *and* environment in a particular way. We thus find a stochastic Schrödinger equation for the 'system' dynamics without any approximation. In particular, we don't need the Markov approximation, nor any form of weak coupling assumption between 'system' and environment. We are thus able to *unravel* the exact evolution of the system's reduced dynamics without even referring to a master equation for its non-Markovian ensemble evolution. Our non-Markovian quantum trajectory approach appears thus more fundamental than the reduced density operator approach. This is in contrast to traditional Markovian quantum trajectory methods which are usually based on the master equation (1). Our result also underlines that, given the total model for 'system' and environment, quantum trajectories give *more* than just the solution of the reduced density operator's master equation.

NON-MARKOVIAN QUANTUM STATE DIFFUSION

Our non-Markovian stochastic Schrödinger equation for the 'system' state vector is based on a standard model of open system dynamics,

$$H_{tot} = H_s + \sum_\lambda \chi_\lambda (L^\dagger a_\lambda + L a_\lambda^\dagger) + \sum_\lambda \omega_\lambda a_\lambda^\dagger a_\lambda. \tag{4}$$

The system Hamiltonian H_s and the coupling (or Lindblad) operator L can be chosen arbitrarily, as can the number of oscillators and the various coupling strengths χ_λ. Our result does, however, rely on this particular environment of harmonic oscillators. It is an open question whether a similarly applicable theory can be established for nonlinear environments.

The results in this paper are valid for a zero temperature environment, finite temperature is equally possible but makes things a bit more complicated[5].

In earlier work[6], we showed that the evolution of the reduced density operator $\rho_t = \mathrm{Tr}_{env}(|\Psi_{tot}\rangle\langle\Psi_{tot}|)$ of the model (4) can be unravelled according to (2) by the linear non-Markovian stochastic equation

$$\frac{d}{dt}\psi_t = -iH\psi_t + L\psi_t z_t - L^\dagger \int_0^t \alpha(t,s)\hat{O}(t,s,z)ds\, \psi_t, \tag{5}$$

where z_t is colored complex Gaussian noise of zero mean and correlations

$$M\left[z_t^* z_s\right] = \alpha(t,s), \quad M\left[z_t z_s\right] = 0. \tag{6}$$

The environment correlation function $\alpha(t,s)$ in (5) and (6), as determined by the model (4), reads[5, 6]

$$\alpha(t,s) = \sum_\lambda \chi_\lambda^2 e^{-i\omega_\lambda(t-s)}. \tag{7}$$

The operator $\hat{O}(t,s,z)$ in (5) is defined through

$$\hat{O}(t,s,z)\psi_t \equiv \frac{\delta}{\delta z_s}\psi_t, \tag{8}$$

and it can in principle be determined from eq.(5). The applicability of our theory rests on the knowledge of an explicit expression for $\hat{O}(t,s,z)$ (see the example and [5]).

The Markov limit arises from an environment with $\alpha(t,s) = \gamma\delta(t-s)$, which leads to $\hat{O} = L$ in (8). Thus, the non-Markovian linear equation (5) reduces to

$$\frac{d}{dt}\psi_t = -iH\psi_t + \sqrt{\gamma}L\psi_t\bar{z}_t - \frac{\gamma}{2}L^\dagger L\psi_t, \tag{9}$$

which we identify as the linear 'quantum state diffusion' equation[2]. In (9), we use delta-correlated complex white noise $\bar{z}_t = z_t/\sqrt{\gamma}$. Clearly, we see from (9) that in this Markov case the density operator (2) evolves according to the master equation (1) with $L_1 = \sqrt{\gamma}L$.

The linear unravelling eq.(9) and equally our non-Markovian generalization (5) are of limited value, since the norm $\|\psi_t(z)\|$ of its solutions tends to 0 with probability 1 and to infinity with probability 0, such that the mean square norm is constant. To be really useful, one should find unravellings in terms of the normalized states

$$\tilde{\psi}_t(z) = \frac{\psi_t(z)}{\|\psi_t(z)\|}, \tag{10}$$

which requires a redefinition of the distribution of the noise ($z_t \to \tilde{z}_t$) so that eq.(2) remains valid for the normalized solutions:

$$\rho_t = M\Big[|\tilde{\psi}_t(\tilde{z})\rangle\langle\tilde{\psi}_t(\tilde{z})|\Big]. \tag{11}$$

Now (11) can be interpreted as an unravelling of the mixed state ρ_t into an ensemble of pure states.

It is remarkable that there is in fact a closed evolution equation for the normalized states $\tilde{\psi}_t$. Its quite tricky derivation[5] leads to

$$\frac{d}{dt}\tilde{\psi}_t = -iH\tilde{\psi}_t + (L - \langle L\rangle_t)\tilde{\psi}_t\tilde{z}_t \qquad (12)$$
$$- \int_0^t \alpha(t,s)\left(\Delta L^\dagger \hat{O}(t,s,\tilde{z}) - \langle \Delta L^\dagger \hat{O}(t,s,\tilde{z})\rangle_t\right) ds\tilde{\psi}_t,$$

which is the central equation of non-Markovian QSD (NMQSD). Here, $\tilde{z}_t$ is the shifted noise

$$\tilde{z}_t = z_t + \int_0^t \alpha^*(t,s)\langle L^\dagger\rangle_s ds, \qquad (13)$$

and for brevity we use $\Delta L^\dagger = L^\dagger - \langle L^\dagger\rangle_t$. The complex noise z_t entering the NMQSD equation (12) through (13) also satisfies the relations (6). Given the operator $\hat{O}$, non-Markovian quantum trajectories can now be computed from (12) in the very same way as in the Markov case - see the example and [5].

In the Markov limit, $\alpha(t,s) = \gamma\delta(t-s)$, and we have $\hat{O} = L$. The NMQSD equation (12) reduces to the widely used non-linear 'quantum state diffusion' (QSD) equation[2], here in Stratonovich form:

$$\frac{d}{dt}\tilde{\psi}_t = -iH\tilde{\psi}_t + \sqrt{\gamma}(L - \langle L\rangle_t)\tilde{\psi}_t(\tilde{z}_t + \sqrt{\gamma}\langle L^\dagger\rangle_t) - \frac{\gamma}{2}(L^\dagger L - \langle L^\dagger L\rangle_t)\tilde{\psi}_t, \qquad (14)$$

where $\langle L\rangle_t \equiv \langle\tilde{\psi}_t|L|\tilde{\psi}_t\rangle$, and again, $\tilde{z}_t$ is delta-correlated complex noise.

ENVIRONMENT DYNAMICS

The central property of an *unravelling* is that we recover the reduced density operator of the 'system' as in (11) by simulating (normalized) quantum trajectories $\tilde{\psi}_t(z)$. Since our NMQSD equation is derived by solving the total model (4), there is more to these trajectories. Without going into details we just want to mention that it is possible to obtain the Q-function of the environment at any time from the knowledge of the shifted processes $\tilde{z}_t$ (13)[7].

AN EXAMPLE - A DAMPED SPIN $\frac{1}{2}$

Here we give a concrete example of NMQSD (12). More examples, including non-Markovian measurement-like situations and extreme non-Markovian cases, where the environment consists of only a single oscillator can be found in[5].

Let us consider a dissipative spin $\frac{1}{2}$ with $H_s = \frac{\omega}{2}\sigma_z$, and coupling $L = \chi\sigma_-$. We choose exponentially decaying correlations $\alpha(t,s) = \frac{\gamma}{2}e^{-\gamma|t-s|-i\Omega(t-s)}$ with an environmental central frequency Ω and memory time γ^{-1}. This choice is just a matter of convenience. For this Hamiltonian and coupling, it is possible to compute trajectories from our NMQSD equation for any $\alpha(t,s)$. Determining the $\hat{O}$ operator[5] for this example, the non-Markovian QSD equation (12) takes the form

$$\frac{d}{dt}\tilde{\psi}_t = -i\frac{\omega}{2}\sigma_z\tilde{\psi}_t - \chi F(t)(\sigma_+\sigma_- - \langle\sigma_+\sigma_-\rangle_t)\tilde{\psi}_t \qquad (15)$$
$$+ \chi(\sigma_- - \langle\sigma_-\rangle_t)\tilde{\psi}_t\left(z_t + \chi\int_0^t \alpha(t,s)^*\langle\sigma_+\rangle_s ds + \langle\sigma_+\rangle_t F(t)\right),$$

with $F(t)$ determined from $\dot{F}(t) = -\gamma F(t) + i(\omega - \Omega)F(t) + \chi F(t)^2 + \frac{\chi\gamma}{2}$ and initial condition $F(0) = 0$. The equation for $F(t)$ can even be solved analytically [5].

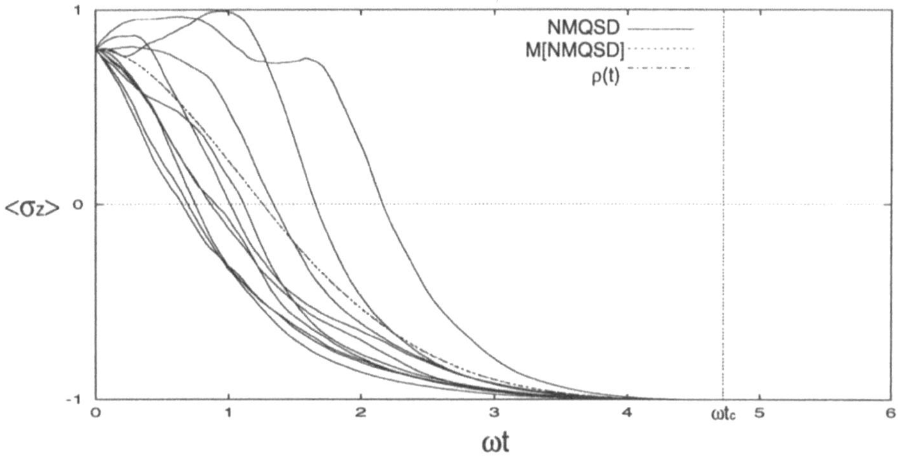

Figure 1. Quantum trajectories of the non-Markovian QSD equation for the dissipative case $H = \frac{\omega}{2}\sigma_z$, $L = \chi\sigma_-$ and an exponentially decaying bath correlation function $\alpha(t,s) = \frac{\gamma}{2}e^{-\gamma|t-s|-i\Omega(t-s)}$. We choose a strong coupling $\chi^2 = \omega$, $\gamma = \omega$ and resonance $\Omega = \omega$. As initial state we use $|\psi_0\rangle = 3|\uparrow\rangle + |\downarrow\rangle$. Displayed is the quantum expectation value $\langle\sigma_z\rangle$ of several solutions of the non-Markovian QSD equation (solid lines), their ensemble average (dashed line) and - almost indistinguishably - the analytical result for the ensemble mean (dashed-dotted line). After the finite time $\omega t_c = \frac{3}{2}\pi \approx 4.71$, all individual trajectories reach the ground state.

It is worth mentioning the case of exact resonance, $\omega = \Omega$. Two regimes should be distinguished. First, when $\gamma > 2\chi^2$ (short memory compared to coupling strength), $F(t)$ tends to $\left(\gamma - \sqrt{\gamma^2 - 2\gamma\chi^2}\right)/(2\chi)$. Hence, for large γ, one recovers Markov QSD (14). For longer memory times or stronger coupling, $\gamma < 2\chi^2$, things are very different: $F(t)$ diverges to infinity when the time t approaches

$$t_c = \frac{\pi + 2\arctan(\gamma/\sqrt{2\chi^2\gamma - \gamma^2})}{\sqrt{2\lambda^2\gamma - \gamma^2}}. \tag{16}$$

All realizations $\psi_t(z)$ reach the 'down' state in this finite time and remain there. The reduction time in this case is $\omega t_c = \frac{3}{2}\pi \approx 4.71$. This is the first example where a continuous quantum state diffusion reaches its asymptotic state in a finite time, which was proven impossible for Markov diffusion [8]. In Fig.1 we show the quantum expectation value $\langle\sigma_z\rangle$ of several solutions of the non-Markovian QSD equation (solid lines), their ensemble average (dashed line) and the analytical result for the ensemble mean (dashed-dotted line), which is almost indistinguishable.

CONCLUSION

In conclusion, we present an unravelling of the dynamics of a quantum system coupled to an environment of harmonic oscillators. In the Markov limit, standard quantum state diffusion[2] is recovered. We emphasize that non-Markovian QSD (12) reproduces the true evolution of the system taking into account the exact unitary dynamics of system and environment [5, 6].

Many features of NMQSD are surprising and have no counterpart in any Markov unravelling. Moreover, our approach represents a new promising tool for the numerical simulation of quantum devices, whenever non-Markovian effects are relevant [9].

Acknowledgments

WTS would like to thank the Deutsche Forschungsgemeinschaft for support through the SFB 237 *Unordnung und große Fluktuationen.* LD is supported by the Hungarian Scientific Research Fund through grant T016047. NG thanks the Swiss National Science Foundation.

REFERENCES

1. J. Dalibard, Y. Castin, and K. Mölmer, *Phys. Rev. Lett.* **68**, 580 (1992); H. Carmichael, *An open system approach to Quantum Optics*, (Springer, Berlin) (1994); M. B. Plenio and P. L. Knight, *Rev. Mod. Phys.* **70**, 101 (1998).
2. N. Gisin and I. C. Percival, *J. Phys. A: Math. Gen.* **25**, 5677 (1992); **26**, 2233 (1993); **26**, 2245 (1993).
3. see e.g. C. Cohen-Tannoudji, B. Zambou, and E. Arimondo, *J. Opt. Soc. Am. B* **10**, 2107 (1993); T. P. Spiller *et. al.*, *J. Low. Temp. Phys.* **101**, 1037 (1995).
4. L. Diósi *et. al.*, *Phys. Rev. Lett.* **74**, 203 (1995); T. A. Brun, *Phys. Rev. Lett.* **78**, 1833 (1997).
5. L. Diósi, N. Gisin, and W. T. Strunz, *Phys. Rev. A* (September issue) (1998), also *quant-ph/9803062.*
6. L. Diósi and W. T. Strunz, *Phys. Lett. A.* **235**, 569 (1997); L. Diósi, *Quantum Semiclass. Opt.* **8**, 309 (1996); W. T. Strunz, *Phys. Lett. A.* **224**, 25 (1996).
7. L. Diósi, N. Gisin, and W.T. Strunz, to be published.
8. P. Pearle, *Phys. Rev. Lett.* **53** 1775 (1984); N. Gisin, *Phys. Rev. Lett.* **53** 1776 (1984); P. Pearle, *Phys. Rev. D* **33**, 2240 (1986); N. Gisin, *Helv. Phys. Acta* **62**, 363 (1990); P. Pearle, in *Experimental Metaphysics*, eds. R.S. Cohen, M. Horne, and J. Stachel, (Kluwer Acad. Press), 143 (1997).
9. A. Imamoglu, *Phys. Rev. A* **50**, 3650 (1994); A. Imamoglu and Y. Yamamoto, *Phys. Lett. A* **191**, 425 (1994); B. M. Garraway, *Phys. Rev. A* **55**, 2290 (1997); S. Bay, P. Lambropoulos, and K. Mölmer, *Phys. Rev. Lett.* **79**, 2654 (1997); G. M. Moy, J. J. Hope, and C. M. Savage, see *quant-ph/9801046* (1998).

MAXIMALLY ROBUST UNRAVELINGS OF QUANTUM MASTER EQUATIONS WITH APPLICATIONS TO THE ATOM LASER

H. M. Wiseman[1] and J. A. Vaccaro[2]

[1]Centre for Laser Science, Department of Physics
The University of Queensland, St. Lucia 4072, Australia
[2]Division of Physics and Astronomy
University of Hertfordshire, Hatfield AL10 9AB, UK

The stationary solution ρ of a quantum master equation can be represented as an ensemble of pure states in a continuous infinity of ways. Here we suggest a definition of the the "most natural" ensemble. The first requirement is that the ensemble be physically realizable through monitoring the system's environment. We call this an 'unraveling' of the master equation. The second requirement is that it be the most robust of such ensembles. Robustness of an ensemble is defined in terms of the survival probability $S(t)$ which is the average probability for one of its elements to remain unchanged after a time t. The most natural ensemble can thus be said to be the one produced by the maximally robust unraveling. An application of interest is the atom laser. We show that the nature of the maximally robust unraveling is closely related to the question of whether the output beam satisfies the conditions required to be a true laser beam.

INTRODUCTION

It is well-known that open quantum systems generally become entangled with their environment. For a suitably weak coupling between the system and the environment, the latter can be ignored (traced over) so as to derive a Markovian master equation for the system evolution. This evolution will in general cause the system to evolve from an initially pure into a mixed state which must be described by a density operator ρ. This density operator can be written in terms of pure states as follows

$$\rho = \sum_i w_i P_i, \tag{1}$$

where the P_i are projection operators and the w_i are positive weights summing to unity. The (possibly infinite) set of ordered pairs,

$$E = \{P_i, w_i\}_i, \tag{2}$$

we will call an ensemble E of pure states. There are continuously infinitely many ensembles E which represent ρ.

Our aim in this paper is to argue that some ensembles are better representations of ρ than other ensembles, and to define the ensemble $E^{\mathcal{R}}$ which is the "most natural" representation for ρ.

PHYSICAL REALIZABILITY

The first criterion which our ensemble must satisfy is that it be *physically realizable.* By this we mean that it should be possible, without altering the average evolution of the system, to prepare it such that we know that its state is one of the pure states P_i, where the probability for that particular state being chosen is w_i. At first sight it is paradoxical to say that the system evolution causes the system to become mixed, but that nevertheless we can know it to be in a pure state. The resolution is that, by monitoring (rather than ignoring) the environment, the system state will be collapsed over time into a pure state. Being simply an example of a quantum measurement, this process (called a quantum trajectory) will be stochastic, and *on average* the system will still obey the master equation.[1] Because of this, the ensemble of quantum trajectories is called an *unraveling* of the master equation.[1] Different unravelings (which we will denote $\mathcal{U}$) will result in different ensembles $E^{\mathcal{U}}$ of pure states which represent ρ.

To make the problem well-defined, we assume that the system obeys a Markovian master equation with a unique stationary state. The most general form of the quantum master equation is[2, 3, 4]

$$\dot{\rho} = -i[H,\rho] + \sum_{k=1}^{K} \mathcal{D}[c_k]\rho \equiv \mathcal{L}\rho, \tag{3}$$

where for arbitrary operators A and B, $\mathcal{D}[A]B \equiv ABA^\dagger - \frac{1}{2}\{A^\dagger A, B\}$. The unique stationary state ρ_∞ is defined by $\mathcal{L}\rho_\infty = 0$. Its existence requires that the Liouvillian superoperator $\mathcal{L}$ be time-independent, which implies that one can remove a Hamiltonian term by moving into its interaction picture if the remainder of the $\mathcal{L}$ is time-independent in that interaction picture.

It is now well recognized that there are many (in fact continuously many) different unravelings for a given master equation.[5] Since any pure state will remain pure under a given unraveling, the unraveling can be represented by a stochastic and nonlinear equation of motion for a state vector. This is often called a stochastic Schrödinger equation (SSE). The set of all SSEs is vast, so to begin this investigation we deliberately narrow our search to those which are continuous and Markovian. These restrictions imply that the SSE must have the form

$$dP = -idt[H,P] + \sum_{k=1}^{K} \{dt\mathcal{D}[c_k] + \mathcal{H}[dW_k(t)c_k]\}\, P \tag{4}$$

Here $P = |\psi\rangle\langle\psi|$ is the projector for the system and $\mathcal{H}[A]$ is a nonlinear superoperator defined (for arbitrary operators A and B) by $\mathcal{H}[A]B \equiv AB + BA^\dagger - \mathrm{Tr}[AB + BA^\dagger]B$. The $dW_k(t)$ are infinitesimal increments of a complex multi-dimensional Wiener process[6] satisfying

$$dW_j(t)dW_k^*(t) = dt\delta_{jk}, \quad dW_j(t)dW_k(t) = dtu_{jk}, \tag{5}$$

where the u_{jk} are arbitrary complex numbers obeying $|u_{jk}| \leq 1$ and $u_{jk} = u_{kj}$.

For a master equation with K Lindblad terms the unraveling (which we will denote $\mathcal{U}$) is thus parametrized by $K(K+1)$ real numbers. Each unraveling $\mathcal{U}$ will give rise to a (not necessarily unique) ensemble of pure states

$$E^{\mathcal{U}} = \{P_i^{\mathcal{U}}, w_i^{\mathcal{U}}\}_i, \tag{6}$$

where $P_i^{\mathcal{U}}$ are the possible pure states of the system at steady state, and $w_i^{\mathcal{U}}$ are their weights. For master equations with a unique stationary state ρ_∞, the SSE (4) is ergodic over $E^{\mathcal{U}}$ and $w_i^{\mathcal{U}}$ is equal to the proportion of time the system spends in state $P_i^{\mathcal{U}}$. The ensemble $E^{\mathcal{U}}$ represents ρ_∞ in that

$$\mathrm{E}[P^{\mathcal{U}}] = \sum_i w_i^{\mathcal{U}} P_i^{\mathcal{U}} = \rho_\infty. \tag{7}$$

This follows from the fact that in Eq. (4) $\mathrm{E}[dP] = \mathcal{L}\mathrm{E}[P]dt$, where E denotes the ensemble average. Note that the projectors $P_i^{\mathcal{U}}$ will not in general be orthogonal.

ROBUSTNESS

The second criterion which our ensemble must obey is that it be *maximally robust* (subject to the first criterion). The idea of robustness has it origins in studies of decoherence and the classical limit.[7, 8, 9, 10] In particular, Zurek[8] introduced the idea of "preferred states" of open systems, defining them as those states which remain relatively pure for a long time. In this work we prefer (see below) to define the robustness of a pure state in terms of its survival probability. This measures how close it remains to its original state after a time t. For a particular member of the ensemble $P_i^{\mathcal{U}}$, the survival probability is given by

$$S_i^{\mathcal{U}}(t) = \mathrm{Tr}[P_i^{\mathcal{U}} \exp(\mathcal{L}t) P_i^{\mathcal{U}}]. \tag{8}$$

In the context of unravelings, this seems a better measure than the purity-measure $\mathrm{Tr}[\left(e^{\mathcal{L}t} P_i^{\mathcal{U}}\right)^2]$. Regardless of whether the system is being observed, $S_i^{\mathcal{U}}(t)$ is the probability for the system to be found in its original state. By contrast, the purity of the state would always be unity under the SSE evolution.

Because we are considering an ensemble of pure states (not just a single one), we next have to consider the *ensemble averaged* survival probability.

$$S^{\mathcal{U}}(t) = \sum_i w_i^{\mathcal{U}} S_i^{\mathcal{U}}(t). \tag{9}$$

The most robust ensemble is that for which the survival probability decays most slowly. This can be quantified in terms of the survival time

$$\tau^{\mathcal{U}} = \min\{t : S^{\mathcal{U}}(t) = \Lambda\}. \tag{10}$$

Here Λ is some suitably chosen constant such that

$$1 = S^{\mathcal{U}}(0) > \Lambda > S^{\mathcal{U}}(\infty) = \mathrm{Tr}[\rho_\infty^2]. \tag{11}$$

The (or perhaps a) most robust ensemble $E^{\mathcal{R}}$ is thus defined by

$$\tau^{\mathcal{R}} \geq \tau^{\mathcal{U}} \ \forall \mathcal{U}. \tag{12}$$

Given that our most robust ensemble is constrained also by the first criterion (Sec. 2), the most succinct way to refer to our proposed "best representation of ρ" is that is the one resulting from the *maximally robust unraveling*, which we have denoted $\mathcal{R}$.

THE ATOM LASER

An atom laser is a device which would (ideally) produce a continuous output beam of atoms which are coherent. Precisely what is meant by 'coherent' in this context was analyzed by one of us in Ref. 11. There, four criteria were identified. The first three are relatively uncontroversial. The fourth is the most interesting as it seems not to have been considered before. In words, it says that

(4) The linewidth must be much less than the output flux (particles per unit time)

Note that there is no mention of the atom laser being in a coherent state.

To model an atom laser we use the following idealized master equation[11] for the laser mode with annihilation operator a :

$$\dot{\rho} = \mu\mathcal{D}[a^\dagger]\mathcal{A}[a^\dagger]^{-1}\rho + \mathcal{D}[a]\rho + N\mathcal{D}[a^\dagger a]\rho - iC[(a^\dagger a)^2, \rho], \tag{13}$$

where $\mathcal{A}[A]B \equiv \frac{1}{2}[A^\dagger AB + BA^\dagger A]$. The first term in Eq. (13) describes the Poissonian addition of atoms to the laser mode at rate μ, appropriate to a laser far above threshold. The second term describes linear loss. This defines the measure of time and is necessary for an output beam to form. The third term is due to excess phase noise (for whatever reasons). The fourth represents the self-interaction of atoms through elastic scattering. The stationary state of this master equation is independent of N and C and is given by

$$\rho_\infty = \sum_{n=0}^{\infty} e^{-\mu}\frac{\mu^n}{n!}|n\rangle\langle n| = \int_0^{2\pi}\frac{d\phi}{2\pi}|\sqrt{\mu}\,e^{i\phi}\rangle\langle\sqrt{\mu}\,e^{i\phi}|. \tag{14}$$

Here $|n\rangle$ and $|\sqrt{\mu}\,e^{i\phi}\rangle$ are eigenstates of $a^\dagger a$ and a respectively.

It can be shown that satisfying condition **(4)** requires

$$\mu \gg 1, N, C^2. \tag{15}$$

When this is satisfied, the output beam can be treated as being a coherent state,[11] despite the fact that the stationary state (14) has no phase information. Evidently the parameters N and C contribute to the phase noise of the laser.

The form of the stationary state (14) suggests the obvious question: which is a better description of the system state, that of being a number state (chosen from a Poissonian distribution), or that of being a coherent state (with a random phase). This is precisely the sort of question the theory of maximally robust unravelings addresses.

To make the problem tractable we linearize the master equation (13) for a state localized about a mean field $\langle a\rangle = \sqrt{\mu}$. For this to be valid obviously requires $\mu \gg 1$, which is the first of the three conditions in Eq. (15). This allows us to make the replacement $a = \sqrt{\mu} + \frac{1}{2}x + i\frac{1}{2}y$. Including the measurement terms for a CM unraveling gives the SSE

$$\begin{aligned} dP = &\left\{\mathcal{D}[\tfrac{1}{2}x + i\tfrac{1}{2}y] + (1+\nu)\mathcal{D}[\tfrac{1}{2}x] + \mathcal{D}[\tfrac{1}{2}y] + \mathcal{H}[i\tfrac{1}{8}(xy+yx)] + \mathcal{H}[-i\chi(\tfrac{1}{2}x)^2]\right\} P dt \\ &+ \left\{\mathcal{H}[dW_1(t)\tfrac{1}{2}(x+iy)] + \mathcal{H}[dW_2(t)\tfrac{1}{2}y] + \sqrt{1+\nu}\,\mathcal{H}[dW_3(t)\tfrac{1}{2}x]\right\} P. \end{aligned} \tag{16}$$

where $\nu = 4N\mu \geq 0$ and $\chi = 4\mu C$. In this linearized approximation the ensemble average stationary state $\rho_\infty = \mathrm{E}[P^{\mathcal{U}}]$ has an infinite y variance and an x variance of unity. The ensemble $E^{\mathcal{U}}$ is parametrized by 6 complex numbers, the u_{ij} as in Eq. (5).

The stationary ensemble in this linearized limit is found by solving Eq. (16) for $t \to \infty$. This can be done analytically and it is found that the solution consists of Gaussian pure states (i.e., squeezed states[12]) which differ (within one ensemble) only by the mean values for x and y. That is to say, the index i in Eq. (6) is replaced by the ordered pair $(\bar{x}, \bar{y})$, and

$$P^{\mathcal{U}}_{(\bar{x},\bar{y})} = \text{ squeezed state with } \langle a \rangle = \alpha = \tfrac{1}{2}[2\sqrt{\mu} + \bar{x} + i\bar{y}]. \quad (17)$$

The two parameters (commonly denoted[12] by the complex number ϵ) which define the orientation and degree of squeezing are the same for all members of a given ensemble $E^{\mathcal{U}}$ and are complicated functions of the unraveling parameters u_{ij}. The weighting function $w^{\mathcal{U}}_{(\bar{x},\bar{y})}$ is a Gaussian function of $\bar{x}, \bar{y}$ such that

$$\int d\bar{x}\, d\bar{y}\, w^{\mathcal{U}}_{(\bar{x},\bar{y})} P^{\mathcal{U}}_{(\bar{x},\bar{y})} = \text{ the linearized } \rho_\infty \quad (18)$$

Because of the size of the parameter space and because the functional dependence of ϵ on u_{ij} can only be determined numerically, the search for the maximally robust unraveling is quite difficult. Work is still underway in completing this project, but some asymptotic results are clear. First of all, for small excess phase noise $\nu, |\chi| \leq 1$, the most robust ensemble consists approximately of coherent states. This coincides with the most common intuition about lasers, that they are "really" in a coherent state but with unknown phase. However, as the parameters ν or χ are increased the elements of $E^{\mathcal{R}}$ become increasingly squeezed. In particular, the variance of the phase quadrature is found from numerical studies to increase as

$$\text{Tr}[(y - \bar{y})^2 P_{(\bar{x},\bar{y})}] \sim \nu^{1/2}, |\chi|^{2/3}. \quad (19)$$

For very large ν or χ, the linearization used to derive Eq. (16) will break down. Specifically, the linearization will only be valid when $\langle (y - \bar{y})^2 \rangle \ll \mu$. Substituting this into Eq. (19) shows that for the linearization to be valid we require

$$\nu \ll \mu^2 \,, \quad |\chi| \ll \mu^{3/2}. \quad (20)$$

From the definitions of ν and χ, these can be seen to be identical to the second and third condition from Eq. (15). In other words the linearization is valid if and only if the device is operating in the regime in which it can be considered a laser.

This result can be interpreted as follows. In the limit where the device acts as a laser, the maximally robust ensemble $E^{\mathcal{R}}$ consists of members which have a well-defined coherent amplitude. For a laser with little excess phase noise these states actually approximate coherent states and the second representation in Eq. (14) is the most robust ensemble. For larger values of ν and χ they do not approximate coherent states, but they are nevertheless states in which the phase variance is small in an absolute sense. In the opposite limit the linearization approximation used in our maximally robust unraveling analysis breaks down. The reason it breaks down is because the members of the maximally robust ensemble are no longer states with well-defined coherent amplitude. They must therefore be states which are closer to number states than they are to coherent states. In the limit where the output of the device utterly fails to satisfy condition (4), the first representation in Eq. (14) must be the more natural one.

CONCLUSION

The above results show that the same parameter regime governs whether an atom laser gives a coherent output and whether its most robust unraveling generates states which have a well-defined coherent amplitude. It must be emphasized that the two results are completely independent in conception and definition. The correspondence between the two is therefore strong evidence that the definition we have used for the maximally robust ensemble is a good one, and that the idea is a useful one.

The techniques we have used are easily generalizable to any system which can be linearized, or which can be solved exactly. However, with more than one irreversible term in the master equation, a partly numerical solution may be necessary, as in the atom laser case. Some other models for which the maximally robust unraveling should be computable include other dissipative Bose-Einstein condensates (see, e.g., Ref. 13), and fluorescent atoms or cavity QED systems.[1] Like the atom laser model, these are systems with rich dynamics, so that the most robust states cannot be guessed *a priori*. Determining the maximally robust unraveling should lead to a better understanding of these and other open quantum systems.

ACKNOWLEDGMENTS

H.M.W. was supported by the Australian Research Council. J.A.V. thanks Profs. S.M. Barnett and K. Burnett for initial discussions.

REFERENCES

1. H. J. Carmichael, *An Open Systems Approach to Quantum Optics* Springer-Verlag, Berlin (1993).
2. G. Lindblad, Commun. math. Phys. 48:199 (1976).
3. M. Rigo and N. Gisin, p. 255 of Ref. 5 (1996).
4. H.M. Wiseman, p. 205 of Ref. 5 (1996).
5. Quant. Semiclass. Opt. 8 (1), special issue on "Stochastic quantum optics", edited by H. J. Carmichael (1996).
6. C. W. Gardiner, *Handbook of Stochastic Methods* Springer, Berlin (1985).
7. J. Gea-Banacloche, in: *New Frontiers in Quantum Electrodynamics and Quantum Optics*, A. O. Barut, ed., Plenum, New York (1990).
8. W. H. Zurek, Prog. Theor. Phys. 89:281 (1993).
9. W. H. Zurek, S. Habib and J. P. Paz, Phys. Rev. Lett 70:1187 (1993).
10. S. M. Barnett, K. Burnett and J. A. Vaccaro, J. Res. Natl. Inst. Stand. Technol. 101:593 (1996).
11. H. M. Wiseman, Phys. Rev. A 56:2068 (1997).
12. D. F. Walls and G. J. Milburn, *Quantum Optics* Springer, Berlin (1994).
13. J. Anglin, Phys. Rev. Lett. 79:6 (1997).

QUANTUM NOISE AND DYNAMIC CHAOS IN OPTO-ELECTRONIC LASER LIGHT TRANSFORMER

S. V. Larin, A. V. Masalov, A. N. Oraevsky and A. A. Putilin

Lebedev Physical Institute
Leninsky prospect 53, 117924 Moscow, Russia

Optical amplifier based on modulation has been implemented. Its noise has been measured in the range 40÷60% above quantum limit for gain up to 1000. Lasing of the amplifier with positive feedback is used to transform laser light. Transition of regimes of laser light transformer from stable oscillation to dynamic chaos has been observed.

INTRODUCTION

New opto-electronic device — optical amplifier based on modulation[1] — proved to be suitable instrument for laser light transformation. Amplification process in the modulation-based amplifier consists of converting the optical signal into radio frequency through heterodyning, and converting the amplified RF-signal back into optical region via sound wave in acousto-optical modulator. Amplification gain up to 10^7 has been achieved. In the present paper the experimental data on noise of modulation-based amplifier are discussed. Due to narrow amplification bandwidth limited by acousto-optic modulator, the modulation-based amplifier does not compete with conventional (stimulated emission) amplifiers in optical signal processing. We used it in oscillation regime (positive feedback added) as a laser light transformer. Transition of operation regimes of our laser light transformer from stable operation to periodic pulsation, then to doubling the period, and finally to chaos has been predicted theoretically.[2] In this paper these regimes are demonstrated experimentally.

NOISE IN MODULATION-BASED AMPLIFIER

The scheme of modulation-based amplifier is shown in Fig.1. It consists of a laser with a beam of high spatial and temporal quality, a photodiode (PD) with high quantum efficiency followed by electronic amplifier, and an acousto-optical modulator (AOM) in the regime of Bragg diffraction on a traveling acoustic wave. The amplification process may be separated into two stages. At first, the input optical signal is transformed into radio frequency by means of heterodyne photoreceiver. Then the RF wave (amplified or not) is converted back into light by acousto-optic modulator. When the intensity of

Quantum Communication, Computing, and Measurement 2
Edited by Kumar et al., Kluwer Academic / Plenum Publishers, New York, 2000.

light at the AOM input is high enough or/and the RF wave is amplified (by conventional electronics), the output light wave exceeds the input one. The amplification band of the amplifier is shifted from the laser frequency over the operation frequency of AOM, and the bandwidth coincides with that of AOM.

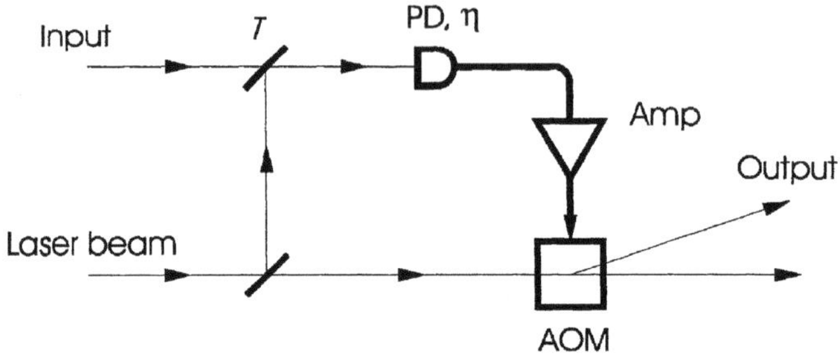

Figure 1. Schematic of modulation-based amplifier: PD — photodiode with quantum efficiency η, AOM — acousto-optical modulator, Amp — electronic amplifier.

To implement the modulation-based amplifier we used an Ar-ion laser in the regime of lowest-transverse single-longitudinal mode at 514 nm. The input signals at shifted frequency were produced from the laser light by additional AOM (not shown in Fig. 1). The input beam splitter transmitted 94.6% of input light intensity towards *p-i-n* photodiode with 80% quantum efficiency. Low noise electronic amplifiers were used to rise the RF signal with a gain in the range 50-100 dB. Some technical noise was introduced by electronics, however it was much less than the Poissonian noise of local oscillator. In our experiments an AOM with 26.6 MHz operation frequency and about 200 kHz bandwidth was used. In our experiments[1] amplification up to 10^7 has been achieved.

The principal noise of modulation based amplifier originates from RF shot noise produced in heterodyne detection. Other noise sources (excess noise of laser intensity, noise of electronics) can be decreased to negligible level. Experimental measurements with modulation-based amplifier had shown that its noise figure is close to quantum limit. Results of these measurements are summarized in this chapter.

To characterize the amplifier noise we had measured the number of photons N in the amplifier output with blocked ("vacuum") input. This value enters the phenomenological model of linear phase-insensitive phase-preserving amplifier:

$$\hat{a}_{out} = \sqrt{G}\hat{a}_{in} + \sqrt{N}\hat{b}_1^+ + \sqrt{N-(G-1)}\hat{b}_2, \tag{1}$$

where $\hat{a}_{in}$, $\hat{a}_{out}$ are the photon annihilation operators of input and output modes, $\hat{b}_1$, $\hat{b}_2$ are these operators for two "vacuum" modes which simulate noise, and G is an amplifier gain. Eq.(1) allows to calculate signal-to-noise transfer in any scheme of amplifier usage. For the ideal (minimal noise) amplifier $N = G - 1$,[3] while for an amplifier based on inversion of populations $(n_1 < n_2)$: $N = (G-1)n_2/(n_2 - n_1)$. The noise of ideal amplifier calculated for input equals $N/G = 1 - 1/G$, which is "one photon per mode" in the limit of large G.

Calculations of the shot-noise contribution in the modulation-based amplifier[4] show that the noise is proportional to gain: $N = G/\eta T$, where T is the transmittance of input beam-splitter for signal and η is the photodiode quantum efficiency. This noise exceeds the standard quantum limit $G - 1$, however, it can be made rather close to it. Only at low gain $(G \to 1)$ the difference is essential: in the ideal amplifier noise is close to zero ($N \to 0$) and it only slightly disturbs the quantum state of amplified light, while modulation based amplifier does disturb the state of light at $G = 1$.

The absolute levels of noise at the output of modulation-based amplifier have been monitored by means of another heterodyne photoreceiver (not shown in Fig. 1) and spectrum analyzer. Measured values of noise versus gain are presented in Fig. 2. Noise calculated for input N/G is displayed, where the scale means "number of photons per mode." According to the theory of modulation based amplifier[4] the principal input noise does not depend on G and equals $1/\eta T$. In our measurements $T = 94.6\%$, $\eta = 80\%$, and $1/\eta T = 1.32$; taking into account the 9% contribution of electronic noise together with shot noise, one should expect $N/G = 1.44$ (marked by dashed line). Our measurements are in agreement with this expectation value. Hence, the noise (at input) of our optical amplifier is within a range $1.40 \div 1.62$ "photons per mode."

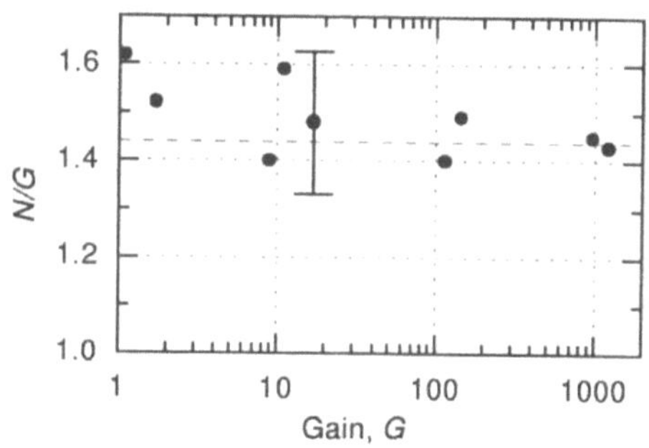

Figure 2. Input noise of modulation-based amplifier (photons per mode) versus gain; expected level is marked by dashed line.

Narrow amplification bandwidth (≈400 MHz, limited by acousto-optic modulator in our case) is unfavorable feature for optical signal processing. However, it is advantages in "lasing" regime.

REGIMES OF LASER LIGHT TRANSFORMER

Oscillation of modulation-based amplifier can be achieved when the part of output beam diffracted by AOM is sent to the input (Fig. 3). While an oscillation in a conventional laser starts from the spontaneous noise of active medium, in our device the process of lasing starts from the shot noise of the heterodyne photodiode, which is a result of Poissonian fluctuations of local wave intensity. Energy of oscillating wave in our oscillator is supplied by laser beam with good spatial quality. Hence, our device is nothing but laser light transformer. Our first results on laser light transformer are presented in Ref. 1. Two analogues of the modulation-based amplifier should be noted: heterodyne interferometer (with acousto-optical modulator) as a microdisplacement sensor;[5] microwave oscillator with electro-optical modulator for generation of stable microwave signals.[6]

Theory

To understand regimes of operation of laser light transformer the stationary theory is enough. In the stationary case these regimes can be described by two equations. The

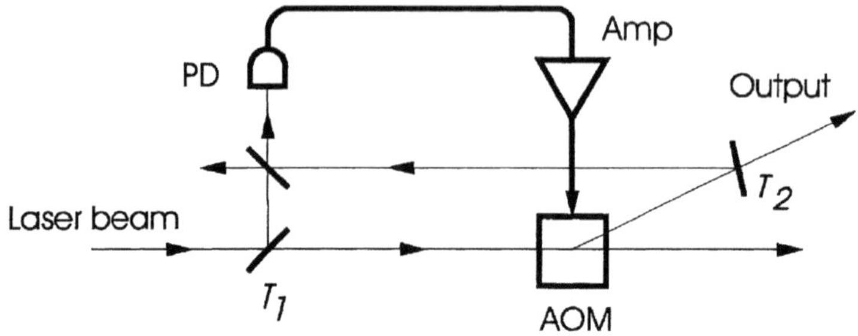

Figure 3. Schematic of laser light transformer (see notations in Fig. 1).

first one relates the amplitude of diffracted light E_{dif} at AOM output with the light amplitude E_{AOM} at AOM input and the amplitude of amplified RF voltage U:

$$E_{dif} = E_{AOM} \sin\left(\frac{\pi}{2} \cdot \frac{U}{U_{AOM}}\right) \quad , \tag{2}$$

where U_{AOM} is the voltage of 100% efficiency of AOM. The second equation relates the RF voltage U produced by heterodyne photoreceiver with the amplitude of local oscillator wave E_{LO} and that of diffracted beam:

$$U = g\rho\left(\frac{\eta e}{\hbar\omega}\right)\sqrt{2(1-T)(1-T_2)}E_{LO}E_{dif} \quad , \tag{3}$$

where e is electron charge, $\hbar\omega$ is energy of light quanta, g is electronic gain, and ρ is RF impedance. We take into account that E_{LO} and E_{AOM} are proportional to the amplitude laser wave, and Eqs.(2),(3) yield for the amplitude of the "lasing" wave E_{dif}:

$$E_{dif} = p\left|\sin E_{dif}\right| \quad , \tag{4}$$

where E_{dif} is made dimensionless, while all parameters are collected in dimensionless value p. Note that p is proportional to laser power. This equation allows many solutions and is known as dynamic chaos productive. We consider here solutions and regimes of only lowest branch $p = 1 \div 3.14$. At $p < 1$ there is no solution at all: $p = 1$ is a threshold of oscillation. Up to $p \approx 2.262$ Eq.(4) has a stable solution, and at $p = \pi/2 \approx 1.571$ conversion efficiency reaches its maximal value T_1T_2 provided the maximal conversion efficiency of AOM is close to 100%. Solutions lose their stability at $p > 2.262.$, and the well known scenario of the transition from stable operation to pulsations and chaos takes place:

$p = 2.262 \div 2.618$ — pulsation of a system with a primary period;
$p = 2.618 \div 2.696$ — pulsation with a doubled period;
$p \geq 2.715$ — chaotic pulsation.

More details on the theory of laser light transformer are given in Ref. 2. Experiments with laser light transformer are described in the next section.

Experiments

In our experiments with laser light transformer the same Ar-ion laser was used (see previous section). Acousto-optical modulator operated at frequency 84.5 MHz with about 400 kHz bandwidth and $U_{AOM} = 0.9$ V. Threshold of oscillation P_{th} at $g \sim 100$ dB was about 10 mW of input laser power. In the range of input power up to $2P_{th}$ the transformer exhibited stable operation. Efficiency as high as 90% have been observed. The ability of light transformer to stabilize laser power at the top of this range is demonstrated in Fig. 4. Input light from Ar-ion laser was tuned to 300 Hz power pulsations with an amplitude about 10%. At the output of the transformer these pulsation were reduced up to 1%, conversion efficiency being about 70%. However, this stabilization reduces only classical variations of intensity (excess noise) while the quantum fluctuations remain unchanged.

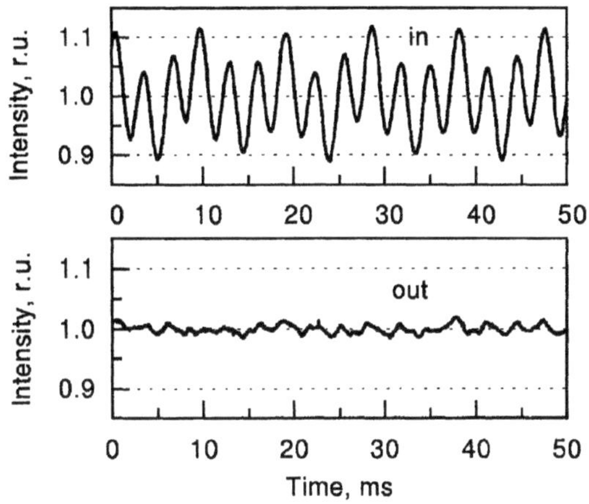

Figure 4. Temporal variations of input laser power (*in*) and that at the transformer output (*out*).

At higher input laser power ($> 2P_{th}$) non-stable regimes have been observed. In Fig. 5. the data on temporal and spectral behavior of RF-amplitude in several regimes of oscillation are collected. Regimes in diagrams from 1 to 4 follow one after another with increase of input laser power. Spectrum of RF-amplitude just after threshold is presented in diagram *1s* (time dependence here is *const*). The main peak has a carier frequency within the frequency range of acousto-optic modulator (84.5 MHz in this case). Two weak sidebands at 340 kHz distance from the main peak are longitudinal modes of oscillation; mode separation interval is defined by acoustic wave delay in AOM. These modes are far below threshold and do not play any role in further regimes. Pulsation of amplitude with period 10 μs is shown in diagram *2t*. The relevant period about 100 kHz is clearly seen in the spectrum *2s*. Further increase of input laser power allows doubling of pulsation period *3t*: two different amplitude maxima follow one another. Weak intermediate peaks at 50 kHz distance appears in the spectrum *3s*. Diagram *4t* shows chaotic pulsation of light transformer. In this case spectrum broadens up to 1 MHz (note change in scale).

The data on temporal behavior of RF-amplitude in chaotic regime have been studied for the value of correlation dimension to verify genuine dynamic chaos. These data have been embedded into a space of dimension from 2 to 10, and the probability P to have point separation higher than ϵ have been calculated. Fig. 6 shows an example of such probability for space dimension 5. Descending part of log-log plot has a slope about 2.2 which is essentially less than the space dimension (=expected value for random noise). Data on slope calculated for various space dimensions are presented in Fig. 7. Values of slope tend to 2.3 which is treated as the correlation dimension of dynamic chaos in the light transformer.

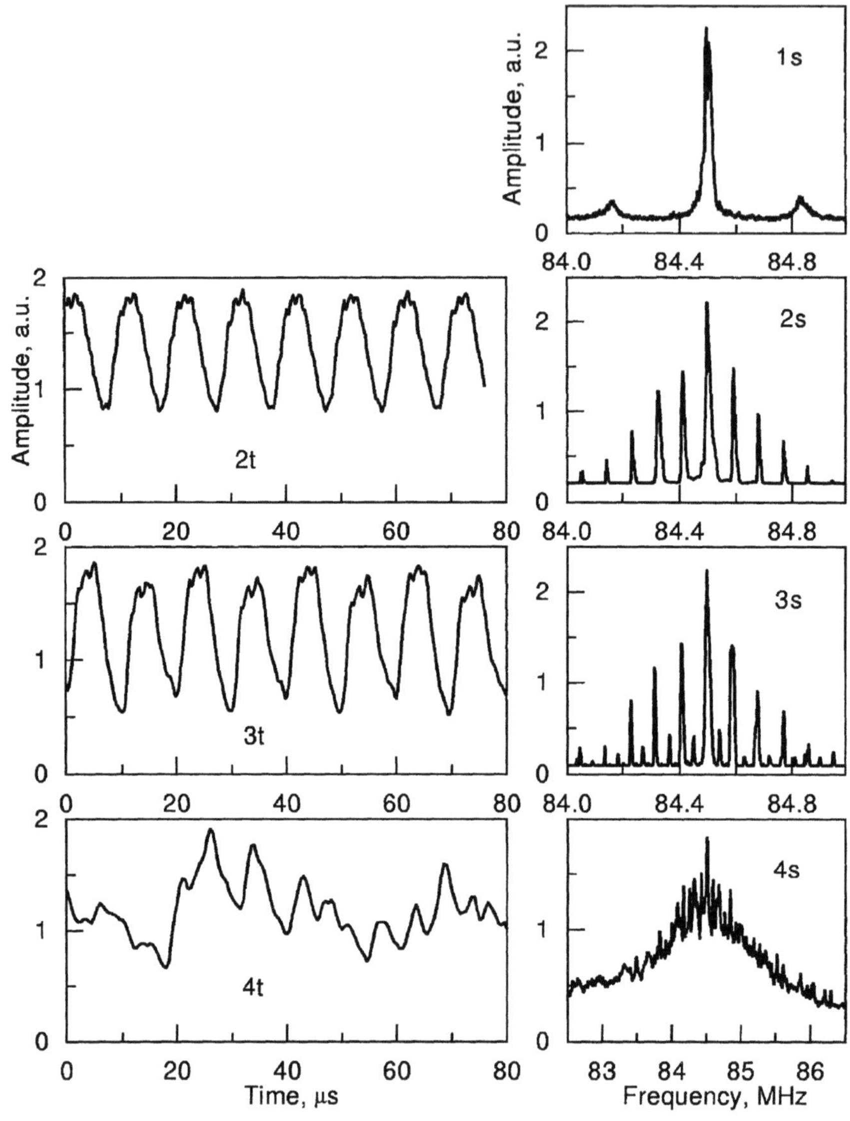

Figure 5. Temporal and spectral diagrams of RF-amplitude in different operation regimes of laser light transformer: *1s* — stable oscillation, *2t,s* — periodic pulsation, *3t,s* — double period pulsation, *4t,s* — dynamic chaos.

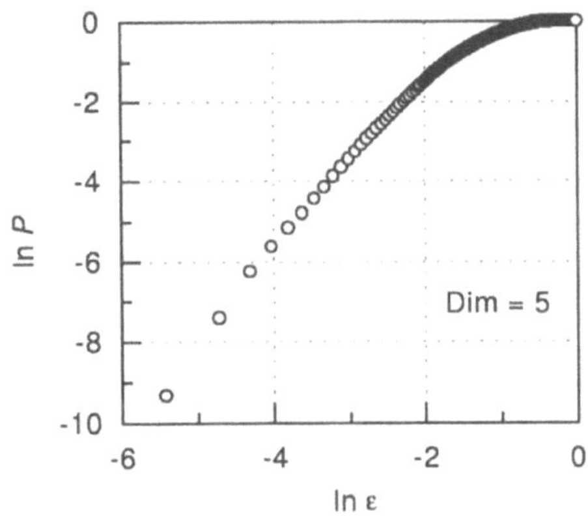

Figure 6. Probability to have point separation higher than ϵ for experimental data embedded into space of dimension 5.

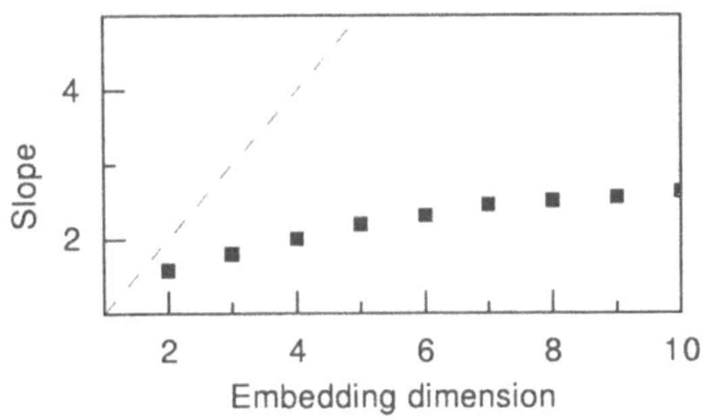

Figure 7. Slope of probability log-log plots versus embedded dimension.

CONCLUSIONS

We have implemented a modulation-based optical amplifier which is able to amplify light waves coherent with pump laser. Noise figure of this amplifier can be made close to the quantum limit; in our device the noise has been measured in the range of 1.4÷1.6 input photons per mode.

Lasing of the modulation-based amplifier has been achieved. Such an oscillator can be used as a laser light transformer; it may operate as an external intensity or/and frequency stabilizer of laser light, which requires no feedback to the laser source. Regimes of the transformer undergo scenario of transition to dynamic chaos: stable operation $\rightarrow$ periodic pulsation $\rightarrow$ double period pulsation $\rightarrow \ldots \rightarrow$ dynamic chaos.

ACKNOWLEDGMENTS

The work was supported by Grant N98-02-16835 of Russian Foundation for Basic Research.

REFERENCES

1. V. N. Konopsky, A. V. Masalov, A. A. Putilin and M. V. Vasilyev, Optical amplifier and oscillator based on modulation, *in*: "Coherence and Quantum Optics VII," J. H. Eberly, L. Mandel, and E. Wolf, Eds., Plenum Press, New York and London (1996).
2. A. V. Masalov and A. N. Oraevsky, A contribution to the theory of a modulation light converter, *Quantum Electronics* 26:909 (1996).
3. C. M. Caves, Quantum limit of noise in linear amplifier, *Phys. Rev. D* 26:1817 (1982).
4. A. V. Masalov and A. A. Putilin, Quantum noise of a modulation optical amplifier, *Optics and Spectroscopy* 82:823 (1997).
5. T. V. Babkina, V. V. Grigoryants, Yu. B. Il'in and A. A. Lobanov, The use of a laser generating interferometer as an optical sensor of microdisplacements, *Quantum electronics* 21:1384 (1991).
6. X. S. Yao and L. Maleki, Electronic microwave oscillator, *JOSA B* 13:1725 (1996).

QUANTIZED PHASE-DIFFERENCE

Jonas Söderholm, Alexei Trifonov[1], Tedros Tsegaye, and Gunnar Björk

Department of Electronics, Royal Institute of Technology (KTH)
Electrum 229, S-164 40 Kista, Sweden

[1]Permanent address: Ioffe Physical Technical Institute
26 Polytekhnicheskaya, 194021 St. Petersburg, Russia

From the experimentalist's point of view, it is natural to think of a phase measurement as a measurement that involves a reference time. As a part of the system, the reference time generating "clock" must be taken into account in the quantum treatment of the measuring process. This could be argued to be the reason why there does not exist any observable corresponding to the absolute phase in the unrestricted Hilbert space. Luis and Sánchez-Soto [Phys. Rev. A **48**, 4702 (1993)] have investigated the case where the "clock" is another harmonic oscillator and defined a Hermitian phase-difference operator, which is valid in the infinite Hilbert space.

In the present work we propose and experimentally demonstrate a setup for direct measurement of the phase-difference. We derive the phase-difference distributions for some two-mode states and compare them with the experimental results.

THE PHASE-DIFFERENCE OPERATOR

The phase problem in quantum mechanics has a long history[1]. It is almost as old as the quantum theory itself. The original phase operator suggested by Dirac[2] was found to have some peculiarities[3,4]. For example it was non-Hermitian[4], while a measurable quantity should have a corresponding Hermitian operator according to the quantum theory. To obtain a Hermitian phase operator several approaches have been made. These include considering extended[6,7] and finite[8] Hilbert spaces. Recently Fujikawa[9] has proved that there does not exist any Hermitian single mode phase operator in the ordinary infinite Hilbert space. The lack of such an operator can be identified with the fact that when actually measuring the phase, a reference system is needed.

In line with this, Luis and Sánchez-Soto[10] have defined a phase-difference operator for a two-mode field. They required the phase-difference operator $\hat{\Phi}_{12}$ to fulfill the decomposition relation

$$\hat{a}_1\hat{a}_2^\dagger = e^{i\hat{\Phi}_{12}}\sqrt{\hat{n}_1\left(\hat{n}_2+1\right)}, \tag{1}$$

and the commutator

$$\left[e^{i\hat{\Phi}_{12}}, \hat{n}_1+\hat{n}_2\right] = 0, \tag{2}$$

Quantum Communication, Computing, and Measurement 2
Edited by Kumar et al., Kluwer Academic / Plenum Publishers, New York, 2000.

which is obtained from correspondence with the Poission bracket in classical mechanics. It turns out that the conjugate observable to phase-difference is the number-difference $\hat{n}_{12} = (\hat{n}_1 - \hat{n}_2)/2$. We therefore define the equipartition states as

$$|\phi^{(N)}\rangle = \frac{1}{\sqrt{N+1}} \sum_{n=0}^{N} e^{i\theta_n} |n, N-n\rangle, \tag{3}$$

where θ_n are arbitrary real numbers. This makes all the two-mode number states in the N:th excitation manifold equally probable. The two-mode Pegg-Barnett states

$$|\varphi^{(N)}\rangle = \frac{1}{\sqrt{N+1}} \sum_{n=0}^{N} e^{in\varphi^{(N)}} |n, N-n\rangle, \tag{4}$$

are special cases of the more general equipartition states (3). A generalization of the Luis and Sánchez-Soto Hermitian phase-difference operator can now be written

$$\hat{\Phi}_{12} = \sum_{N=0}^{\infty} \hat{\Phi}_{12}^{(N)} = \sum_{N=0}^{\infty} \sum_{r=0}^{N} \phi_r^{(N)} e^{-i\phi_r^{(N)}\hat{n}_{12}} |\phi^{(N)}\rangle\langle\phi^{(N)}| e^{i\phi_r^{(N)}\hat{n}_{12}}. \tag{5}$$

The phases defining the orthogonal set of phase-difference states are given by

$$\phi_r^{(N)} = \phi_0^{(N)} + \frac{2\pi r}{N+1}, \qquad r = 0, 1, \ldots, N, \tag{6}$$

where $\phi_0^{(N)}$ is a reference phase-difference, as in the treatment by Pegg and Barnett[8].

Since the phase-difference operator commutes with the operator for the total photon number, it naturally splits into operators operating in different excitation manifolds. We therefore define the (continuous) phase-difference distribution function in the N:th manifold for the state $\hat{\rho}$ as

$$P_{\hat{\rho}}^{(N)}(\phi) = \langle\phi^{(N)}| e^{i\phi\hat{n}_{12}} \hat{\rho} e^{-i\phi\hat{n}_{12}} |\phi^{(N)}\rangle. \tag{7}$$

With this choice of normalization (which is different from the one originally used[10]) we find

$$\sum_{N=0}^{\infty} \sum_{r=0}^{N} P_{\hat{\rho}}^{(N)}\left(\phi_r^{(N)}\right) = \sum_{N=0}^{\infty} p_N = 1, \tag{8}$$

where p_N is the probability of detecting a total photon number $N = n_1 + n_2$ in the state $\hat{\rho}$. Hence, the *discrete* distribution $P_{\hat{\rho}}^{(N)}\left(\phi_r^{(N)}\right)$, including all manifolds N, can be interpreted as a phase-difference probability distribution. This interpretation is confirmed by using (5) to compute the moments of the phase-difference. The mean value of the M:th moment of the phase-difference operator is then given by

$$\langle\hat{\Phi}_{12}^M\rangle = \sum_{N=0}^{\infty} \sum_{r=0}^{N} \left(\phi_r^{(N)}\right)^M P_{\hat{\rho}}^{(N)}\left(\phi_r^{(N)}\right). \tag{9}$$

In the case of a two-mode number state, the phase-difference distribution (7) is flat, and the distribution can be measured at every point by a measurement setup corresponding to the appropriate reference phase $\phi_0^{(N)}$. However, as a consequence of the finite number of orthogonal states in each manifold, the phase-difference variance for a two-mode photon number state is given by the manifold N as

$$\langle n, N-n| \left(\Delta\hat{\Phi}_{12}\right)^2 |n, N-n\rangle = \frac{\pi^2 N(N+2)}{3(N+1)^2}, \tag{10}$$

independent of the chosen reference phase-difference, where $n = 0, 1, \ldots, N$. In the limit $N \to \infty$ the phase-difference states become dense, and we obtain $\langle\left(\Delta\hat{\Phi}_{12}\right)^2\rangle = \pi^2/3$, as expected from a flat distribution function in a 2π-interval.

MEASURING THE PHASE-DIFFERENCE DISTRIBUTIONS

When doing precise measurements in the quantum regime of light, one usually uses single photon detectors. To be able to use single photon detectors in phase-difference measurements we have to unitarily transform the phase-difference states (4) onto the two-mode number states. After such a unitary transformation $\hat{U}$, the number states, and therefore also the phase-difference states, can be distinguished by a dual photon number measurement. As we limit ourselves to direct measurements, which do not involve any additional mode, the unitary transformation is energy-conserving. Since photon number measurements in principle can be done as quantum non-demolition (QND) measurements[11], phase-difference QND-measurements are also allowed by physics. We then make use of the inverse transformation $\hat{U}^{\dagger}$, as sketched in Figure 1.

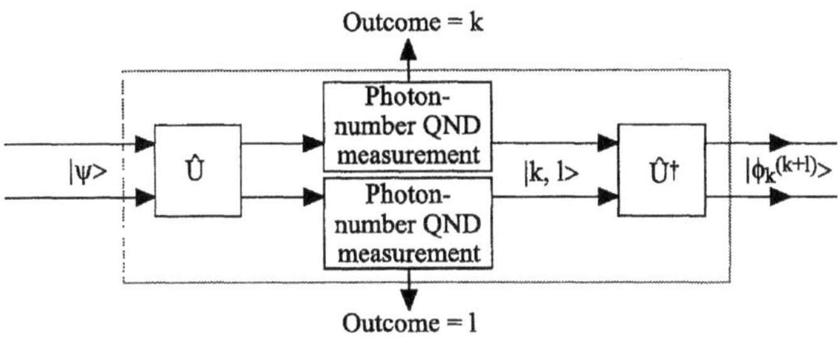

Figure 1. A schematic phase-difference QND-setup.

As we have seen, the phase-difference operator divides into operators in the different manifolds. This makes it possible to treat each manifold individually. The zeroth manifold is of no interest since it just consists of the vacuum state, which is simultaneously a phase-difference state and a two-mode number state. In the first manifold the necessary transformation $\hat{U}$ can be realized with a differential phase-shift and a 50/50 beam-splitter. Different settings of the differential phase-shift will make the measurement correspond to phase-difference operators with different reference phases $\phi_0^{(N)}$. In this way the complete continuous phase-difference distribution in the first manifold can be determined. Such measurements were done a long time ago.

In the second manifold it is impossible to transform the two-mode Pegg-Barnett states to two-mode number states by using just a differential phase-shift and a beam-splitter. However, there is one particular equipartition state that can be unitarily transformed to a two-mode number-state by using a (21/79 %) beam-splitter. The inverse transformation will transform the state $|1, 1\rangle$ onto the equipartition state

$$|\xi\rangle = \frac{1}{\sqrt{3}} \left(|0, 2\rangle + i|1, 1\rangle + |2, 0\rangle \right). \tag{11}$$

This makes it possible to determine the complete phase-difference distribution functions. To do this we let the phase-shift sweep through a 2π-interval, while a $2\pi/3$ sweep would have been sufficient if we could realize a transformation of all three equipartition states specifying the phase-difference operator onto two-mode number states.

In this way we have been able to measure the phase-difference distribution functions for some states in the second manifold, which is the lowest non-trivial manifold. Figure 2 shows our experimental setup for measuring the phase-difference distribution

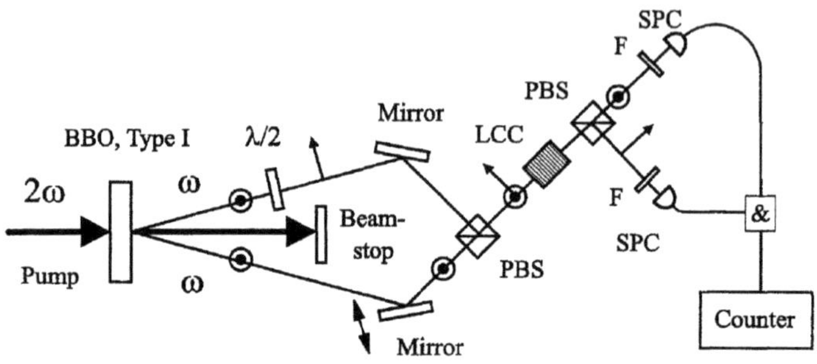

Figure 2. The experimental setup for measuring the phase-difference distribution function of an equipartition state.

function of the equipartition state (11) using a Ti:Sapphire laser, pumped by an Ar-laser, as light source. The 100 fs long pulses of 780 nm IR light were first frequency doubled and the blue pulses were subsequently down-converted. Irises were used to create two modes consisting of frequency degenerate photon-pairs. Since BBO nonlinear crystals with Type I phase matching were used both for the frequency doubling and the spontaneous down-conversion the generated photons had parallel (linear) polarization.

To minimize problems with mechanical drift the photons were made to travel in the same spatial mode, but with perpendicular polarizations. This was accomplished by letting the polarization of one of the modes be rotated by a $\lambda/2$-plate, and using a polarizing beam-splitter (PBS). The differential phase-shift ϕ was realized by a variable bi-refringent liquid crystal cell (LCC). By applying a voltage to the LCC it is possible to change the difference in refractive index, and therefore also the phase-difference, between the orthogonal polarization modes. Rotating the polarization basis, compared to the basis defined by the PBS, is equivalent to the effect of a beam-splitter. Therefore, by just orienting the fast axis of the LCC properly to the impinging polarization modes a 21/79 % beam-splitter could be realized, and an equipartition state could be created and subsequently phase-shifted. To analyze the state a second PBS, parallel to the first PBS, was placed after the LCC. The outputs of the second PBS were detected with a pair of EG&G single photon counters (SPC) preceded by 10 nm interference filters. The correlation between the photon detections was determined by a home-built correlation AND circuit and a HP53131A counter. The measured coincidence as a function of the imposed differential phase-shift is plotted in Figure 3. The pulse repetition rate was 80 MHz and the typical average counting rate for the detectors was 10,000 per second. With two uncorrelated Poissonian sources this would correspond to 1.25 coincidence counts/second, which is well below the obtained values. The dark current of the detectors was 20 counts/second.

Ideally, with perfectly overlapping modes, the correlation would be proportional to

$$P^{(2)}_{|\xi\rangle}(\phi) = \frac{\sin^2(3\phi/2)}{9\sin^2(\phi/2)}, \tag{12}$$

which is the dashed line plotted in Fig. 3. Since the coincidence window of the detectors was 3 ns, two strictly non-overlapping pulses could still cause a double-click. A calculation shows that the probability for such coincidences is

$$\frac{1}{2}\left(1+\frac{\sin^2(3\phi/2)}{9\sin^2(\phi/2)}\right). \tag{13}$$

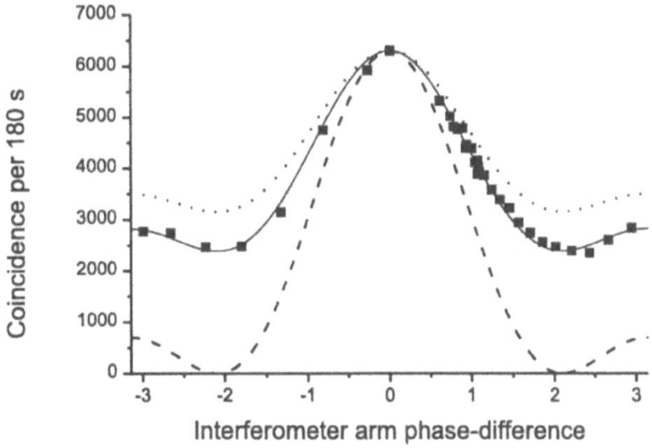

Figure 3. The measured phase-difference distribution for a phase-difference state in the second manifold, compared to the ideal one (dashed) and the result in the case of two non-interfering photons (dotted).

This function is also the one found using the classical wave description of the experiment, and has been plotted dotted in Figure 3. We see that our results fall into the regime that cannot be explained by a classical description. Since the experimental curve has a higher visibility than the function (13), the experiment clearly shows effects of quantum interference. The relatively weak deviation of the measured correlation from the classical result (compared to the ideal one) is believed to be due to non-perfect mode overlap. Partial overlap would give rise to a phase-difference distribution between the ideal and classical one, as obtained in the experiment. That this is the cause of the faint quantum features in this experiment is supported by phase-difference measurements of two-mode coherent states. These results are presented in Figure 4 and are seen to be in good agreement with the curves given by theory. The reason is that it is much simpler to prepare two-mode coherent states with nearly perfect mode overlap.

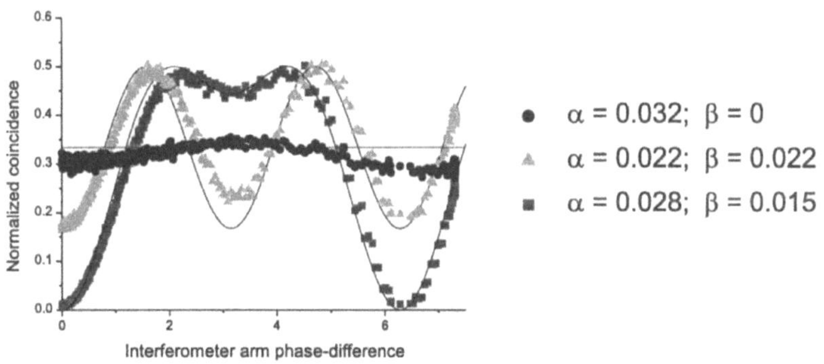

Figure 4. The measured phase-difference distribution functions for three two-mode coherent states $|\alpha\rangle \otimes |\beta\rangle$.

DISCUSSION

We have realized measurements of the phase difference operator in the lowest non-trivial manifold. The measurements on two-mode coherent states exhibits better agreement with theory than the measurements of equipartition states. We think that use of longer pulses may yield a better mode overlap, and therefore a better agreement with theory, for the latter.

ACKNOWLEDGEMENT

This work was supported by Swedish Technical Science Research Council, The Royal Swedish Academy of Science and by INTAS grant 96/167.

REFERENCES

1. R. Lynch, Phys. Rep. **256**, 367 (1995).
2. P. A. M. Dirac, Proc. R. Soc. London Ser. A **114**, 243 (1927).
3. W. H. Louisell, Phys. Lett. **7**, 60 (1963).
4. L. Susskind and J. Glogower, Physics **1**, 49 (1964).
5. P. Carruthers and M. M. Nieto, Rev. Mod. Phys. **40**, 411 (1968).
6. R. G. Newton, Ann. Phys. **124**, 327 (1980).
7. M. Ban, Phys. Rev. A **48**, 3452 (1993).
8. D. T. Pegg and S. M. Barnett, Europhys. Lett. **6**, 483 (1988).
9. K. Fujikawa, Phys. Rev. A **52**, 3299 (1995).
10. A. Luis and L. L. Sánchez-Soto, Phys. Rev. A **48**, 4702 (1993), Opt. Comm. **105**, 84 (1994), Phys. Rev. A **53**, 495 (1996), A. Luis and J. Peřina, Phys. Rev. A **54**, 4564 (1996).
11. V. B. Braginsky and F. A. Khalili, *Quantum Measurement*, (Cambridge University Press, Cambridge, 1992). Chap. 4 and 11.

RELATIVISTIC QUANTUM THEORY WITH A PHYSICAL STATE VECTOR AND HYPOTHETICAL LAWS OF REDUCTION

Bernd A. Berg[1,2]

[1]Department of Physics
[2]Supercomputer Computations Research Institute
Florida State University
Tallahassee, FL, 32306

Evolution of a Physical State Vector (PSV) is described as governed by two distinct physical laws: Continuous, unitary time evolution and a relativistically covariant reduction process. Non-local measurements, proposed by Aharanov and Albert, are excluded when the reduction is attributed to measurement devices which are included in the PSV. The existence of a PSV allows to formulate phenomenological laws of reduction. A proposal is made for which the collapse time is found to be $\tau^c = b\hbar/\triangle E$, where $\triangle E$ is a difference in energy distribution between alternative world branches. Experiments yield the bounds $1.35 \cdot 10^{11} < b < 4.5 \cdot 10^{18}$.

INTRODUCTION

For a relativistic Quantum Field Theory (QFT) the idea of a Physical State Vector faces difficulties due to the well-known fact that spacelike measurements which are instantaneous in one inertial frame are no longer instantaneous in a different inertial frame. Aharanov and Albert[1] conclude that no consistent PSV assignment exists, because no relativistically satisfactory version of the collapse postulate of non-relativistic quantum theory can be found. Their analysis relies on constructing local measurement devices which allow to monitor a non-local state history.

In the following I present an introduction to my work [2], where it is the shown that measurements of the Aharonov-Albert type are excluded, once the measurement device is considered to be part of the PSV. The assumption that unitary, continuous evolution and state vector reduction (collapse) are distinct physical processes enters my argument. In a covariant way the PSV evolution process is defined in the next section. The relativistic spacetime picture proposed by Hellwig and Kraus[3], and criticized in ref.[1], remains inconsistent. Subsequently, ref.[4] is sketched where I have formulated phenomenological laws of reduction as mentioned in the abstract.

Quantum Communication, Computing, and Measurement 2
Edited by Kumar et al., Kluwer Academic / Plenum Publishers, New York, 2000.

PHYSICAL STATE VECTOR AND REDUCTION

In this paper we deal with states in the Schrödinger picture. The question may arise, whether statistical mixtures ought better to be considered. Let us first have a look at classical physics. There, the existence of a single state is guaranteed by definition and statistics is an essential tool in view of the practical impossibility to control a large number of initial conditions. Here I take a similar stance on QFT, I assume that a PSV of the world $|\Psi_W\rangle$ exists, although a complete set of measurements to determine the initial state cannot be performed. The essential feature of QFT is that even complete knowledge of the state does only allow stochastic predictions.

General state vector evolution

In the following we introduce the notion Lorentz Covariant Spacelike Hypersurface (LCSH) for spacelike hypersurfaces which will allow for a covariant formulation of the reduction process. We call two spacetime points $x_0 = (ct_0, \vec{x}_0)$ and $x_1 = (ct_1, \vec{x}_1)$ spacelike when

$$(x_0 - x_1)^2 = c^2 (t_0 - t_1)^2 - (\vec{x}_0 - \vec{x_1})^2 \leq 0 \tag{1}$$

holds. In case of the equal sign, we assume that x_0 and x_1 are defined as the limit of points

$$x_i = \lim_{\delta\to 0, \delta>0} x_{i,\delta}, \ (i = 0, 1) \ \text{ such that } \ (x_{0,\delta} - x_{1,\delta})^2 < 0 \ \text{ holds for all } \ \delta > 0\,.$$

A LCSH is now defined to be a hypersurface S which fulfills two conditions: (1) Equation (1) holds for any two points $x_0 \in S$ and $x_1 \in S$. (2) To insure Lorentz covariance of the collapse process, S is build from light cone sections. Let S_0 and S_1 be LCSHs. The hypersurface S_1 is said to be in the *future* of S_0, when for any two points $x_0 \in S$ and $x_1 \in S_1$ with $\vec{x}_0 = \vec{x}_1$ we have $t_0 \leq t_1$, and there exist some such points with $t_0 < t_1$. Similarly single spacetime points or sets of such points are defined to be in the future (or past) of a LCSH. Finally, for each LCSH S hypersurfaces S_+ and S_- are defined by taking the time component of all their spacetime points in accordance with the limits

$$t_+ = \lim_{\epsilon\to 0,\epsilon>0} (t + \epsilon) \ \text{ and } \ t_- = \lim_{\epsilon\to 0,\epsilon>0} (t - \epsilon) \tag{2}$$

in the sense that the indicated limits have to be taken outside of all other operations.

Assume a state vector is initially defined on some LCSH S_{0+}. Its relativistic evolution follows two laws:

(a) Unitary and causal evolution to some LCSH S_{1-} in the future of S_{0+}

$$|\Psi(S_{1-})\rangle \ = \ U(S_{1-}, S_{0+}) \, |\Psi(S_{0+})\rangle \tag{3}$$

when no reduction happens on any hypersurface in-between S_{1-} and S_{0+}.

(b) A stochastic transformation

$$|\Psi(S_{1-})\rangle \ \to \ |\Psi(S_{1+})\rangle \tag{4}$$

describes reduction on S_1. From S_{1+} the state vector evolves to a next LCSH S_{2-} by unitary and causal evolution. On S_2 reduction takes place, and so on.

The process is graphically depicted in figure 1. A light cone section becomes added to the initial LCSH S_0 such that the created surface S_1 is in the future of S_0. On S_1 reduction takes place from S_{1-} to S_{1+}. The $S_{0\pm}$ and $S_{1\pm}$ surfaces are not explicitly indicated in figure 1, their locations are obvious. In the next subsection we consider the reduction (b) in detail and come to a process where the addition of a single new light cone section is typical. Note, for the speed of light $c \to \infty$ a LCSH becomes an instantaneous hyperplane.

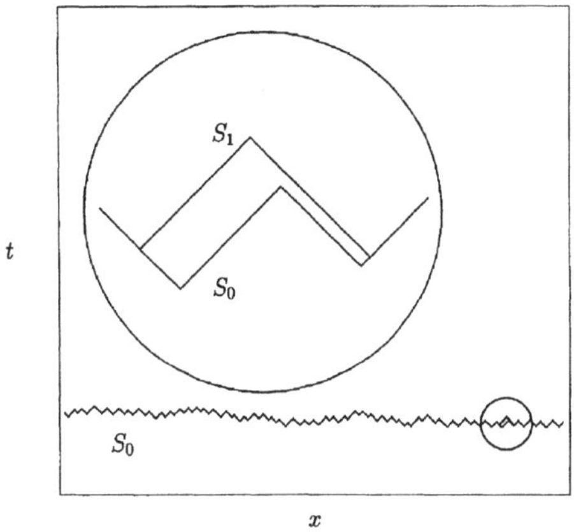

Figure 1. One step in the evolution of a LCSH. In the encircled (and enlarged) region one light cone section is added to S_0, which creates S_1 in the future of S_0.

Measurement and Reduction

From the fact that experimental measurements are made, it follows that it is possible to construct bound states $|\Psi_D\rangle$, called detectors, which have the ability to perform them. Eventually the detector may project some quantum state on an eigenstate of certain operators, or onto a state out of the space of eigenstates in case that the eigenvalue in question is degenerate. I proceed with a description in terms of the PSV: A detector $|\Psi_D\rangle$ can be constructed and a Quantum Object (QO) $|\psi\rangle$ (for instance a single electron) can be prepared, such that over some time[5] period the factorization

$$|\Psi_W\rangle = |\psi\,\Psi_D\rangle\;|\Psi_R\rangle\;, \tag{5}$$

where $|\Psi_R\rangle$ describes the rest of the world, is meaningful in the following sense: Although, due to interactions, this factorization can never be exact, the evolution of

$$|\Psi\rangle = |\psi\,\Psi_D\rangle\;, \tag{6}$$

considered in isolation, describes nevertheless the features we are interested in correctly and proceeds as follows. Initially there is very little overlap between the wave functions of the QO and of the detector, such that approximately

$$|\Psi\rangle = |\psi\rangle\;|\Psi_D\rangle\;. \tag{7}$$

Subsequently, $|\psi\rangle$ and $|\Psi_D\rangle$ interact and this is expressed by the notation $|\psi\,\Psi_D\rangle$, which is already used in equations (5) and (6). At some point in its spacetime the detector performs a transition

$$|\psi\,\Psi_D\rangle \to |\Psi_D^{\prime n}\rangle \tag{8}$$

such that the (macroscopic) state $|\Psi'^{n}_{D}\rangle$ constitutes a measurement of an observable or set of observables $\mathcal{O}_n$, according to the purpose for which the detector was constructed. The measurement (8) is here interpreted as a physical ability of the detector. In the non-relativistic limit complete measurements, as discussed in quantum theory books, can be performed to the extent that it is possible to design a detector which measures a complete set of operators $\mathcal{A}$ such that the state $|\Psi'^{n}_{D}\rangle$ factorizes in the form

$$|\Psi'^{n}_{D}\rangle = |\psi_n\rangle\, |\Psi^{n}_{D}\rangle \tag{9}$$

where $|\psi_n\rangle$ is the eigenstate of the operators $\mathcal{A}$ and the state $|\Psi^{n}_{D}\rangle$ allows us to access this information (through the neglected interactions with $|\Psi_R\rangle$). For real detectors the factorization (9) is the exception and not the rule.

We are now ready to define reduction. For the measurement process (8) we distinguish two cases:

(1) $|\psi\rangle = |\psi_n\rangle$ holds for the initial QO. Then the transformation (8) appears to be consistent with the time evolution (a) and we assume that this is indeed the case.

(2) $|\psi\rangle = c_n\,|\psi_n\rangle + c_{\bar{n}}\,|\psi_{\bar{n}}\rangle$ with $|c_n|^2 + |c_{\bar{n}}|^2 = 1$, both $|c_n|^2 > 0$ and $|c_{\bar{n}}|^2 > 0$, and $\langle\psi_n|\psi_{\bar{n}}\rangle = 0$ holds for the initial QO. Then the transformation (8) appears to be inconsistent with the time evolution (a). We assume that this is indeed the case and understand reduction (b) as a physical law distinct from (a). In the following we limit the definition of reduction strictly to the situation where it constitutes an interruption of the time evolution (a).

Case (1) is measurement without reduction. When, in addition, also the factorization (9) holds, we have a non-demolition measurement. Although not every measurement constitutes now a reduction, every detector has to be attributed the ability to perform reductions. This follows from the fact that a detector is a device constructed for performing measurements of some observable(s) on certain QOs. We can confront the detector with a QO which is prepared to have some component orthogonal to the measured eigenstate(s). According to the standard rules of QFT, the detector has then to make a decision between the eigenstate(s) and the orthogonal complement, *i.e.* performs a reduction in the sense just defined.

The reduction process in spacetime

We consider non-overlapping detectors $A, B, C, \ldots$, each localized in a well-defined spacetime region. The basic idea is that these detectors perform reductions in a well-defined order, which in the following is called *reduction order* and indicated by the symbol r. The labeling of the detectors may be chosen such that the reduction order

$$r_A < r_B < r_C < \ldots \tag{10}$$

holds. If detectors measure at *timelike* positions with respect to one another, their reduction order is requested to agree with the time order of the measurements: $r_A < r_B \Leftrightarrow t_A < t_B$. For detectors at *spacelike* positions the reduction order is still assumed to exist, but does no longer correspond to a well-defined time order. We use the reduction order to construct a PSV $|\Psi\rangle = |\Psi_W\rangle$ in spacetime. As spacelike operators commute, measurable effects are supposed to be independent of the chosen reduction order.

By assumption, the PSV is initially defined on a LCSH S_{0+}: $|\Psi\rangle = |\Psi(S_{0+})\rangle$ and all reductions happen in the future of S_{0+}. Detector A is first in reduction order. Some section of its Backward Light Cone (BLC) is in the future of S_{0+} and denoted C_1^{blc}. The next LCSH S_{1-} is obtained by replacing the part of S_{0+} which is in the past of C_1^{blc} by C_{1-}^{blc}. By unitary and causal time evolution of law (a) the PSV becomes defined on S_{1-}: $|\Psi\rangle = |\Psi(S_{1-})\rangle$. This is always possible, because the reduction order makes sure that no reduction process can be in the way. For this causal evolution it is sufficient to know the PSV on the section of S_{0+} which is within the BLC of the detector's reduction position. On S_1 detector A performs then the stochastic reduction of law (b) and the PSV becomes defined on S_{1+}: $|\Psi\rangle = |\Psi(S_{1+})\rangle$. When the initial hypersurface S_{0+} is at $t = -\infty$ for all its space points, then S_1 is just the BLC of A.

Involving detector B, the promotion of the PSV from S_{1+} to S_{2-} and from there to S_{2+} follows precisely the same scheme. By induction we proceed from detector to detector, always promoting the LCSH, with the PSV defined on it, forward in time: $S_0 < S_1 < S_2 < \ldots$ in the sense of time-ordering of LCSHs. Should there be a last reduction, we can promote the PSV by unitary and causal time evolution to $t = +\infty$ for all space points.

Despite the fact that the reduction order cannot be verified experimentally, this has some interesting consequences. Namely, the non-local measurements envisioned by Aharonov and Albert[1] are not possible in such a formulation, whereas the proposal by Hellwig and Kraus[3] remains inconsistent. In particular, the point that the fundamental understanding of measurements has implication on what measurements are possible or not deserves to be emphasized. An experimentalist actually succeeding in performing Aharonov-Albert measurements would refute the here proposed concept of quantum measurement. For details the interested reader is referred to [2], where many examples are given too.

HYPOTHETICAL LAWS OF REDUCTION

In the previous section the process of reduction has explicitly been defined as distinct from unitary and causal time evolution. Based on this, I introduce in ref.[4] hypothetical laws to understand the Ability To Perform Reductions (ATPR) as an elementary property of microscopic matter. One major law is already a well-established part of QFT: When reduction is known to take place, QFT yields the probabilities for the alternatives to happen. A central open question is, which conglomerates of matter have under which circumstances the ATPR, *i.e.* do act as detectors? The tentative answer is that every boundstate has this ability when particle (*e.g.* photon) absorption or emission takes place. Ruled by stochastic laws, superpositions of Fock space sectors with distinct particle numbers collapse into particle number eigenstate sectors. The other question, not answered by standard QFT, is then about the frequency with which such reductions happen. Assuming a frequency which depends only on the difference $\triangle E$ in energy distribution between emerging, alternative world branches, consistency arguments lead to a collapse time

$$\tau^c(\triangle E) = \frac{b\hbar}{\triangle E}, \tag{11}$$

where b is a dimensionless constant. This equation has the phenomenologically attractive feature that the collapse time becomes large for small energy differences. Especially, superpositions of states degenerate in energy will not collapse. A measurement device is then an apparatus which speeds up the collapse by increasing the difference in energy distribution between quantum branches. Before the distinct branches become macroscopically visible, the energy difference becomes so large that collapse happens with (practical) certainty. The introduced rules leave the most commonly observed interference effects intact: Namely, all those which rely on the wave character of particles in a Fock space sector with fixed particle number. This includes photon or other particle waves passing through double slits and so on.

Superpositions between distinct Fock space sectors do eventually collapse, but for small $\triangle E$ the collapse time is large. To get a good lower bound on b of equation (11) on has to look out for relatively large differences in energy distribution. Bordé atomic beams spectroscopy yields[4] $b_{\min} = 1.35 \cdot 10^{11} < b$. That the constant b has to be large is no surprise, as the action $b\hbar$ marks the transition from quantum to classical physics. The bound can easily be improved by estimating conventional effects which contribute to diminishing the contrast. Beyond, a direct measurement of a non-zero τ^c requires that all other effects can convincingly be controlled and that still a gap between the estimated and measured contrast remains. An upper bound which relies on an analysis[4] of avalanche photodiodes is $b < b_{\max} = 3.8 \cdot 10^{21}$. A similar analysis of channel electron multipliers[6] gives $b_{\max} = 4.5 \cdot 10^{18}$. The remarkable improvement of upper bounds appears to be related to the fact that in the latter case the incoming electrons provide already quite some energy, such that amplification becomes more economical in the sense of providing a better upper bound. The lower and upper bounds leave a wide range open. An analysis of existing experiments should allow to narrow the range further.

CONCLUSIONS

The central point of my approach is that it is possible to understand reduction in QFT as a *fundamental physical process.* Each reduction is performed *locally* by a detector, whereas the consequences can be *global.* Essential for such an interpretation of the reduction process is the existence of a consistent spacetime picture. This was shown by explicit construction of a PSV. In contrast to the non-relativistic limit the PSV can no longer be monitored, because the order in which detectors at spacelike separations perform their reductions cannot be verified experimentally.

The framework was then used as the foundation for a discussion of the possibility of attributing to microscopic matter the ability to perform wave function reductions. It is of interest to improve the bounds for the collapse time $\tau^c(\triangle E)$ of equation (11). From this viewpoint, I would like to argue in favor of a paradigm shift concerning experiments. It is no longer of central interest to demonstrate the existence of one or another exotic interference effect. We know, they are there. Most interesting is to control that interference happens for every single, participating particle. This puts the focus on experiments with a high contrast.

REFERENCES

1. Y. Aharonov and D. Z. Albert, Phys. Rev. **D21** (1980) 3316.
2. B.A. Berg, quant-ph/9807046.
3. K.-E. Hellwig and K. Kraus, Phys. Rev. **D1** (1970) 566.
4. B.A. Berg, hep-ph/9609232 (revised July 1998).
5. To be definite the proper time of the detector may be chosen. By construction a detector is massive and has a well-defined rest frame.
6. B.A. Berg, in preparation, based on *Channeltron, Electron Multiplier Handbook for Mass Spectrometry Applications*, Galileo Electro-Optics Corporation, 1991.

PART III: QUANTUM COMPUTING

TELEPORTATION OF GENERAL QUANTUM STATES, SQUEEZING AND THE DECOHERENCE PROBLEM IN QUANTUM COMPUTERS*

Luigi Accardi

Graduate School of Polymathematics,
Nagoya University, Chikusa-ku, Nagoya, 464-01, Japan,
and Centro Vito Volterra,
Università degli Studi di Roma "Tor Vergata" - 00133 Rome, Italy,
email: accardi@volterra.mat.uniroma2.it, accardi@math.nagoya-u.ac.jp

In sections (1), (2) a channel theoretical formulation of the teleportation problem for quantum states, is given and a general solution, valid in arbitrary finite dimensions is proposed. The problem of uniqueness of the hidden key is discussed and a classification is obtained. The results in these sections survey recent joint work with Masanori Ohya.

It is shown that the stochastic limit of quantum theory suggests the existence of two types of squeezing states of the quantum EM field and correspondingly of two types of squeezing white noises. These states are stationary but usually non equilibrium states of the field and the stochastic limit evidentiates some qualitatively new phenomenon that arises in connection with these states. As a particular case we shall illustrate how one could take advantage of these new physical phenomena to prevent decoherence in quantum computers. The results in the second part of the paper survey recent joint work with S. Kozyrev, Y.G. Lu and I. Volovich.

TELEPORTATION OF GENERAL QUANTUM STATES

Quantum teleportation has been by introduced by Benett and Brassard [BeBr84] and discussed by several authors (cf. [Eke91], [BBCJPW93], [BBPSSW96] and references therein) in the framework of the singlet state. Our starting point shall be the following channel theoretical generalization of the Benett and Brassard scheme proposed by [IOS98]:

- STEP 0: Alice has a unknown quantum state $\rho^{(1)}$ on a Hilbert space $\mathcal{H}_1$, and she wants to teleport it to Bob.
- STEP 1: Prepare a state

  $$\sigma_{23} \in \mathcal{S}\left(\mathcal{H}_2 \otimes \mathcal{H}_3\right)$$

  Hilbert spaces attached to Alice and to Bob, respectively.

- STEP 2: After choosing a projection valued measure
$$\left\{F_k^{(12)}\right\}$$
on the Hilbert space $\mathcal{H}_1 \otimes \mathcal{H}_2$, Alice performs the incomplete measurement corresponding to the k-th projection $F_k^{(12)}$. After Alice's measurement, Bob's state becomes $\rho^{(3)}$.

- STEP 3: Bob is informed which projection (the number k) was used by Alice. This information is transmitted from Alice to Bob without disturbance.

- STEP 4: Bob reconstructs the unknown Alice's state $\rho^{(1)}$ from $\rho^{(3)}$ using a key agreed in advance.

The above procedure can be realized by a channel

$$\Lambda^{*[k]} : \mathcal{S}(\mathcal{H}_1) \longrightarrow \mathcal{S}(\mathcal{H}_1)$$

constructed by the following four maps (channels):

(1) a trivial (i.e. product) lifting in the sense of [AcOh92]

$$\gamma^* : \mathcal{S}(\mathcal{H}_1) \longrightarrow \mathcal{S}(\mathcal{H}_1 \otimes \mathcal{H}_2 \otimes \mathcal{H}) \tag{1}$$
$$\gamma^*\left(\rho^{(1)}\right) = \rho^{(1)} \otimes \sigma^{(23)} \quad \forall \rho^{(1)} \in \mathcal{S}(\mathcal{H}_1)$$

expressing a free coupling of an initial state $\rho^{(1)}$ with the entangled state $\sigma^{(23)}$.

(2) The second step is described by a channel of the form

$$\pi_k^* : \mathcal{S}(\mathcal{H}_1 \otimes \mathcal{H}_2 \otimes \mathcal{H}_3) \longrightarrow \mathcal{S}(\mathcal{H}_1 \otimes \mathcal{H}_2 \otimes \mathcal{H}_3) \tag{2}$$

where, $\forall \rho^{(123)} \in \mathcal{B}(\mathcal{H}_1 \otimes \mathcal{H}_2 \otimes \mathcal{H}_3)$,

$$\pi_k^*\left(\rho^{(123)}\right) := \frac{\left(F_k^{(12)} \otimes I^{(3)}\right)\rho^{(123)}\left(F_k^{(12)} \otimes I^{(3)}\right)}{\mathrm{tr}_{123}\left(F_k^{(12)} \otimes I^{(3)}\right)\rho^{(123)}\left(F_k^{(12)} \otimes I^{(3)}\right)}$$

corresponding to an incomplete first kind measurement describing the state change due to the nonclassical effect determined by Alice's measurement:

(3) the third step is defined by the channel $a^* : \mathcal{S}(\mathcal{H}_1 \otimes \mathcal{H}_2 \otimes \mathcal{H}_3) \longrightarrow \mathcal{S}(\mathcal{H}_3)$ defined by

$$\rho^{(3)} = a^*\left(\rho^{(123)}\right) = \mathrm{tr}_{12}\rho^{(123)} \quad \forall \rho^{(123)} \in \mathcal{S}(\mathcal{H}_1 \otimes \mathcal{H}_\in \otimes \mathcal{H}_\ni)$$

Here tr_{12} is the partial trace on the Hilbert space $\mathcal{H}_1 \otimes \mathcal{H}_2$ defined by

$$< \Phi_1, \mathrm{tr}_{12}\, Q\Phi_2 > \equiv \sum_n < \Psi_n \otimes \Phi_1,\, Q\Psi_n \otimes \Phi_2 >, \;\; Q \in B(\mathcal{H}_1 \otimes \mathcal{H}_2 \otimes \mathcal{H}_3)$$

for any CONS $\{\Psi_n\}$ of $\mathcal{H}_1 \otimes \mathcal{H}_2$ and any $\Phi_1, \Phi_2 \in \mathcal{H}_3$. This channel a^* is a reduction from the state $\rho^{(123)}$ to Bob's state $\rho^{(3)}$ after Alice's measurement.

(4) In the teleportation scheme of Benett and Brassard , there exists a unitary channel

$$u_k^* : \mathcal{S}(\mathcal{H}_3) \longrightarrow \mathcal{S}(\mathcal{H}_1)$$
$$u_k^*\left(\rho^{(3)}\right) = U_k\rho^{(3)}U_k^* \quad \forall \rho^{(3)} \in \mathcal{S}(\mathcal{H}_3)$$

where u_k^* is Bob's unitary operation (key) corresponding to Alice's measurement $F_k^{(12)}$.

Thus the whole teleportation process is written by the channel

$$\Lambda^{*[k]} : \mathcal{S}(\mathcal{H}_1) \longrightarrow \mathcal{S}(\mathcal{H}_1)$$

$$\Lambda^{*[k]} = u_k^* \circ \Lambda_{A\to B}^{*[k]} = u_k^* \circ a^* \circ \pi_k^* \circ \gamma^*$$

The above subscript "$[k]$" means that the channels $\Lambda^{*[k]}$ and $\Lambda_{A\to B}^{*[k]}$ depend on the choice of Alice's measurement $F_k^{(12)}$. More precisely,

$$\Lambda^{*[k]}\rho^{(1)} \equiv U_k \left[\mathrm{tr}_{12} \left[\frac{\left(F_k^{(12)} \otimes I^{(3)}\right)\left(\rho^{(1)} \otimes \sigma^{(23)}\right)\left(F_k^{(12)} \otimes I^{(3)}\right)}{\mathrm{tr}_{123}\left(F_k^{(12)} \otimes I^{(3)}\right)\left(\rho^{(1)} \otimes \sigma^{(23)}\right)\left(F_k^{(12)} \otimes I^{(3)}\right)} \right] \right] U_k^* \tag{3}$$

$$\forall \rho^{(1)} \in \mathcal{S}(\mathcal{H}_1)$$

Note that the channel $\Lambda^{*[k]}$ is generally non linear.

PROBLEM I.

Let

$$N = \dim \mathcal{H}_1 = \dim \mathcal{H}_2 = \dim \mathcal{H}_3 < +\infty$$

we want to find unit vectors $\psi \in \mathcal{H}_2 \otimes \mathcal{H}_3$ and $\xi \in \mathcal{H}_1 \otimes \mathcal{H}_2$ such that, denoting

$$F = |\xi\rangle\langle\xi| \otimes 1_3 \in \text{ Proj } (\mathcal{H}_1 \otimes \mathcal{H}_2 \otimes \mathcal{H}_3) \tag{1}$$

there exists a unitary transformation $U : \mathcal{H}_3 \to \mathcal{H}_1$ such that for any rank one density matrix $\rho \in$ Dens $(\mathcal{H}_1)$ one has

$$U \cdot \frac{\overline{Tr}_{1,2}(F(\rho \otimes |\psi\rangle\langle\psi|)F)}{Tr(F(\rho \otimes |\psi\rangle\langle\psi|)F)} U^* = \rho \tag{2}$$

A possible solution is the following. By our assumption there exists a unit vector $\eta \in \mathcal{H}_1$ such that

$$\rho = |\eta\rangle\langle\eta| \tag{3}$$

Let (ε_α) be an arbitrary orthonormal basis of $\mathcal{H}_3$ and let $(\lambda_{\gamma\alpha})$ be any $N \times N$ complex unitary matrix, so that als o

$$\left(\sum_\alpha \lambda_{\gamma\alpha}\varepsilon_\alpha\right)_\gamma =: (\varphi_\gamma) \tag{4}$$

is an o.n. basis of $\mathcal{H}_3$. Let (ε'_h) be an arbitrary o.n. basis of $\mathcal{H}_2$ and define

$$\psi := \sum \lambda_{h\alpha} |\varepsilon'_h\rangle \otimes |\varepsilon_\alpha\rangle \in \mathcal{H}_2 \otimes \mathcal{H}_3 \tag{5}$$

then

$$\|\psi\|^2 = \sum_{h,\alpha} |\lambda_{h\alpha}|^2 = \sum_h 1 = \dim \mathcal{H}_2 = \dim \mathcal{H}_3 \tag{6}$$

It is at this point that we use in a crucial way the finite dimensionality of $\mathcal{H}_2$ and $\mathcal{H}_3$. Let (ε''_n) be an arbitrary o.n. basis of $\mathcal{H}_1$ and choose

$$\xi = \frac{1}{N^{1/2}} \sum_\mu \varepsilon''_\mu \otimes \varepsilon'_\mu$$

Thus, for F as in (1), one has, with the convention that we sum over repeated indices:

$$F(\rho\otimes|\psi\rangle\langle\psi|)F = \lambda_{h\alpha}\overline{\lambda}_{k\beta}F(|\eta\rangle\langle\eta|\otimes|\varepsilon'_h\rangle\langle\varepsilon'_k|)F\otimes|\varepsilon_\alpha\rangle\langle\varepsilon_\beta| =$$

$$= \frac{1}{N}\lambda_{h\alpha}\overline{\lambda}_{k\beta}|\varepsilon''_\mu\rangle\langle\varepsilon''_\mu,\eta\rangle\langle\eta,\varepsilon''_{\mu'}\rangle\langle\varepsilon''_{\nu'}|\otimes|\varepsilon'_\mu\rangle\langle\varepsilon'_\nu,\varepsilon'_h\rangle\langle\varepsilon'_k,\varepsilon'_{\mu'}\rangle\langle\varepsilon'_{\nu'}|\otimes|\varepsilon_\alpha\rangle\langle\varepsilon_\beta| =$$

$$= \frac{1}{N}\lambda_{h\alpha}\overline{\lambda}_{k\beta}\delta_{\nu h}\delta_{k\mu'}\langle\varepsilon''_\nu,\eta\rangle\langle\eta,\varepsilon''_{\mu'}\rangle|\varepsilon''_\mu\rangle\langle\varepsilon''_{\nu'}|\otimes|\varepsilon'_\mu\rangle\langle\varepsilon'_{\nu'}|\otimes|\varepsilon_\alpha\rangle\langle\varepsilon_\beta|$$

$$= \frac{1}{N}\lambda_{h\alpha}\overline{\lambda}_{k\beta}\langle\varepsilon''_h,\eta\rangle\langle\eta,\varepsilon''_k\rangle|\varepsilon''_\mu\rangle\langle\varepsilon''_{\nu'}|\otimes|\varepsilon'_\mu\rangle\langle\varepsilon'_{\nu'}|\otimes|\varepsilon_\alpha\rangle\langle\varepsilon_\beta|$$

Taking trace over $\mathcal{H}_1\otimes\mathcal{H}_2$ this gives

$$\frac{1}{N}\lambda_{h\alpha}\overline{\lambda}_{k\beta}\delta_{\mu,\nu'}\langle\varepsilon''_h,\eta\rangle\langle\eta,\varepsilon''_k\rangle|\varepsilon_\alpha\rangle\langle\varepsilon_\beta| = \langle\varepsilon''_h,\eta\rangle\langle\eta,\varepsilon''_k\rangle|\varphi_h\rangle\langle\varphi_k| \tag{7}$$

Notice that by taking the trace over $\mathcal{H}_3$ of the above expression one finds

$$\langle\eta,\varepsilon''_k\rangle\langle\varepsilon''_h,\eta\rangle = \langle\eta,\eta\rangle = 1$$

Now, if $U:\mathcal{H}_3\to\mathcal{H}_1$ is the unitary operator such that

$$U\varphi_h = \varepsilon''_h$$

then we have

$$UF(\rho\otimes|\psi\rangle\langle\psi|)FU^* = U\langle\varepsilon''_h,\eta\rangle\langle\eta,\varepsilon''_k\rangle|\varphi_h\rangle\langle\varphi_k|U^* = |\varepsilon''_h\rangle\langle\varepsilon''_h,\eta\rangle\langle\eta,\varepsilon''_k\rangle\langle\varepsilon''_k| = |\eta\rangle\langle\eta|$$

This solves Problem I.

The stochastic limit of quantum theory

In the following I shall quickly report on joint work done with several authors in the past ten years, in particular: Irina Arefeva, Alberto Frigerio, Lu Yun Gang, Igor Volovich, Sergei Kozyrev,...

Both in classical and in quantum theory, two main techniques are used to deal with non integrable systems:

i) scattering theory, dealing with long times ($t\to\infty$)

ii) perturbation theory, dealing with small parameter ($\lambda\to 0$)

By putting together these two techniques one obtains the *stochastic limit of QT* which deals with:

iii) long times & small parameter, more precisely:

$$t\to\infty\ ;\quad \lambda\to 0\ ;\quad \lambda^2 t=\tau\ =\ \text{const}\ \Leftrightarrow t=\ \text{cost}/\lambda^2$$

The physical intuition behind the stochastic limit is:

– the effects of the interaction are "small" ($\lambda\to 0$)

– but they have a lot of time to cumulate ($t\to\infty$) and to produce a nontrivial effect.

As a bonus of the theory one finds that the dominating contribution to the interaction *resum to give a new unitary evolution.* Although the mathematics may become quite involved, the basic prescriptions to apply the stochastic limit of quantum theory are very simple and they can be condensated into the following **stochastic golden rule** (a generalization of the Fermi golden rule):

1.) Start from Schrödinger's equation in interaction representation

$$\partial_t U_t^{(\lambda)} = -i\lambda H_I(t) U_t^{(\lambda)} \ ; \quad U_0^{(\lambda)} = 1$$

$$H_{\text{tot}} = H_0 + H_I \ ; \quad H_I(t) = e^{itH_0} H_I e^{-itH_0}$$

2.) By a simple change of variables one verifies that to rescale time $t \to t/\lambda^2$ is equivalent to rescale energy as follows:

$$\partial_t U_{t/\lambda^2}^{(\lambda)} = -i\frac{1}{\lambda} H_{I\lambda^2}(t) U_{t/\lambda^2}^{(\lambda)} \tag{1}$$

3.) One can prove that, in an appropriate meaning of the convergence, $U_{t/\lambda^2}^{(\lambda)} \to U_t$ unitary

4.) Moreover the limit satisfies the following white noise equation

$$\partial_t U_t = -iH_t U_t \tag{2}$$

5.) The normally ordered form of the white noise equation (4) is a quantum stochastic differential equation.

Summing up: the stochastic limit of QT explains the microscopic (Hamiltonian) origins of many phenomenological (stochastic) equations studied in the past years. It also provides, apparently for the first time, the microscopic structure of the so called *quantum noises*.

The Physical picture of the stochastic limit of quantum theory is the following:

One starts from a "Large" quantum system where "Large" is meant in the sense that the time correlations for the free dynamics are integrable, i.e. $\int dt |\langle A(0)B(t)\rangle| < +\infty$ (purely discrete spectrum excluded)

Inside such a system we distinguish two classes of motions: the slow ones and the fast ones. If we look at the fast degrees of freedom with the clock of the slow degrees of freedom then what we perceive is *pure noise* i.e. maximum chaotic behaviour which is described mathematically by an independent increment process. The simplest example of such a process is white noise!

Typical example

Consider a discrete system S (e.g. an oscillator or a 2–level system) interacting with a continuous one R (e.g. a field or an infinite gas) by means of an Hamiltonian H_I. In the stochastic limit the system remains unaltered while the field converges to a quantum white noise.

Summing up: just like statistical mechanics gives a microscopic interpretation of the thermodynamical quantities ("Those kind of motions we call heat"), the stochastic limit of quantum theory gives a microscopic interpretation of the quantum noises ("Those kind of fields we call noise").

Dissipation and decoherence

A typical explicit form for the white noise equation (2) is

$$\partial_t U_t = -(iDb_t^+ + D^+ b_t)U_t \tag{3}$$

where b_t, b_t^+ is a Boson white noise satysfying $[b_t, b_{t'}^+] = \gamma\delta(t-t')$. The normally ordered form of the white noise equation (3) can be proved to be

$$\partial_t U_t = -iDb_t^+ U_t - iD^+ U_t b_t - YU_t \tag{4}$$

where

$$Y = \begin{cases} \text{Ito correction term (mathematical interpretation)} \\ \text{operator valued transport coefficient (physical interpretation)} \end{cases}$$

The physical meaning of Y is best illustrated by the simplest example: the vacuum transition amplitude associated to the evolution defined by equation (4). Let $\langle\cdot\rangle$ denote vacuum expectation. Then from (4) we deduce:

$$\partial_t \langle U_t\rangle = -\langle YU_t\rangle = -Y\langle U_t\rangle$$

$$\langle U_t\rangle = e^{-tY} = e^{-t\{\text{Re }(Y)+i\text{ Im }(Y)\}}$$

Here Re (Y) can be proved to be a positive operator, so it describes damping, while Im (Y), is a self–adjoint operator describing energy shift. Since

$$|\langle U_t\rangle|^2 = \text{Prob }\{\text{vacuum at } t|\text{ vacuum at } 0\}$$

is the conditional probability of remaining in the vacuum at time t if one was in the vacuum at time zero, we conclude that:

– The real part of the operator transport coefficient drives decoherence.

– The imaginary part descrives: oscillatory behaviour.

How to control decoherence

Basic idea:

Since U_t collects the dominating contributions of the Hamiltonian dynamics, if we are able to control a quantum system so that Re $(Y) = 0$, then

$$\langle U_t\rangle = e^{-it\text{ Im }(Y)}$$

so we have a purely oscillatory behaviour i.e.: *no decoherence*! – The stochastic limit scheme is quite general but: the specific form of Y depends on:

i) the state of the field (Fock, temperature, squeezing,...)

ii) H_I

iii) The free field dynamics

iv) The free system dynamics

However it does not depend on the state of the system!

The simplest example is described by the following ingredients:

– Boson field a_k, a_k^+ interacting with a system

- Fock state of this field
- interaction

$$H_I(t) = \int_{\mathbf{R}^3} dk D^+ \otimes \overline{g}_k e^{it(\omega_k - \omega_0)} a_k + h.c.$$

- D^+ arbitrary system operator (e.g. $D^+ = a^+, \sigma_+, \sigma_x, \ldots$)
- $\overline{g}_k$ form factor $\equiv$ control parameter
- (ω_k) Field energy
- (ω_0) resonant frequency (of the system)
- here we are in Dipole + RWA

Under these assumptions one can prove that

$$Y = \gamma_- D^+ D$$

where γ_- is given by:

$$\gamma_- = \int_{-\infty}^{0} dt \int_{\mathbf{R}^3} e^{-it(\omega_k - \omega_0)} |g_k|^2 dk$$

Using the identity

$$\int_{-\infty}^{0} e^{-it\omega} dt = \pi\delta(\omega) - iP.P.\frac{1}{\omega}$$

we get

$$\gamma_- = \int_{\mathbf{R}^3} dk |g_k|^2 \delta(\omega_k - \omega_0) - iP.P. \int \frac{|g_k|^2}{\omega_k - \omega_0} dk$$

which is a generalization of the Kramers–Krönig dispersion relations. Summing up:

$$\text{Re } Y = D^+ D \int_{\mathbf{R}^3} dk |g_k|^2 \delta(\omega_k - \omega_0) = D^+ D \text{ Re } \gamma_-$$

$\omega_k = \omega_0$ resonant surface and from this condition we see that Re $Y = 0 \Leftrightarrow$ supp $g_k \cap \{\omega_k = \omega\} = \emptyset$ (off–resonance condition) where supp g_k denotes the support of the form factor. This in principle can be controlled by controlling the frequency b and of the original Boson field.

But the general theory of stochastic limit proves that:

$$\text{Re } \gamma_- = \text{ covariance of the noise}$$

Therefore

$$\text{Re } \gamma_- = 0 \Leftrightarrow \text{ no noise}$$

Now, recalling that, in the stochastic limit

$$\text{noise } = \lim_{\lambda \to 0} \text{ field}$$

we conclude that: no noise $\Leftrightarrow \lim_{\lambda \to 0}$ field $= 0$. But physical effects of the field remain under the form of an (operator) energy shif in the system so that the limit white noise equation is in fact deterministic and looks like.

$$\partial_t U_t = i(\text{Im } \gamma_-) D^+ D U_t$$

In the paper [AcKoVo97] this was called a: *Cheshire cat effect*.

SQUEEZING AND THE DECOHERENCE PROBLEM IN QUANTUM COMPUTERS

Let a_k, a_k^+ be a quantum field and let $\langle\cdot\rangle$ be a mean zero gauge invariant gaussian state on the $a_k, a_{k'}^+$–algebra with covariance

$$\begin{pmatrix} \langle a_k^+ a_{k'}\rangle & 0 \\ 0 & \langle a_k a_{k'}^+\rangle \end{pmatrix} = \begin{pmatrix} n(k) & 0 \\ 0 & m(k) \end{pmatrix} \delta(k-k') \tag{1}$$

Then defining

$$b_k := \overline{c}_k a_k + s_k a_{-k}^+ \quad ; \quad b_k^+ := c_k a_k^+ + \overline{s}_k a_{-k} \tag{2}$$

the state $\langle\cdot\rangle$ also gives a mean zero gauge invariant gaussian state on the $b_k, b_{k'}^+$–algebra with covariance matrix

$$\begin{pmatrix} \langle b_k^+ b_{k'}\rangle & \langle b_k b_{k'}\rangle \\ \langle b_k^+ b_{k'}^+\rangle & \langle b_k b_{k'}^+\rangle \end{pmatrix} = \tag{3}$$

$$= \begin{pmatrix} (|c_k|^2 n(k) + |s_k|^2 m(k))\delta(k-k') & (c_k\overline{s}_{-k} n(k) + \overline{s}_k c_{-k} m(k))\delta(k+k') \\ (\overline{c}_k s_{-k} n(k) + s_k \overline{c}_{-k} m(k))\delta(k+k')) & (|c_k|^2 m(k) + |s_k|^2 n(k))\delta(k-k') \end{pmatrix}$$

Notice the $\delta(k+k')$ in the off–diagonal elements. This means that in this state one can have simultaneous creation (resp. annihilation) of pairs, but momentum has to be conserved. The positivity of the state is guaranteed by construction.

We suppose that the state $\langle\cdot\rangle$, a fixed quantum state (Fock, equilibrium, ...), and that it is mean zero gaussian and invariant under the second quantized evolution associated to a given one–particle dynamics defined by a function of momentum $\omega(k)$ (for example $\omega(k) = |k|$)

$$a_k \mapsto a(t,k) = e^{-it\omega_k} a_k \tag{4}$$

In this case the process

$$b(t,k) := \overline{c}_k e^{-it\omega_k} a_k + s_k e^{it\omega_{-k}} a_{-k}^+ \quad ; \quad b^+(t,k) := c_k e^{it\omega_k} a_k^+ + \overline{s}_k e^{-it\omega_{-k}} a_{-k} \tag{5}$$

is mean zero, gaussian and time translation invariant, because it is a linear combination of processes with these properties. The time dependent field $b(t,k), b^+(t,k)$ is called a *squeezed stationary field.* In particular

$$\begin{pmatrix} \langle b^+(t,k)b(t',k')\rangle & \langle b(t,k)b(t',k')\rangle \\ \langle b^+(t,k)b^+(t',k')\rangle & \langle b(t,k)b^+(t',k')\rangle \end{pmatrix} = \begin{pmatrix} \langle b^+(0,k)b(t'-t,k')\rangle & \langle b(0,k)b(t'-t,k')\rangle \\ \langle b^+(0,k)b^+(t'-t,k')\rangle & \langle b(0,k)b^+(t'-t,k')\rangle \end{pmatrix}$$

Explicitly the diagonal part of the covariance is equal to

$$\begin{pmatrix} \langle b^+(0,k)b(t,k')\rangle & 0 \\ 0 & \langle b(0,k)b^+(t,k')\rangle \end{pmatrix} = \delta(k-k') \cdot \tag{6}$$

$$\cdot \begin{pmatrix} |c_k|^2 e^{-it\omega_{-k}} n(k) + |s_k|^2 e^{it\omega_k} m(k) & 0 \\ 0 & |c_k|^2 e^{it\omega_{-k}} m(k) + |s_k|^2 e^{-it\omega_k} n(k) \end{pmatrix}$$

while the off diagonal part of the covariance is equal to

$$\begin{pmatrix} 0 & \langle b(0,k)b(t,k')\rangle \\ \langle b^+(0,k)b^+(t,k')\rangle & 0 \end{pmatrix} = \delta(k+k') \cdot \tag{7}$$

$$\cdot \begin{pmatrix} 0 & c_k\overline{s}_{-k} e^{-it\omega_k} n(k) + c_{-k}\overline{s}_k e^{it\omega_{-k}} m(k) \\ \overline{c}_k s_{-k} e^{it\omega_k} n(k) + \overline{c}_{-k} s_k e^{-it\omega_{-k}} m(k) & 0 \end{pmatrix}$$

Remark. Instead of the b–fields (2), one could also define the fields

$$d_k := \bar{c}_k a_k + s_k a_k^+ \quad ; \quad d_k^+ := c_k a_k^+ + \bar{s}_k a_k \tag{8}$$

leading to the covariance

$$\begin{pmatrix} \langle d_k^+ d_{k'}\rangle & \langle d_k d_{k'}\rangle \\ \langle d_k^+ d_{k'}^+\rangle & \langle d_k d_{k'}^+\rangle \end{pmatrix} = \begin{pmatrix} (|c_k|^2 n(k) + |s_k|^2 m(k)) & (c_k \bar{s}_k n(k) + \bar{s}_k c_k m(k)) \\ (\bar{c}_k s_k n(k) + s_k \bar{c}_k m(k))\rangle & (|c_k|^2 m(k) + |s_k|^2 n(k)) \end{pmatrix} \delta(k-k') \tag{9}$$

and also in this case, by the same argument as above, one would obtain a mean zero, gaussian time translation invariant, field. However it is not clear if these states can be physically realized since for them simultaneous pair creation (resp. annihilation) would take place without conservation of momentum.

Now suppose that the field interacts with a system S whose free Hamiltonian H_S has a discrete spectrum

$$H_S = \sum_n \varepsilon_n P_n \tag{10}$$

where ε_n are the energy levels and P_n the spectral projections. Let the interaction Hamiltonian (x.1) evolved with the free evolution (interaction representation) be:

$$H_I(t) := \int dk[D^+ \otimes \bar{g}_k e^{-it\omega(k)} a_k + D \otimes g_k e^{it\omega(k)} a_k^+] =: D^+ a_t + D a_t^+ \tag{1}$$

where in (4), and often in the following, we omit the symbol $\otimes$ and where we have introduced the shorthand notations:

$$a_t := \int dk \bar{g}_k e^{-it\omega(k)} a_k = A(S_t g) \tag{12}$$

$$S_t := S_t^1 e^{it\omega_o} = e^{-it(\omega_1(k)-\omega_o)} =: e^{-it\omega(k)} \tag{13}$$

$$\omega(k) := \omega_1(k) - \omega_o \tag{14}$$

The stochastic limit studies the rescaled Schrödinger equation

$$\partial_t U_{t/\lambda^2}^{(\lambda)} = -iH_I^{(\lambda)}(t) U_{t/\lambda^2}^{(\lambda)} \tag{15}$$

where the *rescaled interaction Hamiltonian* is defined by:

$$H_I^{(\lambda)}(t) := \frac{1}{\lambda} H_I(t/\lambda^2) = D^+ \left(\frac{1}{\lambda} a_{t/\lambda^2}\right) + D \left(\frac{1}{\lambda} a_{t/\lambda^2}^+\right) =: D^+ a_t^{(\lambda)} + D a_t^{(\lambda)+} \tag{16}$$

and the *rescaled fields* are

$$a_t^{(\lambda)} = \frac{1}{\lambda} a_{t/\lambda^2} = \frac{1}{\lambda} \int_{\mathbf{R}^d} dk \bar{g}_k e^{-i(t/\lambda^2)\omega(k)} a_k \tag{17}$$

The rescaled fields (17) converge in the sense of correlators to a quantum white noise b_t, b_t^+ whose covariance is determined by the reference state of the a–field. In symbols

$$a_t^{(\lambda)} = \frac{1}{\lambda} a_{t/\lambda^2} \to b_t \quad ; \quad a_t^{(\lambda)+} = \frac{1}{\lambda} a_{t/\lambda^2}^+ \to b_t^+ \tag{18}$$

$$\lim_{\lambda \to 0} \frac{1}{\lambda} H_I\left(\frac{t}{\lambda^2}\right) = H_t = D^+ b_t + D b_t^+ \tag{19}$$

and the white noise Hamiltonian equation associated to the right hand side of (19) is

$$\partial_t U_t = -iH_I(t) U_t = -i(D^+ b_t + D b_t^+) U_t \tag{20}$$

Equation (20) can be given a meaning and admits a unique unitary solution U_t. Moreover $\lim_{\lambda\to 0} U^{(\lambda)}_{t/\lambda^2} = U_t$ where convergence is meant in the sense of the matrix elements with respect to the collective vectors. Now suppose that the *generalized rotating wave approximation* condition is satisfied, i.e. there exists a fixed strictly positive real number:

$$\omega_o > 0 \tag{21}$$

such that for any $t \in \mathbf{R}$ one has

$$e^{itH_S} D e^{-itH_S} = e^{-i\omega_o t} D \tag{22}$$

Then, if the reference state of the a–field is the Fock vacuum, then the normally ordered form of the white noise equation (20) is the following equation:

$$\partial_t U_t = -i(D^+ U_t b_t + D b_t^+ U_t) - Y U_t \tag{23}$$

where

$$Y = \gamma_- D^+ D$$

is the operator transport coefficient and where γ_- is given by the following integral

$$\gamma_- := \int_{-\infty}^{0} \langle g, S_s g\rangle ds = \int_{-\infty}^{0} dt \int_{\mathbf{R}^d} e^{-it(\omega_1(k)-\omega_o)} \mid g(k) \mid^2 = \int_{-\infty}^{0} ds e^{-it\omega_o} G_{-+}(s) \tag{24}$$

The normally ordered form of the above white noise equation is equivalent to the quantum stochastic differential equation obtained by multiplying both sides of (23) by dt and by the identification

$$b_t^{\pm} = dB_t^{\pm}$$

and $dB_t^{\pm}$ is the usual quantum Brownian motion.

Using the normally ordered equation (23) for U_t one can easily calculate the matrix elements $\langle\psi, U_t\phi\rangle$ for a large class of states ψ, ϕ in the limit space (number or coherent vectors, ...). As an illustration we compute some vacuum–to–vacuum amplitude of the evolution operator U_t (we mean the vacuum for the operators b_t, b_t^+). The equation for this amplitude is

$$\frac{d}{dt}\langle U_t\rangle = -\gamma_- D^+ D \langle U_t\rangle \tag{25}$$

The explicit form of the solution is

$$\langle\psi, e^{-t\gamma_- D^+ D}\psi\rangle = \langle\psi\otimes\Phi, U_t\psi\otimes\Phi\rangle \tag{26}$$

Since both the operator D^+D and the real part of γ_- are positive we see that the generic behavior one should expect from the amplitude (26) is of an **exponential decay driven by the real part of γ_- and modulated by its imaginary part**. The operator $\gamma_- D^+ D$ could be interpreted as an *operator inverse life–time of the vacuum in the limit space* in the sense that, on each spectral subspace of D^+D, the inverse lifetime is uniquely defined. So the generic behaviour is

$$\langle\alpha, U_t\alpha\rangle \sim e^{it\Delta\varepsilon_\alpha/\hbar} \tag{27}$$

where $\Delta\varepsilon_\alpha$ is a complex number with a **positive imaginary part**:

$$\Delta\varepsilon_\alpha = -\frac{1}{2i}\Gamma_\alpha + \Omega_\alpha\ ; \quad \Gamma_\alpha > 0\ ; \quad \Omega_\alpha \in \mathbf{R} \tag{28}$$

Replace now the vacuum by an exponential vector $\psi(f)$, D and D^+ by a, a^+ (1–mode creation and annihilation) and denote

$$\beta := \xi \otimes \psi(f) \; ; \quad \alpha = \xi \otimes \psi(g)$$

where $\xi = |z>$ is alos a coherent vector for the a, a^+ field.

Then equation (25) becomes

$$\frac{d}{dt} \langle \beta, U_t\alpha \rangle = (-i\overline{z}g(t) - iz\overline{f}(t) - \gamma_-|z|^2)\langle \beta, U_t\alpha \rangle$$

So, if we could choose f, g and z so that

$$\mathrm{Im}\;(\overline{z}g(t) + z\overline{f}(t)) = (\mathrm{Re}\;\gamma_-)|z|^2 \tag{29}$$

we would obtain a purely oscillatory behaviour of the amplitude $\langle \beta, U_t\alpha \rangle$. Such a choice cannot be done for any time t, because it is physically impossible to realize *coherent vectors with constant test functions* (plane waves), however we can obtain reasonably good approximations by taking f and g constant in an arbitrarily large interval of time, say $[0, T]$ and in this interval the transition amplitude $\langle \beta, U_t\alpha \rangle$ shall exhibit a purely oscillatory behaviour.

Another example of purely oscillatory behaviour can be obtained from the stochastic limit for an Hamiltonian of a two–state system coupled to an environment and not satisfying the generalized rotating wave approximation, Models of this kind arise in several contexts in particular in connection with *macroscopic quantum coherence phenomena*, related to the SQUID model [AcKoVo96]. Also here one can prove the emergence of two interesting regimes for the system: the *purely oscillating* and the *purely damping* one. By experimentally controlling the spectral function of the field one should be able to bring the system into the purely oscillatory non decaying regime and this suggests a way to solve the decoherence problem.

REFERENCES

[AcOh92] Accardi, L., and Ohya, M., 1991, "Compound channels, transition expectations and liftings", Volterra Preprint n. 75.

[AcKoVo96] Accardi, L., Kozyrev, S.V., and Volovich, I.V., 1996, Dynamics of dissipative two–state systems in the stochastic approximation, *Phys. Rev.* A 56 N. 3.

[AcOh98] Accardi, L., and Ohya, M., (1998), Teleportation of general quantum states, In preparation.

[AcLuVo99] Accardi, L., Lu, Y.G., and Volovich, I.V., (1999), "Quantum Theory and its Stochastic Limit", Oxford University Press, to appear.

[BeBr84] Bennett, C.H., and Brassard, G., in *Proc. of the IEEE Int. Conf. on Computers, Systems, and Signal Processing, Bangalore, India*, (IEEE, New York, 1984), pp. 175–179.

[BBCJPW93] Bennett, C.H., Brassard, G., Crepeau, C., Josza, R., Peres, A., and Wotters, W.K., (1993), Teleporting an unknown quantum state via dual classical and Einstein–Podolsky–Rosen channels, *Phys. Rev. Lett.* 70, 1895–1899.

[BBPSSW96] Bennett, C.H., Brassard, G., Popescu, S., Smolin J.A., and Wotters, W.K., (1996), Purification of noisy entanglement and faithful teleportation via noisy channels, *Phys. Rev. Lett.* 76, 722–725.

[Eke91] Ekert, A.K., (1991), [Eke91], [BBCJPW93], [BBPSSW96] Quantum criptography based on Bell's theorem, *Phys. Rev. Lett.* 67, 661.

[IOS98] Inoue, K., Ohya, M., and Suyari, H., (1998) Characterization of quantum teleportation process by nonlinear quantum channels and quantum mutual entropy *Physica D*.

QUANTUM TURING MACHINES: LOCAL TRANSITION, PREPARATION, MEASUREMENT, AND HALTING

Masanao Ozawa

School of Informatics and Sciences
Nagoya University
Chikusa-ku, Nagoya 4648601, Japan

Foundations of the theory of quantum Turing machines are investigated. The protocol for the preparation and the measurement of quantum Turing machines is discussed. The local transition functions are characterized for fully general quantum Turing machines. A new halting protocol is proposed without augmenting the halting qubit and is shown to work without spoiling the computation.

INTRODUCTION

The Church-Turing thesis[1, 2] states that to be computable is to be computable by a Turing machine and the modern discipline in computational complexity theory[3] states that to be efficiently computable is to be computable by a Turing machine within polynomial steps in the length of the input data. However, Feynman[4] pointed out that a Turing machine cannot simulate a quantum mechanical process efficiently and suggested that a computing machine based on quantum mechanics might be more powerful than Turing machines. Deutsch introduced quantum Turing machines[5] and quantum circuits[6] for establishing the notion of quantum algorithm exploiting "quantum parallelism." A different approach to quantum Turing machines was investigated earlier by Benioff.[7] Bernstein and Vazirani[8] instituted quantum complexity theory based on quantum Turing machines and showed constructions of universal quantum Turing machines. Yao[9] showed that a computation by a quantum Turing machine can be simulated efficiently by a quantum circuit. Deutsch's idea of quantum parallelism was realized strikingly by Shor,[10] who found efficient quantum algorithms for the factoring problem and the discrete logarithm problem, for which no efficient algorithms have been found for classical computing machines. The purpose of this paper is to discuss foundations of quantum Turing machines and to propose a computational protocol for quantum Turing machines.

A precise formulation of quantum Turing machines is given along with Deutsch's formulation[5] and the computational protocol is discussed for the preparation and the measurement of quantum Turing machines.

The characterization of the transition functions of quantum Turing machines is also discussed. Deutsch[5] required that the transition function should be determined

Quantum Communication, Computing, and Measurement 2
Edited by Kumar et al., Kluwer Academic / Plenum Publishers, New York, 2000.

by local configurations. Bernstein and Vazirani[8] found a simple characterization of the transition functions for the restricted class of quantum Turing machines in which the head must move either to the right or to the left at each step but a general characterization remains open. This problem is discussed and a solution is given.

The computational protocol for the halting of quantum Turing machines is discussed. In order to signal the completion of computation, Deutsch[5] introduced the halt flag by augmenting the halt qubit. Myers[11] pointed out a difficulty in this halting of quantum Turing machines. With improving the preceding work,[12] a halting protocol is proposed without augmenting the halting qubit and it is shown that the monitoring of the halt flag does not spoil the computation.

QUANTUM TURING MACHINES

A *quantum Turing machine (QTM)* $\mathcal{Q}$ is a quantum system consisting of a *processor*, a bilateral infinite *tape*, and a *head* to read and write a symbol on the tape. Its configuration is determined by the *processor configuration* q from a finite set Q of symbols, the *tape configuration* T represented by an infinite string from a finite set Σ of symbols, and the discretized *head position* ξ, taking values in the set $\mathbf{Z}$ of integers. The tape consists of *cells* numbered by the integers and the head position ξ is the place of the cell numbered by ξ. We assume that Q contains two specific symbols q_0 and q_f representing the initial configuration and the final configuration of the processor and that Σ contains the symbol B representing the blank cell in the tape. For any integer m the symbol at the cell m on the tape is denoted by $T(m)$. We assume that the possible tape configurations are such that $T(m) = B$ except for finitely many cells m. The set of all the possible tape configurations is denoted by $\Sigma^{\#}$. The set $\Sigma^{\#}$ is a countable set. Thus, any configuration C of $\mathcal{Q}$ is represented by a triple $C = (q, T, \xi)$ in the *configuration space* $Q \times \Sigma^{\#} \times \mathbf{Z}$. The state of $\mathcal{Q}$ is represented by a unit vector in the Hilbert space $\mathcal{H}$ generated by the configuration space $Q \times \Sigma^{\#} \times \mathbf{Z}$. The complete orthonormal basis canonically in one-to-one correspondence with the configuration space is called the *computational basis.* Thus, the computational basis is represented by $|C\rangle = |q\rangle|T\rangle|\xi\rangle$ for any configuration $C = (q, T, \xi) \in Q \times \Sigma^{\#} \times \mathbf{Z}$; we shall write also $|q, T, \xi\rangle = |q\rangle|T\rangle|\xi\rangle$.

We shall denote by $|X|$ the number of the elements of a set X; for an indexed set the number of elements is understood as the number of indices. In order to define the observables quantizing the configurations, we assume the numbering of the sets Q and Σ such that $Q = \{q_0, \ldots, q_{|Q|-1}\}$ and $\Sigma = \{\sigma_0, \ldots, \sigma_{|\Sigma|-1}\}$. We define observables $\hat{q}$, $\hat{T}(m)$ for $m \in \mathbf{Z}$, and $\hat{\xi}$ as follows.

$$\hat{q} = \sum_{n=0}^{|Q|-1} n|q_n\rangle\langle q_n|, \quad \hat{T}(m) = \sum_{n=0}^{|\Sigma|-1} n|\sigma_n\rangle\langle\sigma_n|, \quad \hat{\xi} = \sum_{\xi\in\mathbf{Z}} \xi|\xi\rangle\langle\xi|.$$

We assume that we have a device to prepare the quantum Turing machine in the state $|q, T, \xi\rangle$ for any configuration $C = (q, T, \xi)$ and that we have a measuring device to measure sufficiently many $\hat{T}(m)$s simultaneously.

Let Γ be a finite set of symbols and Γ^* the set of finite strings from Γ. In this paper, we shall consider computations which are probabilistic transformations on Γ^*, or precisely functions from Γ^* to the set of probability distributions on Γ^*. The set Γ is called the *alphabet* of the computation. A finite string from the set Γ is called a Γ*-string.* The length of a Γ-string x is denoted by $|x|$. When $|x| = 0$, x is called the empty string. We shall identify any Γ-string $x = (x_0, \ldots, x_{|x|-1})$ with a function x from $\{0, \ldots, |x| - 1\}$ to Γ such that $x(m) = x_m$ for all m with $0 \le m \le |x| - 1$.

The computation by a QTM consists of encoding, preparation, time evolution, measurement, and decoding. The encoding transforms the *input* Γ*-string* to the *input tape string.* The preparation prepares the *initial state* of the quantum Turing machine with the input tape string, and the time evolution transforms the initial state to the *final state.* The measurement of the tape string in the final state gives a probability distribution of the *output tape string.* The decoding transforms the output tape string to the output Γ-string and hence transforms the probability distribution of the output tape string to the probability distribution of the output Γ-string. Therefore, the initial Γ-string is transformed to the *output probability distribution* of the Γ-string.

The *encoding* e of the QTM $\mathcal{Q}$ is a polynomial time computable function from Γ^* to $\Sigma^\#$. Thus, the encoding e transforms any Γ-string x to a tape configuration denoted by $e(x)$; if $T = e(x)$ we shall write $T \sim x$ and T is said to *represent* the Γ-string x. Inversely, the *decoding* d of $\mathcal{Q}$ is a polynomial time computable function from $\Sigma^\#$ to Γ^* satisfying $d(e(x)) = x$ for all $x \in \Gamma^*$.

In this paper, we assume that $B \notin \Gamma$ and $\Sigma = \Gamma \cup \{B\}$. We assume that there is an infinite subset $S \subset \mathbf{N}$ of the set of tape cells, called the *data slot*, with polynomial time numbering $S = \{m_1, m_2, \ldots\}$ and that the encoding is such that

$$e(x)(m) = \begin{cases} x(n) & \text{if } m = m_n \in S \text{ and } 0 \leq n < |x|, \\ B & \text{otherwise,} \end{cases} \tag{1}$$

for any $x \in \Sigma^*$, and the decoding is given by

$$|d(T)| = \min\{m_n \in S |\ T(m_n) = B\}, \tag{2}$$

$$d(T)(n) = T(m_n) \tag{3}$$

for $0 \leq n < |d(T)|$, where $T \in \Sigma^\#$.

The computation begins at $t = 0$. At this time $\mathcal{Q}$ is prepared in an *initial state* $|C_0\rangle$ such that

$$|C_0\rangle = |q_0\rangle |T_{in}\rangle |0\rangle, \tag{4}$$

where T_{in} represents some Γ-string x. In this case, T_{in} is called the *input tape*, x is called the *input*, and $|x|$ is called the *input length.*

The computation proceeds in steps of a fixed unit duration τ. Since the position of the head is discretized, the wave function $|\psi(t)\rangle$ may not stay within $\mathcal{H}$ at any time t other than integer multiples of τ. We assume therefore that the time t is discretized to be an integer multiple of τ. We also take the normalized unit of time in which the time t is assumed to take values in $\mathbf{Z}$. The dynamics of $\mathcal{Q}$ are described by a unitary operator U on $\mathcal{H}$ which specifies the evolution of any state $|\psi(t)\rangle$ during a single *computational step* so that we have

$$U^\dagger U = UU^\dagger = I, \tag{5}$$

$$|\psi(t)\rangle = U^t |\psi(0)\rangle \tag{6}$$

for all positive integer t.

Since the number of all the possible tape strings in the data slot is countable, we assume them to be indexed as $\{T_1, T_2, \ldots\}$. Thus, the observable $\hat{T}(S)$ describing the tape string in the data slot can be represented by

$$\hat{T}(S) = \sum_{j=1}^{\infty} \lambda_j\, I_1 \otimes |T_j\rangle\langle T_j| \otimes I_2 \otimes I_3$$

where $\{\lambda_1, \lambda_2, \ldots\}$ is a countable set of positive numbers in one-to-one correspondence with $\{T_1, T_2, \ldots\}$ by a polynomial time function and where I_1 is the identity on the state space spanned by the processor configurations Q, I_2 is the identity on the state space spanned by the tape strings outside the data slot, and I_3 is the identity on the state space spanned by the head positions $\mathbf{Z}$.

We assume that the measurement to obtain the output is allowed only for the computational basis or more specifically the observable $\hat{T}(S)$ describing directly the output symbol string on the tape, while in Deutsch's formulation[5, 6] and in later work no such restriction has been taken place. However, it is an unavoidable assumption in the definition of quantum Turing machine. In fact, if this assumption is dropped, any function would computable without any computational time. To see this, suppose that the tape strings are encoded by the natural numbers. Let $|T_n\rangle$ be the computational basis state, ignoring the inessential degeneracy, in which the output tape string is the one encoded by n and let $\hat{T}$ be the observable such that $\hat{T}|n\rangle = n|T_n\rangle$. Only such $\hat{T}$ is allowed to measure for obtaining the output. Otherwise, given any function f of the natural numbers and a natural number n, if one prepares the tape in the state $|T_n\rangle$ and measures the observable $f(\hat{T})$, one gets $f(n)$ surely without any computation. This contradicts the Church-Turing thesis. Thus, we cannot allow even the measurement of $f(\hat{T})$ unless f is a polynomial time computable function.

LOCAL TRANSITION FUNCTIONS

Deutsch[5] requires that the QTM operate finitely, i.e., (i) only a finite system is in motion during any one step, (ii) the motion depends only on the state of a finite subsystem, and (iii) the rule that specifies the motion can be given finitely in the mathematical sense. To satisfy the above requirement, the matrix elements of U takes the following form*:

$$\begin{aligned}\langle q', T', \xi' | U | q, T, \xi \rangle = [&\delta_{\xi'}^{\xi+1} D(q, T(\xi), q', T'(\xi), 1) + \delta_{\xi'}^{\xi} D(q, T(\xi), q', T'(\xi), 0) \\ &+ \delta_{\xi'}^{\xi-1} D(q, T(\xi), q', T'(\xi), -1)] \prod_{m \neq \xi} \delta_{T(m)}^{T'(m)} \end{aligned} \tag{7}$$

for any configurations (q, T, ξ) and (q', T', ξ'). The continued product on the right ensures that the tape is changed only at the head position ξ at the beginning of each computational step. The terms $\delta_{\xi'}^{\xi\pm1}$, $\delta_{\xi'}^{\xi}$ ensure that during each step the head position cannot change by more than one unit. The function $D(q, T(\xi), q', T'(\xi), d)$, where $q, q' \in Q$, $T(\xi), T'(\xi) \in \Sigma$, and $d \in \{-1, 0, 1\}$, represents a dynamical motion depending only on the local observables $\hat{q}$ and $\hat{T}(\xi)$. We call D the *local transition function* of the QTM $\mathcal{Q}$.

The function D can be arbitrarily given except for the requirement (5) that U be unitary. Each choice defines a different QTM $\mathcal{Q}[D]$. Thus, if we have an intrinsic characterization of the local transition function D, QTMs can be defined formally without referring to the unitary operator U as a primitive notion.

From (7), the time evolution operator U is determined conversely from the local transition function D

$$U|q, T, \xi\rangle = \sum_{p, \tau, d} D(q, T(\xi), p, \tau, d) |p, T_{\xi}^{\tau}, \xi + d\rangle. \tag{8}$$

*This condition is a natural extension of Deutsch's condition[5] to the case where the head is not required to move.

for any configuration (q, T, ξ), where T_ξ^τ is the tape string defined by

$$T_\xi^\tau(m) = \begin{cases} \tau & \text{if } m = \xi, \\ T(m) & \text{if } m \neq \xi. \end{cases} \tag{9}$$

It follows that the relation $D(q, \sigma, q', \tau, d) = c$ can be interpreted as the following instruction of the operation of $\mathcal{Q}$: if the processor is in the configuration q and if the head reads the symbol σ, then it follows with amplitudes c that the processor's state turns to q', the head writes the symbol τ, and that the head moves one cell to the right if $d = 1$, to the left if $d = -1$, or does not move if $d = 0$.

Now we can formulate the characterization problem of local transition functions of QTMs: *Let D be a complex-valued function on $Q \times \Sigma \times Q \times \Sigma \times \{-1, 0, 1\}$ and let U be the operator on $\mathcal{H}$ defined by (8). Then, what conditions ensure that the operator U is unitary?*

This problem is solved by the following theorem.[13]

Theorem 1 *The operator U is unitary if and only if D satisfies the following conditions.*

(a) *For any $(q, \sigma) \in Q \times \Sigma$,*

$$\sum_{p,\tau,d} |D(q, \sigma, p, \tau, d)|^2 = 1.$$

(b) *For any $(q, \sigma), (q', \sigma') \in Q \times \Sigma$ with $(q, \sigma) \neq (q', \sigma')$,*

$$\sum_{p,\tau,d} D(q', \sigma', p, \tau, d)^* D(q, \sigma, p, \tau, d) = 0.$$

(c) *For any $(q, \sigma, \tau), (q', \sigma', \tau') \in Q \times \Sigma^2$, we have*

$$\sum_{p \in Q} D(q', \sigma', p, \tau', 1)^* D(q, \sigma, p, \tau, -1) = 0.$$

(d) *For any $(q, \sigma, \tau), (q', \sigma', \tau') \in Q \times \Sigma^2$, we have*

$$\sum_{p \in Q, d=0,1} D(q', \sigma', p, \tau', d-1)^* D(q, \sigma, p, \tau, d) = 0.$$

If it is assumed that the head must move either to the right or to the left at each step, the condition (d) is automatically satisfied. In this case, the above statement is reduced to the result due to Bernstein and Vazirani.[8]

In order to maintain the Church-Turing thesis, we need to require that the unitary operator U is constructive, or that the matrix elements of U in the computational basis are computable complex numbers; otherwise, we cannot show the existence of the algorithm by the constructive language. From the complexity theoretical point of view, we need also to require that matrix elements are polynomially computable complex numbers. Thus, we require that the range of the transition function δ is in the polynomially computable complex numbers.

HALTING PROTOCOL

The result of a computation is obtained by measuring the tape string after the computation has been completed. Unlike the classical case, the machine configuration cannot be monitored throughout the computation because of the inevitable disturbance caused by measurement. Thus, the machine needs a specific halt scheme to signal actively when the computation has been completed.

Deutsch[5] introduced an additional single qubit, called the halt qubit, together with an observable $\hat{n}_0$, called the halt flag, with the eigenstates $|0\rangle$ and $|1\rangle$, so that the processor configuration q is represented by the state vector $|q\rangle|1\rangle$ if q is the final state in the classical picture or by $|q\rangle|0\rangle$ otherwise. The halt qubit is initialized to $|0\rangle$ before starting the computation, and every valid quantum algorithm sets the halt qubit to $|1\rangle$ when the computation has been completed but does not interact with the halt qubit otherwise. Deutsch claimed that *the observable $\hat{n}_0$ can then be periodically observed from the outside without affecting the operation of the machine.*

Myers[11] argued that the state entangles the non-halt qubits with the halt qubits so that the measurement of the halt flag changes the state and concluded that the halt scheme spoils the computation.

In the preceding work,[12] Deutsch's halt scheme is reformulated precisely and it is shown that, even though it changes the state of the quantum Turing machine, the measurement of the halt flag does not change the probability distribution of the outcome of the computation so that it does not spoil the computation. It is also shown[12] that the halt scheme is equivalent to the quantum nondemolition monitoring of the output observable.

In what follows, we shall give a new formulation of the halt scheme in which the additional halt qubit is not augmented.

The *halt flag* $\hat{n}_0$ is defined to be the observable corresponding to the projection on the final configuration of the processor, i.e.,

$$\hat{n}_0 = |q_f\rangle\langle q_f|. \tag{10}$$

We assume that we have a measuring apparatus to measure $\hat{n}_0$ precisely after each step instantaneously in the manner satisfying the projection postulate. Thus the $\hat{n}_0$-measurement gives surely the outcome 1 if and only if the processor is in $|q_f\rangle$. We shall denote by $[\![\hat{A} = a]\!]$ the spectral projection onto the eigenspace of an observable $\hat{A}$, considered as an operator on $\mathcal{H}$, corresponding to the eigenvalue a. The product $[\![\hat{A} = a]\!][\![\hat{B} = b]\!]$, if commutable, will be denoted by $[\![\hat{A} = a, \hat{B} = b]\!]$.

The precise formulation of the halting protocol is given as follows.

(I) The halt flag $\hat{n}_0$ is measured instantaneously after every step. This measurement is a precise measurement of the observable $\hat{n}_0$ satisfying the projection postulate. (Note that the above measurement is different from the procedure that one measures $\hat{q}$ and checks if the outcome is q_f because this does not satisfy the projection postulate.)

(II) Once the halt flag is set to $\hat{n}_0 = 1$, the QTM no more changes the halt flag nor the result of computation. Thus, we require

$$U[\![\hat{n}_0 = 1, \hat{T}(S) = T_j]\!]U^t|C\rangle = [\![\hat{n}_0 = 1, \hat{T}(S) = T_j]\!]U[\![\hat{n}_0 = 1, \hat{T}(S) = T_j]\!]U^t|C\rangle \tag{11}$$

for any initial configuration C, time $t \geq 0$, and tape string T_j over the data slot S.

(III) After the measurement of the halt flag $\hat{n}_0$ gives the outcome 1, the tape string $\hat{T}(S)$ in the date slot is measured and the outcome of this measurement is defined to be the *output* of the computation.

Now we shall show that the halting protocol does not affect the result of the computation. For that purpose, it suffices to prove that the probability distribution of the output is not affected by monitoring of the halt flag.

Let $\Pr\{\text{output} = T_j|\text{monitored}\}$ be the probability of finding the output T_j up to N steps by the halting protocol. Let $\Pr\{\text{output} = T_j|\text{not-monitored}\}$ be the probability of finding the output T_j by the single measurement after N steps. We shall prove

$$\Pr\{\text{output} = T_j|\text{monitored}\} = \Pr\{\text{output} = T_j|\text{not-monitored}\} \tag{12}$$

Let $P = [\![\hat{n}_0 = 1]\!]$ and $Q_j = [\![\hat{T}(S) = T_j]\!]$. Let $|C\rangle$ be an arbitrary initial state. If $|C\rangle$ is the state of the machine before the computation, We have

$$\Pr\{\text{output} = T_j|\text{not-monitored}\} = \|PQ_jU^N|C\rangle\|^2. \tag{13}$$

By the projection postulate, the joint probability of obtaining the outcome $\hat{n}_0 = 0$ at the times $1, \ldots, K-1$ and obtaining the outcomes $\hat{n}_0 = 1$ and $\hat{T} = \lambda_j$ at the time K is given by

$$\|PQ_j(UP^{\perp})^K|C\rangle\|^2, \tag{14}$$

and hence we have

$$\begin{aligned}&\Pr\{\text{output} = T_j|\text{monitored}\}\\ &\quad = \|PQ_j|C\rangle\|^2 + \|PQ_jUP^{\perp}|C\rangle\|^2 + \cdots + \|PQ_j(UP^{\perp})^N|C\rangle\|^2.\end{aligned} \tag{15}$$

Thus, it suffices to prove the relation

$$\|PQ_jU^N|C\rangle\|^2 = \|PQ_j|C\rangle\|^2 + \|PQ_jUP^{\perp}|C\rangle\|^2 + \cdots + \|PQ_j(UP^{\perp})^N|C\rangle\|^2 \tag{16}$$

for any N and any initial state $|C\rangle$.

Let $\psi = U^t|C\rangle$ where $t \geq 0$. We first consider the relation

$$\|PQ_jU\psi\|^2 = \|PQ_j\psi\|^2 + \|PQ_jUP^{\perp}\psi\|^2. \tag{17}$$

From (11), we have

$$PQ_jUPQ_j\psi = UPQ_j\psi. \tag{18}$$

It follows that

$$PQ_jUPQ_j^{\perp}\psi = \sum_{k \neq j} PQ_jUPQ_k\psi = 0. \tag{19}$$

From (18) and (19), we have

$$\begin{aligned}PQ_jU\psi &= PQ_jUPQ_j\psi + PQ_jUPQ_j^{\perp}\psi + PQ_jUP^{\perp}\psi\\ &= UPQ_j\psi + PQ_jUP^{\perp}\psi.\end{aligned} \tag{20}$$

From (18), we have

$$\begin{aligned}\langle UPQ_j\psi|PQ_jUP^{\perp}\psi\rangle &= \langle PQ_jUPQ_j\psi|UP^{\perp}\psi\rangle\\ &= \langle UPQ_j\psi|UP^{\perp}\psi\rangle\\ &= 0.\end{aligned} \tag{21}$$

From (20) and (21), we have

$$\begin{aligned}\|PQ_jU\psi\|^2 &= \|UPQ_j\psi + PQ_jUP^{\perp}\psi\|^2\\ &= \|UPQ_j\psi\|^2 + \|PQ_jUP^{\perp}\psi\|^2\\ &= \|PQ_j\psi\|^2 + \|PQ_jUP^{\perp}\psi\|^2.\end{aligned} \tag{22}$$

Thus, we have proved (17).

The proof for general N runs as follows. We use mathematical induction and assume that (16) holds for $N-1$. By replacing ψ by $U^{N-1}|C\rangle$ in (17), we have

$$\|PQ_jU^N|C\rangle\|^2 = \|PQ_jU^{N-1}|C\rangle\|^2 + \|PQ_jUP^{\perp}U^{N-1}|C\rangle\|^2. \tag{23}$$

From (18), we have $P^{\perp}UP\psi = \sum_j P^{\perp}UPQ_j\psi = 0$, and hence $P^{\perp}U\psi = P^{\perp}UP^{\perp}\psi$ so that $P^{\perp}U^{N-1}|C\rangle = P^{\perp}(UP^{\perp})^{N-1}|C\rangle$. It follows that

$$\|PQ_jUP^{\perp}U^{N-1}|C\rangle\|^2 = \|PQ_j(UP^{\perp})^N|C\rangle\|^2. \tag{24}$$

By induction hypothesis, we have

$$\|PQ_jU^{N-1}|C\rangle\|^2 = \|PQ_j|C\rangle\|^2 + \|PQ_jUP^{\perp}|C\rangle\|^2 + \cdots + \|PQ_j(UP^{\perp})^{N-1}|C\rangle\|^2. \tag{25}$$

Therefore, from (23), (24), and (25), we obtain (16).

It is concluded that the the probability of finding the output T_j up to N steps by the halt protocol is equal to the probability of finding the output T_i by the single measurement of $\hat{T}(S)$ after N steps. It follows that the halting protocol does not affect the result of the computation.

Recently, Linden and Popescu[14] claimed that the halt scheme given previously[12] is not consistent with unitarity of the evolution operator. However, their argument applies only to the special case in which the whole tape is required not to change after the halt. As suggested in a footnote, the conclusion in the previous work[12] can be obtained from the weaker condition for the general case where the tape is allowed to change except for the date slot. Linden and Popescu[14] disregarded this case and hence their conclusion is not generally true. In this paper, the halting protocol with such a general formulation is treated explicitly and it is proved that even in this case the computation is not affected by the measurement of the halt flag. Moreover, contrary to Linden and Popescu,[14] this general formulation is consistent with the unitarity. In fact, it can be shown that any unidirectional QTMs and stationary QTMs[8] can be simulated by QTMs obeying this halting protocol with constant slowdown.[13] Thus, there is a universal QTM obeying the halting protocol.

REFERENCES

1. A. Church, Am. J. Math. **58**, 345 (1936).
2. A. M. Turing, Proc. Lond. Math. Soc. Ser. 2 **42**, 230 (1936).
3. C. H. Papadimitriou, *Computational Complexity* (Addison-Wesley, Reading, MA, 1994).
4. R. P. Feynman, Inter. J. Theor. Phys. **21**, 467 (1982).
5. D. Deutsch, Proc. R. Soc. Lond. A **400**, 97 (1985).
6. D. Deutsch, Proc. R. Soc. Lond. A **425**, 73 (1989).
7. P. Benioff, J. Stat. Phys. **22**, 563 (1980).
8. E. Bernstein and U. Vazirani, SIAM J. Comput. **26**, 1411 (1997).
9. A. Yao, in *Proceedings of the 34th Annual Symposium on Foundations of Computer Science*, edited by S. Goldwasser, p. 352 (IEEE Computer Society Press, Los Alamitos, CA, 1993).
10. P. W. Shor, in *Proceedings of the 35th Annual Symposium on Foundations of Computer Science*, edited by S. Goldwasser, p. 124 (IEEE Computer Society Press, Los Alamitos, CA, 1994).
11. J. M. Myers, Phys. Rev. Lett. **78**, 1823 (1997).
12. M. Ozawa, Phys. Rev. Lett. **80**, 631 (1998).
13. H. Nishimura and M. Ozawa, "Computational complexity of uniform quantum circuit families and quantum Turing machines," (in preparation).
14. N. Linden and S. Popescu, "The halting problem for quantum computers," (e-print: quant-ph/9806054).

A NEW APPROACH TO QUANTUM COMPUTATION

Horace P. Yuen

Department of Electrical and Computer Engineering,
Department of Physics and Astronomy,
Northwestern University, Evanston, IL 60208

A general quantum machine is characterized to capture all the possible physical structures that may be employed for effective or efficient computation. As an illustration, a specific machine is described for the permanent evaluation problem. The contrast between quantum and classical parallelism is also discussed, as is the effect of quantum error correction against decoherence and loss.

INTRODUCTION

The power of computation devices, as modeled by Turing machines or random access machines, is characterized by the number of machine steps required to solve a given problem — the problem is *effectively* solvable if the solution is provided by the machine in a finite number of steps, *efficiently* solvable if the number of steps is a polynomial function of the size of the problem instance. This number of steps, or (time) complexity of the problem, is clearly machine model dependent, but it turns out that all the standard machine models are polynomially equivalent: one leads to at most polynomial speedup compared to another. However, the possibility that physical structures different from the standard machine models may lead to more powerful machines, both in regard to the scope of effective computation and especially in regard to the complexity of effective computation, has never been systematically studied, although isolated instances of such possibility are long known as in the case of optical Fraunhoffer diffraction computation of Fourier transforms. I believe that the recent development of quantum computation and DNA computation compels us to consider seriously the possible potential of *physical computation* in general, i.e., computation that takes advantages of any possible physical process in nature rather than just Turing machine processes. In the language of computer science, one would want to identify the *oracles*, one-step solvers or more generally efficient solvers of various problems, that can be realized according to the laws of physics. In this paper I'll try to indicate one approach to such issues in the context of quantum computation, with specific consideration on a #P complete problem, the solution of which implies the solutions of all NP problems.

It may be observed that the issues to be addressed are of extreme importance both theoretically and technologically. Whether the class of NP problems can be solved efficiently has long been regarded as the most outstanding problem in theoretical computer

Quantum Communication, Computing, and Measurement 2
Edited by Kumar et al., Kluwer Academic / Plenum Publishers, New York, 2000.

science. The fast solutions of many computation problems, including many in the NP class, would have enormous financial and other impacts on the world. The connection between computation and physics would provide some basic fabrics in the web of our knowledge of the world.

The power of quantum computation compared to classical computation, as exemplified in the Shor algorithm [1], is usually taken to lie in the quantum parallelism first emphasized by Deutsch [2]. This will be analyzed in the next section. Then, the effect of quantum error correction for combating decoherence will be discussed, with the conclusion that decoherence may be harder to suppress than is often supposed. This motivates the one-step quantum machines next introduced. A specific such quantum machine for the permanent evaluation problem is then described. Due to space limitation, only a sketch of the results can be given.

QUANTUM VERSUS CLASSICAL PARALLELISM

The power of quantum computation is often described as the possibility of linearly superposing bits (or the qubits), while of course one can superpose classical bits also. The only relevant point here is that the bits form a linear space. Thus, one does not even need "waves" for superposition, say a space of classical signals $\{s(t)\}$ will do. The essential point is whether the number of linearly independent bits, the dimension of the linear space, can grow exponentially with only polynomial growth in "resources." Quantum mechanically, the tensor product state space of n qubits has dimension 2^n, thus an exponential number of degrees of freedom can be linearly processed simultaneously. To do this classically, one would need a space of dimension 2^n for an n bit input. This cannot be directly obtained as a consequence of the following Dimensionality Theorem.

A function $s(t)$ with Fourier transform $S(f)$ cannot be both strictly time-limited and strictly band-limited, i.e., $s(t)$ and $S(f)$ cannot both have compact support, due to analyticity relations. However, various approximate time-duration T and approximate bandwidth W can be defined that are experimentally meaningful, and for all such definitions it can be shown that the dimension D of the linear space spanned by such signals is about

$$D \sim 2TW. \tag{1}$$

Perfectly precise and rigorous version of this Dimensionality Theorem can be found in [3] and references cited therein.

I believe the Dimensionality Theorem is an extremely important fundamental result, it may even be regarded as a law of nature. It reduces an otherwise infinity to a finite, physically meaningful number. Since the "resource" T or W cannot be increased exponentially, the Dimensionality Theorem prohibits the direct classical construction of a linear space that would otherwise enable one to solve the integer factorization problem in a simplified version of the Shor algorithm. In my opinion, a covariant formulation of the Dimensionality Theorem for Minkowski manifolds would renormalize quantum field theories, for pseudo-Riemannian manifolds would renormalize quantum gravity, as it controls the localization in time (space) and frequency (wavevector).

Nevertheless, the Dimensionality Theorem alone does not preclude the possibility of building classical Hilbert spaces with exponential growth in dimension. Similar to the quantum case, one may try to obtain an exponential growth in dimension via tensor product spaces. In an otherwise very insightful paper, Jozsa [4] indicated that tensor product space cannot be constructed classically, which is literally not true as is evident, for example, in the solution of Laplace equation by separation of variables. However, there are only a total of four space-time dimensions. For $n \geq 4$, it is not clear whether it is possible to construct an n-tensor product space from individual spaces with polynomial growth in resources. The obvious approach of chopping up one dimension does not work, at least not directly. This is clearly an important issue to settle: whether quantum parallelism can be similarly implemented classically. Indeed, other than quantum parallelism, the physical interpretation of a quantum Hilbert space actually introduces various obstacles to the implementation of computation as compared to a classical Hilbert space.

EFFECT OF QUANTUM ERROR CORRECTION

Currently, quantum algorithms are to be realized via quantum networks that are to be constructed from simple quantum gates in analogy with classical computers. This leads to the serious decoherence problem unique to quantum networks that involve the loss of the correct quantum superposition of states through dephasing and energy loss as a result of coupling of the gates to the environment, or simply the vacuum, during the computation process. A large body of work on quantum error correction and fault tolerant quantum computing have been carried out for suppressing decoherence. Typically, fairly optimistic conclusions are drawn under a variety of assumptions on the nature of decoherence, while little were done to analyze decoherence models to test such assumptions. In this section, a few analytical results are described that appear to throw some doubt on the validity of such optimistic conclusions in realistic systems. They are different from the other criticisms one can already find in the literature.

Consider first the case of single quantum error correction for a state $a|0\rangle + b|1\rangle$ represented by a coded state $|\psi\rangle_c = a|0\rangle_c + b|1\rangle_c$, with state fidelity between a given $|\psi\rangle$ and a ρ measured by $F = \langle\psi|\rho|\psi\rangle$. Assuming that the qubits have independent environments each represented by

$$\begin{aligned} |0\rangle &\rightarrow |0\rangle|f_1\rangle + |1\rangle|f_2\rangle\,, \\ |1\rangle &\rightarrow |0\rangle|f_3\rangle + |1\rangle|f_4\rangle\,, \end{aligned} \tag{2}$$

it is often claimed [5, 6] that if ϵ is the single error event probability then the effect of single error correction is to increase $F = 1 - O(\epsilon)$ before decoding to $F' = 1 - O(\epsilon^2)$

$$F = 1 - O(\epsilon) \longrightarrow F' = 1 - O(\epsilon^2)\,. \tag{3}$$

This conclusion (3) was first drawn by Shor [7] under the assumption that the environment produces an effect on each qubit independently with probability ϵ. However, the environment satisfies this Shor assumption only in the special case when the total environmental CP map factorizes into a product $\otimes_i T_i$ with each T_i having the decomposition [8]

$$T_i = (1-\epsilon)I + \epsilon\, T_i' \tag{4}$$

for some CP map T_i'. Condition (4) obtains when $|f_2\rangle = |f_3\rangle = 0$ with $|f_1\rangle = |f_4\rangle =$ real, but is evidently not true in general. The only other proof of (3) provided is Steane's 3-qubit dephasing example [5].

By analyzing the error correction action in the general case, it can be shown that (3) is true under both the general pure dephasing case

$$|f_2\rangle = |f_3\rangle = 0 \tag{5}$$

and the pure loss case

$$|f_2\rangle = 0, \quad \langle f_1|f_3\rangle = \langle f_4|f_3\rangle = 0\,. \tag{6}$$

However, in the case where

$$\begin{aligned} \langle f_1|f_1\rangle &\sim \langle f_4|f_4\rangle \sim 1-\epsilon\,, \\ \langle f_2|f_2\rangle &\sim \langle f_3|f_3\rangle \sim \epsilon\,, \\ \langle f_1|f_{2/3}\rangle &\sim \langle f_4|f_{2/3}\rangle \sim \sqrt{\epsilon(1-\epsilon)}\,, \end{aligned} \tag{7}$$

it is found that F before decoding may be of order $1 - O(\sqrt{\epsilon})$ and that even when $F = 1 - O(\epsilon)$ before decoding, it is possible to have $F' = 1 - O(\sqrt{\epsilon}^3)$ after decoding. Thus, (3) is *not* valid in general. The effect of (7) on multiple error correction and fault tolerant computing remains to be determined. In any case, the *crucial* point of this discussion is that error event probability is not sufficient in dealing with quantum error correction — one needs error probability amplitude in general.

The assumption that the environment acts on each qubit independently does not appear to hold in many typical situations. In the case of spontaneous emission for which a number of radiation modes are involved with wavelengths covering all the qubits in the code, it can be easily shown that the total environmental CP map does not factorize into a product of individual qubit CP maps even in first order perturbation theory. In this case of a general environment, one can readily construct analytic models of small ϵ decoherence for which no quantum error correction can have an F' with a smaller order of error because F before decoding contains only linear ϵ terms which cannot all be removed. However, the decoherence problem could be severely curtailed, if not entirely eliminated, in a one-step quantum device that does not realize a unitary transform by a quantum gate network.

GENERAL QUANTUM MACHINE

Apart from the question of the scope of effective computation which we are not concerned with here, there are several limitations on Turing machine models, classical or quantum, as general representation of computation structures in regard to efficient computation or complexity. The sequential memory limitation is readily removed by random access memory, but the basic machine step count for complexity is way too restrictive for a general physical structure. This is clearly illustrated by the one-step optical calculation of Fourier transforms, which can be regarded as a physical oracle. Other than one-step oracles, one may have a physical structure that processes the data according to a certain physical mechanism with 'steps' totally different from a digital network. Such a device would probably be a special purpose computer. Note that we are not concerned with the invention of a new physical computer to replace current general purpose computers. Rather, a special purpose computer, especially one that solves a complete problem in a class, is sought as an aid or oracle to an ordinary computer. Thus, if we have an (efficient) NP complete problem solver, we can adjoint it to a usual computer to (efficiently) solve all problems in the NP class.

One way to describe a general quantum machine is as follows. Let $|\psi_0\rangle \in \mathcal{H}_S$ be an initial state for L qubits. With an input x of size $n \leq L$, a unitary transformation U_x dependent on x is applied to $|\psi_0\rangle$ and a measurement is made to extract some information j, $j \in \{1, \cdots, M\}$. The most general such measurement is described by a POM, which can be realized [9] by adjoining an "apparatus" in initial state $|\Phi_0\rangle \in \mathcal{H}_A$ and measuring on $\mathcal{H}_S \otimes \mathcal{H}_A$ the observable with orthogonal spectral projections Π_j,

$\sum_j \Pi_j = I$. All the quantum processing is done upon extracting j, and subsequent classical processing on j is used to produce the output. It can be shown from the results of Benioff [10] that there is no loss of generality in this description even if more elaborate and general intermediate processing is used. A quantum Turing machine (QTM) would obtain if appropriate restrictions are placed on U_x and Π_j, as well as on the effective computability of various numerical quantities involved. For efficient computation, all the quantities and operations involved would be required to be "*efficiently realizable*," i.e., they are known to be obtainable in principle in an empirically reasonable amount of time, which for example is to be measured by a polynomial function of the input size. Note that the notion of "efficiently realizable" here is not just a logical one. Even if there is no logical path to obtain the required quantity or operation, it is fine if one just hits upon a way to get it. This is true not only for $|\psi_0\rangle$ or $|\Phi_0\rangle$, but also for U_x. Also, we are obviously talking here about "efficiently realizable" in principle, i.e., that a theoretically efficient procedure is available to specify the quantity or operation, not that an empirical known procedure is available which is a separate issue.

The decoherence problem associated with the U_x in the above machine description can be suppressed in the following compression of the machine. The U_x can be absorbed in the quantum measurement, as can the classical processing. Although it may be desirable to leave the classical processing in an actual device implementation, for simplicity it will be absorbed in the following description. Thus, a *quantum machine* is defined to be a device with L qubits state $|\psi_0\rangle \in \mathcal{H}_S$, apparatus state $|\Phi_0\rangle \in \mathcal{H}_A$, and a map from input x to orthogonal measurement projectors Π_{jx}, $j \in \{1, \cdots, M\}$,

$$x \mapsto \Pi_{jx}, \quad \sum_{j=1}^{M} \Pi_{jx} = I\,, \tag{8}$$

where the readout j is the desired output. From a theorem of Chan [11], it is sufficient that the dimension of $\mathcal{H}_A$ be equal to M,

$$\dim \mathcal{H}_A \leq M\,. \tag{9}$$

For efficient realization, both the map (8) and the actual measurement represented by $\{\Pi_{jx}\}$ for all possible x must be efficiently realizable. Rather than giving many possible sufficient conditions for such efficient realization, the $\#P$ complete problem example in the next section would provide a concrete example. Here it may be noted that both the Shor algorithm and the Grover algorithm [12] can be put in this form, which provides an alternative way for their realization avoiding the decoherence problem, but without any guarantee of efficiency. Analogous definition for a general classical machine may be given. Although it appears strange at first sight, upon reflection it will be clear that a usual computer can indeed be described in this apparent one-step or zero-step (not counting read-in read-out) way.

This characterization of a general quantum machine seems remarkably powerful. Apparently it includes all possible physical structures and physical oracles obtainable in accordance with the laws of quantum physics. Effectively it transforms a quantum computation problem to a quantum measurement problem, albeit with the problem of efficient realization. If the quantum measurement can be carried out without a quantum network, as in the following $\#P$ example, the decoherence problem appears at only one point, between the state preparation and the measurement. Furthermore, when exponential search or computation is accomplished in one overall step, NP complete problems may be efficiently solved if the corresponding measurement can be efficiently realized. To the attention of this problem we now turn.

THE PERMANENT EVALUATION PROBLEM

Consider the problem of evaluating the permanent of an $n \times n$ matrix A

$$\text{perA} = \sum_{\pi} \text{A}_{1\pi(1)} \cdots \text{A}_{n\pi(n)}\,, \tag{10}$$

where the sum is over the $n!$ permutations π on $\{1, \cdots, n\}$, and $\text{A}_{ij} \in \{0,1\}$. This is a #P complete problem [13], the solution of which would provide efficient solutions to all NP problems. A full polynomial randomized approximation scheme (FPRAS) to a counting problem is an algorithm which provides an estimate of the correct number to within a fraction $\epsilon > 0$ with probability $\geq 1 - \delta$, with running time a polynomial function of input size n, $1/\epsilon$ and $\log 1/\delta$. If the running time is a polynomial in $\log \frac{1}{\epsilon}$, an actual solution in the probabilistic sense (BQP) would be obtained. No FPRAS is known for the above perA evaluation problem on a usual computer [14], and of course none has been provided for a quantum computer. In the following, a quantum machine in the sense of the previous section is provided that reduces the FPRAS or BQP solution of this problem to a corresponding accurate quantum measurement problem.

Define an orthonormal basis $\{|e_i\rangle\}$ on $\mathcal{H}$, $\dim \mathcal{H} = n$, so that $\text{A}_{ij} = \langle e_i|A|e_j\rangle$ where A is the operator corresponding to A. Let $|\psi\rangle$ be the symmetrized state

$$|\psi\rangle = \frac{1}{\sqrt{n!}} \sum_{\pi} |e_{\pi(1)}\rangle \cdots |e_{\pi(n)}\rangle \tag{11}$$

on $\otimes^n \mathcal{H}$. It is easily shown that

$$\langle\psi| \otimes^n A|\psi\rangle = \text{perA}, \qquad \otimes^n A \equiv A \otimes \cdots \otimes A\,. \tag{12}$$

Let $A^h = A_1^h \otimes \cdots \otimes A_n^h$ on $\otimes_{i=1}^n (\mathcal{H}_{i1} \oplus \mathcal{H}_{i2}) \equiv \mathcal{H}'$ where

$$A_i^h \equiv \begin{pmatrix} 0 & A^\dagger \\ A & 0 \end{pmatrix} \quad \text{on} \quad \mathcal{H}_{i1} \oplus \mathcal{H}_{i2}\,. \tag{13}$$

It is readily shown that states $|\psi'\rangle \in \mathcal{H}'$ can be simply obtained from $|\psi\rangle$ with

$$\langle\psi| \otimes^n A|\psi\rangle = \langle\psi'|A^h|\psi'\rangle\,. \tag{14}$$

The input of the perA problem is the (0,1) matrix A of size n. The output perA is obtained by the measurements of the hermitian A^h in $|\psi'\rangle$ from (12)–(14) to form the average $\langle\psi'|A^h|\psi'\rangle = \text{perA}$. This measurement is efficiently realizable because each A^h is specified by $2n^2$ known (0,1) coefficients, so that A^h is specified by $2n^3$ bits. Indeed, one can write down explicit operator expressions for A^h with either spin, atomic level, or photon operators. Let a_i be the annihilation operator of state $|e_i\rangle$, $a_i|e_i\rangle = |0\rangle$. Then A can be written $A = \sum_{i,j} \text{A}_{ij} a_i^\dagger a_j$. The state $|\psi\rangle$ can be generated efficiently from the vacuum [15]. If it is taken to be the ground state of a Bose-Einstein condensate, there would be additional advantages that there can be no loss, and no decoherence since the state symmetry would be then destroyed.

As the output is in the form of a quantum average, measurements of N independent systems are needed to form a good estimate. From standard central limit theorem estimates, it is easily shown that the above provides a FPRAS for the perA problem to within a fraction ϵ when N grows quadratically with $\frac{1}{\epsilon}$ if the means m of (14) for all A have corresponding variances σ^2 such that each σ/m is polynomially bounded in n. An exponential number of measurements $N \sim n!$ would be needed to find perA in the sense of BQP, unless σ itself is polynomially bounded in n for all A. Thus, to make this quantum machine useful, a way has to be found to reduce σ to the desired level, or some powerful accurate measurement scheme has to be devised to obtain the required accuracy which is closely related to the content of ref. [16].

To write this machine as a quantum machine of the previous section, we may identify $|\Phi_0\rangle$ to be $|\psi'\rangle$ with $|\psi_0\rangle$ irrelevant, or $|\psi_0\rangle$ as $|\psi'\rangle$ with no need for $|\Phi_0\rangle$, and then extend the system to N identical independent systems, identifying j as the eigenvalue average of the A^h measurements on the N systems. It is also possible to get $\langle\psi| \otimes^n A|\psi\rangle$ in the form $\langle\psi| \otimes^n \mathsf{U}|\psi\rangle$, for some unitary transformation U, that can be measured via probability measurements. Either this approach or (14) can also be directly implemented by an efficient QTM, but the point here is to avoid quantum networks to sidestep the decoherence problem.

It may also be pointed out that the determinant of a real $n \times n$ matrix A can be obtained as the eigenvalue of $\otimes^n A$ in the eigenstate $|\phi\rangle$, the antisymmetrized state on $\{|e_i\rangle\}$,

$$\otimes^n A|\phi\rangle = \det \mathrm{A}|\phi\rangle \tag{15}$$

Equation (15) is equivalent to the well-known determinant representation theorem in multilinear algebra [18]. Currently, determinant evaluation has time complexity $0(n^{2.376})$, and the best known randomized parallel algorithm has complexity $0(\log^2 n)$ using $0(n^{4.376})$ processors [14]. According to (15), a determinant can be evaluated by making measurements on n sets of $\log n$ qubits in a single step, which is quite an improvement.

The above description of a general quantum machine, as exemplified by the specific example here, calls for some sort of universal measurement scheme that would provide a uniform, efficient way to realize classes of quantum measurements. It appears that one promising approach, as far as the evaluation of quantum averages or probabilities goes, is some form of generalized quantum tomography. In particular, in a tensor product system it should be possible to measure a fixed small set of operators on each $\mathcal{H}$ to get all possible operator averages efficiently, as in homodyne tomography [17]. In such schemes, there is no need to go through the space extension (13) since $\langle A\rangle = \langle A_1\rangle + i\langle A_2\rangle$ and the self adjoint averages $\langle A_1\rangle, \langle A_2\rangle$ can be evaluated separately. Also, all the tomographic measurements could be made any time well in advance of the computation.

In conclusion, we may observe that (11)–(14) can be interpreted classically in a classical Hilbert space. If the dimension of such a space is large $\sim n^n$, as for example when arbitrary time resolution is available, the perA problem would be solved exactly since $\langle\psi| \otimes^n A|\psi\rangle$ is just a classical signal correlation, not the average of a random quantity. This underlines the importance of our concern on the realizability of classical parallelism.

REFERENCES

1. P. W. Shor, SIAM J. Comput. **26**, 1474 (1997).
2. D. Deutsch, Proc. Roy. Soc. Lond. **A400**, 97 (1985).
3. D. Slepian, Proc. IEEE **64**, 292 (1976).
4. R. Jozsa, in *The Geometric Universe*, S. A. Huggett, ed. (Oxford University Press, 1998); p. 367.
5. A. Steane, Proc. Roy. Soc. Lond. **A452**, 255 (1996).
6. J. Preskill, quant-ph/9705031.
7. P. W. Shor, Phys. Rev. A **52**, 2493 (1995).
8. E. Knill and R. Laflamme, Phys. Rev. A **55**, 900 (1997); theorem (V.5).

9. A. S. Holevo, *Probabilistic and Statistical Aspects of Quantum Theory* (North Holland, 1982); p. 67.
10. P. A. Benioff, J. Math Phys. **13**, 908 (1972).
11. V. W. S. Chan, Tech. Rep. 489, MIT RLE (1975); theorem 9.
12. L. Grover, Phys. Rev. Lett. **78**, 325 (1997); *ibid* **79**, 4709 (1997).
13. C. H. Papadimitriou, *Computational Complexity* (Addison Wesley, 1994); Ch. 18.
14. R. Motwani and P. Ragharan, *Randomized Algorithms*, (Cambridge University Press, 1995); ch. 11 and 12.
15. A. Barenco, A. Berthiaume, D. Deutsch, A. Ebert, R. Jozsa, and C. Macchiavello, SIAM J. Comput. **26**, 1541 (1997).
16. H. P. Yuen, "The Ultimate Quantum Limits on the Accuracy of Measurements," in *Proceedings of the Workshop on Squeezed States and Uncertainty Relations*" NASA Conference Publications 3135, 1991, pp. 13-22.
17. G. M. D'Ariano, U. Leonhardt, and H. Paul, Phys. Rev. A 52, 1801 (1995).
18. S. MacLane and G. Birkhoff, *Algebra*, (MacMillan, 1979); ch. XVI.

HYPERSENSITIVITY TO PERTURBATION ON AN NMR QUANTUM COMPUTER

Rüdiger Schack[a] and Todd A. Brun[b]

[a]Department of Mathematics, Royal Holloway,
University of London, Egham, Surrey TW20 0EX, UK
E-mail: r.schack@rhbnc.ac.uk

[b]Institute for Theoretical Physics,
University of California, Santa Barbara, CA 93106, USA
E-mail: brun@itp.ucsb.edu

By numerically simulating an implementation of the quantum baker's map on a 3-qubit NMR quantum computer based on the molecule trichloroethylene, we demonstrate the feasibility of quantum chaos experiments on present-day quantum computers. In particular, we describe an experiment investigating the phenomenon of hypersensitivity to perturbation.

INTRODUCTION

Hilbert space is large, but most of it is not directly accessible: To specify the D complex amplitudes of a typical state in D-dimensional Hilbert space to an accuracy of $2^{-\alpha}$, one needs approximately $2\alpha D$ bits of information, but the maximum information about the state that can be obtained in a measurement is $\log_2 D$ bits. This difference is at the heart of many applications of quantum information theory; e.g., in quantum cryptography it is impossible for an eavesdropper to gain enough information to determine the transmitted state without destroying the state and thus alerting the correspondents.

Another fundamental consequence is the potential for *hypersensitivity to perturbation*[1, 2, 3] in quantum systems. Hypersensitivity to perturbation is a quantitative measure of unpredictability in an open classical or quantum system, i.e., a system that cannot be completely isolated from environment perturbations. For classical systems, hypersensitivity to perturbation has been shown to be equivalent to a standard definition of classical chaos under general assumptions.[2] Hypersensitivity to perturbation has therefore been proposed to define quantum chaos.

Recently, one of us has shown that if the dimension of Hilbert space is a power of 2, $D = 2^N$, the quantum baker's map has a simple representation on a quantum computer.[4] The *classical* baker's map is a simple map on the unit square which embodies the two hallmarks of chaos, stretching and folding. The *quantum* baker's map[5] was

Quantum Communication, Computing, and Measurement 2
Edited by Kumar et al., Kluwer Academic / Plenum Publishers, New York, 2000.

invented for the theoretical investigation of quantum chaos. Its mathematical properties have been studied extensively, but no experimental quantum systems are known which embody it. To define the map, we start from position eigenstates $|q_j\rangle$ labeling the discrete positions j/D, $j = 0, \ldots, D-1$ on the unit interval. On a quantum computer, these are represented in the product space of N qubits in such a way that

$$|q_j\rangle = |a_{N-1}\rangle \otimes |a_{N-2}\rangle \otimes \cdots \otimes |a_0\rangle \ , \tag{1}$$

where $j = \sum a_k 2^k$, $a_k \in \{0, 1\}$ $(k = 0, \ldots N-1)$, and where each qubit has basis states $|0\rangle$ and $|1\rangle$. The quantum baker's map can now be written as[4]

$$T = F_N^{-1}(I \otimes F_{N-1}) \ , \tag{2}$$

where F_{N-1} acts on the $N-1$ least significant qubits, and I is the identity operator.

In an earlier paper[6] we described realistic simulations of a simplified version of the quantum baker's map, as implemented on 3-bit NMR quantum computer.[7,8] With this simple map, we were able to demonstrate both hypersensitivity to perturbations and linear growth of entropy in the presence of noise, a signature of quantum chaos[9] which is closely related to the concept of quantum dynamical entropy.[10] Hypersensitivity remained detectable even in the presence of noise at the level of current experiments. In the present paper, we investigate the possibility of studying hypersensitivity to perturbation for the full quantum baker's map as proposed originally in Ref. 5.

THE PROPOSED EXPERIMENT

We consider this map for a $D = 8 = 2^3$ dimensional Hilbert space, implemented on a 3-qubit NMR quantum computer. Since decoherence cannot be neglected for this experiment, we model the NMR system by a master equation of Lindblad form,[11] including the Hamiltonian time evolution, the radio-frequency (RF) pulses, and phase noise due to the environment.

We choose for our physical system the molecule trichloroethylene used in Ref. 12, in which the nuclear spins of the hydrogen (H) and two carbons (C_1 and C_2) serve as our qubits. The spins on a single molecule weakly interact with each other, but are effectively shielded from the environment by rapid tumbling. The molecules are placed in a strong, uniform magnetic field and subjected to RF pulses of various frequencies and durations.

We denote by $\hat{X}$, $\hat{Y}$, and $\hat{Z}$ the σ_x, σ_y, and σ_z Pauli matrices, respectively, and indicate with a subscript (e.g., $\hat{X}_H$) to which spin they apply. The Hamiltonian of the three spins in the interaction picture is approximately[6]

$$\hat{H} \approx \frac{j_1}{4}\hat{Z}_H\hat{Z}_{C_1} + \frac{j_2}{4}\hat{Z}_{C_1}\hat{Z}_{C_2} + \frac{\delta}{2}\hat{Z}_{C_2}, \tag{3}$$

where $j_1 \approx 203$ Hz, $j_2 \approx 102$ Hz, and $\delta \approx -905$ Hz.[13]

In addition to these interactions, we can apply RF pulses which rotate the nuclear spins about the x and y axes, equivalent to operators of the form $\exp(i\theta\hat{X}/2)$ and $\exp(i\theta\hat{Y}/2)$. By controlling the pulse frequencies, these can be selectively applied to single spins, or in some cases to two or three spins at once. The duration of the pulse determines the angle by which the spin is rotated, but in all cases these durations are very short (by about a factor of 100) compared to the timescale of the Hamiltonian $\hat{H}$. It is therefore reasonable to approximate them as instantaneous. For the purposes of our simulations we also assume that these pulses are timed with perfect accuracy.

The general form of a quantum algorithm in an NMR computer is a sequence of pulses, causing rotations of the individual bits, interspersed by precisely-timed delays during which the undriven Hamiltonian couples the neighboring spins.[7, 8] The exact sequence of RF pulses and delays corresponding to the quantum baker's map T is given in Ref. 6. It requires 52 pulses and 20 delays with a total time of $25\pi/2j_1$. Bit 0 is identified with C_1, bit 1 with H, and bit 2 with C_2.

Having defined a quantized version of a classically chaotic map, we need to define a regular map for comparison. This map should take the same time per iteration as the map T, so that it is affected equally by the intrinsic noise. The simplest possibility would be a map that does essentially nothing. This implies that the effects of the Hamiltonian in coupling the spins must be suppressed. We do this by *refocusing*, as described in Ref. 6, where we show that a sequence of evenly spaced x rotations by π on the C_1 spin over a total time t is equivalent to

$$T_{\mathcal{R}} \approx \exp(-it\delta \hat{Z}_{C_2}/2). \tag{4}$$

The total time t should equal the iteration time of the map T, so $t = 25\pi/2j_1$.

The effects of the environment are described by two timescales T_1 and T_2.[14] T_1 is the timescale on which the spins relax to the thermal state and T_2 is the timescale of phase decoherence. For the molecule trichloroethylene, T_1 exceeds T_2 by an order of magnitude, and can be safely neglected for this problem. We model the effect of T_2 by replacing the state vector $|\psi\rangle$ with a density matrix ρ and the Schrödinger equation with a Markovian master equation in Lindblad form,[11]

$$\frac{d\rho}{dt} = -\frac{i}{\hbar}[\hat{H}, \rho] + \frac{1}{2}\sum_k 2\hat{L}_k \rho \hat{L}_k^\dagger - \hat{L}_k^\dagger \hat{L}_k \rho - \rho \hat{L}_k^\dagger \hat{L}_k, \tag{5}$$

where $\hat{H}$ is the Hamiltonian (3) and the $\hat{L}_k$ are a set of operators chosen to model the effects of decoherence. The simplest choice of $\hat{L}_k$'s which capture the essential physics are proportional to the $\hat{Z}$ operator for each of the three spins:

$$\hat{L}_1 = \sqrt{\Gamma_H}\hat{Z}_H \ , \ \hat{L}_2 = \sqrt{\Gamma_{C_1}}\hat{Z}_{C_1} \ , \ \hat{L}_3 = \sqrt{\Gamma_{C_2}}\hat{Z}_{C_2}. \tag{6}$$

The dephasing rate Γ is proportional to $1/T_2$ for each spin. To solve the master equation (5), we use quantum trajectory techniques.[15, 16]

The presence of noise (as described above) can cause a pure state to evolve into a mixture. This is measured by the entropy of the state, $S = -\mathrm{tr}\rho \log_2 \rho$. For a 3-bit system, the maximum entropy is 3.

In many situations, however, the environment is not completely outside the experimenter's control. E.g., by placing detectors around a system, some information about the environment state may be obtained. It seems natural to ask the question *How much can be learned about the system state by making measurements on the environment.* It turns out that this question leads to a definition of quantum chaos.

To simplify the discussion, we assume that the part of the environment noise that is accessible to measurement is modeled by a stochastic perturbation. Suppose for instance that, in addition to the intrinsic phase noise, the system is perturbed by being acted on by an unknown force with a known distribution. Averaging over all possible perturbations also causes the entropy of the state to increase. However, one can reduce *this* part of the entropy growth by obtaining information about the perturbation, such as the actual value of the force to some precision. By having more information about the perturbation, the uncertainty in the state (and hence its entropy) decreases. To reduce the entropy growth by an average amount ΔS requires information $I \geq \Delta S$ about the perturbation. In particular, we want to know the minimum amount of information $I_{\min}$ needed to produce a given entropy reduction ΔS. A system is *hypersensitive to perturbation* if the information $I_{\min}$ needed to lower the system entropy increase by an average amount ΔS is very large compared to ΔS. Precise definitions of the quantities $I_{\min}$ and ΔS are given in Ref. 2. For a general introduction, see Ref. 3.

We will now show that hypersensitivity to perturbation can be detected in our NMR realization of the 3-qubit quantum baker's map. We define perturbed versions of the regular and chaotic maps, $T_{\mathcal{R}}^{+} = e^{i\pi\hat{Z}_{C_1}/2}T_{\mathcal{R}}$ and $T^{+} = e^{i\pi\hat{Z}_{C_1}/2}T$. Applying randomly at each step either the perturbed or the unperturbed map leads, after n steps, to 2^n possible different *perturbation histories*.

The proposed experiment is to apply all $N = 8 = 2^3$ different 3-step perturbation histories, each time starting from the initial state $|\psi_0\rangle = |0_y\rangle \otimes |0_y\rangle \otimes |0_y\rangle$ where $|0_y\rangle = (|0\rangle + i|1\rangle)/\sqrt{2}$, to obtain the list of final density operators $\mathcal{L} = \{\rho_1, \ldots, \rho_N\}$ by quantum tomography,[17, 12] and finally to analyze the distribution of the N density operators in density operator space.

NUMERICAL RESULTS AND CONCLUSIONS

We assume that, in a random trial, all $N = 8$ perturbation histories would occur with the same probability $q_1 = \cdots = q_N = 1/N$. We can find the entropy $\overline{S}_{\max} = -\mathrm{tr}\rho\log_2\rho$ of the *average* density operator $\rho = \sum q_j\rho_j$. This entropy grows quickly with the number of iterations n. We have simulated this experiment by numerically solving the master equation (5) for the 8 perturbation histories. For the relaxation times, we used roughly twice the current experimental values for trichloroethylene (see Figure). The simulation gave $\overline{S}_{\max} = 2.99$ bits in the chaotic case and $\overline{S}_{\max} = 2.84$ bits in the regular case after three steps ($n = 3$).

The values for the entropy increase alone do not reveal much about the distribution of the density operators in the ensemble. In particular, they do not reveal whether the ensemble is *orthogonal* (in which case the entropy increase corresponds to purely classical information) or *nonorthogonal* (corresponding to quantum information). Obtaining information about which perturbation history has been realized can reduce the entropy from $\overline{S}_{\max}$ to a lower value $\overline{S}$; analyzing the dependence of $\Delta\overline{S} = \overline{S}_{\max} - \overline{S}$ on the information needed gives a measure of how nonorthogonal the ensemble is.[2]

We could obtain the total possible information by determining exactly which perturbation history occurred. This corresponds to $\log_2 N = 3$ bits of acquired information, and would reduce us from considering the average density operator ρ to considering only a single final density operator ρ_j. However, we could also obtain *partial* information about the perturbation history by partitioning the N final density operators into $R < N$ groups, and determining only which group the operator was in. Since we are actually interested in the minimum information $I_{\min}$ needed to produce a given entropy reduction, we would like to choose groupings which maximize the entropy reduction.

More precisely, consider a partitioning of the list $\mathcal{L}$ into R groups, labeled by $r = 1, \ldots, R$. We denote by N_r the number of density operators in the rth group, and by $\rho_1^r, \ldots, \rho_{N_r}^r$ the N_r density operators in the rth group. A history lies in the rth group with probability $p_r = N_r/N$. The knowledge that the system state is in group r is described by the density operator $\rho^r = N_r^{-1}\sum\rho_i^r$. Then the entropy conditional on being in group r is $S_r = -\mathrm{tr}\rho^r\log_2\rho^r$. This makes the average conditional entropy

$$\overline{S} = \sum_r p_r S_r = \overline{S}_{\max} - \Delta\overline{S} \leq \overline{S}_{\max} , \tag{7}$$

and the average information

$$I = -\sum_r p_r \log_2 p_r \geq \Delta\overline{S} \ . \tag{8}$$

The information $I_{\min}$ needed about the perturbation to reduce the system entropy by an amount ΔS is now defined as the minimum of I over all groupings for which $\Delta\overline{S} \geq \Delta S$, i.e., all groupings for which the system entropy is reduced by at least ΔS.

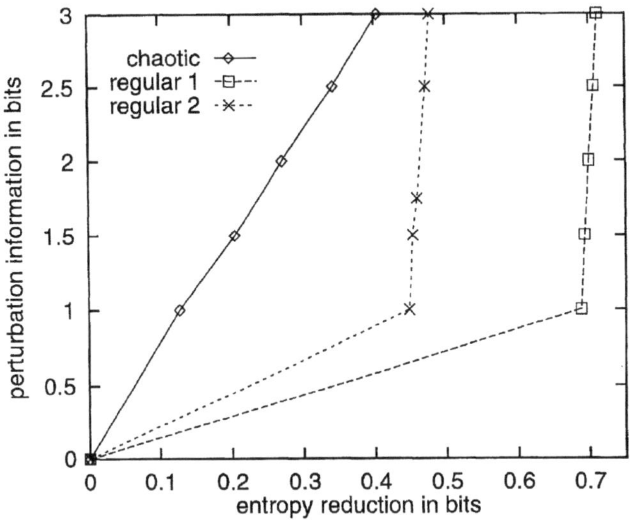

Figure 1. Information needed about the perturbation versus entropy reduction for the baker's map T and the regular map $T_{\mathcal{R}}$. The decoherence times are $1/\Gamma_{C_1} = 1\text{s}$, $1/\Gamma_H = 10\text{s}$ (4s for the curve labeled "regular 2"), and $1/\Gamma_{C_2} = 1\text{s}$.

We obtain an approximation to the optimal grouping numerically; for the details, see Ref. 6. The Figure shows $I_{\min}$ versus ΔS, both for the quantum baker's map and the regular map. The relaxation times we have used for the chaotic map are longer than have been achieved in current experiments by roughly a factor of two. It is our hope that these levels may be achievable in future experiments. We have also plotted a second curve, labeled "regular 2," with a faster relaxation time to illustrate that the form of the curve is unchanged. The slope of the chaotic curve is roughly equal to 8, i.e., about $8n$ bits of information about the perturbation are needed to reduce the system entropy increase by n bits. In the regular case, 1 bit of information about the perturbation is sufficient to reduce the system entropy increase by almost 0.7 bits (0.45 bits in the regular 2 case). The criterion of hypersensitivity to perturbation thus differentiates well between chaotic and regular behavior. Furthermore, the slope of 8 in the chaotic case happens to be equal to the dimension of Hilbert space, $D = 8$. A slope close to D is characteristic for a random distribution of pure states in Hilbert space, and has been conjectured to hold for chaotic quantum systems.[18] The steep slope indicates that the ensemble is highly nonorthogonal.

We have shown that hypersensitivity to perturbation can be detected in the original version[4, 5] of the 3-qubit quantum baker's map in the presence of the noise levels within a factor of 2 of current experiments for trichloroethylene. Using realistic estimates for the experimental parameters, our simulations show that this criterion differentiates very well between chaotic and regular behavior.

The authors would like to acknowledge discussions with C. M. Caves, I. Chuang, J. A. Jones, E. Knill, R. Laflamme, M. Mosca, M. Nielsen, and W. Zurek. Funding for RS came partially from the UK's EPSRC. TAB was funded in part by NSF Grant No. PHY94-07194.

REFERENCES

1. C. M. Caves, in *Physical Origins of Time Asymmetry*, edited by J. J. Halliwell, J. Pérez-Mercader, and W. H. Zurek (Cambridge University Press, Cambridge, England, 1993), p. 47.
2. R. Schack and C. M. Caves, Phys. Rev. E **53**, 3387 (1996); Phys. Rev. E **53**, 3257 (1996).
3. C. M. Caves and R. Schack, Complexity **3**, No. 1, 46 (1997).
4. R. Schack, Phys. Rev. A **57**, 1634 (1998).
5. N. L. Balazs and A. Voros, Ann. Phys. **190**, 1 (1989); M. Saraceno, Ann. Phys. **199**, 37 (1990).
6. T. A. Brun and R. Schack, submitted to Phys. Rev. A, eprint quant-ph/9807050.
7. N. A. Gershenfeld and I. L. Chuang, Science **275**, 350 (1997).
8. D. Cory, A. Fahmy, and T. Havel, Proc. Nat. Acad. Sci. USA **94**, 1634 (1997).
9. W. H. Zurek and J. P. Paz, Phys. Rev. Lett. **72**, 2508 (1994).
10. R. Alicki and M. Fannes, Lett. Math. Phys. **32**, 75 (1994).
11. G. Lindblad, Commun. Math. Phys. **48**, 119 (1976).
12. R. Laflamme, E. Knill, W.H. Zurek, P. Catasti, S.V.S. Mariappan, eprint quant-ph/9709025.
13. E. Knill, private communication.
14. M. Goldman, *Quantum Description of High-Resolution NMR in Liquids* (Clarendon Press, Oxford, 1988).
15. H. J. Carmichael, *An Open Systems Approach to Quantum Optics* (Springer, Berlin, 1993).
16. R. Schack and T. A. Brun, Comp. Phys. Comm. **102**, 210 (1997).
17. M. G. Raymer, M. Beck, and D. F. McAlister, Phys. Rev. Lett. **72**, 1137 (1994).
18. R. Schack and C. M. Caves, in *Quantum Communication, Computing, and Measurement*, eds. O. Hirota, A. S. Holevo, and C. M. Caves, (Plenum Press, New York, 1997), p. 317.

QUANTUM NETWORKS FOR DISTRIBUTED QUANTUM INFORMATION PROCESSING

H. J. Kimble

Norman Bridge Laboratory of Physics 12-33
California Institute of Technology
Pasadena, CA 91125

Quantum networks for the processing and distribution of quantum information could consist of quantum "nodes" for the manipulation and storage of quantum information, with these nodes linked by quantum "channels" for the transmission of quantum states. The leading proposal for the realization of such networks utilizes atoms within optical cavities as the "nodes." Conversion of internal atomic state information into optical photons that propagate from one cavity to another forms the basis for the "channels." Unfortunately, even in terms of laboratory "proof-of-principle" demonstrations, technical capabilities are far short of those required to implement such quantum networks. However, along the road to these ends, there lies a rich set of scientific problems that are the subject of this paper, including nonlinear optics with single atoms and photons, trapping of individual atoms within high finesse optical cavities, and teleportation of quantum states.

INTRODUCTION

Although formal discussions of quantum computation and communication involve abstract unitary transformations in Hilbert space, implementations of quantum logic require attention to the dynamical processes of particular physical systems. In the Quantum Optics Group at Caltech, we are attempting to lay the foundations for quantum information science by way of advances on several fronts in optical physics.

An essential ingredient in this endeavor is the ability to bring two components of a complex quantum system together in a controlled fashion. Qualitatively, assume that the off-diagonal components of the interaction Hamiltonian are characterized by $\langle H_{int} \rangle \sim \hbar\chi$, where χ is the rate of coherent, reversible evolution.[1] If the interaction proceeds for time T (which must be under our external control), then a necessary requirement for conditional quantum dynamics whereby one component has an appreciable impact on another is for the quantity $\theta = \chi T \sim 1$. Certainly, in any real world situation, there will be irreversible interactions with the external environment, which lead to dissipative decay at rate Γ for any individual constituent (e.g., decay of a quantum spin). In addition to the requirement $\theta \sim 1$, we must also then have $\chi \gg \Gamma$.

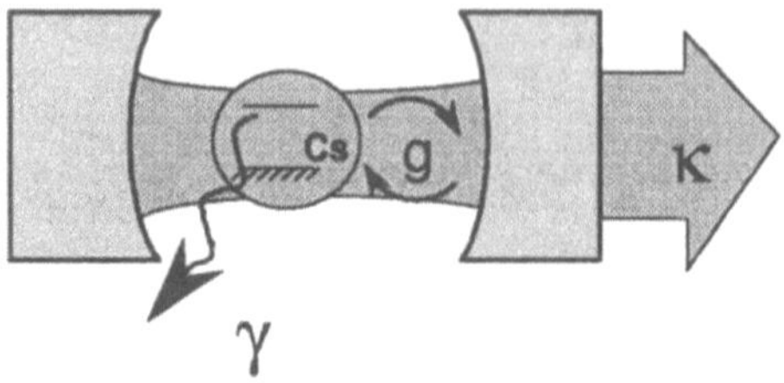

Figure 1. Illustration of a single Cesium atom interacting with the quantized field of an optical cavity. g gives the rate for reversible coherent interaction of atom and cavity field, while (γ, κ) denote irreversible decay channels.

The quest is thus to isolate and manipulate the dynamical processes of individual quantum systems, with interactions studied quantum by quantum. In optical physics, one example of such research is cavity quantum electrodynamics with single atoms and photons,[2] while in condensed matter physics, a notable example is Coulomb blockade with discrete electron energies.[3] Note that the emphasis is not simply to observe event by event as in traditional scattering experiments, but rather to *control* quantum evolution of a single system in *real time.*

Within this context, the program of research that is being pursued in the Quantum Optics Group at Caltech is a broad investigation of the physics of strong coupling in optical physics and its realization, principally within the area of cavity quantum electrodynamics (CQED).[4] As illustrated in Figure 1, if the coupling frequency of a single atom to the mode of an optical resonator is g (i.e., $2g$ is the one-photon Rabi frequency), then our experiments explore strong coupling with $g \gg (\gamma, \kappa)$, where γ is the atomic decay rate to modes other than the cavity mode and κ is the decay rate of the cavity mode itself. Relative to the preceding discussion, the rates χ and Γ are now replaced by $\chi \to g$ and $\Gamma \to (\gamma, \kappa)$. In this circumstance, the number of photons required to saturate an intracavity atom is $n_0 \sim \frac{\gamma^2}{g^2} \ll 1$ and the number of atoms required to have an appreciable effect on the intracavity field is $N_0 \sim \frac{\kappa\gamma}{g^2} \ll 1$.

Given this brief overview of the context and objectives of our research program, in the following sections I present thumbnail sketches of several specific projects.

A QUANTUM PHASE GATE (QPG) AND NONLINEAR SPECTROSCOPY IN CAVITY QED

Several years ago, we carried out the first demonstration of conditional dynamics at the single photon suitable for the implementation of quantum logic.[5] Our measurements utilized the circular dichroism of an atom strongly coupled to the field of a high finesse optical cavity to rotate the polarization state of a linearly polarized "probe" beam (i.e., a one-atom wave plate). Because the rotation angle of the probe beam could be controlled by the intensity of a circularly polarized "pump" beam for intracavity fields with average photon number much less than one, our observations demonstrated conditional dynamics between pump and probe fields at the level of single quanta. The "truth table" for our quantum-phase gate was experimentally verified, with the measurements indicating that the transformation affected by the atom-cavity system is "nontrivial" in that it could serve as a universal element for quantum computation. Here the quantum carriers of information (the "qubits") are fields which propagate in two frequency offset (and hence functionally distinct) channels, with the internal state in each case specified by the circular polarization states $\sigma_\pm$.

Beyond the context of quantum logic and computation, the large values recorded for the dispersive nonlinear interaction between intracavity fields represented a unique achievement within the field of nonlinear optics (with $n_0 \approx 0.02$ photons). Following this theme, we pursued a more general program of investigation of the nonlinear "spectroscopy" of the atom-cavity system in a regime of strong coupling,[6] with excitation spectra obtained by pump-probe techniques. Here, pump and probe fields of independent amplitudes and frequencies were generated by acousto-optic and electro-optic (traveling-wave) modulators. The basic idea was to excite the system to some initial state in the Jaynes-Cummings ladder of states with the pump field (for example, to the state at $+g_0$ for $n = 1$) and then to probe for excitation to higher lying states (for example, from $n = 1$ and $+g_0$ to the state $n = 2$ and $-\sqrt{2}g_0$, resulting in a resonance at $-(\sqrt{2}+1)g_0$ relative to the common frequency $\omega_A = \omega_C$). Pump-probe spectra were recorded as the average number of intracavity atoms varied from $\langle N \rangle = 4.2$ down to $\langle N \rangle = 0.8$ atoms with nonlinear effects observed for as few as 0.1 intracavity pump photons. A deviation from semiclassical predictions was recorded for $\langle N \rangle \lesssim 1$ atom, while the same semiclassical analysis gave reasonable agreement with the data for $\langle N \rangle \gtrsim 2$ atoms. Although not a full dynamical theory, the semiclassical simulation procedure accounted for fluctuations in atom-field coupling which had important effects on both the linear and nonlinear spectra. Indeed, these fluctuations have prevented us from identifying unambiguous quantum aspects of the measured spectra[7] and largely motivated our efforts to work with a system of cold atoms.

COLD ATOMS AND CAVITY QED

To move beyond initial proof-of-principle demonstrations toward more powerful capabilities for quantum measurement and information processing, we have abandoned atomic beams altogether, and have invested a considerable effort toward the marriage of cavity QED with laser cooled and trapped atoms. A significant milestone in this quest to attain strong coupling for a single, stationary atom in an optical cavity has been the real-time detection of single atoms transiting through a high-finesse optical cavity.[8-10] For these experiments, Cesium atoms are dropped from a magneto-optical trap located a few millimeters above a Fabry-Perot cavity. By recording the reduction of the cavity transmission as an atom enters the cavity mode, we can monitor the "trajectory" of an individual atom as it transits through the cavity. Indeed, a rather complex technical infrastructure for cavity locking and frequency stabilization has enabled us to record the full complex susceptibility for single atoms coupled to the cavity field.[10] These measurements are already remarkably close to the standard quantum limit for sensing the atom's position (within a factor of roughly 3).

Given the ability to monitor the motion of a single atom in a cavity in real time with quantum-limited sensitivity, we are attempting to exploit this capability to implement a "quantum servo." The basic plan that we are pursuing involves a magneto-optical trap (MOT) situated a few millimeters above a high finesse optical cavity. When the MOT is switched off, the atoms fall between the cavity mirrors with some small fraction of the atoms actually transiting through the cavity. We have overcome the first hurdle on the path to trapping a single atom with our demonstration of the ability to monitor in real time the transit of individual atoms through the cavity.[8-10] The next step is to switch the intracavity field to trap the atom. Here we make use of the fact that the lower peak of the vacuum-Rabi splitting corresponds to an attractive pseudo-potential, with the average "well depth" increasing with increasing probability for occupation of the lower dressed state, at least in the limit of intracavity fields with photon number $n <$

1. Thus the strategy is first to monitor for the presence of an atom with an intracavity field of photon number $n_1 \ll 1$ which has a small effect on the atomic center-of-mass motion. Having detected an atom, we then switch an external drive tuned to the lower peak of the vacuum-Rabi spectrum to produce an intracavity photon number $n_2 \lesssim 1$ and a corresponding population in the lower dressed-state manifold, thereby creating a confining potential sufficient to trap the atom.

Over the past year, we have had some success in implementing this strategy. Here, by way of this rudimentary "quantum" servo, we have been able to localize a single atom within the cavity for a few hundred microseconds. Note that the atom would have otherwise fallen through the cavity in only about 80 microseconds. Although we are striving to trap the atom "permanently," these measurements already indicate a separation of time scales between the internal (atomic dipole coupled to cavity field) and external (center of mass motion) degrees of freedom of roughly 10^4.

As for longer trapping times, theoretical work in collaboration with the group of Professor D. F. Walls suggests that the atom is being heated by the trapping field and that we need to add subsequent steps to our protocol. Namely, having captured the falling atom, we should then detune the frequency of the trapping laser to avoid (near-resonance) heating. We are also working to develop an active cooling strategy, either by use of quantum feedback to damp the atomic motion or via a Sisyphus-type mechanism within the setting of cavity QED.[11]

A second avenue for cooling and trapping within the cavity mode involves the use of a far-off-resonance trap (FORT).[12, 13] As in the preceding discussion, the presence of an atom in the cavity will first be sensed, and then employed to switch on a second field, which in this case would be a red-detuned diode laser which excites a second longitudinal mode of the cavity. For this strategy the functions of cooling and trapping would be essentially classical in character (and would serve as the "actuator" for control of atomic motion) and would be separate from the quantized atom-field interaction for the resonant mode (which would provide the "sensor").

More generally, some sense of the advance that this work represents relative to other experiments in the area of cavity QED is obtained by noting that the product of coherent coupling constant g_0 with the transit time T_0 is $g_0 T_0 \sim \pi \cdot 10^4$, whereas for all other experiments (which employ atomic beams), $g_0 T_0 \sim \pi$.[2]

CAVITY QED FOR WHISPERING GALLERY MODES OF QUARTZ MICROSPHERES

For a number of years, we have pursued an effort to achieve high quality factors Q for the whispering gallery modes (WGMs) of quartz microspheres. Recently, this program came to fruition with the demonstration of record Q values for three different wavelengths in the visible and near infra-red (NIR), including the highest finesse yet obtained for an optical cavity.[14] Perhaps more importantly, we have begun to understand the basic mechanisms that limit the Q. More explicitly, from the wavelength dependence of the Q, it appears as though some mechanism other than absorption in the bulk silica does not allow access to the theoretical absorption limit in the NIR. Our measurements with atomic force microscopy (AFM) together with a simple model for surface scattering that we have developed suggest that Q is limited by residual surface inhomogeneities. As well, adsorption of water on the surface of the sphere degrades Q over times greater than about one minute. We are now working to overcome these loss mechanisms and to extend our measurements further into the infrared, with a goal of

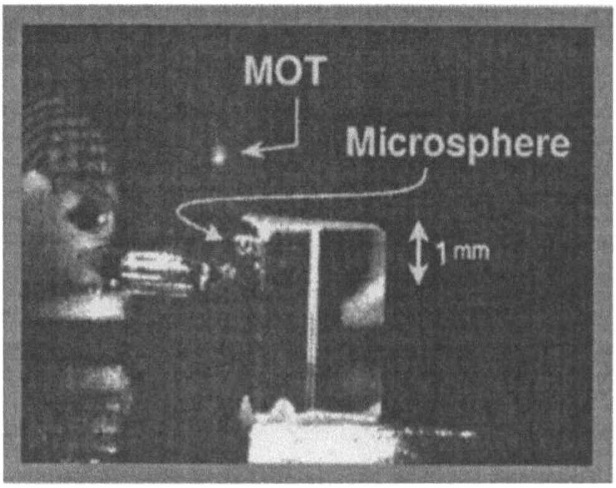

Figure 2. Photograph showing a quartz microsphere situated near a magneto-optic trap (MOT). The goal is to transfer single atoms from the MOT to near the surface of the microsphere for interactions with the whispering gallery modes (WGMs).

reaching ultimate $Q \approx 10^{11}$ values around 1.55μm.

In an initial attempt to exploit these high Q WGMs for cavity QED, we have made the first measurements of cavity-QED effects for the radiative coupling of atoms to the external evanescent field of a whispering-gallery mode (WGM).[15] More specifically, we have recorded modifications of cavity transmission due to the coupling of $\bar{N}_T \sim 1$ Cesium atoms in a thermal gas with a single resonant WGM at the level of a few photons in the mode. That we could achieve sensitivity to $\bar{N}_T \sim 1$ atoms in the microsphere's evanescent field in the face of Doppler broadening of roughly 100 times the natural linewidth is a consequence of the novel features of the microspheres, namely small size (with radius $a < 60$ μm and mode volume $V_m \sim 10^{-8}$ cm^3) and narrow linewidth (with $Q_1 = 5 \times 10^5 \lesssim Q \lesssim Q_2 = 5 \times 10^7$).

Beyond this initial step towards realizing the potential of WGMs for long-lived coherent dynamics in cavity QED, we are working to reduce the Doppler broadening by coupling to cold atoms, as in our work with Fabry-Perot resonators discussed above.[8-10] Figure 2 illustrates the apparatus in which our preliminary (unsuccessful) attempts have been made to couple cold atoms with the WGMs of a quartz microsphere.

QUANTUM TELEPORTATION

Beyond quantum information processing with internal atomic states and photons serving as *qubits*, we are also investigating algorithms for continuous quantum variables. Following the work of Professor L. Vaidman,[16] we have developed a theory for quantum teleportation of continuous quantum variables in an infinite dimensional Hilbert space,[17] including a theory for broad bandwidth teleportation.[18] The same formalism can be applied to achieve super-dense quantum coding.

Motivated by these and other prospects for quantum information processing with continuous variables, we have undertaken an experimental program involving entangled EPR beams generated by parametric down conversion. Quite recently, these experiments have led to the first *bona fide* demonstration of quantum teleportation.[19] As illustrated in schematic form in Fig. 3, an unknown quantum state input to "Alice's sending station" is destroyed, and recreated at "Bob's receiving terminal." The quantum nature of the protocol is confirmed by measurements to determine the fidelity F, which quantifies the match between input and output fields. In our experiment, a fi-

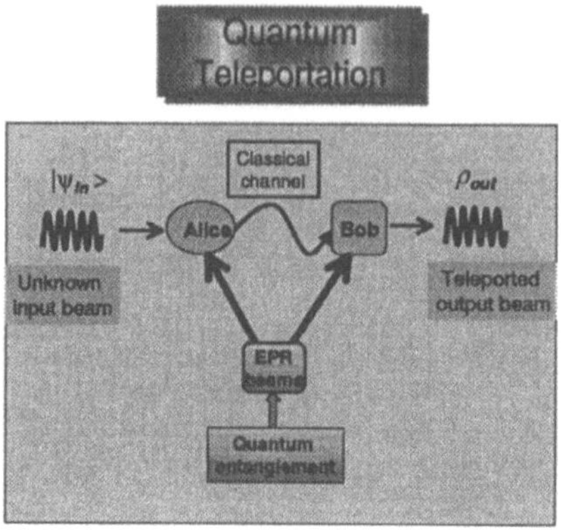

Figure 3. Illustration of quantum teleportation as achieved in Ref. 19.

delity $F_{\text{exp}} = 0.58 \pm 0.02$ is achieved for the teleportation of coherent states. Note that a fidelity greater than 0.5 cannot be achieved for the teleportation of coherent states in the absence of shared entanglement.

Full teleportation is thus realized by our experiment, where "full" means that the following three criteria have been met.

1. An unknown quantum state enters Alice's sending station.
2. A "recreation" of this quantum state emerges from Bob's receiving terminal.
3. The fidelity of input and output quantum states is higher than that which could have been achieved if Alice and Bob shared only a classical communication channel.

Note that prior experiments did not meet criterion (1),[20] or (2),[21] and indeed, because of low efficiencies, could not achieve a nonclassical fidelity for the fields involved.

QUANTIZED CENTER-OF-MASS MOTION

With few exceptions in cavity QED, the atomic center of mass motion has been treated classically, with the spatial variation in coupling coefficient $g(\boldsymbol{r})$ occurring on scales much larger than the size of the atomic wave packet. As the atoms in our experiments become yet even colder and better localized within the cavity mode, it becomes necessary to consider the full, non-perturbative wave-packet dynamics including bound states for the system. We have thus undertaken an investigation of structure and dynamics for an atom strongly coupled to a cavity mode in the domain for which the atomic kinetic energy $E_k < \hbar g$.[22] Beginning with the spectrum of eigenvalues, we have extended the familiar dressed states for the Jaynes-Cummings Hamiltonian to include bound CM states that arise either because of the intrinsic spatial variation of $g(\boldsymbol{r})$ or because of an externally applied atomic potential $V_{ext}(\boldsymbol{r})$, as for example in an RF Paul trap. Spatially localized eigenstates for both the external motion in a potential well and for the internal atom-field interaction are termed "well-dressed" states with the spectrum of eigenvalues illustrated in Ref. 22.

Our analysis explores the interplay of the finite spatial extent of a CM wave packet $\psi(\mathbf{r})$ with the quantum field mode structure $g(\mathbf{r})$. To illustrate the type of dynamical phenomena that can arise, we have considered an atom initially localized in the mode

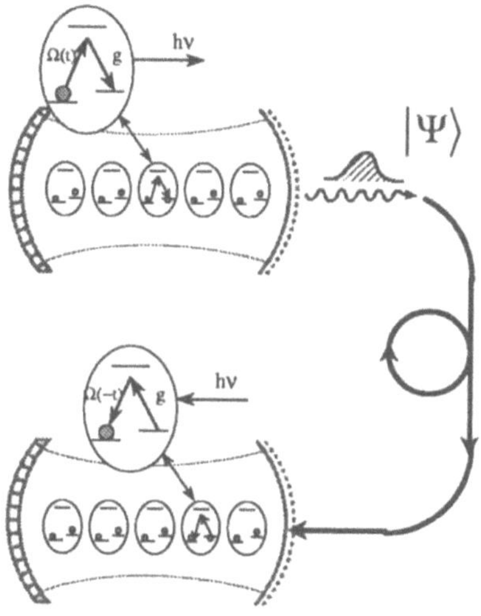

Figure 4. Illustration of the protocol of Refs. 24-27 whereby a component of an entangled state for a set of atoms at one site can be transferred to an atom at another site at a remote location. By extension, more complex quantum networks can be realized.

$g(\mathbf{r})$ with one photon and find substantial modifications to the familiar Rabi nutation rate $2g(\mathbf{r})$. More generally, already implicit in the eigenvalues of the well-dressed states are new CM-dependent spatial and temporal scales that can lead to novel "collapses" and "revivals" of internal atom-field coherence, as well as to modifications of spontaneous emission.

In another calculation, we have investigated the bound state structure and dynamics for an atom trap formed from the whispering gallery modes (WGMs) of a dielectric microsphere in an extensive analytical and numerical treatment.[23] The coupling of the quantized internal and external atomic degrees of freedom is found to play an essential role in the quantum dynamics of the resulting *atom gallery.* In particular, the radiative processes for a cold atom near a microsphere are modified due to the special symmetry of the atom gallery, the WGM mode structure, and the finite extent of the center-of-mass (CM) wave packet. While it is well-known that radiative processes are fundamentally modified for an atom outside of a dielectric sphere, previous calculations have not included the quantum mechanical nature of the CM state, whereas our work is a fully quantum calculation for a realistic *3D* configuration.

QUANTUM MEASUREMENT AND INFORMATION

There are diverse applications within the realms of quantum measurement and information processing for single atoms strongly coupled to the fields of high finesse optical cavities. An important example to which we have been devoting considerable attention is that of quantum networks. Here, as illustrated in Fig. 4, multiple atom-cavity systems, located at spatially separated "nodes" could be interconnected via optical fibers to create quantum networks (QNs) whose unique and powerful properties have been anticipated by recent advances in quantum information theory. Together with Professors I. Cirac and P. Zoller, we have put forth the first realistic scheme for the realization of such a quantum network,[24] which would be capable of a complete set of elementary network operations, including local processing of quantum information, transmission of quantum states from one node to another, and the distribution of quantum entanglements.[24-27] The technical advances described in the preceding sections are important steps toward enabling such quantum networks.

ACKNOWLEDGMENTS

The experiments described herein have been carried out in the Quantum Optics Laboratory at Caltech, with the personnel responsible for the research including graduate students J. Buck, N. Georgiades, C. Hood, H. Mabuchi (now an assistant professor of physics at Caltech), T. Lynn, J. Sorensen (visitor from Aahrus University), Q. Turchette (now at NIST, Boulder), and D. Vernooy, and undergraduate E. Streed. Senior members of the group include Drs. M. Chapman (now an assistant professor of physics at Georgia Tech), C. Fuchs, A. Furusawa (Nikon Advanced Research Labs), S. van Enk, and J. Ye. We have benefited greatly from ongoing collaborations with the groups of Professors S. L. Braunstein, E. S. Polzik, D. F. Walls, and P. Zoller. This work is supported by DARPA via the QUIC Institute which is administered by ARO, by the National Science Foundation, and by the Office of Naval Research. For more information, please visit our web site at *http://www.cco.caltech.edu/~qoptics.*

REFERENCES

1. H. J. Kimble, Physica Scripta, (1998).
2. *Cavity Quantum Electrodynamics,* ed. P. Berman (Academic Press, San Diego, 1994).
3. See, for example, D. C. Ralph, C. T. Black, and M. Tinkham, Phys. Rev. Lett. **78**, 4087 (1997).
4. H. J. Kimble, in *Cavity Quantum Electrodynamics,* ed. P. Berman (Academic Press, San Diego, 1994).
5. Q. A. Turchette, C. J. Hood, W. Lange, H. Mabuchi, and H. J. Kimble, Phys. Rev. Lett. **75**, 4710 (1995).
6. R. J. Thompson, Q. A. Turchette, O. Carnal, and H. J. Kimble, Phys. Rev. A **57**, 3084 (1998).
7. B. C. Sanders, H. J. Carmichael, and B. F. Wielinga, Phys. Rev. A **55**, 1358 (1997).
8. H. Mabuchi, Q. A. Turchette, M. S. Chapman, and H. J. Kimble, Opt. Lett. **21**, 1393 (1996).
9. C. J. Hood, M. S. Chapman, T. W. Lynn, and H. J. Kimble, Phys. Rev. Lett. **80**, 4157 (1998).
10. H. Mabuchi, J. Ye, and H. J. Kimble, Appl. Phys. B (submitted, 1998).
11. P. Horak, G. Hechenblaikner, K. M. Gheri, H. Stecher, and H. Ritsch, Phys. Rev. Lett. **79**, 4974 (1997).
12. H. J. Lee, C. S. Adams, M. Kasevich, and S. Chu, Phys. Rev. Lett. **76**, 2658 (1996).
13. J. D. Miller, R. A. Cline, D. J. Heinzen, Phys. Rev. **A47**, R4567 (1993).
14. D. W. Vernooy, V. S. Ilchenko, H. Mabuchi, E. W. Streed, and H. J. Kimble, Opt. Lett. **23**, 247 (1998).
15. D. W. Vernooy, A. Furusawa, N. Ph. Georgiades, V. S. Ilchenko, and H. J. Kimble, Phys. Rev. A **57**, R2293 (1998).
16. L. Vaidman, Phys. Rev. **A49**, 1473 (1994).
17. S. L. Braunstein and H. J. Kimble, Phys. Rev. Lett. **80**, 869 (1998).
18. P. van Loock, S. L. Braunstein, and H. J. Kimble, in preparation.
19. A. Furusawa, J. Sorensen, S. L. Braunstein, C. Fuchs, H. J. Kimble, and E. S. Polzik, Science **282**, 706 (1998).
20. D. Boschi, S. Branca, F. Demartini, L. Hardy, S. Popescu, Phys. Rev. Lett. **80**, 1121 (1998).
21. D. Bouwmeester *et al.*, Nature **390**, 575 (1997).
22. D. W. Vernooy and H. J. Kimble, Phys. Rev. A **56**, 4287 (1997) and references therein.
23. D. W. Vernooy and H. J. Kimble, Phys. Rev. A **55**, 1239 (1997).
24. J. I. Cirac, P. Zoller, H. J. Kimble, and H. Mabuchi, Phys. Rev. Lett **78**, 3221 (1997).
25. S. J. Van Enk, J.-I. Cirac, and P. Zoller, Phys. Rev. Lett. **79**, 5178 (1997).
26. S. J. Van Enk, J.-I. Cirac and P. Zoller, Phys. Rev. Lett. **78**, 4293 (1997).
27. S. J. Van Enk, J.-I. Cirac, and P. Zoller, Science **279**, 5348 (1998).

QUANTUM PHENOMENA OF TRAPPED IONS

H. Walther

Sektion Physik der Universität München and
Max-Planck-Institut für Quantenoptik
85748 Garching, Germany

In this paper recent experiments with single trapped ions are reviewed. They deal with the spectrum of the resonance fluorescence of a single ion, the ion trap laser and with experiments towards a new frequency standard on the basis of a single In^+ ion. The experiments demonstrate essential ingredients necessary to realize quantum logic gates using trapped ions.

INTRODUCTION

In the followig recent single ion experiments performed at Garching will be discussed. We will start the discussion with the description of experiments on the resonance fluorescence of a single ion.

RESONANCE FLUORESCENCE

Resonance fluorescence of an atom is a basic process in radiation-atom interactions, and has therefore always generated considerable interest. The methods of experimental investigation have changed continuously due to the availability of new experimental tools. A considerable step forward occurred when tunable and narrow band dye laser radiation became available. These laser sources are sufficiently intense to easily saturate an atomic transition. In addition, the lasers provide highly monochromatic light with coherence times much longer than typical natural lifetimes of excited atomic states. Excitation spectra with laser light using well collimated atomic beam lead to a width being practically the natural width of the resonance transition, therefore it became possible to investigate the frequency spectrum of the fluorescence radiation with high resolution. However, the spectrograph used to analyze the reemitted radiation was a Fabry-Perot interferometer, the resolution of which did reach the natural width of the atoms, but was insufficient to reach the laser linewidth, see e.g. Hartig et al.[1] and Cresser et al.[2] A considerable progress in this direction was achieved by investigating the fluorescence spectrum of ultra-cold atoms in an optical lattice in a heterodyne experiment. [3] In these measurements a linewidth of 1 kHz was achieved, however, the quantum aspects of the resonance fluorescence such as antibunched photon statistics

Quantum Communication, Computing, and Measurement 2
Edited by Kumar et al., Kluwer Academic / Plenum Publishers, New York, 2000.

cannot be investigated under these conditions since they wash out when more than one atom is involved.

Thus the ideal experiment requires a single atom to be investigated. Since some time it is known that ion traps allow to study the fluorescence from a single laser cooled particle practically at rest, thus providing the ideal case for the spectroscopic investigation of the resonance fluorescence. The other essential ingredient for achievement of high resolution is the measurement of the frequency spectrum by heterodyning the scattered radiation with laser light as demonstrated with many cold atoms.[3] Such an optimal experiment with a single trapped Mg^+ ion is reviewed in this paper. The measurement of the spectrum of the fluorescent radiation at low excitation intensities is presented. Furthermore, the photon correlation of the fluorescent light has been investigated under practically identical excitation conditions. The comparison of the two results shows a very interesting aspect of complementarity since the heterodyne measurement corresponds to a "wave" detection of the radiation whereas the measurement of the photon correlation is a "particle" detection scheme. It will be shown that under the same excitation conditions the wave detection provides the properties of a classical atom, i.e. a driven oscillator, whereas the particle or photon detection displays the quantum properties of the atom. Whether the atom displays classical or quantum properties thus depends on the method of observation.

The spectrum of the fluorescence radiation is given by the Fourier transform of the first order correlation function of the field operators, whereas the photon statistics and photon correlation is obtained from the second order correlation function. The corresponding operators do not commute, thus the respective observations are complementary. Present theory on the spectra of fluorescent radiation following monochromatic laser excitation can be summarized as follows: fluorescence radiation obtained with low incident intensity is also monochromatic owing to energy conservation. In this case, elastic scattering dominates the spectrum and thus one should measure a monochromatic line at the same frequency as the driving laser field. The atom stays in the ground state most of the time and absorption and emission must be considered as one process with the atom in principle behaving as a classical oscillator. This case was treated on the basis of a quantized field many years ago e.g. by Heitler.[4] With increasing intensity upper and lower states become more strongly coupled leading to an inelastic component, which increases with the square of the intensity. At low intensities, the elastic part dominates since it depends linearly on the intensity. As the intensity of the exciting light increases, the atom spends more time in the upper state and the effect of the vacuum fluctuations comes into play through spontaneous emission. The inelastic component is added to the spectrum, and the elastic component goes through a maximum where the Rabi flopping frequency $\Omega = \Gamma/\sqrt{2}$ (Γ is the natural linewidth) and then disappears with growing Ω. The inelastic part of the spectrum gradually broadens as Ω increases and for $\Omega > \Gamma/2$ sidebands begin to appear.[2, 5]

The experimental study of the problem requires, as mentioned above, a Doppler-free observation. In order to measure the frequency distribution, the fluorescent light has to be investigated by means of a high resolution spectrometer. The first experiments of this type were performed by Schuda et al.[6] and later by Walther et al.,[7] Hartig et al.[1] and Ezekiel et al.[8] In all these experiments, the excitation was performed by single-mode dye laser radiation, with the scattered radiation from a well collimated atomic beam observed and analyzed by a Fabry-Perot interferometer. Experiments to investigate the elastic part of the resonance fluorescence giving a resolution better than the natural linewidth have been performed by Gibbs et al.[9] and Cresser et al.[2]

The first experiments which investigated antibunching in resonance fluorescence were also performed by means of laser-excited collimated atomic beams. The initial results obtained by Kimble, Dagenais, and Mandel [10] showed that the second-order correlation function $g^{(2)}(t)$ had a positive slope which is characteristic of photon antibunching. However, $g^{(2)}(0)$ was larger than $g^{(2)}(t)$ for $t \to \infty$ due to number fluctuations in the atomic beam and to the finite interaction time of the atoms.[11, 12] Further refinement of the analysis of the experiment was provided by Dagenais and Mandel.[12] Rateike et al.[13] used a longer interaction time for an experiment in which they measured the photon correlation at very low laser intensities (see Cresser et al.[2] for a review). Later, photon antibunching was measured using a single trapped ion in an experiment which avoids the disadvantages of atom number statistics and finite interaction time between atom and laser field.[14]

As pointed out in many papers photon antibunching is a purely quantum phenomenon (see e.g. Cresser et al.[2] and Walls.[15]) The fluorescence of a single ion displays the additional nonclassical property that the variance of the photon number is smaller than its mean value (i.e. it is sub-Poissonian).[14, 16]

The trap used for the present experiment was a modified Paul-trap, called an endcap-trap.[17] The trap consists of two solid copper-beryllium cylinders (diameter 0.5 mm) arranged co-linearly with a separation of 0.56 mm. These correspond to the cap electrodes of a traditional Paul trap, whereas the ring electrode is replaced by two hollow cylinders, one of which is concentric with each of the cylindrical endcaps. Their inner and outer diameters are 1 and 2 mm, respectively and they are electrically isolated from the cap electrodes. The fractional anharmonicity of this trap configuration, determined by the deviation of the real potential from the ideal quadrupole field is below 0.1 % (see Schrama et al.[17]). The trap is driven at a frequency of 24 MHz with typical secular frequencies in the xy-plane of approximately 4 MHz. This required a radio-frequency voltage with an amplitude on the order of 300 V to be applied between the cylinders and the endcaps.

The measurements were performed using the $3^2S_{1/2} - 3^2P_{3/2}$ transition of the 24 Mg^+-ion at a wavelength of 280 nm. The heterodyne measurement is performed as follows. The dye laser excites the trapped ion while the fluorescence is observed in a direction of about 54° to the exciting laser beam. However, both the observation direction and the laser beam are in a plane perpendicular to the symmetry axis of the trap. A fraction of the laser radiation is removed with a beamsplitter and then frequency shifted (by 137 MHz with an acousto-optic modulator (AOM)) to serve as the local oscillator. An example of a heterodyne signal is displayed in Fig. 1. The signal is the narrowest optical heterodyne spectrum of resonance fluorescence reported to date. Thus our experiment provides the most compelling confirmation of Weisskopf's prediction of a coherent component in resonance fluorescence. The linewidth observed implies that exciting laser and fluorescent light are coherent over a length of 400 000 km. Further details on the experiment are given by Höffges et al.[18],[19] Investigation of photon correlations employed the ordinary Hanbury-Brown and Twiss setup. The setup was essentially the same as described by Diedrich and Walther.[14] The results are shown and discussed in Ref. 18 also.

The presented experiment describes the first high-resolution heterodyne measurement of the elastic peak in resonance fluorescence of a single ion. At identical experimental parameters we have also measured antibunching in the photon correlation of the scattered field. Together, both measurements show that, in the limit of weak excitation, the fluorescence light differs from the excitation radiation in the second-order correlation but not in the first order correlation. However, the elastic component of resonance fluorescence combines an extremely narrow frequency spectrum with antibunched pho-

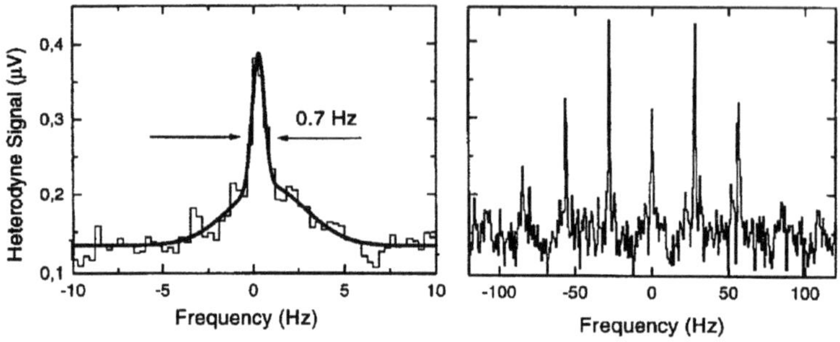

Figure 1. Heterodyne spectrum of a single trapped 24 Mg^+-ion. Left side: Resolution bandwidth 0.5 Hz. The solid line is a Lorentzian fit to the experimental data; the peak appears on top of a small pedestal being 4 Hz wide. The latter signal is due to random phase fluctuations in the spatially separated sections of the light paths of local oscillator and fluorescent light; they are generated by variable air currents in the laboratory. Right side: Heterodyne spectrum of the coherent peak with sidebands generated by mechanical vibrations of the mount holding the trap. For details see Ref. 18 and 19.

ton statistics, which means that the fluorescence radiation is not second-order coherent as expected from a classical point of view.[20] The heterodyne and the photon correlation measurement are complementary since they emphasize either the classical wave properties or the quantum properties of resonance fluorescence, respectively.

THE ION-TRAP LASER

There have been several theoretical papers on one-atom lasers in the past.[21, 22, 23, 24, 25] This system provides a testing ground for new theoretical concepts and results in the quantum theory of the laser. Examples are atomic coherence effects[26] and dynamic (i.e. self-generated) quantum-noise reduction.[27, 28, 24] All these aspects are a consequence of a pump process whose complex nature is not accounted for in the standard treatment of the laser. So far there is one experiment where laser action could be demonstrated with one atom at a time in the optical resonator.[29] A weak beam of excited atoms was used to pump this one-atom laser.

A formidable challenge for an experiment is to realize a laser with a trapped ion in the cavity. Mirrors with an ultrahigh finesse are required, and a strong atom-field coupling is needed. After the emission of a photon, the ion has to be pumped before the next stimulated emission can occur. Similar as in the resonance fluorescence experiments which show antibunching,[10, 14] there is a certain time gap during which the ion is unable to add another photon to the laser field. It has been shown[24] that this time gap plays a significant role in the production of a field with sub-Poissonian photon statistics.

We have investigated the theoretical basis for an experiment of this type. Our analysis takes into account details such as the multi-level structure, the coupling strengths and the parameters of the resonator. It has been a problem to find an ion with an appropriate level scheme. We could show that it is possible to produce a laser field with the parameters of a single Ca^+ ion. This one-atom laser displays several features, which are not found in conventional lasers: the development of two thresholds, sub-Poissonian statistics, lasing without inversion and self-quenching. The details of this

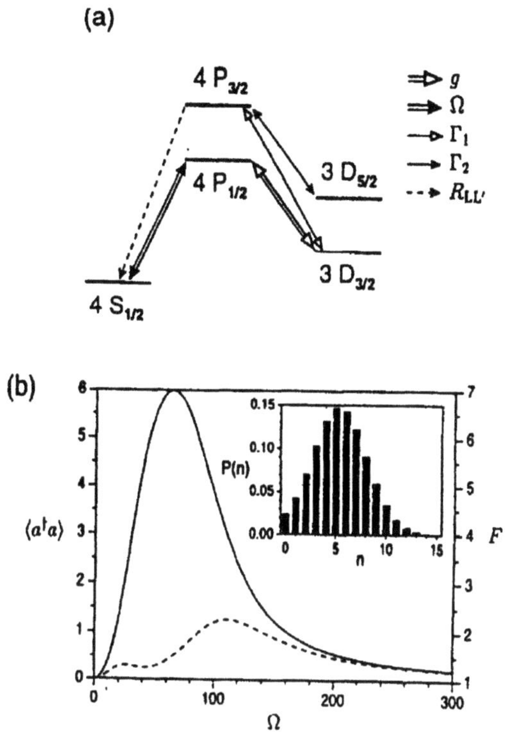

Figure 2. (a) Schematic representation of the Ca^+ scheme for the ion-trap laser. (b) Mean photon number $|a^+a\rangle$ (solid) and Fano factor F (dashed) versus the coherent pump strength Ω. The parameters are A = 1, g = 14.8, $\Gamma_1 = 40$, and $\Gamma_2 = 100$. The inset shows the photon distribution for $\Omega = 50$. All rates are in MHz.

work are reported by Meyer et al.[30, 31] In a subsequent paper[32] also the center-of-mass motion of the trapped ion was quantized. This leads to additional features of the ion trap laser, especially a multiple vacuum Rabi-splitting is observed.

The Ca^+ level scheme is sketched in Fig.2(a). It contains a Λ-type subsystem: the ion is pumped coherently from the ground state to the upper laser level $4P_{1/2}$, stimulated emission into the resonator mode takes place on the transition to $3D_{3/2}$ at a wavelength of 866 nm. Further pump fields are needed to close the pump cycle and to depopulate the metastable levels.

Although spontaneous relaxation from the upper laser level to the ground state takes place at a relatively large rate of 140 MHz and suppresses the atomic polarization on the laser transition, laser light is generated for realistic experimental parameters due to atomic coherence effects within the Λ subsystem. The occurrence of laser action is demonstrated in Fig. 2(b) for a resonator with a photon damping rate A = 1 MHz and a vacuum Rabi frequency g = 14.8 MHz on the laser transition. For the numerical calculation of the realistic scheme, the Zeeman substructure and the polarizations of the fields have to be taken into account. With increasing coherent pump Ω, the mean photon number inside the resonator first increases and then decreases. Both the increase and decease of the intensity are accompanied by maxima in the intensity fluctuations, which can be interpreted as thresholds. Laser action takes place in between these two thresholds. This is confirmed by the Poissonian-like photon distribution given in the

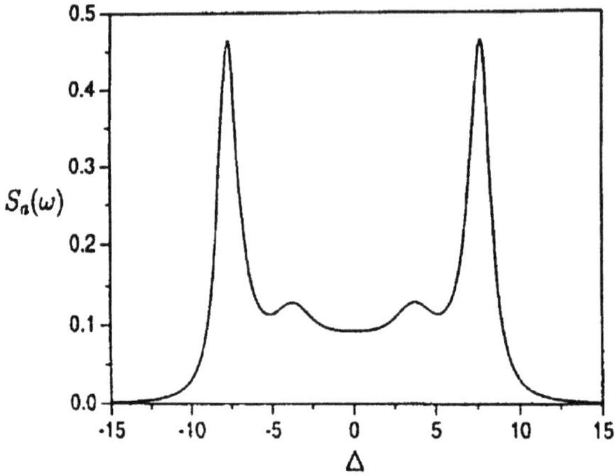

Figure 3. Multiple vacuum Rabi splitting in the output spectrum $S_a(\omega)$ for the two-level atom with quantized CM motion. The parameters are A = 0.1, B = 0.05, $\mu = 0.5$, $R_{AB} = 0.1$, $R_{BA} = 0.001$, and $\eta = 0.7$. All rates are in units of g_0.

inset of Fig. 2(b). In addition, the linewidth of the output spectrum is in the laser region up to ten times smaller than below the first and beyond the second threshold.[31] Note that for a thermal distribution the solid and dashed curves in Fig. 2(b) for the intensity and the intensity fluctuations would coincide.

For a nonvanishing Lamb-Dicke parameter η, higher vibrational states will be excited during the pump and relaxation processes; the amplitude of the atomic motion will increase. Therefore, the ion will in general not remain at an antinode of the resonator mode, and the strength of the atom-field coupling will decrease. However, the atom can be prevented from heating up by detuning a coherent pump field. The coupling strength is given by the product of a constant g_0 and a motion-dependent function[32] that is determined by an overlap integral involving the motional wave function of the atom and the mode function of the field.

In a simple two-level laser model with decay rate R_{AB} and pump rate R_{BA}, the cooling process may be incorporated by coupling the atomic motion to a thermal reservoir with cooling rate B and thermal vibron number μ. Already in such a simple model, the discrete nature of the quantized motion shows up below threshold in a multiple vacuum Rabi splitting of the output spectrum.[32] This is illustrated in Fig. 3. The pairs of peaks correspond to different vibrational states with different atom-field coupling.

The cooling mechanism is most transparent in the special case of resolved-sideband cooling. The coherent pump may be detuned to the first lower vibrational sideband so that with each excitation from $4S_{1/2}$ to $4P_{1/2}$ one vibron is annihilated and the CM motion is cooled. Eventually, all the population will collect in the motional ground state of the atomic ground state $4S_{1/2}$ and cannot participate in the lasing process. The coherent pump strength is now given by Ω_0 times a motion-dependent function. In order to maintain laser action in the presence of the cooling, an additional broadband pump field Γ may be applied to the cooling transition. Figure 4 indicates that a field with a mean photon number $\langle a^+a\rangle = 2.3$ is generated while the mean vibron number is restricted to a value of $\langle b^+b\rangle = 0.5$. If a larger mean vibron number is acceptable, the pump rate Γ can be increased and more population takes part in the laser action. This leads to considerably larger mean photon numbers. The calculation shows that it

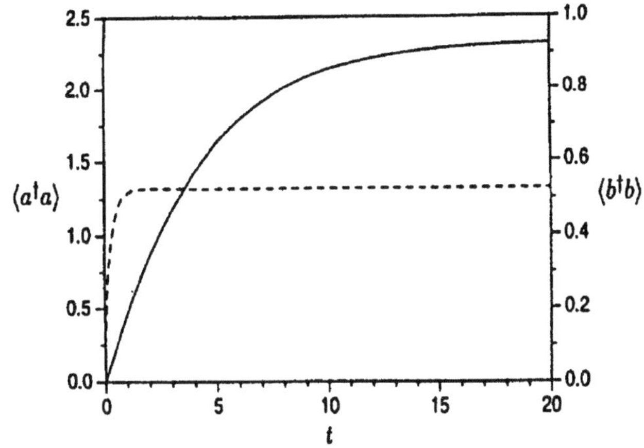

Figure 4. Time evolution of the mean photon number (solid) and the mean vibron number (dashed) in the Ca^+ ion-trap laser with sideband cooling. The parameters are A $= 0.5$, $g_0 = 14.8$, $\Omega_0 = 100$, $\Gamma = \Gamma_1 = 40$, $\Gamma_2 = 100$, and $\eta = 0.1$ on the laser transition. Initially, the atom is in the ground state and the vibronic distribution is thermal with $\langle b^+b\rangle = 0.1$. All rates are in MHz.

is possible to incorporate a cooling mechanism in a multilevel one-atom laser scheme and to obtain significant lasing also for nonperfect localization of the atom. Although it is difficult to reach the resolved-sideband limit in an experiment, cooling may still be achieved in the weak-binding regime by detuning a coherent pump field.

WORK TOWARDS A FREQUENCY STANDARD ON THE BASIS OF A SINGLE IN$^+$ ion

We report on progress towards an optical frequency standard in the spirit of the group-III monoion-oscillator, based on the $5s^2\,{}^1S_0 \to 5s5p\,{}^3P_0$ transition of the In^+ ion at a wavelength of 236.5 nm. This type of forbidden transition between two levels with vanishing electronic angular momentum is favorable for a frequency standard because line shifts due to external stray fields are small. For isotopes with nonzero nuclear spin the hyperfine interaction leads to small admixtures of levels with $J = 1$ to the ${}^3P_0$ state leading to a small electronic dipole decay rate of this level. For the case of In^+ the natural linewidth of the ${}^1S_0 \to {}^3P_0$ line becomes 1.1 Hz [33].

Besides the clock transition the ion also has to provide a sufficiently fast optical transition for laser cooling and fluorescence detection. In this respect indium is the most promising candidate among the group-III ions since for this ion the wavelength of the $5s^2\,{}^1S_0 \to 5s5p\,{}^3P_1$ transition at 230.6 nm can still be reached by frequency doubled light of a blue dye laser or by using the fourth harmonic of a high power near infrared diode laser. The use of a relatively narrow intercombination line limits in principle the initial cooling power and the single-ion fluorescence signal, but it also results in a very low temperature of the ion. The temperature achieved by sideband cooling is about 100 μK. Another advantage is that the In^+ ion can be brought rather easily in the strong binding regime, that means that the oscillation frequencies of the ion in the trap are larger than the linewidth of the cooling transition being 360 kHz.[33] The clock transition at 236.5 nm has the advantage that it coincides with the fourth harmonic of the 946 nm Nd:YAG laser line. So this intrinsically frequency stable solid-state laser can be used as the oscillator for the frequency standard.[34]

Among the foreseeable line shifts the most severe one is the Zeeman shift resulting from the small admixtures of J = 1 levels into $^3P_0$ which make the g-factor larger than the purely nuclear g-factor of the $^1S_0$ ground state. For the $m_F = 1/2 \rightarrow m_F = 1/2$ component of the clock transition this resutls in a shift of 240 Hz/G.[33]

Details of our experimental setup are given by Peik et al.[33, 34] and Hollemann et al.[35]. The ion is stored in a Paul-Straubel trap. [17] The trap consists of just a ring electrode to which the radiofrequency voltage is applied. This design can easily be miniaturized to submillimeter size and provides a large observation angle, because the endcaps are replaced by shield electrodes, which are located far away from the trap center. The ring used for our experiments had a hole diameter of 1 mm and was made of copper-beryllium. The applied AC voltage was 1000 V at 10 MHz and in addition a DC voltage of 30 V was used resulting in a secular frequency of 1 MHz.

The laser radiation at 230.6 nm for cooling of the ion is generated by frequency doubling blue light from a stilbene-3 ring dye laser with a BBO crystal in an external enhancement cavity. Typically 30 μW of UV light is focused to a spot size of about 50 μm at the position of the ion. The fluorescence signal from the ion is observed using a solar-blind photomultiplier.

The initial search for the forbidden clock transition required a laser source with high output power. Our clock laser system[35] therefore consists of two diode-pumped Nd:YAG lasers: a master laser at 946 nm contains the tuning elements and is frequency locked to a Zerodur cavity with a finesse of 70 000. It produces 30 mW of IR light with a linewidth in the range of 10 Hz. This light is injected into the second diode pumped Nd:YAG ring laser containing a $KNb0_3$ crystal for intracavity frequency doubling to 473 nm. This light is then doubled again in another external cavity, finally about 0.5 mW of frequency stable UV light are obtained.

To control and measure the absolute wavelengths of both laser systems we use reference lines of the $^{130}Te_2$ molecule in the blue spectral region around 461 nm and 473 nm. In a first series of experiments we measured the wavelength, hyperfine constants, isotope shift and lifetime of the $5s^2\,{}^1S_0 \rightarrow 5s5p\,{}^3P_1$ cooling transition.[33] Recent results on a double resonance experiment[36] on the cooling and clock transition are shown in Fig. 5[37]. Fig. 5(a) shows the cooling transition. For this scan the frequency doubled dye laser was set to the fourth motional sideband of the transition on the long wavelength side for cooling and a weak sideband of the laser was scanned across the central line. The line observed has a width being slightly larger than the expected value of 360 kHz resulting from a small power broadening. A careful analysis shows that secular sidebands are practically not present so that the ion is predominantly in its vibrational ground state. The measurement with the clock laser gives results of the type shown in Fig. 5(b). The number of induced quantum jumps are plotted versus the laser frequency. The plot shows the result of a run of about 3 min. The clock and cooling laser are chopped and excite the ion alternatively in order to avoid power broadening. The observed linewidth of the clock transition is in the moment still much broader than the natural linewidth being due to the linewidth of the clock laser. It is expected that this deficiency can be eliminated quite soon. Presently also measurements are under way to determine the absolute frequency of the clock transition with respect to the Cs-clock. The preliminary results presented here are very promising. There is the hope that the natural linewidth of the clock transition can be achieved quite soon.

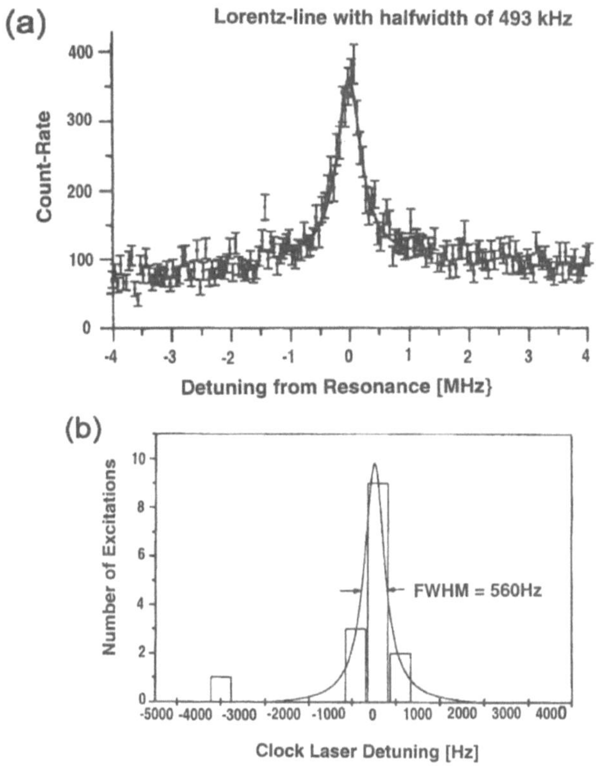

Figure 5. Excitation spectra of a single indium ion. Part (a) shows the result for the cooling transition $^1S_0 \rightarrow {}^3P_1$ and (b) the one for the clock transition $^1S_0 \rightarrow {}^3P_0$ measured in an optical-optical double resonance experiment.

CONCLUSION

In this paper recent experiments performed in Garching in order to study quantum phenomena of single trapped ions have been described. The experiments demonstrate that the available techniques provide the ingredients necessary to demonstrate quantum logic gates on the basis of trapped ions.

REFERENCES

1. W. Hartig, W. Rasmussen, R. Schieder, H. Walther, Study of the frequency distribution of the fluorescent light induced by monochromatic excitation, *Z. Physik* A278:205 (1976).
2. J. D. Cresser, J. Häger, G. Leuchs, F.M. Rateike, H. Walther, Resonance fluorescence of atoms in strong monochromatic laser fields, *Topics in Current Physics* 27:21 (1982).
3. P.S. Jessen, C. Gerz, P.D. Lett, W.D. Phillipps, S.L. Rolston, R.J.C. Spreuuw, and C.I. Westbrook, Observation of quantized motion of Rb atoms in an optical field, *Phys. Rev. Lett.* 69:49 (1992).
4. W. Heitler, *in:* "The Quantum Theory of Radiation," Oxford University Press (1954), Third Edition, pp. 196-204
5. B.R. Mollow, Power spectrum of light scattered by two-level systems, *Phys. Rev.* 188:1969 (1969).
6. F. Schuda, C. Stroud, Jr., M. Hercher, Observation of the resonant Stark effect at optical frequencies, *J. Phys.* B7:L198 (1974).

7. H. Walther, Atomic fluorescence induced by monochromatic excitation, *Lecture Notes in Physics* 43:358 (1975).
8. F. Y. Wu, R.E. Grove, S. Ezekiel, Investigation of the spectrum of resonance fluorescence induced by a monochromatic field, *Phys. Rev. Lett.* 35:1426 (1975); R.E. Grove, F.Y. Wu, S. Ezekiel, Measurement of the spectrum of resonance fluorescence from a two-level atom in an intense monochromatic field, *Phys. Rev.* A 15:227 (1977),
9. H.M. Gibbs and T.N.C. Venkatesan, Direct observation of fluorescence narrower than the natural linewidth, *Opt. Comm.* 17:87 (1976).
10. H.J. Kimble, M. Dagenais, and L. Mandel, Photon antibunching in resonance fluorescence, *Phys. Rev. Lett.* 39:691 (1977).
11. E. Jakeman, E.R. Pike, P.N. Pusey, and J.M. Vaugham, The effect of atomic number fluctuations on photon antibunching in resonance fluorescence, *J. Phys.* A 10:L257 (1977).
12. H.J. Kimble, M. Dagenais, and L. Mandel, Multiatom and transit-time effects in photon correlation measurements in resonance fluorescence, *Phys. Rev.* A 18:201 (1978); M. Dagenais, L. Mandel, Investigation of two-atom correlations in photon emissions from a single atom, *Phys. Rev.* A 18:2217 (1978).
13. F.M. Rateike, G. Leuchs, and H. Walther, results cited in Ref. 2.
14. F. Diedrich, H. Walther, Non-classical radiation of a single stored ion, *Phys. Rev. Lett.* 58:203 (1987).
15. D.F. Walls, Evidence for the quantum nature of light, *Nature* 280:451 (1979).
16. R. Short and L. Mandel, Observation of sub-Poissonian photon statistics, *Phys. Rev. Lett.* 51:384 (1983).
17. C.A. Schrama, E. Peik, W.W. Smith, and H. Walther, Novel miniature ion traps, *Opt. Comm.* 101:32 (1993).
18. J.T. Höffges H.W. Baldauf, T. Eichler, S.R. Helmfrid, and H. Walther, Heterodyne measurement of the fluorescent radiation of a single trapped ion, *Opt. Com.* 133:170 (1997).
19. J.T. Höffges H.W. Baldauf, W. Lange, and H. Walther, Heterodyne measurement of the resonance fluorescence of a single ion, *Journal of Modern Optics* 44:1999 (1997).
20. R. Loudon, Non-classical effects in the statistical properties of light, *Rep. Progr. Phys.* 43:913 (1980).
21. Y. Mu and C.M. Savage, One-atom lasers, *Phys. Rev.* A 46:5944 (1982).
22. C. Ginzel, H.J. Briegel, U. Martini, B.-G. Englert, and A. Schenzle, Quantum optical master equations: the one-atom laser, *Phys. Rev.* A 48:732 (1993).
23. T. Pellizzari and H.J. Ritsch, Photon statistics of the three-level one-atom laser, *Mod. Opt.* 41:609 (1994); Preparation of stationary Fock states in a one-atom Raman laser, *Phys. Rev. Lett.* 72:3973 (1994); P. Horak, K.M. Gheri, and H. Ritsch, Quantum dynamics of a single-atom cascade laser, *Phys. Rev.* A 51:3257 (1995).
24. H.-J. Briegel, G.M. Meyer, and B.-G. Englert, Dynamic noise reduction in multilevel lasers: nonlinear theory and the pump-operator approach, *Phys. Rev.* A 53:1143 (1996); Pump operator for lasers with multi-level excitation, *Europhys. Lett.* 33:515 (1996).
25. M. Löffler, G.M. Meyer, and H. Walther, Spectral properties of the one-atom maser, *Phys. Rev* A 55:3923 (1997).
26. For a recent review see E. Arimondo, Coherent population trapping in laser spectroscopy, *Progress in Optics*, E. Wolf, ed., Elsevier, Amsterdam (1996), vol. **XXXV**, pp. 257-354.
27. A.M. Khazanov, G.A. Koganov, and E.P. Gordov, Macroscopic squeezing in three-level laser, *Phys. Rev.* A 42:3065 (1990); T.C. Ralph and C.M. Savage, Squeezed light from a coherently pumped four-level laser, *Phys. Rev.* A 44:7809 (1991); H. Ritsch, P. Zoller, C.W. Gardiner, and D.F. Walls, Laser light by dynamic pump-noise suppression, *Phys. Rev.* A 44:3361 (1991).
28. K.M. Gheri and D.F. Walls, Squeezed lasing without inversion or light amplification by coherence, *Phys. Rev.* A 45:6675 (1992); H. Ritsch and M.A.M. Marte, Quantum noise in Raman lasers: effects of pump bandwidth and super- and sub-Poissonian pumping, *Phys. Rev.* A 47:2354 (1993).
29. K. An J.J. Childs R.R. Dasari M.S. Feld Microlaser: a laser with one atom in an optical resonator, *Phys. Rev. Lett.* 73:3375 (1994).
30. G.M. Meyer H.-J. Briegel and H. Walther Ion-trap laser, *Europhys. Lett.* 37:317 (1997).
31. G.M. Meyer M. Löffler and H. Walther, Spectrum of the ion-trap laser, *Phys. Rev.* A 56:R1099 (1997).
32. M. Löffler G.M. Meyer and H. Walther, One atom laser with quantized centre-of-mass motion, *Europhys. Lett.* 40:263 (1997).

33. E. Peik, G. Hollemann, H. Walther, Laser cooling and quantum jumps of a single indium ion, *Phys. Rev.* A 49:402 (1994).
34. G. Hollemann, E. Peik, H. Walther, Frequency-stabilized diode-pumped Nd:YAG laser at 946 nm with harmonics at 473 nm and 237 nm, *Opt. Lett.* 19:192 (1994).
35. G. Hollemann, E. Peik, A. Rusch, H. Walther, Injection locking of a diode-pumped Nd:YAG laser at 946 nm, *Opt. Lett.* 20:1871 (1995).
36. E. Peik, G. Hollemann, H. Walther, Single-ion spectroscopy of indium: towards a group-III monoion oscillator, *in:* Proceedings Fifth Symposium on Frequency and Metrology, J. Berquist, ed., World Scientific, Singapore (1996), pp. 376-379.
37. E. Peik, J. von Zanthier, T. Becker, J. Abel, M. Fries, H. Walther, *to be published.*
38. S. Chu, Three-dimensional viscous confinement and cooling of atoms by resonance radiation pressure, *Phys. Rev. Lett.* 55:48 (1985).

PROGRESS TOWARDS USING A CALCIUM ION TRAP TO PERFORM QUANTUM LOGIC OPERATIONS

M. S. Gulley, J. J. Gomez, M. H. Holzscheiter, D. F. V. James, P. G. Kwiat, S. K. Lamoreaux, C. G. Peterson, V. Sandberg, M. M. Schauer, C. Simmons, D. Tupa, P. Wang, A. G. White, R. J. Hughes.

Los Alamos National Laboratory
Los Alamos, NM 87545

We briefly review the development and theory of an experiment to investigate quantum computation with trapped calcium ions. The ion trap, laser and ion requirements are determined, and the parameters required for simple quantum logic operations are described.

INTRODUCTION

In the last 15 years various authors have considered the generalization of information theory concepts to allow the representation of information by quantum systems. The introduction into computation of *quantum mechanical* concepts, in particular the superposition principle, opened up the possibility of new capabilities, such as quantum cryptography [1], that have no classical counterparts. One of the most interesting of these new ideas is quantum computation, first proposed by Benioff [2]. Feynman [3] suggested that quantum computation might be more powerful than classical computation, a notion which gained further credence through the work of Deutsch [4]. However, until quite recently quantum computation was an essentially academic endeavor because there were no quantum algorithms that exploited this power to solve useful computational problems, and because no realistic technology capable of performing quantum computations had been envisioned. This changed in 1994 when Shor discovered quantum algorithms for efficient solution of integer factorization and the discrete logarithm problem [5, 6], two problems that are at the heart of the security of much of modern public key cryptography [7]. Later that same year Cirac and Zoller proposed that quantum computational hardware could be realized using known techniques in the laser manipulation of trapped ions [8]. Since then interest in quantum computation has grown dramatically, and remarkable progress has been made: a single quantum logic gate has been demonstrated with trapped ions [9]; quantum error correction schemes have been invented [10, 11]; several alternative technological proposals have been made [12, 13, 14, 15, 16, 17] and quantum algorithms for solving new problems have been discovered [18, 19, 20, 21]. In this paper we will review our development of an experiment to investigate the potential of quantum computation using trapped calcium ions [22].

The three essential requirements for quantum computational hardware are: (1) the ability to isolate a set of two-level quantum systems from the environment for long enough to maintain coherence throughout the computation, while at the same time being able to interact with the systems strongly enough to manipulate them into an arbitrary quantum state; (2) a mechanism for performing quantum logic operations: in other words a "quantum bus channel" connecting the various two-level systems in a quantum mechanical manner; and (3) a method for reading out the quantum state of the system at the end of the calculation. All three of these requirements are in principle met by the cold trapped ion quantum computer. In this scheme each qubit consists of two internal levels of an ion trapped in a linear configuration. In order to perform the required logic gates, a third atomic state known as the auxiliary level is required. The quantum bus channel is realized using the phonon modes of the ions' collective oscillations. These quantum systems may be manipulated using precisely controlled laser pulses.

CHOICE OF ION

Various ions used in atomic frequency standards work satisfy the requirements to be a qubit. Of these ions, Ca^+ offers the advantages of transitions that can be accessed with titanium-sapphire or diode lasers and a reasonably long-lived metastable state to allow computations to take place. The relevant energy levels of the $A = 40$ isotope are shown in Fig.1.

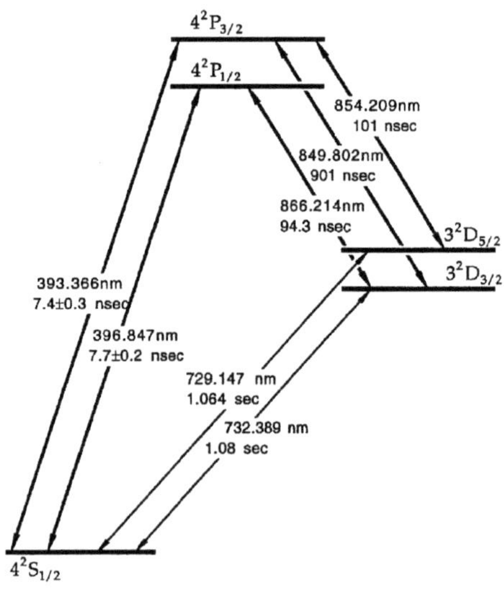

Figure 1. The lowest energy levels of $^{40}Ca^+$ ions, with transition wavelengths and lifetimes listed.

The dipole-allowed transition from the $4\,^2S_{1/2}$ ground state to the $4\,^2P_{1/2}$ level with a wavelength of 397 nm can be used for Doppler cooling and quantum jump readout; The 732 nm electric quadrupole transition from the $4\,^2S_{1/2}$ ground state to the $3\,^2D_{3/2}$ metastable level (lifetime $\approx 1.08 sec.$) is suitable for sideband cooling. In the single laser computation scheme, the qubits and auxiliary level can be chosen as the electronic states

$$|0\rangle = |4\,^2S_{1/2},\, M_j = 1/2\rangle,$$
$$|1\rangle = |3\,^2D_{5/2},\, M_j = 3/2\rangle,$$
$$|aux\rangle = |3\,^2D_{5/2},\, M_j = -1/2\rangle.$$

This ion can also be used for Raman type qubits, with the two Zeeman sublevels of the $4\,^2S_{1/2}$ ground state forming the two qubit states $|0\rangle$ and $|1\rangle$, with one of the sublevels of the $4\,^2P_{1/2}$ level being the upper level $|2\rangle$. A magnetic field of 200 Gauss should be sufficient to split these two levels so that they can be resolved by the lasers. The pump and Stokes beams would be formed by splitting a 397nm laser into two, and shifting the frequency of one with respect to the other by means of an acousto-optic or electro-optic modulator. This arrangement has an advantage in that any fluctuations in the phase of the original 397nm laser will be passed on to both the pump and Stokes beams, and will therefore be canceled out, because the dynamics is only sensitive to the difference between the pump and Stokes phases.

EXPERIMENTAL APPARATUS

In our experiment ions are confined in a linear RF quadrupole trap [22]. Radial confinement is achieved by a quadrupole RF field provided by four 1 mm diameter rods in a rectangular arrangement. Axial confinement is provided by DC voltages applied to conical endcaps at either end of the RF structure; the endcap separation is 10 mm. The design of the trap used in these experiments is shown in diagrammatically in Fig.2.

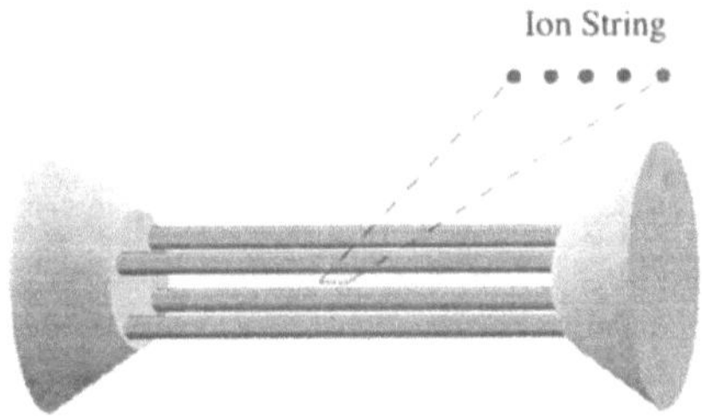

Figure 2. Side view diagram of the linear RF trap used to confine Ca^+ ions in these experiments. The endcap separation is 10 mm and the gap between the RF rods is 1.7 mm.

The first stage of cooling is to cool a small number of ions to their Doppler limit in the ion trap. This requires a beam at 397 nm, the $4\,^2S_{1/2} - 4\,^2P_{1/2}$ resonant transition. Tuning the laser to the red of the transition causes the ions to be Doppler cooled. Having carefully selected the trap parameters, many cycles of absorption and re-emission will bring the system to the Lamb-Dicke regime, leaving the ions in a string-of-pearls geometry. We have recently found ion crystals of up to thirteen Ca^+ ions. Fig.3. shows a typical string of ions. The fluorescence signal from the ions is detected perpendicular to the trap axis by an image intensified CCD camera and independently by a photomultiplier mounted on the opposite side of the trap. The photomultiplier is used to observe the 397 nm fluorescence spectra both for detuning of the 397 nm and the 866 nm lasers from resonance to establish the exact line positions.

In order to Doppler cool the ions, we use a Titanium:Sapphire (Ti:Sapphire) laser (Coherent CR 899-21) with an internal frequency doubling crystal to produce the 397

Figure 3. Image of a string of four trapped calcium ions. The total length is roughly 80 μm.

nm light. We are also developing a diode laser with a frequency doubling cavity to produce 397 nm light, which will be a smaller, less expensive alternative to the Ti:Sapphire laser. During the Doppler cooling, the ions may decay from the $4\,^2P_{1/2}$ state to the $3\,^2D_{3/2}$ state, whose lifetime is $\sim$ 1sec. To empty this metastable state, we use a diode laser at 866 nm.

Once the string of ions is Doppler cooled to the Lamb-Dicke regime, the second stage of optical cooling, sideband cooling, will be used to reduce the collective motion of the string of ions to its lowest vibrational level [25]. In this regime, a narrow optical transition, such as the 732 nm $4\,^2S_{1/2} - 3\,^2D_{3/2}$ dipole forbidden transition, develops sidebands above and below the central frequency by the vibrational frequencies of the ions. The sidebands closest to the unperturbed frequency correspond to the CM vibrational motion. If ω_0 is the optical transition frequency and ω_x the frequency of the CM vibrational motion, the phonon number is increased by one, unchanged, or decreased by one if an ion absorbs a photon of frequency $\omega_0 + \omega_x$, ω_0 or $\omega_0 - \omega_x$, respectively. Thus, sideband cooling is accomplished by optically cooling the string of ions with a laser tuned to $\omega_0 - \omega_x$.

The need to resolve the sidebands of the transition implies a much more stringent requirement for the laser linewidth; it must be well below the CM mode vibrational frequency of $\sim (2\pi) \times 200$ kHz. The laser power must also be greater in order to pump the forbidden transition. We plan to use a Ti:Sapphire laser locked to a reference cavity to meet the required linewidth and power. The transitions required for realization of quantum logic gates and for readout can be performed with the same lasers used in the Doppler and sideband cooling procedures.

QUBIT ADDRESSING OPTICS

In order for the Ca^+ ion qubits to be useful for actual calculations, it will be necessary to address the ions in a very controlled fashion. Our optical system for qubit addressing is shown schematically in Fig.4.

There are two aspects to be considered in the design of such a system: the precise interactions with a single ion; and an arrangement for switching between different ions in the string. In addition to the obvious constraints on laser frequency and polarization, the primary consideration for making exact π- or 2π-pulses is control of the area (over time) of the driving light field pulse. The first step toward this is to stabilize the intensity of the laser, as can be done to better than 0.1%, using a standard "noise-eater". Switching the light beam on and off can be performed with a similar (or even the same) device. Because such switches can possess rise/fall times on the scale of nanoseconds, it should be possible to readily control the area under the pulse to within $\sim 0.1\%$, simply by accurately determining the width of the pulse.

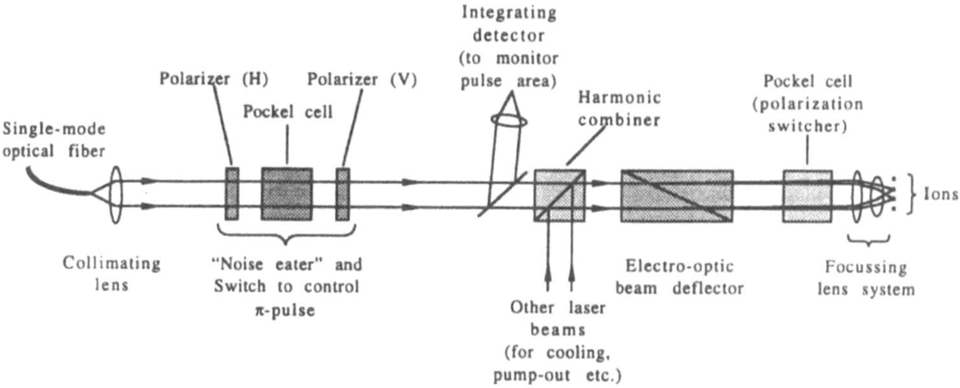

Figure 4. Illustration of the laser beam control optics system.

The controls for switching between ions must be fast, reproducible, display very precise aiming and low "crosstalk" (i.e. overlap of the focal spot onto more than one ion), and be as simple as possible. In particular, it is desirable to be able to switch between different ions in the string in a time short compared to the time required to complete a given π-pulse on one ion. This tends to discount any sort of mechanical scanning system. As a tentative solution, we propose to use an electro-optic beam deflector, basically a prism whose index of refraction, and consequently whose deflection angle, is varied slightly by applying a high voltage across the material; typical switching times for these devices is 10 nanoseconds, adequate for our purposes. We have also constructed a prototype system for testing the focussing lens system. With this test system, we can achieve a crosstalk of less than 0.2 percent for an ion spacing of 20 μm.

SUMMARY

In this paper we have described in some detail the experiment we are currently developing to investigate the feasibility of cold trapped ion quantum computation. We should emphasize that our intentions are at the moment exploratory: we have chosen an ion on the basis of current laser technology, rather than on the basis of which ion which will give the best performance for the quantum computer. Other species of ion may well give better performance: In particular Beryllium ions do have the potential for a significantly lower error rate due to spontaneous emission, although it is also true that lighter ions may be more susceptible to heating. Other variations, such as the use of Raman transitions in place of single laser transitions, or the use of standing wave lasers need to be investigated. Our choice of Calcium will allow us to explore these issues. Furthermore, calculations suggest that it should be possible to trap 20 or more Calcium ions in a linear configuration and manipulate their quantum states by lasers on short enough time scales that many quantum logic operations may be performed before coherence is lost. Only by experiment can the theoretical estimates of performance be confirmed [26, 27]. Until all of the sources of experimental error in real devices are thoroughly investigated, it will be impossible to determine what ion and addressing scheme enables one to build the best quantum computer or, indeed, whether it is possible to build a useful quantum computer with cold trap ions at all.

ACKNOWLEDGMENTS

This research was funded by the National Security Agency.

REFERENCES

1. R. J. Hughes et al. Contemp. Phys. **36** (1995) 149-163.
2. P. A. Benioff, Int. J. Theor. Phys.**21** (1982) 177-201.
3. R. P. Feynman, Foundations of Physics **16** (1986) 507-531.
4. D. Deutsch, Proc. R. Soc. Lond.**A 425** (1989) 73-90.
5. P. W. Shor, *Proceedings of the 35th Annual Symposium on the Foundations of Computer Science*, S. Goldwasser ed., IEEE Computer Society Press, Los Alamitos CA, 1994.
6. A. Ekert and R. Jozsa, Rev. Mod. Phys. **68** (1996) 733-753.
7. R. J. Hughes, "Crptography, Quantum Computation and Trapped Ions", submitted to Phil. Trans. Roy. Soc. (London), 1997; quant-ph/9712054.
8. J. I. Cirac and P. Zoller, Phys. Rev. Lett. **74**, (1995) 4094-4097.
9. C. Monroe et al., Phys. Rev. Lett. **75** (1995) 4714-4717.
10. E. Knill, R. Laflamme and W. Zurek, "Accuracy threshold for quantum computation"; submitted to *Science* (1997).
11. J. Preskill, "Reliable quantum computers", preprint (1997), quant-ph/9705031.
12. Q. A. Turchette et al., Phys. Rev. Lett.**75** (1995) 4710-4713.
13. D. G. Cory, A. F. Fahmy and T. F. Havel, Proc. Natl. Acad. Sci. USA **94** (1997) 1634-1639.
14. J. R. Friedman et al., Phys. Rev. Lett. **76** (1996) 3830-3833.
15. V. Privman, I. D. Vagner and G. Kventsel, "Quantum computation in quantum-Hall systems", preprint (1997), quant-ph/9707017.
16. M. F. Bocko, A. M. Herr and M. J. Feldman, "Prospects for quantum coherent computation using superconducting electronics", preprint (1997).
17. D. Loss and D. P. DiVincenzo, "Quantum computation with quantum dots" preprint (1997), cond-mat/9701055.
18. L. K. Grover *Proceedings of the 28th Annual ACM Symposium on the Theory of Computing*, ACM Press, New York, 1996 p.212
19. B. M. Terhal and J. A. Smolin, " Superfast quantum algorithms for coin weighing and binary search problems", preprint (1997) quant-ph/9705041.
20. D. Boneh and R. Lipton, "Quantum cryptanalysis of hidden linear functions," Proc. CRYPTO'95 (Springer, New York, 1995)
21. A. Kitaev, "Quantum measurements and the Abelian stabilizer problem," preprint (1995) quant-ph 9511026.
22. R. J. Hughes, et al., "The Los Alamos Trapped Ion Quantum Computer Experiment", Fortschritte der Physik, **46** (1998) 329-361; quant-ph/9708050.
23. T. Pellizzari et al., Phys. Rev. Lett.**75** (1995) 3788-3791.
24. F. Diedrich et al., Phys. Rev. Let. **62** (1989) 403-407.
25. D. J. Wineland and W. M. Itano, Phys. Rev.**A 20** (1979) 1521-1540.
26. R. J. Hughes, D. F. V. James, E. H. Knill, R. Laflamme and A. G. Petschek, Phys. Rev. Lett. **77** (1996) 3240-3243.
27. D. F. V. James, R. J. Hughes, E. H. Knill, R. Laflamme and A. G. Petschek, *Photonic Quantum Computing*, S. P. Hotaling, A. R. Pirich eds, *Proceedings of SPIE* **3076** (1997) 42-50.

QUANTUM COMPUTATION WITH "HOT" TRAPPED IONS

Daniel F. V. James,[1], Sara Schneider,[1,2], and Gerard J. Milburn[2]

[1] University of California, Los Alamos National Laboratory, Los Alamos, NM 87545, USA
[2] Centre for Laser Science, Department of Physics, The University of Queensland, St. Lucia, QLD 4072, Australia

We describe two methods that have been proposed to circumvent the problem of heating by external electromagnetic fields in ion trap quantum computers. Firstly the higher order modes of ion oscillation (i. e. modes other than the center-of-mass mode) have much slower heating rates, and can therefore be employed as a reliable quantum information bus. Secondly we discuss a recently proposed method combining adiabatic passage and a number-state dependent phase shift which allows quantum gates to be performed using the center-of-mass mode as the information bus, regardless of its initial state.

INTRODUCTION

Ion trap quantum computers, first proposed by Ignacio Cirac and Peter Zoller of the University of Innsbruck[1], and demonstrated experimentally shortly afterwards by Dave Wineland and his collaborators at NIST Boulder[2], are, arguably, the most promising quantum computation technology for realizing systems of dozens of qubits in the foreseeable future. An ion trap quantum computer consists of a linear array of ions confined in a radio-frequency trap. Two internal states of the ion compose each qubit and the center-of-mass (CM) vibrational mode of the ions' collective oscillations acts as a quantum information bus, by means of which quantum logic gate operations can be performed between pairs of ions. The computer is controlled by a series of appropriately tuned laser pulses, which can change the internal states of each ion and, at the same time, exciting or destroying a quantum of the ions' collective oscillations.

The most daunting problem to be overcome in the development of this technology is the very fragile nature of the quantum mechanical ground-state of the CM mode. Cooling and maintaining the ions in this ground state are required for performing quantum logic gates in the manner proposed by Cirac and Zoller. Any excitation by external fields will diminish the accuracy of logic gates, thus leading to unreliable performance of the quantum computer as a whole, and maybe making the implementation of even simple algorithms impossible.

Elimination of all possible causes of heating is an incredibly demanding task. Thus necessity of being in the ground state of the CM mode. In this paper, we will briefly describe two possible methods by which this problem can be circumvented (fuller accounts can be found elsewhere[3,4]). Other schemes for tackling this problem have also been investigated by different authors[5,6].

SUPRESSION OF HEATING OF 'HIGHER' OSCILLATION MODES

As mentioned above, ion trap quantum computers transfer quantum information between different qubits by means of quanta of the ions' collective oscillation modes. Because of the Coulomb interaction between the different ions, the oscillations of a string of ions will be strongly coupled, and is best considered in terms of *normal modes.* Of these modes, which have been described in detail elsewhere[7], the most convenient mode to use for quantum computation is the CM mode, in which all of the ions oscillate as if they are rigidly clamped together. This mode has two advantages: it is well resolved in frequency space and all of the ions couple to it with equal strength. However, the transfer of information from one ion to another relies on these oscillatory modes being carefully prepared in their quantum ground states. If this were not the case, and the ions were oscillating randomly, quantum logic could not be performed reliably using Cirac and Zoller's scheme. The degradation of these modes due to various influences has been a subject of considerable study[3,8].

The different oscillatory modes are numbered $p = 1, 2, \ldots N$ in order of increasing oscillation frequency, N being the total number of ions in the string. The heating time (i. e. the time for the phonon excitation number to increase by one) for the p-th mode, $\tau_{N,p}$, is given by[3]:

$$\frac{\tau_1}{\tau_{N,p}} = \frac{\sum\limits_{m,n=1}^{N} b_m^{(p)} b_n^{(p)} \gamma_{mn}}{\sqrt{\mu_p}}, \tag{1}$$

where τ_1 is the heating time for a single ion, $b_n^{(p)}$ is the normalized eigenvector and μ_p the eigenvalue, both of the p-th mode[7], and γ_{mn} is the degree of cross-coherence of the external fields acting on the n-th and m-th ions. The relative heating rates, calculated using eq. (1), when degree of cross-coherence is assumed to be a exponential function (i. e. $\gamma_{mn} = \exp\left[-|x_n - x_m|/\ell_{coh}\right]$, x_n being the position of the n-th ion and ℓ_{coh} the coherence length) are shown in fig. 1, for the case $N = 5$. One can easily calculate the heating time of the p-th mode in the limit of a coherent field ($\gamma_{mn} = 1$) and the limit of incoherent fields ($\gamma_{mn} = \delta_{n,m}$, where $\delta_{n,m}$ is the Kronecker delta):

$$\frac{\tau_1}{\tau_{N,p}} = \begin{cases} N & p = 1 \\ 0 & p > 1. \end{cases} \quad \text{coherent} \tag{2}$$

$$\frac{\tau_1}{\tau_{N,p}} = \frac{1}{\sqrt{\mu_p}} \quad \text{incoherent.} \tag{3}$$

The heating times in the coherent limit is a very important result: Only the lowest ($p = 1$) CM mode will be heated up by spatially coherent fields, the modes other than the CM mode remaining in their ground states. The separations of the ions is of the order of a few microns, while the wavelength of the external exciting fields (in resonance with the ions oscillation frequency) is hundreds of meters; thus the coherent case is the experimentally important case. This relatively rapid heating of the CM mode compared to the other modes has also been demonstrated experimentally[9].

Since (with a few exceptions) each ion couples with each phonon mode, any one mode. These modes can be selectively excited or de-excited by tuning the addressing laser to the appropriate mode frequency, and controlling the pulse duration by the appropriate amount, although care must be exercised in this procedure because of the variations in coupling strength between different ions to the different modes[7].

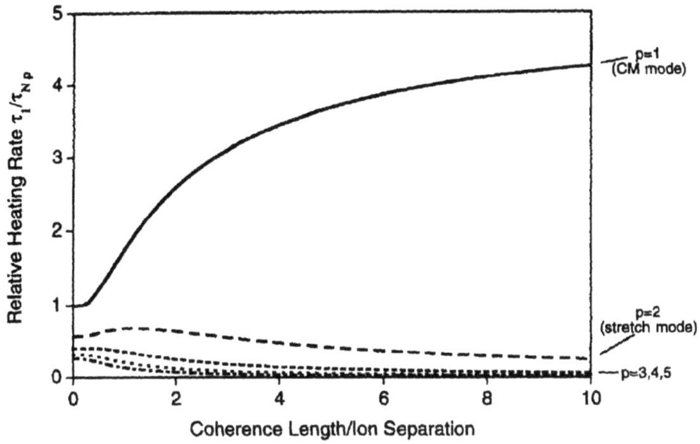

Figure 1. Relative heating rates of different oscillation modes for $L = 5$ ions. For heating fields with long coherence lengths, the $p = 1$ center of mass mode is heated strongly, while the heating rates of the 'higher' modes is significantly reduced.

Drawbacks of this scheme are: the strength of the coupling of each ion to those higher order modes varies from ion to ion, making it harder to adjust the pulse durations; the higher modes are closer together in frequency space, thus making them harder to resolve; and also there is a 'Debye-Waller' factor due to heating of the CM mode which alters the strength of the laser interaction by an undetermined factor.

ADIABATIC PASSAGE AND THE D'HELON-MILBURN PHASE SHIFT

We will now discuss an entirely different approach to the problem of performing quantum logic with 'hot' trapped ions. We make use of the fact that although the ions are not necessarily in the CM mode ground state, they all share the same CM mode, thus enabling them to interact with each other. Our scheme performs a controlled-rotation (CROT) gate between a pair of ions designated control (c) and target (t). This gate operation consists of a conditional sign change which takes place only if both ions are in the excited state. It can be realized by a sequence of four laser pulses, illustrated symbolically in fig. 2.

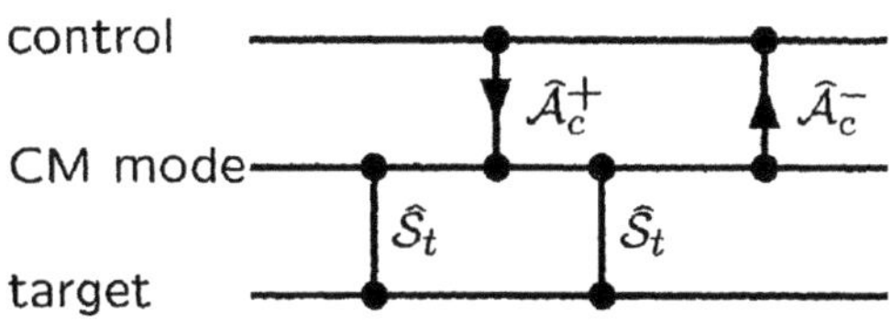

Figure 2. Schematic illustration of the steps involved in the CROT gate with hot ions. The

We will assume that the CM phonon mode is in a pure state* given by the following formula:

$$|\phi\rangle = \sum_n a_n |n\rangle \ , \tag{4}$$

where a_n are a set of unknown complex coefficients and $|n\rangle$ is the Fock state of occupation number n. It will be convenient in what follows to introduce the odd and even parts of this wavefunction, viz. :

$$\begin{aligned} |+\rangle &= \sum_n a_{2n} |2n\rangle \ , \\ |-\rangle &= \sum_n a_{2n+1} |2n+1\rangle \ . \end{aligned} \tag{5}$$

We will also use the following notation for phonon states to which a single quantum has been added:

$$\begin{aligned} |-'\rangle &= \sum_n a_{2n} |2n+1\rangle \ , \\ |+'\rangle &= \sum_n a_{2n+1} |2n+2\rangle \ . \end{aligned} \tag{6}$$

The first step is a conditional phase change between odd phonon number states and the excited internal state of an ion, which can be carried out using an effect first considered by D'Helon and Milburn[11]. They introduced a Hamiltonian for a two-level ion at the node of a detuned standing wave laser field. In the limit of large detuning and for interaction times much greater than the vibrational period of the trap, this Hamiltonian for the jth ion is

$$\hat{H}^{(j)} = \hbar \hat{a}^\dagger \hat{a} \chi (\hat{\sigma}_z^{(j)} + 1/2) \ , \tag{7}$$

where $\hat{\sigma}_z^{(j)}$ is the population inversion operator for the jth ion, $\hat{a}$ and $\hat{a}^\dagger$ are the annihilation and creation operators of the CM mode, and $\chi = \eta^2 \Omega^2 / (N\delta)$ (η is the Lamb-Dicke parameter, Ω is the Rabi frequency for the transitions between the two internal states of the ions and δ the laser detuning). If we choose the duration τ of this interaction to be $\tau = \pi/\chi$, the time evolution is represented by the operator

$$\hat{S}_j = \exp[-i\hat{a}^\dagger \hat{a} (\hat{\sigma}_z^{(j)} + 1/2)\pi] \ . \tag{8}$$

This time evolution flips the phase of the ion when the CM mode is in an odd state and the ion is in its excited state, thus providing us with a conditional phase shift for an ion and the CM mode.

The adiabatic passage[10] which we require for the next step of our gate operation can be realized as follows. We use two lasers, the pump (which couples the qubit state $|1\rangle_c$ to some second auxiliary state $|3\rangle_c$ and is detuned by an amount Δ) and the Stokes (which couples to the red side band transition $|2\rangle_c|n+1\rangle \leftrightarrow |3\rangle_c|n\rangle$, with the same detuning Δ) (see fig. 3). If the population we want to transfer adiabatically is initially in the state $|1\rangle_c|n\rangle$, we turn on the Stokes field and then slowly turn on the pump field until both lasers are turned on fully. Then we slowly turn off the Stokes laser: this is the famous "counter-intuitive" pulse sequence used in adiabatic passage[10]. The adiabatic passage has to be performed very slowly. The condition in our scheme is that $T \gg 1/\Omega_{p,n}, 1/\Omega_{S,n}$, where T is the duration of the adiabatic passage and $\Omega_{p,n}$ ($\Omega_{S,n}$)

*The gate can be shown to work when the CM mode is in a mixed state: we assume a pure state for

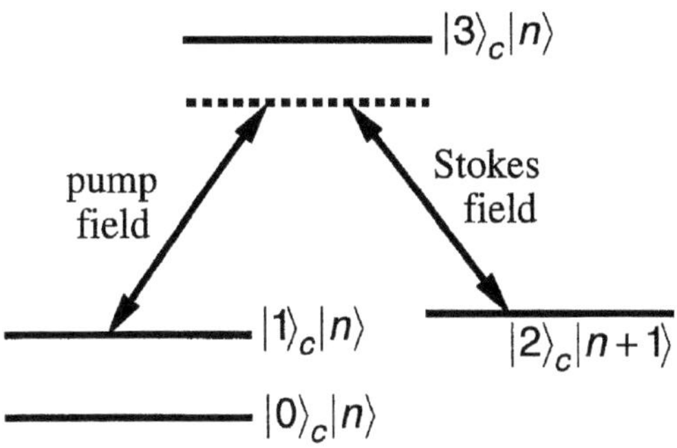

Figure 3. Schematic illustration of the level scheme of the control ion used to realize the adiabatic passage operations $\hat{\mathcal{A}}_c^+$ and $\hat{\mathcal{A}}_c^-$.

are the effective Rabi frequencies for the pump and the Stokes transition, respectively. Using the adiabatic passage we can transfer the population from $|1\rangle_c|n\rangle$ to $|2\rangle_c|n+1\rangle$. To invert the adiabatic passage, we just have to interchange the roles of the pump and the Stokes field. We will denote the adiabatic passage by operators $\hat{\mathcal{A}}_j^+$ and $\hat{\mathcal{A}}_j^-$ defined as follows:

$$\begin{aligned} \hat{\mathcal{A}}_j^+ &: |1\rangle_j|n\rangle \rightarrow |2\rangle_j|n+1\rangle \\ \hat{\mathcal{A}}_j^- &: |2\rangle_j|n+1\rangle \rightarrow |1\rangle_j|n\rangle \ . \end{aligned} \tag{9}$$

Putting these two operations together in the sequence shown in fig. 2, we can write down the step-by-step states for our gate:

$$\begin{aligned} &|00\rangle(|+\rangle+|-\rangle) \xrightarrow{\hat{S}_t} |00\rangle(|+\rangle+|-\rangle) \xrightarrow{\hat{\mathcal{A}}_c^+} |00\rangle(|+\rangle+|-\rangle) \xrightarrow{\hat{S}_t} \\ &\qquad\qquad |00\rangle(|+\rangle+|-\rangle) \xrightarrow{\hat{\mathcal{A}}_c^-} |00\rangle(|+\rangle+|-\rangle) \\ &|01\rangle(|+\rangle+|-\rangle) \xrightarrow{\hat{S}_t} |01\rangle(|+\rangle-|-\rangle) \xrightarrow{\hat{\mathcal{A}}_c^+} |01\rangle(|+\rangle-|-\rangle) \xrightarrow{\hat{S}_t} \\ &\qquad\qquad |01\rangle(|+\rangle+|-\rangle) \xrightarrow{\hat{\mathcal{A}}_c^-} |01\rangle(|+\rangle+|-\rangle) \\ &|10\rangle(|+\rangle+|-\rangle) \xrightarrow{\hat{S}_t} |10\rangle(|+\rangle+|-\rangle) \xrightarrow{\hat{\mathcal{A}}_c^+} |20\rangle(|+'\rangle+|-'\rangle) \xrightarrow{\hat{S}_t} \\ &\qquad\qquad |20\rangle(|+'\rangle+|-'\rangle) \xrightarrow{\hat{\mathcal{A}}_c^-} |10\rangle(|+\rangle+|-\rangle) \\ &|11\rangle(|+\rangle+|-\rangle) \xrightarrow{\hat{S}_t} |11\rangle(|+\rangle-|-\rangle) \xrightarrow{\hat{\mathcal{A}}_c^+} |21\rangle(|-'\rangle-|+'\rangle) \xrightarrow{\hat{S}_t} \\ &\qquad\qquad |21\rangle(-|-'\rangle-|+'\rangle) \xrightarrow{\hat{\mathcal{A}}_c^-} -|11\rangle(|+\rangle+|-\rangle). \end{aligned} \tag{10}$$

Thus we end up with a controlled rotation gate between the ions c and t. A controlled-NOT (CNOT) gate can be realized by performing $\pi/2$ rotation pulses on the target qubit both before and after this series of operations.

A possible source of error in performing gate operations using this scheme is the heating during gate operations. To perform logic operations, effectively the quantum information stored in the two levels of the control qubit is transferred to the even and odd states of the CM mode. Heating mixes these two states, thereby degrading the information stored. Other experimental issues are the phase change induced during the

the state $|1\rangle_c$; also the Stokes and pump fields cause A. C. Stark shifts to other levels, resulting in phase changes. These effects can be compensated by judicious choices of free parameters such as the time taken for the adiabatic passage or the detuning. The D'Helon- Milburn conditional phase operator $\hat{S}_t$ relies on the assumption of the Lamb-Dicke regime as well as a standing wave field acting on target ion. However, standing waves are difficult to achieve in practice, but it may be be possible to obtain similar effects using traveling waves; further for the low trap frequencies required to resolve spatially individual ions with a focused laser, the Lamb-Dicke regime may not be valid. All of these questions, as well as the study of heating and decoherence mechanisms will be the subject of future papers.

The authors would like to thank Ignacio Cirac, Richard Hughes, Brian King, Paul Kwiat, Hideo Mabuchi, Jörg Steinbach, Gil Toombes and Dave Wineland for useful discussions and comments. This work was supported by the U. S. National Security Agency and the Australian Research Council International Program.

REFERENCES

1. J. I. Cirac and P. Zoller, *Phys. Rev. Lett.* **74**, 4091 (1995).
2. C. Monroe *et al.*, *Phys. Rev. Lett.* **75**, 4714 (1995).
3. D. F. V. James, *Phys. Rev. Lett.* **81**, 317 (1998).
4. S. Schneider, D. F. V. James and G. J. Milburn, "Method of quantum computation with hot trapped ions", submitted to *Phys. Rev. Lett.* 1998.
5. J. F. Poyatos, J. I. Cirac, and P. Zoller, *Phys. Rev. Lett.* **81**, 1322 (1998).
6. A. Sørensen and K. Mølmer, "Quantum computation with ions in thermal motion", preprint, 1998.
7. D. F. V. James,*Appl. Phys. B* **66**, 181 (1998).
8. S. K. Lamoreaux, *Phys. Rev. A* **56**, 4970 (1997); T. A. Savard, K. M. O'Hara and J. E. Thomas, *Phys. Rev. A* **56**, R1095 (1997); S. Schneider and G. J. Milburn, *Phys. Rev. A* **57**, 3748 (1998); M. Murao and P. L. Knight, *Phys. Rev. A* **58**, 663 (1998).
9. B. E. King *et al.* "Cooling the collective motion of trapped ions to initialize a quantum register," *Phys. Rev. Lett.* (in press); quant-ph/9803023.
10. K. Bergmann and B. W. Shore, "Coherent Population Transfer", in: *Molecular Dynamics and Spectroscopy by Stimulated Emission Pumping*, H. L. Dai and R. W. Field, eds. (World Scientific, Singapore, 1995).
11. C. D'Helon and G. J. Milburn, *Phys. Rev. A* **54**, 5141 (1996).

NOVEL QUANTUM EFFECTS IN LIGHT SCATTERING FROM COLD TRAPPED ATOMS

A. Orłowski[1], M. Gajda[1], P. Krekora[1], R. J. Glauber[2], and J. Mostowski[1]

[1]Instytut Fizyki PAN & Szkoła Nauk Ścisłych
Aleja Lotników 32/46, 02-668 Warszawa, Poland
[2]Lyman Laboratory, Harvard University
Cambridge, MA 12138, USA

INTRODUCTION

Trapping and cooling of neutral atoms is a mature but still rapidly developing area of contemporary atomic and optical physics. Progress in experimental techniques provided a possibility of trapping single atoms in magnetic and magneto-optical traps. This gives a unique opportunity to study properties of single quantum systems subjected to repeatable measurements and provides a tool for creating the quantum states of the atomic center-of-mass motion[1, 2]. On the other hand, it is now possible to trap and to cool down millions of atoms to the regime where quantum statistics become very important. A peculiar phenomenon of Bose-Einstein condensation has been observed experimentally [3]. There is a natural question: how can we learn about properties of trapped atoms? It is especially important to have methods that are not destructive. The lesson from actual development of physics is that in most cases scattering is the only way we can gain some information about the structure and behavior of investigated system[4].

In this work a detailed analysis of both far-off-resonance and resonant scattering of light from a single neutral atom trapped in a harmonic potential is given. Although the real magnetic and magneto-optical traps are not ideally harmonic, this assumption is a very good approximation to most of the existing configurations. For off-resonance scattering an exact closed-form expression for the differential scattering cross section is derived from a general S matrix theory via an effective Hamiltonian[5]. Physical meaning of the obtained result is elaborated and potential applications to non-destructive measurements of ultra-low temperatures are suggested. The atom is assumed to be initially in a thermal state.

For the resonant case there are several scattering scenarios possible. Complete quantum expressions for scattering cross sections are derived and analyzed. A detailed physical discussion is performed. As expected, especially interesting features of the scattering process are found when the atomic recoil energy is much larger than the

Quantum Communication, Computing, and Measurement 2
Edited by Kumar et al., Kluwer Academic / Plenum Publishers, New York, 2000.

difference between the subsequent energy levels of the harmonic trap[6]. An extensive analysis of the most realistic Compton-like regime in resonant scattering is performed. Possibility of more exotic quantum effects is mentioned. For simplicity we assume that the center-of-mass of the atom is initially in a ground state of the trapping potential.

FAR-OFF-RESONANCE SCATTERING

Our model of the far-off-resonance scattering consists of an atom of mass M which can move in the isotropic harmonic potential of oscillatory frequency Ω. The atomic center-of-mass motion is governed by the following Hamiltonian:

$$H_{CM} = \frac{\mathbf{P}^2}{2M} + \frac{1}{2}M\Omega^2\mathbf{R}^2. \tag{1}$$

The Hamiltonian describing the dipole interaction of the atom with the electromagnetic field is given by $H_I(t) = -\mathbf{d}(t)\mathbf{E}(t)$. Actually, due to the fact that for low-frequency electromagnetic field the dipole moment is $\mathbf{d}(t) = \alpha\mathbf{E}(t)$ (α being low-frequency polarizability), we can introduce an *effective* Hamiltonian $H_I^{eff}(t) = -\frac{1}{2}\alpha\mathbf{E}^2(t)$.

The scattering process can be easily described within the framework of the S matrix theory. First it is convenient to define the S matrix using T matrix: $S_{FI} = \delta_{FI} - 2\pi i\delta(E_F - E_I)\langle F|T(0)|I\rangle$, where T matrix can be expressed by the interaction Hamiltonian as a perturbative expansion:

$$T(t) = H_I(t) + \frac{1}{i\hbar}H_I(t)\int_{-\infty}^{t} \mathrm{d}t' H_I(t') + \cdots \tag{2}$$

Initial $|I\rangle = |\mathbf{k}_L, i\rangle$ and final $|F\rangle = |\mathbf{k}, f\rangle$ states of the total system consist of the states of photons and center-of-mass ones. For low frequency fields we have:

$$\langle F|T(0)|I\rangle \simeq \langle F|H_I^{eff}|I\rangle = \frac{\alpha\hbar}{2}\sqrt{\omega\omega_L}(\epsilon_f^*\epsilon_i)\langle F|\exp\left[i(\mathbf{k}_L - \mathbf{k})\mathbf{R}\right]|I\rangle. \tag{3}$$

To get the differential scattering cross section we divide $|S_{FI}|^2$ by the flux of incoming photons. As a result we have:

$$\frac{\mathrm{d}\sigma_{fi}}{\mathrm{d}\Omega_k \mathrm{d}E_f} = \frac{\alpha^2}{(4\pi)^2}\frac{k^3 k_L}{\hbar}|\epsilon_f^*\epsilon_i|^2|\langle f|\exp[i(\mathbf{k}_L - \mathbf{k})\mathbf{R}]|i\rangle|^2\delta(\omega + \omega_f - \omega_L - \omega_i). \tag{4}$$

In the case of nonpolarized light we get: $|\epsilon_f^*\epsilon_i|^2 = (1 + \cos^2\vartheta)/2$, where ϑ is the angle of scattering. Now we are able to compute the scattering cross section for the initial center-of-mass state being the thermal state:

$$\frac{\mathrm{d}\sigma_{fi}}{\mathrm{d}\Omega_k \mathrm{d}E_f} = F(k, k_L; \alpha)\frac{1}{2\pi\hbar}\int_{-\infty}^{\infty} \mathrm{d}t \exp[-i(\omega - \omega_L)t]K(\mathbf{q}, t). \tag{5}$$

The integral in the above expression is known as a response function and its Fourier transform can be written as

$$K(\mathbf{q}, t) = \langle \exp[-i\mathbf{q}\mathbf{R}(0)]\exp[i\mathbf{q}\mathbf{R}(t)]\rangle_T =$$

$$= \exp\left[-\frac{2\varepsilon}{\Omega}(1 - \cos\vartheta)\coth(\hbar\Omega\beta/2)\right]\exp\left[\frac{2\varepsilon}{\Omega}(1 - \cos\vartheta)\frac{\cosh(i\Omega t + \hbar\Omega\beta/2)}{\sinh(\hbar\Omega\beta/2)}\right], \tag{6}$$

where the first term is called the Debye-Weller factor. In the above formula we have $\beta = 1/(k_B T)$ and $\hbar\varepsilon = (\hbar^2 k_L^2)/(2M)$ is a recoil energy. For the very low temperatures

($\hbar\Omega\beta \gg 1$) the Debye-Weller factor is $\exp[-(2\varepsilon/\Omega)(1-\cos\vartheta)]$. In the opposite case it is given by $\exp[-(2\varepsilon/\Omega)(1-\cos\vartheta)/(\hbar\Omega\beta)]$. We also have $F(k,k_L;\alpha) = \alpha^2 k^3 k_L(1+\cos^2\vartheta)/(32\pi^2)$ and $\hbar\mathbf{q} = \hbar\mathbf{k}_L - \hbar\mathbf{k}$ is the momentum transfer from the field to the atomic center of mass.

The final result can be presented as a series expansion of the modified Bessel functions of the first kind:

$$\frac{\mathrm{d}\sigma_{fi}}{\mathrm{d}\Omega_k \mathrm{d}E_f} = F(k,k_L;\alpha)\exp\left[-\frac{2\varepsilon(1-\cos\vartheta)}{\Omega}\coth(\hbar\Omega\beta/2)\right]$$
$$\times \sum_{n=-\infty}^{n=\infty} \exp[n\hbar\Omega\beta/2] I_n\left(\frac{2\varepsilon(1-\cos\vartheta)}{\Omega\sinh(\hbar\Omega\beta/2)}\right)\delta(\hbar(\omega-\omega_L)-n\hbar\Omega). \qquad (7)$$

It follows from the above expressions that at zero temperature the scattering is of the Stokes type (in the scattering spectrum there are only lines with frequency lower than that of incident light). At higher temperatures there can be some energy transfer from the center-of-mass degrees of freedom to the electromagnetic field and scattering spectrum gets more symmetric with respect to the frequency of the incident light. This is well illustrated in Fig 1. The asymmetry of the spectrum can be used as a measure of the temperature of the system.

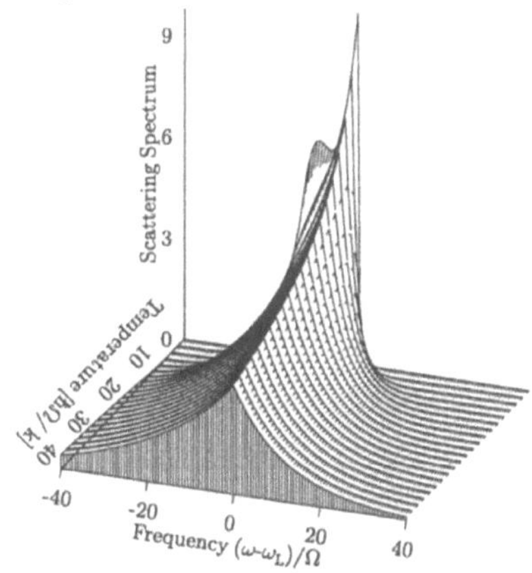

Figure 1. Spectrum of the scattered light for various values of temperature in the range from 0 to $40\hbar\Omega/k$.

RESONANT SCATTERING

In the regime of resonant scattering the Hamiltonian of the system can be divided into three parts $H = H_A + H_F + H_I$, where:

$$H_A = \hbar\omega_0|e\rangle\langle e| + \frac{\mathbf{P}^2}{2M} + \frac{1}{2}M\Omega^2\mathbf{R}^2 \qquad (8)$$

is the free-atom Hamiltonian, $H_F = \sum_{\mathbf{k}\mu}\hbar\omega\, a^\dagger_{\mathbf{k}\mu}a_{\mathbf{k}\mu}$ is the free-field Hamiltonian, and the interaction is given by

$$H_I = -\sum_{\mathbf{k}\mu} i\sqrt{\frac{2\pi\hbar\omega}{V}}\left(|e\rangle\langle g|(\boldsymbol{\epsilon}_{\mathbf{k}\mu}\mathbf{d})\, e^{i\mathbf{kR}}\, a_{\mathbf{k}\mu} - \text{h.c.}\right). \qquad (9)$$

The interaction Hamiltonian describes coupling of the electric field to both internal and center-of-mass degrees of freedom of the atom; it is proportional to the field at the position of the atom as well as to the transition dipole moment. Because we restrict our analysis to very weak incident electromagnetic fields (single photon states only) the most convenient way of studying the scattering process is the perturbation theory. Straightforward calculations lead to analytical expression for the transition amplitude from the initial to some final state of the system. We study the scattering in the stationary limit, i.e., we assume that atom-field interaction had been switched on in the remote past, lasted for a very long (infinite) time until it was switched off in the remote future. In this way asymptotic eigenstates of the system are those of the free Hamiltonian. The probability per unit of time of transition between the initial and the final state of the free Hamiltonian can be related to the scattering cross section in the standard way. In the infinite interaction-time limit the total energy of the system has to be conserved in the scattering process.

The obtained scattering cross section is a function of the final state vibrational quantum numbers n and n_z. Straightforward calculations based on the second order perturbation theory lead to the following expression:

$$\frac{\mathrm{d}\sigma_{n,n_z}}{\mathrm{d}\Omega_k} = \sigma_0(1+\cos^2\vartheta)\frac{1}{n!}\left(\frac{\varepsilon}{\Omega}\right)^n \binom{n}{n_z}(\sin^2\vartheta)^{n-n_z}\left|\int_0^T g_{n_z}(\vartheta,\tau)\mathrm{d}\tau\right|^2, \tag{10}$$

where we have introduced the unit of the cross section $\sigma_0 = (d\omega_L/c)^4/(\hbar\Gamma/2)^2$ and differential solid angle $\mathrm{d}\Omega_k = \sin\vartheta\mathrm{d}\vartheta\mathrm{d}\varphi$; ϑ and φ being angles of propagation of the scattered photon. Using the quantity σ_0 we can express the polarazibility mentioned in the previous section: $\alpha^2 = (2\pi\Gamma/\Delta)^2\sigma_0/(k_L^3 k)$. $T = 2\pi/\Omega$ is the period of the center-of-mass oscillations in the trap and $g_{n_z}(\vartheta,\tau)$ is time dependent amplitude defined as follows:

$$g_{n_z}(\vartheta,\tau) = \mathcal{N}\exp[(\mathrm{i}\,\Delta-\Gamma/2)\tau]\left(\mathrm{e}^{-\mathrm{i}\Omega\tau}-\cos\vartheta\right)^{n_z}\exp\left[-(\varepsilon/\Omega)\left(1-\mathrm{e}^{-\mathrm{i}\Omega\tau}\cos\vartheta\right)\right], \tag{11}$$

where we have introduced the notion of the detuning:

$$\Delta = \omega_L - \omega_0, \tag{12}$$

and a normalization factor:

$$\mathcal{N} = \frac{\Gamma/2}{1-\exp\left[2\pi\mathrm{i}(\Delta+\mathrm{i}\Gamma/2)/\Omega\right]}. \tag{13}$$

Because we are not interested in a particular final state of the center-of-mass but in the spectral and angular characteristics of the scattered light we sum up all contributions from different n_z-components introducing the cross section $\mathrm{d}\sigma_n/\mathrm{d}\Omega_k$:

$$\frac{\mathrm{d}\sigma_n}{\mathrm{d}\Omega_k} = \sum_{n_z=0}^{n}\frac{\mathrm{d}\sigma_{n,n_z}}{\mathrm{d}\Omega_k}. \tag{14}$$

The index n distinguishes between different final frequencies of the scattered photon according to the energy balance. At this point we want to comment that although the second order differential cross section has been obtained without any other but model approximations, in writing formulas describing it we have substituted final photon frequency ω by the frequency of the incident photon ω_L.

A new and interesting features of the scattering are related to the center-of-mass dynamics which becomes not trivial if the recoil energy is sufficiently large, $\varepsilon/\Omega \gg 1$.

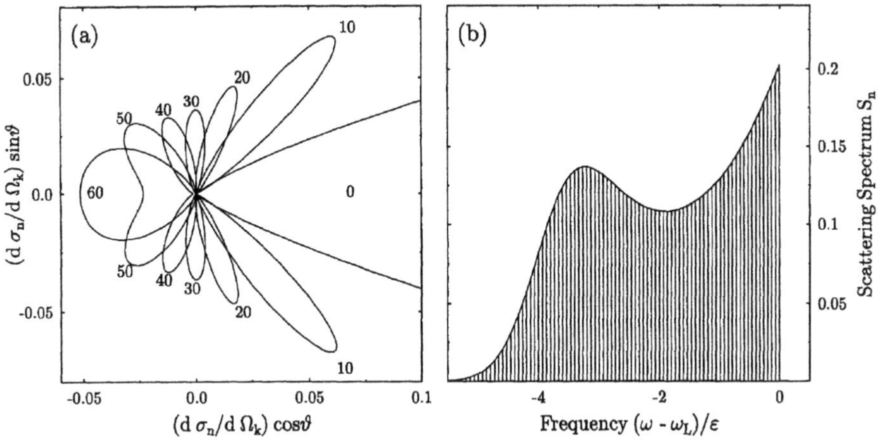

Figure 2. (a) The polar plot of the angular pattern of the differential cross section in the CS regime; $\varepsilon/\Gamma = 1.5 \times 10^{-4}$, $\Omega/\Gamma = 10^{-5}$, $\Delta = 0$. Each component corresponds to a different frequency of the scattered light and is numbered by the vibrational quantum number of a final center-of-mass state n. The component of $n = 0$ which corresponds to the elastic line is very large and is out of the scale of the figure. The inelastic lines corresponds to $n = 10, 20, \ldots, 60$. The cross section is measured in the natural unit σ_0. Up-down symmetry of the figure reflects the axial symmetry of the scattering; (b) The spectrum of the scattered radiation in the CS regime; $\varepsilon/\Gamma = 1.5 \times 10^{-4}$, $\Omega/\Gamma = 10^{-5}$, $\Delta = 0$. The spectrum consist of series of delta-like spikes separated by the vibrational frequency Ω. The line connecting tops of spikes is drown in order to lead an eye. The spectrum is measured in the natural unit σ_0.

If, in addition, the atomic lifetime is much smaller than the oscillation period, $\Omega/\Gamma \ll 1$ the emission follows the absorption act 'almost immediately'. The amplitude $g_{n_z}(\vartheta, \tau)$ (see Eq.(11)) is large only in the neighborhood of $\tau = 0$. Therefore, an approximate expression for the cross section can be obtained by expanding it into a power series of τ. The order of the expansion depends on a ratio of relevant parameters. The first interesting consequences of the recoil experienced by the atom can be found already in the limit of very small Doppler shift, $\varepsilon/\Gamma \ll 1$. It is the case when atomic velocity is so small that a distance traveled by the excited atom during its lifetime is much smaller than the light wavelength and zero order expansion is sufficient to approximate the function $g_{n_z}(\vartheta, \tau)$. The scattering cross section can be expressed in the following form:

$$\frac{d\sigma_n}{d\Omega_k} = \sigma_0(1 + \cos^2\vartheta)\, \mathcal{P}_n(\vartheta)\, \frac{(\Gamma/2)^2}{\Delta^2 + (\Gamma/2)^2}, \tag{15}$$

where $\mathcal{P}_n(\vartheta)$ can be interpreted as an occupation probability of the final vibrational states:

$$\mathcal{P}_n(\vartheta) = \exp[-2\varepsilon/\Omega(1 - \cos\vartheta)][2\varepsilon/\Omega(1 - \cos\vartheta)]^n/n!. \tag{16}$$

The recoil effect introduces the first non-vanishing correction as compared to the infinitely heavy atom limit; it leads to a substantial redistribution of populations of the center-of-mass energy levels. The cross section Eq.(15) is a product of two terms. The first one, $\mathcal{P}_n(\vartheta)$, originates from the Frank-Condon factor and has a sharp maximum for these values of frequencies and scattering angles which are mutually related by the momentum conservation law, i.e., for

$$n\Omega = 2\varepsilon(1 - \cos\vartheta). \tag{17}$$

This leads to some angular-frequency correlation in scattered radiation (Fig. 2a) which is very similar to that observed in the Compton effect so the scattering in this regime can be called the *Compton scattering.* The bandwidth of the scattered radiation is limited by 4ε, the amount which is much smaller then the natural broadening of the electronic excited state, Fig. 2b, and this corresponds to the bandwidth of the spectrum of X-rays scattered on electrons in Compton effect. Two maxima in Fig. 2b correspond to the elastic, $n = 0$, and deeply inelastic component, $n \approx 4\varepsilon/\Omega$, what is the result of our choice of polarization.

BRIEF SUMMARY

Both far off-resonance and resonant scattering of light from single atoms trapped by 3D harmonic potentials has thoroughly been studied. Novel effects are predicted for different physical regimes. We have shown that dynamics of the atomic center-of-mass strongly influences the scattering cross section. Possibility of using spectrum of the scattered light in far-off-resonance regime to nondestructively measure the temperature of ultracold atoms is advocated: off-resonance scattering can be used as an 'optical thermometer'. The realistic Compton-like regime in resonant scattering has been investigated in detail. Another interesting quantum effect in resonant regime, which has not been discussed here due to the lack of space, is the time resolved scattering, showing up when the atom can remain in the excited state long enough to make many trips back and forth in the trap before emitting a photon. The possibility of the experimental observation of the predicted effects is now being scrutinized.

This work was supported in part by the Polish Committee for Scientific Research (KBN) under Grant No. 2 P03B 130 15.

REFERENCES

1. F. Ruschewitz, D. Bettermann, J. L. Peng, and W. Ertmer, Statistical investigations on single trapped neutral atoms, *Europhys. Lett.* **34** (9), 651-656 (1996); D. Haubrich, H. Schadwinkel, F. Strauch, B. Ueberholz, R. Wynands, and D. Meschede, Observation of individual neutral atoms in magnetic and magneto-optical traps, *Europhys. Lett.* **34** (9), 663-668 (1996).
2. D. M. Meekhof, C. Monroe, B. E. King, W. M. Itano, and D. J. Wineland, Generation of nonclassical motional states of a trapped atom, *Phys. Rev. Lett.* **76** (11), 1796-1799 (1996); D. Leibfried, D. M. Meekhof, B. E. King, C. Monroe, W. M. Itano, and D. J. Wineland, Experimental determination of the motional quantum state of a trapped atom, *Phys. Rev. Lett.* **77** (21), 4281-4285 (1996); C. Monroe, D. M. Meekhof, B. E. King, and D. J. Wineland, A "Schroedinger cat" superposition state of an atom, *Science* **272** (5265), 1131-1136 (1996).
3. The first observation of Bose-Einstein condensation (BEC) was reported for rubidium: M. H. Anderson, J. R. Ensher, M. R. Matthews, C. E. Wieman, and E. A. Cornell, Observation of Bose-Einstein condensation in a dilute atomic vapor, *Science* **269** (5221), 198-201 (1995). The same year BEC was observed also for lithium: C. C. Bradley, C. A. Sackett, J. J. Tollett, and R. G. Hulet, Evidence of Bose-Einstein condensation in an atomic gas with attractive interactions, *Phys. Rev. Lett.* **75** (9), 1687-1690 (1995), and for sodium: K. B. Davis, M. -O. Mewes, M. R. Andrews, N. J. van Druten, D. S. Durfee, D. M. Kurn, and W. Ketterle, Bose-Einstein condensation in a gas of sodium atoms, *Phys. Rev. Lett.* **75** (22), 3969-3973 (1995). Up to now several other groups have also produced BEC.
4. R. J. Glauber, Scattering of neutrons by statistical media, *in:* "Lectures in Theoretical Physics," Vol. IV, edited by W. E. Brittin, B. W. Downs, and J. Downs (Wiley, New York, 1962) pp. 571-615.
5. R. J. Glauber and A. Orłowski, Off-resonance light scattering from single atom trapped by harmonic potential (unpublished).
6. M. Gajda, P. Krekora, and J. Mostowski, Light scattering by an ultracold trapped atom, *Phys. Rev. A* **54** (1), 928-942 (1996).

PREPARATION OF HIGHLY EXCITED MOTIONAL ENERGY EIGENSTATES WITH A TRAPPED ATOM

Sascha Wallentowitz,[1] Werner Vogel,[1] and Peter E. Toschek[2]

[1]Fachbereich Physik, Universität Rostock
Universitätsplatz 3, D-18051 Rostock, Germany
[2]Institut für Laser-Physik, Universität Hamburg
Jungiusstraße 9, D-20355 Hamburg, Germany

Highly excited eigenstates of the motion of an individual spatially confined atom may be prepared making use of a "trapping" mechanism based on the nonlinearity of the laser-mediated coupling of internal excitation and center-of-mass vibration of the atom. In contrast to micromaser-type trapping states, these states are insensitive to laser-phase fluctuations, and a long sequence of laser pulses may be applied to the atom with little degradation only of the atomic motion. Thus, vibrational number states of high quantum numbers are achievable.

INTRODUCTION

The preparation of quantum states of the motion of trapped atoms has substantially progressed in the past years. Starting from a single laser-cooled trapped ion[1-4], the preparation of various motional quantum states has been experimentally achieved[5,6]. Energy eigenstates of the ion's center-of-mass motion in the trap potential for small numbers of vibrational quanta have been prepared by the application of a limited number of laser pulses, exploiting the coherent dynamics of the selective excitation of vibrational sidebands of the ion. Since the number of the required laser pulses grows with the desired quantum number of the energy eigenstate, the preparation of highly excited number states is eventually limited by the electronic dephasing due to the fluctuating phase of the laser field.

An alternative way of preparing vibrational number states is the use of a micromaser-type dynamics[7,8] for the center-of-mass motion of the trapped atom.[9,10] The ideal situation where the desired number state is a stationary trapping state of the atomic motion is impaired by both laser-phase fluctuations and recoil effects of spontaneously emitted photons. Nevertheless, for the incoherent regime where the bandwidth of the applied laser field exceeds the Rabi frequency an interesting dynamics exists where the populations of vibrational number states reveal sub-Poissonian statistics.[10]

A completely different mechanism for the generation of trapping states is based on the interference-type effects in the interaction of the laser wave with the atomic center-of-mass wavefunction, that lead to a nonlinear modification of the vibronic Jaynes–

Quantum Communication, Computing, and Measurement 2
Edited by Kumar et al., Kluwer Academic / Plenum Publishers, New York, 2000.

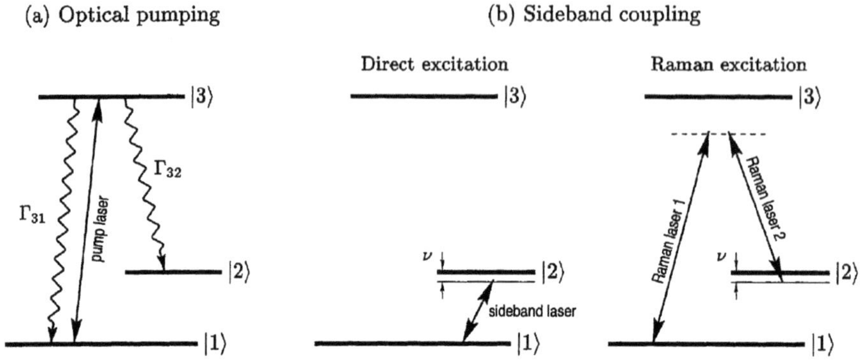

Figure 1. Excitation scheme of the electronic three-level system of the trapped atom. (a) The two-level system $|1\rangle \leftrightarrow |2\rangle$ is inverted by optical pumping. (b) The first lower vibrational sideband (vibrational frequency ν) is excited by direct laser excitation or in a Raman-laser configuration.

Cummings interaction.[11] This effect can usually be neglected in the Lamb–Dicke regime where the spatial width of the atomic wavefunction is much smaller than the laser wavelength. This approximation fails, however, for highly excited states of large spatial extension. The "laser-atom interference" can be used for the manifestation of a trapping mechanism which is advantageous for preparing highly excited number states of the motion of the trapped ion.[12] We show in this paper that this type of trapping mechanism is insensitive to fluctuations of the laser phase, and a large number of laser pulses will not degrade the state.

EXCITATION SCHEME

The excitation scheme for the electronic three-level system of the trapped atom is shown in Fig. 1. The preparation process consists of several cycles where first the two-level system $|1\rangle \leftrightarrow |2\rangle$ is inverted by optical pumping, see Fig. 1a, and then the first lower vibrational sideband is coupled to the electronic transition by either directly exciting the sideband by a resonant laser field, or by using Raman-laser excitation resonant to the vibrational sideband (Fig. 1b). Let us consider the two steps of a cycle in more detail.

Optical Pumping

Optical pumping is achieved by application of a laser field resonant with the strong (dipole) transition $|1\rangle \leftrightarrow |3\rangle$. The propagation direction of the laser beam is set perpendicular to the direction of motion under consideration, such that the vibrational quantum state is left undisturbed by the laser. In the interaction picture the atom-laser interaction is described by the pump (p) Hamiltonian

$$\hat{H}_p = \frac{1}{2}\hbar\left(\Omega_p \hat{A}_{31} + \Omega_p^* \hat{A}_{13}\right), \tag{1}$$

with Ω_p being the Rabi frequency and $\hat{A}_{ij} = |i\rangle\langle j|$ being the electronic flip operator which induces the electronic-state transition $|j\rangle \rightarrow |i\rangle$. The master equation reads

$$\frac{\partial}{\partial t}\hat{\varrho}(t) = \frac{1}{i\hbar}\left[\hat{H}_p, \hat{\varrho}(t)\right] + \hat{\mathcal{L}}(t)\,\hat{\varrho}(t), \tag{2}$$

where the effects of the spontaneous emissions are described by the Liouvillean operator

$$\hat{\mathcal{L}}(t)\,\hat{\varrho} = \Gamma_{31}\hat{A}_{13}\hat{\mathcal{R}}_1(t)\,\hat{\varrho}\,\hat{A}_{31} + \Gamma_{32}\hat{A}_{23}\hat{\mathcal{R}}_2(t)\,\hat{\varrho}\,\hat{A}_{32} - \frac{\Gamma_{31}+\Gamma_{32}}{2}\left[\hat{A}_{33}\,\hat{\varrho} - \hat{\varrho}\,\hat{A}_{33}\right]. \quad (3)$$

Here, Γ_{3i} denote the spontaneous emission rates from the state $|3\rangle$ to the states $|i\rangle$ $(i=1,2)$, and the operators $\hat{\mathcal{R}}_i(t)$ describe the recoil effects of spontaneous emission,

$$\hat{\mathcal{R}}_i(t)\,\hat{\varrho} = \int_{-1}^{1} ds\, w(s)\, \exp\left\{is\eta_i\left[\hat{a}^\dagger(t)+\hat{a}(t)\right]\right\}\,\hat{\varrho}\,\exp\left\{-is\eta_i\left[\hat{a}^\dagger(t)+\hat{a}(t)\right]\right\}, \quad (4)$$

with $w(s)=\frac{3}{8}(1{+}s^2)$ being the radiation characteristics of the strong (dipole) transitions $|3\rangle \leftrightarrow |i\rangle$ $(i=1,2)$ and $\hat{a}(t)=\hat{a}e^{i\nu t}$ includes the oscillation with vibrational frequency ν. The Lamb–Dicke parameters occurring in Eq. (4) are defined as $\eta_i=2\pi\Delta x/\lambda_{3i}$ $(i=1,2)$ with Δx being the spatial width of the ground-state wavepacket in the trap potential, and λ_{3i} being the wavelengths of the electronic transitions $|3\rangle \leftrightarrow |i\rangle$ $(i=1,2)$. In the Lamb–Dicke limit the center-of-mass wavepacket is strongly localised with respect to the wavelengths, and therefore the Lamb–Dicke parameters η_i attain very small values. The vibrational transitions during spontaneous emissions can then be neglected. Nevertheless, for higher motional excitation the recoil effects play a crucial role, since they broaden the statistics of the populations of the vibrational number states. For the present purpose we describe the recoil effects by rate equations for the number-state populations, since motional coherences are strongly damped with a large laser bandwidth.

Sideband Coupling

While for the direct excitation of the vibrational sideband (Fig. 1b) the Lamb–Dicke parameter occurring in the laser-atom interaction is determined by the vibrational and electronic frequencies, the Lamb–Dicke parameter can be varied with Raman excitation by changing the geometry of propagation of the laser beams. In both cases the vibronic coupling can be written in the interaction picture and in rotating-wave approximation as[11]

$$\hat{H}_{sb} = \frac{1}{2}\,\hbar\Omega_{sb}\,\hat{A}_{21}\,\hat{f}(\hat{a}^\dagger\hat{a};\eta)\,\hat{a} + \text{h.c.}, \quad (5)$$

with Ω_{sb} being the Rabi frequency on the vibrational sideband (sb). The function $\hat{f}(\hat{a}^\dagger\hat{a};\eta)$ depends solely on the number of vibrational quanta $\hat{n}=\hat{a}^\dagger\hat{a}$, and its diagonal elements $f(n;\eta)=\langle n|\hat{f}(\hat{a}^\dagger\hat{a};\eta)|n\rangle$ read

$$f(n;\eta) = \frac{1}{n+1}\,L_n^{(1)}(\eta^2)\,e^{-\eta^2/2}, \quad (6)$$

with $L_n^{(k)}(x)$ being Laguerre polynomials. The nonlinear modification (6) represents the overlap integral $\int dp\,\psi^*_{n+1}(p)\,\psi_n(p+\hbar k)$ of two motional eigenfunctions in momentum representation, shifted against each other by the momentum of the laser photon $\hbar k$. These factors, for the vibration of a trapped ion, are counterparts to molecular Franck–Condon factors and may be interpreted as interference between the laser wave field and the center-of-mass wavefunctions of the trapped atom. It plays an important role when the spatial extension of the atomic wavepacket is of the order of the laser wavelength. Moreover, since this situation occurs for highly excited motional wavepackets, this nonlinearity is especially useful for our purpose. This is easily seen from the coefficients (6). At certain numbers N of vibrational quanta the factor of nonlinear coupling becomes zero. At these quantum numbers the laser interaction collapses, and no further vibrational quanta are generated in the trapped atom, i.e. the atom has entered a motional trapping state.

TRAPPING IN HIGH NUMBER STATES

In order to prepare trapping states of high numbers of vibrational quanta, we have to take into account the phase fluctuations of the laser that excites the vibrational sideband. We quantify these fluctuations by the laser bandwidth γ. The master equation for the sideband coupling reads then

$$\frac{\partial}{\partial t}\,\hat{\varrho}(t) = \frac{1}{i\hbar}\left[\hat{H}_{sb}\,,\,\hat{\varrho}(t)\right] - \frac{\gamma}{2}\left[\hat{A}_{11}\,\hat{\varrho}(t)\,\hat{A}_{22} + \hat{A}_{22}\,\hat{\varrho}(t)\,\hat{A}_{11}\right], \tag{7}$$

where the second term accounts for damping of the electronic coherences. With a typical vibrational frequency $\nu/2\pi \approx 10\text{MHz}$, Rabi frequency $|\Omega|/2\pi \approx 10\ldots500\text{kHz}$, and bandwidth $\gamma/2\pi \approx 1\text{MHz}$, an incoherent regime is reached where the bandwidth of the laser field exceeds the Rabi frequency, but the vibrational sideband is well resolved. In this situation the electronic coherences can be adiabatically eliminated in order to finally obtain the solution for the motional statistics at times t_k after the pump/sideband-excitation cycles,

$$P_n(t_{k+1}) = w(n|n)\,P_n(t_k) + w(n|n-1)\,P_{n-1}(t_k), \tag{8}$$

with the transition probabilities being defined as

$$w(n+1|n) = \frac{1}{2}\left\{1 - \exp\left[-\gamma_n(\eta)\,t_{sb}\right]\right\}, \qquad w(n|n) = \frac{1}{2}\left\{1 + \exp\left[-\gamma_n(\eta)\,t_{sb}\right]\right\}. \tag{9}$$

Here, t_{sb} is the interaction time of the laser field, and the excitation-dependent damping coefficients $\gamma_n(\eta)$ are given by

$$\gamma_n(\eta) = \frac{4|\Omega_{sb}|^2}{\gamma}\,(n+1)\,f^2(n;\eta); \tag{10}$$

they reduce in the Lamb–Dicke approximation to $\lim_{\eta\to0}\gamma_n(\eta) = 4|\Omega_{sb}|^2(n+1)/\gamma$. Note, that in order to obtain the recurrence relation (8), we have neglected the weak perturbation of the motional quantum state during the optical pumping.

Equation (8) shows that the trapping state $|N\rangle$ emerges when the transition probability $w(N+1|N)$ vanishes,

$$w(N+1|N) = \frac{1}{2}\left\{1 - \exp\left[-\gamma_N(\eta)\,t_{sb}\right]\right\} = 0. \tag{11}$$

Thus, the zeroes of $w(N+1|N)$ are determined by the zeroes of the damping coefficient $\gamma_N(\eta)$, that are determined by the zeros of the Nth-order Laguerre polynomial,

$$L_N^{(1)}(\eta_0^2) = 0. \tag{12}$$

Therefore if, according to (12), some value η_0 is acquired either by direct or by Raman excitation of the sideband, the number state $|N\rangle$ is approached after several cycles by the interference-induced trapping mechanism.

Figure 2 shows the time evolution of the trapping-state population P_N (a) and the relative motional variance (b) for the trapping state $|N=160\rangle$ and the corresponding Lamb–Dicke parameter $\eta=0.151$. Here, the disturbance of the motional statistics during the optical pumping is included for the chosen values of the Lamb–Dicke parameters η_{3i} $(i=1,2)$ of the strong dipole transitions. Larger Lamb–Dicke parameters of optical-pumping transitions reduce the maximum population in the trapping state. This reduction is caused by the population in vibrational levels below the trapping

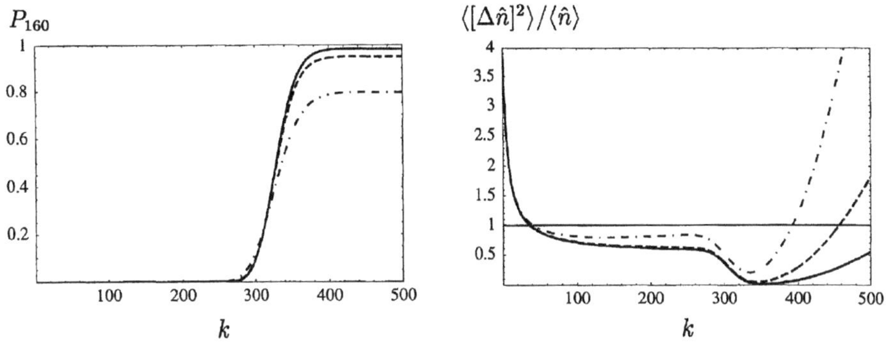

Figure 2. (a) Population P_{160} in the trapping state $|N=160\rangle$ for $\eta=0.151$ and $|\Omega_{sb}|^2 t_{sb}/\gamma=10$ versus the number k of cycles (laser pulses) for an initial thermal state with 25% population in the ground state. The Lamb–Dicke parameters for the optical pumping transitions are $\eta_{31}\approx\eta_{32}=0.01$ (solid), 0.02 (dashed), 0.05 (dot-dashed). (b) Relative variance versus k for the same conditions as in (a), with Lamb–Dicke parameters $\eta_{31}\approx\eta_{32}=0.01$ (solid), 0.02 (dashed), 0.05 (dot-dashed), and $\Gamma_{31}=\Gamma_{32}=5\nu$, $\Omega_p=\nu$. While the results for $\eta_p=0.01, 0.02$ are shown to visualise the recoil effects, the combination $\eta_p=0.05$ and $\eta=0.151$ can be realised experimentally.

state that is scattered during the optical pumping to number states beyond the trapping state. This effect is also seen in Fig. 2b where the values of relative variance are almost identical up to the point where the trapping state is reached at $k\approx 350$. At this stage, lower-lying number states are scattered across the trapping-state boundary by the recoil effects during the optical pumping. It is therefore advantageous to stop the sequence of laser interactions at an optimum number of pulses where both the trapping-state population is maximum and the relative variance is below the classical limit of unity, shown as a line in Fig. 2b.

SUMMARY AND CONCLUSIONS

In summary, it has been shown that the laser-induced nonlinear vibronic coupling causes a trapping mechanism where the trapping states are selected by adjusting the Lamb–Dicke parameter of the laser-atom interaction. In contrast to the micromaser-type trapping mechanism, this mechanism is insensitive to laser-phase fluctuations, since it is based on the vanishing of the atom-laser coupling strength. Due to the insensitivity of this mechanism with respect to decoherence a large number of laser pulses can be applied for the preparation of high number states of the motion of a trapped atom.

Acknowledgments

This research was supported by the Deutsche Forschungsgemeinschaft.

REFERENCES

1. W. Neuhauser, M. Hohenstatt, P.E. Toschek, and H.G. Dehmelt, *Phys. Rev. Lett.* 41:233 (1978).
2. D. Wineland, R. Drullinger, and F. Walls, *Phys. Rev. Lett.* 40:1639 (1978).
3. F. Diedrich, J.C. Bergquist, W.M. Itano, and D.J. Wineland, *Phys. Rev. Lett.* 62:403 (1989).
4. C. Monroe, D.M. Meekhof, B.E. King, S.R. Jefferts, W.M. Itano, and D.J. Wineland, *Phys. Rev. Lett.* 75:4011 (1995).

5. D.M. Meekhof, C. Monroe, B.E. King, W.M. Itano, and D.J. Wineland, *Phys. Rev. Lett.* 76:1796 (1996).
6. C. Monroe, D.M. Meekhof, B.E. King, and D.J. Wineland, *Science* 272:1131 (1996).
7. P. Filipowicz, J. Javanainen, and P. Meystre, *Phys. Rev. A* 34:3077 (1986).
8. G. Rempe, F. Schmidt-Kaler, and H. Walther, *Phys. Rev. Lett.* 64:2783 (1990).
9. R. Blatt, J.I. Cirac, and P. Zoller, *Phys. Rev. A* 52:518 (1995).
10. S. Wallentowitz, W. Vogel, I. Siemers, and P.E. Toschek, *Phys. Rev. A* 54:943 (1996).
11. W. Vogel and R.L. de Matos Filho, *Phys. Rev. A* 52:4214 (1995).
12. This effect seems to have been observed as multistability in the vibrational stationary state of a trapped and cooled ion, cf. T. Sauter, H. Gilhaus, W. Neuhauser, R. Blatt, P.E. Toschek, *Europhys. Lett.* 7:317 (1988).

OPTICAL IMPLEMENTATION OF GROVER'S ALGORITHM: IT'S ALL DONE WITH MIRRORS

Paul Kwiat, Jay Mitchell, Peter Schwindt, and Andrew White

Physics Div., P-23, MS-H803
Los Alamos National Laboratory
Los Alamos, NM, 87545

The essential operations of a quantum computer can all be accomplished using only standard linear optical elements (e.g., beamsplitters, waveplates, polarizers, etc.), with the individual bits represented by different spatial or polarization degrees of freedom. The difference from a genuine quantum computer with *distinct* entangleable registers is that the optical implementation requires a number of elements which grows exponentially with the number of bits. However, by "compiling" the code, and making multiple use of redundant components, the required number of elements can be substantially reduced, allowing the realization of quantum algorithms involving several bits. Here we present a simple optical implementation of Grover's algorithm for efficiently searching a database. In our example a database of four elements is searched with a single query, in contrast to the classical expected value of 2.25 queries. It is seen that the "quantum" computer in this case is no more than an interferometer, albeit a complicated one.

INTRODUCTION

Quantum computation, if implementable, enables tremendous improvements over traditional computing methods for certain types of problems, such as efficiently searching a database [1]. Thus far, the implementation of actual algorithms has been limited to schemes employing bulk NMR methods [2, 3], although basic gate operations have also been performed using other "hardware". In the past several years, a number of works have discussed the possibility of encoding quantum information in one or more degrees of freedom of light [4], some specifically focusing on the possible implementation of simple quantum circuits [5-8]. The central idea in these schemes is that individual bits can be represented by different polarization or spatial-mode degrees of freedom of light. It is then very easy to obtain superpositions, interference, and even entanglement. The difference from a genuine quantum computer with *distinct* entangleable registers is that the optical implementation requires a number of elements which grows exponentially with the number of bits. Nevertheless, as we shall see, it is possible to "compile" the circuits, allowing one to easily realize and test nontrivial algorithms involving several

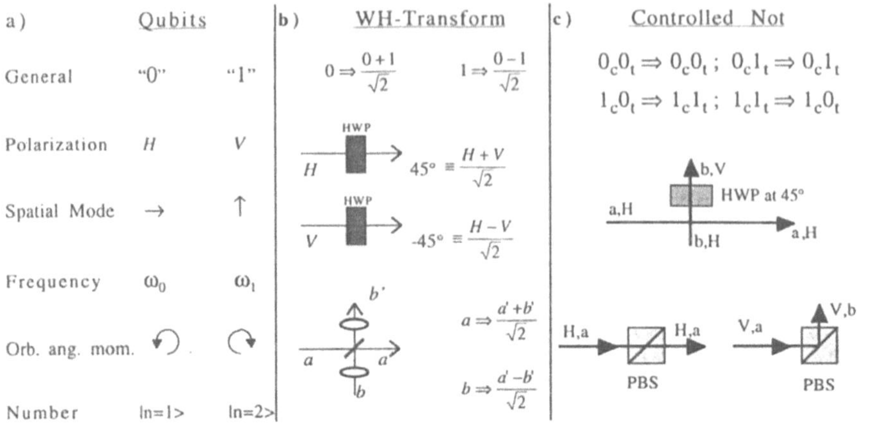

Figure 1. Basic optical representation of quantum elements (a) Qubits encoded in polarization, spatial mode, frequency, orbital angular momentum and photon number; (b) WH-transformation for polarization and spatial-mode qubits; (c) Controlled-Not Gates, with spatial- (polarization-)mode as control (target) [top], and vice versa [bottom].

bits. One conclusion of this approach is that a quantum circuit is essentially a (complicated) interferometer. Here we present the experimental realization of an all-optical implementation of Grover's database-search algorithm [9].

As shown in Figure 1, all the basic circuit elements of quantum computation can be accomplished using only linear passive optical elements. In particular, we can use two orthogonal polarizations (e.g., H and V) to represent a bit. We can also use two different spatial modes to represent another bit. The most straightforward extension to more bits is simply to add extra spatial modes (via beamsplitters), so that, for example, whether the photon is transmitted at the first BS defines the first bit, whether it is transmitted at the next BS becomes the second bit, and so on. We are also exploring the use of frequency as another independent bit. For example, one could use acousto-optic modulators to make various transformations between a base frequency ω_o and a shifted value $\omega_o + \delta\omega$. And because there already exists a sizable technology devoted to frequency multiplexing, much of the development work has already been done. Another possibility, at least for classical beams, is to use orthogonal states of *orbital* angular momentum, i.e., different helical wavefront modes; transformations between them are enabled by computer-generated holograms. Finally, we note that one could also employ the photon-occupation number as yet another degree of freedom. However, this last has the disadvantage that, unlike all the other optical representations of bits, it is susceptible to loss, which transforms one bit into another; furthermore, there are no simple optical elements whose operation *depends* on photon number.

Optical implementations of the Walsh-Hadamard (WH) transforms, which take $0 \rightarrow (0+1)/\sqrt{2}$ and $1 \rightarrow (0-1)/\sqrt{2}$, are shown in Fig. 1b. The WH transformation on a polarization qubit can be accomplished with a half waveplate (HWP) oriented at 22.5° to the horizontal*; and a simple 50-50 beamsplitter (BS) performs the WH transform on a right-propagating spatial mode ($\equiv 0$) and an upward-propagating spatial mode ($\equiv 1$), after two extra $-\pi/2$ phase shifters are included to compensate for the relative $\pi/2$ phase shift upon reflection from any beamsplitter. Finally, although we will not

*The effect of a HWP is to *reflect* the linear polarization about the direction of the optic axis, effectively rotating the polarization by twice the angle between the incident polarization and the optic axis.

need them for what follows, it is easy to perform Controlled Not between various qubits, as indicated in Fig. 1c. For example, the polarization can "control" the "target" spatial mode direction simply by passing the light through a polarizing beam splitter. Similarly, a waveplate (at 45°) in the "1" spatial mode will flip the polarization qubit only if the photon is in that path. We will now describe how to construct a nontrivial quantum circuit using these elements.

GROVER'S ALGORITHM

The specifics of Grover's algorithm are described in detail elsewhere;[1, 3] a schematic representation is shown in Fig. 2a. A number of qubits (n qubits suffice to encode a 2^n-element database) are input, each in its "0" state. The first WH transform on each individual bit prepares an equal superposition of all the database elements. The "Oracle" then simultaneously evaluates all the database elements, and "marks" that one which satisfied some desirable characteristic (e.g., was the only prime number in a list), by imposing a $-\pi$ phase shift on it. At this stage a measurement on the state would still be equally likely to yield any element of the database. The following sequence then transfers some of the amplitude from the non-marked elements to the marked one: A WH transform is performed on each of the bits; a π phase shift is applied to all but the first element; and another WH transform is performed on each of the bits. In general for a database of N elements, running the circuit $\sim \pi\sqrt{N}/8$ times will lead to a final state in which the magnitude of the marked element's amplitude is greater than $1/\sqrt{2}$ – a measurement on the system will then yield this element more than half the time [10]. Remarkably, for $N = 4$ (i.e., a 2-bit database), *all* of the amplitude is transfered to the desired element in a *single* run of the quantum circuit. For our proof-of-principle experiment we focused on this simple case; below we discuss the extension to more bits.

Optical implementation

Fig. 2b shows a direct optical implementation of Grover's algorithm for searching a 4-element, 2-bit database; each step has been implemented individually, to demonstrate the process. A practical realization of the Oracle will be discussed below, but conceptually it is achievable by simply inserting a HWP in path a or b, with optic axis horizontal or vertical. In this way, the Oracle can flip the sign of one of the datebase elements aH, aV, bH, or bV. The two outputs of the second interferometer are analyzed with polarizing beamsplitters (PBS), in the H-V basis. One can show that for each setting of the Oracle, an incident photon will with certainty exit to a specific port, due to complete constructive interference of the four trajectories through the system [9]. Hence, with the all-optical methods, we see explicitly that Grover's algorithm (and, we speculate, *all* quantum computer algorithms) can be understood as an interference effect, albeit a complicated one.

Because many of the optical transformations "commute" with each other, it is possible to do considerable simplification of the optical arrangement of Fig. 2b. For example, by moving the second WH polarization-transform to *after* the spatial-mode transform (i.e., moving the HWPs after the second BS), one can then combine and eliminate some elements (e.g., the HWP-π_V-HWP combination in path a' is equivalent to a single HWP oriented at 45°). The resulting "compiled" circuit is shown in Fig. 2c.

Our actual experimental layout is shown in Fig. 3a. There are a number of significant modifications/improvements to the system of Fig. 2c. Instead of using some

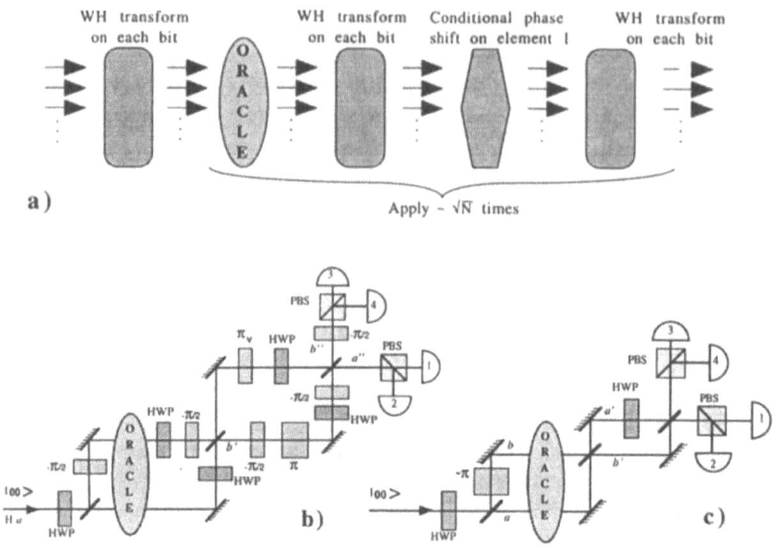

Figure 2. Grover's technique for efficiently searching a database. (a) A block diagram indicating the necessary steps of the algorithm. (b) All-optical implementation of Grovers algorithm for a 2-bit, 4-element database. (All half waveplates (HWP) are oriented at 22.5°). (c) A "compiled" version of the algorithm. (The first and second HWPs are at 22.5° and 45°, respectively.)

sort of active stabilization, we instead employed a displaced Sagnac interferometer design – the great advantage is relative insensitivity to changes in path lengths (but not to tilts of the beamsplitter or mirrors), because both clockwise and counterclockwise paths are affected similarly. Nevertheless, because the paths are displaced from each other by $\sim$ 4mm, one can situate different elements in each path. The right-angle prisms simplify alignment considerably, but have the disadvantage of undesirable polarization-dependent phase shifts from total internal reflections. For this reason, the HWP in the second interferometer intersects three of the paths, instead of just one. This causes each of the four possible trajectories through the system to be horizontally-polarized while passing through one prism, and vertically-polarized in the other, so that any birefringent phase-shifts factor out. Finally, although not shown in the figure, we found it preferable to use the beamsplitter and two mirrors at closer to normal incidence, to again reduce polarization-dependent effects.

For our Oracle we used the system shown in the inset of Fig. 3a, consisting of a Pockels cell (PC) and a liquid crystal (LC) variable phase retarder in both paths, and a quartz rotator (QR) in path b. The PC and LC may each be viewed as a two-state device. In the "off" state, they are equivalent to pieces of glass; in the "on" state, they are equivalent to HWPs with optic axes oriented horizontally, and thus take $H \to H$ and $V \to -V$. The QR rotates any input linear polarization by 90°. The net effect of the Oracle is to apply a relative π phase shift to one of the states aH, aV, bH, or bV, depending on the settings of the PC and LC. For example, when both are acting like HWPs, the state bH is "marked".

Figure 3b shows the results for our system when we injected the light from a 670-nm laser diode (after passing it through a single-mode fiber to give a clean gaussian spatial mode), and measured the intensities at each output using calibrated power meters. As expected, we can clearly distinguish among the four states of the oracle.

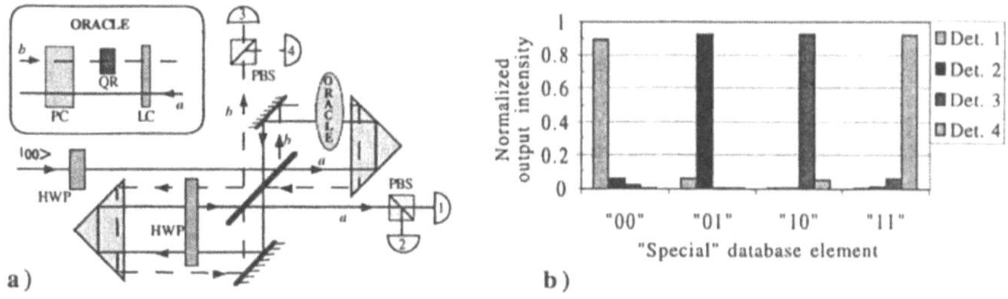

Figure 3. (a) Experimental setup using self-stabilized displaced Sagnac interferometers. (The first and second HWPs are at 22.5° and 45°, respectively.) Inset: Our implementation of the Oracle, consisting of a Pockels cell (PC), liquid crystal (LC) phase retarder, and quartz rotator (QR). (b) Experimental results.

The average probability of identifying a specific non-marked state was about 2.8%, and was largely due to the detrimental effects of the Oracle components on the wavefront quality – in the absence of the LC and QR, the noise could be reduced to less than 1%. Note that although we used a laser intensity ($\sim 1\mu W$) consisting of very many photons, the identical results are predicted if we were to employ single photons instead. Just as the individual molecules in the liquid NMR schemes are essentially isolated quantum computers [2], each photon in our system makes an independent "run" of the code.

DISCUSSION

It is natural to ask what features of quantum computation are embodied in our all-optical implementation of Grover's algorithm. In particular, initially it does not seem that there are any operations (e.g., CNOT gates) to generate entanglement, even of the formal kind between different degrees of freedom. However, the Oracle itself serves to entangle the bits: while $00+01+10+11$ can be factored into a product state for each of the bits, $-00+01+10+11$ cannot, for example. That one can achieve an effective entanglement *without* using CNOT gates is one of the advantages of the all-optical implementations of quantum circuitry, related to the fact that the optical implementation is much easier for operations that transform a given element of the database, because each element corresponds to a different physical mode of the system. Conversely, operations that transform a given *bit* become very difficult, requiring $\sim 2^{n-1}$ optical components (e.g., to flip the sign of one bit requires a HWP in half of the paths). Contrast this with the standard multi-particle implementations of quantum computers, in which operations at the bit-level are more natural, while those on the elements are among the most difficult. Another major advantage of the all-optical methods is the virtual lack of decoherence, stemming from the fact that there are no preferred bases (e.g., the H/V polarization basis is not preferred to the 45/-45 basis).†

The major limitation of these all-optical schemes is that the number of optical elements grows exponentially with the number of bits.‡ However, as we have seen, this

†However, the effective decoherence may be *adjusted* simply by varying the coherence length of the light relative to the interferometer path-length imbalance. This technique was recently used to prepare an arbitrary mixed state of polarization [11].

‡Curiously, implementations using bulk NMR effectively have a similar limitation, because the signal to noise for n spins (bits) drops as $1/2^n$, due to the decreasing likelihood that all bits (which are in a thermal distribution) start off in the ground state [2].

may be mitigated to some extent by "compiling" the algorithm, so that it should be possible to extend the current work to a greater number of bits and a correspondingly larger database.§ For example, while a naïve implementation of Grover's algorithm to search a 3-bit, 8-element database requires nearly 60 elements, this may be readily reduced to only 33 by combining various operations, and further consolidated to only 17 by multiple use of the same components. It is an open question how to achieve the maximal reduction, and to what extent reduction in the optical circuit corresponds to reduction with other "hardware".

Because these methods are not scalable to large numbers of bits, they are not intended as replacements of conventional quantum computer hardware. Rather, we view them as a convenient test bench on which to investigate nontrivial quantum circuits involving several bits. It is our hope that they will lead to a deeper understanding of the essential nature of quantum computation, and to improvements in algorithm design and implementation.

REFERENCES

1. L. K. Grover, Quantum mechanics helps in searching for a needle in a haystack, *Phys. Rev. Lett.* 79:25 (1997).
2. N. Gershenfeld and I. L. Chuang, Bulk spin-resonance quantum computation, *Science* 275:350 (1997); D. G. Cory, A. F. Fahmy, and T. F. Havel, Ensemble quantum computing by NMR-spectroscopy, *Proc. Natl. Acad. Sci., USA* 94: 1634 (1997).
3. I. L. Chuang, N. Gershenfeld, and M. Kubinuc, Experimental implementation of fast quantum searching, *Phys. Rev. Lett.*, 80:3408 (1998); I. L. Chuang *et al.*, Experimental realization of a quantum algorithm; *Nature* 393:143 (1998); J. A. Jones, M. Mosca, and R. H. Hansen, Implementation of a quantum search algorithm on a quantum computer, *Nature* 393:344 (1998).
4. M. Reck, A. Zeilinger, H. J. Bernstein, and P. Bertani, Experimental realization of any discrete unitary operator, *Phys. Rev. Lett.* 73:58 (1994); S. Stenholm, Polarization coding of quantum information, *Opt. Comm.* 123:287 (1996); R.J.C. Spreeuw, A classical analogy of entanglement, *Found. of Phys.*, in press, (1998).
5. S. Takeuchi, A simple quantum computer: experimental realization of the Deutsch-Jozsa algorithm with linear optic, *in*: "Proc. of the 4th Workshop on Physics and Computation" (PhysComp96), p. 299 (1996).
6. J. F. Clauser and J. P. Dowling, Factoring integers with Young's N-slit interferometer, *Phys. Rev. A* 53:4587 (1996); J. Summhammer, Factoring and Fourier transformation with a Mach-Zehnder interferometer, *Phys. Rev. A* 56: 4324 (1997).
7. N. J. Cerf, C. Adami, and P. G. Kwiat, Optical simulation of quantum logic, *Phys. Rev. A* 57:R1477 (1998).
8. S. L. Braunstein, Quantum error correction for communication with linear optics, *Nature* 394:47 (1998).
9. P. G. Kwiat, J. R. Mitchell, P. D. D. Schwindt, and A. G. White, All-optical implementation of Grover's search algorithm, submitted to *Phys. Rev. Lett.*.
10. M. Boyer, G. Brassard, P. Hoyer, and A. Tapp, Tight bounds on quantum searching, *Fortschr. Phys.* 46:493 (1998).
11. P. D. D. Schwindt, P. G. Kwiat, and B.-G. Englert, Quantitative wave-particle duality and non-erasing quantum erasure, submitted to *Nature*.
12. P. G. Kwiat, Hyper-entangled states, *J. Mod. Opt.* 44:2173 (1997).

§Another interesting possibility for extending the capabilities of the present techniques is to incorporate the true entanglement (and even *hyper*-entanglement in several degrees of freedom) that exists between correlated photons produced via spontaneous parametric down-conversion.[12]

QUANTUM STATE PROTECTION IN OPTICAL CAVITIES

Paolo Tombesi,[1,2] David Vitali,[1,2]

[1]Dipartimento di Matematica e Fisica, University of Camerino, Italy
[2]Istituto Nazionale per la Fisica della Materia, Italy

We show how an initially prepared quantum state of a radiation mode in a cavity can be preserved for a long time using a feedback scheme based on the injection of appropriately prepared atoms. We also discuss the possibility of applying these methods for decoherence control in quantum information processing.

INTRODUCTION

What makes quantum information processing much more attractive than its classical counterpart is its capability of using entangled states and of processing generic linear superpositions of input states. The entanglement between a pair of systems is capable of connecting two observers separated by a space-like interval, it can neither be copied nor eavesdropped on without disturbance, nor can it be used by itself to send a classical message [1]. The possibility of using linear superposition states has given rise to quantum computation, which is essentially equivalent to have massive parallel computation [2]. However all these applications crucially rely on the possibility of maintaining quantum coherence, that is, a defined phase relationship between the different components of linear superposition states, over long distances and for long times.

The possibility of an experimental control of decoherence is important also from a more fundamental point of view. In fact decoherence is the practical explanation of why linear superposition of macroscopically distinguishable states, the states involved in the famous Schrödinger cat paradox [3], are never observed and how the classical macroscopic world emerges from the quantum one [4]. The progressive decoherence of a mesoscopic Schrödinger cat has been observed for the first time in the experiment of Brune *et al.* [5], where the linear superposition of two coherent states of the electromagnetic field in a cavity with classically distinct phases has been generated and detected. We have already shown [6] that a stroboscopic feedback mechanism could be devised to slowdown the decoherence of the experiment of Brune *et al.* [5]. Here we propose a similar feedback scheme, based on direct photodection. The idea is very simple: whenever the cavity looses one photon, a feedback loop supplies the cavity mode with another photon, through the injection of an appropriately prepared atom. The proposed feedback almost instantaneously "cures" the effect of a quantum jump and is able therefore to minimize the destructive effects of dissipation on the quantum state of the cavity mode.

THE MASTER EQUATION

The master equation for continuous feedback has been derived by Wiseman and Milburn [7], and, in the case of perfect detection via a single loss source, is given by

$$\dot{\rho} = \gamma\Phi(a\rho a^\dagger) - \frac{\gamma}{2}a^\dagger a\rho - \frac{\gamma}{2}\rho a^\dagger a \,, \tag{1}$$

where γ is the cavity decay rate and $\Phi(\rho)$ is a generic superoperator describing the effect of the feedback atom on the cavity state ρ. Eq. (1) assumes perfect detection, i.e., all the photons leaving the cavity are absorbed by a unit-efficiency photodetector and trigger the cavity loop. It is practically impossible to realize such an ideal situation and therefore it is more realistic to generalize this feedback master equation to the situation where only a fraction $\eta < 1$ of the photons leaking out of the cavity is actually detected and switches on the atomic injector. It is immediate to see that (1) generalizes to

$$\dot{\rho} = \eta\gamma\Phi(a\rho a^\dagger) + (1-\eta)\gamma a\rho a^\dagger - \frac{\gamma}{2}a^\dagger a\rho - \frac{\gamma}{2}\rho a^\dagger a \,. \tag{2}$$

Now, we have to determine the action of the feedback atom on the cavity field $\Phi(\rho)$; this atom has to release exactly one photon in the cavity, possibly regardless of the field state in the cavity. In the optical domain this could be realized using *adiabatic transfer of Zeeman coherence* [8]. Using dark state adiabatic transfer of Zeeman coherence in which the atom passage realizes the adiabatic transformation

$$|g_1, n\rangle \rightarrow |g_2, n+1\rangle \,,$$

where g_i refers to the Zeeman shifted lower states and n is the number of photons in the cavity, it is possible to release always the photon in the cavity, regardless the cavity mode state, i.e.,

$$\Phi(\rho) = a^\dagger (aa^\dagger)^{-1/2}\rho(aa^\dagger)^{-1/2}a \,. \tag{3}$$

The feedback master equation can be rewritten as

$$\dot{\rho} = \frac{(1-\eta)\gamma}{2}\left(2a\rho a^\dagger - a^\dagger a\rho - \rho a^\dagger a\right) - \frac{\eta\gamma}{2}\left[\sqrt{\hat{n}}, \left[\sqrt{\hat{n}}, \rho\right]\right] \tag{4}$$

that is, a standard vacuum bath master equation with effective damping $(1-\eta)\gamma$, supplemented with a sort of "square root of phase diffusion". This phase diffusion is due to the fact that the photon injected by feedback has no phase relationship with the photons already present in the cavity and this means that feedback does not guarantee perfect state protection for a generic state, even in the ideal condition $\eta = 1$ However this does not mean that the proposed feedback scheme is good for preserving number states only, because the unconventional "square-root of phase diffusion" is much slower than the conventional one (described by a double commutator with the number operator). A semiclassical estimation of the diffusion constant can be obtained from the representation of the master equation in terms of the Wigner function

$$-\left[\sqrt{n}, \left[\sqrt{n}, \rho\right]\right] \leftrightarrow \frac{1}{4\bar{n}}\frac{\partial^2}{\partial\theta^2}W(r,\theta) \,, \tag{5}$$

showing that (at least at large photon number) in the case of the feedback-induced unconventional phase diffusion, the diffusion constant is scaled by a factor $4\bar{n} \gg 1$.

The above arguments completely neglect dissipative effects due to cavity losses and atomic spontaneous emission. For example, cavity dissipation couples a given manifold $|g_1, n\rangle$, $|e, n\rangle$, $|g_2, n+1\rangle$ with those with a smaller number of photons. Since ideal adiabatic transfer occurs when the passage involves a single manifold, optimization is obtained when the photon leakage through the cavity is negligible during the atomic crossing, that is

$$T_{cross}^{-1} \gg \bar{n}\gamma \;, \tag{6}$$

where $\bar{n}$ is mean number of photons in the cavity. On the contrary, the technique of adiabatic passage is robust against the effects of spontaneous emission as, in principle, the excited atomic state $|e\rangle$ is never populated.

SCHEME FOR THE PROTECTION OF SUPERPOSITIONS OF NUMBER STATES

A clear example of how the quantum coherence associated to nonclassical superposition states of the radiation field inside the cavity is well preserved by the feedback scheme based on the adiabatic passage, is given by the study of the evolution of linear superpositions of two Fock number states

$$|\psi(0)\rangle = \alpha|n\rangle + \beta|m\rangle \;. \tag{7}$$

These states have not been experimentally generated in optical cavities yet, but there are now a number of proposals for their generation [9, 10].

For a quantitative characterization of how the feedback scheme is able to protect an initial pure state we study the fidelity $F(t)$

$$F(t) = \mathrm{Tr}\left\{\rho(0)\rho(t)\right\} \tag{8}$$

i.e., the overlap between the final and the initial state $\rho(0)$ after a time t. In general $0 \le F(t) \le 1$. For an initially pure state $|\psi(0)\rangle$, $F(t)$ is in fact the probability to find the system in the initial state at a later time. In the present case the fidelity function $F(t)$ can be easily evaluated analytically ($m > n$)

$$\begin{aligned} F(t) = {} & |\alpha|^4 e^{-n(1-\eta)\gamma t} + |\beta|^4 e^{-m(1-\eta)\gamma t} + 2|\alpha|^2|\beta|^2 e^{-\gamma t\left[(m+n)/2 - \eta\sqrt{nm}\right]} \\ & + |\alpha|^2|\beta|^2 e^{-n(1-\eta)\gamma t}\left(1 - e^{-(1-\eta)\gamma t}\right)^{m-n}\frac{m!}{(m-n)!n!} \end{aligned} \tag{9}$$

and when this expression is plotted for different values of η and compared with that in absence of feedback ($\eta = 0$), we see a significant increase of the "lifetime" of the state (7). This comparison is shown in Figure 1, which refers to the initial state $(|2\rangle + \sqrt{2}|4\rangle)/\sqrt{3}$ and where the notation is as follows: the full line refers to the feedback model in the ideal case $\eta = 1$; the dotted line to the feedback case with $\eta = 0.75$, small dashes refer to the case $\eta = 0.5$; big dashes refer to $\eta = 0.25$ and the dot-dashed line to the evolution in absence of feedback ($\eta = 0$).

A qualitative confirmation of how well an initial superposition of number states is protected by feedback is given by Figure 2: (a) shows the Wigner function of the initial superposition state $(|2\rangle + \sqrt{2}|4\rangle)/\sqrt{3}$ considered in Fig. 1; (b) shows the Wigner function of the same state evolved for a time $t = 1/\gamma$ in the presence of feedback ($\eta = 1$) and (c) the Wigner function of the same state again after a time $t = 1/\gamma$, but evolved in absence of feedback. This elapsed time is about three times the decoherence time $t_{dec} = (\gamma\bar{n})^{-1}$. As it is shown by (c), this means that after this time the state has lost

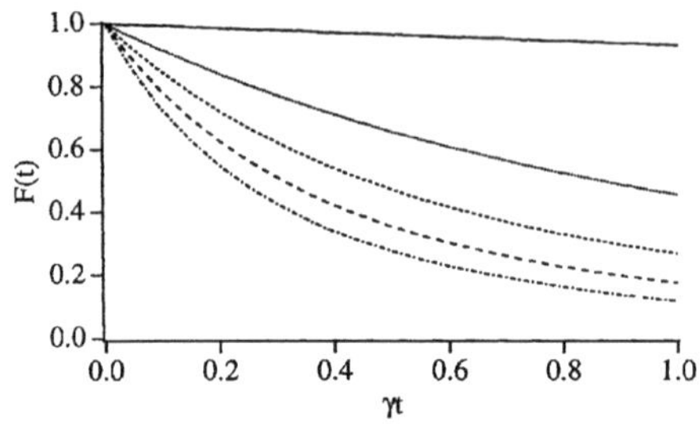

Figure 1. Time evolution of the fidelity $F(t)$ for the initial superposition state of two Fock states $(|2\rangle + \sqrt{2}|4\rangle)/\sqrt{3}$; full line: $\eta = 1$; dotted line: $\eta = 0.75$; small dashes: $\eta = 0.5$; big dashes: $\eta = 0.25$; dot-dashed line: evolution in absence of feedback ($\eta = 0$).

the oscillating part of the Wigner function associated to quantum interference; however this is no longer true in the presence of our feedback scheme: (b) shows in fact that, after $t \sim 3t_{dec}$, the state is still almost indistinguishable from the initial one and that the quantum wiggles of the Wigner function are still well visible.

SCHEME FOR THE PROTECTION OF QUBITS

Photon states are known to retain their phase coherence over considerable distances and for long times and for this reason high-Q optical cavities have been proposed as a promising example for the realization of simple quantum circuits for quantum information processing. To act as an information carrying quantum state, the electromagnetic fields must consist of a superposition of few distinguishable states. The most straightforward choice is to consider the superposition of the vacuum and the one photon state $\alpha|0\rangle + \beta|1\rangle$. However it is easy to understand that this is not convenient because any interaction coupling $|0\rangle$ and $|1\rangle$ also couples $|1\rangle$ with states with more photons and this leads to information losses. Moreover the vacuum state is not easy to observe because it cannot be distinguished from a failed detection of the one photon state. A more convenient and natural choice is *polarization coding*, i.e., using two degenerate polarized modes and qubits of the following form

$$|\psi\rangle = \left(\alpha a_+^\dagger + \beta a_-^\dagger\right)|0\rangle = \alpha|0,1\rangle + \beta|1,0\rangle \,, \tag{10}$$

in which one photon is shared by the two modes [11]. In fact this is a "natural" two-state system, in which the two basis states can be easily distinguished with polarization measurements; moreover they can be easily transformed into each other using polarizers. Polarization coding has been already employed in one of the few experimental realization of a quantum gate, the quantum phase gate realized at Caltech [12].

In the preceding section we have shown that the proposed feedback scheme is able to increase the "lifetime" of linear superpositions of Fock states. Therefore it is quite natural to look if our scheme can be used to protect qubits like those of the Caltech experiment, against the destructive effects of cavity damping. To be more specific, here we shall not be concerned with the protection of the quantum gate dynamics, but we shall focus on a simpler but still important problem: protecting an unknown input state for the longest possible time against decoherence. For this reason we shall not consider the two interacting fields, but a single frequency mode with a generic polarization, i.e.,

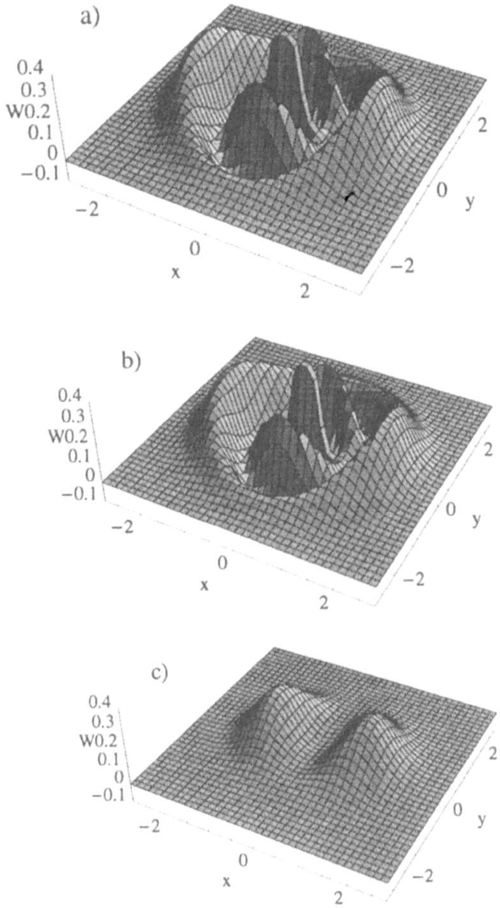

Figure 2. (a) Wigner function of the initial superposition state of two Fock states $(|2\rangle + \sqrt{2}|4\rangle)/\sqrt{3}$; (b) Wigner function of the same state evolved for a time $t = 1/\gamma$ ($t \simeq 3t_{dec}$), in the presence of feedback ($\eta = 1$); (c) Wigner function of the same state after a time $t = 1/\gamma$, but evolved in absence of feedback.

a single qubit. We shall consider a class of initial states more general than those of Eq. (10), i.e.,

$$|\psi\rangle = \alpha|n, m\rangle + \beta|m, n\rangle \ , \tag{11}$$

where $m + n$ photons are shared by the two polarized modes.

If we want to apply the adiabatic transfer feedback scheme described above for protecting qubits as those of Eq. (11), one has to consider a feedback loop for each polarized mode. This can be done using polarization-sensitive detectors which electronically control the polarization of the classical laser field and the initial state of the injected atoms. In fact one has to release in the cavity a left or right circularly polarized photon depending on which detector has fired and this can be easily achieved when the $|g_1\rangle \rightarrow |e\rangle$ and $|g_2\rangle \rightarrow |e\rangle$ transitions are characterized by opposite angular momentum difference $\Delta m_J = \pm 1$. In this case a left polarized photon, for example, is given back to the cavity with the adiabatic transition $|g_1\rangle \rightarrow |g_2\rangle$, while the right polarized one is released into the cavity through the reversed adiabatic transition $|g_2\rangle \rightarrow |g_1\rangle$ and the two possibilities are controlled by the polarization sensitive detectors.

Since the input state we seek to protect is unknown, the protection capabilities of the feedback scheme are better characterized by the minimum fidelity, i.e., the fidelity of Eq. (8) minimized over all possible initial states. This minimum fidelity can be easily evaluated by solving the master equation (4) for each polarized mode and one gets the following expression

$$F_{min}(t) = \frac{1}{2}\left(e^{-(1-\eta)\gamma t(n+m)} + e^{-\gamma t(n+m-2\eta\sqrt{nm})}\right) . \tag{12}$$

In the absence of feedback ($\eta = 0$), this expression becomes $F_{min}(t) = \exp\{-\gamma t(n+m)\}$ showing that in this case, the states most robust against cavity damping are those with the smallest number of photons, $m + n = 1$, i.e., the states of the form of Eq. (10). Moreover, in a typical quantum information processing situation, one has to consider small qubit "storage" times t with respect to γ^{-1} so to have reasonably small error probabilities in quantum information storage. Therefore the protection capability of an optical cavity with no feedback applied is described by

$$F_{min}(t) = 1 - \gamma t . \tag{13}$$

If we now consider the situation in the presence of feedback (Eq. (12)), the best protected states for a given nonzero efficiency η, may be different from the states with only one photon, $\alpha|0,1\rangle + \beta|1,0\rangle$, and they depend upon the explicit value of the feedback efficiency η. For the determination of the optimal qubit of the form of (11) (i.e., the optimal values for m and n), one has to minimize the deviation from the perfect protection condition $F(t) = 1$. For $\gamma t \ll 1$ one gets

$$\begin{aligned} &\min_{m\neq n}\left[(2-\eta)(n+m) - 2\eta\sqrt{nm}\right] \\ &= \min_{m\neq n}\left[(\sqrt{m}-\sqrt{n})^2 + (1-\eta)(\sqrt{m}+\sqrt{n})^2\right] \\ &= \min_{n\geq 0,\, p\geq 1}\left[\frac{p^2}{(\sqrt{n+p}+\sqrt{n})^2} + (1-\eta)(\sqrt{n+p}+\sqrt{n})^2\right] , \end{aligned} \tag{14}$$

where $p = m - n$. From these expression it can be easily seen that one has to choose $p = 1$, and therefore the optimal qubits are those of the form

$$|\psi\rangle = \alpha|n_{opt}, n_{opt}+1\rangle + \beta|n_{opt}+1, n_{opt}\rangle , \tag{15}$$

where n_{opt} is determined by the minimization condition

$$\min_{n\geq 0}\left[\frac{1}{(\sqrt{n+1}+\sqrt{n})^2} + (1-\eta)(\sqrt{n+1}+\sqrt{n})^2\right] . \tag{16}$$

As long as

$$\eta \leq 2/(1+\sqrt{2}) \simeq 0.83 , \tag{17}$$

one has $n_{opt} = 0$ and therefore the situation is similar to that of the no-feedback case: the states of the form (10) are the best protected states and the corresponding minimum fidelity is given by

$$F_{min}(t) = \frac{1}{2}\left(e^{-(1-\eta)\gamma t} + e^{-\gamma t}\right) \simeq 1 - \gamma t\left(1 - \frac{\eta}{2}\right) . \tag{18}$$

In this case, feedback leads to a very poor qubit protection with respect to the no-feedback case and therefore our scheme proves to be practically useless for the protection of single photon qubits of (10) employed in the Caltech experiment of Ref. [12].

However, when the feedback efficiency η becomes larger than 0.83, the situation can improve considerably. In fact n_{opt} becomes nonzero and can become very large in the limit $\eta \to 1$, and in this case the minimum fidelity decays very slowly. To be more specific, n_{opt} is approximately given by the condition

$$\left(\sqrt{n_{opt}+1}+\sqrt{n_{opt}}\right)^2 = (1-\eta)^{-1/2} \tag{19}$$

and the corresponding small time behavior of $F_{min}(t)$ is given by

$$F_{min}(t) \simeq 1 - \frac{\gamma t}{\left(\sqrt{n_{opt}+1}+\sqrt{n_{opt}}\right)^2} \simeq 1 - \gamma t\sqrt{1-\eta}\,. \tag{20}$$

This means that in the limit of a feedback efficiency very close to one, it becomes convenient to work with a large number of photons per mode, since in this limit the probability of errors in the storage of quantum information can be made very small. This can be easily understood from Eq. (4), because in this limit the square-root of phase diffusion term prevails in the master equation and its quantum state protection capabilities improve for increasing photon number (see Eq. (5)). In the ideal case $\eta = 1$, n_{opt} becomes infinite and therefore the minimum fidelity can remain arbitrarily close to one. It is convenient to work with the largest possible number of photons, that is,

$$|\psi\rangle = \alpha|n, n+1\rangle + \beta|n+1, n\rangle \qquad n \gg 1 \tag{21}$$

and the corresponding minimum fidelity is

$$F_{min}(t) \simeq \frac{1}{2}\left(1+e^{-\gamma t/4n}\right) \simeq 1 - \frac{\gamma t}{8n}\,. \tag{22}$$

The time evolution of the minimum fidelity as a function of the feedback efficiency in the polarization coding case is summarized in Fig. 3.

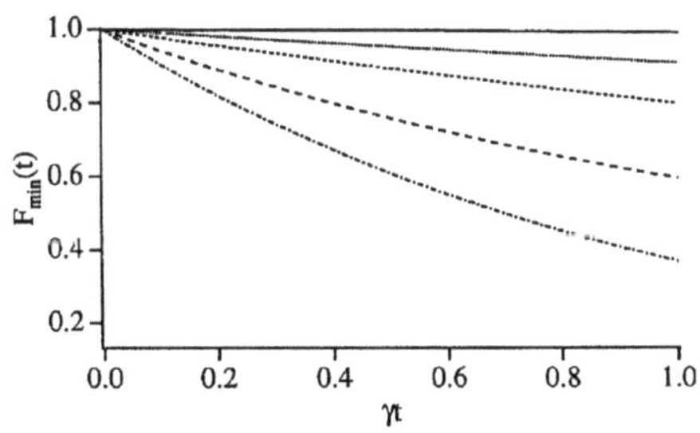

Figure 3. Time evolution of the minimum fidelity $F_{min}(t)$ for the polarization coding case; full line: $\eta = 1$ (Eq. (22) with $n = 10$); dotted line: $\eta = 0.99$ (Eq. (20) with $n_{opt} = 2$); small dashes: $\eta = 0.95$ (Eq. (20) with $n_{opt} = 1$); big dashes: $\eta = 0.8$ (Eq. (18)); dot-dashed line: evolution in absence of feedback ($\eta = 0$).

The feedback method proposed here to deal with decoherence in quantum information processing is different from most of the proposals made in this research field, which are based on the so called quantum error correction codes [13], which are a way to use *software* to preserve linear superposition states.

In our case, feedback allows a physical control of decoherence, through a continuous monitoring and eventual correction of the dynamics and in this sense our approach is similar in spirit to the approach of Ref. [14, 15]. The present feedback scheme is not very useful in the case of one-photon qubits (10) of the quantum phase gate experiment of Ref. [12]; however it predicts a very good decoherence control in the case of high feedback efficiency $\eta > 0.83$ and for larger photon numbers (see Eq. (15)). It is very difficult to achieve these experimental conditions with the present technology, but our scheme could become very promising in the future.

REFERENCES

1. C.H. Bennett, in *Quantum Communication, Computing and Measurement*, edited by O. Hirota, A.S. Holevo and C.M. Caves (Plenum Press, New York, 1997), pag. 25.
2. A. Ekert and R. Josza, Rev. Mod. Phys **68**, 733 (1996).
3. E. Schrödinger, *Naturwissenschaften* **23**, 807, 823, 844 (1935).
4. W.H. Zurek, Phys. Today **44**(10), 36 (1991), and references therein.
5. M. Brune, E. Hagley, J. Dreyer, X. Maitre, A. Maali, C. Wunderlich, J.M. Raimond and S. Haroche, Phys. Rev. Lett. **77**, 4887 (1996).
6. D. Vitali, P. Tombesi, G.J. Milburn, Phys. Rev. Lett. **79** 2442 (1997).
7. H.M. Wiseman and G.J. Milburn, Phys. Rev. Lett. **70**, 548 (1993); Phys. Rev. A **49**, 1350 (1994); H.M. Wiseman, Phys. Rev. A **49**, 2133 (1994).
8. A.S. Parkins, P. Marte, P. Zoller, O. Carnal and H.J. Kimble, Phys. Rev. A **51**, 1578 (1995) and references therein.
9. K. Vogel, V.M. Akulin and W.P. Schleich, Phys. Rev. Lett. **71**, 1816 (1993).
10. C.K. Law and J.H. Eberly, Phys. Rev. Lett. **76**, 1055 (1996).
11. S. Stenholm, Opt. Comm. **123**, 287 (1996); P. Törmä and S. Stenholm, Phys. Rev. A **54**, 4701 (1996).
12. Q.A. Turchette, C.J. Hood, W. Lange, H. Mabuchi and H.J. Kimble, Phys. Rev. Lett. **75**, 4710 (1995).
13. E. Knill and R. Laflamme, Phys. Rev. A **55**, 900 (1997) and references therein.
14. T. Pellizzari, S.A. Gardiner, J.I. Cirac, and P. Zoller, Phys. Rev. Lett. **75**, 3788 (1995).
15. H. Mabuchi and P. Zoller, Phys. Rev. Lett, **76**, 3108 (1996).

CONTROL OF DYNAMICAL LOCALIZATION BY ADDITIONAL QUANTUM DEGREES

K. Riedel,[1] P. Törmä,[1,2] V. Savichev,[1] and W. P. Schleich[1]

[1]Abteilung für Quantenphysik, Universität Ulm, 89069 Ulm, Germany
[2]Institut für Theoretische Physik, Universität Innsbruck, Technikerstraße 25, 6020 Innsbruck, Austria

The phenomenon of localization manifests itself in many quantum mechanical systems ranging from the localization of light in a random medium via Anderson localization of an electronic wave to the motion of atoms in time-dependent laser fields. In all these cases the underlying classical system is chaotic and shows diffusion as a function of time. In contrast, the quantum mechanical counterpart has a localized wave function whose width is governed by the classical diffusion and Planck's constant. In this paper we show that there exists an additional quantum parameter that controls the localization length. In the system of a two-level ion stored in a Paul trap and interacting with a standing wave it is the detuning between the transition frequency and the laser field. We also discuss the effect of decoherence in form of spontaneous emission.

INTRODUCTION

Dynamical localization is one of the most striking finger prints of quantum chaos[1, 2]. It is related to Anderson localization[3] of electronic waves in one-dimensional disordered solids, where randomness of the lattice-site energies leads to classical diffusion. The quantum suppression of this diffusion manifests itself in the localization of eigenstates. In systems capable of showing dynamical localization no real randomness is present; the dynamics is deterministic but chaotic due to the time periodicity and nonlinearity of the potential. Quantum localization is associated with the Floquet states of the system since there are no stationary eigenstates.

The periodically kicked rotator[1], described by the dimensionless Hamiltonian

$$H = \frac{p^2}{2} + \lambda \cos\theta \sum_n \delta(t - nT), \tag{1}$$

is the paradigm of dynamical localization. Here Delta-function-type kicks with period T cause classical diffusion in the action variable p. The interaction strength depends on the angle variable θ and on the control parameter λ. When the motion of the rotator is quantized one observes localization in p. It is interesting that even then the kicks are still 'classical'.

Quantum Communication, Computing, and Measurement 2
Edited by Kumar et al., Kluwer Academic / Plenum Publishers, New York, 2000.

Other systems closely related to the kicked rotator are atoms interacting with a phase-modulated standing wave[4] and an ion[5] stored in an explicitly time dependent Paul trap[6] and interacting with a standing wave. In both examples the internal quantum structure of the atom has been adiabatically eliminated in the atom-light interaction. In the present paper we ask what happens when the kicks have also a quantum character. We therefore take into account the internal stucture in the two-level ion. We observe localization also in this case, but in addition discover another quantum effect: Oscillations in the localization length as a function of the atom-laser detuning. These oscillations do not appear in the classical diffusion rate, unlike in the case of atoms in a phase modulated standing wave, where there is a correspondence between oscillations in the quantum and classical momentum distribution[7, 8, 9].

In our previous work[5] we have already emphasized the importance of the micromotion in the Paul trap for the phenomenon of dynamical localization. For the qualitative description of the localization the reference frequency[10] of the time-dependent oscillator was essential. The reference frequency carries information about the micromotion. In the present paper we show that also the secular frequency, which describes the time-averaged motion in the trap, plays a role in localization: The oscillations in the localization length as a function of the detuning are determined by the secular frequency.

Dynamical localization has been experimentally observed only in two types of systems, namely Rydberg atoms in microwave fields[11] and atoms moving in an amplitude and phase modulated standing wave[4]. We have proposed[5] that dynamical localization can also be observed in the center-of-mass motion of an ion stored in a Paul trap. This system is very promising since nowadays ions in traps can be cooled down to the quantum regime, their dynamics can be controlled using laser fields and the quantum state can be reconstructed from the resonance fluorescence[12, 13, 14].

Dynamical localization is a coherence effect and thus its experimental verification requires a system that is not too sensitive to noise[9, 15, 16, 17]. In this paper we simulate the effect of spontaneous emission on the dynamical localization in the Paul trap. We show that indeed the drastic difference between the classical and quantum behavior is preserved in the limit of far-detuning.

We first describe the system and consider the localization for finite detunings. We then discuss the oscillations in the localization length. The effect of decoherence is considered before turning to the conclusions.

THE SYSTEM

We consider the standard Paul trap set-up realized[12, 13, 14] experimentally in many labs: A standing electromagnetic wave of frequency ω_L and wave number k aligned along the x-axis couples the internal states of a single two-level ion of mass m to the center-of-mass motion. The resulting dynamics of the ion follows from the time-dependent Schrödinger equation with the Hamiltonian

$$\tilde{\hat{H}} = \frac{\tilde{\hat{p}}^2}{2m} + \frac{1}{2}\frac{m\omega^2}{4}\left[a + 2q\cos\left(\omega\tilde{t}\right)\right]\tilde{\hat{x}}^2 + \frac{1}{2}\hbar\omega_0\hat{\sigma}_z + \hbar\tilde{\Omega}_0\hat{\sigma}_x\cos(k\tilde{\hat{x}} + \phi)\cos(\omega_L\tilde{t}). \quad (2)$$

Here the parameters a and q denote[6] the DC and AC voltages applied to the trap and the quantities ω, ω_0 and $\tilde{\Omega}_0$ correspond to the frequency of the AC field, the atomic transition frequency and the Rabi-frequency, respectively, and ϕ is the phase of the standing wave. We describe the internal structure of the ion by the Pauli matrices $\hat{\sigma}_x$, $\hat{\sigma}_y$ and $\hat{\sigma}_z$.

We introduce the interaction picture and make the rotating wave approximation. In dimensionless variables of position $\hat{x} \equiv k\hat{\tilde{x}}$, time $t \equiv \omega\tilde{t}/2$, momentum $\hat{p} \equiv 2k\hat{\tilde{p}}/(m\omega)$, and with the coupling $\Omega_0 \equiv \tilde{\Omega}_0/\omega$ and detuning $\Delta \equiv (\omega_a - \omega_L)/\omega$ the dimensionless Hamiltonian then reads

$$\hat{H} \equiv \frac{4k^2}{m\omega^2}\hat{\tilde{H}} = \frac{1}{2}\hat{p}^2 + \frac{1}{2}(a + 2q\cos 2t)\hat{x}^2 - k\!\!\!{}^{-}\Delta\hat{\sigma}_z + k\!\!\!{}^{-}\Omega_0\cos(\hat{x}+\phi)\hat{\sigma}_x. \tag{3}$$

The effective Planck constant $k\!\!\!{}^{-} \equiv 2k^2\hbar/(m\omega)$ is consistent with the commutation relation

$$[\hat{x},\hat{p}] = \frac{2k^2}{m\omega}[\hat{\tilde{x}},\hat{\tilde{p}}] = \frac{2k^2}{m\omega}i\hbar \equiv ik\!\!\!{}^{-},$$

and the dynamics of the ion is described by the Schrödinger equation[18]

$$ik\!\!\!{}^{-}\frac{\partial}{\partial t}\begin{pmatrix}\psi_e(x,t)\\ \psi_g(x,t)\end{pmatrix} = \hat{H}\begin{pmatrix}\psi_e(x,t)\\ \psi_g(x,t)\end{pmatrix}. \tag{4}$$

Here the indices e and g refer to excited and ground states, respectively.

In order to investigate localization we calculate the time evolution of a Gaussian wave-packet centered initially at the origin of $x-p$ phase space. For the integration we use the split-operator method[19] with a grid of 8192 points. We control numerical errors using an adaptive time step-size algorithm[20].

To make a comparison between the quantum and classical dynamics we calculate 4096 trajectories of a classical Gaussian ensemble centered initially at the origin and having the same widths in the position and momentum as the quantum wavepacket[21]. We use the equations[22]

$$\dot{x} = p \tag{5}$$
$$\dot{p} = -(a + 2q\cos(2t))x + k\!\!\!{}^{-}\Omega_0\sin(x+\phi)r_1 \tag{6}$$

for the center-of-mass position $x(t) \equiv \mathrm{Tr}[\hat{x}\hat{\rho}(t)]$ and momentum $p(t) \equiv \mathrm{Tr}[\hat{p}\hat{\rho}(t)]$ of the ion and

$$\dot{r}_1 = -2\Delta r_2 \tag{7}$$
$$\dot{r}_2 = 2\Delta r_1 + 2\Omega_0\cos(x+\phi)r_3 \tag{8}$$
$$\dot{r}_3 = -2\Omega_0\cos(x+\phi)r_2 \tag{9}$$

for the components $r_1(t) \equiv \mathrm{Tr}[\hat{\sigma}_x\hat{\rho}(t)]$, $r_2(t) \equiv \mathrm{Tr}[\hat{\sigma}_y\hat{\rho}(t)]$ and $r_3(t) \equiv \mathrm{Tr}[\hat{\sigma}_z\hat{\rho}(t)]$ of the Bloch vector. Here $\hat{\rho}$ is the density operator of the system. Throughout the paper we choose the phase ϕ of the standing wave to be zero.

The effect of spontaneous emission is simulated by the quantum Monte Carlo method[23] using the effective non-hermitian Hamiltonian $\hat{H}_{eff} \equiv \hat{H} - ik\!\!\!{}^{-}\gamma/2\hat{\sigma}_+\hat{\sigma}_-$, where γ is the spontaneous decay rate scaled by ω. The times when a spontaneous emission takes place are chosen at random. Then the wave-function is projected on the ground-state and renormalized $|\Psi\rangle \to \hat{\sigma}_-|\Psi\rangle/\langle\Psi|\Psi\rangle$. The recoil $p \in [-k\!\!\!{}^{-}, k\!\!\!{}^{-}]$ is taken into account by choosing p randomly according to the probability distribution[24] $N(p) = 3\left(1 + (p/k\!\!\!{}^{-})^2\right)/(8k\!\!\!{}^{-})$. When the results given by single runs are averaged, one obtains the same result as given by a master equation[25].

DYNAMICAL LOCALIZATION

Dynamical localization occurs [5, 26, 27] in our system both in position and momentum (c.f. also Fig.3). This is in contrast to the kicked rotator [1, 4], where localization appears only in the momentum. The Floquet theory of the Paul trap[10] provides us with a self-consistent explanation of the phenomena we observe. The motion in a time dependent quadratic potential is described by the solution $\epsilon(t)$ of the classical Mathieu equation $\ddot{\epsilon}(t) + [a + 2q\cos 2t]\,\epsilon(t) = 0$, subjected to the initial conditions $\epsilon(0) = 1$ and $\dot{\epsilon}(0) = i\omega_r$, where ω_r is a reference frequency. The solution becomes the Floquet solution

$$\epsilon(t) \equiv e^{i\omega_s t}\phi(t) \equiv e^{i\omega_s t} \sum_{n=-\infty}^{\infty} c_n e^{i2nt}, \tag{10}$$

that is a periodic function $\phi(t)$ multiplied by a time-dependent phase, when the reference frequency is chosen to be

$$\omega_r = \omega_s + \sum_{n=-\infty}^{\infty} 2nc_n. \tag{11}$$

Here the so-called secular frequency ω_s determined by the trap parameters a and q describes the time-averaged motion in the trap.

In the explanation of dynamical localization the reference frequency plays the main role, since it contains information about the micro-motion and the time-periodicity of the potential. They are responsible for classical chaos and dynamical localization. The secular frequency, being related to a time-independent device, is insufficient to describe the dynamical phenomena.

The classical diffusion becomes apparent when we transform the Hamiltonian of the system utilizing the Floquet solution (10)–(11) of the Mathieu equation. The time, position, and momentum variables are scaled with $|\epsilon(t)|$ as follows

$$\tau(t) = \int_0^t \frac{dt'}{|\epsilon(t')|^2}, \quad \chi(\tau) = \frac{x(t)}{|\epsilon(t)|}, \quad \Pi(\tau) = \frac{d\chi(\tau)}{d\tau}. \tag{12}$$

When we use the action-angle variables I and θ of the harmonic oscillator with frequency ω_r defined by

$$\chi = \sqrt{\frac{2I}{\omega_r}}\sin\theta \qquad \Pi = \sqrt{2I\omega_r}\cos\theta, \tag{13}$$

the equations of motion for the center-of-mass motion described by the action I and the angle θ become

$$\dot{I} = \hbar\Omega_0\sqrt{\frac{2I}{\omega_r}}\cos\theta|\epsilon(t)|^3 \sin\left(|\epsilon(t)|\sqrt{\frac{2I}{\omega_r}}\sin\theta\right) r_1 \tag{14}$$

and
$$\dot{\theta} = \omega_r - \frac{\hbar\Omega_0\sin\theta}{\sqrt{2I\omega_r}}|\epsilon(t)|^3 \sin\left(|\epsilon(t)|\sqrt{\frac{2I}{\omega_r}}\sin\theta\right) r_1. \tag{15}$$

Here the dot means differentiation with respect to the new time τ.

Hence the time evolution of the angle θ is governed by a constant frequency ω_r, like for a harmonic oscillator, plus a time-dependent perturbation. The dynamics described by these equations leads[28, 29] to fast diffusion of the angle variable θ and slow diffusion of the action variable I. This in turn causes diffusion in both position and momentum, which is visible in our numerical results. Furthermore, if diffusion in I is approximately linear in time and proportional to a diffusion constant D, we have

$$\Delta\chi_{classical} = D_\chi\tau^{\alpha_\chi} = \sqrt{D/\omega_r}\tau^\alpha \tag{16}$$

$$\Delta\Pi_{classical} = D_\Pi\tau^{\alpha_\Pi} = \sqrt{D\omega_r}\tau^\alpha, \tag{17}$$

where $\alpha = 1/4$. This leads to

$$\frac{\Delta\Pi_{classical}}{\Delta\chi_{classical}} = \omega_r. \tag{18}$$

Quantum mechanically the action variable corresponds to the vibrational quantum number of the reference oscillator, which should be the variable showing dynamical localization — this leads to localization in both position and momentum, which we observed. The widths of the localized distributions should consequently be related to the localization length l via the simple relations

$$\Delta\chi^2 = \frac{\hbar}{\omega_r}\left(l + \frac{1}{2}\right) \quad ; \quad \Delta\Pi^2 = \hbar\omega_r\left(l + \frac{1}{2}\right), \tag{19}$$

which implies

$$\frac{\Delta\Pi}{\Delta\chi} = \omega_r. \tag{20}$$

We have numerically confirmed the relations (18) and (20) for the variables (x, p) rather than χ and Π. We observe that the values $D_p/D_x \simeq 0.475$ and $\Delta p/\Delta x \simeq 0.415$ are constant when the detuning Δ is varied from zero to ten and $\Omega = 2.24$. Moreover, these ratios are indeed close to the reference frequency $\omega_r = 0.436$ and not to the secular frequency $\omega_s = 0.293$. For detunings in the range $\Delta \in \{0.1, 2.0\}$ we found that $\alpha_x = 0.26$ and $\alpha_p = 0.23$ with the standard deviation of 0.01.

Quantum oscillations in the localization length

The widths of the quantum mechanical position and momentum distributions, that is the localization lengths, show characteristic oscillations as a function of the detuning Δ. These oscillations do not appear in the widths of the corresponding classical distributions as shown in Fig.1 for the position distributions.

The classical curve decays monotonously. This is easily understood when we recall that the classical width is determined by the classical diffusion constant, which is proportional to the perturbing potential caused by the laser field. Moreover, we note that in the limiting case of large detuning the effective potential becomes[18] proportional to the effective coupling constant Ω_0^2/Δ. Hence, as Δ increases, the perturbation is smaller and the diffusion slower.

The quantum widths show a distinctly different behavior, in particular two sharp minima and maxima for $\Delta < 1$. The first minimum is around $\Delta = 0$ and the next is at $\Delta \approx \omega_s$, where $\omega_s = 0.29$ is the secular frequency. The origin of these resonances becomes clear, when we consider the quantum equations and write the initial Hamiltonian in the interaction picture where the dynamics of the trap as well as the internal state energy are transformed away by unitary transformations. The interaction Hamiltonian describing transitions between the trap states becomes[30]

$$H_{int}(t) = \sum_{n=0}^{\infty} \sum_{k=-[n/2]}^{\infty} \sum_{l=-\infty}^{\infty} \hbar\omega_l^{(n,n+2k)} e^{2i(l-k\omega_s+\Delta)t} \hat{\sigma}_+ |n\rangle\langle n+2k| + H.c., \tag{21}$$

with $\eta \equiv [\hbar/(2\omega_r)]^{1/2}$ and

$$\omega_l^{(n,n+2k)} \equiv \Omega_0 \sqrt{\frac{n!}{(n+2k)!}} [i\eta]^{2k} \frac{1}{\pi} \int_{-\pi/2}^{\pi/2} dt [\phi^*(t)]^{2k} e^{-\frac{1}{2}\eta^2|\phi(t)|^2} L_n^{2k}(\eta^2|\phi(t)|^2) e^{-2ilt}, \tag{22}$$

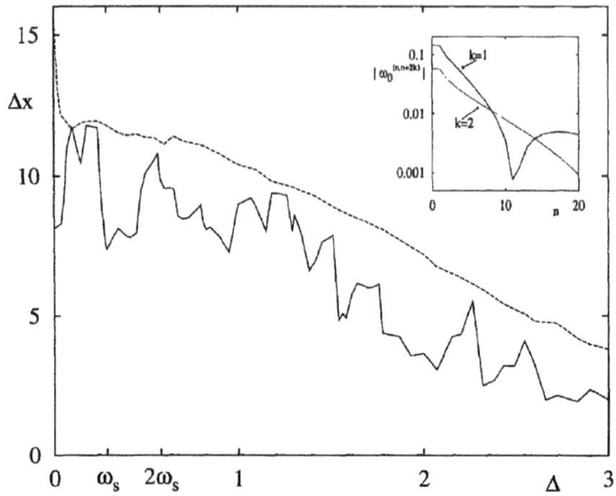

Figure 1. The widths Δx of the classical (dashed line) and quantum mechanical (solid line) position distributions as a function of the detuning Δ averaged from $t = 200\pi$ to 500π. The classical curve decays monotonously as a function of detuning. In contrast, the quantum one displays a lot of structure. The inset displays the transition amplitudes $|\omega_0^{(n,n+2)}|$ and $|\omega_0^{(n,n+4)}|$ between the reference oscillator states as a function of the oscillator excitation number n. Similar curves arise for the momentum variable.

Here $|n\rangle$ denotes the n-th energy eigenstate of the time independent reference oscillator.

From the term $e^{2i(l-k\omega_s+\Delta)t}$ we can expect resonance effects, when Δ obeys the condition $l - k\omega_s + \Delta = 0$. For $l = 0, k = 1$ we have $\Delta = \omega_s$, which corresponds to the two-phonon resonance transitions. One could expect enhancement of the quantum diffusion rate on resonance. Instead we have a deep minimum. In order to understand this we have calculated the characteristic frequencies $\omega_l^{(n,n+2k)}$ for $l = 0$ and $k = 1, 2$ at different vibrational quantum numbers n. From the inset of Fig.1 we note that $|\omega_0^{(n,n+2)}|$ has a deep minimum around $n = 10$, which explains suppression of the diffusion over the vibrational states. For $|\omega_0^{(n,n+4)}|$ we observe a normal behavior, this causes the quantum diffusion at $\Delta = 2\omega_s$ (resonance for $k = 2$) to turn back to its maximum near the classical value.

Oscillations in the momentum width appear also in another example of a system showing classical chaos and quantum localization, namely that of atoms in a phase-modulated standing wave[7, 8]. There the oscillations, however, are of classical origin and appear both in the classical and quantum widths. This is easy to understand when we recall that for the kicked rotator the localization length is related to the classical diffusion constant D and the Planck's constant by $l \sim D/\hbar^2$. Thus any oscillations in the rate of the classical diffusion reflect themselves in the quantum localization length as well. However the oscillations we see here are a quantum effect. This has remarkable consequences as we show now.

In Fig.2 we show the localization lengths obtained from the quantum widths as well as the estimate based on Eqs.(16)–(19) and the assumption $l \sim t_*$. Here t_* is the quantum break time when the classical and quantum results start to differ. The estimate gives from the position diffusion $l \sim (D_x\omega_r/\hbar)^{1/(1-\alpha_x)}$ or from the momentum diffusion $l \sim (D_p/(\hbar\omega_r))^{1/(1-\alpha_x)}$. In our case these estimates based on classical diffusion are not able to completely describe the quantum localization; they give some order of magnitude estimate, but do not explain the prominent oscillations in the quantum localization length. Unlike in the previous examples of dynamical localization with only one quantum degree of freedom, we have now another one, the two-level structure, which leaves its traces on the quantum localization length but not on the classical diffusion rate.

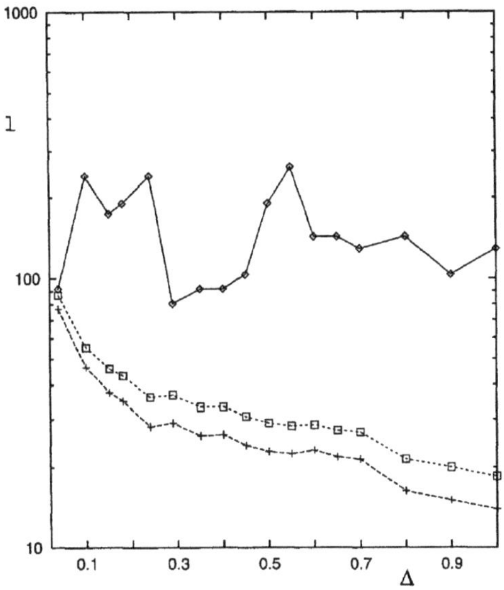

Figure 2. The localization length (full line) and the estimates given by $l \sim D/k\!\!\!^-{}^2$ from the position and momentum distributions (dashed and dotted lines, respectively) as a function of the detuning Δ.

THE EFFECT OF DECOHERENCE

In this section we consider the effect of noise on our system by simulating spontaneous emission using the quantum Monte Carlo technique[23]. We show that in the limit of large detuning the effect of noise is negligible, the phenomenon of dynamical localization remains and can be experimentally observed.

In Fig.3 we show simulations with a realistic rate of spontaneous emission $\gamma = 2$ corresponding to the decoherence rate[31] of the $S \rightarrow P$ transition at 19.4 MHz in ${}^9Be^+$ and for a detuning[5] $\Delta = 1000$ corresponding to 10 GHz. The small oscillations in the localized Δx and Δp are destroyed, but the main phenomenon, the substantial quantum suppression of classical diffusion is still visible. Note that the small difference between the results with and without spontaneous emission in Fig.3 is of the same order of magnitude that was found for the experiments on localization of atomic momentum transfer from a phase modulated standing wave[16]. Since dynamical localization appears for a large range of parameters[5], there is a lot of room for optimizing the parameters. Thus the phenomenon of dynamical localization in a Paul trap could be experimentally observed. The results of Ref.[16] also support this.

We have shown[27] that on resonance and for the above mentioned parameters, dynamical localization is completely destroyed, as it should be in presence of strong decoherence. In order to observe the oscillations in the localization length discussed in the previous section, we suggest a configuration where decoherence is weaker even

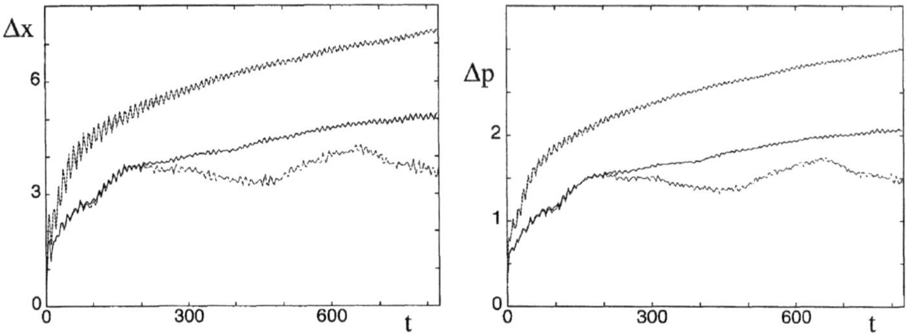

Figure 3. The position and momentum widths as a function of time in the case of the detuning $\Delta = 1000$. The quantum mechanical curve with spontaneous emission (middle line) lies above the corresponding curve without spontaneous emission (bottom line) but well below the classical result (top line). Spontaneous emission does not destroy the localization. An average over 49 runs has been made.

for small detunings: For example, we can envision a different set of parameters in the present system, or a different system such as a two-level Raman transition.

CONCLUSIONS

Conventional wisdom knows basically two ways of controlling dynamical localization: Namely through the classical diffusion and through the quantized nature of the variable showing localization. This expresses itself in the estimate $l \sim D/\hbar^2$ for the localization length. The classical diffusion coefficient D is determined by the perturbing potential; the coupling to this potential is called the control parameter, since it is a direct way to control the localization length. The scaled Planck constant $\hbar$ describes how important the quantization of the system is with respect to the perturbation. In the present paper we have shown that, when the perturbing potential has a quantum character as well, the relation $l \sim D/\hbar^2$ is not exactly true any more: We observe oscillations in the localization length which do not result from the classical diffusion rate. We therefore call the parameter determining the oscillations a quantum control parameter.

We have considered a two-level ion stored in an explicitly time dependent Paul trap and interacting with a standing wave. We remove the time dependence from the center-of-mass motion using Floquet theory and therefore arrive at a static Floquet reference oscillator affected by an explicitly time-dependent perturbing potential. We observe classical diffusion and quantum localization in the reference oscillator action variable, leading to localization in both position and momentum. When we vary the atom-field detuning Δ, we find oscillations in the quantum position and momentum widths, which are not present in the classical ones. These resonances appear when the quantum control parameter Δ is an integer multiple of the secular frequency plus an integer multiple of the AC trap frequency.

We have simulated the effect of spontaneous emission on the system and showed, that for large enough detunings it becomes negligible. The quantum interference effect of dynamical localization is preserved.

Acknowledgements We thank B. Kneer and M. El Ghafar for many discussions and useful comments. One of us (W. P. S.) expresses his deep appreciation to P. Kumar for organizing a splendid conference. P. T. and V. S. acknowledge the support of the Deutsche Forschungsgemeinschaft. P. T. also acknowledges the support by the TMR Programme of the European Commission (ERBFMBICT983061). We thank the Rechenzentrum Ulm and the Rechenzentrum Karlsruhe for their technical support.

REFERENCES

1. F. Haake, *Quantum Signatures of Chaos* (Springer-Verlag, Berlin 1992).
2. *Quantum Chaos*, Eds. G. Casati and B. Chirikov (Cambridge University Press, Cambridge 1995).
3. P. W. Anderson, Phys. Rev. **109**, 1492 (1958).
4. F. L. Moore, J. C. Robinson, C. Bharucha, P. E. Williams and M. G. Raizen, Phys. Rev. Lett. **73**, 2974 (1994); theoretical proposal in R. Graham, M. Schlautmann, and P. Zoller, Phys. Rev. A **45**, R19 (1992).
5. M. El Ghafar, P. Törmä, V. Savichev, E. Mayr, A. Zeiler, and W. P. Schleich, Phys. Rev. Lett. **78**, 4181 (1997).
6. W. Paul, Rev. Mod. Phys. **62**, 531 (1990).
7. J. C. Robinson, C. Bharucha, F. L. Moore, R. Jahnke, G. A. Georgaki, M. G. Raizen and B. Sundaram, Phys. Rev. Lett. **74**, 3963 (1995).
8. P. J. Bardroff, I. Bialynicki-Birula, D. S. Krähmer, G. Kurizki, E. Mayr, P. Stifter, and W. P. Schleich, Phys. Rev. Lett. **74**, 3959 (1995).
9. In [R. Graham and S. Miyazaki, Phys. Rev. A **53**, 2683 (1996)] the problem of atoms moving in a phase modulated standing wave was considered without the adiabatic elimination of the upper state. The emphasis of that paper was, however, on the study of spontaneous emission, not on the effect of the additional quantum degree of freedom.
10. R. J. Glauber, *Laser Manipulation of Atoms and Ions*, Proc. Int. School of Physics 'Enrico Fermi' Course 118, Eds. E. Arimondo et al. (North Holland, Amsterdam 1992); see also G. Schrade, P. J. Bardroff, R. J. Glauber, C. Leichtle, V. Yakovlev and W. P. Schleich, Appl. Phys. B **64**, 181 (1997).
11. For a review see for example P. M. Koch and K. A. H. van Leeuwen, Phys. Rep. **255**, 289 (1995); G. Casati, Phys. Rev. A **45**, 7670 (1992).
12. D. M. Meekhof, C. Monroe, B. E. King, W. M. Itano and D. J. Wineland, Phys. Rev. Lett. **76**, 1796 (1996); C. Monroe, D. M. Meekhof, B. E. King and D. J. Wineland, Science **272**, 1131 (1996); D. Leibfried, D. M. Meekhof, B. E. King, C. Monroe, W. M. Itano, and D. J. Wineland, Phys. Rev. Lett. **77**, 4281 (1996).
13. G. Birkl, J. A. Yeazell, R. Rückerl, and H. Walther, Europhys. Lett. **27**, 197 (1994); H. Katori, S. Schlipf, and H. Walther, Phys. Rev. Lett. **79**, 2221 (1997).
14. B. Appasamy, Y. Stalgies, and P. E. Toschek, Phys. Rev. Lett. **80**, 2805 (1998)
15. M. Arndt, A. Buchleitner, R. N. Mantegna and H. Walther, Phys. Rev. Lett. **67**, 2435 (1991); F. L. Moore, J. C. Robinson, C. Bharucha, B. Sundaram and M. G. Raizen, Phys. Rev. Lett. **75**, 4598 (1995); R. Blümel, A. Buchleitner, R. Graham, L. Sirko, U. Smilansky, and H. Walther, Phys. Rev. A **44**, 4521 (1991).
16. P. Goetsch and R. Graham, Phys. Rev. A **54**, 5345 (1996).
17. For an experimental study of the effect of noise and dissipation on dynamical localization see H. Ammann, R. Gray, I. Shvarchuck, and N. Christensen, Phys. Rev. Lett. **80**, 4111 (1998); B. G. Klappauf, W. H. Oskay, D. A. Steck, and M. G. Raizen, Phys. Rev. Lett. **81**, 1203 (1998).
18. A. P. Kazantsev, G. I. Surdutovich and V. P. Yakovlev, *Mechanical Action of Light on Atoms* (World Scientific, Singapore 1990).
19. M. D. Feit, J. A. Fleck, Jr. and A. Steiger, J. of Comput. Phys. **47**, 412 (1982).
20. K. Riedel, *Quanteneffekte in Paul-Fallen, Dynamische Lokalisierung und Dekohärenz*, Diplomarbeit, Universität Ulm, 1997.
21. For more information about the classical dynamics of this system see R. Chacón and J. I. Cirac, Phys. Rev. A **51**, 4900 (1994); M. El Ghafar, E. Mayr, V. Savichev, P. Törmä, A. Zeiler, and W. P. Schleich, J. Mod. Opt. **44**, 1985 (1997).
22. In deriving these equations, we have assumed that the quantum mechanical position distribution is very narrow and localized near the classical trajectory $x(t)$. This property allows us to take the slowly varying cosine potential out of the integrals.
23. R. Dum, A. S. Parkins, P. Zoller, and C. W. Gardiner, Phys. Rev. A **46**, 4382 (1992); K. Mœlmer, Y. Castin, and J. Dalibard, J. Opt. Soc. Am. B **10**, 523 (1993).

24. J. Javanainen and S. Stenholm, Appl. Phys. **21**, 35 (1980).
25. H. Carmichael, *An Open System Approach to Quantum Optics* (Springer–Verlag, Berlin 1991).
26. M. El Ghafar, K. Riedel, P. Törmä, V. Savichev, E. Mayr, A. Zeiler, and W. P. Schleich, Acta Physica Slovaca **47**, 291 (1997).
27. K. Riedel, P. Törmä, V. Savichev, and W. P. Schleich, submitted to Phys. Rev. A, (1998).
28. G. M. Zaslavsky, *Chaos in Dynamic Systems* (Harwood Academic Publishers, Chur 1985).
29. Of course, not any potential is able to cause classical diffusion; for instance for a running wave we have found neither classical nor quantum diffusion, because we can define a frame where the ion does not see the potential caused by the running wave.
30. P. J. Bardroff, C. Leichtle, G. Schrade, and W. P. Schleich, Phys. Rev. Lett. **77**, 2198 (1996).
31. S. R. Jefferts, C. Monroe, E. W. Bell, and D. J. Wineland, Phys. Rev. A **51**, 3112 (1995).

TWO-PHOTON INTERACTIONS FOR USE IN QUANTUM COMPUTING

J.D. Franson and T.B. Pittman

Applied Physics Laboratory
The Johns Hopkins University
Laurel, MD 20723

Nonlinear interactions between two photons would have many potential applications, including the implementation of quantum logic gates for use in quantum computing. We are investigating a new mechanism for nonlinear optics at the two-photon level that is based on the exchange of photons by pairs of atoms in a medium, such as an atomic vapor. The effect can be understood as being due to a factor of two increase in the probability of having two excited atoms in the medium compared to the case in which the two photons propagate in two separate media. These effects are expected to be sufficiently large that cavities and traps should not be required, which may eventually allow the construction of large numbers of simple quantum logic gates.

INTRODUCTION

Nonlinear optics at the two-photon level would have many potential applications, including logic devices for quantum computing, quantum non-demolition experiments, and quantum teleportation. Unfortunately, nonlinear optical effects usually require high-intensity fields containing many photons, since the electric field associated with a single photon is normally very weak. The Caltech group (Turchette et al., 1995) has demonstrated nonlinear phase shifts using high-Q cavities and atomic beams or traps, but that approach may not be practical for the construction of full-scale quantum computers due to the size and cost of the cavities and traps. We are investigating a new approach to nonlinear optics at the two-photon level that is based on the exchange of photons by pairs of atoms in the media (Franson, 1997). This approach does not require cavities or traps and is somewhat analogous to the exchange forces that exist between two indistinguishable particles.

The effect of interest corresponds to a nonlinear phase shift (Kerr effect), which can be used to construct an XOR quantum logic gate (Milburn, 1989; Franson and Pittman, 1997). Two photons with different frequencies ω_1 and ω_2 are incident upon an atomic medium in the same transverse mode. The origin of the nonlinear phase shift can be most easily understood by comparing the case in which the two photons propagate through the same medium with the case in which they propagate through

Quantum Communication, Computing, and Measurement 2
Edited by Kumar et al., Kluwer Academic / Plenum Publishers, New York, 2000.

two separate but identical media. Subtracting the phase shifts obtained in the two cases eliminates the usual linear phase shift and leaves only the nonlinear phase shift $\Delta\phi_{non}$ of interest.

This approach is based on the fact that the number of atoms in an atomic vapor cell can readily be increased to the point that photon 1 has a 50% chance of interacting with one of the atoms, for example. Photon 2 will also have a 50% chance of interacting with one of the atoms, so that the interaction is large in some sense. However, the probability that both photons will interact with the *same* atom is negligibly small, and this a requirement for most (if not all) other mechanisms for nonlinear optics. Our approach allows the two photons to interact with two different atoms, which does occur with a reasonably large probability under these conditions, while still giving a nonlinear phase shift. The effect can be understood as being due to a factor of two increase in the probability that there are two excited atoms in the medium, which is due to quantum interference effects similar to photon bunching (Hanbury-Brown and Twiss, 1956). The application of a laser pulse produces a phase shift in the excited states of the atoms, which gives a different overall phase shift when the two photons pass through the same medium than when they pass through two different media.

TWO-PHOTON DRESSED STATES

The two incident photons are described by localized single-photon wave packets created by the operators

$$
\begin{aligned}
a_1^\dagger &= \sum_{\mathbf{k}} f_1^*(\mathbf{k})\, a_{\mathbf{k}}^\dagger \\
a_2^\dagger &= \sum_{\mathbf{k}} f_2^*(\mathbf{k}) a_{\mathbf{k}}^\dagger
\end{aligned}
\tag{1}
$$

Here the operator $a_{\mathbf{k}}^\dagger$ creates a plane-wave photon with wave vector $\mathbf{k}$ and $f(\mathbf{k})$ are the Fourier coefficients of a gaussian wave packet. In the experiments in progress, these wave packets have a width of approximately 1 ns. The matrix elements for the absorption and emission of these photons is relatively large, since their electric field is localized at the position of the atoms. The matrix elements for the absorption and emission of plane-wave photons is much smaller, since their fields are not localized, and those matrix elements can therefore be neglected to a good approximation. This corresponds to the neglect of scattering and dispersion, which are both small for large detunings, as well as the neglect of small radiative corrections such as the Lamb shift. With this approximation and the rotating-wave-approximation, the only states of the field that can be excited are:

$$
\begin{aligned}
|\gamma_1, \gamma_2(t)\rangle &= a_1^\dagger(t) a_2^\dagger(t)|0\rangle \\
|\gamma_1(t)\rangle &= a_1^\dagger(t)|0\rangle \\
|\gamma_2(t) > &= a_2^\dagger(t)|0\rangle \\
&|0\rangle \\
|\gamma_1, \gamma_1(t) > &= a_1^\dagger(t) a_1^\dagger(t)|0\rangle \\
|\gamma_2, \gamma_2(t) > &= a_2^\dagger(t) a_2^\dagger(t)|0\rangle
\end{aligned}
\tag{2}
$$

where $|0\rangle$ is the vacuum state. Thus the state of the system can be represented by a six-dimensional effective state vector. It should be emphasized that the photon wave packets are not assumed to be eigenstates of the Hamiltonian.

As the two photons propagate into the medium, the state vector is perturbed by the coupling with the atoms:

$$|\psi\rangle = \begin{pmatrix} 1 \\ 0 \\ 0 \\ 0 \\ 0 \\ 0 \end{pmatrix} \Rightarrow \begin{pmatrix} c(\gamma_1,\gamma_2) \\ c(\gamma_1) \\ c(\gamma_2) \\ c(0) \\ c(\gamma_1,\gamma_1) \\ c(\gamma_2,\gamma_2) \end{pmatrix} = |\psi\rangle_{dressed} \tag{3}$$

Here $c(\gamma_1, \gamma_2)$ is the probability amplitude that photons 1 and 2 are both present with no excited atoms, $c(\gamma_1)$ is the probability amplitude that only photon 1 is present with one excited atom, $c(0)$ is the probability amplitude that no photons are present with two excited atoms, etc. Such a perturbed state can be thought of as a two-photon dressed state. We assume that the frequency spread of the photon wave packets is much less than the detunings, in which case the magnitude of the electric field is slowly varying as the photons approach the atoms. The above probability amplitudes can then be calculated by using the adiabatic theorem and by calculating the eigenstates of the corresponding slowly-varying Hamiltonian, which was done numerically.

If the two photons were totally non-interacting, it might be expected that the two-photon dressed state would be the tensor product of two single-photon dressed states, in which case there would be no nonlinear phase shift. That is not the case, as can be seen by considering the probability that the excited atomic states are populated. Let P_{1S} be the probability that exactly one atom is excited when the photons are in the same medium, let P_{2S} be the probability that two atoms are excited when the photons are in the same medium, and let P_{1D} and P_{2D} be similarly defined for the case in which the two photons are in different media. There are two distinct ways in which the two-excited-atom state can be created when the photons are in the same medium. For any pair of atoms A and B, photon 1 can excite atom A while photon 2 excites atom B, or photon 1 can excite atom B while photon 2 excites atom A. These two processes will interfere constructively provided that $\Delta\mathbf{k} \cdot \Delta\mathbf{x} \ll \pi$, where $\Delta\mathbf{k}$ is the difference in the wave vectors of the two photons and $\Delta\mathbf{x}$ is the difference in the position of the two atoms; this condition is well satisfied in the experiments in progress. As a result, we may expect that $P_{2S}/P_{2D} = 2$ in analogy with the Hanbury-Brown Twiss effect. The results of the numerical calculations show that this is indeed the case in the limit of large numbers of atoms.

Although it may not be as apparent, the mean number $< N_e >$ of excited atoms is the same in the two cases:

$$\begin{aligned} < N_e >= P_{1S} + 2P_{2S} &= P_{1D} + 2P_{2D} \\ P_{1S} - P_{1D} &= -P_{2S} \end{aligned} \tag{4}$$

The probability of having exactly one excited atom is reduced when the two photons are in the same medium, since the creation of a two-excited-atom state comes at the expense of depleting the one-excited-atom states. The differences in the atomic energy levels are the same in the two cases in the limit of large numbers of atoms.

EFFECTS OF A LASER PULSE

The effects of applying a short laser pulse can now be understood. The frequency of the laser pulses is tuned near the transition frequency of the excited atomic state of interest and a third, higher-energy level. The detuning is sufficiently large that negligible population transfer occurs. This produces a shift in the energy of the excited atomic states and a corresponding phase shift, which can be chosen to be $\pi/2$ by adjusting the intensity of the laser pulses. The effects of such an impulsive phase shift are then given by

$$|\psi_0\rangle = \begin{pmatrix} c(\gamma_1,\gamma_2) \\ c(\gamma_1) \\ c(\gamma_2) \\ c(0) \\ c(\gamma_1,\gamma_1) \\ c(\gamma_2,\gamma_2) \end{pmatrix} \Rightarrow \begin{pmatrix} c(\gamma_1,\gamma_2) \\ e^{i\pi/2}c(\gamma_1) \\ e^{i\pi/2}c(\gamma_2) \\ e^{i\pi}c(0) \\ c(\gamma_1,\gamma_1) \\ c(\gamma_2,\gamma_2) \end{pmatrix} = |\psi\rangle' \tag{5}$$

The two components corresponding to one excited atom are subjected to a $\pi/2$ phase shift, while the component corresponding to two excited atoms receives a phase shift of π, which just reverses its sign.

We are interested in the overall phase shift of the original state, while the generation of any orthogonal component will correspond to an error that must be corrected in some way for quantum computing applications. We therefore express the new state vector in terms of the original state vector and an orthogonal component:

$$|\psi'\rangle = re^{i\phi}|\psi_0\rangle + |\psi_\perp\rangle \tag{6}$$

where $|\psi_\perp\rangle$ is orthogonal to the state vector before the pulse. The coefficient in front of $|\psi_0\rangle$ is determined by the projection of $|\psi'\rangle$ onto $|\psi_0\rangle$, or

$$re^{i\phi} = \langle\psi_0|\psi'\rangle = r' + i(|c(\gamma_1)|^2 + |c(\gamma_2)|^2) + (-1)\,|c(0)|^2 \tag{7}$$

Here r' is the contribution from the terms with no phase shift, which is real and on the order of unity for large detunings, while the other terms are much less than 1. In that limit, the value of ϕ is determined by the imaginary part of the above equation, which gives

$$i\phi_S \doteq iP_{1S} \qquad (for \quad r' \sim 1) \tag{8}$$

The nonlinear phase shift is obtained by subtracting off the corresponding expression for the case in which the two photons propagate in two different media:

$$\Delta\phi_{non} = P_{1S} - P_{1D} = -P_{2S} \tag{9}$$

Thus the nonlinear phase shift is proportional to the probability that two atoms were excited at the time of the laser pulse, which can be on the order of unity with relatively small losses due to scattering or absorption in the atomic vapor cell.

After the laser pulse has been applied, the photon wave packets will propagate out of the medium and the perturbation that produced the two-photon dressed state will gradually decrease. Once again, the adiabatic theorem can be applied when the frequency spread of the gaussian wave packets is small compared to the detuning, and the original "bare" photon states are left with an overall phase factor of $\Delta\phi_{non}$ imparted to the quantum state of the system.

Density matrix calculation were performed to compute the nonlinear phase shift produced by a random sequence of laser pulses. The results of those calculations were in reasonable agreement with the simplified derivation given above. The density matrix calculations included the effects of collisions and are not limited to the case of small couplings.

The generation of orthogonal components $|\psi_\perp\rangle$ will produce errors in a quantum computation. These errors could be controlled using quantum error correction techniques, but it is also possible to apply a sequence of laser pulses in such a way that the contributions to $|\psi_\perp\rangle$ from all of the pulses cancel out. A suitable sequence of laser pulses can be conveniently produced by passing a single laser pulse through a series of beam splitters. The use of such a sequence of laser pulses can eliminate any decoherence due to the laser pulses themselves, while decoherence due to scattering and absorption can be reduced to a small level by the use of large detunings.

SUMMARY

Nonlinear optical effects using only two photons are difficult to achieve because the probability that both of the photons will interact with the same atom in medium (without cavities or traps) is extremely small. In our approach, the two photons are allowed to interact with two different atoms in the medium, which can occur with probability on the order of unity. Even though the two photons never interact with the same atom, their presence in the same medium does increase the probability of having two excited atoms by a factor of two, in analogy with the Hanbury-Brown and Twiss effect (photon bunching). The application of a laser pulse then produces phase shifts that depend on the populations of the atomic states and are different when the photons propagate in the same medium, which produces a nonlinear phase shift that can be on the order of unity. This effect is roughly analogous to the "exchange forces" that exist between indistinguishable particles even when there is no physical interaction between them.

The practical advantages of this approach for the construction of a full-scale quantum computer have been discussed elsewhere and include the ability to fabricate large numbers of simple quantum logic devices that can be connected using optical fibers. The preliminary results (Franson and Pittman, 1997) from an experiment using a sodium vapor cell and a buffer gas have shown small nonlinear phase shifts in rough agreement with what would be expected from the theory, but the observed phase shifts were too small to allow systematic investigations. An improved experiment employing picosecond laser pulses is in progress.

This work was supported in part by the Army Research Office, the National Security Agency, the Office of Naval Research, and by IR&D funds from the Applied Physics Laboratory.

REFERENCES

Brown, R. Hanbury, and Twiss, R.Q., 1956, Correlations between photons in two coherent beams of light, *Nature* 177:27.

Franson, J.D., 1997, Cooperative enhancement of Optical Quantum Gates, *Phys. Rev. Lett.* 78:3852.

Franson, J.D., and Pittman, T.B., 1997, Nonlocality in Quantum Computing, Proceedings of the Fundamental Problems in Quantum Theory Workshop, Baltimore, MD, to appear in Fort. der Physik.

Milburn, G.J., 1989, Quantum optical Fredkin gates, *Phys. Rev. Lett.* 62:2124.

Turchette, Q.A., Hood, C.J, Lange, W., Mabuchi, H. and Kimble, H.J., 1995, Measurement of conditional phase shifts for quantum logic, *Phys. Rev. Lett.* 75:4710.

DECOHERENCE AND QUANTUM ERROR CORRECTION IN FREQUENCY STANDARDS

C. Macchiavello,[1] S. F. Huelga,[2] J.I. Cirac, [3] A.K. Ekert,[4] M.B. Plenio[2]

[1] Dipartimento di Fisica "A. Volta" and INFM - Unità di Pavia, Via Bassi 6, 27100 Pavia, Italy
[2] Optics Section, The Blackett Laboratory, Imperial College, London SW7 2BZ, U.K.
[3] Institute fuer Theoretische Physik, Universitaet Innsbruck, Technikerstrasse 25, A-6020, Innsbruck, Austria
[4] Clarendon Laboratory, Department of Physics, University of Oxford, Oxford OX1 3PU, U. K.

The precision of frequency measurements performed on trapped ions in the presence of decoherence is analysed. In particular, standard Ramsey spectroscopy on uncorrelated ions and optimal measurements on maximally entangled states are proved to lead to the same resolution, while the best precision is achieved using partially entangled preparations. The use of symmetrisation procedures is proposed and it is shown how this allows to overcome even the optimal precision achievable when both the initial preparation and the final measurement are optimized.

INTRODUCTION

The goal of frequency standards is to stabilize the frequency of a reference oscillator to a given atomic transition frequency. The development of laser cooling and trapping techniques has allowed new possibilities for improved frequency standards based on laser cooled ions. Limitations due to the second order Doppler effect present in traditional atomic beam devices can now be overcome and accuracies in the range of 1 part in $10^{14}-10^{18}$ are expected[1]. Recently accuracies of the order 10^{-13} have been reported[2]. In this paper we discuss the limits to the maximum precision achievable in the spectroscopy of n two level atoms in the presence of decoherence. This question is particularly timely in view of current efforts to improve high precision spectroscopy by means of quantum entanglement. We also present a symmetrization method that allows to improve the uncertainties that can be achieved by optimising the initial preparation of the n-ions as well as the final measurement on them. Such a method is based on a symmetrisation procedure which was originally proposed as a technique for error correction in quantum computation[3]. The result shows that techniques developed in quantum computation can have convenient applications in several contexts.

Quantum Communication, Computing, and Measurement 2
Edited by Kumar et al., Kluwer Academic / Plenum Publishers, New York, 2000.

The paper is organised as follows. We first review the standard Ramsey spectroscopic technique implemented with trapped ions and derive the shot noise limit. We later show how the shot noise limit can be overcome by exploiting entanglement among the ions, and describe the corresponding scheme in terms of elementary quantum gates. We analyse the effects of decoherence in the above mentioned schemes and show that they lead to the same best precision. We then apply an optimisation procedure to find out the most convenient initial preparation for the state of the ions and the most efficient measurement scheme that lead to the best precision in the presence of decoherence. Finally, we briefly review a stabilisation technique based on symmetrisation and apply it to frequency standards. In the end we summarise the results shown and draw some conclusions.

STANDARD RAMSEY SPECTROSCOPY

We will briefly review here the implementation of an optical frequency standard in an ion trap using the method of separated oscillatory fields due to Ramsey. The ion trap is loaded with n ions initially prepared in the same internal state $|0\rangle$ (we denote by $|0\rangle$ and $|1\rangle$ the ground and the excited states of each ion). A Ramsey pulse of frequency ω is applied to all ions. The pulse shape and duration are carefully chosen so that it drives the atomic transition $|0\rangle \leftrightarrow |1\rangle$ of natural frequency ω_0 and prepares an equally weighted superposition of the two internal states $|0\rangle$ and $|1\rangle$ for each ion. In the basis $\{|0\rangle, |1\rangle\}$ this transformation is given by

$$R = \frac{1}{\sqrt{2}} \begin{pmatrix} 1 & -1 \\ 1 & 1 \end{pmatrix} . \tag{1}$$

It is assumed here that the Rabi frequency of the local oscillator is much larger than the detuning $\Delta = \omega - \omega_0$ between the classical driving field and the atomic transition. Next the system evolves freely for a time t. In a reference frame rotating at the oscillator frequency ω, the free evolution is governed by the Hamiltonian

$$H = -\hbar\Delta|1\rangle\langle 1| \tag{2}$$

where $\Delta = \omega - \omega_0$ denotes the detuning between the classical driving field and the atomic transition. The evolution of the basis atomic states can then be represented as a conditional phase shift on the atomic basis states, described by the transformation

$$F = \begin{pmatrix} 1 & 0 \\ 0 & e^{i\Delta t} \end{pmatrix} . \tag{3}$$

and the frequency difference between the atomic transition and the reference oscillator leads to the accumulation of a relative phase between the components $|0\rangle$ and $|1\rangle$. The following step is to apply a second Ramsey pulse of the same characteristics as the first one. The probability that an ion is found in the state $|1\rangle$ after this pulse is given by

$$P = \frac{1}{2}(1 + \cos\Delta t). \tag{4}$$

This basic scheme is repeated yielding a total duration T of the experiment. The aim is to estimate Δ as accurately as possible for a given T and a given number of ions n. The two quantities T and n are the physical resources we consider when comparing the performance of different schemes.

The statistical fluctuations associated with a finite sample yield an uncertainty ΔP in the estimated value of P given by

$$\Delta P = \sqrt{\frac{P(1-P)}{N}} \,, \tag{5}$$

where $N = nT/t$ denotes the actual number of experimental data (we assume that N is large). Hence the uncertainty in the estimated value of ω_0 is given by

$$|\delta\omega_0| = \frac{\Delta P}{|dP/d\omega|} = \frac{1}{\sqrt{nTt}}. \tag{6}$$

This value is often referred to as the *shot noise limit*[4].

MAXIMALLY ENTANGLED STATES IN FREQUENCY STANDARDS

The theoretical possibility of overcoming this limit has been put forward recently[5]. The basic idea is to prepare the ions initially in an entangled state. To see the advantage of this approach, let us consider the case of two ions prepared in the maximally entangled state

$$|\psi\rangle = \frac{1}{\sqrt{2}}(|00\rangle + |11\rangle). \tag{7}$$

We note the example we will give does not represent exactly the original proposal[5]. We will illustrate it in terms of a quantum network, shown in Fig. 1, however, the two schemes are equivalent as far as the resulting precision is concerned. The state (7) can be generated by the initial part of the network. A Ramsey pulse (1) is first applied to the first ion. The Ramsey pulse is followed by a controlled–NOT gate[6] (see Fig. 1 for a schematic picture of the protocol). The controlled-NOT gate is an elementary quantum gate which performs the following operation on the two ions

$$|\,\epsilon_1\rangle\,|\,\epsilon_2\rangle \rightarrow |\,\epsilon_1\rangle\,|\,\epsilon_1 \oplus \epsilon_2\rangle \,, \tag{8}$$

where $\epsilon_1, \epsilon_2 = \{0, 1\}$ and $\oplus$ denotes addition modulo two. After a free evolution period of time t the state of the composite system in the interaction picture rotating at the driving frequency ω reads

$$|\psi\rangle = \frac{1}{\sqrt{2}}(|00\rangle + e^{2i\Delta t}\,|11\rangle)\,. \tag{9}$$

In the second part of the network another controlled-NOT gate is applied in order to to disentangle the ions after the free evolution period. The population in state $|1\rangle$ of the first ion will now oscillate at a frequency 2Δ

$$P_2 = \frac{1}{2}(1 + \cos 2\Delta t). \tag{10}$$

This scheme can be easily generalised to the n ion case by applying a Ramsey pulse to the first ion and then a sequence of $n-1$ controlled–NOT gates linking the first ion with each of the remaining ones. In this way, a maximally entangled state of n ions of the form

$$|\psi\rangle = \frac{1}{\sqrt{2}}(|00...0\rangle + |11...1\rangle) \tag{11}$$

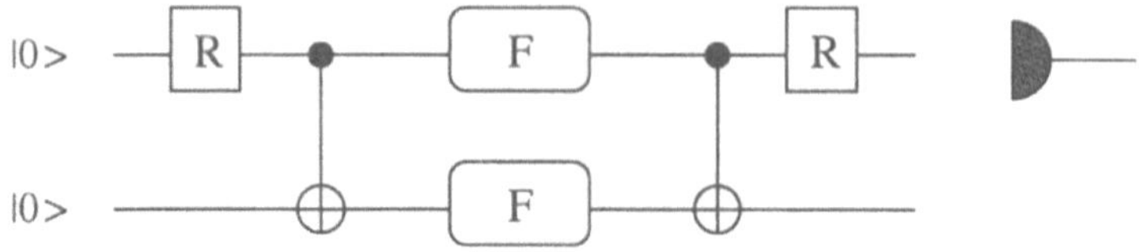

Figure 1. Schematic illustration of spectroscopy with two maximally entangled particles by means of a quantum network. The particles are entangled and disentangled by means of "controlled–NOT" gates.

is generated. The final measurement on the first ion, after the free evolution period and the second cascade of $n-1$ controlled–NOT gates, gives the signal

$$P_n = \frac{1}{2}(1 + \cos n\Delta t) . \tag{12}$$

The advantage of this scheme is that the oscillation frequency of the signal is now amplified by a factor n with respect to the case of uncorrelated ions and the corresponding frequency uncertainty is

$$|\delta\omega_0| = \frac{1}{n\sqrt{Tt}} . \tag{13}$$

Note that this result represents an improvement of a factor $1/\sqrt{n}$ over the shot noise limit (6) by using the same number of ions n and the same total duration of the experiment T, and it was argued that this is the best achievable precision[7].

EFFECTS OF DECOHERENCE

In this section we will examine the same situation in a realistic experimental scenario, where decoherence effects are inevitably present. The main type of decoherence in an ion trap is dephasing due to processes that cause random changes in the relative phase of quantum states while preserving the population in the atomic levels. Important mechanisms that result in dephasing effects are collisions, stray fields and laser instabilities. We model the time evolution of the reduced density operator for a single ion ρ in the presence of decoherence by the following master equation[8]:

$$\dot{\rho}(t) = -i\Delta\left(\rho|1\rangle\langle 1| - |1\rangle\langle 1|\rho\right) + \gamma\left(\sigma_z\rho\sigma_z - \rho\right) \tag{14}$$

[Actually, our analysis is not restricted to this particular model but holds for any process where off-diagonal elements decay exponentially with time.] Equation (14) is written in a frame rotating at the frequency ω. By $\sigma_z = |0\rangle\langle 0| - |1\rangle\langle 1|$ we denote a Pauli spin operator. Here we have introduced the decay rate $\gamma = 1/\tau_{dec}$, where τ_{dec} is the decoherence time. For the case of independent particles this will give rise to a broadening of the signal (4):

$$P = \frac{1}{2}(1 + e^{-\gamma t}\cos\Delta\, t). \tag{15}$$

As a consequence the corresponding uncertainty in the atomic frequency is no longer detuning-independent. We now have

$$|\delta\omega_0| = \sqrt{\frac{e^{-2\gamma t}1 - \cos^2(\Delta t)}{nTte^{-2\gamma t}\sin^2(\Delta t)}} . \tag{16}$$

In order to obtain the best precision it is necessary to optimise this expression as a function of the duration of each single measurement t. The minimal value is attained for

$$\Delta t = k\pi/2 \ (k \ \ odd) \qquad t = \tau_{dec}/2 \tag{17}$$

provided that $T > \tau_{dec}/2$. Thus the minimum frequency uncertainty reads

$$|\delta\omega_0|_{opt} = \sqrt{\frac{2\gamma e}{nT}} = \sqrt{\frac{2e}{n\tau_{dec}T}}. \tag{18}$$

For maximally entangled preparation the signal (12) in the presence of dephasing is modified as follows:

$$P_n = \frac{1}{2}(1 + e^{-n\gamma t}\cos n\Delta\, t) \tag{19}$$

namely the oscillating term picks up a damping factor which is also amplified by n with respect to the independent ions' signal (15). The resulting uncertainty for the estimated value of the atomic frequency is now minimal when

$$\Delta t = k\pi/2n \ (k \ \ odd) \qquad t = \tau_{dec}/2n \ . \tag{20}$$

Interestingly, we recover exactly the same minimal uncertainty (18) as for standard Ramsey spectroscopy. In the presence of decoherence both preparations reach the same precision.

OPTIMISATION OF THE INITIAL PREPARATION AND THE MEASUREMENT

We will now treat the problem of the best estimate of small detunings from a more general point of view and investigate the possibility of overcoming the limit (18) in the presence of decoherence. We will conclude that the limit (18) is not a fundamental bound on precision but is only related to the particular initial preparations and experimental setups of the previous sections. The precision can actually be improved by the preparation of different initial states and the implementation of more general measurement schemes[9]. For this purpose, we note that the problem addressed in precision spectroscopy (i.e. the optimal measurement of small atomic phase shifts) maps onto that of statistical distinguishability of nearby states, analyzed by Wootters[10] and generalized for the case of mixed states by Braunstein and Caves[11]. By finding measurements that optimally resolve neighbouring states, they have provided an upper bound for the precision in the estimation of a given variable that parametrizes a family of quantum states. In our case this variable is the detuning Δ. Moreover, the optimal measurements always correspond to a set of orthogonal projectors in the n ions' Hilbert space. If we apply the Braunstein and Caves optimisation procedure to find the best final measurement starting from uncorrelated or maximally entangled states we recover the limit (18) for the optimal precision. In this sense, the schemes shown in the previous sections are optimal for independent or maximally entangled states. The Braunstein and Caves procedure can also be applied to optimise the experiment with respect to both the initial state preparation and the final measurement. The initial state preparation which leads to the best precision is of the form

$$|\psi_n\rangle = \sum_{k=0}^{\lfloor \frac{n}{2} \rfloor} a_k |k\rangle \ , \tag{21}$$

where $|k\rangle$ denotes an equally weighted superposition of all states of n ions which contain either a number k or a number $n-k$ of excited states. By $\lfloor\cdots\rfloor$ we denote the corresponding integer part. The coefficients a_k can be chosen to be real.

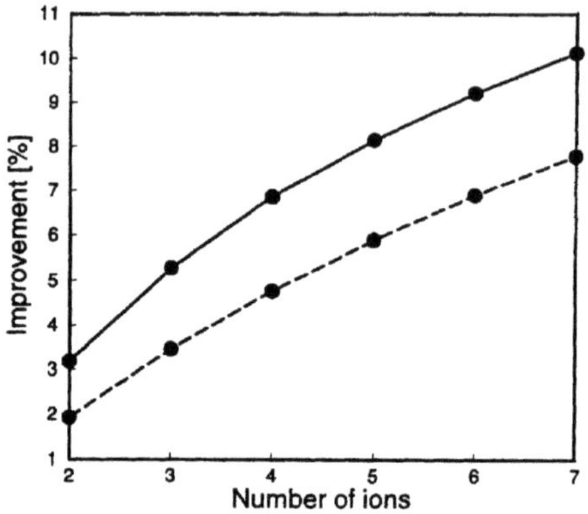

Figure 2. The optimum percentage of improvement in the precision relative to the limit (18) as a function of the number of ions n. Solid line: Numerical optimization with respect to the initial preparation and application of the Braunstein and Caves algorithm for determining the optimal measurement. Dashed line: Optimized initial preparation and generalized Ramsey spectroscopy as the final measurement[9].

Note that this family of states exhibits a high degree of symmetry: it is completely symmetric under permutations of the n ions and under exchange of the excited and the ground state for each ion. The optimum percentage of improvement in the precision relative to the limit (18) as a function of the number of ions n is shown in Fig. 2. The solid curve shows the improvement obtained by optimizing both the initial preparation and the final measurement using the algorithm of Braunstein and Caves. The dashed line exhibits the improvement obtained by optimizing only the initial preparation and performing generalized Ramsey spectroscopy[9].

STABILISATION BY SYMMETRISATION

We will now show a new method to improve the optimal precision in frequency standards in the presence of decoherence. Such a method is based on the use of a symmetrisation procedure[3], originally proposed as a technique to suppress errors in quantum computation. We briefly summarise here the basic idea. Suppose you have a quantum system, you prepare it in some initial state $|\psi_i\rangle$ and you want to implement a prescribed unitary evolution $|\psi(t)\rangle$. Now, suppose that instead of a single system you can prepare M copies of $|\psi_i\rangle$ and subsequently you can project the state of the combined system on the symmetric subspace i.e. the subspace containing all states which are invariant under any permutation of the sub-systems. The claim is that frequent projections on the symmetric subspace will reduce decoherence or any kind of errors induced by the environment. The intuition behind this concept is based on the observation that a prescribed error-free storage or evolution of the M independent copies starts in the symmetric sub-space and should remain in that sub-space. Therefore, since the error-free component of any state always lies in the symmetric subspace, upon successful projection it will be unchanged and part of the error will have been removed. Note however that the projected state is generally not error–free since the symmetric subspace contains states which are not of the simple product form $|\psi\rangle\,|\psi\rangle\ldots|\psi\rangle$. Nevertheless it has been shown that the error probability will be suppressed by a factor of $1/M^{12}$.

We illustrate here this effect in the simplest case of two ions. The projection into the symmetric subspace is performed in this case by introducing the symmetrisation operator:

$$S = \frac{1}{2}(P_{12} + P_{21}) , \tag{22}$$

where P_{12} represents the identity and P_{21} the permutation operator which exchanges the states of the two ions. The symmetric–projection of a pure state $|\Psi\rangle$ of two ions is just $S|\Psi\rangle$, which is then renormalised to unity. It follows that the induced map on mixed states of two ions (including renormalisation) is:

$$\rho_1 \otimes \rho_2 \longrightarrow \frac{S(\rho_1 \otimes \rho_2)S^\dagger}{\mathrm{Tr}[S(\rho_1 \otimes \rho_2)S^\dagger]} \tag{23}$$

The state of either ion separately is then obtained by partial trace over the other ion.

Let us assume that each copy is initially prepared in pure state $\rho_0 = |\psi\rangle\langle\psi|$ and that the two copies interact with independent environments. After some short period of time δt the state of the two copies ρ will have undergone an evolution

$$\rho(0) = \rho_0 \otimes \rho_0 \quad \longrightarrow \quad \rho(\delta t) = \rho_1 \otimes \rho_2 \tag{24}$$

where $\rho_i = \rho_0 + \varrho_i$ for some Hermitian traceless ϱ_i. We will retain only terms of first order in the perturbations ϱ_i so that the overall state at time δt is

$$\rho = \rho_0 \otimes \rho_0 + \varrho_1 \otimes \rho_0 + \rho_0 \otimes \varrho_2 + O(\varrho_1\varrho_2) . \tag{25}$$

We can calculate the average purity of the two copies before symmetrisation by calculating the average trace of the squared states:

$$\frac{1}{2}\sum_{i=1}^{2} \mathrm{Tr}((\rho_0 + \varrho_i)^2) = 1 + 2\mathrm{Tr}(\rho_0\bar{\varrho}), \tag{26}$$

where $\bar{\varrho} = \frac{1}{2}(\varrho_1 + \varrho_2)$. Note that $\mathrm{Tr}(\rho_0\bar{\varrho})$ is negative, so that the expression above does not exceed 1. After symmetrisation each ion is in state

$$\rho_s = [1 - \mathrm{Tr}(\rho_0\bar{\varrho})]\rho_0 + \frac{1}{2}\bar{\varrho} + \frac{1}{2}(\rho_0\bar{\varrho} + \bar{\varrho}\rho_0) \tag{27}$$

and has purity

$$\mathrm{Tr}(\rho_s^2) = 1 + \mathrm{Tr}(\rho_0\bar{\varrho}). \tag{28}$$

Since $\mathrm{Tr}\rho_s^2$ is closer to 1 than (26), the resulting symmetrised system ρ_s is left in a purer state.

Let us now see how the fidelity changes by applying the symmetrisation procedure. The average fidelity before symmetrisation is

$$F_{bs} = \frac{1}{2}\sum_i \langle\psi|\rho_0 + \varrho_i|\psi\rangle = 1 + \langle\psi|\bar{\varrho}|\psi\rangle , \tag{29}$$

while after successful symmetrisation it takes the form

$$F_{as} = \langle\psi|\rho_s|\psi\rangle = 1 + \frac{1}{2}\langle\psi|\bar{\varrho}|\psi\rangle . \tag{30}$$

The state after symmetrisation is therefore closer to the initial state ρ_0.

We have applied this technique for frequency standards based on standard Ramsey spectroscopy with two initially uncorrelated states. Repeated symmetrisation steps are applied during the free evolution region. After each symmetrisation step the ions are kept only if the symmetrisation is successful, namely if the surviving state corresponds to the symmetric component of the state before symmetrisation. Otherwise the ions are discarded and reset to state $|0\rangle$ to start the scheme from the beginning. Although this reduces the number of experimental data available for statistics, this has proven to be a convenient strategy to improve the overall precision of the experiment.

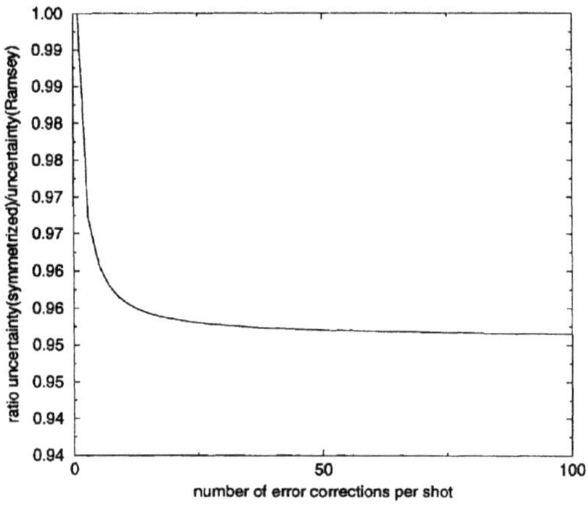

Figure 3. Ratio of the uncertainty for two initially uncorrelated particles in standard Ramsey spectroscopy with and without symmetrisation, as a function of the number of symmetrisation steps performed during the free evolution region.

In Fig. 3 we report the ratio of the uncertainty for two initially uncorrelated ions with and without symmetrisation, as a function of the number of symmetrisation steps performed during the free evolution region. Notice that for frequent repetitions of symmetrisation during the free evolution period the improvement obtained over the shot noise (18) for uncorrelated particles is larger than the one achieved by optimising the initial preparation and the final measurement according to the Braunstein and Caves procedure (compare Fig. 2 with Fig. 3 at $n = 2$). Note also, that the symmetrization procedure outlined here is compatible with the optimized initial states Eq. (21) that have been found in[9]. This suggests to use these two methods simultaneously to obtain even larger improvements. The limits to the precision achievable with symmetrisation procedures for arbitrary n and arbitrary initial preparation of the state of the ions are still under investigation

CONCLUSIONS

We have shown that standard Ramsey spectroscopy is optimal for uncorrelated particles both in the presence and in the absence of decoherence effects. We have also shown that for experiments with a duration longer than the decoherence time, e.g. high precision optical frequency standards, the use of maximally entangled states does not provide higher resolution as compared to using independent particles. The best sensitivity is obtained by optimising both the initial preparation of the ions and the final measurement. The optimal initial states are highly symmetric but only partially entangled. We have presented a method to further improve the best precision achievable in the presence of decoherence based on a symmetrisation procedure originally proposed to stabilise quantum computation against noise. This technique seems to be very promising for improving precision in frequency standards and shows how ideas of quantum computation can be profitably applied also in other contexts.

Acknowledgments

We acknowledge useful discussions with David Deutsch.

This work was supported in part by the European TMR Research Network ERB 4061PL95-1412, the European TMR Research Network ERBFMRXCT96066, the Royal Society, Hewlett-Packard and Elsag-Bailey.

REFERENCES

1. D. J. Wineland et al., *IEEE Trans. on Ultrasonics, Ferroelectrics and Frequency Control A* 37:515 (1990).
2. D.J. Berkeland *et al.*, Laser-cooled mercury ion frequency standard, *Phys. Rev. Lett.* 80:2089 (1998).
3. D. Deutsch, (1993) talk presented at the Rank Prize Funds Mini–Symposium on Quantum Communication and Cryptography, Broadway, England; A. Berthiaume, D. Deutsch R. and Jozsa, in *Proceedings of Workshop on Physics and Computation — PhysComp94*, IEEE Computer Society Press, Dallas, Texas, (1994).
4. W. H. Itano et al., Quantum projection noise: population fluctuations in two-level systems, *Phys. Rev. A* 47:3554 (1993).
5. D. J. Wineland et al., Spin squeezing and reduced quantum noise in spectroscopy, *Phys. Rev. A* 46:R6797 (1992); D. J. Wineland et al., Squeezed atomic states and projection noise in spectroscopy, *Phys. Rev. A* 50:67 (1994).
6. A. Barenco *et al.*, Conditional quantum dynamics and logic gates, *Phys. Rev. Lett.* 74:4083 (1995).
7. J. J. Bollinger et al., Optimal frequency measurements with maximally correlated states, *Phys. Rev. A* 54:R4649 (1996).
8. C. W. Gardiner, "Quantum Noise", Springer–Verlag, Berlin (1991).
9. S. F. Huelga, C. Macchiavello, T. Pellizzari, A. K. Ekert, M. B. Plenio and J. I. Cirac, Improvement of frequency standards with quantum enanglement, *Phys. Rev. Lett.* 79:3865 (1997).
10. W. K. Wootters, Statistical distance and Hilbert space, *Phys. Rev. D* 23:357 (1981).
11. S. L. Braunstein and C. Caves, Statistical distance and the geometry of quantum states, *Phys. Rev. Lett.* 72:3439 (1994).
12. A. Barenco, A. Berthiaume, D. Deutsch, A. Ekert, R. Jozsa and C. Macchiavello, Stabilization of quantum computations by symmetrisation, *SIAM J. Comput.* 26:1541 (1997).

TRANSITION PHENOMENA IN NETWORKS OF QUANTUM SYSTEMS

B. Kraus and P. Törmä

Institut für Theoretische Physik, Universität Innsbruck, Technikerstraße 25/2, A-6020 Innsbruck, Austria

The advances in experimental manipulation and theoretical understanding of single quantum systems has fed the interest in *quantum networks*: quantum computers, quantum communication networks and multimode interferometers. According to statistical physics, a set of probabilistically behaving individual systems can exhibit critical behaviour when connected. In this paper we show that transition phenomena exist also in networks which behave probabilistically not because of finite temperature but due to their quantum nature [1].

We define a model which carries in its structure a formal analogy to the two-dimensional Ising-model. Such networks can be experimentally realized by various active (non-unitary) and passive (unitary) components. We give physical meaning to the transitions found. In the case of active components, such as amplifiers, the transition point coincides with the fundamental cloning limit given by quantum mechanics. We indicate how the transition phenomena could be observed, and discuss their significance and connection to quantum information related issues such as quantum computation and quantum cellular automata.

INTRODUCTION

Quantum information processing requires not only the understanding and control of single quantum systems but also knowledge about the global properties and funtioning of quantum networks. A quantum computer[2] is a network where any two of the nodes (qubits) can interact with each other; for a certain computation selected qubits have to be connected in a predefined manner, which can be expressed by network diagrams. Because of the presence of noise such as decoherence the optimization of computational networks provides a further challenge[3]. Apart from quantum computation and communication[4], quantum networks have been used to study higher dimensional quantum interference effects, c.f. the experiments on photons in beam splitter networks[5] and excitations in multilevel systems with a network-like energy spectrum[6, 7].

In this paper we consider the question whether quantum networks can exhibit transition phenomena[1]. In statistical physics a set of individual systems which behave probabilistically because of finite temperature can show critical behaviour. We ask whether this is possible in quantum networks which behave probabilistically due to

Quantum Communication, Computing, and Measurement 2
Edited by Kumar et al., Kluwer Academic / Plenum Publishers, New York, 2000.

their quantum nature. By a transition we mean in this context that a global parameter describing the network becomes singular as a function of a local parameter characterizing the individual nodes. As a simple model system we consider a network which carries in its structure a formal analogy to the two-dimensional Ising-model.

Below, we consider a network described by a unitary transform which has a connection to the Ising-model. We present the derivation of the eigenvalues in a more detailed manner as in[1], and discuss briefly the physical interpretation of the results, as well as the connection to quantum computation and quantum cellular automata.

NEAREST NEIGHBOUR QUANTUM NETWORK

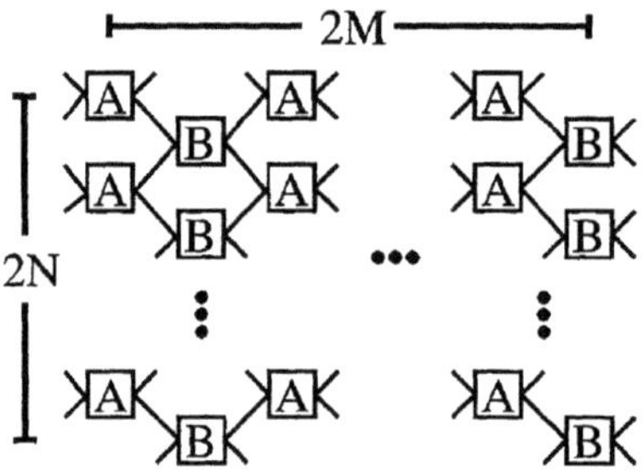

Figure 1. A quantum network where the nodes are connected to their nearest neighbours. The boxes A and B denote the transformations performed at the nodes.

The Ising model describes a set of two-state systems which interact with their nearest neighbours; a quantum analogy of such a setup can be experimentally realized in various ways, as will be explained below. Fig.1 shows schematically a 2-D quantum network with nearest neighbour interactions. To define the building blocks of this network we now take a closer look at the Ising model.

The two-state systems in the Ising model, such as spins, are on a 2-D lattice of the size $N \times M$. Since only nearest neighbour interactions are taken into account, the total energy of the system can be expressed using the energy $E(\mu)$ of one column (with the spin configuration μ) and the energy $E(\mu, \mu')$ between two columns. Let $s_k = \pm 1$ denote the values of individual spins and ϵ be the absolute value of the energy of a spin-spin interaction. Then the energies can be written as $E(\mu) = -\epsilon \sum_{k=1}^{N} s_k s_{k+1}$ and $E(\mu, \mu') = -\epsilon \sum_{k=1}^{N} s_k s'_k$. The partition function $Q(T)$ can be expressed in a simple form by defining a $2^N \times 2^N$ matrix $\mathcal{P}$ whose matrix elements are the thermal weight factors corresponding to a particular spin configuration of two neighbouring columns $\langle \mu | \mathcal{P} | \mu' \rangle \equiv e^{-\beta[E(\mu,\mu')+E(\mu)]}$. With this notation $Q(T) = Tr\mathcal{P}^M = \sum_{\alpha=1}^{2^N} (\lambda_\alpha)^M$. The eigenvalues λ_α thus determine the thermodynamics of the system. The $2^N \times 2^N$ matrix $\mathcal{P}$ is a spinor representation of a set of plane rotations in $2N \times 2N$ dimensional space 8. The eigenvalues of $\mathcal{P}$ are uniquely determined by the eigenvalues of the corresponding plane rotation matrix P, which is

$$P=\begin{bmatrix} \boxed{A} & \begin{matrix}0&0\\0&0\end{matrix} & \cdots & \\ \begin{matrix}0&0\\0&0\end{matrix} & \boxed{A} & & \\ \vdots\ \vdots & & \ddots & \\ & & & \boxed{A}\end{bmatrix}\begin{bmatrix} B_{22} & 0\ \ 0 & \cdots & & B_{21}\\ 0 & \boxed{B} & & & \\ 0 & & & & \\ \vdots & & \boxed{B} & & \\ & & & \ddots & \\ B_{12} & & & & B_{11}\end{bmatrix} \equiv \mathcal{AB}$$

$$\boxed{A}=\begin{bmatrix}\cosh\theta & i\sinh\theta\\ -i\sinh\theta & \cosh\theta\end{bmatrix}\quad \boxed{B}=\begin{bmatrix}\cosh\phi & i\sinh\phi\\ -i\sinh\phi & \cosh\phi\end{bmatrix}$$

$$\coth\theta=\cosh\phi\ ,\quad \phi=2\epsilon/kT. \tag{1}$$

The form of the matrix P suggests a quantum-network analogue, as it turns out, actually the one shown in Fig.1. The matrices A and B can be interpreted to describe the evolution of a two-state or two-mode system. The matrix P is then the evolution operator over a period in the network of Fig.1. The inputs of the network are mixed pairwise according to the transformation B, and then the pairs are let to interact with the neighbouring ones by applying the shifted set of operations A. By repeating this M times, a $2N \times 2M$ dimensional network can be constructed. The networks should be either large enough, or imposed to periodic boundary conditions — in the case of optical components the latter can be easily realized.

The physical realizations of the quantum network P can be divided into two groups. When the angle ϕ is real, A and B are SU(1,1)-type matrices describing active, i.e. energy-consuming operations. Imaginary ϕ leads to SU(2) matrices, which correspond to energy-preserving (passive) manipulations of the two modes or two states. Parametric amplifiers, four-wave mixers and phase-conjugating mirrors are SU(1,1) devices which can operate also in the quantum regime[9]: they could be used to build a network of active (quantum) optical components. The corresponding passive networks could be realized, for example, with beam splitters or fibre couplers. Also a network of intersecting energy levels can be described by a network of the type in Fig.1: the avoided crossings between the levels are identified with the operations A and B. Corresponding physical systems are for instance Rydberg atoms[10] and longitudinal electro-magnetic modes in a cavity[6]. One can also consider the matrix P as a set of operations in the computational space of a quantum computer[11].

Now we transform the matrix P into a form that gives directly the exact eigenvalues. The derivation given here differs slightly from the solution of the Ising model given in many textbooks; the one given here is more intuitive in the context of quantum networks. In P, using a simple shift operation, the matrix $\mathcal{B}$ can be written in the same kind of block diagonal form as $\mathcal{A}$. Mathematically, $\mathcal{B}=S_{2N}^T(I_N\otimes B)S_{2N}$, where $(S_{2N})_{kl}=\delta_{k,l+1}+\delta_{0,N-1}$. The matrix P becomes

$$P=(I_N\otimes A)S_{2N}^T(I_N\otimes B)S_{2N}. \tag{2}$$

The key to diagonalization of the above matrix is the fact that an amplitude shift, S_{2N}, corresponds to a relative phase shift in its Fourier-conjugate space. That is, Fourier transform of a shift matrix produces a diagonal matrix with phase factors (roots of unity) on the diagonal. By inserting $I_{2N}=(F_N^\dagger\otimes I_2)(F_N\otimes I_2)$, where $(F_N)_{kl}=e^{i2\pi kl/N}/\sqrt{N}\equiv\omega^{kl}/\sqrt{N}$ is the N-dimensional discrete Fourier transform, and noting that it leaves the block diagonal $I_N \otimes A = \mathcal{A}$ and $I_N \otimes B$ unaffected leads to

$$\begin{aligned} P &= I_{2N}\mathcal{A}I_{2N}S_{2N}^T I_{2N}(I_N \otimes B)I_{2N}S_{2N}I_{2N} \\ &= (F_N^\dagger \otimes I_2)\mathcal{A}(F_N \otimes I_2)S_{2N}^T(F_N^\dagger \otimes I_2)(I_N \otimes B)(F_N \otimes I_2)S_{2N}(F_N^\dagger \otimes I_2)(F_N \otimes I_2). \end{aligned}$$

Because of the above mentioned relation between amplitude and phase changes, $(F_N \otimes I_2)S_{2N}(F_N^\dagger \otimes I_2)$ is a block diagonal matrix with the kth block being

$$\begin{pmatrix} 0 & \omega^k \\ 1 & 0 \end{pmatrix} \equiv \boxed{L_k}. \tag{3}$$

Thus we finally have

$$\begin{aligned} P &= (F_N^\dagger \otimes I_2)(I_N \otimes A)\begin{bmatrix} \boxed{L_0} & 0_2 & \cdots \\ 0_2 & \ddots & \\ \vdots & & \boxed{L_{N-1}} \end{bmatrix}(I_N \otimes B)\begin{bmatrix} \boxed{L_0} & 0_2 & \cdots \\ 0_2 & \ddots & \\ \vdots & & \boxed{L_{N-1}} \end{bmatrix}(F_N \otimes I_2) \\ &\equiv (F_N^\dagger \otimes I_2)\begin{bmatrix} \boxed{K_0} & 0_2 & \cdots \\ 0_2 & \ddots & \\ \vdots & & \boxed{K_{N-1}} \end{bmatrix}(F_N \otimes I_2), \end{aligned} \tag{4}$$

where

$$\begin{aligned} \boxed{K_n} &= A\boxed{L_n}B\boxed{L_n} \\ &= \begin{bmatrix} \cosh\theta\cosh\phi - \sinh\theta\sinh\phi\omega^{-n} & i\cosh\phi\sinh\theta - i\cosh\theta\sinh\phi\omega^n \\ -i\cosh\phi\sinh\theta + i\cosh\theta\sinh\phi\omega^{-n} & \cosh\theta\cosh\phi - \sinh\theta\sinh\phi\omega^n \end{bmatrix} \end{aligned} \tag{5}$$

From K_n one obtains the eigenvalues $e^{\pm\gamma_n}$, where γ_n are determined via

$$\begin{aligned} \cosh\gamma_n &= \cosh\theta\cosh\phi - \cos(2\pi n/N)\sinh\theta\sinh\phi \\ &= \coth\phi\cosh\phi - \cos(2\pi n/N). \end{aligned} \tag{6}$$

The latter equality comes from the relation $\coth\theta = \cosh\phi$ given in Eq.(1). Note that the form of (4) brings into mind an interferometer — here the modes (the componenents of the input vector) are mixed in a balanced way by the transforms F_N and $F_N^\dagger$[12], and inbetween a two-mode rotation is performed to pairs of modes (c.f. a phase shift to single modes in a Mach-Zehnder interferometer). Utilizing this diagonalized form one can calculate the Mth power of P and describe the whole network in Fig.1.

An explicit expression for γ_n is given via the integral representation[8] $\gamma_n = \int_0^\pi d\nu/\pi \log[2(\coth\phi\cosh\phi - \cos(2\pi n/N) - \cos\nu)]$. There is a singularity in γ_0, and it defines the transition point. For active systems it is $\cosh\theta = \sqrt{2}$, and coincides with the cloning limit in quantum optics[13] above which an initially squeezed input loses its quantum properties. For passive systems the relation $\cosh\theta = \coth\phi$ given in Eq.(1) can be true for only one trivial choice of ϕ and θ, thus we have to define the network differently. For example if we fix $\theta = i\pi/4$ and let ϕ vary, then the transition will be between the adiabatic and diabatic regimes of the network, see[1]. Note that both the active and passive networks are maximally symmetric at the transition point.

Imagining a quantum computer to perform the transformation P helps to understand why collective phenomena can happen in the network even when only the nearest neighbours interact. The matrix between the Fourier transforms in (4) makes a rotation to the first qubit conditional on the states of *all* other qubits.

Considering the Mth power of P one can derive the observable output corrensponding to a certain input. For example for an symmetric input of the form $[1,0,1,0,...,1,0]/\sqrt{N}$ the output amplitudes are[1] $\cosh M\gamma_0$, or $\cos M\gamma_0$ for passive networks. The global amplification coefficient $M\gamma_0$, or the "phase" in the SU(2) case (c.f. an interferometer), has a discontinuity in its derivative $Md\gamma_0/d\phi$ at the transition point.

CONNECTION TO QUANTUM COMPUTATION AND QUANTUM CELLULAR AUTOMATA

The $2N \times 2N$ dimensional transformation P can be viewed also as a unitary operation applied in a $2N$-dimensional computational space spanned by $\log_2(2N)$ qubits — an operation that can be done by a quantum computer. Imagining a quantum computer to perform the transformation P helps to understand why collective phenomena can happen in the network even when only the nearest neighbours interact. The matrix between the Fourier transforms in (4) makes a rotation to the first qubit conditional on the states of *all* other qubits. Because, as shown clearly by Eq.(4), the transformation P is of interferometric type, it becomes close to some quantum algoritms such as the Grover's search algorithm[14] which is basically composed of mixing transformations such as the Hadamard transform (special case of the Fourier transform) and phase shifts.

Quantum cellular automata as treated in[15, 16] describe the quantum evolution of a particle by means of discretized space-time. In one-dimensional cellular automata, for example, the particle can move at each time step either left or right. The network of Fig.1, and described by P, corresponds to this when we interprete the vertical axis as space and the horizontal as time. In comparing with the definition of quantum cellular automata precented in[16], it is interesting to note that the condition $A \neq B$ corresponds to the existence of mass, and $A = B$, which is the situation at the transition point, is the case of a massless particle.

CONCLUSIONS

We have considered the question whether quantum networks can show transition phenomena, that is, singularities in the global parameters of the network. We explicitely constructed a network which has a mathematical analogy to the solution of the Ising model. Physical realizations of such networks as well as the interpretations of the transition points found were discussed. Connection to quantum computation and quantum cellular automata was pointed out — these are only two examples of the possible directions of related research in future.

Acknowledgements P.T. acknowledges the support by the TMR Programme of the European Commission (ERBFMBICT983061).

REFERENCES

1. P. Törmä, "Transitions in Quantum Networks", quant-ph/9802024, to appear in Phys. Rev. Lett. September (1998).
2. R. Feynman, Int. J. Theor. Phys. **21**, 467 (1982); P. W. Shor, in *Proc. of the 35th Ann. Symp. on the Foundations of Computer Science*, ed. S. Goldwasser (IEEE Computer Soc. Press, Los Alamitos, CA, 1994), p.124.
3. A. Barenco, A. Ekert, K-A. Suominen, and P. Törmä, Phys. Rev. A **54**, 139 (1996).
4. A. Ekert, Phys. Rev. Lett. **67**, 661 (1991); C. H. Bennett, Phys. Rev. Lett. **68**, 3121 (1992).

5. M. Reck, A. Zeilinger, H. J. Bernstein, and P. Bertani, Phys. Rev. Lett. **73**, 58 (1994); K. Mattle, M. Micheler, H. Weinfurter, A. Zeilinger, and M. Zukowski, Appl. Phys. B **60**, S111 (1995).
6. D. Bouwmeester, I. Marzoli, G. Karman, W. P. Schleich, and J. P. Woerdman, submitted to Phys. Rev. A.
7. D. A. Harmin, Phys. Rev. A **56**, 232 (1997).
8. L. Onsager, Phys. Rev. **65**, 117 (1944); B. Kaufmann, Phys. Rev. **76**, 1232 (1949); K. Huang, *Statistical Mechanics* (John Wiley & Sons, New York, 1987).
9. C. M. Caves, Phys. Rev. D **26**, 1817 (1982), and references therein.
10. *Rydberg States of Atoms and Molecules*, edited by R. F. Stebbings and F. B. Dunning (CUP, Cambridge, 1983).
11. The first two-qubit gate has been realized with trapped ions [C. Monroe, D. M. Meekhof, B. E. King, W. M. Itano, and D. J. Wineland, Phys. Rev. Lett. **75**, 4714 (1995)] according to the theoretical proposal in [J. I. Cirac and P. Zoller, Phys. Rev. Lett. **74**, 4091 (1995)]. Other proposals utilize for example cavity QED techniques [T. Pellizzari, S. A. Gardiner, J. I. Cirac, and P. Zoller, Phys. Rev. Lett. **75**, 3788 (1995); Q. A. Turchette, C. J. Hood, W. Lange, H. Mabuchi, and H. J. Kimble, Phys. Rev. Lett. **74**, 4710 (1995)], solid state structures [A. Barenco, D. Deutsch, A. Ekert, and R. Josza, Phys. Rev. Lett. **74**, 4083 (1995)] and NMR [N. A. Gerschenfeld and I. L. Chuang, Science **275**, 350 (1997)].
12. The Fourier transform matrices F_N can be viewed as generalizations of balanced beam splitters (F_2) into higher dimensions, and can be efficiently constructed from two-dimensinal components with FFT-type procedures [P. Törmä, S. Stenholm, and I. Jex, Phys. Rev. A **52**, 4853 (1995); H. Paul, P. Törmä, T. Kiss, and I. Jex, Phys. Rev. Lett. **76**, 2464 (1996)].
13. S. Friberg and L. Mandel, Opt. Comm. **46**, 141 (1983); R. Loudon and T. J. Shepherd, Optica Acta **31**, 1243 (1984); S. Stenholm, Opt. Comm. **58**, 177 (1986); U. Leonhardt, Phys. Rev. A **49**, 1231 (1994).
14. L. K. Grover, Phys. Rev. Lett. **79**, 324 (1997).
15. G. Grössing and A. Zeilinger, Complex Systems **2**, 197 (1988); I. Bialynicki-Birula, Phys. Rev. D **49**, 6920 (1994) and references therein.
16. D. A. Meyer, Phys. Rev. E **55**, 5261 (1997).

EXPERIMENTAL DEMONSTRATION OF COHERENT QUANTUM FEEDBACK

Richard J. Nelson,[1] Yaakov Weinstein,[1] Seth Lloyd,[1] and David G. Cory[2]

[1]d'Arbeloff Laboratory for Information Systems and Technology
Department of Mechanical Engineering
[2]Department of Nuclear Engineering
Massachusetts Institute of Technology
Cambridge, Massachusetts, 02139

In the conventional picture of quantum feedback control, sensors make measurements on a quantum system, a classical controller processes the results of the measurements, and semi-classical actuators act back on the system to alter its behavior. Quantum measurements are well known to yield probabilistic results and to introduce unavoidable disturbances in the system measured. Accordingly, quantum feedback is traditionally considered to be both probabilistic and destructive. This paper proposes and provides an experimental demonstration of an alternative method for quantum feedback control, in which sensors, controller, and actuators are themselves quantum systems that interact coherently with the system to be controlled. The resulting feedback loop preserves quantum coherence. Such coherent quantum feedback control is deterministic, non-destructive, and can be used to perform actions such as transferring quantum entanglement that are not possible using conventional feedback control of quantum systems.

INTRODUCTION

Feedback control of classical dynamic systems involves measuring system variables, computing inputs, and applying those inputs back to the system to be controlled. These three main tasks are carried out by elements in the feedback loop usually called sensors, controller, and actuators, respectively. (See Fig. 1.) Traditionally, when this control scheme is applied to a quantum mechanical system, the result is a feedback loop that necessarily destroys quantum coherence because some or all of the elements are semi-classical.[1, 2, 3, 4, 5, 6, 7, 8, 9, 10, 11, 12] For example, if the sensor is a semi-classical device that measures the spin state of a spin-1/2 quantum system, the measurement causes the wave function to collapse probabilistically to one of two possible eigenstates, irreversibly destroying the original wave function in the process. In addition, if the controller is a digital computer, then a von Neumann type (*i.e.*, wave function collapse–inducing) measurement must be made to digitize the measurement to classical data that can be processed by the controller.

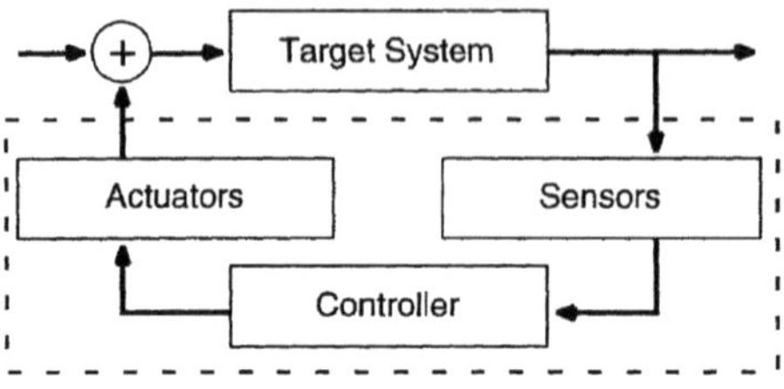

Figure 1. A Feedback System. The solid lines and arrows represent signals, or the flow of information. Traditional feedback schemes for quantum mechanical systems use semi-classical measuring devices as sensors, necessarily destroying quantum phase information. However, if all the subsystems within the dotted line are themselves quantum mechanical, the information in the feedback loop can coherently processes quantum phase information in addition to classical information.

This paper proposes a new method for controlling quantum mechanical systems in which the sensors, controller, and actuators are themselves quantum mechanical systems that interact coherently with the quantum system to be controlled. By eliminating the necessity of a measurement, the fully quantum mechanical feedback loop can drive the system to states that are impossible to achieve via traditional (semi-classical) feedback. Such a feedback loop is then completely deterministic and reversible. To demonstrate this approach, a fully quantum feedback loop has been designed to perform a task no semi-classical controller can perform, and then implemented in a nuclear magnetic resonance (NMR) setting.

CORRELATION TRANSFER VIA FULLY QUANTUM FEEDBACK

One consequence of the quantum theory is the propensity to create quantum states that are more correlated than classically allowed.[13, 14, 15, 16, 17] Creation and manipulation of such correlations is necessary for many of the emerging applications in quantum mechanical engineering today, including quantum computation,[18, 19, 20, 21] quantum teleportation, and quantum cryptography.[22]

Two quantum systems may exhibit a range of correlations, including classical correlations, quantum correlations that do not violate Bell's inequalities (yet cannot be created classically), and entangled states. Specifically, if the density matrix for two spin-1/2 systems (denoted spins B and C) can be written

$$\rho^{BC} \propto \mathbf{1} + bI^B + cI^C + dI^B I'^C \tag{1}$$

for some generalized Pauli matrices I^B, I^B, and I'^C, then the correlation between B and C can be transferred by a classical controller. A composite system whose density matrix satisfies Eq. 1 is separable[23] and can be transferred by a classical controller because it contains only classical information, while a composite system whose density matrix cannot be written as in Eq. 1 contains quantum phase information that would necessarily be destroyed by a semi-classical measurement.

Let A be the spin system to be controlled, and suppose the desired final state of A is a quantum correlation unseparable from an ancillary, possibly remote spin C. It is possible to use another quantum system, spin B, with which system A can interact coherently, to close the quantum feedback loop. One way this can be accomplished is by selecting as the controller spin B that is in the desired final state of the target system

A, and performing a transfer via three conditional rotations (or controlled-NOT's). The first controlled-NOT, where spin B is flipped conditioned on the state of A, acts as the quantum sensor: B acquires quantum information about A. The last two controlled-NOT's allow the information acquired by B during the quantum sensing to be fed back to system A in a coherent fashion, thus allowing B to also act as the controller and quantum actuator.

EXPERIMENTAL RESULTS USING NMR

Demonstration of a fully quantum feedback loop that puts an unknown initial state into a desired final state is accomplished using the three ^{13}C spins on an enriched alanine molecule. As above, denote spin A the target spin, B the quantum controller, and C an ancilla spin. The ensemble of spins begins in a highly mixed state at thermal equilibrium at room temperature. The initial density matrix for the systems together is approximately

$$\rho_o \approx \frac{1}{8}\mathbf{1} + \frac{1}{4}\beta\left(I_z^A + I_z^B + I_z^C\right) , \tag{2}$$

where β is the Boltzmann factor (of order 10^{-6} at room temperature), and where $I_z^k = \frac{1}{2}\sigma_z^k$, the Pauli spin matrix around z. The deviation density matrix—the part of ρ that varies from the identity—represents the surplus or deficit populations in energy levels (compared to complete unpolarization) caused by the background field, and is denoted by $\bar{\rho}$. Only these populations permit a detectable signal, with the spectral lines at the Larmor frequencies (including the effects of chemical shifts), split by the scalar couplings in the system's natural Hamiltonian:

$$\mathcal{H} = \omega_A I_z^A + \omega_B I_z^B + \omega_C I_z^C + 2\pi J_{AB} I_z^A I_z^B + 2\pi J_{BC} I_z^B I_z^C + 2\pi J_{AC} I_z^A I_z^C \,. \tag{3}$$

Using a Bruker AMX400 spectrometer (background field of 9.6 Tesla) and a transmitter frequency of $\omega = 100.6049426$ MHz, the spectral lines for A, B, and C appear at $\omega_A = \omega + 14298.4$ Hz, $\omega_B = \omega + 1720.9$ Hz, and $\omega_C = \omega - 1720.9$ Hz, respectively, split by the coupling constants $J_{AB} = 53.4$ Hz, $J_{BC} = 35.3$ Hz, and/or $J_{AC} = 1.4$ Hz.

Correlation Preparation

A quantum correlation between the controller spins B and the ancilla spins C is produced by the following sequence:

$$\rho_0 \longrightarrow \left[\frac{\pi}{2}\right]_y^{BC} \longrightarrow \frac{1}{4J_{BC}} \longrightarrow [\pi]_x^A \longrightarrow \frac{1}{4J_{BC}} \longrightarrow \left[\frac{\pi}{2}\right]_x^B \longrightarrow \rho_i \,. \tag{4}$$

In this notation, the bracketed expressions refer to radio frequency (rf) pulses that rotate the superscripted spins by the given angle, around the subscripted axis; terms involving the scalar coupling constants indicate a delay for the indicated time to allow natural Hamiltonian evolution. The state of the system after the above sequence has a deviation density matrix $\bar{\rho}_i \propto -I_z^A + 2I_z^B I_z^C - 2I_y^B I_y^C$. This density matrix represents a mixture of Einstein-Rosen-Podolsky (EPR) entangled states involving spins B and C; if the same sequence were applied to a perfectly polarized sample, B and C would be in a strongly-entangled, pure EPR state. Note that this density matrix cannot be written in the form given in Eq. 1; hence, the correlation between B and C cannot be transferred to A by semi-classical means.

The state of the system can be determined experimentally using state tomography: rf pulses can be applied to tip selected spins around known axes, the resulting spectral lines integrated, and the data compiled to map all elements of the density matrix. An example of a trace is given in Fig. 2, which shows the spectra after a $\left[\frac{\pi}{2}\right]^B_{-x}$ rotation is applied to ρ_i. The trace shows a correlation between the B and C spins, but the B and C spins are each completely uncorrelated with A. Using the tomographic technique, the actual deviation density matrix is experimentally determined to be

$$\begin{aligned}\bar{\rho}_i \approx &+ 0.01\left(I_x^A\right) - 1.00\left(I_z^A\right)\\ &- 0.10\left(I_y^B\right) - 0.02\left(2I_y^B I_z^C\right) + 0.04\left(2I_z^B I_x^C\right) + 1.00\left(2I_y^B I_y^C\right) - 0.99\left(2I_z^B I_z^C\right)\ ,\end{aligned}$$

which verifies the correlated state between B and C.

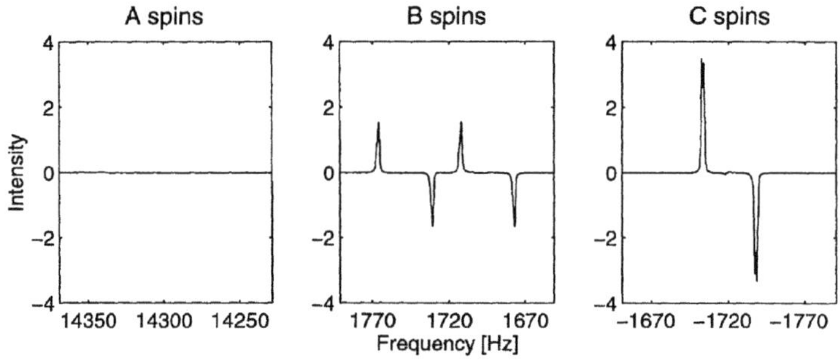

Figure 2. Quantum Correlation Spectrum in Alanine. The traces are the Fourier–transformed signals from the free–induction decay of the sample, as registered by pick-up coils in the $x-y$ plane. The A spins lie along z, and produce no signal. In the trace for B, the frequency difference between the peaks of the doublet on the left (or on the right) is J_{BC}; the frequency difference between the first and third peak (or second and fourth peak) equals J_{AB}. The spectrum shows a correlation with the C spins: the spectral lines are "up" ("down") whenever the state of C is aligned (anti-aligned) with the external magnetic field. There is no correlation between B and A, since half of the spectral lines for B are "up" and half "down" for each possible state of A. The trace for C shows the same correlations as in B (C correlated with B, but not with A), although the small value of J_{AC} makes it difficult to fully resolve both peaks in the left and right doublets.

Quantum Feedback Loop

The B spins are now prepared in the state into which it is desired to put the A spins. Although in this particular experiment the A spins are in a state with deviation density matrix proportional to I_z^A, there is nothing special about this initial state, and indeed the feedback loop outlined here functions for any initial state of A.

The first sequence in the quantum feedback, implemented as

$$\rho_i \longrightarrow \left[\frac{\pi}{2}\right]^B_{-x} \longrightarrow \frac{1}{4J_{AB}} \longrightarrow [\pi]^C_x \longrightarrow \frac{1}{4J_{AB}} \longrightarrow \left[\frac{\pi}{2}\right]^B_y \longrightarrow \rho_j\ , \tag{5}$$

allows B to acquire information about the state of A by means of coherent sensing—becoming coherently correlated with A without making a measurement. The deviation density matrix $\bar{\rho}_j$ is now proportional to $-I_z^A - 4I_z^A I_z^B I_z^C - 2I_x^B I_y^C$.

The controller now processes the information acquired during the coherent sensing stage and feeds information back into the system via coherent interactions (actuation). Thus, the sequence implemented as

$$\rho_j \longrightarrow \left[\frac{\pi}{2}\right]^A_x \longrightarrow \frac{1}{4J_{AB}} \longrightarrow [\pi]^C_x \longrightarrow \frac{1}{4J_{AB}} \longrightarrow \left[\frac{\pi}{2}\right]^A_y \longrightarrow \rho_k\ , \tag{6}$$

flips the A spins conditioned on the state of the B spins, producing a deviation density matrix $\bar{\rho}_k \propto 2I_z^A I_z^B - 2I_z^A I_z^C + 4I_x^A I_y^B I_y^C$. Then the final sequence

$$\rho_k \longrightarrow \left[\frac{\pi}{2}\right]_x^B \longrightarrow \frac{1}{4J_{AB}} \longrightarrow [\pi]_x^C \longrightarrow \frac{1}{4J_{AB}} \longrightarrow \left[\frac{\pi}{2}\right]_y^B \longrightarrow \rho_f \tag{7}$$

removes the correlation the coherent sensor/controller/actuator has with A, leaving as the final deviation density matrix $\bar{\rho}_f \propto -I_z^B + 2I_z^A I_z^C - 2I_y^A I_y^C$.

The final density matrix, determined experimentally using the tomography method outlined above reveals

$$\begin{aligned}\bar{\rho}_f \approx & + 0.12\left(I_x^B\right) - 0.96\left(I_z^B\right) \\ & - 0.05\left(I_y^A\right) + 0.06\left(I_z^A\right) + 0.03\left(I_y^C\right) - 0.02\left(2I_x^A I_x^C\right) - 0.04\left(2I_x^A I_y^C\right) \\ & - 0.02\left(2I_y^A I_x^C\right) - 0.02\left(2I_z^A I_y^C\right) + 0.95\left(2I_y^A I_y^C\right) - 0.98\left(2I_z^A I_z^C\right) .\end{aligned}$$

A spectrographic trace indicative of the final state of A, B, and C is shown in Fig. 3. As expected, the trace shows A and C perfectly correlated (along 2 sets of axes), while the B spins are uncorrelated with both A and C.

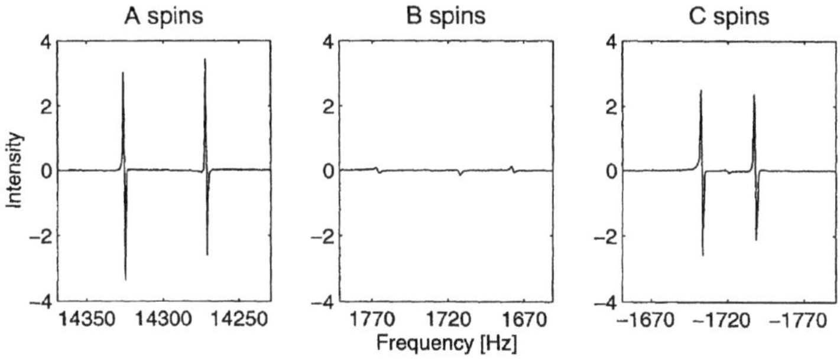

Figure 3. Results After Quantum Feedback. Now the A and C spins are correlated with each other, but both are completely uncorrelated with B. The B spins are along the z axis and produce almost no observable signal. Comparison with Fig. 2 shows that B has successfully transferred its correlation with C to the A spins—something that no semi-classical feedback loop can accomplish.

Using $F(\bar{\rho}_f, \bar{\rho}_i) \equiv 1 - [\mathrm{Tr}(\bar{\rho}_f - \bar{\rho}_i)^2/\mathrm{Tr}\bar{\rho}_i^2]$ as a measure of the fidelity of the feedback loop, the transfer of the quantum correlation occurred with 0.915 fidelity. The fidelity for the transfer of the initial state of A to the final state of B was 0.915; the fidelity for the transfer from the initial correlated state between B and C to the final correlated state between A and C was 0.885. These data demonstrate the unitarity of transformation: the feedback loop preserves quantum coherence. The main errors came from imperfect flip angles, noise in the data, and numerical errors. Note that decoherence processes were negligible since the entire execution lasted only 42.3 ms—much shorter that the smallest T_2 for the spins (about 450 ms for the C spins). Finally, no direct transfer of quantum information between A and C could have occurred since the only interaction allowing such communication is much slower ($1/2J_{AC} \approx 360$ ms) than the program duration.

CONCLUSIONS

The NMR methods used to construct the quantum feedback loop in the above experiment are not new. However, fully coherent quantum feedback is not restricted to NMR applications only. The explicit construction of the feedback loop demonstrated here serves to underscore the important potentials quantum feedback offers. Specifically, coherent quantum feedback control is deterministic and non-destructive: initial quantum correlations are preserved, and no probabilistic measurement is ever performed. Fully quantum feedback can be used to perform actions (such as transferring quantum correlations, even entanglement) that are not possible using conventional feedback control of quantum systems. Finally, applications for fully quantum feedback include, but are not limited to, stabilizing quantum dynamics,[24] quantum trajectory tracking, and quantum error correction.[18, 19, 20, 21]

REFERENCES

1. G. Huang, T. Tarn, and J. Clark, On the controllability of quantum-mechanical systems, *J. Math. Phys.* 24:2608-2618 (1983).
2. A. Blaquiere, S. Diner, and G. Lochak, eds. "Information, Complexity and Control in Quantum Physics," Springer-Verlag, New York (1987).
3. A. Blaquiere, "Modeling and Control of Systems in Engineering, Quantum Mechanics, Economics and Biosciences," Springer-Verlag, New York (1989).
4. A. Butkovskiy and Y. Samoilenko, "Control of Quantum-Mechanical Processes and Systems," Kluwer Academic, Dordrecht (1990).
5. H. Ezawa and Y. Murayama, eds., "Quantum Control and Measurement," North-Holland, Amsterdam (1993).
6. A. Peirce, M. Dahleh, and H. Rabitz, Optimal control of quantum-mechanical systems: existence, numerical approximation, and applications, *Phys. Rev. A* 37:4950-4964 (1988).
7. M. Dahleh, A. Peirce, and H. Rabitz, Optimal control of uncertain quantum systems, *Phys. Rev. A* 42:1065-1079 (1990).
8. W. Warren, H. Rabitz, and M. Dahleh, Coherent control of quantum dynamics: the dream is alive, *Science* 259:1581-1589 (1993).
9. V. Ramakrishna, M. Salapaka, M. Dahleh, H. Rabitz, and A. Peirce, Controllability of molecular systems, *Phys. Rev. A* 51:960-966 (1995).
10. H. Wiseman and G. Milburn, Squeezing via feedback, *Phys. Rev. A* 49:1350-1366 (1994).
11. H. Wiseman, Quantum theory of continuous feedback, *Phys. Rev. A* 49:2133-2150 (1994).
12. M. Keller and G. Mahler, Nanostructures, entanglement and the physics of quantum control, *J. Mod. Opt.* 41:2537-2555 (1994).
13. A. Einstein, B. Podolsky, and N. Rosen, Can quantum- mechanical description of reality be considered complete?, *Phys. Rev.* 47:777-780 (1935).
14. D. Bohm, "Quantum Theory," Prentice-Hall, Englewood Cliffs (1951).
15. J. Bell, On the Einstein Podolsky Rosen paradox, *Physics* 1:195-200 (1964).
16. C. Bennett, D. DiVincenzo, J. Smolin, and W. Wootters, Mixed-state entanglement and quantum error correction, *Phys. Rev. A* 54:3824-3851 (1996).
17. J. Schlienz and G. Mahler, Description of entanglement, *Phys. Rev. A* 52:4396-4404 (1995).
18. P. Shor, Scheme for reducing decoherence in quantum computer memory, *Phys. Rev. A* R2493-R2496 (1995).
19. A. Calderbank and P. Shor, Good quantum error-correcting codes exist, *Phys. Rev. A* 54:1098-1105 (1996).
20. A. Steane, Error-correcting codes in quantum theory, *Phys. Rev. Lett.* 77:793-797 (1996).
21. R. Laflamme, C. Miquel, J. Paz, and W. Zurek, Perfect quantum error correcting code, *Phys. Rev. Lett.* 77:198-201 (1996).
22. C. Bennett, Quantum information and computation, *Physics Today* 48(10):24-30 (1995).
23. A. Peres, "Quantum Theory: Concepts and Methods," Kluwer, Dordrecht (1993).
24. S. Lloyd and J.-J. Slotine, Analog quantum error correction, *Phys. Rev. Lett.* 80:4088-4091 (1998).

CLASSICAL BEHAVIOR OF A MACROSCOPIC SCHRÖDINGER CAT

Carlo Presilla

Dipartimento di Fisica, Università di Roma "La Sapienza"
Piazzale A. Moro 2, Rome, Italy, 00185
and INFM, Unità di Ricerca di Roma "La Sapienza"

We study the dynamics of classical and quantum systems linearly interacting with a classical environment represented by an infinite set of harmonic oscillators. The environment induces a dynamical localization of the quantum state into a generalized coherent state for which the $\hbar \to 0$ limit always exists and reproduces the classical motion. We describe the consequences of this localization on the behavior of a macroscopic system by considering the example of a Schrödinger cat.

INTRODUCTION

The problem of how classical behavior is regained from quantum mechanics in the macroscopic limit can be conceptually solved by recognizing that a macroscopic system is never completely isolated by the external world. It has been argued[1, 2, 3] that the interaction with an environment can, after a transient whose duration presumably depends on the coupling strength, drive the totality of the admissible states of the Hilbert space into those having classical limit, formally $\hbar \to 0$.

In a recent paper in collaboration with R. Onofrio and M. Patriarca[4], we substantiated this conjecture by analyzing the dynamics of general classical and quantum systems linearly interacting with an infinite set of degrees of freedom. Here, we briefly review the main results of this model and describe in detail how the pathologies of a simple Schrödinger cat are cured by the presence of the environment.

DYNAMICS OF SYSTEMS INTERACTING WITH AN ENVIRONMENT

Let us consider a system described by the classical Hamiltonian

$$H(p,q,t) = \frac{p^2}{2m} + V(q,t) \tag{1}$$

We model its interaction with a classical environment by a linear coupling to an infinite set of degrees of freedom $\{P_n, Q_n\}$. The Hamiltonian for the total system is

$$H_{tot} = H(p,q,t) + H_m(P, Q - q), \tag{2}$$

Quantum Communication, Computing, and Measurement 2
Edited by Kumar et al., Kluwer Academic / Plenum Publishers, New York, 2000.

where

$$H_m(P, Q-q) = \sum_n \left[\frac{P_n^2}{2M} + \frac{M\omega_n^2}{2}(Q_n - q)^2 \right]. \tag{3}$$

The classical dynamics of the system modified by the environment is described in terms of equations obtained by formally solving the harmonic motion of $\{P_n, Q_n\}$. For an environment having frequencies $\{\omega_n\}$ distributed with density $dN/d\omega = \theta(\Omega - \omega)2m\gamma/\pi M\omega^2$, we get, for times $\gg \Omega^{-1}$, the Markovian evolution

$$dp(t) = -\left[\gamma p(t) + \partial_q V(q(t), t)\right] dt + \sqrt{2m\gamma k_B T}\xi(t)dt \tag{4}$$

$$dq(t) = \frac{p(t)}{m} dt \tag{5}$$

with initial conditions p' and q' at time t'. If the the initial conditions $\{P_n', Q_n'\}$ of the environment are chosen as a realization of the equilibrium Gibbs measure at temperature T, then $\xi(t)$ is a realization of a stochastic process in time with respect to the same measure with properties $\overline{\xi(t)} = 0$ and $\overline{\xi(t)\xi(s)} = \delta(t-s)$. Therefore, Eqs. (4,5) are stochastic Langevin equations.

In alternative to the detailed stochastic description, we may be interested to determine the average behavior of the system obtained by considering all the possible realizations of the initial conditions of the environment. In this case, the system is described by a probability density $W(p, q, t)$ solution of the Fokker-Plank equation associated to (4-5)

$$\partial_t W(p, q, t) = \left[-\frac{p}{m}\partial_q + \partial_q V(q, t)\partial_p + \partial_p \left(\gamma p + m\gamma k_B T \partial_p\right) \right] W(p, q, t) \tag{6}$$

with initial conditions $W(p, q, t') = \delta(p - p')\delta(q - q')$.

The classical analysis can be repeated at quantum level. Besides obvious technical modifications, there is now a conceptual difference. Since we do not know how to describe the coupling of classical and quantum degrees of freedom, we must start with a quantum description of both the system and the environment. The condition of classical behavior of the environment can be reintroduced later by asking that the thermal energy $k_B T$ is much larger than the energy spacing of the highest-frequency oscillators $\hbar\Omega$. If this high temperature condition is satisfied, for times $\gg \Omega^{-1}$ the system is described by the nonlinear stochastic Schrödinger equation

$$\begin{aligned} d|\psi_{[\xi]}(t)\rangle &= -\frac{i}{\hbar}\left[\hat{H}(\hat{p}, \hat{q}, t) + \frac{\gamma}{4}(\hat{p}\hat{q} + \hat{q}\hat{p}) \right] |\psi_{[\xi]}(t)\rangle dt \\ &\quad -\frac{1}{2}\left[\hat{A}^\dagger \hat{A} + a(t)^* a(t) - 2a(t)^* \hat{A} \right] |\psi_{[\xi]}(t)\rangle dt + \left[\hat{A} - a(t) \right] |\psi_{[\xi]}(t)\rangle \xi(t) dt, \end{aligned} \tag{7}$$

where $\hat{A} = \sqrt{2m\gamma k_B T/\hbar^2}\,\hat{q} + i\sqrt{\gamma/8mk_B T}\,\hat{p}$, $a(t) = \langle\psi_{[\xi]}(t)|\hat{A}|\psi_{[\xi]}(t)\rangle$, and $\xi(t)$ is a real white noise.

A direct characterization of the average properties of the quantum system is also possible. By introducing the reduced density matrix operator

$$\hat{\rho}(t) = \overline{|\psi_{[\xi]}(t)\rangle\langle\psi_{[\xi]}(t)|} \tag{8}$$

and the associated Wigner function

$$W(p, q, t) = \frac{1}{2\pi\hbar} \int dz \, \exp\left(\frac{i}{\hbar} pz\right) \langle q - \frac{z}{2}|\hat{\varrho}(t)|q + \frac{z}{2}\rangle, \tag{9}$$

from Eq. (7) we obtain

$$\partial_t W(p,q,t) = \Big[-\frac{p}{m}\partial_q + \sum_{n=0}^{\infty} \left(\frac{\hbar}{2i}\right)^{2n} \frac{1}{(2n+1)!} \partial_q^{2n+1} V(q,t) \partial_p^{2n+1} + \partial_p \left(\gamma p + m\gamma k_B T \partial_p\right) + \frac{\hbar^2 \gamma}{16 m k_B T} \partial_q^2 \Big] W(p,q,t). \tag{10}$$

Note that (7) and (10) reduce to the corresponding quantum equations for an isolated system when $\gamma = 0$.

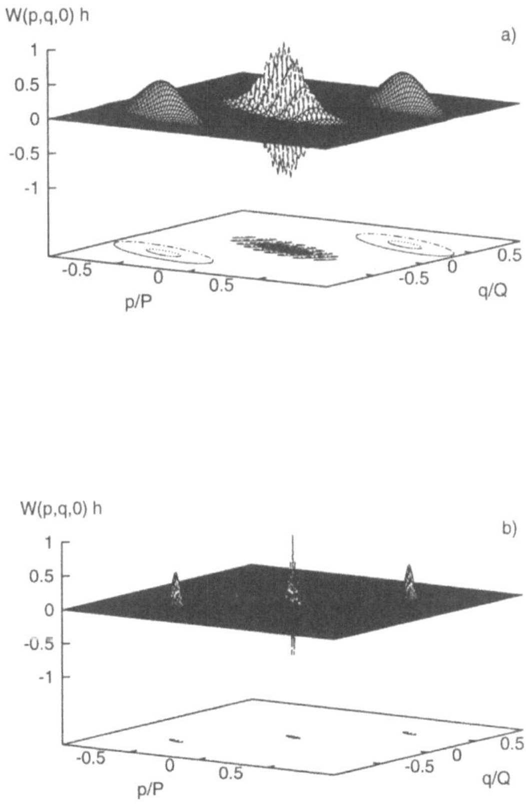

Figure 1. Wigner function at time 0 corresponding to the cat state (16) for $k_B T/\hbar\gamma = 100$. In case a) we have $P = 20\sqrt{\hbar m \gamma}$ and $Q = 2\sqrt{m\gamma/\hbar}$ while in b) $P = 100\sqrt{\hbar m\gamma}$ and $Q = 10\sqrt{m\gamma/\hbar}$.

DYNAMICAL LOCALIZATION INTO A COHERENT STATE

In the previous section, we have described the equations which govern the dynamics of classical and quantum systems in interaction with a classical environment. Now, we show that the quantum dynamics reduces to the classical one when the limit $\hbar \to 0$

is taken. Note that this is not always possible in the case of an isolated system where well known pathological limits are encountered[5].

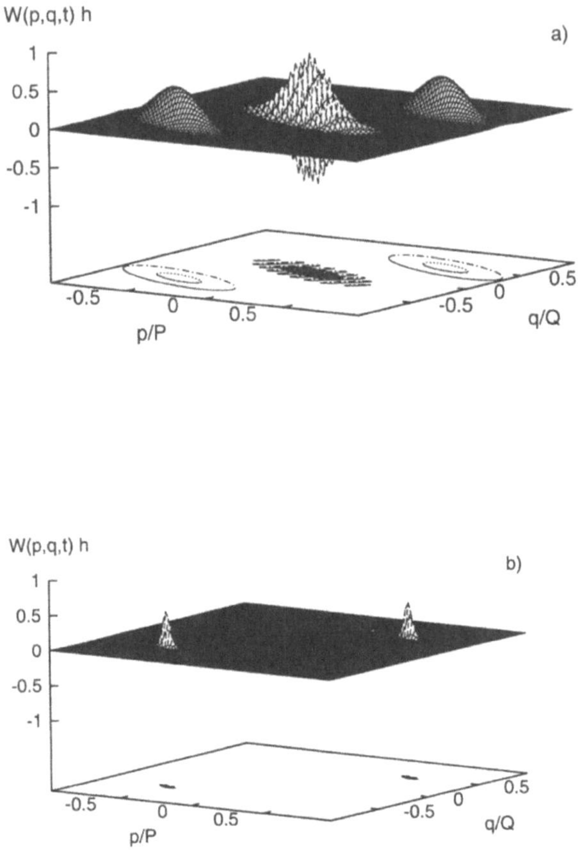

Figure 2. As in Fig. 1 but at the adimensional time $\gamma t = \frac{1}{2}10^{-3}$. Note that the adimensional characteristic time (20) is about $\frac{1}{4}10^{-2}$ in case a) and 10^{-4} in case b).

It is possible to demonstrate rigorously[6,4] that for a system with potential

$$V(q,t) = v_0(t) + v_1(t)q + \frac{1}{2}m\omega_0^2 q^2 \tag{11}$$

after a time which, in the worst case, is of the order of $\sqrt{\hbar/\gamma k_B T}$ the solutions of Eq. (7) become of the form

$$|\psi_{[\xi]}(t)\rangle = \exp\left[-\frac{i}{\hbar}\varphi(t)\right] \; |p(t)q(t)\rangle, \tag{12}$$

where $\varphi(t)$ is real, $p(t) = \langle\psi_{[\xi]}(t)|\hat{p}|\psi_{[\xi]}(t)\rangle$, $q(t) = \langle\psi_{[\xi]}(t)|\hat{q}|\psi_{[\xi]}(t)\rangle$, and

$$\langle q|p(t)q(t)\rangle = (2\pi\sigma_q^2)^{-1/4}\exp\left\{-\frac{1-\frac{2i}{\hbar}\sigma_{pq}^2}{4\sigma_q^2}\,[q-q(t)]^2+\frac{i}{\hbar}p(t)[q-q(t)]\right\}. \tag{13}$$

The state (13) defined in terms of the parameters

$$\sigma_q^2 = \frac{\hbar}{m}\sqrt{\frac{\gamma^2 - \omega_0^2 + \sqrt{(\gamma^2-\omega_0^2)^2 + 16\,(\gamma k_B T/\hbar)^2}}{32\,(\gamma k_B T/\hbar)^2}} \tag{14}$$

$$\sigma_{pq}^2 = \sqrt{m^2(\gamma^2-\omega_0^2)\sigma_q^4 + \frac{\hbar^2}{4}} - m\gamma\sigma_q^2 \tag{15}$$

is a generalized coherent state which admits the $\hbar \to 0$ limit. Moreover, the convergence into this coherent state takes place in a time which vanishes for $\hbar \to 0$. Therefore, a linear system like (11) always has classical limit at any time $t > t'$ even if the $\hbar \to 0$ limit does not exist at the initial time t'. Since the contribution to the Green function of Eq. (7) due the system-environment coupling is of the form $q^2\hbar^{-1}$, these results apply in the limit $\hbar \to 0$ also to nonlinear systems.

We illustrate the consequences of the localization into a coherent state induced by the coupling with the environment by analyzing the evolution of a cat state. For simplicity, consider a free quantum particle which at time $t' = 0$ is in the superposition state

$$|\psi(0)\rangle = N\left(|\frac{1}{2}P - \frac{1}{2}Q\rangle + |-\frac{1}{2}P\frac{1}{2}Q\rangle\right) \tag{16}$$

where $|\pm\frac{1}{2}P \pm \frac{1}{2}Q\rangle$ are coherent states (13-15) with $\omega_0 = 0$ and N is a normalization factor. The state (16) has no classical counterpart. Indeed, its corresponding Wigner function

$$W(p,q,0) = N^2\left\{ W_{\frac{1}{2}P-\frac{1}{2}Q}(p,q) + W_{-\frac{1}{2}P\frac{1}{2}Q}(p,q) + W_{00}(p,q)\,2\cos\left[\frac{p}{\hbar}Q + \frac{q}{\hbar}P\right]\right\} \tag{17}$$

does not have $\hbar \to 0$ limit due to the presence of the last oscillating term. In Eq. (17), we indicated with $W_{p_0q_0}(p,q)$ the Wigner function of the coherent state $|p_0q_0\rangle$ having limit $W_{p_0q_0}(p,q) \stackrel{\hbar\to 0}{\longrightarrow} \delta(p-p_0)\delta(q-q_0)$. In more physical terms, if we consider a macroscopic limit, for instance increase the values of P and/or Q, the oscillating interference term in (17) never disappear contrarily to common sense. An example of this pathological behavior is shown in Fig. 1.

At a later time, the situation is different. By solving Eq. (10) with the initial condition (17), we find the following expression for the Wigner function

$$W(p,q,t) = N^2\Big\{ W_{\frac{1}{2}P-\frac{1}{2}Q}(p,q,t) + W_{-\frac{1}{2}P+\frac{1}{2}Q}(p,q,t) + W_{00}(p,q,t)$$
$$\times e^{-C_{P-Q}+\Sigma_{P-Q}(t)}\,2\cos\left[p\Upsilon_{P-Q}(t) + q\Phi_{P-Q}(t)\right]\Big\}. \tag{18}$$

The definitions of $W_{p_0q_0}(p,q,t)$, C_{P-Q}, $\Sigma_{P-Q}(t)$, $\Upsilon_{P-Q}(t)$, and $\Phi_{P-Q}(t)$ can be found in Ref. 4. The functions $W_{p_0q_0}(p,q,t)$ are solution of Eq. (10) with initial condition $W_{p_0q_0}(p,q)$. In the $\hbar \to 0$ limit, they reduce to phase-space probability densities $W^{\text{cl}}_{p_0q_0}(p,q,t)$ solution of the classical Fokker-Plank equation (6) with initial condition $\delta(p-p_0)\delta(q-q_0)$. The exponential term $\exp[-C_{P-Q}+\Sigma_{P-Q}(t)]$ vanishes for $\hbar \to 0$ at any $t > 0$ and

$$\lim_{\hbar\to 0} W(p,q,t) = \frac{1}{2}\left[W^{\text{cl}}_{\frac{1}{2}P-\frac{1}{2}Q}(p,q,t) + W^{\text{cl}}_{-\frac{1}{2}P\frac{1}{2}Q}(p,q,t)\right]. \tag{19}$$

The classical limit is equivalently reached for large values of P and/or Q. An estimate of the critical values of P and Q for which the Wigner function changes from (18) to (19) can be obtained by observing that $\Sigma_{P-Q}(t) = C_{P-Q} - t/\tau + \mathcal{O}(t^2)$ with the characteristic time τ given by

$$\gamma\tau \simeq \left(\frac{k_B T}{\hbar\gamma} \frac{\hbar Q^2}{m\gamma} + \frac{1}{16} \frac{\hbar\gamma}{k_B T} \frac{P^2}{\hbar m\gamma} \right)^{-1} . \tag{20}$$

Note that $\tau(P,Q) \to 0$ when P and/or Q diverge. At a chosen time $t > 0$, we have quantum behavior for $\tau(P,Q) \gg t$ (microscopic system) and classical behavior for $\tau(P,Q) \ll t$ (macroscopic system). An example of this quantum-to-classical transition is shown in Fig. 2.

Finally, we note that $\Sigma_{P-Q}(t)$, $\Upsilon_{P-Q}(t)$, and $\Phi_{P-Q}(t)$ vanish for $t \to \infty$ so that in this limit the Wigner function becomes

$$W_\infty(p,q,t) = N^2 \left\{ W^{\rm cl}_{\frac{1}{2}P-\frac{1}{2}Q}(p,q,t) + W^{\rm cl}_{-\frac{1}{2}P\frac{1}{2}Q}(p,q,t) + W^{\rm cl}_{00}(p,q,t) e^{-C_{P-Q}2} \right\}. \tag{21}$$

The classical limit, formally $\hbar \to 0$, is reached only for a macroscopic system, not in the long time limit of a microscopic one.

REFERENCES

1. E. Joos and H. D. Zeh, *Z. Phys. B* 59: 223 (1985).
2. W. H. Zurek, *Phys. Rev. D* 24: 1516 (1981); 26: 1862 (1982).
3. M. Cini, *Nuovo Cimento B* 73: 27 (1983).
4. C. Presilla, R. Onofrio, and M. Patriarca, *J. Phys. A* 30:7385 (1997).
5. E. Schrödinger, *Naturwissenschaften* 23: 807 (1935); 23: 823 (1935); 23: 844 (1935) [English translation by J. P. Trimmer, *Proc. Am. Philos. Soc.* 124: 323 (1980)].
6. J. Halliwell and A. Zoupas, *Phys. Rev. D* 55: 4697 (1997).

PART IV: QUANTUM TELEPORTATION AND CRYPTOGRAPHY

FREE-SPACE QUANTUM CRYPTOGRAPHY

R. J. Hughes, W. T. Buttler, P. G. Kwiat, S. K. Lamoreaux,
G. L. Morgan, J. E. Nordholt, and C. G. Peterson

University of California, Los Alamos National Laboratory
Los Alamos, New Mexico 87545, USA

An experimental free-space quantum key distribution (QKD) system has been tested over an outdoor optical path of $\sim$ 1 km under nighttime conditions at Los Alamos National Laboratory. This system employs the Bennett 92 protocol; here we give a brief overview of this protocol, and describe our experimental implementation of it. An analysis of the system efficiency is presented, as well as a description of our error detection protocol, which employs a two-dimensional parity check scheme. Finally, the susceptibility of this system to eavesdropping by various techniques is determined. Possible applications include the rekeying of satellites in low earth orbit.

INTRODUCTION

Quantum cryptography was introduced in the mid-1980s[1] as a new method for generating the shared, secret random number sequences, known as cryptographic keys, that are used in crypto-systems to provide communications security. The appeal of quantum cryptography is that its security is based on laws of Nature, in contrast to existing methods of key distribution that derive their security from the perceived intractability of certain problems in number theory,[2] or from the physical security of the key distribution process.

Since the introduction of quantum cryptography, several groups have demonstrated quantum communications[3,4] and key distribution[5–10] over multi-kilometer distances of optical fiber. Free-space QKD (over an optical path of 32 cm) was first introduced in 1991,[11] and recent advances have led to demonstrations of QKD over free-space indoor optical paths of 205 m,[12] and outdoor optical paths of 75 m.[13] These demonstrations increase the utility of QKD by extending it to line-of-site laser communications systems. Indeed there are certain key distribution problems in this category for which free-space QKD would have definite practical advantages (for example, it is impractical to send a courier to a satellite). Here we report our results of free-space QKD over outdoor optical paths of up to 950 m under nighttime conditions.[14]

Quantum Communication, Computing, and Measurement 2
Edited by Kumar et al., Kluwer Academic / Plenum Publishers, New York, 2000.

Table 1. Observation Probabilities

Alice's Bit Value	"0"	"0"	"1"	"1"
Bob Tests With	"1"	"0"	"1"	"0"
Observation Probability	p= 0	p= $\frac{1}{2}$	p= $\frac{1}{2}$	p= 0

The Bennett 92 Protocol

A QKD procedure starts with the sender, "Alice," generating a secret random binary number sequence. For each bit in the sequence, Alice prepares and transmits a single photon to the recipient, "Bob," who measures each arriving photon and attempts to identify the bit value Alice has transmitted. Alice's photon state preparations and Bob's measurements are chosen from sets of non-orthogonal possibilities. For example, using the B92 protocol[18] Alice agrees with Bob (through public discussion) that she will transmit a horizontal-polarized photon, $|h\rangle$, for each "0" in her sequence, and a right-circular-polarized photon, $|r\rangle$, for each "1" in her sequence. Bob agrees with Alice to randomly test the polarization of each arriving photon with vertical polarization, $|v\rangle$, to reveal "1s," or left-circular polarization, $|\ell\rangle$, to reveal "0s." In this scheme, Bob will never detect a photon for which he and Alice have used a preparation/measurement pair that corresponds to different bit values, such as $|h\rangle$ and $|v\rangle$, which happens for 50% of the bits in Alice's sequence. However, for the other 50% of Alice's bits the preparation and measurement protocols use non-orthogonal states, such as $|h\rangle$ and $|\ell\rangle$, resulting in a 50% detection probability for Bob, as shown in Table 1. Thus, by detecting single-photons Bob identifies a random 25% portion of the bits in Alice's random bit sequence, assuming a single-photon Fock state with no bit loss in transmission or reception. This 25% efficiency factor, η_Q, is the price that Alice and Bob must pay for secrecy.

Bob and Alice reconcile their common bits through a public discussion by revealing the locations, but not the bit values, in the sequence where Bob detected photons; Alice retains only those detected bits from her initial sequence. The resulting detected bit sequences comprise the raw key material from which a pure key is distilled using classical error detection techniques. The single-photon nature of the transmissions ensures that an eavesdropper, "Eve," can neither "tap" the key transmissions with a beam splitter (BS), owing to the indivisibility of a photon,[19] nor copy them, owing to the quantum "no-cloning" theorem.[20] Furthermore, the non-orthogonal nature of the quantum states ensures that if Eve makes her own measurements she will be detected through the elevated error rate she causes by the irreversible "collapse of the wavefunction.[21]"

Quantum-Key Transmitter: Alice

The faithful transmission of polarized single photons through a turbulent medium (the atmosphere), receiving them with non-negligible probability and detecting them against a high ambient background, appear to be serious obstacles to free-space QKD. However, these obstacles can be overcome by exploiting sub-nanosecond timing techniques, narrow wavelength filters,[15,16] spatial filtering,[12,14] and adaptive optics.[17]

The QKD transmitter for our experiments (Fig. 1) consisted of a temperature-controlled single-mode (SM) fiber-pigtailed diode laser, a fiber to free-space launch system, a 2.5-nm bandwidth interference filter (IF), a variable optical attenuator, a polarizing beam splitter (PBS), a low-voltage Pockels cell, and a 27× beam expander. The diode laser wavelength is temperature adjusted to 772 nm, and the laser is con-

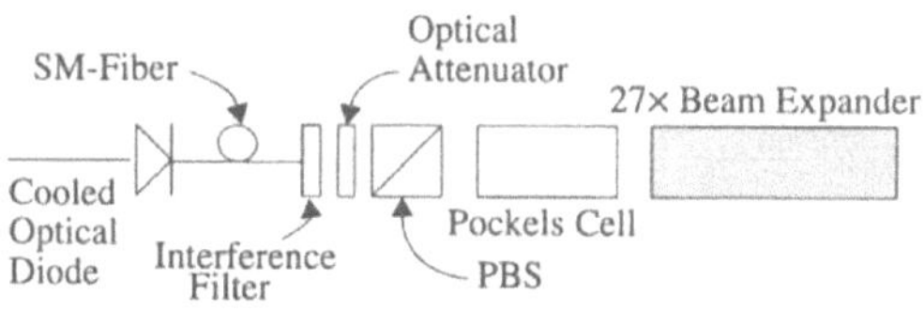

Figure 1. QKD Transmitter.

figured to emit a short, coherent pulse of approximately 1-ns length, containing $\sim 10^5$ photons.

A computer control system (Alice) starts the QKD protocol by pulsing the diode laser at a rate previously agreed upon between herself and the receiving computer control system (Bob). Each laser pulse is launched into free-space through the IF, and the ~ 1 ns optical pulse is then attenuated to an average of less than one photon per pulse, based on the assumption of a statistical Poisson distribution.[22] (The attenuated pulse only approximates a "single-photon" state; we tested the system with averages down to less than 0.1 photon per pulse. This corresponds to a 2-photon probability of $< 0.5\%$ and implies that less than 6 of every 100 detectable pulses will contain 2 or more photons, i.e., for a Poisson distribution with an average photon number of $\bar{n} = 0.1$, for every 1000 pulses there will be ~ 905 empty pulses, ~ 90 pulses of 1 photon, ~ 5 pulses of 2 photons, and ~ 1 pulse of 3 or more photons.) The photons that are transmitted by the optical attenuator are then polarized by the PBS, which transmits an average of less than one $|h\rangle$ photon to the Pockels cell. The Pockels cell is randomly switched to either pass the "single-photon" unchanged as $|h\rangle$ (zero-wave retardation) or change it to $|r\rangle$ (quarter-wave retardation). The random switch setting is determined by discriminating the voltage generated by a white noise source.

Quantum-Key Receiver: Bob

The free-space QKD receiver (Fig. 2) comprised a 8.9 cm Cassegrain telescope followed by the receiver optics and detectors. The receiver optics consisted of a 50/50 BS that randomly directs collected photons onto either of two distinct optical paths. The lower optical path contained a polarization controller (a quarter-wave retarder and a half-wave retarder), adjusted as an effective quarter-wave retarder, followed by a PBS to test collected photons for $|h\rangle$ (at first glance this may be confusing, but the effective quarter wave retarder converts $|h\rangle$ to $|r\rangle$ leading to a 50% probability an $|h\rangle$ photon will be detected); the upper optical path contained a half-wave retarder followed by a PBS to test for $|r\rangle$. The output port along each optical path was coupled by multi-mode (MM) fiber to a single-photon counting module (SPCM: EG&G part number: SPCM-AQ 142-FL). [Although the receiver did not include IFs, the spatial filtering provided by the MM fibers effectively reduced noise caused by the ambient background during nighttime operations to negligible levels (the background was ~ 1.1 kHz).]

Bit values are determined in the following fashion: a single $|r\rangle$ photon traveling along the lower path encounters the polarization controller, and is converted to $|v\rangle$ and reflected away from the SPCM by the PBS, but a single $|h\rangle$ photon traveling the same path is converted to $|r\rangle$ and transmitted toward or reflected away from the SPCM in this path with equal probability; in contrast, a single $|h\rangle$ photon traveling the upper path is converted to $|v\rangle$ and reflected away from the SPCM in this path, but a single

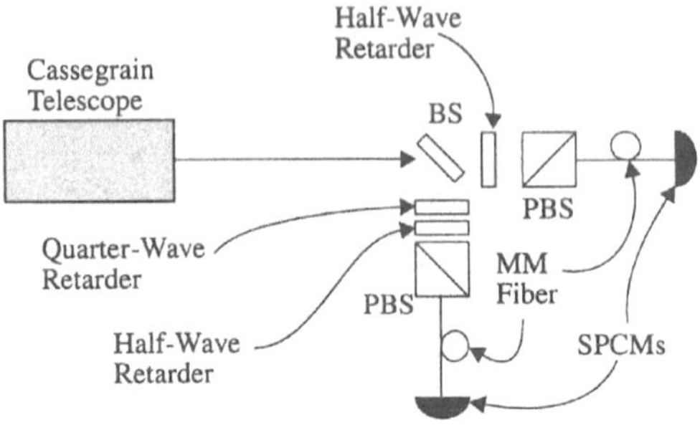

Figure 2. QKD receiver.

$|r\rangle$ photon traveling this path is converted to $|\ell\rangle$ and transmitted toward or reflected away from the SPCM with equal probability.

In this detection scheme, there are a total of four possible optical paths through the receiver, but only two of the paths, those which terminate upon the detectors seen in Fig. 2, contain definite polarization information (definite in the sense that Bob can know what polarization Alice has transmitted if one of these detectors fire). However, while the remaining two paths contain indeterminate polarization information (indeterminate in the sense that Bob cannot know with certainty whether Alice has transmitted $|h\rangle$, or $|r\rangle$ if a detector placed in either of these paths fires), this information remains important for the secure implementation of B92.

OUTDOOR FREE-SPACE EXPERIMENTS

The transmitter and receiver optics were operated over 240-, 500-, and 950-m outdoor optical paths, with the transmitter and receiver collocated in order to simplify data acquisition. The various total optical path lengths were determined by positioning a 25.4 cm diameter mirror at the transmission distance half way point that reflected the transmitted beam back to the receiver. All measurements were made at night.

System Efficiency

The optical coupling efficiency between the transmitter and receiver for the 950-m path was $\eta \sim 14\%$, which accounts for losses between the transmitter and the MM fibers at the receiver. Bob's detection probability,

$$P_B = e^{-\bar{n}} \sum_{n=1}^{\infty} \frac{\bar{n}^n}{n!} [1 - y^n] = 1 - e^{-\bar{n}\,\eta_B} , \tag{1}$$

is the convolution of the Poisson probability distribution of photons in Alice's transmitted weak pulse with average photon number $\bar{n}$, and the probability that Bob detects at least 1 photon. Here, $y = (1 - \eta_B)$, where $\eta_B = \eta \cdot \eta_D \cdot \eta_Q$, and $\eta_D = 65\%$ is Bob's detector efficiency. When the transmitter was pulsed at a rate of 20 kHz with an average of 0.1 photon per pulse for the 950-m path, Eq. 1 gives $\bar{n} \cdot \eta_B = 0.1 \cdot (0.14 \cdot 0.25 \cdot 0.65) \sim 2.3 \times 10^{-3}$, and hence a bit rate in agreement with the experimental result of $\sim$ 50 Hz.

The bit error rate (BER, defined as the ratio of the bits received in error to the total number of bits received) for the 950-m path was $\sim$ 1.5% when the system was

Table 2. A 200-Bit Sample of Alice's (A) and Bob's (B) Raw Key Material Generated by QKD over 1 km.

A	0000010101	1101101001	0000000000	0110010101
B	0000010101	1101101001	0000000000	0110010101
A	0011100010	0111011101	1110111000	0100100011
B	0011100010	0111011101	1110111000	0100100011
A	1110000000	0101101111	1001001010	0010000011
B	1110000000	0101101111	1001001010	0010000011
A	0000010111	0000111111	1111000000	1010101101
B	0000010111	0000111111	11**0**1000000	1010101101
A	1111100111	1110111101	0100110100	1011101111
B	11111000**0**11	1110111101	0100110100	1011101111

operating down to < 0.1 photon per pulse level. (A BER of $\sim 0.7\%$ was observed over the 240-m optical path and a BER of $\sim 1.5\%$ was also observed over the 500-m optical path.) A sample of raw key material from the 950-m experiment, with errors, is shown in Table 2.

Bit errors caused by the ambient background were minimized to less than ~ 1 every 9 s by narrow gated coincidence timing windows (~ 5 ns) and spatial filtering. Further, because detector dark noise (~ 80 Hz) contributed only about 1 dark count every 125 s, we believe that the observed BER was mostly caused by misalignment and imperfections in the optical elements (wave-plates and Pockels cell).

Error Detection

Our experiments implement a two-dimensional (2D) parity check scheme that allows the generation of error-free key material. Error detection is accomplished by Bob and Alice organizing their reconciled bits into 2D square matrices in the order that they were detected. Once organized, the parities of the rows and columns are determined and openly exchanged between Alice and Bob, and any column or row in which Bob and Alice possess different parities is discarded. To ensure privacy, Alice and Bob also discard the bits oriented along the diagonals of their matrices. This guarantees the elimination of two bits for each row and column of the matrix, even when no errors are detected, eliminating knowledge revealed during the parity exchange.

EAVESDROPPING BY EVE

The original form of the B92 protocol has a weakness to an opaque attack by Eve. For example, Eve could measure Alice's photons in Bob's basis and only send a dim photon pulse when she identifies a bit. However, if Eve retransmits each observed bit as a single-photon she will noticeably lower Bob's bit-rate. To compensate for the additional attenuation to Bob's bit-rate Eve could send on a dim photon pulse of an intensity appropriate to raise Bob's bit-rate to a level similar to her own bit-rate with Alice. [In fact, if Eve sends a bright classical pulse (a pulse of a large average photon number) she guarantees that Bob's bit-rate is close to her own bit-rate with Alice.] However, this type of attack would be revealed by our two SPCM system through an increase in "dual-fire" errors, which occur when both SPCMs fire simultaneously. In a perfect system dual-fire errors would not exist, regardless of the average photon number per pulse, but in a real experimental system, where bit-errors occur, dual-fire errors will occur. (We have used the dual-fire information to estimate the average number of photons per pulse reaching the SPCMs.) Our system could also be modified to operate under the BB84 protocol[1] which also protects against an opaque attack.

Eve could also passively, or translucently, attack the the system using a BS and a receiver identical to Bob's (perhaps of even higher efficiency) to identify some of the bits for which Alice's weak pulses contain more than 1 photon, i.e., Eve receives pulses reflected her way by the BS which has reflection probability R, whereas Bob receives the transmitted pulses, and the BS has transmission probability $T = 1 - R$. Introducing a coupling and detection efficiency factor η_E, for Eve, analogous to Bob's η_B, we find that Eve's photon detection probability is $P_E = 1 - e^{-\bar{n}\,\eta_E\,R}$, whereas Bob's detection probability becomes $P_B = 1 - e^{-\bar{n}\,\eta_B\,T}$. (Note: we do not explicitly consider any eavesdropping strategy, with or without guessing, in which Eve might use more than 2 detectors.)

The important quantity in a BS attack is the ratio of the number of bits Eve shares with Bob to the number of bits Bob and Alice share. We find that the probability that Eve and Bob will both observe a photon on the same pulse from Alice is[23,24]

$$P_{B\wedge E} = [1 - e^{-\bar{n}\,\eta_E\,R}]\,[1 - e^{-\bar{n}\,\eta_B\,T}]\,. \tag{2}$$

To take an extreme case, if Eve's BS has $R = 0.9999$, her efficiency is perfect (i.e., $\eta_E = 0.25$), and Alice transmits pulses of $\bar{n} = 0.1$, then Eve's knowledge $P_{B\wedge E}/P_B$ of Bob and Alice's common key will never be more than 2.5%. Thus, Alice and Bob have an upper bound on the amount of privacy amplification[25] needed to protect against a BS attack. Of course, such an attack would cause Bob's bit-rate to drop to near zero; for smaller reflection coefficients, R, Eve's information on Bob and Alice's key is reduced. For example, if Alice transmits pulses of $\bar{n} = 0.1$, and $R = T = 0.5$, then for every 250 key bits Alice and Bob acquire, Eve will know ~ 3 bits.

CONCLUSIONS

The results in this paper demonstrate free-space QKD through a turbulent medium under nighttime conditions. We have described a system that provides two parties a secure method to secretly communicate with a simple system based on the B92 protocol. This system was operated at a variety of average photon number per pulse down to an average of < 0.1 photon per pulse. The results were achieved with low BERs, and the 240-m experiment demonstrated that BERs of 0.7% or less are achievable with this system. This protocol could be implemented with classical signature authentication[2] and privacy amplification procedures to ensure the security of private information.

As a final discussion, we consider the feasibility to transmit the quantum states required in QKD between a ground station and a satellite in a low earth orbit. To that end, we designed our QKD system to operate at 772 nm where the atmospheric transmission from surface to space can be as high as 80%, and where single-photon detectors with efficiencies as high as 65% are commercially available; at these optical wavelengths atmospheric depolarizing effects are negligible, as is the amount of Faraday rotation experienced on a surface to satellite path.

To detect a single QKD photon it is necessary to know when it will arrive. The photon arrival time can be communicated to the receiver by using a bright precursor reference pulse. Received bright pulses allow the receiver to set a 1-ns time window within which to look for the QKD photon. This short time window reduces background photon counts dramatically, and the background can be further reduced by using narrow bandwidth filters.

Atmospheric turbulence impacts the rate at which QKD photons would arrive at a satellite from a ground station transmitter. Assuming 30-cm diameter optics at both the transmitter and satellite receiver, the diffraction-limited spot size would be $\sim$ 1.2-m diameter at a 300-km altitude satellite. However, turbulence induced beam-wander can vary from $\sim 2.5 - 10$ arc-seconds leading to a photon collection efficiency at the satellite of $10^{-3} - 10^{-4}$. Thus, with a laser pulse rate of 10 MHz, an average of one photon-per-pulse, and atmospheric transmission of $\sim$ 80%, photons would arrive at the collection optic at a rate of $800 - 10,000$ Hz. Then, with a 65% detector efficiency, the 25% intrinsic efficiency of the B92 protocol, IFs with transmission efficiencies of $\sim$ 70%, and a MM fiber collection efficiency of $\sim$ 40%, we find a key generation rate of $35 - 450$ Hz is feasible. With an adaptive beam tilt corrector the key rate could be increased by about a factor of 100 leading to a key rate of $3.5 - 45$ kHz; these rates will double using the BB84 protocol.

Errors would arise from background photons collected at the satellite. The night-time earth radiance observed at 300 km altitude at the transmission wavelength is ~ 1 mW m^{-2} str^{-1} μm^{-1}, or $\sim 4 \times 10^{16}$ photons s^{-1} m^{-2} str^{-1} μm^{-1}, during a full moon, dropping to $\sim 10^{15}$ photons s^{-1} m^{-2} str^{-1} μm^{-1} during a new moon. Assuming a 5 arc-seconds receiver field of view, and 1-nm IFs preceding the detectors, a background rate of $\sim$ 800 Hz (full moon), and $\sim$ 20 Hz (new moon) would be observed (with a detector dark count rate of $\sim$ 50 Hz, the error rate will be dominated by background photons during full moon periods, and by detector noise during a new moon). We infer a BER from background photons of $\sim 9 \times 10^{-5} - 10^{-3}$ (full moon), and $\sim 2 \times 10^{-6} - 3 \times 10^{-5}$ (new moon).

During daytime orbits the background radiance would be much larger ($\sim 10^{22}$ photons s^{-1} m^{-2} str^{-1} μm^{-1}), leading to a BER of $\sim 2 \times 10^{-2} - 3 \times 10^{-1}$, if an atomic vapor filter[26] of $\sim 10^{-3}$ nm bandwidth was used instead of the IF. (Note: it would also be possible to place the transmitter on the satellite. In this situation, the beam wander is similar, but it is only over the lowest $\sim$ 2 km of the atmosphere. In this situation, the bit-rate would improve by $\sim$ 150, decreasing the BER by the same amount.)

Because the optical influence of turbulence is dominated by the lowest $\sim$ 2 km of the atmosphere, our experimental results and this simple analysis show that QKD between a ground station and a low-earth orbit satellite should be possible on nighttime orbits and possibly even in full daylight. During the several minutes that a satellite would be in view of the ground station there would be adequate time to generate tens of thousands of raw key bits, from which a shorter error-free key stream of several thousand bits would be produced after error correction and privacy amplification. From these results we believe that it will be feasible to use free-space QKD for re-keying satellites in low-earth orbit from a ground station.

REFERENCES

1. C. H. Bennett and G. Brassard, "Quantum cryptography: public key distribution and coin tossing," *Proc. of IEEE Int. Conf. on Comp., Sys., and Sig. Proc.*, Bangalore, India, p. 175 (1984).
2. A. J. Menezes, P. C. van Oorschot, and S. A. Vanstone, "Handbook of Applied Cryptography," CRC Press, New York (1997).
3. A. Muller, J. Breguet, and N. Gisin, "Experimental demonstration of quantum cryptography using polarized photons in optical fibre over more than 1 km," Europhys. Lett. **23**, p. 383

(1993).
4. P. D. Townsend, J. G. Rarity, and P. R. Tapster, "Enhanced single-photon fringe visibility in a 10 km-long prototype quantum cryptography channel," *Elec. Lett.* **29**, p. 634 (1993).
5. J. D. Franson and H. Ilves, "Quantum cryptography using optical fibers," *Appl. Opt.* **33**, p. 2949 (1994).
6. C. Marand and P. D. Townsend, "Quantum key distribution over distances as long as 30 km," *Opt. Lett.* **20**, p. 1695 (1995).
7. R. J. Hughes, D. M. Alde, P. Dyer, G. G. Luther, G. L. Morgan, and M. Schauer, "Quantum cryptography," *Contemp. Phys.* **36**, p. 149 (1995).
8. R. J. Hughes, G. G. Luther, G. L. Morgan, C. G. Peterson, and C. M. Simmons, "Quantum cryptography over underground optical fibers," *Lecture Notes In Computer Science* **1109**, p. 329 (1996).
9. A. Muller, H. Zbinden, and N. Gisin, "Quantum cryptography over 23 km in installed under-lake telecom fiber," *Europhys. Lett.* **33**, p. 335 (1996).
10. R. J. Hughes, W. T. Buttler, P. G. Kwiat, G. G. Luther, G. L. Morgan, J. E. Nordholt, C. G. Peterson, and C. M. Simmons, "Secure communications using quantum cryptography," *Proc. of SPIE* **3076**, p. 2 (1997).
11. C. H. Bennett, F. Bessette, G. Brassard, L. Salvail, and J. Smolin, "Experimental quantum cryptography," *Lecture Notes In Computer Science* **473**, p. 253 (1991).
12. W. T. Buttler, R. J. Hughes, P. G. Kwiat, G. G. Luther, G. L. Morgan, J. E. Nordholt, C. G. Peterson, and C. M. Simmons, "Free-space quantum-key distribution," *Phys. Rev. A* **57**, p. 2379 (1998).
13. B. C. Jacobs and J. D. Franson, "Quantum cryptography in free space," *Opt. Lett.* **21**, p. 1854 (1996).
14. W. T. Buttler, R. J. Hughes, P. G. Kwiat, S. K. Lamoreaux, G. G. Luther, G. L. Morgan, J. E. Nordholt, C. G. Peterson, and C. M. Simmons, "Practical free-space quantum key distribution over 1 km," *Submitted to Phys. Rev. Lett.* (1998).
15. J. G. Walker, S. F. Seward, J. G. Rarity, and P. R. Tapster, "Range measurement photon by photon," *Quant. Opt.* **1**, p. 75 (1989).
16. S. F. Seward, P. R. Tapster, J. G. Walker, and J. G. Rarity, "Daylight demonstration of a low-light-level communication system using correlated photon pairs," *Quant. Opt.* **3**, p. 201 (1991).
17. C. A. Primmerman, D. V. Murphy, D. A. Page, B. G. Zollars, and H. T. Barclay, "Compensation of atmospheric optical distortion using a synthetic beacon," *Nature (London)* **353**, p. 141 (1991).
18. C. H. Bennett, "Quantum cryptography using any two nonorthogonal states," *Phys. Rev. Lett.* **68**, p. 3121 (1992).
19. J. F. Clauser, "Experimental distinction between quantum and classical field-theoretic predictions for photoelectric effect," *Phys. Rev. D* **9**, p. 853 (1974).
20. W. K. Wooters and W. H. Zurek, "A single quantum cannot be cloned," *Nature (London)* **299**, p. 802 (1982).
21. A. K. Ekert, B. Huttner, G. M. Palma, and A. Peres, "Eavesdropping on quantum cryptosystems," *Phys. Rev. A* **50**, p. 1047 (1994).
22. B. E. A. Saleh and M. C. Teich, "Fundamentals of Photonics," Jon Wiley and Sons, Inc., New York (1991).
23. W. T. Buttler, R. J. Hughes, P. G. Kwiat, S. K. Lamoreaux, G. G. Luther, G. L. Morgan, J. E. Nordholt, C. G. Peterson, and C. M. Simmons, "Free-space quantum key distribution at night," *Proc. of SPIE* **3385**, p. 14 (1998).
24. P. D. Townsend, "Quantum Cryptography over multi-user optical fibre networks," *Nature (London)* **385**, p. 47 (1997).
25. C. H. Bennett, G. Brassard, C. Crepeau, C., and U. Maurer, "Generalized privacy amplification," *IEEE Trans. Inf. Th.* **41** p. 1915 (1995).
26. H. Zhilin, X. Sun, and X. Zeng, "Rb 780 nm Faraday anomalous dispersion optical filter in a strong magnetic field," *Opt. Communications* **101**, p. 175 (1993).

PARAMETRIC SCHROEDINGER CAT WITH NO DECOHERENCE

S. Branca, F. De Martini, G. Di Giuseppe and F. Mariani

Dipartimento di Fisica, Universit "La Sapienza"
Istituto Nazionale di Fisica della Materia
00185, Roma, Italia.

The new process of *quantum-injection* into a parametric amplifier *amplifies* into a large dimensionality spin-$\frac{1}{2}$ Hilbert space the quantum superposition properties of the photon-couples generated by parametric down conversion. The structure of the Wigner function and of the field's correlation functions shows a multiphoton Schroedinger-cat behaviour of the emitted field which is largely detectable against the squeezed-vacuum noise.

INTRODUCTION

The generation of classically distinguishable quantum states, a major endeavor of modern physics, has long been the object of extensive theoretical studies. In recent times important experimental investigation with atoms has been carried out in this field by various research groups.[1,2,3,4,5] In this context it has been proved that the realization of the *Schroedinger-cat* program is generally challenged by an extremely rapid "decoherence" process due to the stochastic interactions of any freely evolving mesoscopic system with the environment.[6,7] Within the close framework of *quantum computation* the same process has also been recognized to represent a major limitation toward the coherent interaction of the set of particles carrying the quantum information.[8] In the domain of quantum optics several strategies have been proposed to partially overcome the problem, e.g., the back-action evasion[9] and the cavity control by optical feedback.[10,11] In the present letter we present a new approach to the problem based on the amplifying/squeezing operation of the optical parametric amplifier (OPA) when initiated by a process of *quantum injection*, e.g., provided by the sub-Poissonian character of a single photon in the Fock state $n = 1$.[12] This photon may belong to a couple generated by spontaneous parametric down-conversion (SPDC) in a nonlinear (NL) crystal cut for Type II phase-matching, in a Φ – *phase tunable* entangled state of some degree of freedom, e.g., of *linear polarization* π, defined within a Hilbert space of dimensionality 2×2.[13] This process is at the basis of the recent demostrations of the violation of Bell inequalities,[13] of quantum state-teleportation[14] and of all processes belonging to the general chapter of *nonlocal entangled interferometry.*[15,16] The key idea of the present work consists of "amplifying" the superposition character of the generated particles to a higher dimensionality spin-$\frac{1}{2}$ Hilbert space, i.e., involving a large number of photon

Quantum Communication, Computing, and Measurement 2
Edited by Kumar et al., Kluwer Academic / Plenum Publishers, New York, 2000.

couples, by taking advantage of the *unitary* character of the amplification process. Precisely we show that this can be realized by a novel optical device, the *quantum-injected, entangled-state* OPA leading to a new virtually *decoherence-free, Schroedinger-cat* (S-cat) configuration.

EXPERIMENT

Consider the diagram shown in Figure 1: two NL crystals, e.g., BBO ($\beta-$barium borate) cut for Type II phase-matching, are excited by two beams derived from a common UV laser at wavelength λ_p. Crystal 1 is the SPDC source of couples of $\pi-$entangled photons with $\lambda = 2\lambda_p$, emitted over the two output modes $\mathbf{k}_i$ ($i = 1, 2$) determined by two fixed pinholes according to the standard phenomenology.[13,15] We found that the entanglement phase $|\Phi|$ can be easily tuned over the range $0 - \pi$ by rotating by an angle ψ the crystal around the excitation axis $\mathbf{k}_p$, $\Phi(\psi)$ being a linear function.[13] The photon emitted over $\mathbf{k}_1$ provides the *quantum injection* into the OPA, physically consisting of the NL crystal-2 cut and oriented for collinear operation over the two linear polarization modes $\mathbf{k}_{1\perp}$ and $\mathbf{k}_{1||}$, respectively parallel and orthogonal to the horizontal. The photon associated with the output mode $\mathbf{k}_2$ is filtered by a polarization analyzer with axis at 45° and then detected by D_2. The $\pi -$ *analyzer* prevents the nonlocal, i.e., EPR type, state reduction on the correlated mode $\mathbf{k}_1$ that spoils the superposition character of the overall process. A detection event at D_2 opens a gate that selects the registered outcomes thus providing the *conditional* character of the experiment.[17]

An alternative solution for quantum injection is provided by a Type I NL crystal-1 feeding the OPA by a single photon with π oriented at 45°, the other photon exciting D_2 without any $\pi-$*selection*. In order to investigate the properties of the output beam after filtering by the filter F against UV, a birefringent plate $D(\Psi)$ and a Fresnel-rhomb $R(\varphi)$ induce respectively a field's phase delay $\Psi = (\psi_\perp - \psi_{||})$ and a $(45° + \varphi)$ $\pi-$*rotation*. Then the two orthogonal π components are detected by $D_\perp$ and $D_{||}$ after separation by a polarizing beam splitter PBS. This realizes a $\pi-$interferometer as we shall see.

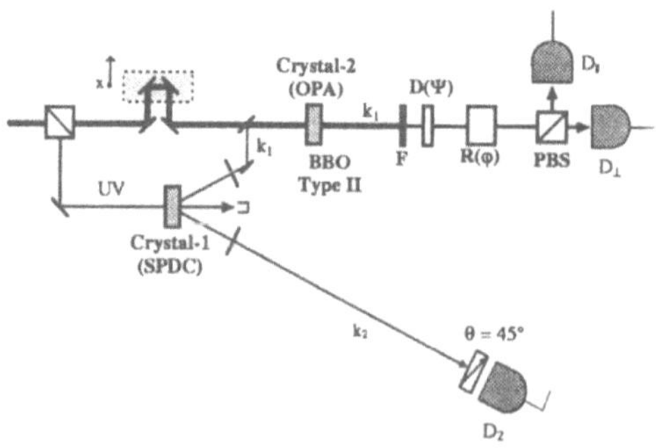

Figure 1. Optical configuration of the *quantum injected* optical parametric amplifier realizing the process of multiphoton quantum superposition. The SPDC *quantum injector* (crystal 1) is provided by a Type II $\Phi -$ *phase tunable* generator of *linear polarization* (π)-entangled photon couples. Crystal-2, realizing the OPA action is cut for *collinear* Type II phase matching. By a "trombone" device the time superposition in the OPA of the excitation UV pulse and of the injection photon wavepacket is obtained. After filtering (F) against UV, a birefringent slab (D) and a Fresnel-rhomb (R) determine a phase-shift Ψ and a rotation of π by an angle $\varphi + 45°$. Then the orthogonal $\pi-$omponents are directed toward two detectors by a polarizing beam splitter PBS. For simplicity we keep the the same denomination $\mathbf{k}_1$ for the injection mode involving both SPDC and OPA processes.

Let us analyze the amplification process in details. The input state is:

$$|\Psi\rangle_{IN} = \frac{1}{\sqrt{2}} \left[|1\rangle_{1\perp} \otimes |0\rangle_{1\|} + (\exp i\Phi) |0\rangle_{1\perp} \otimes |1\rangle_{1\|} \right] \tag{1}$$

The OPA process induces a unitary transformation on the couple of time dependent field operators: $\hat{a}(t) \equiv \hat{a}_{1\perp}$, and $\hat{b}(t) \equiv \hat{a}_{1\|}$ for which, at the initial time of the NL interaction, $t = 0$, is $\left[\hat{a}(0), \hat{a}(0)^\dagger\right] = \left[\hat{b}(0), \hat{b}(0)^\dagger\right] = 1$ and $\left[\hat{a}(0), \hat{b}(0)^\dagger\right] = 0$. A quantum analysis of the dynamics leads to a linear dependence of the field operators on the corresponding input quantities:[12]

$$\begin{bmatrix} \hat{a}(t) \\ \hat{b}(t)^\dagger \end{bmatrix} = \begin{bmatrix} C & S \\ S & C \end{bmatrix} \begin{bmatrix} \hat{a}(0) \\ \hat{b}(0)^\dagger \end{bmatrix} \tag{2}$$

being: $C \equiv \cosh g, S \equiv \sinh g$, $g \equiv \chi t$ = *amplification gain*, χ the coupling term proportional to the product of the 2nd-order NL susceptibility of the crystal and of the *pump* field, assumed classical and undepleted by the interaction. The evolution operator is then expressed in the form of the *squeeze operator*: $U_A(t) = \exp[g(\widehat{A}^\dagger - \widehat{A})]$ being: $\widehat{A}^\dagger \equiv \hat{a}(t)^\dagger \hat{b}(t)^\dagger, \widehat{A} \equiv \hat{a}(t)\hat{b}(t)$. By use of the *disentangling theorem*,[18] the output state is found:

$$|\Psi\rangle_{OUT} = G\left[\left|\overline{n}+1_\perp \otimes \overline{n}_\|\right\rangle + \exp(i\Phi)\left|\overline{n}+1_\| \otimes \overline{n}_\perp\right\rangle\right] \tag{3}$$

where $G \equiv (\frac{1}{2}C)^{-2}$. There the two mutually orthogonal, interfering *pure* states are given in the form:

$$\left|\overline{n}+1_\perp \otimes \overline{n}_\|\right\rangle \equiv \sum_{n=0}^{\infty} \Gamma^n \sqrt{n+1}\, |n+1\rangle_{1\perp} \otimes |n\rangle_{1\|} \tag{4}$$

with $\Gamma \equiv S/C$.[19]

In order to inspect the superposition status of our S-cat, let's determine the Wigner function of the output field state by first evaluating the *symmetrically* ordered *characteristic function* of the set of complex variables $(\eta, \eta^*, \xi, \xi^*) \equiv \{\eta, \xi\}$: $\chi_S\{\eta, \xi\} \equiv_{IN} \langle\Psi| D[\eta(t)] D[\xi(t)] |\Psi\rangle_{IN}$ expressed in terms of the *displacement* operators $D[\eta(t)] \equiv \exp[\eta(t)\hat{a}(0)^\dagger - \eta^*(t)\hat{a}(0)]$ and $D[\xi(t)] \equiv \exp[\xi(t)\hat{b}(0)^\dagger - \xi^*(t)\hat{b}(0)]$ where: $\eta(t) \equiv (\eta C - \eta^* S)$; $\xi(t) \equiv (\xi C - \xi^* S)$.[12] The Wigner function of the complex *phase-space variables* $(\alpha, \alpha^*, \beta, \beta^*) \equiv \{\alpha, \beta\}$ is the fourth dimensional Fourier transform of $\chi_S\{\eta, \xi\}$. By evaluating analytically in *closed form* both functions $\chi_S\{\eta, \xi\}$ and $W\{\alpha, \beta\}$ the *exact* result is obtained:

$$W\{\alpha, \beta\} = -(2/\pi)^2 \exp\left(-|\Delta|^2\right) \left[1 - \left|\Delta_A\, e^{+i\frac{\Phi}{2}} + \Delta_B\, e^{-i\frac{\Phi}{2}}\right|^2\right] \tag{5}$$

where the quantities:

$$|\Delta|^2 \equiv \left[|\gamma_{A+}|^2 + |\gamma_{A-}|^2 + |\gamma_{B+}|^2 + |\gamma_{B-}|^2\right]/2 \tag{6}$$

$$\Delta_A \equiv (\gamma_{A+} - i\gamma_{A-})/\sqrt{2} \qquad \Delta_B \equiv (\gamma_{B+} - i\gamma_{B-})/\sqrt{2} \tag{7}$$

are expressed in terms of the *squeezed variables*:

$$\gamma_{A+} \equiv (\alpha + \beta^*)\exp(-g) \qquad \gamma_{A-} \equiv i(\alpha - \beta^*)\exp(+g) \tag{8}$$

$$\gamma_{B+} \equiv (\beta + \alpha^*)\exp(-g) \qquad\qquad \gamma_{B-} \equiv i(\beta - \alpha^*)\exp(+g) \tag{9}$$

Inspection of Eq.5 shows precisely the *quantum superposition* character of the S-cat, with the same phase Φ affecting the injected state, Eq.1. Note in Eq.5 and in Fig. 2 the non definite positivity of $W\{\alpha,\beta\}$ over its definition space, which assures the overall quantum character of our multiparticle, quantum-injected amplification scheme.[12,20] Note also, most important, the absence of *definite positivity* of $W\{\alpha,\beta\}$ over the overall phase space $\{\alpha,\beta\}$. This assures the quantum character of the multiparticle system generated by our *quantum injected* amplification scheme.[7] A tridimensional representation of the Wigner function showing this relevant feature is given in Figure 2. There the two positive peaks of the distribution, corresponding to the two interfering multiparticle states, are clearly distinguishable, i.e., well resolved in phase space. All this reproduces the expected behaviour of any system realizing the condition of macroscopic quantum coherence.[2,7,20]

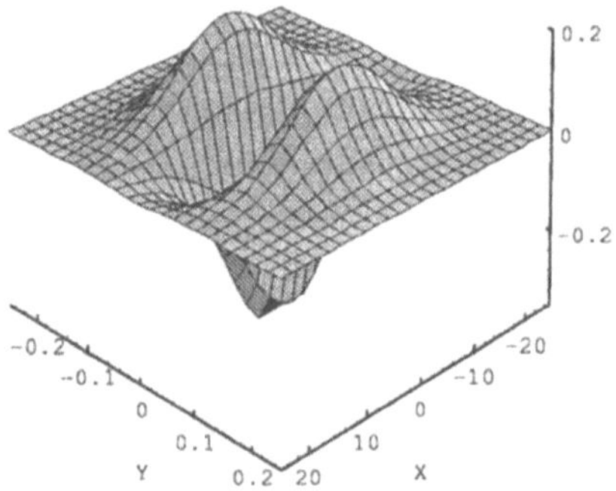

Figure 2. Tridimensional plot of the Wigner Function of the amplified field on mode $\mathbf{k}_1$ at the output of the *quantum injected* OPA as function of the *squeezed* variables: $X = (\alpha + \beta^*)\exp(-g)$; $Y = i(\beta - \alpha^*)\exp(+g)$, for an input superposition angle: $\Delta\Phi = \pi$ and a parametric *gain* g=2.5.

The quantum features of the system are also revealed by the *correlation functions* of the output fields.

The $1^{st}-$ *order* functions $G^{(1)}(\varphi) \equiv_{IN} \langle\Psi|[\widehat{N}(t)]_\varphi|\Psi\rangle_{IN}$ are ensemble averages of the number operators $\widehat{N}(t) \equiv \widehat{c}^\dagger(t)\widehat{c}(t)$ written in terms of the field exciting $D_{||}$ after phase-shifting and $\pi -$ *rotation*: $\widehat{c}(t) \equiv [\xi_\perp \hat{a}(t) + \xi_{||}\widehat{b}(t)]$, $[\widehat{c}(t), \widehat{c}^\dagger(t)] = 1$, $\xi_\perp \equiv 2^{-\frac{1}{2}}(\cos\varphi - \sin\varphi)\exp i\psi_\perp$, $\xi_{||} = 2^{-\frac{1}{2}}(\cos\varphi + \sin\varphi)\exp i\psi_{||}$. The general result is expressed by:

$$G^{(1)}(\varphi) = \overline{n} + (\overline{n} + \frac{1}{2})[1 + \cos(2\varphi)\cos\Delta\Phi \tag{10}$$

with $\Delta\Phi \equiv (\Psi - \Phi)$, and it leads to an *interference fringe visibility*

$$V = (G^{(1)}_{max} - G^{(1)}_{min})/(G^{(1)}_{max} + G^{(1)}_{min}) \geq \frac{1}{2} \tag{11}$$

for $\Delta\Phi \equiv 0$. V is limited by the unavoidable *squeezed-vacuum* noise, with average photon number $\overline{n} \equiv S^2$, but is large enough to allow for a clear detection of the quantum superposition implying a Schroedinger - cat process. This is presently being investigated experimentally in our laboratory by detection of the 1st -order inteference fringes as function of the angles φ and/or Ψ.

The 2nd – *order* correlation functions $G^{(2)}_{\varphi,\varphi'} \equiv_{IN} \langle\Psi| : [\widehat{N}(t)]_{\varphi}[\widehat{N}(t)]_{\varphi'} : |\Psi\rangle$ may also be measured by use of the detectors $D_{\perp}$ and $D_{||}$. Their expressions are given here for completeness, by postponing further comments to a future work: $G^{(2)}_{\varphi,\varphi} = 6\overline{n}^2 + 2\overline{n} + 3\overline{n}(\overline{n}+1)\cos^2(2\varphi) + 2\overline{n}(3\overline{n}+2)\cos(2\varphi)\cos\Delta\Phi$; $G^{(2)}_{\varphi,\overline{\varphi}} = 2\overline{n}(3\overline{n}+2)$ - $3\overline{n}(\overline{n}+1)\cos^2(2\varphi)$; $\overline{\varphi} \equiv \varphi + 90^0$.

The absence of *decoherence* within our ideal, non-dissipative multiparticle system is due to its nature of nonlinearly *driven* excitation. As such, it is coupled with a continuously re-phasing environment here provided by the parametric NL polarization. Similar situations are encountered in physics of the nonlinear dynamical systems, e.g., in nonlinear surface plasmon generation in solid state NL spectroscopy.[21] Of course any single photon loss event, mainly contributed in our case by stray reflections, implies an elementary decoherence process. In our experiment two equal 1mm thick, BBO crystals are excited by $0.2ps$ pulses at $\lambda_p = 400nm$ second-harmonic-generated by a mode-locked Ti:Sa laser at a 76 MHz rep-rate with an average power $P \approx 0.3W$. The detection system is equal to the one shown in Fig. 1. The initial phase is: $\Phi = 0$. All surfaces are treated by special AR coatings at $\lambda = 800nm$ with an overall transmittivity: $T \approx 99.60$. This figure implies the loss of a single photon every $20 \geq$ pulses with the generation of $\overline{n} \sim 10$ per pulse. This would make our S-cat experiment quite feasible.

CONCLUSION

In summary, we have given the quantum analysis of a novel *multi-particle* nonlinear system showing macroscopic, nearly decoherence free, quantum superposition features that can be detectable. This result is reached by a smart interplay of the fundamental paradigms of modern quantum optics, i.e., *quadrature-squeezing, multiparticle state entanglement* and *quantum nonseparability in parametric correlations*. From a foundational perspective, our *quantum injection* method and the simple application reported in the present work could prove useful within the realization of nonlocality and noncontestuality tests of quantum mechanics that have not yet been performed, e.g., the ones proposed by GHZ, Mermin and others.[22] In addition, within the fields of *quantum information* and *computation* it may represent a new way to amplify quantum coherence over large systems providing at the same time an elegant way to beat decoherence.[23] As a final comment, and referring to Fig. 1, a very interesting scheme would correspond to considering a OPA crystal cut and oriented for Type II *non-collinear* phase matching, i.e., emitting π*-entangled* photon couples over two *distinct* output modes, say $\mathbf{k}_1$ and $\mathbf{k}_3$.[13] This OPA may be fed by a single-photon *quantum-injection* over the mode $\mathbf{k}_1$ within a conditional experiment, similar to the present one. In this case an interesting combination of multi-photon $\pi-$*entanglement* and quantum superposition should affect the amplified output field.[24] A further, interesting and more complex 3-crystal variant of the above scheme consists of two independent SPDC devices feeding in a symmetrical fashion the OPA on *both* imput modes $\mathbf{k}_j$, by two distinct single-photon quantum-injection processes and within a double-conditional experiment. We thank M.D'Ariano, D.P.Di Vincenzo, G.Ghirardi, L.Hardy, M. Paris, A. Peres for enlightening discussions, the CEE-TMR Program (Contract N.ERBMRXCT96-066), MURST, and INFM (Contract PRA97-cat) for funding..

REFERENCES

1. E. Schroedinger, *Naturwissenshaften* 23, 807 (1935); 23, 823 (1935).
2. A. O. Caldeira and A. J. Leggett, *Physica A* 121, 587 (1983).
3. C. Monroe, D. M. Meekhof, B. E. King and D. J. Wineland, *Science* 272, 1131 (1996).
4. M. Brune, E. Hagley, J. Dreyer, X. Maistre, A. Maali, C. Wunderlich, J. M. Raimond and S. Haroche, *Phys. Rev. Lett.* 77, 4887 (1996) and: *Phys. Rev. Lett.* 79, 1964 (1997) (1983); *Ann. Phys.* (N.Y.) 149, 374 (1983); *Phys. Rev.A* 31, 1059 (1985).
5. M. W. Noel, C. R. Stroud, *Phys. Rev. Lett.* 77, 1913 (1996).
6. D. F. Walls and G. J. Milburn, *Phys. Rev. A* 31, 2403 (1985); *Phys. Rev. D* 32, 3208 (1985).
7. W. H. Zurek, *Phys. Rev. D* 26, 1862 (1982), *Physics Today*, October 1991 and April 93.
8. I. L. Chuang, R. Laflamme, P. W. Shor and W. H. Zurek, *Science*, 270, 1633 (1995).
9. S. Song, C. M. Caves and B. Yurke, *Phys. Rev. A*, 41, 5261 (1990).
10. H. M. Wiseman and G. I. Milburn, *Phys. Rev. A*, 49, 4110 (1994).
11. P. Tombesi and D. Vitali, *Phys. Rev. A*, 50, 4253 (1994).
12. D. F. Walls and G. I. Milburn, *Quantum Optics* (Springer-Verlag, Berlin, 1995) Ch. 5. In the present Letter *Quantum injection* is defined as the one provided by an input field whose $P-function$ doesn't exist as a tempered distribution.
13. D. Boschi , F. De Martini and G. Di Giuseppe in *Quantum Interferometry*, Ed. by F. De Martini, G. Denardo and Y. Shih (VCH, Weinheim, 1996). P. G. Kwiat, K. Mattle, H. Weinfurter, A. Zeilinger, A. V. Sergienko and Y. H. Shih, *Phys. Rev. Lett.*, 75, 4337 (1995); D. Boschi (Laurea Thesis, Università di Roma, June 1995).
14. D. Bouwmester, J. W. Pan, K. Mattle, M. Eibl, H. Weinfurter and A. Zeilinger, *Nature*, 390, 575 (1997); D. Boschi, S. Branca, F. De Martini, L. Hardy and S. Popescu, *Phys. Rev. Lett.* 80, 1121 (1998).
15. D. Boschi, F. De Martini and G. Di Giuseppe in *Fundamental Problems in Quantum Theory*, UMBC Baltimore, August 1997 (*Fortschritte der Physik*, 1998). Cf. also: K. Mattle, H. Weinfurter, P. G. Kwiat and A. Zeilinger, *Phys. Rev. Lett.* 76, 4656 (1996); G. Di Giuseppe, L.Heiberger, F. De Martini, A. V. Sergienko, *Phys. Rev. A* 56, R21 (1977).
16. S. Branca, D. Boschi, F. De Martini and G. Di Giuseppe, submitted to *Phys. Rev. A*.
17. A. Einstein, B. Podolsky and N. Rosen, *Phys. Rev.* 47, 777 (1935).
18. M. J. Collett, *Phys. Rev. A* 38, 2233 (1988).
19. The n-state distribution functions accounting for quantum injection are equal to zero at $n+1 = 0$ and their peak value is increasingly shifted toward higher values of n for an increasing number of injected photons in Fock state. These distributions are nonzero over the complete set of n-states, thus justifying the appeal to the *macroscopic* quantum coherence.
20. C. W. Gardiner, *Quantum Noise* (Springer, Berlin, 1991), Ch. 8.
21. F. De Martini and Y. R. Shen, *Phys. Rev. Lett.* 36, 216 (1976); F. De Martini, P. Mataloni, E. Palange and Y. R. Shen, *Phys. Rev. Lett.* 37, 440 (1976); F. De Martini, M. Colocci, S. Kohn and Y. R. Shen, *Phys. Rev. Lett.* 38, 1223 (1977).
22. D. M. Greenberger, M. A. Horne and A. Zeilinger, *Am. J. Phys.* 58, 1131 (1990); N. D. Mermin, *Phys. Rev. Lett.* 65,1838 (1990), *Revs. Mod. Physics* 65, 803 (1993); G. Alcaine, *Phys. Rev. Lett.* 80, 1797 (1998).
23. The realization of *nonlinear* universal logic gates is now being considered as a sensible route toward experimental quantum computation: D.Di Vincenzo, private comm. Our present results support this trend.
24. F. De Martini, *Phys. Rev. Lett.* 81, 2619, (1998).

QUANTUM CRYPTOGRAPHY VIA PARAMETRIC DOWNCONVERSION

Gilles Brassard,[1] Tal Mor,[1,2] and Barry C. Sanders,[3]

[1]Département IRO, Université de Montréal
C.P. 6128, succ. centre-ville, Montréal (Québec), Canada H3C 3J7
E-mail: {brassard,mor}@iro.umontreal.ca

[2]Electrical Engineering, University of California at Los Angeles
Los Angeles, CA 90095-1594, USA
E-mail: talmo@EE.UCLA.EDU

[3]Department of Physics, Macquarie University
Sydney, New South Wales, Australia 2109
E-mail: barry@mpce.mq.edu.au

The use of quantum bits (qubits) in cryptography holds the promise of secure cryptographic quantum key distribution schemes. It is based usually on single-photon polarization states. Unfortunately, the implemented "qubits" in the usual weak pulse experiments are not true two-level systems, and quantum key distribution based on these imperfect qubits is totally insecure in the presence of high (realistic) loss rate. In this work, we investigate another potential implementation: qubits generated using a process of parametric downconversion. We find that, to first (two-photon) and second (four-photon) order in the parametric downconversion small parameter, this implementation of quantum key distribution is equivalent to the theoretical version.

Once realistic measurements are taken into account, quantum key distribution based on parametric downconversion suffers also from sensitivity to extremely high (nonrealistic) losses. By choosing the small parameter of the process according to the loss rates, both implementations of quantum key distribution can in principle become secure against the attack studied in this paper. However, adjusting the small parameter to the required levels seems to be impractical in the weak pulse process. On the other hand, this can easily be done in the parametric downconversion process, making it a much more promising implementation.

INTRODUCTION

Quantum information theory suggests the possibility to accomplish tasks that are beyond the capability of classical computer science, such as information-secure cryptographic key distribution.[1] While theoretical quantum key distribution (QKD) schemes are proven secure against very sophisticated attacks,[2] the experimental QKD

Quantum Communication, Computing, and Measurement 2
Edited by Kumar et al., Kluwer Academic / Plenum Publishers, New York, 2000.

schemes are not yet proven secure even against very simple attacks. In this work, we analyse the effect of losses on the security of experimental quantum key distribution. We investigate a novel implementation, qubits produced by a process of parametric downconversion (PDC), and we compare it to the more common implementation based on weak coherent pulses (WCP).

A protocol is considered secure if the adversary is restricted only by the rules of quantum mechanics, and yet cannot obtain any information on the final key. In the four-state scheme[1] usually referred to as BB84, the sender (Alice) and the receiver (Bob) use two conjugate bases (say, the rectilinear basis, $+$, and the diagonal basis, $\times$) for the polarization of single photons. In basis $+$ (resp. $\times$), they use the two orthogonal basis states $|0_+\rangle$ and $|1_+\rangle$ (resp. $|0_\times\rangle$ and $|1_\times\rangle$) to represent "0" and "1" respectively. The basis is revealed later on, which enables Bob to decode the bit whenever he used the same basis as Alice; otherwise, they throw the bit away. Finally, they use error-correction and privacy amplification to obtain a potentially secure final key.[3, 2]

All the experiments done so far to demonstrate protocols for secure quantum key distribution use pulses of light containing (on average) much less than one photon. We approximate the state of the modified qubit created by this process to be in single mode, which we call a "weak coherent pulse" (WCP). [For an explanation regarding a description of a pulse, see Blow *et al.*[4]] We analyse the security of WCP-based schemes while paying special attention to the losses. The channel causes huge loss rate (whether a fiber, which causes attenuation, or free space, which causes beam broadening). In the experimental literature, it is usually assumed that the only effect of losses is to reduce the bit rate. We show that there are two different types of losses, channel losses and losses due to the state ("state losses"). The state losses have impact on the bit rate. The channel losses have a vital impact on security, in addition to their impact on the bit rate. A careful analysis of channel losses shows that schemes that were assumed secure are in fact totally insecure even against a simple intercept-resend attack. In intercept-resend attacks, an eavesdropper (Eve) performs a complete measurement on the input qubit, and she prepares and sends to Bob a state of her own, according to the outcome of her measurement. When Alice and Bob are using linearly independent states, Eve can *sometimes* get full information by performing a "positive operator value measure" (POVM) that conclusively distinguishes such states. This is fatal in presence of high channel losses between Alice and Bob because Eve can recreate the state near Bob and send it to him without loss whenever she measured it conclusively, whereas she forwards nothing to Bob otherwise! We shall refer to this attack as the *conclusive-measurement attack.* This was discussed when the two-state scheme[5] was invented, and its power against the four-state scheme was realized by Yuen.[6]

Recently, parametric downconversion has been used to generate a polarization singlet state[7] to test Bell's inequalities, and it is believed that it can be used as a much better single-photon source for quantum key distribution. Here, we explain the potential experiment and we present the modified singlet state resulting from this PDC process. Then, we calculate the state sent to Bob, including two-photon and four-photon terms, assuming dispersion-free devices, no dark counts and perfect detectors. [A different use of a PDC for QKD was previously suggested,[8] based on Franson-type uncertainties, but the polarization encoding we suggest here allows for a much simpler analysis].

We find that PDC-QKD is much more secure than WCP-QKD: The security of WCP-QKD is destroyed in the presence of high channel loss rate due to the linear independence obtained when adding the second-order terms. The crucial advantage of the PDC-QKD is that the second-order terms do not affect the fact that the states in one basis are linearly dependent on the states in the other basis. Thus, the attack that destroys the security of WCP-QKD in the presence of high losses has no impact on PDC-QKD (when second-order calculation and perfect detection are considered).

When imperfections in the process are taken into account, this euphoric picture changes, and the second-order states sent to Bob are not linearly dependent anymore. Fortunately, PDC-QKD becomes totally insecure against the conclusive-measurement attack only in the presence of such extremely high loss rate that more serious practical problems would have already arisen, such as the importance of dark counts, or errors due to various inaccuracies in the devices. As we explain in the discussion, it is probably impossible to make the WCP implementation secure against the conclusive measurement attack, thus we suggest that the experimental effort should be directed towards the implementation of PDC-QKD.

SECURITY OF WCP-BASED QKD

Experimental QKD is mainly based on the use of weak pulses of coherent light. By definition, a pulse consists of a linear superposition of many frequency contributions, but the laser pulse itself can be considered to be in a single, localized mode provided that dispersion is not significant in any of the optical elements.[4]

Using Fock state notation, $|0,0\rangle$ denotes the vacuum state, and the state $|n_{\updownarrow}, m_{\leftrightarrow}\rangle$, which describes n photons with vertical polarization and m photons with horizontal polarization, is denoted more simply by $|n,m\rangle$. Ideally, the four BB84 states should be $|\updownarrow\rangle = |0_+\rangle = |1,0\rangle$ and $|\leftrightarrow\rangle = |1_+\rangle = |0,1\rangle$ in the $+$ basis, and $|0_\times\rangle = (1/\sqrt{2})[|1,0\rangle + |0,1\rangle]$ and $|1_\times\rangle = (1/\sqrt{2})[|1,0\rangle - |0,1\rangle]$ in the $\times$ basis.

Consider now a weak coherent pulse with parameter α, meaning that a photon would be detected with probability α^2 if the pulse were measured by a perfect detector. If this pulse is polarized in the $+$ basis, the two states are simply, to second order in α,

$$\begin{aligned}
|0_+^{wcp}\rangle &\approx \left(1 - \frac{\alpha^2}{2}\right)|0,0\rangle + \alpha|1,0\rangle + \frac{\sqrt{2}\alpha^2}{2}|2,0\rangle \\
|1_+^{wcp}\rangle &\approx \left(1 - \frac{\alpha^2}{2}\right)|0,0\rangle + \alpha|0,1\rangle + \frac{\sqrt{2}\alpha^2}{2}|0,2\rangle \ .
\end{aligned}$$

However, the two states in the $\times$ basis, when expressed as Fock states in terms of the $+$ basis, are more complicated:

$$\begin{aligned}
|0_\times^{wcp}\rangle &\approx \left(1 - \frac{\alpha^2}{2}\right)|0,0\rangle + (\alpha/\sqrt{2})\big[|1,0\rangle + |0,1\rangle\big] + \frac{\sqrt{2}\alpha^2}{4}\big[|2,0\rangle + \sqrt{2}|1,1\rangle + |0,2\rangle\big] \\
|1_\times^{wcp}\rangle &\approx \left(1 - \frac{\alpha^2}{2}\right)|0,0\rangle + (\alpha/\sqrt{2})\big[|1,0\rangle - |0,1\rangle\big] + \frac{\sqrt{2}\alpha^2}{4}\big[|2,0\rangle - \sqrt{2}|1,1\rangle + |0,2\rangle\big] \ .
\end{aligned}$$

We call those four states the *modified qubits*. Note that they are not two-level systems anymore but six-level systems, or *qu-hexits*.

If we considered only the first order in α, as is usually done, the four states would behave very much like the ideal BB84 states leading us to the wrong conclusion that the protocol is secure! However, when the second order is considered, the two states in one basis are no longer linear combinations of the two states in the other basis. As noted by Yuen,[6] this linear independence in the six-dimensional Hilbert space creates a fatal flaw for BB84 in the presence of high losses. These states can be distinguished conclusively by an appropriate POVM. Such measurement yields no information about the state most of the time, but sometimes it identifies it unambiguously. As explained in the introduction, this allows for a successful conclusive-measurement attack provided the loss rate expected by Alice and Bob is sufficiently high. To provide numerical analysis, one must find the states that form the POVM. This is a cumbersome calculation and we leave it for the final paper. However, it is clear that the success probability is of order α^2 (relative to the one-photon counts). Therefore, with Eve getting a conclusive result with relative probability of order α^2, and with $\alpha^2 = 0.1$ as in the current experiments, it seems that a channel loss rate of 90%–95% is fatal. With current channel loss rates, there is no escape from decreasing α by more than one order of magnitude if reasonable security is to be achieved, and by more than two orders of magnitudes if we expect to have secure key distribution to distances required for practical purposes.

CREATING A MODIFIED SINGLET STATE IN THE PROCESS OF PDC

In this section, we present the parametric downconversion process and we give the output state to second order in the PDC parameter. The PDC process provides a source of photons for Bob and Alice with important advantages over the weak coherent pulse discussed in the previous section. A classical pump field with vertical polarization drives a PDC crystal below threshold, thereby producing photon pairs from a two-mode vacuum state input field $|0,0,0,0\rangle$. The two output fields from the parametric downconverter are correlated in time of emission as well as polarization, and conservation laws apply to the sum of energies and momenta of the photons in the two fields. The quantum field input to the parametric downconverter is assumed to be in the vacuum state. We consider the field emitted by the PDC process and channeled through a polarization rotator and a beam splitter, which creates entanglement between them. One arm of the resulting output goes to Alice and the other arm goes to Bob.

We denote by $|k_{a_\updownarrow}, l_{a_\leftrightarrow}, n_{b_\updownarrow}, m_{b_\leftrightarrow}\rangle$, or more simply $|k,l,n,m\rangle$, the state in which there are k photons with vertical polarization and l photons with horizontal polarization going into Alice's arm "a", and n photons with vertical polarization and m photons with horizontal polarization going into Bob's arm "b". The PDC small parameter χ, which is proportional to the strength of the pump field, the interaction time between the field and the crystal and the nonlinearity of the medium, is so that a photon pair would be detected with probability χ^2 if the output of the interaction were measured by perfect detectors. The state created by this process is an entangled state, and it is usually assumed to be a singlet $|\psi_-\rangle = (1/\sqrt{2})[|0,1,1,0\rangle - |1,0,0,1\rangle]$, but we show in the final paper how to calculate it more precisely, to obtain the *modified singlet* $|\chi\rangle = |\psi_-^{(mod)}\rangle$ to second order in χ:

$$\begin{aligned}|\chi\rangle = \left(1-\frac{\chi^2}{2}\right)|0,0,0,0\rangle &+ \frac{\chi}{2}\Big[|0,1,1,0\rangle + |1,1,0,0\rangle - |0,0,1,1\rangle - |1,0,0,1\rangle\Big] \\ &+ \frac{\chi^2}{4}\Big[|0,2,2,0\rangle + |2,2,0,0\rangle + |0,0,2,2\rangle + |2,0,0,2\rangle - 2|1,1,1,1\rangle \\ &\quad + \sqrt{2}|1,0,1,2\rangle - \sqrt{2}|0,1,2,1\rangle + \sqrt{2}|1,2,1,0\rangle - \sqrt{2}|2,1,0,1\rangle\Big]\,. \qquad (1)\end{aligned}$$

CREATING A MODIFIED QUBIT IN A PDC PROCESS

In order to use PDC for performing the BB84 four-state scheme, we need to consider the state sent from Alice to Bob. This is produced by Alice measuring her arm in a basis ($+$ or $\times$) of her choice and letting the other arm, which is the modified qubit, go to Bob through the quantum channel. More precisely, Alice directs her arm to an adjustable rotator (to choose the basis of measurement: angle 0 for $+$ and $\pi/4$ for $\times$) followed by a polarization-dependent beam splitter that sends the horizontal mode to one direction and the vertical mode to another spatial direction. Each of these spatial modes is now subjected to a measurement, which in the limit of perfect efficiency provides an exact count of the number of photons that reached each detector. In this section, we analyse to orders χ and χ^2 the modified qubit thus sent to Bob resulting from the modified singlet state.

Considering $|\chi\rangle$ to order χ and perfect detectors (used by Alice), the modified singlet is projected to yield a perfect qubit that is sent towards Bob in one of the BB84 states. With imperfect detection, but not allowing dark counts, Alice might send the vacuum, while she thinks she sent a single photon, but this causes only state losses and it has no effect on security as far as we could see.

When we consider $|\chi\rangle$ to order χ^2 and perfect detectors, this process yields a modification of the four BB84 states, but surprisingly still results in a perfect BB84 scheme! With perfect measurements, only the terms with exactly one photon at Alice's site will not be discarded, so that we need only consider the terms $\frac{\chi}{2}[|0,1,1,0\rangle - |1,0,0,1\rangle]$ and $\frac{\chi^2}{2\sqrt{2}}[|1,0,1,2\rangle - |0,1,2,1\rangle]$. In case Alice decides to rotate her mode by angle $\pi/4$ in order to send Bob a qubit in the $\times$ basis, the above terms change to $\frac{\chi}{2\sqrt{2}}[|0,1,1,0\rangle + |1,0,1,0\rangle - |1,0,0,1\rangle + |0,1,0,1\rangle]$ and $\frac{\chi^2}{4}[|1,0,1,2\rangle - |0,1,1,2\rangle - |0,1,2,1\rangle - |1,0,2,1\rangle]$.

With ideal detectors and Alice measuring without rotation, the state of Eq. 1 is projected onto $|0,1\rangle$ or $|1,0\rangle$ (in Alice's arm), yielding respectively

$$|0_+^{pdc}\rangle \approx \frac{\chi}{2}|1,0\rangle - \frac{\chi^2}{2\sqrt{2}}|2,1\rangle$$
$$|1_+^{pdc}\rangle \approx -\frac{\chi}{2}|0,1\rangle + \frac{\chi^2}{2\sqrt{2}}|1,2\rangle$$

(since Alice used the $+$ basis). When Alice uses the $\times$ basis, the rotated terms (calculated as before) provide the relevant contribution, yielding

$$|0_\times^{pdc}\rangle \approx \frac{\chi}{2\sqrt{2}}\big[|1,0\rangle + |0,1\rangle\big] - \frac{\chi^2}{4}\big[|1,2\rangle + |2,1\rangle\big]$$
$$|1_\times^{pdc}\rangle \approx \frac{\chi}{2\sqrt{2}}\big[|1,0\rangle - |0,1\rangle\big] + \frac{\chi^2}{4}\big[|1,2\rangle - |2,1\rangle\big] .$$

The modified qubit is not a two-level system but a four-level system. Yet, all four states lie in a two-level system spanned by any two of them. Furthermore, they satisfy the same conditions as the theoretical BB84 states; each one in the $\times$ basis is an equal superposition of the states in the $+$ basis. Thus, all theoretical security analyses apply to these states.

DISCUSSION

We have seen that PDC-QKD has a crucial advantage over WCP-QKD due to the fact that the four states created in the PDC process are equivalent to the theoretical states. However, the calculation so far assumed that Alice uses perfect measuring devices. A calculation taking account of realistic measurements will contain also other corrections. Then the states *will be* linearly independent, so that Eve can find a POVM to distinguish between them conclusively.

Nevertheless, let us show a vital advantage of the more realistic PDC-QKD over WCP-QKD. Even though both schemes are insecure in principle in the presence of high channel losses, the use of PDC as a source of qubits is potentially much preferable: For PDC qubits, the controlled parameter χ is usually smaller than 10^{-3}, thus the probability of having more than one photon is 10^{-6}, conditional to having at least one photon, and seems to be negligible when the channel losses are 99% or even much more. Furthermore, the small parameter can be easily further decreased according to the loss rate to potentially solve the problem, perhaps while increasing the pulse frequency to keep the same bit rate. In WCP, the corresponding parameter α is usually around 0.3. Unfortunately, this parameter cannot be adjusted so easily because it plays a dual role. Decreasing it immediately increases the state losses, which are $1 - \alpha^2$. Although these are state losses and not channel losses—hence we didn't see any effect of these losses on security—they are crucial in this implementation: with much smaller α it is impossible to achieve any reasonable bit rate since the state loss rate is $1 - \alpha^2$. Increasing the number of pulses to overcome this problem is not an appropriate solution since Alice needs to write down the polarization of the states in all pulses, and change the polarization for each one.

Another important advantage of PDC-QKD is that it solves a problem usually left unnoticed: Eve can attack WCP-QKD by eavesdropping *into Alice's lab*; this can be done by finding the setting of Alice's polarizers using a strong pulse sent to, and reflected from the polarizers[9] in between Alice's pulses. We are not aware of any such attack that can be used against the PDC-QKD implementation.

Our work is only an initial step. Analysis of more realistic scenarios and of other attacks might show that PDC-QKD is not as superior to WCP-QKD as this preliminary study indicates.

ACKNOWLEDGEMENTS

We are very thankful to Amiram Ron for helpful questions and remarks. T. Mor is thankful to Eugene Polzik for providing the initial motivation for this work, and to the AQIP'98 conference, organized by BRICS in Denmark. B. Sanders is thankful to S. Warburton for checking calculations and for useful comments.

REFERENCES

1. C.H. Bennett and G. Brassard, in *Proc. of IEEE Inter. Conf. on Computers, Systems and Signal Processing, Bangalore, India* (IEEE, New York, 1984) p. 175.
2. E. Biham, M. Boyer, G. Brassard, J. van de Graaf and T. Mor, "Security of quantum key distribution against all collective attacks", Los Alamos Archive: quant-ph 9801022.
3. C.H. Bennett, F. Bessette, G. Brassard, L. Salvail and J. Smolin, *J. Crypto.* **5**, 1 (1992).
4. K.J. Blow, R. Loudon, S.J.D. Phoenix and T.J. Sheperd, *Phys. Rev. A* **42**, 4102 (1990).
5. C.H. Bennett, *Phys. Rev. Lett.* **68**, 3121 (1992).
6. H. P. Yuen, *Quant. Semiclass. Opt.* **8**, 939 (1996).
7. Z.Y. Ou and L. Mandel, *Phys. Rev. Lett.* **61**, 50 (1988).
8. A.K. Ekert, J.G. Rarity, P.R. Tapster and G.M. Palma, *Phys. Rev. Lett.* **69** , 1293 (1992).
9. Such an attack was suggested by several people independently, such as Adi Shamir, Charles H. Bennett and others.

DIM COHERENT STATES AS SIGNAL STATES IN THE BB84 PROTOCOL: IS IT SECURE?

Norbert Lütkenhaus

Helsinki Institute of Physics
PL 9
00014 Helsingin yliopisto
Helsinki, Finland

Practical realizations of quantum key distribution employ dim coherent states as an approximation to single photon states. The advantage is that dim coherent states are convenient to use. The draw-back is that the security proofs for quantum key distribution have to be adapted. Difficulties arising from the use of signal states containing multi-photon contributions in combination with lossy channels have been pointed out earlier by Huttner et al.[1]. In this paper I describe a first numerical bound allowing quantum key distribution using dim coherent signal states in the presence of noise.

INTRODUCTION

The first scheme of quantum key distribution was proposed by Bennett and Brassard[2] in 1984 (BB84). Recent proofs by Mayers and Yao[3], and Lo and Chau[4] prove the security of this scheme in the presence of noise. Since some time systems are being designed to implement quantum key distribution. The experimental situation deviates in some points from the BB84 protocol. The proof of security for these systems needs therefore to be modified to reach the goal of a provable secure key distribution which can be efficiently implemented. Such a modification must

- involve statistical issues (not only statements about average values),
- incorporate the actual experimental set-up (signal states),
- allow for efficient classical protocols during the key establishment (bi-directional error correction codes).

In the effort to reach this goal the above mentioned approaches sacrifice practicability to obtain a security proof valid against all eavesdropping attacks allowed by quantum mechanics. In my approach I sacrifice the claim of total security by replacing it with security against all eavesdropping attacks on *individual signals*. This allows me to incorporate practicability into the proof. After outlining the protocol, I describe the security situation for single-photon states, mixtures of Fock states and for coherent states.

Quantum Communication, Computing, and Measurement 2
Edited by Kumar et al., Kluwer Academic / Plenum Publishers, New York, 2000.

PRACTICAL PROTOCOL FOR QUANTUM KEY DISTRIBUTION

The practical protocol for quantum key distribution consists of four parts: 1. Generation of a *sifted key* using a quantum set-up. 2. Error correction using redundant bits, thereby creating the *reconciled key*. 3. Application of privacy amplification to create a shorter secure key, the *final key*. 4. Authentication of the final key to ensure that the key is shared.

It should be noted that in practice it is not possible to establish a secret key without Alice and Bob sharing at least some short secret key beforehand. The reason is that a public channel, which can not be manipulated by an eavesdropper, is an idealized concept which in practice can be realized only by the classical method of *authentication* which uses shared secret bits. With the use of secret bits we change from quantum key distribution to quantum key growing. We take additional advantage of initially shared secret bits and use them to control the flow of information to an eavesdropper during error correction. We will now briefly describe the main steps of the practical protocol. More details can be found in a longer article.[5]

Generation of the Sifted Key: The principles of quantum key distribution (better known as quantum cryptography) are described in great detail in many papers and introductory articles[6,7]. The basic scenario of the four-state Bennett Brassard protocol[2] (BB84) is that Alice sends non-orthogonal quantum states to Bob. In the language of spin-1/2 particles the signal states are, for example, two eigenstates of σ_z and two eigenstates of σ_x. These four states are sent at random with equal probability in a signal slot. Bob chooses with equal probability to measure the arriving system in the x or the z basis. Whenever the signal state was prepared as an eigenstate of the measurement device, Bob's measurement result is deterministic, while it gives a random result otherwise. Communication over a public channel allows Alice and Bob to select the subset of deterministic signals as starting point for the remainder of the protocol, which is purely classical. The subset is called "sifted key". The power to select the sifted key from the random signals measured in randomly selected bases gives Alice and Bob the advantage over an eavesdropper.

Error Correction: Before Alice and Bob can proceed, they have to reconcile (correct) their sifted key. During this process they have to exchange redundant bits. To avoid extra flow of information to reach Eve, these bits can be encoded with a previously shared secret key.* This makes it necessary to use the minimum number of exchanged redundant bits, since the protocol is successful only if more secret key bits are generated in the output than are invested in the process.

The minimum number of redundant bits to be exchanged is given by the Shannon limit. If the amount of Shannon information shared between Alice and Bob is $I_{AB} = 1 - \bar{\epsilon}\log\bar{\epsilon} - (1-\bar{\epsilon})\log(1-\bar{\epsilon})$ for an bit error rate $\bar{\epsilon}$, then the number of redundant bits N_{rec} is bounded by $N_{rec} \geq 1 - I_{AB}$. In practice it is hard to get to that limit with the usual linear error correction codes. However, by allowing an interactive protocol with information flow going forth and back between Alice and Bob, one can come close to that limit (up to a factor of 5/4), as shown by Brassard and Salvail[9]. With this protocol we correct the errors rather than discarding them. Other existing proofs of security do not allow to use this efficient reconciliation protocol.

Privacy Amplification: Bennett et al.[10] introduced the method of privacy amplification. It allows to construct a short secret key from a longer key on which an eavesdropper has some information. For this to work it is necessary to bound the *col-*

*An alternative approach by Cachin and Mauer[8] takes the extra flow of information into account during privacy amplification.

lision probability of the eavesdropper on the long key as a function of the transmission error rate. Then Alice and Bob can shorten the key by a fraction τ_1 which guarantees that the eavesdropper's average Shannon information $\overline{I}_E$ on the total short key is below 1 bit. By shortening the key by extra n_S bits, this remaining information can be reduced exponentially to $\overline{I}_E \leq \frac{2^{-n_S}}{\ln 2}$. It should be pointed out, that this step reduces the *average* Shannon information of the final key from the knowledge of the *average* error rate. The statistical issue arising here is discussed in detail elsewhere[5]. There it is described how to bound the eavesdropper's Shannon information on a *specific* key from the *measured* error rate of the sifted key it is generated from.

Authentication: The authentication of the final key is necessary to ensure that an eavesdropper can not perform a "separate world attack" where she would establish different keys with Alice and Bob, pretending to each of them to be the other one. In practice authentication of the final key means that Alice and Bob hash the key in a short sequence (for example by selecting random parity bits) and exchange this hashed key encoded with previously shared secret bits. The probability that an eavesdropper could fake this authentication decreases exponentially with the length of the hashed key.

Security Statement: The protocol described is not successful in all realizations of a sifted key. Three different kinds of failure can occur: With some probability α the security is compromised without Alice and Bob knowing. This probability arises, for example, during the estimation of the average error rate from the measured error rate, the estimation of the specific Shannon information from the average Shannon information and during authentication[5]. With probability β Alice and Bob fail to reach a shared key from the initial sifted key while knowing that they failed. This can be caused by insufficient redundant bits for error correction which becomes apparent by failing of authentication. The probability γ refers to keys which are neither shared nor secret without Alice and Bob knowing that. This situation typically arises from the small probability that a key is successfully authenticated although two local copies of Alice and Bob differ. All three probabilities are under control of Alice and Bob and can be made exponentially small provided the sifted key is long enough.

With probability β Alice and Bob *succeed* to reach a point where they can issue the following certificate:
This key of length n_{rec} *is* shared *between Alicc and Bob with a probability* $1 - \gamma$. *It is* secure *in the sense that no eavesdropper attacking individual signals can possibly possess more Shannon information about the key than* I_{tol} *except with a probability* α.

For the design of an experimental set-up it is of interest to know the secure bit rate for the length of the sifted key going to infinity and reconciliation along the Shannon limit. The relevant quantities are I_{AB} for the Shannon limit and τ_1 to describe the fraction discarded during privacy amplification. (The additional n_S discarded bits can be neglected in the presented limit.) Secure bits are obtained if the number of invested and gained bits satisfy

$$\underbrace{n_{sif}\,(1-\tau_1)}_{\text{out}} \geq \underbrace{n_{sif}\,(1-I_{AB})}_{\text{in}} \tag{1}$$

which simplifies to $I_{AB} \geq \tau_1$. The balance of these quantities is subject of the following discussion for different signal states. The secure key rate is given by $n_{sif}\,(I_{AB} - \tau_1)$.

SINGLE-PHOTON SIGNALS

The first example is that of signal states which can be represented in a form motivated by single-photons in different polarizations. They allow the representation of $|1,0\rangle \pm |0,1\rangle$ and $|1,0\rangle \pm \mathrm{i}\,|0,1\rangle$ with $|1,0\rangle$ and $|0,1\rangle$ as states of one photon in one of two orthogonal polarization states. In a detailed analysis paper[5] I derived the estimate (see figure 1)

$$\tau_1 \leq \log\left(1 + 4\bar{\epsilon} - 4\bar{\epsilon}^2\right) . \tag{2}$$

From the above estimation we see that practical key distribution should work up to an

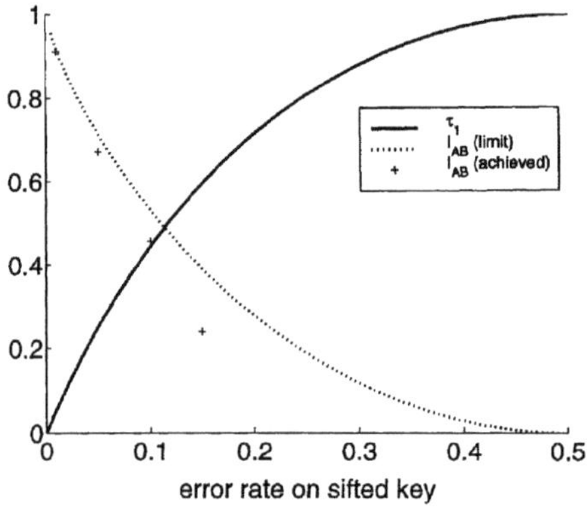

Figure 1. For error rates below 11% we can gain secure bits in the limit of infinite length of the sifted key and ideal error correction. For realistic error correction the crosses point to values achieved with the protocol by Brassard and Salvail.

error rate of 10% using available error correction techniques. The use of single photon states means that loss along the quantum channel reduces the rate with which the sifted key is generated, but it does not affect the security of the final key.

MIXTURES OF FOCK STATES

Next we consider input states which use the same polarization encoding as for single states. However, the states are now represented as a mixture of Fock number states in the polarization modes. They are given by $\rho_i = \sum_n P_n |n\rangle\langle n|$ with the Fock state $|n\rangle = \left(a_i^\dagger\right)^n |0\rangle$ composed with the creation operators $a_{1/2}^\dagger = \frac{1}{\sqrt{2}}\left(b_1^\dagger \pm b_2^\dagger\right)$ and $a_{3/4}^\dagger = \frac{1}{\sqrt{2}}\left(b_1^\dagger \pm \mathrm{i}\, b_2^\dagger\right)$ in terms of the creation operators $b_1^\dagger, b_2^\dagger$ of two orthogonal modes (e.g. polarization modes). Thermal light is an example of those states. Here we can assume without loss of generality that an eavesdropper performs a quantum non-demolition measurement on the total photon number of the signal states and then acts depending on that result. If she does not find a photon, she will not give any photon to Bob. If she finds a photon, then she can eavesdrop on it using the results of the previous section. In case she does find more than one photon she can split off one photon from the two polarization modes without changing the polarization of the remaining or the extracted signal. If the channel between Alice and Bob is lossy, then she can in addition to that suppress more of the single photon signals in favor of those originating from multi-photon signals. Using a superior ideal quantum channel, she still will be able to make sure that Bob detects the expected number of signals in his detectors. She therefore exploits the gap between the probability $1 - P_0$ of non-vacuum signals and the observed detection probability P_{det} to her advantage. In the limit of large loss the transmission becomes insecure if the probability of multi-photon signals $P_{>1}$ exceeds the expected detection probability, that is if $P_{>1} > P_{det}$. In this scenario all errors are assumed to come from eavesdropping on single-photon signals only. If we can bound the number of multi-photon signals contributing to the sifted key of n_{sif} to be less than n_m, we can bound $\tau_1^{(mult)}$ with the help of the single-photon bound τ_1 as

$$\tau_1^{(mult)}\left[\bar{\epsilon}\right] \leq \frac{n_m}{n_{rec}} + \left(1 - \frac{n_m}{n_{rec}}\right)\tau_1\left[\bar{\epsilon}\frac{n_{rec}}{n_{rec} - n_m}\right] .$$

An example for the resulting numbers is given in figure 2.

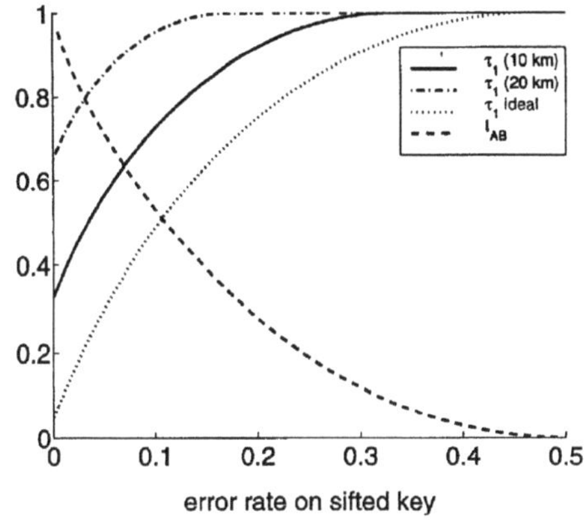

Figure 2. Example of a fiber with a loss coefficient of 0.3 dB/km, a detection efficiency of 30% and an average photon number of $\bar{n} = 0.1$ with Poisson distribution. The gain domain decreases as signal losses increase. Nevertheless, secure transmission with existing experiments is possible in this estimate. For ideal quantum devices the effect is negligible.

COHERENT STATES

With coherent states as input states in four different polarizations as before, the equations become more difficult. It is no longer possible to separate the multi-photon contributions from the vacuum and the single-photon contribution. Instead, we have to start again from the beginning. It is possible to formulate the desired estimate as an inequality for τ_1 with two constraints: the average error rate $\bar{\epsilon}$ and the average detection probability P_{det}. The two constraining numbers can be estimated from *measured* quantities. The inequality and the two constraints can be formulated using 12 real parameters with the help of symmetries. The resulting formula have not led to an analytical result yet. However, I performed a numerical optimization using the GAMS optimization program. An example is shown in figure 3. We find that this estimate gives much higher values for τ_1 than the comparable quantum non-demolition attack from the previous section. It is not clear yet, whether this is an artifact from the estimate leading to the inequality for τ_1 as a function of the 12 real parameters or whether it really means that coherent states are more susceptible to eavesdropping attacks than their Fock state mixtures with the same photo number distribution.

REFERENCES

1. B. Huttner and N. Imoto and N. Gisin and T. Mor. Quantum cryptography with coherent states. *Phys. Rev. A*, 51(3):1863–1869, March 1995.

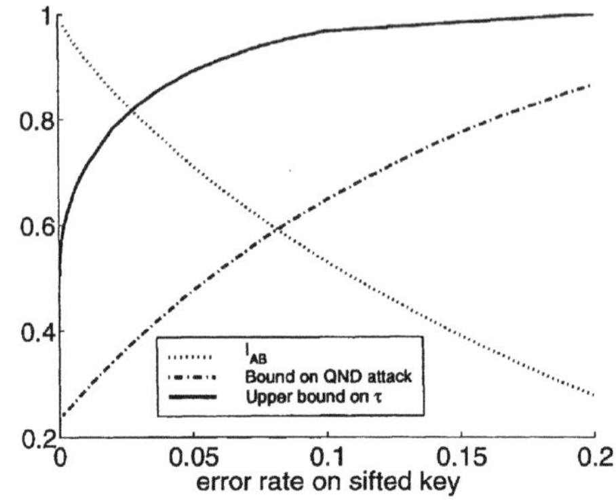

Figure 3. The numerical evaluation of the bound on τ_1 shows that the bound is much higher than the QND attack version. The coherent input states have an average photon number of $\overline{n} = 0.1$ and the detection probability is $P_{det} = 0.02$, which corresponds to ca. 5 km of fiber in the situation of figure 2.

2. C. H. Bennett and G. Brassard. Quantum cryptography: Public key distribution and coin tossing. In *Proceedings of IEEE International Conference on Computers, Systems, and Signal Processing, Bangalore, India*, pages 175–179, New York, December 1984. IEEE.
3. D. Mayers and A. Yao. Unconditional security in quantum cryptography. Report quant-ph/9802025, (1998).
4. H. K. Lo and H. F. Chau. Security of quantum key distribution. quant-ph/9803006, 1998.
5. N. Lütkenhaus. Estimates for practical quantum cryptography. Submitted to Phys. Rev. A, 1998. e-print: quant-ph/9806008.
6. S. J. D. Phoenix and P. D. Townsend. Quantum cryptography - how to beat the code breakers using quantum-mechanics. *Contemp. Phys.*, 36(3):165–195, 1995.
7. R. J. Hughes, D. M. Alde, P. Dyer, G. G. Luther, G. L. Morgan, and M. Schauer. Quantum cryptography. *Contemp. Phys.*, 36(3):149–163, 1995.
8. C. Cachin and U. M. Maurer. Linking information reconciliation and privacy amplification. *J. Crypt.*, 10(2):97–110, 1997.
9. G. Brassard and L. Salvail. Secret-key reconciliation by public discussion. In *Proceedings of Eurocrypt '93, held in Lofthus, Norway, 1993*, 1993.
10. C. H. Bennett, G. Brassard, C. Crépeau, and U. M. Maurer. Generalized privacy amplification. *IEEE Trans. Inf. Theo.*, 41:1915, 1995.

PRACTICAL ASPECTS OF QUANTUM CRYPTOGRAPHY

Miloslav Dušek,[1] Ondřej Haderka,[2,1] and Martin Hendrych[2,1]

[1]Department of Optics, Palacký University
[2]Joint Laboratory of Optics of Palacký Univ.
& Phys. Inst. Czech Acad. Sci.
17. listopadu 50, 772 00 Olomouc, Czech Republic

An apparatus for quantum key distribution using two unbalanced fibre Mach-Zehnder interferometers has been constructed in our laboratory. Physical aspects qualifying good performance of the system were thoroughly studied. The research covered a matter of coherence properties of the light source, a question of losses, noise, polarization, optimization of detection, problems associated with the decrease of visibility caused by imperfections of beam-splitters and unbalanced losses in different arms of interferometers, and active stabilization of interference (the problem of thermal stability). A quantum identification system has been proposed and tested. It combines a simple classical identification procedure and quantum key distribution, where the latter functions to replace used identification sequencies by new ones. Each identification sequence is used only once. The questions of authentication of public discussion have also been studied.

INTRODUCTION

In everyday life there are many situations when it is necessary to conceal the contents of confidential information conveyed over insecure communications lines. Classical cryptographic techniques have proved very helpful for this task. However, nearly all these techniques are merely computationally secure, i.e., they rely on limited advancement of computer power, technologies, and mathematical algorithms in the foreseeable future. The construction of a quantum computer can seriously menace their security. In the recent past, there has been a good deal of research of a new cryptographic method whose security is based on the fundamental laws of quantum physics – quantum cryptography.[1-7] Its main triumph is that it can solve the problem of key distribution. From the practical point of view, it is interesting that quantum cryptography may expediently be realized by means of quantum optics. The alphabet is represented by quantum states of electromagnetic field and optical fibre serves as a transmission channel. To encode information, e.g., polarization or phase can be used. The method based on "one-photon" interferometry has been analyzed by our group.

One of the basic cryptographic tasks is to certify the identities of the legitimate users of a communications line. Existing identification systems are only computationally secure. A quantum identification system was first proposed by C. Crépeau and

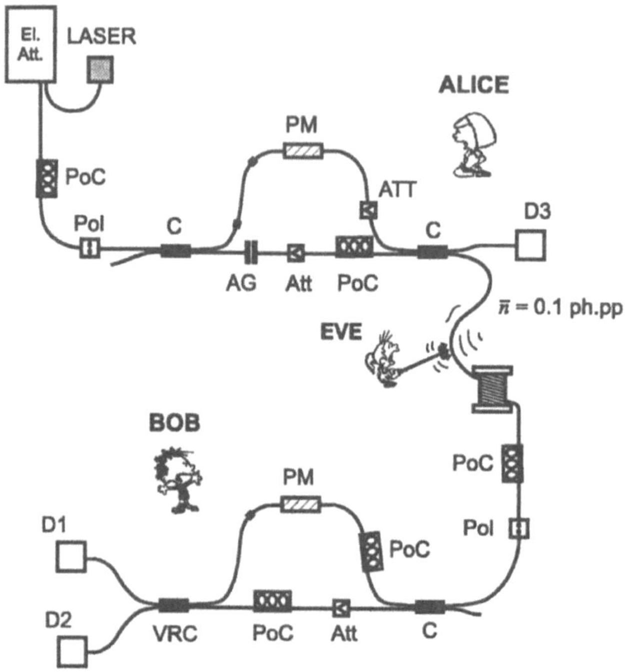

Figure 1. A scheme of optical part of the apparatus. El. Att. – electronic attenuator, PoC – polarization controllers, PM – planar electro-optic phase modulators, Att – attenuators, Pol – polarizers, C – fibre couplers, VRC – variable ratio coupler, AG – air gap.

L. Salvail.[8] Their identification protocol is based on quantum oblivious transfer. Alice and Bob mutually check their knowledge of a common secret string without disclosing it. However, quantum oblivious transfer has been proved insecure against the so-called collective attacks by D. Mayers,[9,10] and H.K. Lo and H.F. Chau.[11] Although to perform collective attacks is not possible with current technology, recent developments suggest that it might be possible in the near future. We propose an identification protocol combining expediently a simple classical identification method with quantum key distribution.[12]

EXPERIMENTAL REALIZATION

Outlines of experimental implementation of our system are given in Fig. 1. The apparatus is based on an interferometric setup with time multiplexing. It consists of two unbalanced fibre Mach-Zehnder interferometers. The path difference (2 m) of the arms of each interferometer is larger than the width of the laser pulse (its duration is 4 ns). Interference occurs at the outputs of the second interferometer for pulses "going through" long-short or short-long paths. These paths are of the same length and they are indistinguishable. Each of these interferometers represents the main part of the "terminals" of both communicating parties (called conventionally Alice and Bob). The terminals are interconnected by a 0.5 km single mode optical fibre acting as a quantum channel and also by a classical channel (local computer network). As a light source, a semiconductor laser operating at 830 nm is used. Laser pulses are attenuated by a computer-controlled attenuator so that the intensity level at the output of the first interferometer is 0.1 photon per pulse. The accuracy of this setting is monitored by detector D3. Polarization properties of light in the interferometers are controlled by polarization controllers PoC. To balance the lengths of the arms, an air gap AG with remotely controlled gap-distance is used. The phase coding is performed by means of two planar electro-optic phase modulators PM (one at each terminal). As the last beam splitter a variable ratio coupler VRC is used. With this setup, it is possible to reach visibilities well above 99.6 %. Detectors D1–D3 are single photon counting modules with Si-avalanche photodiodes. Their output signals are processed by detection electronics based on time-to-amplitude converters and single channel analyzers. Both terminals are fully driven by computers. The interferometers are placed in thermo-isolating boxes. Together with automatic active stabilization of interference, it enables us to reach low error rates below 0.4 % with raw data transmission rates of the order of kilobits per second.

WHAT AFFECTS GOOD PERFORMANCE OF THE APPARATUS

An analysis of various physical influences is important for minimizing the device's error – a necessary condition for effective detection of eavesdropping.

Coherence properties of laser pulses used: The coherence length and the shape of the autocorrelation function determine the precision with which it is necessary to balance the lengths of the arms of the interferometer in order to obtain high visibility of interference. In our particular case, to reach visibilities above 99.5 % the path difference of the arms must not exceed 5 μm.

Unbalanced losses and beam-splitting ratios: Fringe visibility in a Mach-Zehnder interferometer is adversely affected by beam-splitter imperfections and unequal losses in its arms. The effect of these factors may be eliminated by inserting additional losses in one of the arms of the interferometer.[13] However, if unity visibilities are required at both detectors, then the last beam splitter must have an ideal 50:50 splitting ratio. This is the reason why we employ a variable ratio coupler as the output beam splitter in our interferometer. (Real beam splitters are "nonunitary," thus the ratio 50:50 is meant between two inputs of the coupler to each output separately.) The additional losses mentioned should be concentrated in Alice's part of the interferometer since losses in Bob's part decrease transmission rates. In our apparatus the losses of Bob's part are about 4.5 dB.

Polarization of light: Fringe visibility further depends on the degree of polarization of light entering the interferometer and, of course, on congruence of polarization states of the beams combined at the output of the interferometer.[14] In our system the degree of polarization of laser radiation is improved by planar phase modulators (placed in both arms) which also serve as polarizers with extinction coefficient 10^{-6}. The changes of polarization states in optical fibres (due to birefringence caused by bending the fibres, etc.) must be compensated for by polarization controllers. Another problem arises because of distortions of polarization on the fibre connecting Alice's and Bob's parts of the interferometer. A partial solution is to place a polarizer in front of Bob's apparatus – then the polarization changes affect the data rate but do not very effect the error rate.

Thermal fluctuations of phase: Fluctuations of temperature and temperature gradients cause changes of refraction indices of fibers. This is the reason for substantial instability of the interference pattern. Both parts of the interferometer must be thermally isolated (we use polystyrene boxes). The environmental perturbations may further be reduced by means of active stabilization of the interferometer. After certain time intervals (during key distribution), constructive interference at one detector and destructive at the other one is found by scanning the phase difference, and "relative zero" is set on the phase modulators. The combination of the passive and active methods of stabilization gives very good results. In our measurements, the period after which the interferometer was calibrated was usually 3 s (the phase deviations were then smaller than $\pi/100$).

Time multiplexing and synchronization: If only one interconnecting fibre is to be used, time multiplexing is necessary. There are three time-separated peaks, but only the middle one "interferes." The separation of the peaks should be as small as possible since small path difference of the arms of the "sub-interferometers" is advantageous (especially due to lower sensitivity to environmental influences). However, the peaks must not overlap – it would decrease visibility. We use separation 10 ns.

Noise of detectors: The dark counts of detectors represent – together with losses on transmission line – the crucial factor limiting the range of quantum cryptographic transmission. We have used detectors based on Si avalanche photodiodes with less then 60 dark counts per sec.

QUANTUM IDENTIFICATION SYSTEM

In this section we briefly describe two protocols for mutual identification that were implemented in our laboratory quantum cryptographic system. In both these protocols, Alice and Bob check their common secret (random) string in a classical way. To prevent from a later misuse, each identification sequence is used only once and the distribution of a new common secret string is achieved by means of quantum key distribution (QKD) based on the BB84 protocol.[1] QKD has recently been proved secure against any collective attack allowed by quantum mechanics,[15, 16] and thus it offers unconditional protection even against eavesdroppers possessing unlimited computational and technological power.

Protocol I (unjammable open channel)

The protocol consists of a three-pass exchange of identification sequences (ISs) and it can be realized as follows (Alice and Bob already share several secret triads of IS):

- Alice and Bob say each other their ordinal numbers of the first unused IS triads in the stack and choose the higher one if they differ.
-
  - Alice sends the first IS of the triad to Bob.
  - Bob checks whether it agrees with his copy. If not, Bob aborts communication and shifts his pointer to the next triad. Otherwise he sends the second IS of the triad to Alice.
  - Alice compares whether her and Bob's second ISs agree. If not, she aborts communication and shifts her pointer. Otherwise she sends the third IS to Bob. If Bob finds it correct, the identification is successfully finished.
- To replace the used ISs, Alice and Bob "refuel" new ISs by means of QKD and set the pointers to their initial positions.

The three passes are necessary for the following reason: An eavesdropper (Eve) can pretend to be Bob and get the first IS from Alice. Of course, Alice recognizes that Eve is not Bob because Eve cannot send the correct second IS. So Alice aborts connection and discards this triad (i.e., shifts her pointer to the next one). However, later on Eve could turn to Bob and impersonate Alice. She knows the first IS! Bob can recognize a dishonest Eve just only because she does not know the third IS.

Let us note that Alice and Bob can tolerate a certain small number of errors in compared ISs (such that corresponding information leaked to Eve during QKD would not be sufficient for her to succeed in the identification procedure). Thus there is no need to perform error correction and privacy amplification after QKD.

Protocol II (with authenticated public discussion)

In reality, it is difficult to create a physically unjammable communication channel. Therefore unconditionally secure authentication of the messages sent over the open channel is necessary. The authentication of public discussion performed during QKD can be, however, utilized for the identification itself. Three-pass authenticated public discussion can function as the three-pass exchange of ISs described in the Protocol I. Note, that the authentication would anyway require additional "key" material to be prestored and transmitted similarly to ISs. The concrete authentication algorithm employed is based on the so-called orthogonal arrays[17] and is briefly described elsewhere.[12] A more detailed description of the protocol will be given in a separate paper.

- Alice and Bob first perform transmission over the quantum channel according to the BB84 protocol.
- Alice and Bob say each other their addresses in the pool of shared secret information and choose the higher one if they differ. Then a three-pass authenticated public discussion follows:
  - Bob sends to Alice an authenticated message containing the positions of bits randomly selected for error rate estimation.
  - Alice checks authentication and aborts communication if it fails. Otherwise she sends back to Bob an authenticated message containing the bases and bit values of the selected bits.
  - Bob checks authentication and aborts communication if it fails. Next he compares bases of the selected subset and retains only those qubits where his and Alice's bases coincide. At last, he estimates error rate and aborts communication when his result exceeds a certain limit value.[12] If all these three tests are correctly passed, he sends to Alice an authenticated message to inform her that identification was successful. Alice checks authentication and aborts communication if it fails.
- Alice and Bob compare bases of the rest of their raw data and arrive at their sifted keys.
- Then they perform error correction and privacy amplification procedures and arrive at an error-free distilled key.
- Alice and Bob refuel their shared secret information.

The used authentication sequences are always thrown away. The length of the raw quantum transmission must be selected such that the length of the newly obtained distilled key is greater than the number of bits consumed for authentication/identification purposes. It is convenient if it covers several unsuccessful identification procedures.

It is worth mentioning that the extent of authenticated information transmitted during QKD is limited due to the fact that the authentication procedure consumes a considerable part of the distributed key for its next run. In fact, only a fraction of public discussion can be authenticated. As the crucial characteristic of quantum cryptography is that any attempt at eavesdropping inevitably increases the number of errors in the transmitted key, it is necessary to authenticate just the part of public discussion serving error-rate estimation.[12]

Acknowledgement: The authors wish to express their gratitude to Professor Jan Peřina for his help throughout and for encouragement. This research was supported by Czech Ministry of Education (VS 96028), Czech Grant Agency (202/95/0002) and Czech Home Department (19951997007, 19982003012).

REFERENCES

1. C.H. Bennett and G. Brassard, in *Proceedings of the IEEE International Conference on Computers, Systems, and Signal Processing, Bangalore, India* (IEEE, New York, 1984), p. 175.
2. C.H. Bennett, F. Bessette, G. Brassard, L. Salvail, and J. Smolin, J. Cryptology **5**, 3 (1992).
3. A.K. Ekert, Phys. Rev. Lett. **67**, 661 (1991).
4. C.H. Bennett, G. Brassard, and N.D. Mermin, Phys. Rev. Lett. **68**, 557 (1992).
5. C.H. Bennett, Phys. Rev. Lett. **68**, 3121 (1992).
6. A.K. Ekert, J.G. Rarity, P.R. Tapster, and G.M. Palma, Phys. Rev. Lett. **69**, 1293 (1992).
7. C. Crépeau, in *Proc. 1st Intl. Conf. Theory and Applications of Cryptology, Pragocrypt '96, Prague* (CTU Publishing, Prague, 1996), p. 193.
8. C. Crépeau and L. Salvail, in *Advances in Cryptology: Proc. of Eurocrypt '95,* (Springer-Verlag, 1995), p. 133.
9. D. Mayers, Unconditionally secure quantum bit commitment is impossible, available at http://xxx.lanl.gov/abs/quant-ph/9605044.
10. D. Mayers, The Trouble with Quantum Bit Commitment, available at http://xxx.lanl.gov/abs/quant-ph/9603015.
11. H.-K. Lo and H.F. Chau, Is Quantum Bit Commitment Really Possible?, available at http://xxx.lanl.gov/abs/quant-ph/9603004.
12. M. Dušek, O. Haderka, and M. Hendrych, Acta Physica Slovaca, **48**, 169 (1998).
13. M. Hendrych, M. Dušek, and O. Haderka, Acta Physica Slovaca, **46**, 393 (1996).
14. M. Hendrych, M. Dušek, and O. Haderka, in *Proc. 1st Intl. Conf. Theory and Applications of Cryptology, Pragocrypt '96, Prague* (CTU Publishing, Prague, 1996), p. 234.
15. E. Biham, M. Boyer, G. Brassard, J. van de Graaf, and T. Mor, Security of Quantum Key Distribution Against All Collective Attacks, available at http://xxx.lanl.gov/abs/quant-ph/9801022.
16. D. Mayers and A. Yao, Unconditionally Security in Quantum Cryptography, available at http://xxx.lanl.gov/abs/quant-ph/9802025.
17. D.R. Stinson: *Cryptography, Theory and Practice,* CRC Press, Boca Raton, 1995.

QUANTUM VERSUS CLASSICAL NOISE CRYPTOGRAPHY

Horace P. Yuen

Department of Electrical and Computer Engineering,
Department of Physics and Astronomy,
Northwestern University, Evanston, IL 60208

A new quantum cryptographic scheme is introduced that, in contrast to previous ones, can be amplified and regenerated similar to classical noise-based cryptosystems. The essentially classical reasons for the success of various known quantum cryptosystems are indicated.

Quantum cryptography [1]-[5] is an approach for establishing a cryptographic key, a string of information bits for encrypting messages, through an open insecure channel between two users. Its security is derived from quantum effects while the usual public key agreement protocols [6] are based upon computational complexity. In this paper I will distill the principles underlying the success of various known quantum cryptosystems, which all involve nonorthogonal states in an essential way [7] and hence cannot be amplified [8] for application to long distance communications. I will introduce a new quantum cryptosystem that involves transmission in the usual way but detection of eavesdroppers via quantum effects, and that can be amplified and regenerated. It will be shown that the basic reasons for the success of all these quantum cryptosystems can be considered the same as those of a classical noise-based cryptosystem [9], which would have several advantages in practical implementations.

In the two-state (all states in this paper are pure states) quantum cryptosystem [3, 10] two legitimate users, Adam and Babe, establish a key as follows. Adam modulates a random sequence of 0's and 1's by the nonorthogonal states $|u\rangle$ or $|v\rangle$. Babe measures a POM [11] that would yield an inconclusive result with probability $|\langle u|v\rangle|$ but otherwise tells exactly whether $|u\rangle$ or $|v\rangle$ was sent. After Babe informs Adam openly the bit positions where she obtained conclusive results, the values of the corresponding bits would be known only to themselves. An active eavesdropper Eve can attempt to break the system in a variety of ways consistent with the laws of quantum physics. The following attacks can be launched on any quantum cryptosystem. On the one extreme called opaque eavesdropping [10] Eve can interrupt every message, try to make a decision on whether it is 0 or 1, and then re-transmit the corresponding quantum state to Babe. In the opposite extreme called translucent eavesdropping (with entanglement) [10], Eve can try to correlate Babe's future measurement result for each bit with her own through some sort of quantum correlator, such as a photon number duplicator [8] that correlates perfectly the photon counting readings of Eve

Quantum Communication, Computing, and Measurement 2
Edited by Kumar et al., Kluwer Academic / Plenum Publishers, New York, 2000.

and Babe. There are a continuum of possible attacks between these two limits, with the translucent type of attack having the special advantage that Eve can choose her measurement after Adam and Babe's open exchange, including error correction and privacy amplification to be discussed later. A cryptosystem is only successful if the users can always detect the presence of Eve whenever she learns about the key beyond a given tolerance level.

It should be clear that the security of a cryptosystem is a quantitative, not mere qualitative question. Detailed cryptoanalysis of quantum cryptosystems only begin to appear recently [10], [12]-[14]. Nevertheless, one can gauge the potential of such systems by a few basic parameters and considerations, as was done in Ref. [3] and [4]. For all the cryptosystems discussed in this paper, one can characterize the system by three parameters: F_+, the frequency of conclusive results; P_e, the probability that Babe makes an error in cases she makes a decision; and P_e^E, the probability that Eve makes an error in opaque eavesdropping. For the 2-state system, $P_e = 0$ ideally but $P_e \neq 0$ from interference of various kinds in a practical system. Translucent eavesdropping is thwarted because probably no good correlator exists that does not induce a large detectable $\overline{P}_e$. Opaque eavesdropping for all cryptosystems can be described by $\overline{P}_e$, Babe's error probability with respect to Adam's transmission as interfered by Eve,

$$\overline{P}_e = P_e^E(1 - P_e) + P_e(1 - P_e^E). \tag{1}$$

¿From (1), $\overline{P}_e \gg P_e$ for any P_e if $P_e^E \gg P_e$, and the eavesdropping can be detected by Babe openly checking with Adam the error frequency on a test bit string. In the 2-state case, $P_e^E \gg P_e$ when $|\langle u|v\rangle| \lesssim 1$.

This 2-state system has exact classical analogs [9] in which opaque eavesdropping is detected as above and the possibility of translucent eavesdropping is thwarted by having independent noises at the detectors of Eve and Babe, which obtains, e.g., in the case of quantum amplifier noise limited detection of high speed optical signals. For antipodal signals $\{\pm s(t)\}$ of energy S in additive Gaussian noise of variance σ^2, the users can set a decision threshold at $\pm m\sqrt{S}$, $m > 1$, with resulting

$$F_+ = P_e' + P_c', \tag{2}$$

$$P_e = P_e'/(P_e' + P_c'), \tag{3}$$

where $P_e' = Q((m-1)\sqrt{S/\sigma^2})$, $P_c' = Q((m+1)\sqrt{S/\sigma^2})$, $Q(x) \equiv \int_x^\infty e^{-y^2/2}\mathrm{dy}/\sqrt{2\pi}$, while

$$P_e^E = Q(\sqrt{\lambda S/\sigma_E^2}) \tag{4}$$

with λS and σ_E^2 the signal energy and noise variance Eve sees.

In the original 4-state system [2], Adam employs and Babe measures randomly one of two orthonormal bases at $\pi/4$ angle in a two-dimensional state space for each bit transmission. After they openly check the bit positions where their bases match, the corresponding bit values would be known only to themselves. In contrast to the 2-state case, Eve can effectively utilize this basis information. In particular, her P_e^E can be reduced to ≥ 0.15 in opaque eavesdropping [4]. Perfect translucent attack is thwarted by the impossibility of nonorthogonal state duplication, the so-called no-cloning theorem [15]-[16]. Experimentally, the four-coherent-state (4c-state) system [4]-[5] is employed instead for ease of realization, which is in fact closer to the 2-state system. There are other variants [3] among the known quantum cryptosystems equivalent to one of these three. Table 1 summarizes the characteristics of the four systems just described together with a new quantum system, the u-state system.

A major limitation on all the previous quantum cryptosystems is that nonorthogonal states are essentially involved and so amplification cannot be introduced [8] to compensate for possible transmission loss, particularly in long distance fiber communication. Similarly, a regenerator involving detection and signal regeneration cannot be employed. However, as we have seen above, the success of quantum cryptosystems depends only on the possibility of detecting the presence of Eve via quantum effects when she is able to learn too much about the key being established. If Adam transmits states in the *usual* way and Babe measures an observable such that the states are not eigenstates of the observable, Eve can launch a translucent attack with a quantum nondemolition (QND) measurement of the observable or a perfect duplicator of that observable as if the system is classical, but Babe can detect the presence of Eve by monitoring the conjugate observable. In this u-state (usual state) system, opaque eavesdropping is detected by a sufficiently large P_e^E as in all the other systems.

For definiteness, consider a single mode optical field for which the users employ coherent states $|\alpha\rangle$ for real α. The homodyne noise is then Gaussian with variance 1/4. If Eve performs a single quadrature QND measurement (i.e., measurement of the first kind that leaves the system in an eigenstate corresponding to the measured value), the quadrature eigenstate $|\alpha\rangle_1$ would be presented to Babe upon Eve's reading of α. Babe's homodyne measurement result would perfectly correlate with that of Eve. However, $|\alpha\rangle_1$ has infinite fluctuation in the conjugate quadrature that Babe can easily observe at any of the bit positons. Thus, a perfect translucent attack, while not thwarted as in the 4-state case, can be easily detected. In general, Babe can introduce a TCS-heterodyne measurement [17] that introduces a noise variance Δ_1 to the signal quadrature, Δ_2 to the conjugate quadrature, with $\Delta_1\Delta_2 \geq 1/16$. The ratio F_2 of the noise variance and the ratio F_1 of the signal-to-noise ratio (SNR) with and without interference are, for minimum $\Delta_1\Delta_2 = 1/16$,

$$F_1 = (1+4\Delta_1)^{-1}, \quad F_2 = 1 + (4\Delta_1)^{-1}, \quad \Delta_1 \geq 0\,. \tag{5}$$

If $F_2 \sim 2$ is on the boundary of detection by Babe, $\Delta_1 \sim 1/4$ which corresponds to heterodyning by Eve. The resulting degradation of SNR by 1/2 from F_1 gives a SNR advantage to Babe that can be used in various ways. To detect opaque eavesdropping, the users can imply, e.g., antipodal classical amplitudes as in the above classical noise system, i.e., use the two states $|\alpha_0\rangle$, $|-\alpha_0\rangle$ and adjust α_0 so that P_e^E is sufficiently large and P_e sufficiently small via a high enough threshold $\ell = \pm m\alpha_0$. This P_e^E and also P_e can be computed under a variety of receivers employed by Eve and Babe. For the optimum quantum detector for discrimination between two states $|u\rangle, |v\rangle$, the error probability is [18]

$$P_e^E = \frac{1}{4}(2 - \parallel |u\rangle\langle u| - |v\rangle\langle v| \parallel_1)\,, \tag{6}$$

where $\parallel B \parallel_1$ is the trace norm $\mathrm{Tr}(B^\dagger B)^{1/2}$ of a trace class operator B.

The three previous types of quantum cryptosystems, the present u-state system, and the classical noise system are summarized and compared in Table 1. It has often been stated that the no-cloning theorem underlies the success of quantum cryptography. Actually, the no-cloning theorem merely rules out a perfect translucent attack in the 4-state system. Not only does it not rule out the possibility of a sufficiently effective though not perfect translucent attack, it has nothing to do with the 4c-state or the 2-state system. The fundamental reason for the success of quantum cryptosystems that have been introduced thus far does not reside in the no-cloning theorem, it does not even reside in transmitting nonorthogonal states as our u-state system shows. Rather, common to all such systems is the fact that they exploit noncommuting observables

Table 1. Comparison of cryptosystems — the three previous quantum ones, a new quantum one, and a classical one. [a] The P_e is nonzero in practice from various noises and imperfections. [b] These hold assuming a Gaussian noise system is employed with antipodal classical signals or antipodal coherent-state signals. [c] Even nonperfect eavesdropping can be detected subject to Eq. (5). [d] Even nonperfect eavesdropping are ruled out by totally independent noises, while possible up to a certain level for partially correlated noises.

	4-state	4c-state	2-state	u-state	classical noise
F_+	1/2	S	$\lvert\langle u\lvert v\rangle\rvert$	Eq. (2)[b]	Eq. (2)[b]
P_e	0^a	0^a	0^a	Eq. (3)[b]	Eq. (3)[b]
P_e^E	$\gtrsim 0.15$	$\gtrsim 1-S$	Eq. (6)	Eq. (4)[b]	Eq. (4)[b]
Opaque Eaves-dropping	Eq. (1)	Eq. (1)	Eq. (1)	Eq. (1)	Eq. (1)
Perfect Translucent Eaves-dropping	thwarted by no-cloning theorem	thwarted by no perfect correlator for nonorthogonal states	thwarted by no perfect correlator for nonorthogonal states	conjugate observable detection[c]	thwarted by independent noises[d] for Eve and Babe
Amplification and Regeneration	no	no	no	yes	yes

or their corresponding eigenstates to detect Eve's presence and to thwart some of her possible actions. However, the use of quantum effects to this end, while certainly interesting and novel, is not essential. A closely similar classical noise-based system can achieve the same ends as Table 1 shows. In contrast to a quantum law of physics, a fact of nature — the independence of noises in Eve's and Babe's receivers — is used instead, which occurs in many realistic situations.

It has also often been stated that the security of quantum cryptosystem can be proved from the laws of quantum physics, in effect from the no-cloning theorem. The limited role of this theorem has just been discussed above. In general, a quantum cryptosystem can be proved to be secure at a certain level only if one can show that the laws of physics would have to be violated if security is breached beyond that level. While major advances have been recently made [12]-[14] in this direction for the 4-state system, such general proofs remain to be developed.

Common to all the five systems in Table 1 are the needs to eliminate any possible errors in the agreed key string and to make sure that Eve cannot effectively utilize any "partial information" she may be able to obtain. The first need on error correction has been treated [4], [19] but has not been systematically quantified — it is not established how many bits need to be dropped in how many trials in order to get the error probability to a given very low level. The second need has been extensively treated under the name "private amplification," [4], [19]-[20] and hashing techniques are available to reduce Eve's mutual information on the key to any specified level. Here, a simple technique is described that addresses directly on how Eve may use her partial information. With Eve knowing each bit with error probability P_e^E, she can try to guess at the key by running trials in a known-plaintext attack [6]. If N_E is the total number of key trials she may make, it is readily shown that in an optimal trial strategy her probability P_c^E of successfully guessing the key of length $|k|$ is determined by [9]

$$\sum_{n=0}^{n_0} \binom{|k|}{n} \leq N_E \,, \tag{7}$$

$$\begin{aligned} P_c^E &= \sum_{n=0}^{n_0} \binom{|k|}{n} (P_e^E)^n (1-P_e^E)^{|k|-n} \\ &+ \left[N_E - \sum_{n=0}^{n_0} \binom{|k|}{n}\right] (P_e^E)^{n_0+1} (1-P_e^E)^{|k|-n_0-1} , \end{aligned} \tag{8}$$

where n_0 is the maximum integer satisfying (7). Depending on $|k|$, P_e^E and the tolerable level of P_c^E, the users may need to use some form of privacy amplification to increase P_e^E. If a simple mod 2 sum of m bits is used to produce 1 new bit, the probability of Eve's error $\overline{P}_e^E$ for this new bit is given by

$$\overline{P}_e^E = \sum_{n\,(\text{odd})}^{m} \binom{m}{n} (P_e^E)^n (1-P_e^E)^{m-n} . \tag{9}$$

For $m \sim 10$, this would increase $P_e^E \sim 0.1$ to $P_e^E \sim 0.446$, and resulting from (5)–(7) a $P_c^E \sim 5.53 \times 10^{-8}$ for $N_E \sim 2^{20}$. Typically, $|k| \sim 10^3$ after error correction and can be used in an optical stream cipher directly. Even with $P_e^E \sim 0.1$, the resulting P_c^E from (8) is very small [9]. In applications requiring a shorter key $|k| \sim 50$, the above mod 2 compression may be used. Even though this compression throws away a lot of bits that could otherwise be saved, it is efficient not to save them as only $|k| \sim 50$ is needed, particularly in view of the fact that the main rate inefficiency in these systems comes from the whole agreement process, not from the length $|k|$, apart from the intrinsic weak signals. Note that this kind of cryptography is probably too slow if the agreed string is used as one-time pad.

The information that Adam and Babe need to publicly exchange during the key agreement process may be utilized by Eve in her attack, especially for the 4-state system. It is this fact that makes it so difficult to prove the security of a quantum cryptosystem. However, an absolute proof of security may be too much to ask for in any cryptosystem, quantum or classical — none is known except the one-time pad. What is needed is continual cryptoanalysis on various "obvious" attacks on the systems, which we cannot go into in this paper.

Finally, we treat the nontrivial problem of designing a cryptosystem over an attenuation channel where Eve may be near the transmitter and enjoy considerable signal energy advantage over the users. In a classical noise system in which quantum effects are negligible such as radio wave cellular phones, amplifiers with essentially no additive noise can be found and the system can be amplified and regenerated indefinitely to compensate for the attenuation, the only proviso being that Eve's energy advantage does not allow her to break the system. For example, an energy advantage $\lambda = 4$ in (4) for $\sigma_E^2 = \sigma^2$ can be sustained with $S/\sigma^2 \sim 1$, $m \sim 9$ so that $P_e \sim 10^{-2}$ from (3) with $P_e^E \sim 10^{-1}$. In optical implementation of the classical noise system or in the u-state system, quantum noise may be present in both amplifiers and attenuators, and the SNR continues to degrade significantly along the transmission line. Nevertheless, the transmission can still be greatly extended with the use of amplifiers. As an example, consider the use of the standard phase-insensitive linear amplifier. With n amplifiers of gain G each succeeded by an attenuation $\eta = G^{-1}$ in a chain, the output homodyne SNR is degraded from that after the first amplifier by a factor of n [21]. Thus, for coherent states one can achieve a transmission distance L given by $e^{-\alpha L} = G^{-n}$ for attenuation coefficient α, as compared to $e^{-\alpha L'} = \frac{1}{n}$ without the use of amplifiers, where n is the maximum tolerable signal energy advantage λ that Eve is allowed to have. This gives an improvement ratio

$$L/L' = n \log G / \log n . \tag{10}$$

Using the same parameters just presented above so that $n = 4$, (10) indicates a factor of 10 improvement for $G = 32$. The use of more exotic amplifiers [8], [21] and regenerators can, in principle, greatly further extend the ratio (10) which will be treated elsewhere.

There are many possible intermediate combinations of the u-state and classical noise schemes with different classical and quantum signal sets. Detailed analyses of these schemes are yet to be carried out for the proper choice of a cryptosystem in a given application. They all possess the advantage that amplification and regeneration can be introduced compared to the previous quantum cryptosystems.

Acknowledgements

This work was supported by The Office of Naval Research.

REFERENCES

1. S. Wiesner, SIGACT News **15**, 78 (1983).
2. C. H. Bennett, G. Brassard, and A. K. Ekert, Scientific American, October, 50 (1992).
3. C. H. Bennett, Phys. Rev. Lett. **68**, 3121 (1992) and references cited therein.
4. C. H. Bennett, F. Bessette, G. Brassard, L. Salvail, and J. Smolin, J. Cryptol. **5**, 3 (1992).
5. P. D. Townsend, J. G. Rarity, and P. K. Tapster, Electron. Lett. **29**, 1291 (1993).
6. A. J. Menezes, P. C. van Oorschot, and S. A. Vanstone, *Handbook of Applied Cryptography* (CRC, New York, 1997).
7. Nonorthogonal states are essential in the cryptosystem of L. Goldenberg and L. Vaidman, Phys. Rev. Lett. **75**, 1239 (1995). See T. Mor, Phys. Rev. Lett. **80**, 3137 (1998). This system is insecure against an attack involving measuring the signaling time and then re-send, which is to be discussed elsewhere.
8. H. P. Yuen, Quantum Semiclass. Opt. **8**, 939 (1996).
9. H. P. Yuen and A. M. Kim, Phys. Lett. A **241**, 135 (1998), and errata.
10. A. K. Ekert, B. Huttner, G. M. Palma, and A. Peres, Phys. Rev. A **50**, 1047 (1994).
11. A simple description of quantum measurement probabilities via POM (positive operator-valued measure) is given in H. P. Yuen, Phys. Lett. **91A**, 101 (1982).
12. P. Mayers and A. Yao, online preprint quant-ph/9802025.
13. N. Lüetkenhaus and S. M. Barnett, in *Quantum Communication, Computing, and Measurement*, O. Hirota, A. S. Holevo, and C. M. Caves, Eds. (Plenum, New York, 1997); p. 89.
14. E. Biham and T. Mor, Phys. Rev. Lett. **79**, 4034 (1997).
15. W. K. Wootters and W. H. Zurek, Nature (London) **299**, 802 (1982). This proof from linearity applies to a minimum of three states. For two nonorthogonal states, it follows from unitarity as given in Ref. [16].
16. H. P. Yuen, Phys. Lett. A **113**, 405 (1986).
17. H. P. Yuen and J. H. Shapiro, IEEE Trans. Inform. Theory **IT-26**, 78 (1980).
18. C. W. Helstrom, *Quantum Detection and Estimation Theory*, (Academic Press, 1976). This P_e is usually expressed in terms of the sum of the positive or negative eigenvalues of $\rho_1 - \rho_2$ for two arbitrary states ρ_1, ρ_2, each of which is $\frac{1}{2} \parallel \rho_1 - \rho_2 \parallel_1$ from $\mathrm{Tr}(\rho_1 - \rho_2) = 0$.
19. C. H. Bennett, G. Brassard, and J. M. Roberts, SIAM J. Comput. **17**, 210 (1988).
20. C. H. Bennett, G. Brassard, C. Crépeau, and U. M. Mauer, IEEE Trans. Inform. Theory **41**, 1915 (1995).
21. H. P. Yuen, Opt. Lett. **17**, 73 (1992).

QUANTUM CRYPTOGRAPHY WITH FEMTOSECOND PARAMETRIC DOWN CONVERSION

A. V. Sergienko, M. Atature, B. M. Jost, J. Perina Jr., B. E. A. Saleh, and M. C. Teich

Quantum Imaging Laboratory
Department of Electrical and Computer Engineering
Boston University, Boston MA 02215

We experimentally demonstrate a quantum cryptography system using two-photon entangled (EPR) states generated via the nonlinear process of spontaneous parametric down conversion pumped by a femtosecond laser. There are two major approaches in quantum cryptography which historically appeared almost simultaneously. One uses the quantum features of single photon states produced by significant attenuation of original light in a coherent state. The other is based on the quantum nonlocal character of two-photon entangled EPR states. The applicability of the latter one was strongly limited because of low visibility and poor stability of the systems which require synchronous manipulation of two Mach-Zehnder interferometers well separated in space. We developed a new scheme for quantum cryptography which is based on the use of a distributed polarization quantum intensity interferometer. This technique utilizes a double-entangled EPR quantum states generated in the nonlinear process of type-II spontaneous parametric down conversion (SPDC). The high contrast and stability of quantum interference demonstrated in our preliminary experiments promises to bring the performance of this system above the level of the best single-photon polarization techniques, and to do so without their specific limitations. The use of a high-repetition rate femtosecond pulses as a pump source enhances significantly the flux of entangled photon pairs available for the reliable and secure key distribution.

INTRODUCTION

Today's modern communication and information systems transmit a substantial amount of sensitive and financial information through both regular data networks and specialized channels. The level of communication security using traditional encryption tools depends on the computational intractability of mathematical procedures such as factoring large numbers. This approach is intrinsically vulnerable to advances in computer power. The explosion of new information services dictates a need for totally new and unconventional approaches to the problem of security and data authentication

Quantum Communication, Computing, and Measurement 2
Edited by Kumar et al., Kluwer Academic / Plenum Publishers, New York, 2000.

in communication networks. Recent developments in experimental tests of the fundamental problems of quantum mechanics, such as the Einstein-Podolsky-Rosen (EPR) paradox and the violation of Bell's inequalities,[1] have introduced a new paradigm for secure communications — quantum cryptography. The privacy of transmitted information can now be protected by the fundamental laws of nature.

Quantum cryptography has made use of two principal approaches that utilize the quantum nature of the photon state. One approach makes use of near single-photon states prepared from light initially in a coherent state.[2,3] Its major drawback arises from the statistical fluctuations of the number of photons in the original state. This adds the possibility of simultaneously having two photons in the channel; the eavesdropper can use the second one to extract partial information. The other approach is based on the nonlocal character of two-photon entangled (EPR) states generated in the nonlinear optical process of spontaneous parametric down conversion (SPDC).[4,5] The unique correlation of two photons in space, time, energy, and momentum resolves the problem inherent in the first approach. Unfortunately, the applicability of the latter technique has been severely limited because of low visibility and poor system stability inherent in the use of type-I SPDC, as well as the need for the synchronous manipulation of two Mach-Zehnder interferometers that are well separated in space.

Based on our previous experimental results,[1] we have demonstrated that the use of doubly entangled EPR states generated by type-II SPDC provides richer physics than type-I SPDC, and thereby creates a more flexible, robust, and reliable quantum apparatus for cryptographic appliacations. The high contrast and stability of the fourth-order quantum interference patterns demonstrated in our initial experiments promise to bring the performance of EPR-based quantum cryptography systems beyond the level of the best single-photon systems.

The key feature of quantum cryptography, that is, the impossibility of cloning the quantum state or extracting information without destroying it, carries with it a major limitation on the distance of secure information transfer. The limit is the distance that a single-photon state can travel without absorption. The level of signal attenuation in modern fibers would appear to pose a limit of 30-50 km for reliable quantum cryptography. Open-air communication may be more feasible especially when fibers are not available (ship-to-ship or in-field communication). The problem of secure communication to a satellite is also a vital issue in modern telecommunications. Ground-to-satellite, satellite-to-satellite, and satellite-to-ground communication becomes even more important when communication links must go over the horizon. Open-air quantum cryptography is expected to become a crucial tool in these situations. The thickness of the atmospheric layer is several kilometers and its density rapidly decreases with altitude, making ground-to-satellite communication attractive. Satellite-to-satellite communications using our cryptographic method, in the vacuum of open space, has only one problem — how to collimate and point the beam. Finally, the synthesis of both approaches - local distribution over fiber lines and transmittion over the horizon using a satellite-based link, can provide a global secure communication network (see Fig. 1).

Entangled Photons Created by Spontaneous Parametric Downconversion

Correlated (entangled) two-particle states have been known since the early 1920's. Entangled states comprise two or more particles whose state cannot be written as products of single-particle states.[6] These states have played an important role in the study of the basic questions of quantum mechanics such as the Einstein-Podolsky-Rosen (EPR) paradox[7] and tests of Bell's inequalities.[8] Two-photon correlations of the light created in the nonlinear process of SPDC permit the investigation of many fundamental

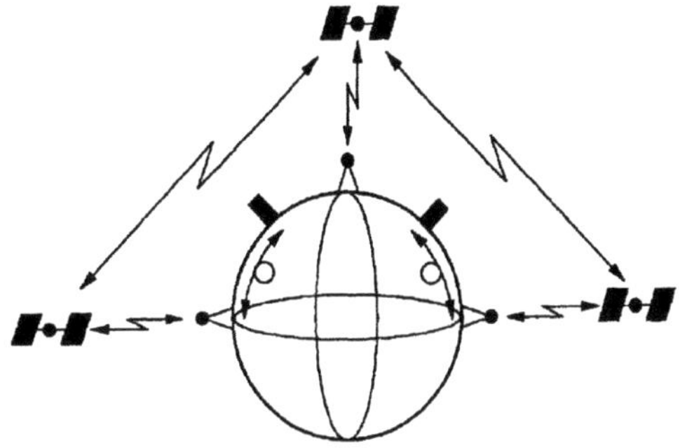

Figure 1. Illustration of combined short-distance optical fiber and long-distance open-air secure communication.

issues of quantum mechanics of photons.[1]

In spontaneous parametric down conversion, a pump laser beam is incident on a birefringent crystal. Nonlinear effects in the crystal lead to the spontaneous emission of pairs of entangled light quanta. The entanglement in frequency-wavenumber space, or equivalently space-time, comes from the frequency- and phase-matching (equivalently energy- and momemtum-conserving) conditions[9-11]

$$\omega_1 + \omega_2 = \omega_p, \qquad \mathbf{k}_1 + \mathbf{k}_2 = \mathbf{k}_p \tag{1}$$

where ω_i is the frequency and $\mathbf{k}_i$ the wave number, linking the input pump (p), and output signal (1) and idler (2). The down conversion is called Type-I or Type-II depending on whether the photons in the pair have parallel or orthogonal polarizations. The photon pair that emerges from the nonlinear crystal may propagate in different directions or may propagate collinearly. The frequency and propagation directions are determined by the orientation of the nonlinear crystal and the phase matching relations. Initially, Type-I SPDC was used extensively as a convenient source of two-photon entangled states.[1]

It was shown recently in our work that Type-II SPDC provides a richer tool due to the two-photon entanglement both in space-time and in polarization (spin).[12] The dispersion of the ordinary and extraordinary waves in a nonlinear crystal lead to a space-time structure of a wave function which is different from that generated in Type-I SPDC. This unique double entanglement of the two-photon state in Type-II SPDC provides us with control of the relative positions of these two photons in space-time.

Experimental attempts to develop quantum cryptography using Type-I entangled-photon pairs (EPR states) was initiated shortly after the notion was introduced by Ekert.[4] This approach requires the use of a Franson-type interferometer.[13] This is a distributed system of two interferometers, well-separated in space, with synchronously varied optical delay. Non-locality of the quantum features imbedded in the EPR pair should lead to an almost 100% visibility of quantum interference observed in coincidence between detectors at the output of each interferometer.

The visibility is the most crucial parameter in this technique. Only undisturbed quantum state will produce $\sim$ 100% visibility. Intervention of any classical measurement apparatus (eavesdropping) will cause an immediate reduction of visibility to 75% providing clear evidence of intrusion.

However, practical attempts to demonstrate the feasibility of quantum cryptography with EPR photons in fibers were not very successful. The applicability of this technique has been severely limited because of low visibility inherent in the need of synchronous manipulation of the two spatially separated Mach-Zehnder interferometers.

CRYPTOGRAPHY WITH POLARIZED ENTANGLED PHOTONS

To demonstrate that the EPR state is a reliable tool for quantum cryptography we designed a new approach using non-local quantum interference of two-photon entangled states (EPR states) generated in Type-II SPDC. It is based on the use of a double, strongly unbalanced, and distributed polarization interferometer as shown in Fig. 2.

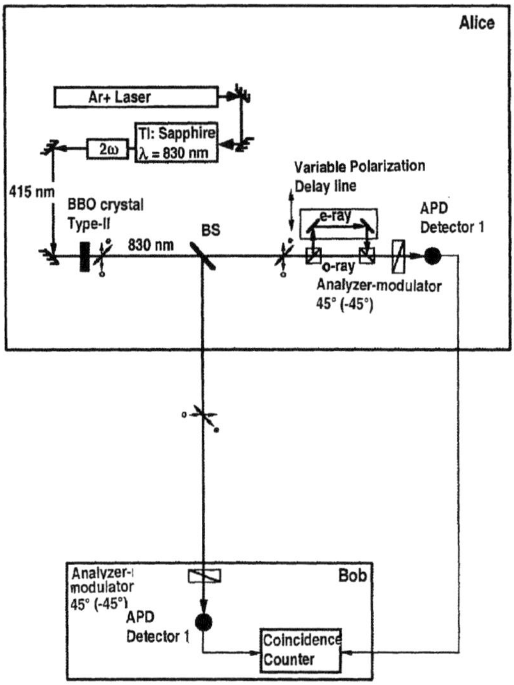

Figure 2. Schematic of the experimental setup for the generation of Type-II entangled photons (with orthogonal polarization) and their registration using quantum interference with coincidence detection.

Polarization-entangled photons are created by sending frequency doubled femtosecond Ti:Sapphire laser pulses through an appropriately oriented Type-II second-order nonlinear BBO crystal. Using the experimental system in Fig. 2, one can accurately manipulate the phase and position of the emitted entangled photon pairs. The photons enter two spatially separated arms via a polarization insensitive 50/50 beamsplitter (BS) allowing both ordinary and extraordinary polarized photons to be reflected and transmitted with equal probability. One arm contains a controllable polarization-dependent optical delay (the e-ray/o-ray loop). The introduction of polarization analyzers oriented at 45 degrees in front of each photon counting detector completes the creation of the polarization interferometer. Signal correlation is registered by detecting the coincidence counts between the two detectors as a function of the polarization delay.

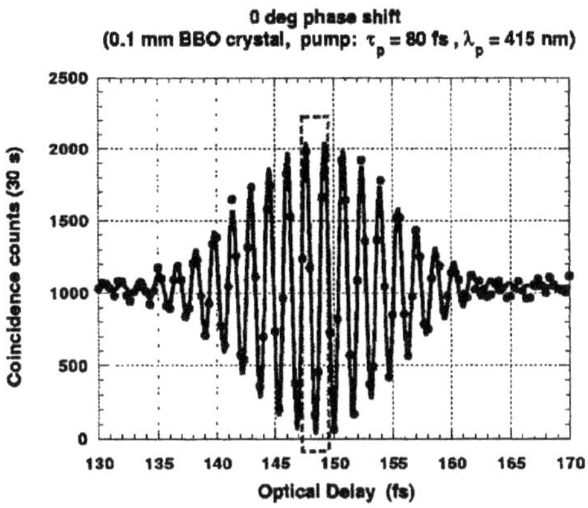

Figure 3. Experimentally obtained intensity correlation function from a 0.1-mm thick BBO crystal. The high-frequency carrier inside the envelope reflects the period of the UV pump wavelength rather than of inividual signal or idler waves. This is due to the two-photon entanglement of the twin beams. This quantum *destructive* interference is observed at **0 deg phase shift** between polarization analyzers.

The crucial features of this quantum interferometer are:

Double - One input beamsplitter (BS) and two output polarization beamsplitters (analyzers at 45°), well-separated in space.

Strongly unbalanced - polarization delay line introduced only in one interferometer.

Distributed - first beamsplitter is with Alice, one of the output beamsplitters is far away with Bob.

Nonlocal quantum interference - a phase shift imposed on one of the entangled photons does work for both of them even though they are well separated in space.

Polarization interferometer - Type-II SPDC and polarization analyzers at the output beamsplitter.

Intensity correlations - measure an intensity correlation function by detecting the variation in the coincidence counting rate.

Results obtained using this experiment are shown in Fig. 3. The pattern in this figure arises from the contributions of two effects. First, the full-width at half-maximum of the envelope defines the coherence time

$$T_{coh} = \left(\frac{1}{u_o} - \frac{1}{u_e}\right) L_c \tag{2}$$

where u_o and u_e are the group velocities of the ordinary and extraordinary waves and L_c is the length of the crystal. Second, the internal modulation has a period that depends only on the pump wavelength.

The 90° shift of the phase in one of the analyzers will change the quantum interference immediately so as to be constructive (rather than destructive) at the central fringe (see Fig. 5) with a very high ($\sim$ 99%) contrast.

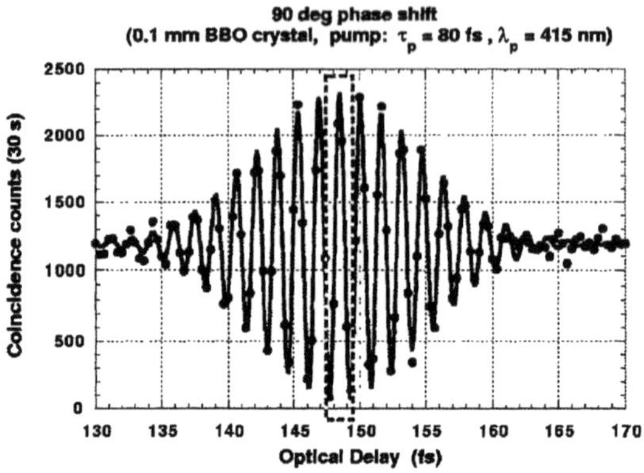

Figure 4. Experimentally observed *constructive* interference at **90 deg phase shift** between polarization analyzers.

In order to complete the procedure of quantum key distribution using our new design, we have to randomly modulate the polarization parameters of the two-photon entangled state by switching each analyzer-modulator between two sets of polarization settings 0°/90° or 45°/135°. This can be accomplished using fast Pockels-cell polarization rotators in front of the detectors.

These two sets of selected angles will force the mutual measurement by Alice and Bob to be a binary "0" (destructive) or "1" (constructive) with 50%-50% probability. Each particular outcome depends on mutual orientation of modulators on both sides. Discussion between Alice and Bob over the public communication channel of which set of orientations was selected in each trial but not what was the outcome will complete the standard quantum key distribution described in a literature.[4, 14, 15]

The use of high-repetition rate femtosecond pulses as a pump source enhances significantly the flux of entangled photon pairs available for reliable and secure key distribution. The downconverted entangled pairs appear only at well-defined times when pump pulses are present. A fixed 12.5-ns timing separation between the pump pulses enhances significantly the performance of single-photon detectors increasing the high-fidelity detection rate. The femtosecond timing will help significantly to develop a daylight operating communication system.

Our study has shown that the phase-sensitive quantum interference of two entangled photons in a strongly unbalanced polarization intensity interferometer delivers robust quantum hardware suitable for practical quantum cryptography applications. The high contrast and stability of quantum interference demonstrated in our preliminary experiments promises to bring the performance of this system above the level of the best single-photon polarization techniques, and to do so without their specific limitations.

REFERENCES

1. Ou, Z. Y. and Mandel, L., *Phys. Rev. Lett.* 1988, 61:50; Shih, Y. H. and Alley, C. O., 1988, *Phys. Rev. Lett.* 61:2921; Hong, C. K., Ou, Z. Y., and Mandel, L., *Phys. Rev. Lett.* 1987, 59:2044; Ou, Z. Y. and Mandel, L., *Phys. Rev. Lett.* 1988, 61:54; Kwiat, P. G., Steinberg, A. M., and Chiao, R. Y., *Phys. Rev. A.*, 1993, 47:2472; Brendel,J., Mohler, E., and Martienssen, W.,*Phys. Rev. Lett.* 1991, 66:1142; Larchuk, T. S., Campos, R. A. , Rarity, J. G., Tapster, P. R., Jake-

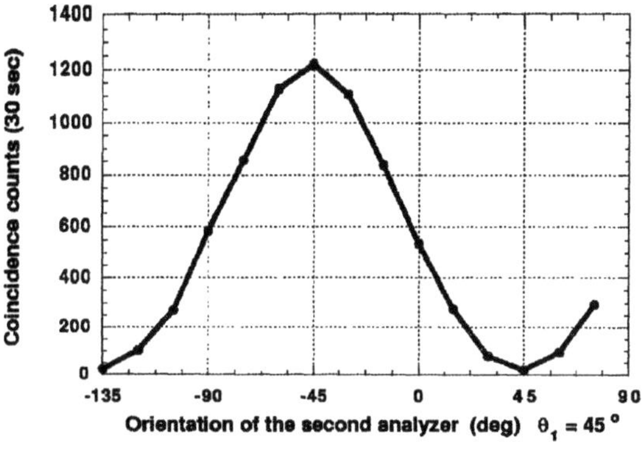

Figure 5. Experimentally observed modulation at the central fringe as a function of relative position of polarization modulators. 99% visibility insures a strong violation of Bell's inequalities.

man, E., Saleh, B. E. A., and Teich, M. C., *Phys. Rev. Lett.* 1993, 70:1603; Steinberg, A. M., Kwiat, P. G., and Chiao, R. Y., *Phys. Rev. Lett.* 1993, 71:708; Hong, C. K., Ou, Z. Y., and Mandel, L., *Phys. Rev. Lett.* 1987, 59:1903; Rarity, J. G., and Tapster, P. R.,*J. Opt. Soc. Am. B.* 1989, 6:1221; Kiess, T. E., Shih, Y. H., Sergienko, A. V., and Alley, C. O., *Phys. Rev. Lett.* 1993, 71:3893; Shih, Y. H. and Sergienko, A. V., *Phys.Lett A.* 1994, 186:29; Sergienko, A. V., Shih, Y. H., and Rubin,M. H., *J. Opt. Soc. Am. B.* 1995, 12:859.

2. Bennett, C. H., Bessette, F., Brassard, G., Savail, L., and Smolin, J., *Journal of Cryptology.* 1992, 5:3; Bennett, C. H., Brassard, G., Mermin, N. D, *Phys. Rev. Lett.* 1992, 68:557; Bennett, C H., Wiesner, S. J., *Phys. Rev. Lett.* 1992, 69:2881; Breguet, J., Muller, A., and Gisin, N., *Journal of Mod. Opt.* 1994, 41:2405, Franson, J. D. and Ilves, H., *J. of Mod. Opt.* 1994, 41:2391; Franson, J. D. and Jacobs, B. C., *Electronic Letters* 1995, 31:232; Muller, A., Herzog, T., Huttber, B., Tittel, W., Zbinder, H., and Gisin, N., *Appl. Phys. Lett.* 1997, 70:793.
3. Hughes, R. J., Alde, D. M., Dyer, P., Luter, G. G., Morgan, G. L., and Schauer, M., *Contemporary Physics* 1995, 36:149.
4. Ekert, A. K., *Phys. Rev. Lett.* 1991, 67:661; Ekert, A. K., and Palma, G. M., *J. of Mod. Opt.* 1994, 41:3413; Rarity, J. G., Owens, P. C. M., and Tapster, P. R., *J. of Mod. Opt.* 1994, 41:2435; Rarity, J. G., and Tapster, P. R., *Phys. Rev. A.* 1992, 45:2052.
5. Ekert, A. K., Palma, G. M., Rarity, J. G., and Tapster, P. R., *Phys. Rev. Lett.* 1992, 69:1293.
6. Schrödinger, E., *Naturwissenschaften* 1935, 23:807, [Translation *in* 1983, "Quantum Theory of Measurement," ed. Weeler, J. A., and Zurek, W. H., Princeton University Press Princeton].
7. Einstein, A., Podolsky, B., and Rosen, N., *Phys. Rev.* 1935, 47:777.
8. Bell, J. S., *Physics* 1964, 1:195.
9. Yariv, A., 1967, "Quantum Electronics," Wiley, New York.
10. Klyshko, D. N., 1988, "Photons and nonlinear optics," Gordon and Breach, New York.
11. Blombergen, N., 1965, "Nonlinear Optics," Benjamin, New York.
12. Rubin,M. H., Klyshko, D. N., Shih, Y. H., and Sergienko, A. V., *Phys.Rev. A.* 1994, 50:5122; Shih, Y. H., and Sergienko, A. V., *Phys.Rev. A.* 1994, 50:2564; Pittman, T. B., Sergienko, A. V., Shih, Y. H., and Rubin,M. H., *Phys.Rev. A.* 1995, 51:3495; Kwiat, P. G., Mattle, K., Weinfurter, H., Zeilinger, A., Sergienko, A. V., and Shih, Y. H., *Phys. Rev. Lett.* 1995, 75:4337.
13. Franson, J. D., *Phys. Rev. Lett.* 1989, 62:2205.
14. Bennett, C. H., Brassard, G., 1984 *in* "Proc. Int. Conf. Computer Systems and Signal Processing," Bangalor, 175.
15. Bennett, C. H., *Phys. Rev. Lett.* 1992, 68:3121.

GENERATION OF NARROW-BAND TWO-PHOTON STATE WITH DOUBLY RESONANT CAVITY IN PARAMETRIC DOWN-CONVERSION

Z. Y. Ou

Department of Physics
Indiana University-Purdue University Indianapolis
402 N. Blackford St.
Indianapolis, IN 46202

Spontaneous parametric down-conversion normally produces two-photon state with a wide bandwidth of Tera Hz, which is too large for optical detector to resolve and will limit its application in quantum information processing. Passive filtering may be used to narrow the bandwidth at the expense of signal rate. Here we report an experiment using active filtering to narrow the bandwidth to sub Giga Hz range and in the meantime greatly enhance the signal level. The observed enhancement factor is 5.6×10^4 per mode.

INTRODUCTION

Recently, there has been renewed interest in the interference between independent sources of light, mainly for the applications in quantum information processing (Bouwmeester et al., 1997; Pan et al., 1998). In these applications, two or more independent quantum sources are often mixed to create quantum entanglement among them. Quantum entanglement is essential in information exchange for processing. More specifically, three or more photon state can be produced indirectly by mixing two or more sources in two-photon state commonly generated by parametric down-conversion. The reason that independent sources have to be exploited here is that entangle quantum states with more than two particles are difficult to produce directly.

Interference between two independent sources was first demonstrated by Magyar and Mandel (1963) in a classic experiment involving two independent lasers. However, because classical sources were used for the optical fields, it does not have any practical application in quantum information processing other than serves as a proof of principle. Detailed quantum theory was later worked out by Ou (1988), who showed that the bandwidth of the independent sources must be much smaller than that of the detectors

Quantum Communication, Computing, and Measurement 2
Edited by Kumar et al., Kluwer Academic / Plenum Publishers, New York, 2000.

in order to achieve high visibility in the interference between independent stationary fields of any kind. Recently, however, Rarity (1995) and Zukowski et al. (1995) proposed a scheme that utilizes pulsed (nonstationary) field to solve the bandwidth problem. Narrow bandpass (passive) filters are still required to achieve high visibility. Thus narrow-band optical fields involved in interference is still preferable for satisfying the condition of high visibility in interference between independent sources. But passive filtering will substantially reduce the signal level as in recent experiments (Rarity, 1996; Bouwmeester et al., 1997; Pan et al., 1998)

In this paper, we will demonstrate experimentally how to produce a narrow-band two-photon state while increasing or at least maintaining the signal level by cavity-enhanced parametric down conversion process. We show a significant reduction in the bandwidth of the source (3 orders of magnitude) with signal level (overall rate) enhanced 20 times resulting in more than 4 orders of magnitude enhancement factor per mode.

GENERAL THEORY

Two-photon state can be produced via spontaneous parametric down-conversion by splitting the pump photon of higher energy into two photons with lower energy (known as signal and idler photons). However, the down-converted photons usually have a wide bandwidth due to an extra degree of freedom for the down-conversion process (two frequency components are produced). The only limitation on bandwidth results from the phase matching condition, which sets a loose constraint on the frequencies of down-converted photons and gives rise to a wide bandwidth of a typical value of 10^{13} Hz. This is much wider than the response bandwidth of the fastest photodetector available (typically 10^{10} Hz). Thus spontaneous parametric down-conversion cannot be used in quantum interference between independent sources. In order to reduce the bandwidth, we may either use a narrow bandpass filter (Hong et al., 1987) or a small pinhole (Rarity and Tapster, 1989). For the reduction of bandwidth to sub-Giga Hz level (to match the bandwidth of the detectors), both methods, when applied, will significantly decrease the flux of the photon pairs because the photon number is proportional to the bandwidth (or number of modes) owing to the spontaneous nature of the process.

Active filtering by placing the source inside an optical resonator can also reduce the bandwidth of the down-conversion. Quite different from passive filtering, it will enhance the down-conversion process by equivalently lengthening the interaction distance. By making both down-conversion components resonant with the cavity (double resonance), the interaction length is effectively increased by $\mathcal{F}$ times with $\mathcal{F}$ being the finesse of the cavity or equivalently the number of round trips in the cavity for the photons before they are coupled out. The enhancement results from a constructive two-photon interference. The pump power has to be carefully controlled to operate at a relatively low level so that no stimulated process occurs and only two-photon states are produced.

In order to see how the enhancement scheme compensates the decrease of signal level due to bandwidth reduction, we recall that the signal from spontaneous down-conversion process is proportional to the bandwidth $\Delta\omega$ and to the square of interaction length d (Yariv, 1989). So we can write the photon rate of the signal as

$$R_1 = CPd^2\Delta\omega, \tag{1}$$

where P is the power of the pump field, C is the proportional constant that depends on the nonlinear coupling. With cavity enhancement, the interaction length d is changed to the effective length $d\mathcal{F}$ and the bandwidth $\Delta\omega$ to $\Delta\Omega$ of the cavity linewidth. It can be shown more rigorously (Ou et al., 1992) that when double resonance condition is satisfied, the enhancement factor for one frequency component is

$$B = \frac{4}{\pi^2}\mathcal{F}_1\mathcal{F}_2 \equiv \frac{R_2/\Delta\Omega}{R_1/\Delta\omega} \tag{2}$$

for doubly resonant parametric down-conversion. Here $\mathcal{F}_{1,2}$ is the finesse of the cavity for the two down-converted fields, respectively. So the photon rate R_2 for the cavity-enhanced down-converted field is then

$$R_2 \approx \frac{4}{\pi^2}CP\mathcal{F}_1\mathcal{F}_2 d^2 \Delta N \Delta\Omega, \tag{3}$$

where $\Delta\Omega$ is the bandwidth of the down-conversion from the enhancement cavity due to the linewidth of the cavity and ΔN is the number of modes from the cavity (see later discussion on mode structure). $\Delta\Omega$ is usually of the order of 10^8 rad/sec. It can be seen from Eq.(3) that the reduction due to bandwidth (10^{-4}) is compensated by the cavity enhancement ($\mathcal{F}_1\mathcal{F}_2 \sim 10^5$)

However, optical cavity allows simultaneous resonance of a number of frequencies (longitudinal modes of the cavity). Double resonance condition for parametric down-conversion will reduce the number. But depending on whether it is type I or II down-conversion, we may still have more than one frequency mode of the cavity. Even for a single mode of the cavity, the field still has multi-frequency components with a bandwidth of $c/L\mathcal{F}$. L is the effective length of the cavity.

MODE STRUCTURE

Although cavity allows simultaneous resonance of many frequency components forming the longitudinal mode of the cavity with equal spacing of $\Delta\nu = c/L$, the free spectral range of the cavity, the two down-converted fields may not be both resonant in the same time. The double resonance condition will further restrict the down-conversion bandwidth.

I. Type-I down-conversion

Two highly correlated fields with same polarization are generated in Type-I parametric down-conversion. For studying interference effect, it is preferable to make them equal in frequency. This corresponds to degenerate parametric down-conversion. The resonance condition requires that

$$\omega_0 = \frac{2\pi N_0 c}{l + n(\omega_0)d} \quad \text{or} \quad [l + n(\omega_0)d]\omega_0 = 2\pi N_0 c \tag{4}$$

for the degenerate frequency ω_0. Here N_0 is some integer, d is the length of the nonlinear crystal providing parametric interaction, $n(\omega_0)$ is the index of refraction, and l is the cavity length excluding the crystal. Because of the degeneracy in frequency, double resonance is automatically satisfied.

On the other hand, because of the linewidth of each cavity mode, some other pairs of down-converted fields with nondegenerate frequencies of $\omega_{1,2} = \omega_0 \pm \Delta\omega$ may also satisfy double resonance approximately to within the linewidth. The conditions for them are

$$[l + n(\omega_0 + \Delta\omega)d](\omega_0 + \Delta\omega) = [2\pi(N_0 + \Delta N) + \delta_+]c \tag{6a}$$

$$[l + n(\omega_0 - \Delta\omega)d](\omega_0 - \Delta\omega) = [2\pi(N_0 - \Delta N) + \delta_-]c \tag{6b}$$

with $\Delta N = 1, 2, \ldots$ and $|\delta_{+,-}| < \pi/\mathcal{F}$. Here $\pi/\mathcal{F}$ is the phase shift corresponding to half linewidth of the cavity mode. When $|\delta_{+,-}| > \pi/\mathcal{F}$, we do not consider the frequency components to be on double resonance. The question is how large is ΔN, which determines the number of modes that are doubly resonant.

Obviously, we have

$$|\delta_+ \pm \delta_-| \leq |\delta_+| + |\delta_-| < 2\pi/\mathcal{F}. \tag{7}$$

This provides a constraint on ΔN. Adding and subtracting Eqs.(6a,b) and using Eqs.(5, 7), we obtain

$$\Delta\omega < \sqrt{2\pi c/\mathcal{F}d|2n' + n''\omega_0|}, \tag{8a}$$

and

$$\Delta N \approx (l + n_0 d + n'\omega_0 d)\Delta\omega/2\pi c. \tag{8b}$$

Here $n_0 \equiv n(\omega_0)$, $n' \equiv \partial n/\partial\omega$ and $n'' \equiv \partial^2 n/\partial\omega^2$. Combining Eqs.(8a,b), we have

$$\Delta N \leq \frac{(l + n_0 d + n'\omega_0 d)}{\sqrt{2\pi c\mathcal{F}d|2n' + n''\omega_0|}}. \tag{9}$$

For a crystal of $KNbO_3$ of 4 mm length and a monolithic standing wave cavity, we have $\Delta N \approx 26$, i.e., about twenty pairs of down-conversion frequencies are on resonance with the cavity.

II. Type-II down-conversion

Two orthogonally polarized subharmonic fields are generated in Type-II parametric down-conversion. Because of the birefringence in nonlinear crystals, indices of refraction for the two fields are not same even at degenerate frequency. The condition for double resonance is even more restrict than Type-I case. Assuming we have double resonance at degenerate frequency:

$$(l + n_1 d)\omega_0 = 2\pi N_1 c, \quad (l + n_2 d)\omega_0 = 2\pi N_2 c, \quad (N_1 \neq N_2) \tag{9}$$

then, the double resonance condition for Type-II case is

$$(l + n_1 d)(\omega_0 + \Delta\omega) = 2\pi(N_1 + \Delta N_1)c, \tag{10a}$$

$$(l + n_2 d)(\omega_0 - \Delta\omega) = 2\pi(N_2 - \Delta N_2)c. \tag{10b}$$

$N_{1,2}$ are integers and $|\Delta N_2 - \Delta N_1| \equiv \Delta N = 1, 2, 3, \ldots$. So we have

$$\Delta\omega = 2\pi c(\Delta N_1 - \Delta N_2)/(n_1 - n_2)d. \tag{11}$$

However, Type-II parametric down-conversion has a bandwidth of

$$\Delta\omega_{PDC} \approx \pi c/|n_1 - n_2|d, \tag{12}$$

derived from phase matching condition:

$$|\Delta k|d = d|2\omega_0 n_0 - (\omega_0 + \Delta\omega)n_1 - (\omega_0 - \Delta\omega)n_2|/c < \pi, \tag{13}$$

where we assume that phase matching is achieved at degenerate frequency ($2n_0 = n_1 + n_2$). So $\Delta\omega$ in Eq.(11) lies outside the bandwidth of down-conversion. We have double resonance only for degenerate frequency and thus achieve the single mode operation in Type-II down-conversion.

EXPERIMENT

Because of its large nonlinearity, we choose $KNbO_3$ as our nonlinear crystal for parametric down-conversion. The pump field is from frequency-doubled Ti:sapphire laser operating around 855 nm, which is also the down-conversion wavelength. We use a semi-monolithic arrangement for the enhancement cavity: the $KNbO_3$ crystal is polished at one end in the geometry of a convex surface with a radius of curvature of 7 mm while at the other end with a flat surface. The length of the crystal is 4 mm. The curved surface is optically coated in such a way that it is highly reflective at subharmonic and highly transmissive at harmonic pump wavelength. The flat surface is anti-reflection coated at subharmonic wavelength. To form a resonator, a flat output coupler with 1% transmissivity is placed very close to the flat side of the crystal. The total length of the cavity is about 5 mm. The crystal is mounted on a thermal electric cooler (TEC) to maintain its temperature for proper phase matching.

To achieve enhancement in the signal, one critical technical problem is to maintain resonance of cavity for the degenerate frequencies of parametric down-conversion. A simple solution is to lock it to the fundamental frequency generated from the Ti:sapphire laser. But since it has exactly the same frequency as the down-converted signal, it will create an enormous background. To avoid the background, the technique of chopped locking (i.e. alternating the periods for cavity locking and the signal detection) can be used. The semi-monolithic design helps to form a rigid cavity so that the cavity stays locked when the locking field is absent (data collecting period). Another technical problem is to match the mode of pump beam to the cavity of down-conversion in order to achieve maximum efficiency of conversion. This is achieved by an auxiliary cavity. First, harmonic field is produced from the down-conversion cavity by reversing the down-conversion process and is mode-matched to the auxiliary cavity. The pump beam from Ti:sapphire laser is then mode-matched to the auxiliary cavity. In this way, the pump beam is matched to the down-conversion cavity.

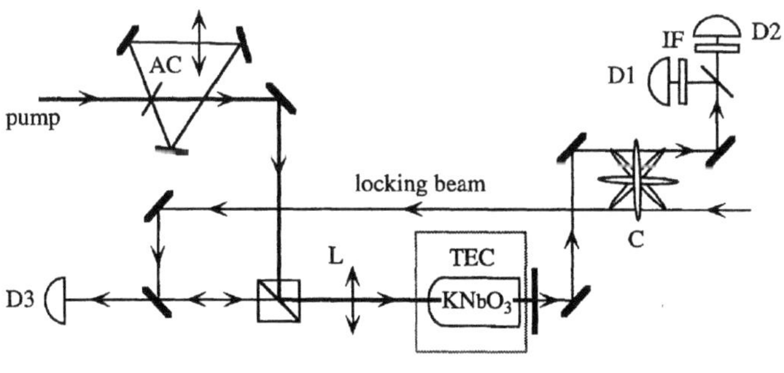

Figure 1. Experimental arrangement: L is a mode match lens, D3 is cavity locking detector, C is the chopper, AC is the auxiliary cavity, and IF is interference filter.

The schematic diagram for the experimental arrangement is shown in Fig.1. The down-converted signal is measured with two avalanche photodetectors (D1, D2). Interference filters with $\Delta\omega = 1.26 \times 10^{12}$ rad/sec are placed in front of the detectors. Count rates from as well as coincidence between the detectors are recorded as the length of the cavity is scanned. Fig.2 shows the result of the measurement. Strong resonant peaks in both count rate and coincidence rate occur whenever the cavity is on resonance. The secondary peaks are due to higher order transverse modes of the cavity and is a result of mode mismatching between the pump field and the down-conversion cavity.

To compare with the theory presented earlier, we also measured signal rate without the coupler resulting in no resonance effect. At 1 mW pumping power, we registered $R_1 = 10^5$ counts per second. At the same pumping power with the coupler in and cavity on resonance, we recorded a count rate of 1.6×10^7 per second. The finesse of the down-conversion cavity is measured to be $\mathcal{F} = 350$ and the free spectral range for the cavity is $\Delta\omega_{FSR} = 9.4 \times 10^{10}$ rad/sec resulting in a bandwidth of $\Delta\Omega = \Delta\omega_{FSR}/\mathcal{F} = 2.7 \times 10^8$ rad/sec for single mode. Because only Type-I phase matching is possible for $KNbO_3$, the down-converted field consists of multimode of the cavity. From previous section, we find that there are more than 20 modes. However, because of the interference filters, there are only $\Delta N = \Delta\omega/\Delta\omega_{FSR} = 13$ modes reaching the detectors so that the count rate for a single cavity mode is $R_2 = 1.6 \times 10^7/13 = 1.2 \times 10^6$ per second. Hence the observed enhancement factor per mode is

$$B \equiv \frac{R_2/\Delta\Omega}{R_1/\Delta\omega} = 5.6 \times 10^4, \tag{14}$$

which is in good agreement with the theoretical value of 5.0×10^4 from Eq.(2).

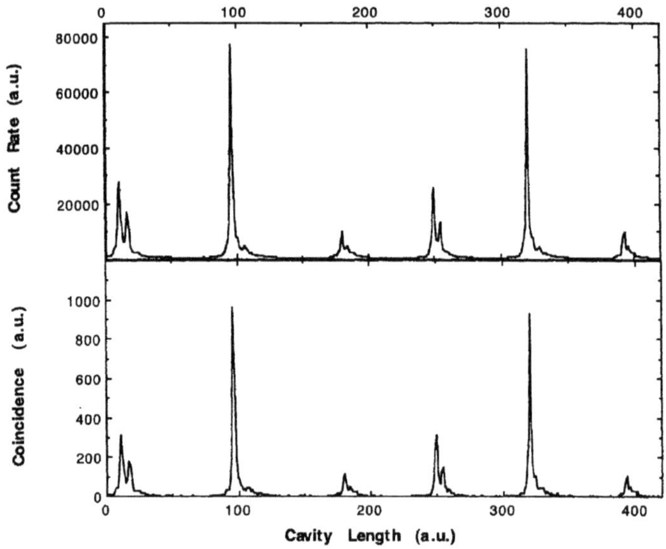

Figure 2. Count rate and coincidence as a function of cavity length

From Fig.2, we can see that the coincidence rate follows exactly the count rate. A simple regression shows a linear relation between the count rate and the coincidence, indicating that we have a two-photon source. Due to Type-I phase matching, the two-photon source is still not a narrow-band source. It contains about 13 longitudinal modes of the cavity with equal spacing of $\Delta\nu_{FSR} = 1.5 \times 10^{10}$ Hz. But this can be easily fixed with a passive Fabre-Perot filter. Because energy is highly concentrated in the 13 modes, elimination of all but one mode will only reduce the signal rate by 13 times. The bandwidth will be substantially reduced to 4.3×10^7 Hz, the cavity linewidth. This bandwidth is much smaller than the response bandwidth of photon detectors. With the high signal rate, we have achieved a truly high flux two-photon source and are ready to employ it to study interesting phenomena in quantum information processing and quantum nonlocality of multi-particle (Greenberger et al., 1989).

Acknowledgments

This work is supported by the Office of Naval Research.

REFERENCES

Bouwmeester, D., Pan, J-W., Mattle, K., Eibl, M., Weinfurter, H., and Zeilinger, A., 1997, Nature (London), **390**:575.
Greenberger, D. M., Horne, M. A. , and Zeilinger, A., 1989, *in*: "Bell's Theorem, Quantum Theory, and Conceptions of the Universe," M. Katafos, ed., Kluwer Academic, Dordrecht, The Netherlands.
Hong, C. K., Ou, Z. Y., and Mandel, L., 1987, *Phys. Rev. Lett.* **59**:2044.
Magyar, G. and Mandel, L., 1963, Nature **198**:255
Ou, Z. Y., 1988, Phys. Rev. A**37**:1607.
Ou, Z. Y., Pereira, S. F., and Kimble, H. J., 1992, Appl. Phys. B**55**:265
Pan, J-W., Bouwmeester, D., Weinfurter, H., and Zeilinger, A., 1998, *Phys. Rev. Lett.* **80**:3891.
Rarity, J. G., 1995, Ann. NY Acad. Sci. **755**:624.
Rarity, J. G., 1996, *in*: "Quantum Interferometry," F. De Martini, G. Denardo, Y. Shih, eds., VCH, Weinheim, Germany.
Rarity, J. G. and Tapster, P. R., 1989, J. Opt. Soc. Am. B**6**:1221.
Yariv, A., 1989, "Quantum Electronics", 3rd ed., Wiley, New York.
Zukowski, M., Zeilinger, A., and Weinfurter, H., 1995, Ann. NY Acad. Sci. **755**:91.

PART V: QUANTUM STATE ENGINEERING

FROM ENTANGLED PHOTONS TO ENTANGLED ATOMS

Eugene S. Polzik, Jan Hald and Jens L. Sørensen

Institute of Physics and Astronomy, Aarhus University
Aarhus 8000, Denmark

Mapping quantum states of light on atoms opens up the possibilities for quantum information storage, teleportation of atoms, and quantum computing. We describe a novel approach to such mapping based on creation of pairwise atomic entanglement in a large macroscopic atomic ensemble. As opposed to the previously demonstrated methods involving strong coupling of a single two-level atom to a cavity field, our strategy utilizes a large sample of three- and more level atoms weakly interacting with a free propagating quantum field. Besides practical advantages offered by the new method (no high-Q cavities involved), it also allows for production of a large number of correlated atoms which can be spatially separated and trapped in magnetic or magneto-optical traps.

INTRODUCTION

Strong interaction of the cavity e.-m. field with atoms traversing the cavity has been a subject of extensive research activity during the last decade. Recently remarkable experiments demonstrating elementary quantum operations [1] and two-atom entanglement [2] have been carried out with such systems. The cavity field has been used as the bus transferring the quantum state of one atom onto another atom.

Recently we started exploring an alternative approach for transferring quantum states of light onto atoms. Our method utilizes free propagating quantum-correlated light to produce correlated atoms. Strong interaction between the cavity field and a single atom is replaced by weak interaction of the propagating field with an atomic ensemble. An extra feature of this method is due to the fact that propagating quantum correlated e.-m. fields can be relatively easy to separate in space. This feature opens up a possibility for creating spatially separated entangled atomic ensembles.

As an example illustrating this idea, consider an Einstein-Podolsky-Rosen pair of photons with polarizations σ_+, σ_- propagating through an ensemble of three level atoms with the initial state $m_F = 0$ and the final states $m_F = -1, 1$ (Fig. 1).

All atoms are initially in the state $m_F = 0$. The atomic ensemble is assumed to be optically thick for the EPR light, so that the light is absorbed completely. As a result a pairwise delocalized atomic excitation of the kind $1/\sqrt{2}(|-1,1\rangle - |1,-1\rangle)$ is created. The necessary requirements for such mapping are therefore, three level atoms with initially empty final states and the complete absorption of light. The fact that

Quantum Communication, Computing, and Measurement 2
Edited by Kumar et al., Kluwer Academic / Plenum Publishers, New York, 2000.

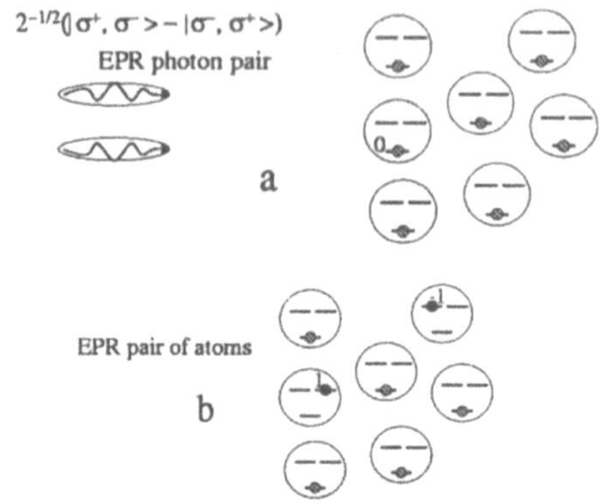

Figure 1. Mapping of EPR correlations of light onto atoms. a - before the interaction, b - after the interaction.

any pair of atoms may become entangled implies that, in order to detect the atomic entanglement, the whole ensemble of the atoms in the final states has to be addressed uniformly. A straightforward way to perform such readout process is to use a probe laser resonant with the $|-1\rangle, |1\rangle$ states but not with the $|0\rangle$ state. Such mapping is a natural development of quantum cryptography with EPR photons [3] towards the realization of the storage of quantum information.

EXCHANGE OF THE ENTANGLEMENT BETWEEN MULTI-PHOTON AND MULTI-ATOM SYSTEMS

Since the initial atomic reservoir ($m_F = 0$ in Fig. 1) may contain a very large number of atoms N_0, the same method can work for a rather large number of quantum correlated photons, which however should not exceed N_0. An example of such multi-photon, multi-atom entanglement exchange is a recent proposal for generation of Spin Squeezed States (SSS) of atoms via complete absorption of the free propagating squeezed light [4]. We consider the effect of *driving* V-atoms with a normal Gaussian beam containing a mixture of orthogonally polarized squeezed light and coherent light (Fig. 2).

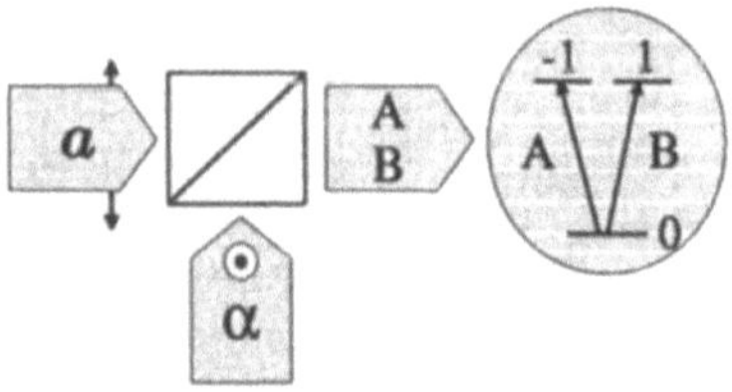

Figure 2. Generation of the spin squeezed state of atoms via interaction with the polatization squeezed light. *a* - squeezed vacuum field, α - coherent field.

The light absorption process creates an entanglement of the field and the individual atoms, and when all the light is absorbed in the sample we realize a steady state multiparticle entanglement, which is only partly (50 %) deteriorated by the effect of atomic spontaneous emission. We stress that considering spin squeezing within the final states of the transition is crucial because the large initial state population of atoms provides a dominant non-squeezed contribution to the collective optical atomic coherence. Spin squeezing within a close pair of atomic states is, in fact, the situation relevant for precision spin measurements in frequency standards, magnetometers, etc. Quantum correlations are created between the two excited states of V-type atoms. Following [4], let us introduce the collective continuous spin operator of the excited state $F_{ij}(r,t) = \frac{1}{\rho\delta V}\sum_\mu \exp\left[i\frac{\omega_{ij}}{c}(z-z_\mu)\right] F_{ij}^\mu$ where $F_{ij}^\mu = |i\rangle\langle j|, i,j = -1, 1$ for μ-th atom, ω_{ij} is the frequency splitting between the two upper substates, z the z-coordinate, ρ the number density of atoms and ΔV the volume. For atomic samples of the laboratory size and ω_{ij} in the MHz range the exponent can be substituted with unity. The spin (pseudo-spin) components of the excited state are then $F_z = \frac{1}{2}(F_{-1,-1} - F_{1,1}), F_x = \frac{1}{2}(F_{-1,1} + F_{1,-1}), F_y = \frac{-i}{2}(F_{-1,1} - F_{1,-1})$. In the lowest order of the perturbation theory one obtains:

$$\begin{aligned} &F_x \propto \alpha\alpha^* \\ &F_z \propto \alpha(a+a^+) \propto X(\varphi = 0) \\ &F_y \propto \alpha(a-a^+) \propto X(\varphi = \frac{\pi}{2}) \end{aligned} \qquad (1)$$

Eq. 1 reads that the mean spin is oriented along x axis since this is the only component which is nonzero on average. This is natural given that the coherent light has its polarization along x axis. The transverse components $F_{z,y}$ have zero mean and their variances $\delta F_{z,y}^2 \propto X^2(\varphi)$ - the variance of the quadrature phase amplitude of the squeezed vacuum $X(\varphi) = \frac{1}{\sqrt{2}}[ae^{i\varphi} + a^\dagger e^{-i\varphi}]$. φ is the phase of the squeezed vacuum relative to the coherent field. Since for the squeezed vacuum one of the quadrature phase amplitudes has the variance less than that for the vacuum field, one of the transverse collective spin components has its variance less than that for the pure coherent spin state which corresponds to the excitation with linearly polarized coherent light . Such a spin state is called squeezed spin state [5,6]. The nature of the entanglement is in pairwise correlations between the individual atomic spins created by light. The connection between individual atomic spin states on the one hand and the collective spin states on the other is illustrated in Fig. 3.

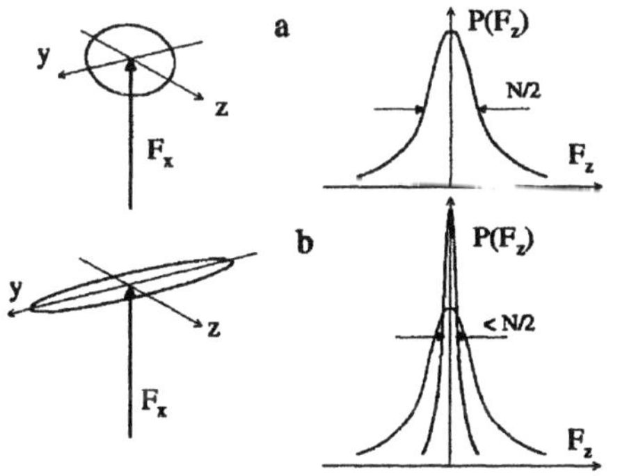

Figure 3. a. Coherent pseudo-spin state. Collective spin projections (left). Poissonian (binomial) probability distribution of atomic spin projections (right). b. Same for the squeezed spin states

Consider for simplicity spin 1/2 atoms. For N uncorrelated atoms oriented along the x axis the collective state projected on the y axis can be written as $|\Psi\rangle_{coh} = 2^{-N/2}(|\frac{1}{2}\rangle + |-\frac{1}{2}\rangle)^N$. The resulting collective spin is characterized by the binomial distribution with the zero mean value and the variance $\delta F_y^2 = \delta F_z^2 = \frac{1}{4}N = \frac{1}{2}\langle F_x\rangle$. A value of the variance less than that suggests correlations between the individual spins of the two ensembles. These correlations lead to the reduced quantum noise of the collective atomic spin of the ensemble which can be experimentally detected.

MAPPING QUANTUM STATES OF LIGHT ONTO ATOMS: EXPERIMENT

The experiment aimed at achieving spin squeezing of atoms via entanglement exchange with non-classical light has been carried out in the Quantum Optics Lab at Aarhus University. About 10^9 Cs atoms in a magneto-optical trap were used to study the spin noise of the $6P$ state using the probe polarization noise technique developed in [7]. The principle idea is to study an additional quantum noise imposed by the spin noise of atoms on the probe light transmitted through the atomic ensemble (Fig. 4).

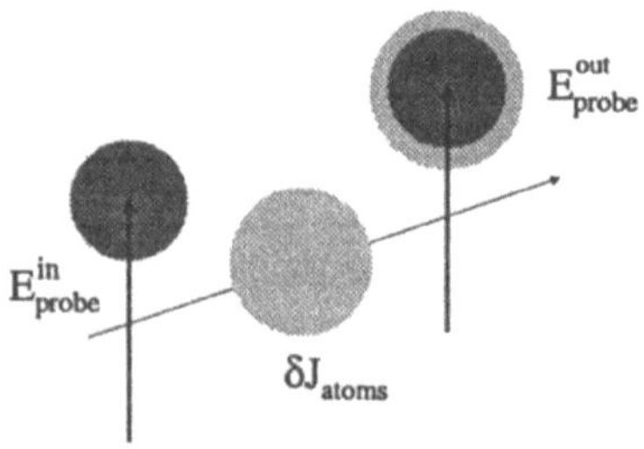

Figure 4. The probe light with its intrinsic quantum noise (shot noise) propagates through an atomic ensemble. The quantum noise of the collective atomic spin can be studied by observing an additional noise imposed on the transmitted probe.

In the experiment we study the spin noise of a magnetically degenerate state $6P_{3/2}$ in Cs with $F = 5$. The theory developed in [4] for $1, -1$ states can be generalized for states with larger angular momenta. For the geometry used in our experiments the three operators obeying the commutations relation $\frac{1}{2i}\left[\hat{J}_z, i\left(\hat{J}_-^2 - \hat{J}_+^2\right)\right] = -\left(\hat{J}_+^2 + \hat{J}_-^2\right)$, are the components whose quantum properties are measured. $\hat{J}_{x,y,z}$ are the atomic physical spin components and $\hat{J}_\pm = \hat{J}_x \pm i\hat{J}_y$. The following uncertainty relation $\delta\left(\hat{J}_z\right)\delta\left(i\left(\hat{J}_-^2 - \hat{J}_+^2\right)\right) \geq \left|\left\langle \hat{J}_+^2 + \hat{J}_-^2 \right\rangle\right|$ allows to introduce the "standard quantum limit" for the J_z and $i\left(\hat{J}_-^2 - \hat{J}_+^2\right)$ noise corresponding to uncorrelated individual atomic spins.

The experimental layout is shown in Fig. 5. The trapping beams and the magnetic field are chopped at 450Hz and the spin state of the $6P$ ensemble is studied only during the 1.2ms "dark" periods. According to [7] we expect the spin noise contribution to the probe noise be small compared to the shot noise of the probe. To observe this contribution we implement the lock-in method. During every second dark period the quantum pump is turned on. In this way the $6P$ state is modulated at 225Hz providing the possibility for lock-in discrimination of the atomic spin noise against the background of the shot noise of the probe and the dark noise.

First we excite the $6P$ collective spin with the coherent pump in order to establish the quantum limit of the spin noise. The pump resonant with the $6S_{1/2}, F = 4 \rightarrow 6P_{3/2}, F = 5$ 852nm transition provides weak excitation for the $6P$ collective spin of about 10^7 atoms. The noise of the probe is analyzed by the balanced polarization interferometer and the spectrum analyzer (SA in Fig. 5) at $\Omega = 1.9$MHz with the zero span [7]. The video bandwidth of the SA is chosen to allow for the detection of the 225Hz modulated component of the noise. The output of the SA is fed into the lock-in amplifier synchronized with the 225Hz frequency of the chopper. In this way the output voltage of the lock-in amplifier is proportional to the modulated part of the

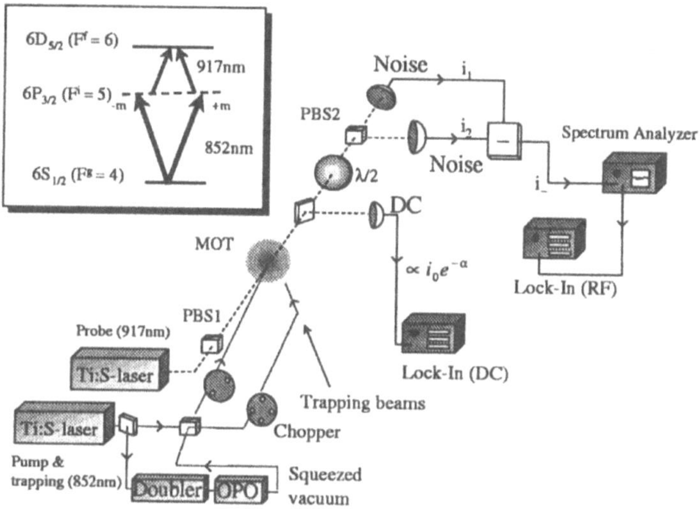

Figure 5. Experimental setup for generating spin squeezed states of cold atoms via excitation with squeezed light.

probe noise $i^2(\Omega) = i^2_{Mod\_shot}(\Omega) - i^2_{Coh\_spin}(\Omega)$ which consists of the quantum spin noise $i^2_{Coh\_spin}(\Omega)$ and the fraction of the probe shot noise $i^2_{Mod\_shot}(\Omega)$ modulated due to the modulation of the $6P$ population. The value of $i^2_{Mod\_shot}(\Omega)$ is measured independently via sending the probe through a cavity with the cavity length modulated at 225Hz providing the same modulated transmission of the probe as the atoms. In order to confirm that we are indeed at the quantum limit of the spin noise we plot $i^2(\Omega)$ (lower curve) and $i^2_{Mod\_shot}(\Omega)$ (upper curve) in Fig. 6 as a function of the optical depth α of the $6P$ ensemble (α is proportional to N, the number of atoms in $6P$). The probe transmission $e^{-\alpha}$ is measured with a dc detector and a second lock-in amplifier (Fig. 5).

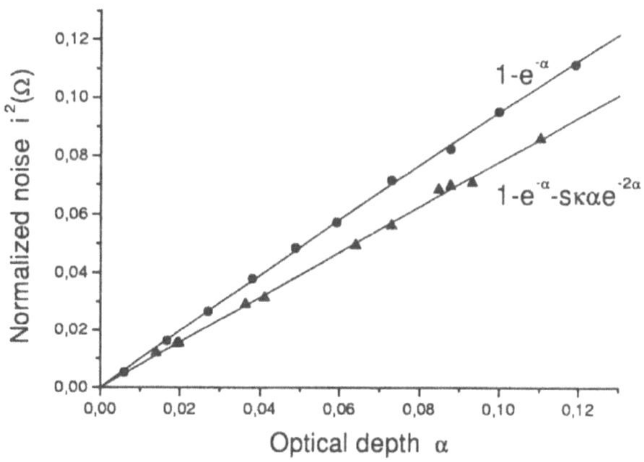

Figure 6. Quantum spin noise versus optical depth. Upper curve (circles): modulated probe shot noise $i^2_{Mod\_shot}(\Omega)$. Lower curve (triangles): modulated probe shot noise and quantum spin noise. $i^2(\Omega)$

According to the theory of [7] $i^2_{Mod\_shot}(\Omega) - i^2_{Coh\_spin}(\Omega) \propto [1 - \exp(-\alpha)] - s\kappa\alpha\exp(-2\alpha)$. s is the probe saturation parameter and κ is a constant depending on atomic parameters. The fit confirms that $i^2_{Coh\_spin}(\Omega) \propto \alpha \propto N$ and therefore we are indeed measuring the quantum limit of the spin noise. Fixing now α at its maximal value, we measure the spectral distribution of the coherent spin noise by scanning the probe detuning. The spectrum of the coherent spin noise is shown in Fig. 7 along with the dc absorption spectrum shown for reference. The spectrum of the spin noise can be well described by the sum of the squared dispersion profile corresponding to the noise of J_z observed off resonance and the Lorentzian corresponding to the noise of the alignment observed on resonance[8]. During the 1.2ms measurement time the weak pump accelerates the atoms. The resulting Doppler broadening gives an observed linewidth which is somewhat larger than the natural linewidth.

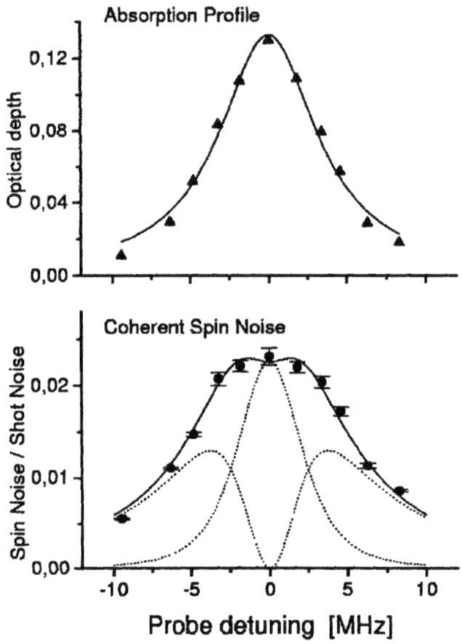

Figure 7. Upper curve: Probe absorption profile with a Lorentzian fit. Lower curve: Atomic spin noise, fitted with the square of a dispersion and a Lorentzian line shape having the same width as in upper curve.

We proceed with exciting the $6P$ ensemble with the quantum correlated pump. Adding suitably phased squeezed vacuum into the other port of the beamsplitter PBS1 (Fig. 5) we map the correlations of the light onto atoms. Squeezed vacuum with the central frequency resonant with the pump transition is generated in the sub-threshold optical parametric oscillator described elsewhere [9]. 2.5dB of the polarization squeezed light at $\Omega = 1.9$MHz is available at the trap site. In order to achieve the best mapping of quantum properties of light onto atoms, the optical depth for the quantum pump is sustained at the highest possible level, $\alpha_{pump} \approx 4$. We concentrate on the measurement of δJ_z because the sensitivity of our polarization measurements to the conjugated variable $i\left(\hat{J}^2_- - \hat{J}^2_+\right)$ is much smaller. When the squeezed vacuum is out-of-phase with the coherent component of the pump, i.e. the squeezing ellipse is elongated perpendicular to the coherent amplitude in the phase space, $X^2_{\pi/2} > 1$. The $\pi/2$ phase corresponds to the excess quantum noise of the spin above the coherent spin noise level: $\delta\left(\hat{J}_z\right)^2_{antisq} = \frac{1}{\eta+1}\delta\left(\hat{J}_z\right)^2_{coh}(1 + \eta X^2_{\pi/2}) > 1$. We introduce η, the mapping-readout efficiency for quantum correlations. $\eta = 1$ corresponds to perfect mapping of squeezing of light onto atoms and perfect readout of the quantum spin noise by the probe within the limits set by spontaneous decay[4]. On the contrary, in case of $\eta = 0$, $\delta\left(\hat{J}_z\right)^2 = \delta\left(\hat{J}_z\right)^2_{coh}$ independent of $X^2_{\pi/2}$. The spectrum of the observed anti-squeezed spin noise defined as $\left(\delta\left(\hat{J}_z\right)^2 - \delta\left(\hat{J}_z\right)^2_{coh}\right) D^2(\Delta)$, where $D^2(\Delta)$ is the spectral shape normalized to take the maximum value of 1, is plotted in Fig. 8 (circles). The data corresponds to 5.5dB of the excess noise in the anti-squeezed quadrature of the pump. Fitting the data with the square of the dispersion function $D^2(\Delta)$ with the width corresponding to the experimental absorption curve (Fig. 7) we obtain the value of the mapping-readout efficiency $\eta = 0.05$ at $\Delta = \pm 3.8$MHz. With the squeezed vacuum in-phase we measure $X_0^2 = 0.56$ corresponding to -2.5dB of squeezing of the pump. With such pump and the efficiency $\eta = 0.05$ the expected spin squeezing spectrum is plotted as a dotted line in Fig. 8. The result of the preliminary measurement is plotted as a triangle, and is in reasonable agreement with the expectations. We conclude that the first evidence of the spin squeezed ensemble with the noise below the quantum limit corresponding to uncorrelated individual atomic spins (dashed line) is observed. More measurements are on the way to confirm this preliminary result.

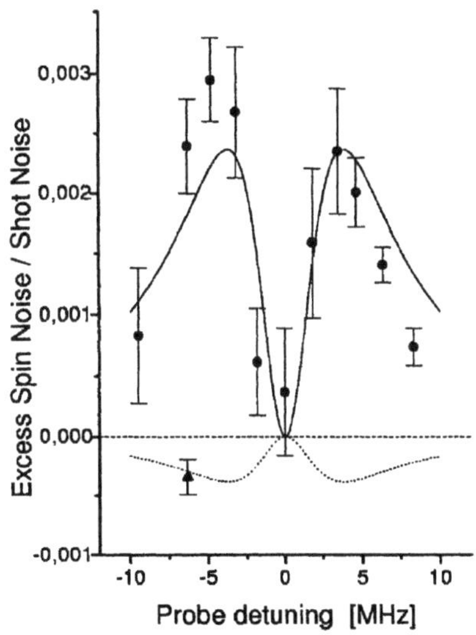

Figure 8. Upper curve (circles): Excess spin noise when exciting atoms with anti-squeezed light. The solid line is the squared dispersion profile from Fig. 7 but rescaled. Lower curve (triangle): dotted line is the expected noise reduction when pumping with squeezed light. The triangle is the preliminary measurement.

DISTANT EPR-CORRELATED ATOMIC ENSEMBLES

Along the same lines creation of two spatially separated macroscopic entangled atomic ensembles also seems feasible [10]. Such entanglement can be achieved, for example, in two magneto-optical traps (MOTs), via interaction with two EPR-correlated beams from an optical parametric oscillator. The proposal is illustrated in Fig. 9.

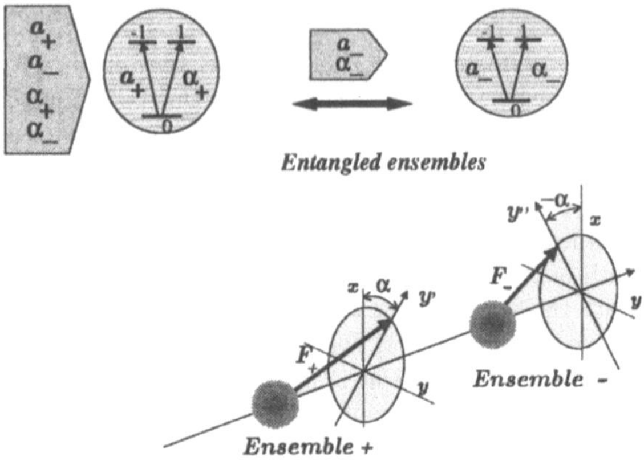

Figure 9. Creating EPR entangled atomic ensembles via interaction of atoms with EPR entangled light.

Two atomic ensembles are excited with the EPR correlated optical fields a_+ and a_- along with two coherent fields α_+ and α_-. The frequencies of the fields and the resonance frequencies of the atomic ensembles are chosen so that the "+" fields interact only with one ensemble whereas the "−" fields, only with the other one. Transfer of the quantum state of light onto atomic spins results in EPR correlations between the z component of the collective spin F_+ of the first ensemble and the same component of the spin F_- of the second ensemble. Similar correlations occur between the y components of the two ensembles. In case of continuous excitation the degree of EPR correlations is limited to 50% by spontaneous emission for the reasons considered in detail in [4].

SUMMARY

A method for mapping of quantum properties of light onto an ensemble of multi-level atoms is proposed and verified experimentally. The method is based on the complete absorption of the light carrying quantum correlations by the atomic medium. As a result, such properties of light as squeezing and EPR correlations may be transferred onto atomic collective spin. The method is readily scalable to large numbers of atoms. The resulting atomic entanglement is not a maximally entangled state. Rather the ensemble bears pairwise partial entanglement between the atoms. This feature explains why the atomic entanglement of this kind has a lifetime comparable with the lifetime of the atomic population, as opposed to the case of maximal entanglement which decays much faster.

In the experiment we have demonstrated observation of the collective atomic spin noise of cold atoms at the quantum level corresponding to the "shot noise" of atoms. Furthermore we observed mapping of the quantum noise of light onto the atomic cloud and corresponding excess (anti-squeezed) quantum noise of atoms. Recently we have succeeded in observing the first preliminary evidence of the spin squeezed state of atoms. The squeezed atomic ensemble contains around 10^7 atoms. We are therefore dealing with a "macroscopic" entangled object containing around 2 femtogram of matter.

The first experimental realization of the entanglement exchange between free propagating light and atoms described above concerns atoms in the excited state with the lifetime of 32ns. We plan to extend this research program towards long lived atomic states. Such states can be created by quantum correlated light via pulsed Raman processes. Besides the work on the storage of entangled light in atoms for quantum information processing, such long lived entangled states are relevant for sub-quantum atom interferometry and frequency standards.

REFERENCES

1. Q. A. Turchette, C. J. Hood, W. Lange, H. Mabuchi, and H. J. Kimble, Phys. Rev. Lett. **75**, 4710 (1995).
2. E. Hagley, X. Maitre, G. Nogues, C. Wunderlich, M. Brune, J. M. Raimond and S. Haroche, Phys. Rev. Lett. **79**, 1 (1997).
3. A. K. Ekert, Phys. Rev. Lett. **67**, 661 (1991).
4. A. Kuzmich, K. Molmer, and E. S. Polzik, Phys. Rev. Lett. **79**, 4782 (1997).
5. M. Kitagawa and M. Ueda, Phys. Rev. A **47**, 5138 (1993).
6. D. J. Wineland, J. J. Bolinger, W. M. Itano, F. L. Moore, D. J. Heinzen, Phys. Rev. **46**, R6797 (1992).
7. J. Sørensen, J. Hald, and E. S. Polzik, Phys. Rev. Lett. **80**, 3487 (1998).
8. C. Cohen-Tannoudji and F. Laloe, J. Phys. **30**, 277 (1969); F. Laloe, M. Leduc and P. Minguzzi, *ibid.* **28**, 505 (1967).
9. J. L. Sørensen, J. Hald, N. Jørgensen, J. Erland and E. S. Polzik, Quantum Semiclass. Opt. **9**, 239 (1997) .
10. E. S. Polzik, submitted, 1998.

SINGLE PHOTON QUANTUM CONTROL VIA HIGH-$\chi^{(3)}$ MEDIA

Michael Dunstan, Stojan Rebic, Sze Tan, Scott Parkins, Matthew Collett, Dan Walls

Physics Department, University of Auckland,
Private Bag 92019, Auckland, New Zealand

We discuss the generation of very large $\chi^{(3)}$ nonlinearities via electromagnetically-induced transparency, and the applications of these to quantum control with single photons.

INTRODUCTION

Given a sufficiently large optical nonlinearity, with low quantum noise, it should be possible to switch optical systems with single photons. In this paper we outline a method for achieving such nonlinearities and also some of their applications.

The simplest $\chi^{(3)}$ nonlinear optical medium is one consisting of an assemblage of far-detuned two-level atoms. In such a medium one can always increase the strength of the coupling by tuning closer to resonance. However, this increases the excitation of the atoms and hence the spontaneous emission rate; the penalty for large nonlinearity is large noise.

Atomic absorption can be reduced by exploiting quantum coherence effects in multilevel atoms. Such effects include coherent population trapping[1], electromagnetically induced transparency[2] (EIT), lasing without inversion[3] and ghost transitions[4]. Quantum non-demolition experiments based on the ghost transition scheme have produced the best quantum correlations to date[5]. The method we consider here is a form of EIT.

We begin by reviewing the basis of the method[6], and a simple application[7]. We then consider nonidealities of such a medium that may limit its usefulness for switching purposes. We note that some of these may be reduced or avoided by the use of a single atom in a high-finesse cavity. Finally we discuss possible applications of a high $\chi^{(3)}$ to state preparation and quantum control.

ELECTROMAGNETICALLY-INDUCED TRANSPARENCY IN AN ATOMIC MEDIUM

EIT is based on the use of a dark resonance: quantum interference cancels the absorption of the signal. This can result in a $\chi^{(3)}$ nonlinearity free of spontaneous emission noise.

Quantum Communication, Computing, and Measurement 2
Edited by Kumar et al., Kluwer Academic / Plenum Publishers, New York, 2000.

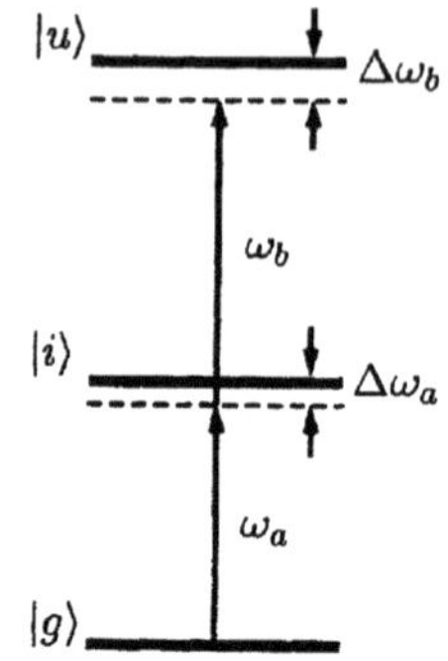

Figure 1. Three level scheme.

Schmidt and Imamoğlu[6] recently suggested an EIT scheme which is effectively a four-level generalisation of a conventional three-level cross-phase modulation scheme. The three-level scheme shown in Fig. 1 requires large detuning from the intermediate level to reduce one-photon loss. Unfortunately the effective nonlinearity also falls off rapidly with the detuning:

$$\mathrm{Re}\left[\chi^{(3)}\right] \propto \frac{1}{\Delta\omega_a^2}\,. \tag{1}$$

The four-level scheme shown in Fig. 2 is based on a Λ system, with ground states levels $|1\rangle$ and $|3\rangle$, and excited state $|2\rangle$. EIT occurs when the probe beam at ω_a and the coupling field at ω_c are at two-photon resonance with the $|1\rangle \leftrightarrow |3\rangle$ transition. The EIT eliminates one-photon loss without a corresponding reduction in the nonlinearity. In fact, the effective nonlinearity depends on the Rabi frequency of the coupling field,

$$\mathrm{Re}\left[\chi^{(3)}\right] \propto \frac{1}{\Omega_c^2}\,, \tag{2}$$

and can evidently be made large. In principle phase shifts of the order of π should be achievable with single photons.

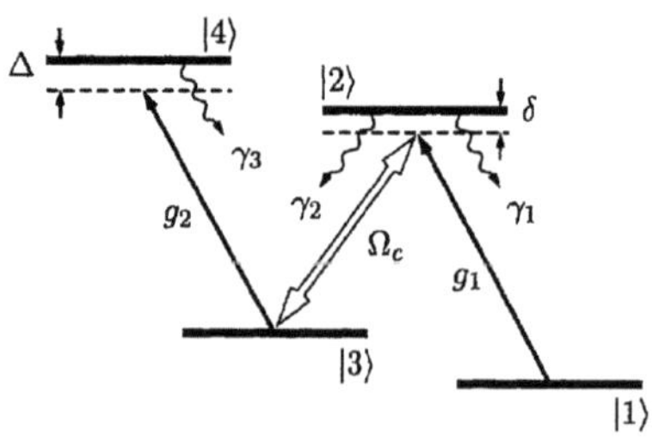

Figure 2. Four level EIT scheme.

Photon Blockade

Imamoğlu et al.[7] have shown that, given a sufficiently large pure $\chi^{(3)}$ nonlinearity in an optical cavity, there is an effect they have called a *photon blockade*.

The first photon to enter the cavity detunes it via the $\chi^{(3)}$ interaction; if this large enough, the driving field is sufficiently far from the new cavity resonance to prevent

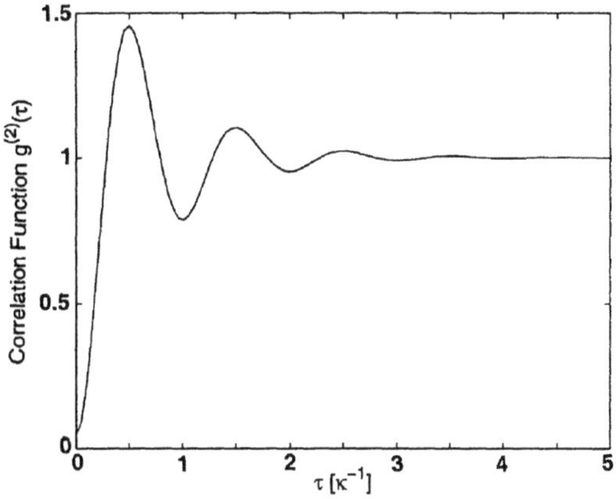

Figure 3. Second order correlation function, as calculated from pure high Kerr nonlinearity.

a second photon entering the cavity. The result is that only one photon at a time can be in the cavity, the next only being able to enter when the first has left. The result is strong antibunching of the transmitted light, as indicated by the second-order correlation function,

$$g^{(2)}(\tau) = \frac{\langle n(\tau)n(0)\rangle}{\langle n\rangle^2}, \tag{3}$$

shown in Fig. 3.

NONIDEALITIES IN THE EIT MEDIUM

The question now arises of what conditions are required for the atomic system to approach the pure $\chi^{(3)}$ nonlinearity that is assumed for the photon blockade:

$$H = \chi(a^\dagger a)^2 . \tag{4}$$

The presence of the atomic medium may add noise; the medium may not be able to respond fast enough to changes in the state of the cavity; the propagation of the field in the cavity may be altered. We consider each of these in turn.

Quantum Noise Analysis

We start from a description including the cavity mode and all of the atoms. We chose the operating conditions to satisfy the EIT condition, i.e. two-photon resonance on the ($|1\rangle \to |3\rangle$) transition. We then perform a standard linearised quantum fluctuation analysis on the atomic variables, and adiabatically eliminate them to obtain an equation for the field mode operator only:

$$\dot{a} = -\kappa\left(1 + \frac{i}{\kappa}\eta' a^\dagger a\right)a - \sqrt{2\kappa}\, a^{\text{in}} + \Gamma_{\text{atomic}}\,, \tag{5}$$

where

$$\eta' = \left(\frac{g_1}{\Omega_c}\right)^2 \left[\frac{g_2^2\Delta}{\left(\frac{\gamma_3}{2}\right)^2 + \Delta^2} - \frac{g_1^2\delta}{\gamma^2 + \delta^2}\right], \text{ and} \tag{6}$$

$$\gamma = \frac{\gamma_1 + \gamma_2}{2}. \tag{7}$$

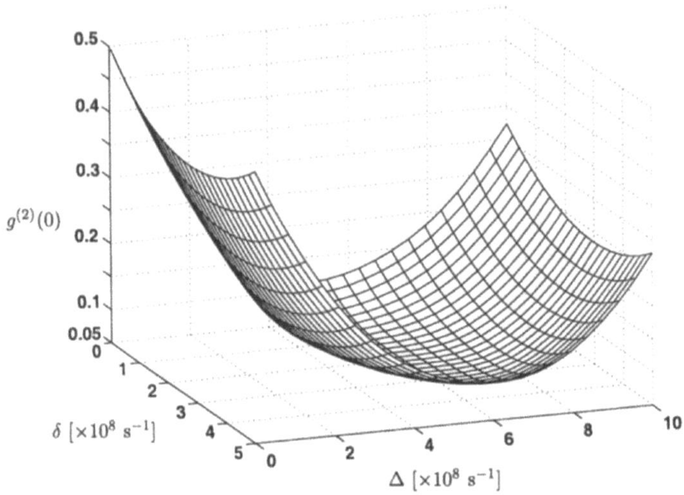

Figure 4. Minimisation of noise in the adiabatic approximation.

This differs from the case of a cavity mode with an ideal $\chi^{(3)}$ medium by the presence of the atomic noise term Γ_{atomic}. We can now evaluate the second-order correlation function for various parameters and look for the optimum choice. Fig. 4 shows $g^{(2)}(0)$ with atomic noise as a function of the detunings, keeping the effective $\chi^{(3)}$ at the same value as in Imamoğlu's work; the minimum $g^{(2)}(0)$ occurs for

$$\frac{\delta}{\gamma} \approx 7\,, \ \frac{\Delta}{\gamma} \approx 16\,, \tag{8}$$

and is close to zero, suggesting that for these parameters it is possible to avoid the blockade being interrupted by atomic noise. The other parameters here are:

$$\begin{array}{ll} \kappa = 0.5\gamma & \gamma_1 = \gamma_2 = \gamma_3 = \gamma \\ \Omega_c = 15\gamma & N = 10^7 \text{ atoms} \\ g_1 = g_2 = g & \\ C = g^2 N/\kappa\gamma = 10^6 & \end{array} \tag{9}$$

Validity of Adiabatic Approximation

Both the ideal effective Hamiltonian and the quantum noise analysis in the previous section assume that the atoms can be adiabatically eliminated.

The validity of this assumption has recently been questioned by Grangier et al.[8] In the adiabatic approximation the response time of the medium is assumed to be much shorter than the typical timescale of the cavity. Direct comparison of the cavity and atomic linewidths would suggest this to be the case.

However, the reason that the noise is kept small, despite the large effective coupling, is that the atomic polarisation is small, the atoms remaining in the ground state. In fact, the atomic polarisation vanishes at first order due to EIT. But EIT requires two-photon resonance and the two-photon resonance has a very narrow linewidth. Slightly off-resonant frequency components see a very high refractive index which switches them out of resonance.

Thus the effective Hamiltonian is only valid close to the central frequency component of the cavity field.

By linearising the single mode light field as well as the atomic variables, it is possible to perform a linearised fluctuation analysis without the adiabatic approximation, including the dynamics of the atoms[8]. The resulting intensity squeezing spectrum Fig. 5 shows linewidth narrowing of factor of ~ 50 due to the narrowness of the two-photon resonance. That is, the quantum noise reduction is restricted to a narrow bandwidth. This implies that the photon blockade does not work well as a switch: while the average transmitted intensity over times long compared to the inverse of the two-photon resonance width is stabilised to well below shot-noise levels, on short (single-photon) time scales the improvement is much less dramatic.

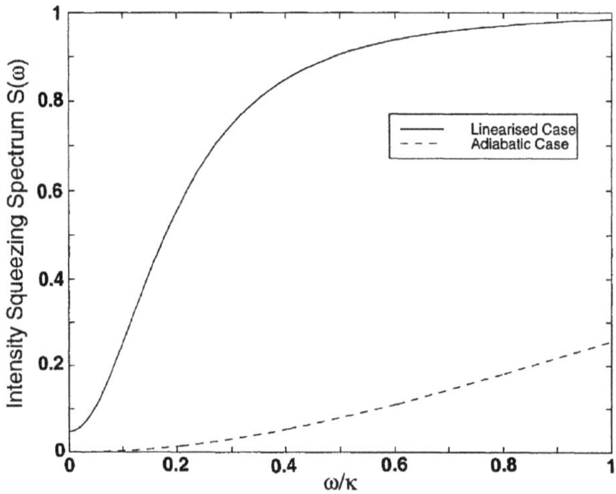

Figure 5. Intensity squeezing spectrum.

Cavity Linewidth Narrowing

Some compensation for the narrowness of the two-photon resonance may be provided by the linear dispersion of the atoms. The classical properties of the optical cavity are modified by the presence of a nonabsorbing highly dispersive medium: in particular, a sharply reduced group velocity leads to substantial linewidth narrowing.

There have been theoretical predictions[9,10] of linewidth narrowing of several orders of magnitude, and a linewidth reduced by a factor of 50 has been experimentally demonstrated[9].

It is conceivable that such narrowing could be enough to restore the validity of the adiabatic approximation, and hence the imperviousness of the photon blockade. However, the response timescales resulting would be much increased over those of the original model, and the usefulness of the system as a switch accordingly reduced.

At the low intensities of interest here, the problem is further complicated by the nonlinearity; the group velocity, and hence the effective cavity loss rate, is dependent on the intensity. That is, the nonlinear dispersion results in an effective nonlinear damping.

EIT WITH A SINGLE ATOM

As an alternative to the atomic medium, we may also consider using a single four-level atom in a high-finesse cavity. We use the relative loss rates achieved by Hood et al.[11] in Cs:

$$\gamma \approx 5\,\text{MHz}\,, \tag{10}$$
$$\kappa \approx 20\,\text{MHz} = 4\gamma\,, \tag{11}$$
$$g \approx 120\,\text{MHz} = 60\gamma\,; \tag{12}$$

and choose detuning and driving to mimic Imamoğlu's effective parameter[7] $\chi = 20\kappa$:

$$\delta = 0\,, \tag{13}$$
$$\Delta = 30\gamma\,, \tag{14}$$
$$\Omega_c = 100\gamma\,. \tag{15}$$

The result (see Fig. 6) is a second-order correlation $g^{(2)}(0)$ very close to zero, and a response time comparable to that of the adiabatic case. We conclude that the single-atom system, unlike the many-atom medium, *can* produce a photon blockade fast enough to be usable as a switch.

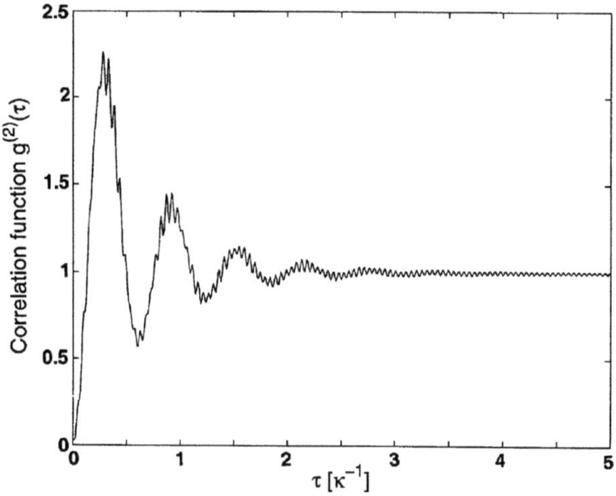

Figure 6. Second order correlation function for a single atom.

APPLICATIONS OF HIGH-$\chi^{(3)}$ NONLINEARITIES

Given a high $\chi^{(3)}$ nonlinearity, perhaps due to a single atom, there are other interesting applications in addition to the photon blockade.

Arbitrary state preparation

Ignoring now the nonidealities of the medium, a single-mode cavity with $\chi^{(3)}$ nonlinearity can be described by the effective Hamiltonian

$$H_0 = \hbar\omega_0 a^\dagger a + \hbar\chi a^{\dagger 2} a^2 \,, \tag{16}$$
$$H_{\text{driving}} = i\hbar(\varepsilon^* a - \varepsilon a^\dagger) \,, \tag{17}$$
$$H_{\text{damping}} = a\Gamma^\dagger + a^\dagger\Gamma \,, \tag{18}$$

where the driving field ε may be detuned from the cavity resonance, or even include a number of different frequencies. The energy eigenvalues of H_0 are

$$E_n = \hbar n \left[\omega_0 + (n-1)\chi\right] \,; \tag{19}$$
$$E_{n+1} - E_n = \hbar(\omega_0 + 2n\chi) \,, \tag{20}$$

as shown in Fig. 7. Driving with an on-resonance driving field corresponds to the photon blockade case: it prepares a one-photon state inside the cavity.

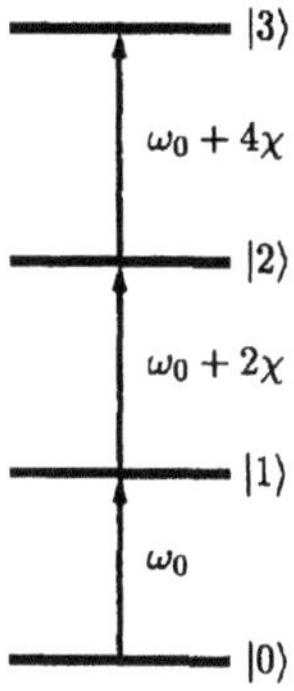

Figure 7. Lower energy levels of the nonlinear cavity.

More general target quantum states of the field,

$$|\psi_{\text{target}}\rangle = \sum_{n=0}^{N} c_n |n\rangle \,, \tag{21}$$

can also be created by driving with with different frequency components. This may be done sequentially, stepping through the photon number states with a series of pulses, each matched to the next transition. Obviously, the resulting superpositions can only survive for times shorter than the cavity lifetime. Alternatively, it may be done by driving with a field containing all required frequency components simultaneously. Even in this case, cavity losses reduce superpositions to mixtures in the steady state.

We can quantify the success of state preparation by the quality of state production

$$\langle\psi_{\text{target}}|\,\rho\,|\psi_{\text{target}}\rangle \,, \tag{22}$$

and also by the state purity

$$\text{Tr}\left(\rho^2\right) \,. \tag{23}$$

Both should be close to 1, and this can indeed be achieved on a short enough timescale. In Fig. 8 we show the results for the production of a number state $|n = 3\rangle$ by a sequence of pulses. The pulse widths (τ_{01}, τ_{12}, and τ_{23}), detunings ($\delta_1 = 0$, $\delta_2 = 2\chi$, and $\delta_3 = 4\chi$), and strengths ($|\varepsilon_1| = |\varepsilon_2| = |\varepsilon_3| = \chi/10$) are chosen to match the single photon transitions, stepping through the number states to reach the target. The effect of increasing the cavity decay rates is also shown.

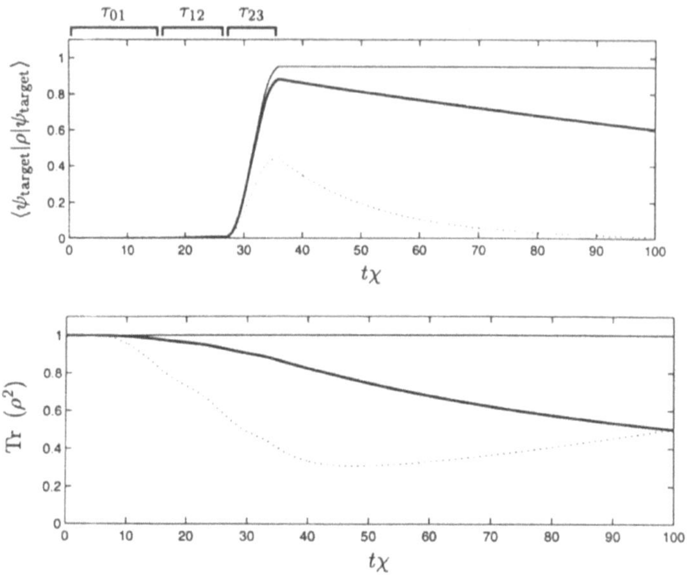

Figure 8. Production of a number state, $|\psi_{\rm target}\rangle = |n=3\rangle$, by a sequence of pulses. Cavity decay rates $\kappa = \chi/1000$ (bold solid line), $\kappa = 0$ (thin solid line), $\kappa = \chi/100$ (dotted line).

Quantum Control with a Two-Mode Field

The possibilities of single-photon switching are better illustrated with a two-mode cavity; we assume here that the two modes are two polarisations at the same frequency, having the same spatial modefunctions. The effective nonlinearity is then described by

$$H_{\rm int} = \hbar\chi\left(a^{\dagger 2}a^2 + b^{\dagger 2}b^2 + 4a^\dagger a b^\dagger b\right) . \tag{24}$$

The energy levels of this are shown in Fig. 9. We choose the detuning of the driving field for mode a to be on resonance, $\delta_a = 0$ and that for mode b to be detuned by $\delta_b = 4\chi$.

Initially no photons will enter mode b. One photon from beam a enters the cavity: a photon blockade is now in effect and no more will enter from beam a. But this one

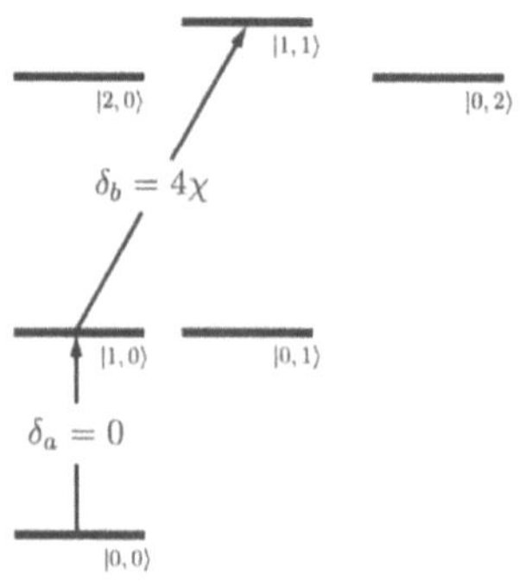

Figure 9. Energy levels of the quantum gate.

photon in mode a detunes the cavity for mode b — into resonance with its driving field, allowing one photon from driving field b to enter the cavity. In Fig. 10 we show populations of the quantum modes in response to the pulse envelopes. The first pair of pulses arrive at the same time ($t\chi = 50$) and each mode receives a single photon. For the latter pair ($t\chi > 400$) mode b is pulsed after mode a. In this case no photon enters mode b. A single photon enters mode a but then exits the cavity before the mode b is pulsed. The pulse widths are chosen to optimise success for the first case shown of equal pulse arrival.

On the experimental side, conditional quantum dynamics similar to this have been demonstrated by Brune et al.[12] and Turchette et al.[13] in microwave and optical cavity QED arrangements, respectively. Both experiments employed Kerr-type nonlinearities involving single atoms in high finesse cavities; conditional phase shifts of $\sim 36°$ per photon in the microwave regime and $\sim 16°$ per photon in the optical regime were observed. The intensity-based switch suggested here would require phase shifts of $\sim \pi/4$ per photon.

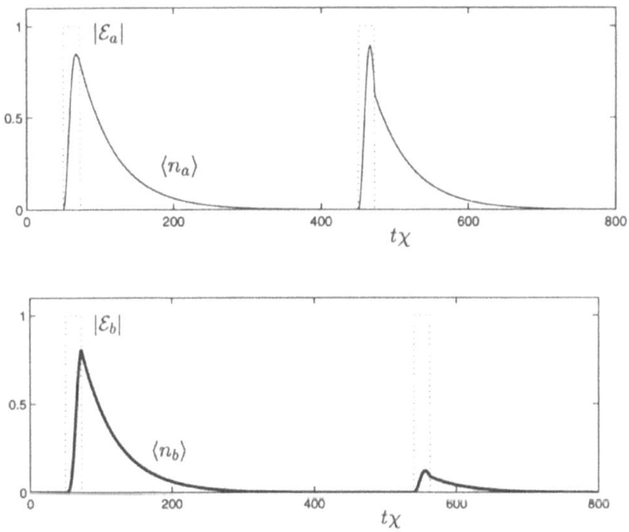

Figure 10. Conditional dynamics of the quantum gate. Pulses strengths are $|\mathcal{E}_a| = |\mathcal{E}_b| = \chi/10$ and the cavity decay rate, $\kappa = \chi/100$, is equal for both modes.

CONCLUSIONS

We have studied how quantum interference effects in four level atoms may be used to generate high $\chi^{(3)}$ optical nonlinearity with low associated quantum noise. We find that for a medium consisting of an ensemble of atoms dispersive effects cause the response time to be too long for an effective photon switch or blockade. We show that a single atom in a high finesse cavity may provide a suitably high $\chi^{(3)}$ with a fast response time. We have demonstrated how a high $\chi^{(3)}$ medium in an optical cavity may be used to generate arbitrary quantum states and to obtain quantum control with single photons.

ACKNOWLEDGEMENTS

This research was supported by the Marsden Fund of the Royal Society of New Zealand. The authors wish to thank Philippe Grangier and Klaus Gheri for useful discussions and their results.

REFERENCES

1. G. Alzetta, A. Gozzini, L. Moi and G. Orriols, An experimental method for the observation of R. F. transitions and laser beam resonances in orientated Na vapour, *Nuovo Cim. B* 36:5 (1976).
2. K.-J. Boller, A. Imamoğlu and S. E. Harris, Observation of electromagnetically induced transparency, *Phys. Rev. Lett.* 67:3062 (1991).
3. M. O. Scully, S.-Y. Zhu and A. Gavrielides, Degenerate quantum-beat laser: lasing without inversion and inversion without lasing, *Phys. Rev. Lett.* 62:2813 (1989).
4. K. M. Gheri, Ph. Grangier, J. P. Poizat, and D. F. Walls, Quantum-nondemolition measurements using ghost transitions, *Phys. Rev. A* 46:4276 (1992).
5. J.-F. Roch, K. Vigneron, Ph. Grelu, A. Sinatra, J.-Ph. Poizat, and Ph. Grangier, Quantum nondemolition measurements using cold trapped atoms, *Phys. Rev. Lett.* 78:634 (1997).
6. H. Schmidt and A. Imamoğlu, Giant Kerr nonlinearities obtained by electromagnetically induced transparency, *Opt. Lett.* 21:1936 (1996).
7. A. Imamoğlu, H. Schmidt, G. Woods and M. Deutsch, Strongly interacting photons in a nonlinear cavity, *Phys. Rev. Lett.* 79:1467 (1997).
8. P. Grangier, D. F. Walls and K. M. Gheri, Comment on "Strongly interacting photons in a nonlinear cavity", *Phys. Rev. Lett.* (in press)
9. G. Müller, M. Müller, A. Wicht, R.-H. Rinkleff and K. Danzmann, Optical resonator with steep internal dispersion, *Phys. Rev. A* 56:2385 (1997).
10. M. D. Lukin, M. Fleischhauer, M. O. Scully and V. L. Velichansky, Intracavity electromagnetically induced transparency, *Opt. Lett.* 23:2951 (1998).
11. C. J. Hood, M. S. Chapman, T. W. Lynn and H. J. Kimble, Real-time cavity QED with single atoms, *Phys. Rev. Lett.* 80:4157 (1998).
12. M. Brune, P. Nussenzveig, F. Schmidt-Kaler, F.Bernardot, A. Maali, J. M. Raimond, and S. Haroche, From Lamb shift to light shifts: vacuum and subphoton cavity fields measured by atomic phase sensitive detection, *Phys. Rev. Lett.* 72:3339 (1994).
13. Q. A. Turchette, C. J. Hood, W. Lange, H. Mabuchi and H. J. Kimble, Measurement of conditional phase shifts for quantum logic, *Phys. Rev. Lett.* 75:4710 (1995).

EXPERIMENTAL REALIZATION OF POPPER'S EXPERIMENT: VIOLATION OF UNCERTAINTY PRINCIPLE?

Yoon-Ho Kim, Rong Yu, and Yanhua Shih

Department of Physics, University of Maryland,
Baltimore County, Baltimore, MD 21250

An entangled pair of photon 1 and 2 are emitted in opposite directions along the positive and negative x-axis. A narrow slit is placed in the path of photon 1 which provides precise knowledge about its position along the y-axis and because of the quantum entanglement this in turn provides precise knowledge of the position y of its twin, photon 2. Does photon 2 experience a greater uncertainty in its momentum, i.e., a greater Δp_y, due to the precise knowledge of its position y? This is the historical thought experiment of Sir Karl Popper which was aimed to undermine the Copenhagen interpretation in favor of a realistic viewpoint of quantum mechanics. This paper reports an experimental realization of the Popper's experiment. One may not agree with Popper's position on quantum mechanics; however, it calls for a correct understanding and interpretation of the experimental results.

INTRODUCTION

Consider the situation where an ensemble of photons is prepared in a pure state of a superposition which contains either continuous or discrete but a considerably large bandwidth of spectrum, i.e., multi-colors. A Geiger counter gives a "click" when triggered by a photon. One simple yet fundamental question has been asked for many years: "Does this photon have a definite color which is labeled by a frequency, ω, in the course of its travel to the Geiger counter from the source, regardless of whether we measure it or not?" Quantum mechanics answers "No!" However, on the other hand, quantum mechanically, a "click" is the result of an annihilation of $\hbar\omega$. A memorable quote from Wheeler: "No elementary quantum phenomenon is a phenomenon until it is a recorded phenomenon"[1] summarizes what Copenhagen has been trying to tell us. Since 1927, most physicists had accepted the Copenhagen interpretation as the standard view of the quantum formalism due to the works of Heisenberg, Bohr and Pauli[2]. However, there are physicists and philosophers refuse to compromise. Among the most notable opponents (Einstein, Podolsky and Rosen; de Broglie; Landé), Karl Popper, a metaphysical realist, had indulged in a life-long argument against the central theses of Copenhagen. In his opinion, the quantum formalism could, and should, be interpreted realistically. Popper invented a thought experiment in the early 1930's with the aim to support the realistic interpretation and undermine Copenhagen[3,4,5]. What Popper's experiment intends to show is that a particle can have both precise position and momentum at the same time which shares the same goal of that of the EPR *gedankenexperiment*[6].

POPPER'S THOUGHT EXPERIMENT

Similar to EPR's *gedankenexperiment*, Popper's experiment is also based on *two-particle entanglement*. Quantum mechanics allows the entangled EPR type state, a state in which if the position or momentum of particle 1 is known the position or momentum of its twin-brother, particle 2, is then 100% determined[6]. Popper's experiment is schematically shown in Fig. 1a. A point source S, positronium for example (assume exact momentum conservation), is placed at the center of the experimental arrangement from which entangled pairs of particles 1 and 2 are emitted in opposite directions along the positive and negative x-axis, towards two screens A and B respectively. There are slits on both screens, parallel to the y-axis, and the slits may be adjusted by varying their width Δy. Beyond the slits on each side an array of Geiger counters are arranged for the coincidence measurements of the particle pairs as shown in the figure. The entangled pair may be emitted in any direction within the 4π solid angles from the point source. However, if particle 1 is detected in a certain direction, particle 2 is known to be in the opposite direction whether we measure it or not.

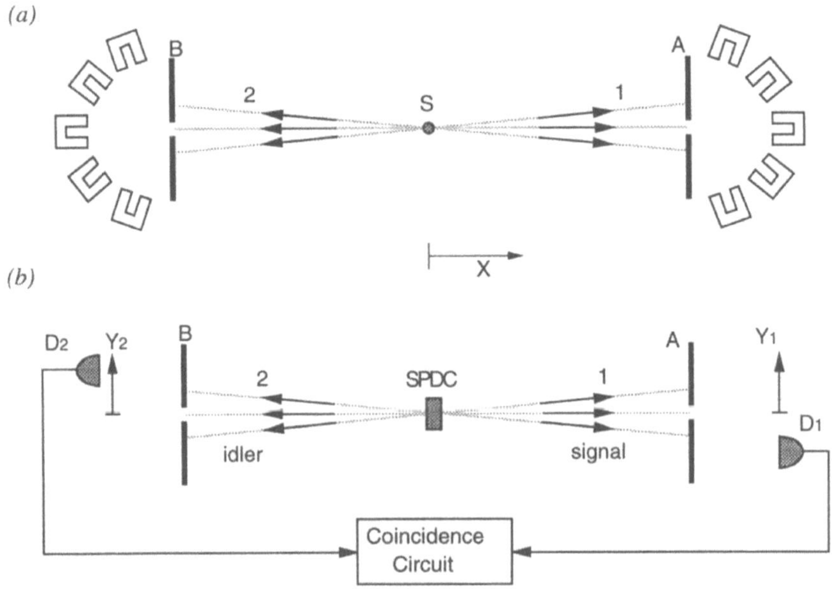

Figure 1. (a) Historical Popper's experiment. (b) Modified version of Popper's experiment. An EPR photon-pair is generated by Spontaneous Parametric Down Conversion. Two detectors D_1 and D_2 are used to scan in y directions for coincidence counts

First let us consider the case in which slits A and B are adjusted both very narrowly, then counters which are higher up and lower down should come into play, as viewed from the slits. The firing of these counters is indicative of the wider scattering angles, i.e., the greater Δp_y, for each particle due to narrower the slits, i.e., smaller Δy. There seems to be no disagreement in this situation between Copenhagen and Popper, and both sides can provide a reasonable explanation according to their philosophy.

Next we keep the slit at A very narrow and leave the slit at B wide open. The use of narrow slit A provides precise knowledge about the position y of particle 1 and in turn gives precise knowledge of the position of its twin (particle 2) on side B, which is possible due to the entanglement. Does particle 2 experience a greater uncertainty of Δp_y due to the precise knowledge of position? If not, it runs into a serious problem: the product of Δy and Δp_y of particle 2 could be smaller than $\hbar$ ($\Delta y \Delta p_y < \hbar$?). It seems that particle 2 going to the left must scatter like its twin-brother which has passed though slit A, even though slit B is wide open 2. The realism point of view is different. A particle can have precise position and momentum, therefore $\Delta y \Delta p_y$ could be smaller than h. From Popper's point of view, particle 2 (twins of these passed narrow slit A) will not experience a greater Δp_y unless a physical narrow slit B is applied.

EXPERIMENTAL REALIZATION

Popper's experiment is realized with the use of the entangled two-photon source of spontaneous parametric down conversion (SPDC)[7,8]. In order to clearly demonstrate all aspects of the historical and modern concerns practically, we have slightly modified Popper's experiment, see Fig. 1b. The two-photon source is a CW Argon ion laser pumped SPDC, which provides a two-photon entangled state preserving exact momentum conservation for the signal-idler photon pair. The laser beam diameter is small enough to be treated as a "point source". The signal-idler is labeled 1 and 2, respectively. Instead of a battery of Geiger counters, only two photon counting detectors, D_1 and D_2, behind silts A and B, respectively, are used for the coincidence detection. Both D_1 and D_2 can be scanned along the y-axis by step motors. Suppose slit A is adjusted to a certain width and slit B is wide open. If detector D_1 receives a "click", we know particle 1 must have passed through slit A. The position y of particle 1 is then known with uncertainty of Δy equals the width of slit A. So is the position of particle 2 at position B, due to the entanglement. Δp_y of particle 2 can be determined by measuring the width of its "diffraction pattern". This is realized by recording the coincidences of D_1 and D_2 while scanning detector D_2 along its y axis. $\Delta y \Delta p_y$ is easily calculated and compared with $\hbar$.

The schematic experimental setup is shown in Fig. 2 with indications of distances which are important for the evaluation of uncertainties. A CW Argon ion laser line of $\lambda_p = 351.1nm$ is used to pump a $3mm$ long beta barium borate (BBO) crystal for type-II SPDC to generate an orthogonally polarized entangled signal-idler photon pair[9,10]. The colinear signal-idler beams, with $\lambda_s = \lambda_i = 702.2nm = 2\lambda_p$ are separated from the pump beam by a fused quartz dispersion prism, P_2, and then split into two orthogonal components (*signal* and *idler*) by a polarization beam splitter, PBS. The signal and idler beams pass through the slits A and B respectively, travel the same distance, X_1 to reach two photon counting detectors D_1 and D_2. The detectors are Geiger mode avalanche photodiodes with $180\mu m$ size in diameter. The output pulses from the detectors are sent to a coincidence circuit. The y_1 coordinate of D_1 is selected by a micrometer and D_2 is scanned in y_2 axis by a step motor.

Measurement 1: we first studied the case in which both slit A and B were adjusted to be $0.16mm$. The diameter of SPDC pump laser was $1mm$. The y coordinate of D_1 was chosen to be $0mm$ (center), while D_2 was allowed to scan along its y axis. Fig. 3 shows the coincidence counting rates against the y coordinate of D_2. It is a typical

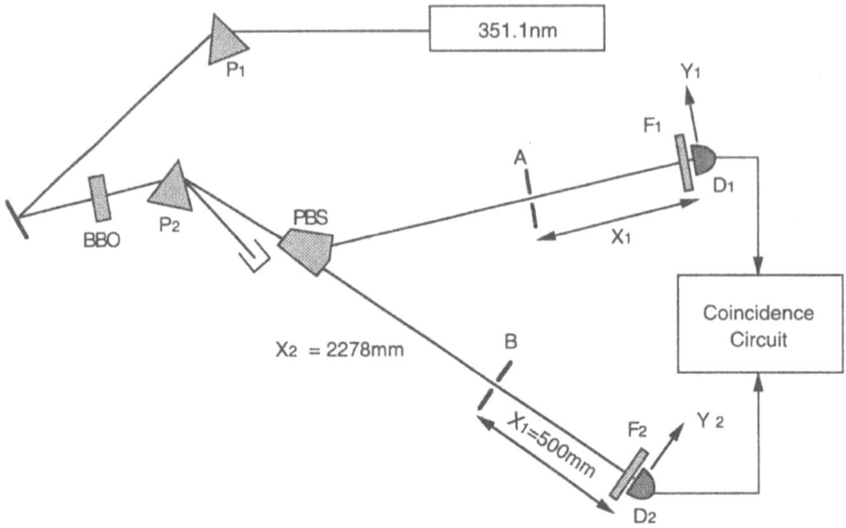

Figure 2. Schematic of the experimental setup. Prism P_1 is to deflect plasma luminescence and prism P_2 is to separate SPDC from the pump. PBS is a polarization beam splitter. F_1 and F_2 are spectral filters centered at $702.2nm$ with bandwidth of $9.4nm$ FWHM. See the text for the definition of X_2.

single-slit diffraction pattern with minimum uncertainty $\Delta y \Delta p_y = \hbar$.

Measurement 2: we kept the same experimental conditions except making slit B wide open. The y coordinate of D_1 was chosen to be $0mm$ and $1mm$, respectively while D_2 was allowed to scan along its y axis. Due to the entanglement nature of the SPDC and the coincidence measurement, only those twins which passed $\Delta y = 0.16mm$ contributed to the coincidence rate through the triggering of D_1 and D_2. Fig. 4 reports the measured coincidence counting rates against the y coordinate of D_2. In each of the two positions of D_1, the measured width of the pattern is narrower than that of the diffraction pattern in Fig. 3. In addition, the width of the pattern is also much narrower then the actual size of the diverging SPDC beam at D_2. The actual size of the "photon 2" beam at D_2 is measured by the single detector counting rate of D_2 (a "unconditional" measurement, it is different then the coincidence measurement "conditionally" determined by D_1,), which is almost a constant in that region. Compare to the diffraction pattern width we have measured in measurement 1, it is clear that the experimental data indicates $\Delta y \Delta p_y < \hbar$ for $\Delta y = 0.16mm$.

QUANTUM MECHANICAL FORMALISM

Before drawing any conclusion, let us first examine what quantum formalism gives. The crucial point is that we are dealing with an entangled two-photon state of SPDC[8,910],

$$|\Psi\rangle = \sum_{s,i} \delta\left(\omega_s + \omega_i - \omega_p\right) \delta\left(\mathbf{k}_s + \mathbf{k}_i - \mathbf{k}_p\right) a^\dagger(\omega(\mathbf{k}_s)) a^\dagger(\omega(\mathbf{k}_i)) \left|0\right\rangle \tag{1}$$

where ω_j, $\mathbf{k}_j$ (j = s, i, p) are the frequency and wavevectors of the signal (s), idler (i), and pump (p), respectively, ω_p and $\mathbf{k}_p$ can be considered as constants, $a^\dagger(\omega(\mathbf{k}_s))$ and $a^\dagger(\omega(\mathbf{k}_i))$ are the creation operators for the signal and the idler, respectively. Even

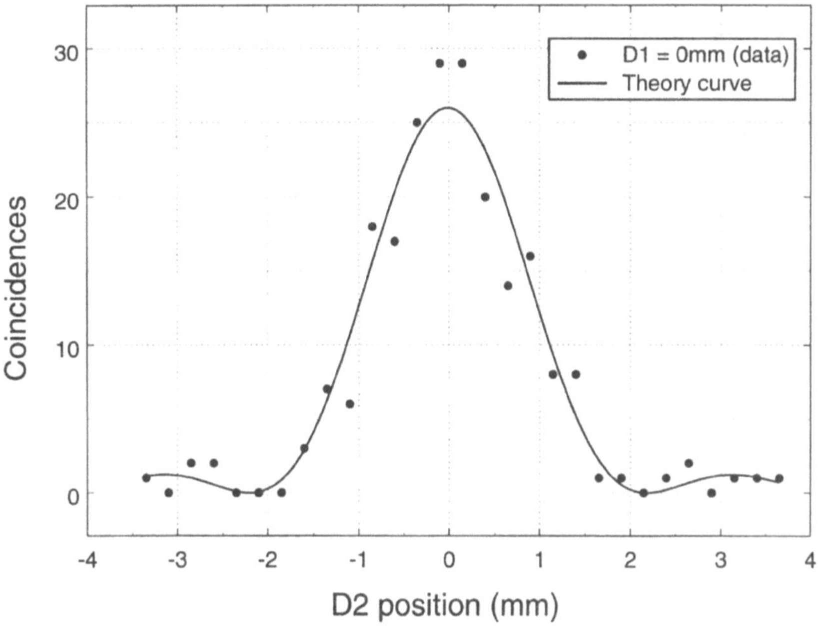

Figure 3. The observed coincidence pattern with slit A and B. Slit-width = $0.16mm$. The distance from the slit to the detector $X_1 = 500mm$. The line curves are calculated from Eq. (2).

though there is no precise knowledge of the momentum for either signal or idler, the state provides precise knowledge of the momentum correlation of the pair. The pair must be emitted in certain directions which satisfy $\mathbf{k}_s + \mathbf{k}_i = \mathbf{k}_p$. In a simplified picture we may "unfold" the SPDC by assuming $\mathbf{k}_s + \mathbf{k}_i = 0$, the pair will always be emitted in exactly opposite directions.

A simplified picture is shown in Fig. 5. We have "unfolded" the SPDC, which is equivalent the assumption $\mathbf{k}_s + \mathbf{k}_i = 0$. Even though there is no precise knowledge about the momentum for either the signal or the idler, the state provides precise knowledge of the momentum correlation of the pair. In this simplified picture the pair will always be emitted in exactly "opposite directions". This important peculiarity selects the possible two-photon paths in Fig. 5 when Popper's diffraction experiment is considered. The solid lines (two-photon amplitudes) of the two-photon paths in Fig. 5 represents the momentum conservation of the signal-idler pair. The coincidence counting rate R_c is determined by the probability P_{12} of detecting a signal-idler pair by detectors D_1 and D_2. $P_{12} \propto \langle \Psi | E_1^+ E_2^+ E_2^- E_1^- | \Psi \rangle$ where E_1 and E_2 are field operators at D_1 and D_2, respectively. Summing up all the possible paths (two-photon amplitudes) as in Fig. 5, which corresponds to measurement 1, the diffraction pattern of the coincidences can be calculated as the follows,

$$R_c \propto P_{12} = \left| \int_{-a/2}^{a/2} \int_{-b/2}^{b/2} dy_3 dy_4 \ \exp[-ikr(y_1, y_2, y_3, y_4)] \right|^2 \cong R_0 \ \mathrm{sinc}^2\left(\frac{\pi a}{\lambda X_1} y_2\right) \tag{2}$$

where a (b) is the width of the slit A (B), λ is the wavelength, y_1 and y_2 are the positions of D_1 and D_2, y_3 and y_4 are the coordinates where a possible path takes at slit A and B, respectively. r is the distance from D_1 to D_2 along each solid line. Moreover, single longitudinal mode and far field approximations have been applied in the evaluations. Eq.(2) shows the "standard" diffraction pattern of a single-slit.

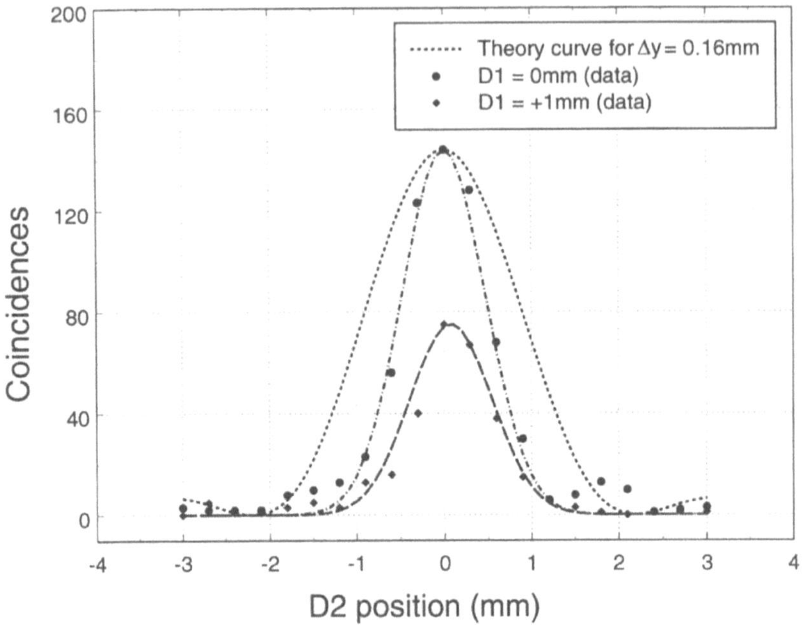

Figure 4. The observed coincidence pattern. slit A = $0.16mm$. The line curves are calculated from Eq. (4). Note: the data curve has a width which is narrower than that of the diffraction pattern in Fig. 3.

When slit B is wide open and an infinite large size of SPDC is considered, see Fig. 5b, the diffraction pattern is calculated to be:

$$R_c \propto P_{12} = \left| \int_{-a/2}^{a/2} dy_3 \ \exp[-ikr(y_1, y_2, y_3)] \right|^2 \cong R_0 \ \mathrm{sinc}^2[\frac{\pi a}{\lambda}(\frac{y_2}{X_2} + \frac{y_1}{X_1})] \qquad (3)$$

However, due to the considerably small size of the two-photon source ($1mm$), see Fig. 5c, the possible two-photon paths are limited within the $1mm$ size of the SPDC. With a Gaussian profile of a pump beam for measurement 2, the coincidence counting rate is approximately,

$$R_c \cong R_0 \ \exp(-\frac{y_2^2}{2\sigma^2}) \ \mathrm{sinc}^2[\frac{\pi a}{\lambda}(\frac{y_2}{X_2} + \frac{y_1}{X_1})] \qquad (4)$$

where $\sigma \equiv \frac{X_2}{X_0}\sigma_p$, with σ_p being the half width of the SPDC pump laser beam at half maximum intensity, and X_0 is the distance from the source to slit A. X_2 is the distance from slit A going backward along the signal beam to SPDC and then going along the idler beam to detector D_2 11.

Based on curve fittings of the experimental data we conclude that Eqs. (2) and (4) have provided accurate solutions within experimental errors. The quantum formalism does provide a solution which gives $\Delta y \Delta p_y < \hbar$ for the above measurement. The only requirement for Eq. (4) is that the small size two-photon source preserves momentum conservation of the photon pair. According to Eq. (4), one would expect $\Delta y \Delta p_y \ll \hbar$ if σ_p, the size of the entangled two-photon source is small enough. In principle, one could even have $\Delta y \Delta p_y \sim 0$ for a idealized point two-photon source and an infinite narrow width of slit A. In this situation, "photon 2" will have definite momentum and position at the same time.

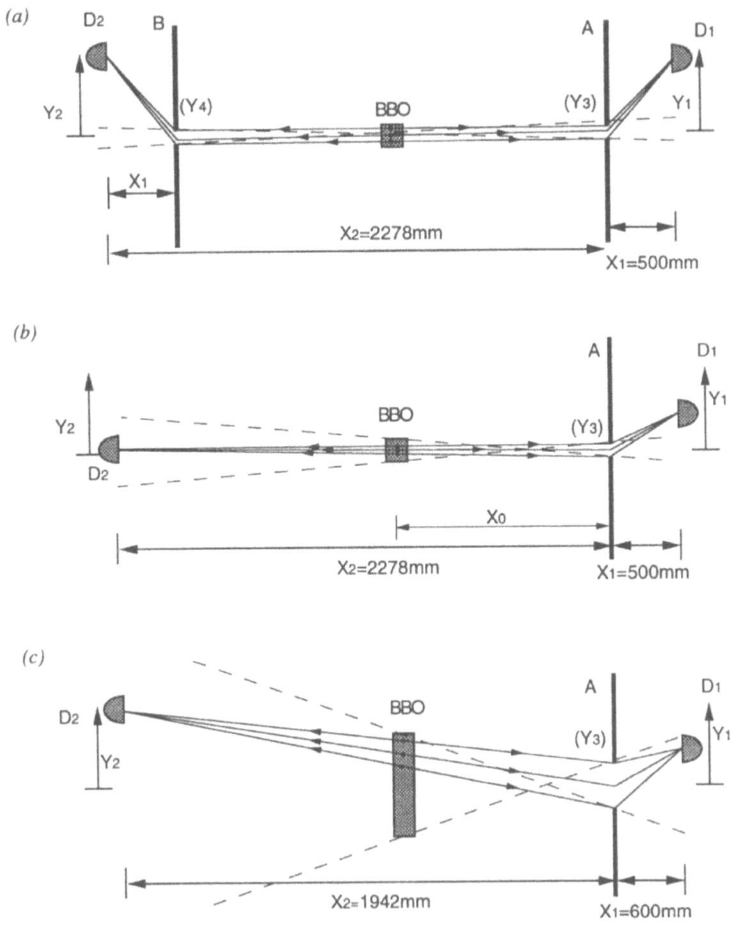

Figure 5. The unfold schematic of the experiment.

DISCUSSIONS

It seems possible to have $\Delta y \Delta p_y \ll h$ experimentally in Popper's experiment and the quantum formalism does allow such a solution. How do we understand this?

First, we would say that this is not a violation of the uncertainty principle. The measurement is a two-particle "joint detection", "coincidence conditional detection" or "correlation measurement" for an entangled two-particle state. The uncertainty which both Popper and Copenhagen refer to is for the physical property of a quantum. Quantum mechanically, an entangled two-particle state provides the knowledge of the correlation of the pair only. Neither of the single subsystem is determined by the state. In the above standard quantum mechanical calculations we did not consider either "photon 1" or "photon 2" *independently* even though the "clicks" of the detectors occur at a considerably large distance. In general, if one is concerned with a two-particle entangled state for a "click-click" "joint detection" or "coincidence measurement", only the two-particle amplitudes are considered in the formalism. In Popper's experiment, it is the superposition of the two-photon amplitudes are considered and each of the two-photon amplitude can be viewed as a "straight line" which passes the slit, source, and D_2. If there is a uncertainty it is to be understood as meaningful only for the two-particle system but not for one of the subsystems. The quantum formalism does predict the two-photon diffraction pattern, see Eq. (3) and experimental data in the Appendix. When the two-photon source is small enough so that the possible two-photon paths are "precisely" constrained to a small region, the straight lines (two-photon amplitudes) will be ended up with the "central counters" only. In an idealized situation, if there is only one line (one amplitude) left, we would have $\Delta y \Delta p_y \sim 0$. This means that photon 2 will go along the "straight lines" predetermined by the entangled two-photon state. In Popper's words: "photon 2 will go on and hit one of the central counters", despite the precise knowledge of its position. However, $\Delta y \Delta p_y \ll h$ here means a "precise" knowledge of the two-particle correlation only. Quantum mechanically it is allowed to have 100% determination for both momentum-momentum and position-position correlation in an entangled two-particle system. As a matter of fact, this was the basic point EPR made in their 1935 paper[6]. What Popper tried to show us is very similar to what EPR did. There is nothing wrong to have precise knowledge of 100% momentum-momentum and position-position determination for an entangled two-particle system and this is what we have learned from Popper's experiment. A 100% momentum-momentum determination does not necessarily destroy the position-position determination. However, having precise momentum-momentum and position-position determination is fundamentally different from that of having precise knowledge of momentum and position for a particle.

In the spirit of the above argument, one would conclude that it was a historical mistake to mix up uncertainty principle with the measurement of an entangled two-particle system.

Second, we would like to note that our experiment is not a complete realization of Popper's experiment since we have not achieved a "point-like" two-particle source yet. One may not be able to realize a point SPDC source because of the lack of "phase matching condition", i.e., $\mathbf{k}_s + \mathbf{k}_i - \mathbf{k}_p \equiv \Delta \mathbf{k} \neq 0$, if the SPDC nonlinear optical interaction is restricted spatially within a "point" volume. Is it possible to have an entangled two-particle source which shall preserve the two-particle momentum conservation for Popper's experiment? Historically, we had different answers. Popper was even criticized for making a "fundamental" mistake. A typical argument was that a point source will cause the uncertainty of momentum for particle 2 (particle 1 as well) so that one can never know where particle 2 is going to "fly". This argument is trivially wrong. Yes, it is true, neither of the momentum of particle 1 nor the momentum of particle 2 is determined so that both of them could be emitted to any direction in the 4π solid angle; however, as long as the momentum conservation is applied in the pair production precess the point source is ready for the test of Popper's experiment. We believe we have made this point clear enough in this paper. Another argument has been that a "point source" (an atom, for example) may not preserve momentum conservation of the pair[12]. The momentum uncertainty (non-zero momentum) of the "point-like source" will contribute to the pair production process. However, as long as the non-zero momentum of the "point-like source" is small enough compare to the momentum of the pair, we still have $\mathbf{p}_1 + \mathbf{p}_2 \cong 0$, there is no problem for the testing of Popper's idea. We believe it is experimentally achievable. A cooled and trapped "point-like source" is definitely a good candidate.

CONCLUSION

In conclusion, we have realized the historical thought experiment of Karl Popper by means of a two-photon source of SPDC. The experiment has clearly demonstrated that, in an entangled two-particle system, the precise knowledge of the position of a particle, which is determined by the position-position entanglement, does not necessarily produce an uncertainty in its momentum. Preliminary discussions are provided for stimulating deeper understanding.

Acknowledgments

The authors acknowledge important suggestions and encouragements from T. Angelidis, A. Garuccio, C.K.W. Ma (LSE), and J.P. Vigier. We also thank T.E. Keller and M.H. Rubin for helpful discussions. This research was partially supported by the U.S. Office of Naval Research and the National Security Agency.

Appendix: Two-photon diffraction measurement

The experimental data of the two-photon diffraction pattern is reported here for reference. In this measurement, the pump laser beam was set to be $2.7mm$ in diameter. The width of slit A was chosen to be $a = 0.5mm$ and slit B was left wide open. The y coordinates of D_1 were chosen to be 0, $0.5mm$, and $1mm$, respectively, while D_2 was scanning along its y axis. The measured coincidence counting rates are indicated in Fig. 6. According to curve fittings we conclude that Eq. (3) has provided accurate solutions within experimental errors. (a): it is clear that a diffraction pattern which is the result of the superposition of these two-photon amplitudes of the entangled two-photon state is observed. (b) the diffraction pattern was shifted considerably to one side of the y axis corresponding to the movement of D_2.

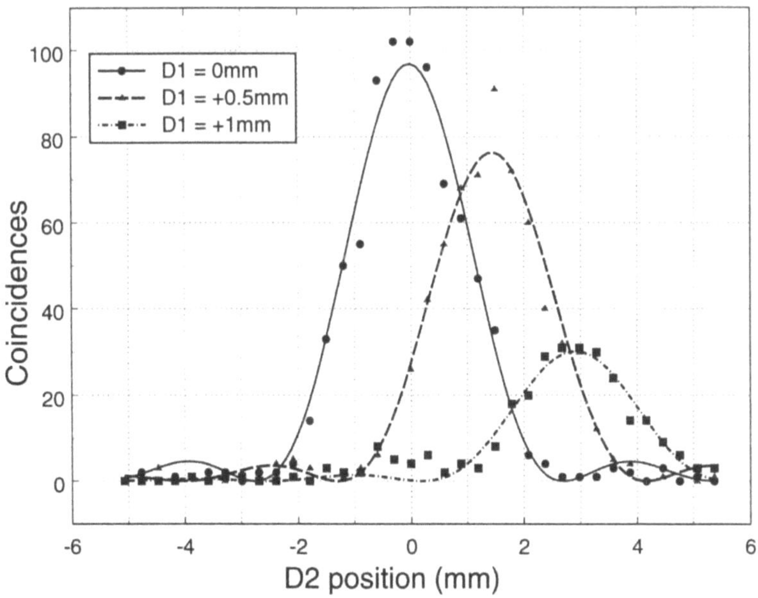

Figure 6. The observed coincidence pattern. Slit A $= 0.5mm$. Slit B is wide open. The line curves are curve fitting using Eq.(3).

REFERENCES

1. J.A. Wheeler, "Niels Bohr in Today's Words" in *Quantum Theory and Measurement*, edited by J.A. Wheeler and W.H. Zurek, Princeton University Press, New York, (1983).
2. For example, W. Heisenberg, *The Physical Principles of the Quantum Theory*, Dover Publications, (1930)
3. K. Popper, *Zur Kritik der Ungenauigkeisrelationen*, Die Naturwissenschaften, 22, Helft, 48, 807 (1934).
4. K. Popper, *From the Postscript to the Logic of Scientific Discovery*, edited by E.I. Bitsakis and N. Tambakis, Gutenberg Publishing, (1984).
5. K. Popper,*Quantum Theory And The Schism In Physics*, edited by W.W. Bartly, Hutchinson, London, 28 (1983).
6. A. Einstein, B. Podolsky, and N. Rosen, Phys. Rev., **47**, 777 (1935).
7. A. Yariv, *Quantum Electronics*, John Wiley and Sons, New York, (1989).
8. D.N. Klyshko, *Photon and Nonlinear Optics*, Gordon and Breach Science, New York, (1988).
9. Y.H. Shih, A.V. Sergienko, and M.H. Rubin, Phys. Rev. A, **50**, 23 (1994).
10. M.H. Rubin, D.N. Klyshko, Y.H. Shih, and A.V. Sergienko, Phys. Rev. A, **50**, 5122 (1994).
11. A special case for $y_1 = 0$ has been demonstrated as a "ghost" diffraction reported in D.V. Strekalov, A.V. Sergienko, D.N. Klyshko,and Y.H. Shih, Phys. Rev. Lett., **74**, 3600 (1995).
12. D.Bedford and F. Selleri, Lettere al Nuovo Cimento, **42**, 325 (1985).

SOLITON PHOTON NUMBER SQUEEZING: AN OVERVIEW

Stephen R. Friberg and Michael J. Werner

NTT Basic Research Laboratories
3-1 Morinosato Wakamiya
Atsugi, Kanagawa, Japan 243

Soliton photon number squeezing of more than 6 dB has recently been reported and reliable theoretical estimates suggest that squeezing of more than 10 dB is to be expected. Squeezing can be obtained either by spectral filtering or by using an unbalanced nonlinear optical loop mirror.

INTRODUCTION

Soliton photon number squeezing is a newly discovered method of photon number noise reduction that is easy to implement and produces substantial amounts of squeezing.[1] It is done most simply by propagating solitons down a length of optical fiber and spectrally filtering them as they emerge.[1,2,3] Theoretical treatments suggest that this can produce about 7 dB of squeezing for near fundamental solitons[4,5,6,7] or up to 8.1 dB for higher order solitons. Experimental demonstrations of spectral filtering squeezing have recorded 6.4 dB of squeezing[3] and non-soliton spectral filter squeezing of 0.9 dB[8] and 2.6 $\pm$ 0.7 dB[9] has been observed. Soliton photon number squeezing can also be implemented using an unbalanced nonlinear optical loop mirror (NOLM).[10,11] 11 $\pm$ 1 db squeezing has been theoretically predicted[10] and demonstrated squeezing of 6.0 dB[12] and 6.2 dB[13] has been reported.

The important implications of these findings are several. First of all, soliton photon number squeezing offers a new method for producing substantial amounts of squeezed light. Simple techniques for noise reduction in excess of 10 dB may become feasible. The squeezing bandwidth for temporal solitons should be that of the Kerr nonlinearity, extending from DC to over 10 THz. If spatial solitons are used, cw light with similar amounts of amplitude squeezing could become available.[8,14] Because the squeezing is due to the nonlinear propagation of solitons, it requires neither precision cavities, expensive optical coatings, active stabilization controls, or injection seeding of $\chi^{(2)}$ parametric amplifier sources.[15] Another implication has to do with the rich set of correlated spectral fluctuations that are obtained. Temporal soliton photon number squeezing is usually done using ultrashort pulses with a broad spectrum and a variety of correlation phenomena in the spectral regime are readily observed. Important examples include quantum correlations between different components of the soliton's spectrum and spectral fluctuation maps that show internal quantum fluctuation structures of propagated

Quantum Communication, Computing, and Measurement 2
Edited by Kumar et al., Kluwer Academic / Plenum Publishers, New York, 2000.

pulses.[1,5,16] These novel maps of the internal quantum structure of solitons offer the possibility of new diagnostics applicable to both optical pulses and other related phenomena such as Bose-Einstein condensates. However, these phenomena have not been observed before and are not yet well understood. Finally, soliton and optical pulse communications are of great technical importance in modern communications technology, and insight into the noise properties of optical pulses may contribute to improved noise properties of systems.[17]

QUANTUM CORRELATIONS AND QUANTUM OPTICS IN FIBERS

Modern quantum optics relies strongly on technologies of photon entanglement, optical squeezing and quantum light generation which derive from parametric processes that create quantum-correlated (entangled) pairs of photons. Typical processes involve the creation of quantum-correlated signal and idler photons from pump photons via three-wave mixing in $\chi^{(2)}$ nonlinear media, or alternatively from pairs of pump photons via four-wave mixing in $\chi^{(3)}$ materials. Quantum-correlated photon pairs were first observed experimentally in $\chi^{(2)}$ media by Burnham and Weinberg in 1970[18] and by Friberg *et al* in 1985.[19] The first application of quantum-correlated photon pairs, by Hong, Friberg and Mandel in 1985, was to quantum communications.[20] In 1986, $\chi^{(2)}$ processes were used to demonstrate 7.0 dB of quadrature-amplitude squeezing[21] and subsequently have been the mainstay of squeezing and quantum entanglement experiments. Four-wave mixing in $\chi^{(3)}$ media was used for the first demonstrations of quadrature-amplitude squeezing[22,23] but has been employed infrequently since. This is in part because $\chi^{(3)}$ nonlinearities are considerably smaller than $\chi^{(2)}$ nonlinearities, so that cavities become too large to be practical. Resonant $\chi^{(3)}$ nonlinearities can be large, but only for signal and idler wavelengths very close to the pump wavelength.[24] Then, scattering from pump beams tends to mask the desired quantum effects. In contrast, $\chi^{(2)}$ nonlinearities are usually nonresonant, very fast, and have signal and idler beams at frequencies widely different from the pump beam, eliminating problems due to scattering.

Many of the usual problems of $\chi^{(3)}$ materials can be overcome – in principle – by the use of single mode optical fibers. Such fibers combine extremely low losses ($\sim$ 0.2 dB/km at 1.5 μm) with strong confinement ($\sim$ 50 μm^2 cross sections) over long distances. Optical fiber nonlinearities are a weak off-resonant $\chi^{(3)}$ Kerr effect with an ultrafast frequency response which extends from DC to well above 10 THz. In concert with long propagation distances and tight confinement, they can provide large nonlinear effects. Shelby, Levenson and co-workers took advantage of these properties to obtain 0.6 dB of quadrature-amplitude squeezed light in the second of the reported squeezing demonstrations, using 625 nm light in a 114 m long optical fiber.[23] This demonstration, while the first to use optical fibers for quantum effects, highlighted the severe difficulties peculiar to fibers. These include stimulated Brillouin backscattering (SBS) that could strongly deplete cw pump beams and guided acoustic wave Brillouin scattering (GAWBS) phase noise that could prevent observation of sensitive phase-dependent phenomena such as quadrature-amplitude squeezing. It was soon recognized that using short optical pulses can eliminate SBS effects, strongly reduce the effect of GAWBS noise, and strongly enhance intensities. The Raman effects which then come into play are usually disadvantageous, but often not overwhelmingly so. Rosenbluh and Shelby in 1991 used solitons – which can propagate in fibers at wavelengths longer than 1300 nm – to obtain 3.2 dB of broadband squeezing.[25,26] Bergman and Haus used non-solitons to eventually obtain 5.1 dB squeezing in a narrow 80-100 kHz window and broadband squeezing of 4 dB.[27] Quantum soliton effects are reviewed by Drummond *et al*[26] and Sizmann.[28]

SOLITON PHOTON NUMBER SQUEEZING USING SPECTRAL FILTERING

Soliton photon number squeezing was first observed experimentally and later explained theoretically.[1,4] In the initial observations, optical solitons of 2.7 ps duration were propagated through a 1.5 km length of optical fiber and spectrally filtered to remove part of the outlying spectral sidebands, leaving 90% to 95% of the soliton's energy. The experimental schematic, Fig. 1(a), shows a diffraction grating and a slit as the spectral filter, but a monochromator works just as well. A maximum of 2.3 dB of photocurrent reduction was observed, corresponding to 3.7 dB of photon number squeezing when losses were accounted for. Soliton energies were about 12 pJ (vs. fundamental soliton energies of 8 pJ). The optical fiber was about 4.5 soliton periods long (a soliton period is the restoration distance for an $N = 2$ breather soliton). When the spectral filter was translated across the soliton's spectrum, a strong variation in the transmitted noise was observed, as shown in Fig. 1(b).

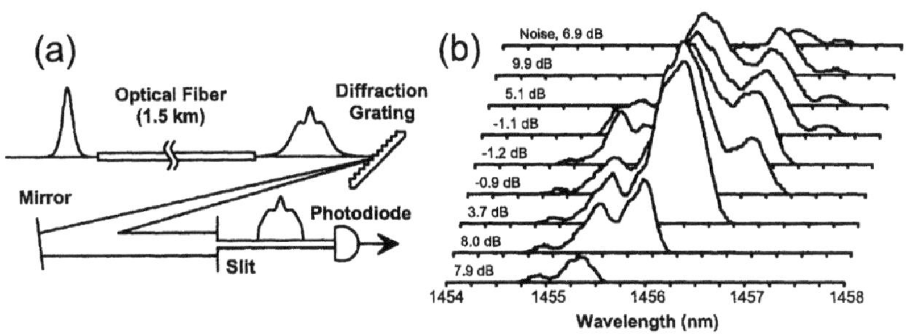

Figure 1. Soliton photon number squeezing using spectral filtering. (a) is a schematic showing the squeezing apparatus and (b) shows the noise reduction and enhancement measured for different filterings of the spectrum.

Spectral filtering of sub-picosecond solitons has produced even better squeezing.[2,3] Using first 160 fs pulses centered at 1506 nm from a chromium YAG laser, a maximum of 3.2 dB of photocurrent reduction was observed, corresponding to squeezing of 4.5 dB when quantum efficiencies were accounted for.[2] Soliton energies were 78 pJ and the squeezing fiber was 10 m long. In a second experiment, 130 - 135 fs pulses were used. The best squeezing results were 3.8 ± 0.2 dB photocurrent reductions and 6.4 ± 0.8 dB squeezing with losses taken into account.[3] In this case, the squeezing was obtained for 85 pJ pulses in fibers 90 meters long, corresponding to a distance of over 100 soliton periods. Raman effects caused a center frequency shift of 80 nm. In the absence of Raman-induced soliton self-frequency shifts, the maximum squeezing was similar to that for the previous experiment.

Soliton spectral filtering has been modeled in several different ways. A simple heuristic model of soliton spectral filter squeezing considers the nonlinear input/output (i/o) response of soliton propagation followed by spectral filtering. Solitons, unless they are exactly fundamental solitons, oscillate temporally and spectrally as they propagate [Fig. 2 (a) and 2(b)] with an oscillation period determined by the soliton amplitude. The transmission of a fixed-width spectral filter at a certain propagation distance accordingly is determined by the spectral profile of the soliton, which in turn is determined

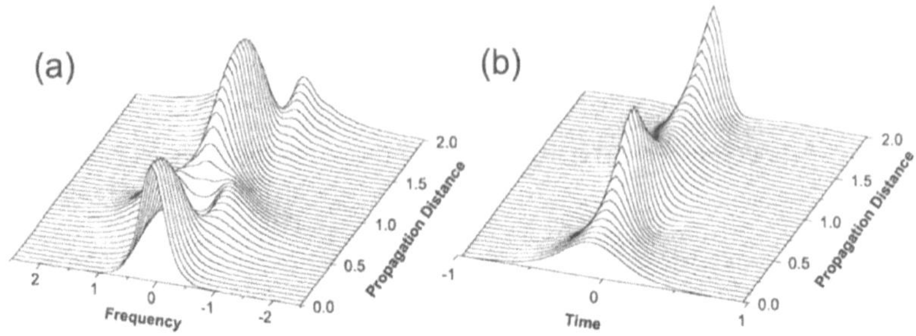

Figure 2. Propagation of an $N = 1.7$ soliton over a distance of two soliton periods showing oscillatory breathing behavior. In the spectral regime (a), the spectrum bifurcates and then recombines. In the temporal regime (b), the pulse narrows and broadens.

by its amplitude, provided the other parameters are held constant. As the amplitude changes, the transmission also changes, mapping out a nonlinear i/o response. A simple estimate of the squeezing can then be obtained by assuming the input pulse to be in a coherent state with shot noise fluctuations due to amplitude variations. The output variance can then be calculated from the i/o curve, with corrections accounting for photons assumed to be randomly deleted by the spectral filter. Comparing the output variances with those for a linear system with the same loss estimates the squeezing. Figure 3(a) shows the calculated nonlinear i/o curve assuming a 4.5 soliton period fiber length and pulse and filter characteristics similar to those used in the experiment of Friberg *et al.* Figure 3(b) shows the calculated squeezing and the corresponding estimates from quantum field theoretic models discussed later. The results are in qualitative agreement with the experimental results, but overestimate the squeezing. They also predict larger amounts of squeezing than do the more exact field theoretic models.

Soliton perturbation theory models should give considerably more accurate estimates of the squeezing.[6,7,14] Starting with the quantum nonlinear Schrödinger equation, the linearization approximation can be applied to yield a linear equation for an amplitude fluctuation operator. This can be solved to obtain the variance of a filtered output soliton emerging from the fiber, leading to an expression that includes terms due to the soliton, the continuum radiation associated with deviations from a perfect fundamental soliton, and beating between the continuum and the soliton. If the filtering is optimized, 6.5 dB of squeezing can be expected for near fundamental solitons, according to these calculations.

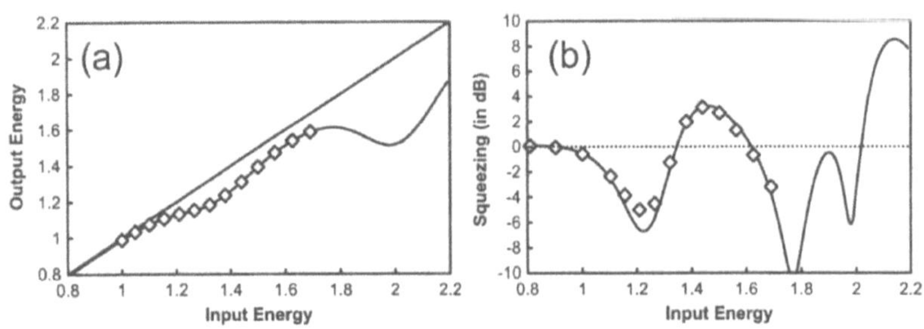

Figure 3. Nonlinear input/output curves. (a) shows the i/o curve calculated for the experiment of Friberg *et al.* In (b), the solid lines are predictions from the i/o curves of the squeezing. The rectangles are the predictions from a quantum field-theoretic model. Input and output energies are normalized to that of a fundamental soliton.

The most exacting estimations of the behavior of soliton spectral filtering systems come from quantum field-theoretic models of soliton propagation.[1,4,5,8,10,11] A coherent state positive-P representation is used to derive a stochastic nonlinear Schrödinger equation from a dispersive nonlinear Lagrangian for pulse propagation in nonlinear fibers.[29] This is used to numerically obtain solutions for specific initial conditions and spectral filter characteristics. Inhomogeneous Raman effects – increasingly important as pulse widths dip below a picosecond – are incorporated, as are other effects.[4,5] These models discard little of the relevant quantum physics, and therefore behave as numerical experiments that can accurately portray quantum optical solitons. They are computationally intensive, however, and cannot readily obtain results when propagation is over very long distances.

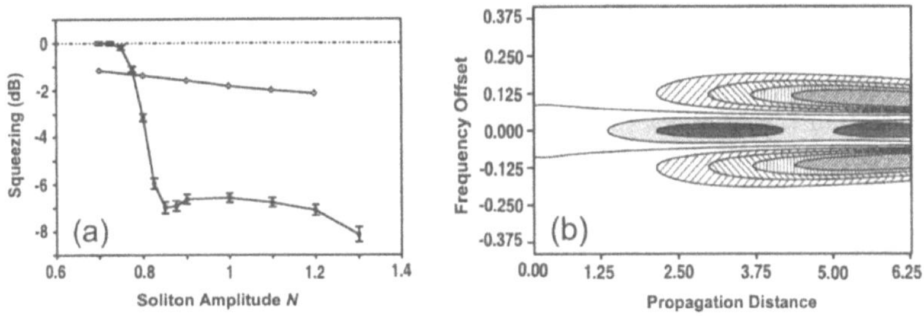

Figure 4. Quantum field theoretic modeling results for soliton spectral filtering. The lower line in (a) shows maximized squeezing in the absence of Raman effects: 8.1 dB ± 0.3 dB squeezing is predicted. The upper line shows the squeezing predictions for non-soliton propagation: 2.4 ± 0.1 dB is the maximum for these values of N. On the right, (b) shows spectral variations relative to the standard quantum limit versus propagation distance (dispersion length units) and the frequency offset (inverse soliton width units). Grey indicates below shot-noise, hatching indicates above shot-noise.

Results of numerical calculations using quantum field-theoretic models are shown in Fig. 3(b), Fig. 4(a), and Fig. 4(b). For chirped solitons in a 4.5 soliton period fiber, a maximum of 5 dB ± 0.2 dB squeezing is predicted for a bandpass cutoff filter, slightly more than the observed 3.7 dB. For non-chirped solitons, about 7 dB of squeezing is predicted,[5] reduced to about 4.8 dB when Raman effects appropriate to 1 ps solitons are included. New results – obtained by varying parameters to produce the best possible squeezing – are presented in Fig. 4(a). They indicate that squeezing of 8.1 db ± 0.3 dB should be obtainable using soliton spectral filtering if Raman effects are made negligible by using relatively wide (> 2 ps) pulses with low energies.

An important feature of the quantum field-theoretic calculations is that they can give mappings of the intensity fluctuations of soliton spectral components.[5] One such mapping, showing intensity variance as a function of center frequency offset and propagation distance, is graphed in Fig. 4(b) for an $N = 1$ soliton. As the soliton propagates, spectral regions both quieter (shown as shades of grey) and noiser (shown cross-hatched) than the standard quantum limit develop and set in. Cutting out spectral regions of excess noise can then produce photon number squeezed solitons. However, fluctuations in some spectral region are often correlated (either positively or negatively) with fluctuations in other spectral regions. This implies that eliminating regions of large spectral fluctuations may not necessarily reduce the overall fluctuation level. Experimental verification of these numerical results has recently been obtained by Spälter et al.[16] Another result, a bit surprising, is that the fluctuations provide a diagnostic that reveals aspects of the internal structure of the solitons. Another interesting result is that the noise properties of solitons can be shown to undergo a phase transition as soliton amplitudes are raised above critical levels and the nonlinear effects turn on.

PHOTON NUMBER SQUEEZING USING NOLMS

A nonlinear optical loop mirror (NOLM) – often called a nonlinear Sagnac interferometer – is a nonlinear Mach-Zehnder interferometer suitable for long optical fibers. Fiber interferometers are long because of the weakness of the Kerr nonlinearity, and their length makes them highly sensitive to temperature, position, acoustic and other fluctuations. Use of a NOLM strongly reduces such sensitivities by having pulses counterpropagate down a single fiber, rather than having two separate fibers. Environmental fluctuations then affect both pulses identically and perturbations cancel out when the pulses interfere. Use of a NOLM makes it possible to implement amplitude squeezing (photon number squeezing for cw light) using a fiber nonlinear Mach-Zehnder interferometer, as first predicted by Ritze and Bandilla.[30] In quadrature-amplitude squeezing demonstrations using fibers, NOLMs have been used as balanced Mach-Zehnder inteferometers to subtract off the classical portions of solitons and optical pulses to produce a squeezed vacuum.[25,26,27]

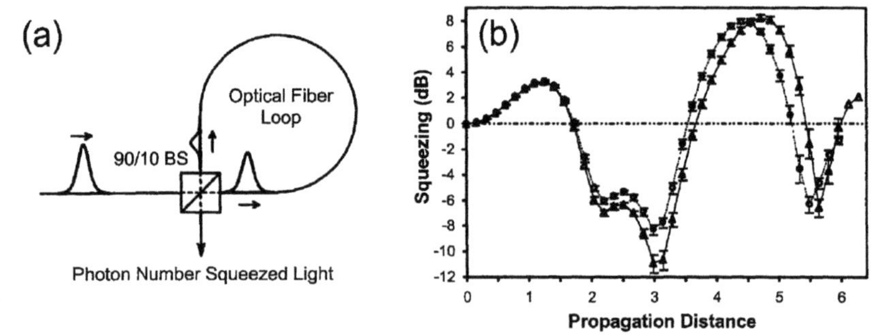

Figure 5. Unbalanced NOLM photon number squeezing. A schematic of an unbalanced NOLM (nonlinear optical loop mirror) is shown in (a). Calculated photon number squeezing predictions for a NOLM with a 90/10 beamsplitter ratio are shown in (b). 11 ± 1 dB of squeezing is predicted in the absence of Raman effects (solid line) and 8.3 ± 0.4 dB in its presence (dotted line). Raman effects are assumed to be those appropriate for 176 fs pulses at room temperature.

It was initially thought that a NOLM would not produce photon number squeezed solitons, as two propagating solitons independently acquire quadrature-amplitude squeezing noise that does not cancel. This was verified for balanced NOLMs: the small amounts of squeezing that could be obtained required very high powers. For unbalanced NOLMs, Werner discovered that the case was quite different and that substantial amounts of photon number squeezing could be produced.[10,11] For an unbalanced NOLM with a 90/10 beamsplitter as diagramed in Fig. 5(a), quantum field theoretic models predict 11.0 ± 1.0 dB squeezing for $N = 1.5$ input solitons. If room temperature Raman effects appropriate for 1.76 ps solitons are taken into account, the squeezing is reduced to 9.0 ± 0.5 dB, which is still large. Results from the model are shown in Fig. 5(b). The more intense of the two counterpropagating pulses propagates as an $N > 1$ soliton, whereas the weak pulse progagates as a linear pulse. Returning to the beamsplitter, the two interfere to provide a substantial amount of photon number noise cancelation, provided the phase shift is correct. Raman effects lower the amount of squeezing that can be obtained, but do not eliminate it.

Werner's predictions have been verified experimentally, although the very large expected squeezing has not yet been observed.[12,13] In the experiment by Schmitt *et al*, 126 fs pulses centered at 1510 nm were propagated through an unbalanced NOLM with a variable beamsplitter splitting ratio.[12] It was found that the squeezing, along with the needed intensities, increased as the beamsplitter ratio became more unbalanced until Raman effects induced by increasing powers degraded the interference. The best results were a photocurrent reduction of 3.9 ± 0.2 dB, corresponding to a squeezing of 6.0 ± 0.9 dB. They were obtained with a splitting ratio of 90/10 and input pulses with pulse energies of 60 pJ. In a similar experiment, Krylov and Bergmann have used 182 fs optical parametric oscillator pulses centered at 1550 nm with a NOLM constructed of a 54 cm length of optical fiber.[13] For an 88/12 beamsplitter ratio and 830 pJ input pulses, they report 5.7 dB of photocurrent reduction, or 6.2 dB of photon number squeezing, taking into account measurement losses.

OTHER PHOTON NUMBER SQUEEZING METHODS

Optical pulses, of course, do not always propagate as solitons, nor are all solitons optical pulses. In fibers, solitons only propagate at wavelengths longer than 1300 nm where the group velocity dispersion can be anomalous. Optical physics, however, is mainly done at shorter wavelengths, suggesting limited applicability for fiber soliton squeezing. Also, most applications of optical squeezing have been for cw light, rather than pulsed sources. The problems that this raises can be addressed with the techniques outlined in this review either by using non-soliton photon number squeezing in fibers at shorter wavelengths, by using solitons propagating in materials other than glass, or by using different types of solitons (spatial solitons, for example) which have a similar behavior.

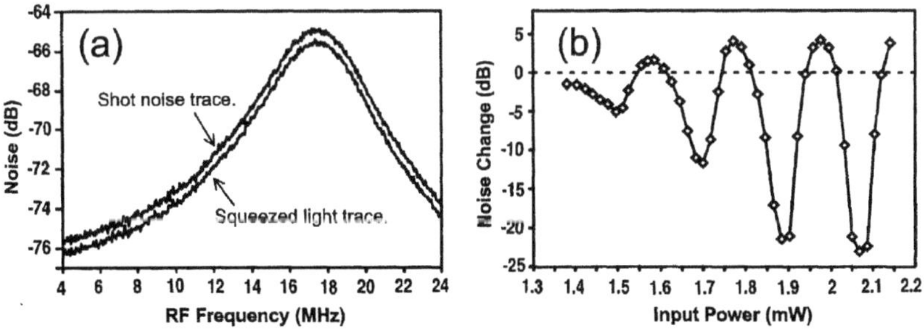

Figure 6. Non-soliton photon number squeezing and classical noise reduction. RF analyzer traces showing the shot noise level and the reduced noise levels of squeezed pulses are graphed in (a). The results of classical noise reduction measurements using spectral filtering are shown in (b).

Non-soliton photon number squeezing was first predicted by Werner and Friberg using quantum field theoretic methods, with results shown in Fig. 4(a).[5] For pulses with energy equivalent to an $N = 3$ soliton, non-soliton squeezing of 5.4 ± 0.1 dB in the absence of Raman effects has been predicted in a 90/10 NOLM.[10] Two experimental demonstrations of non-soliton squeezing have recently been reported.[8,9] Friberg and Werner used 2.4 ps input pulses centered at 1451 nm propagating in 500 meters of dispersion shifted fiber with a normal dispersion of 28.5 ps/nm km.[8] Averaging over RF spectrum analyzer data from 4 to 24 MHz, photocurrent noise was reduced to 0.631 ± 0.006 dB below the shot noise level, corresponding to 0.9 dB of squeezing when losses were taken in account. Figure 6(a) shows the RF spectrum traces for the shot noise calibration and the squeezing measurement. König *et al* have done better, obtaining photocurrent reductions of 1.2 ± 0.2 dB with 29 fs pulses centered at 808 nm from a mode-locked Ti-sapphire laser propagating through a 2 m length of single-mode fiber.[9] This gave squeezing levels of 2.6 ± 0.7 dB when losses were taken into account. Pulse energies were about 300 pJ.

Spatial soliton squeezing is expected to be a cw source of squeezed light. Spatial solitons propagating in a nonlinear planar waveguide are identical mathematically to optical fiber temporal solitons, provided that spatial width is substituted for time and spatial frequency for spectral frequency. The soliton effect is then due to diffraction balanced against Kerr-induced self-focusing, rather than group velocity dispersion balanced against Kerr-induced self-phase modulation. Spatial soliton photon number squeezing was first described by Meccozi and Kumar.[14] Their perturbation theory approach predicts that filtering of spatial solitons can produce up to 6.5 dB of squeezing. Friberg and Werner predict that 7 dB of squeezing should be attainable.[8] They suggest that spatial solitons should be observable in AlGaAs planar waveguides at 1.5 μm.

The noise reduction capabilities of nonlinear spectral filtering and NOLM photon number squeezing mechanisms are not limited to shot noise reduction. Friberg and Machida have used a nonlinear spectral filter to suppress the excess $1/f$ noise that invariably accompanies optical pulse trains from laser sources.[31] With the help of a 1.5 km length of optical fiber, low frequency $1/f$ noise in pulses from a passively mode-locked erbium-doped fiber laser was reduced by 23 dB at 1 MHz frequencies and lower, as shown in Fig. 6(b). The broadband character of the noise reduction – DC to greater than 10 THz – makes it practical for reducing noise for diagnostics, data-acquisition, communications and other applications.

OUTLOOK

Soliton photon number squeezing has rapidly established itself as a viable method to produce squeezed light. More than 6 dB of squeezing has been demonstrated and more than 10 dB appears to be realistic. The basic squeezing mechanism – the localization of fluctuations to particular portions of a pulse's spectrum via nonlinear propagation effects – allows noise reduction using both soliton spectral filtering and unbalanced NOLMs. The counterintuitive lesson learned from the recent predictions and demonstrations of NOLM squeezing is that interference with a weak local oscillator not subject to nonlinear effects can give rise to large amounts of photon number squeezing. I think that this is an insight that may allow the development of yet other methods of squeezing. Both soliton spectral filtering systems and unbalanced NOLMs act as nonlinear optical switching devices, suggesting the generalization that nonlinear optical switches can produce photon number squeezing. This is, of course, only true provided that the switching is not caused by the interference of two beams acquiring equal amounts of SPM noise. We know this to provide mainly amplified quantum noise.

These results also demonstrate that $\chi^{(3)}$ nonlinearities can be as effective as $\chi^{(2)}$ nonlinearities at producing squeezed light. Earlier demonstrations of quadrature-amplitude squeezed light in optical fibers could produce substantial squeezing only in very narrow bandwidth windows because of GAWBS noise. Spectral filtering squeezing bypasses the problem of GAWBS noise altogether, as it is immune to the phase fluctuations that affect quadrature amplitude squeezing. Interferometric devices like NOLMs should show some sensititivity to GAWBS noise, but it is greatly reduced by the use of short pulses. Good design using offset co-propagating pulses should overcome it completely.

The broader implications of photon number squeezing in optical fibers may be that it shows us how to take full advantage of fiber four-wave mixing processes. It was initially hoped that the use of optical fibers would overcome the problems associated with $\chi^{(3)}$ nonlinearities. This did not happen, in large part because of GAWBS noise and SPM noise. Doing photon number squeezing with fibers has shown us ways to deal both with GAWBS noise and SPM noise. This suggests that we may usefully start to consider how to use other fiber four-wave mixing techniques for quantum light generation.

REFERENCES

1. S. R. Friberg, S. Machida, M. J. Werner, A. Levanon, and Takaaki Mukai, Phys. Rev. Lett. 77, 3775 (1996).
2. S. Spälter, M. Burk, U. Strössner, M. Böhm, A. Sizmann, and G. Leuchs, Europhys. Lett. 38, 335 (1997).
3. S. Spälter, M. Burk, U. Strössner, M. Böhm, A. Sizmann, and G. Leuchs, Opt. Express 2, 77 (1998).
4. M. J. Werner, Phys. Rev. A 54, R2567 (1996).
5. M. J. Werner and S. R. Friberg, Phys. Rev. Lett. 79, 4143 (1997).
6. A. Mecozzi and P. Kumar, Opt. Lett. 22, 1232 (1997).
7. D. Levandovsky, M. Vasilyev, and P. Kumar, Paper QWG5, QEC'98 Technical Digest Series, Vol. 7, San Francisco, CA, May 3-8, 1998, p. 131.
8. S. R. Friberg and M. J. Werner, Paper QThH3, IQEC'98 Technical Digest Series, Vol. 7, San Francisco, CA, May 3-8, 1998, p. 197.
9. F. König, S. Spälter, I. L. Shumay, U. Strössner, A. Sizmann, Th. Fauster and G. Leuchs, submitted to J. Mod. Optics.
10. M. J. Werner, Paper ThD1, 1997 Annual Meeting, Optical Society of America, OSA '97, Long Beach, CA, Oct. 1997.
11. M. J. Werner and S. R. Friberg, Paper QWG2, IQEC'98, San Francisco, CA, May 5-8, 1998, p. 130; M. J. Werner, submitted to Phys. Rev. Lett.
12. S. Schmitt, J. Ficker, M. Wolff, F. König, A. Sizmann, and G. Leuchs, to appear in Phys. Rev. Lett.
13. D. Krylov and K. Bergman, to appear in Opt. Lett.
14. A. Mecozzi and P. Kumar, Quantum and Semiclassical Optics, 10, p. L21-26, (1998).
15. K. Schneider, M. Lang, J. Mlynek and S. Schiller, Opt. Express 2, 59 (1998).
16. S. Spälter, N. Korolkova, F. König, A. Sizmann, and G. Leuchs, Phys. Rev. Lett. 81, 786 (1998).
17. A. Hasegawa and Y. Kodama, *Solitons in Optical Communications*, (Oxford University Press, Oxford, 1995).
18. D. C. Burnham and D. L. Weinberg, Phys. Rev. Lett. 25, 84 (1970).
19. S. Friberg, C. K. Hong, and L. Mandel, Phys. Rev. Lett. 54, 84 (1985); S. Friberg, C. K. Hong, and L. Mandel, Opt. Commun. 54, 311 (1985).
20. C. K. Hong, S. R. Friberg, and L. Mandel, Appl. Opt. 24, 3877 (1985).
21. L. A. Wu, H. J. Kimble, J. L. Hall, and H. Wu, Phys. Rev. Lett. 57, 2520 (1986).
22. R. E. Slusher, L. W. Hollberg, B. Yurke, J. C. Mertz, and J. F. Valley, Phys. Rev. Lett. 55, 2409 (1985).
23. R. M. Shelby, M. D. Levenson, S. H. Perlmutter, R. G. DeVoe, and D. F. Walls, Phys. Rev. Lett. 57, 691 (1986).
24. M. Vallet, M. Pinard, and G. Grynberg, Europhys. Lett. 11, 739 (1990).
25. M. Rosenbluh and R. M. Shelby, Phys. Rev. Lett. 66, 153 (1991).
26. P. D. Drummond, R. M. Shelby, S. R. Friberg, and Y. Yamamoto, Nature 365, 307 (1993).
27. K. Bergman and H. A. Haus, Opt. Lett. 16, 663 (1991); K. Bergmann, H. A. Haus, E. P. Ippen and M. Shirasaki, Opt. Lett. 19, 290 (1994).
28. A. Sizmann, Appl. Phys. B. 65, 745 (1997).
29. P. D. Drummond, Phys. Rev. A 42, 6845 (1990).
30. H. -H. Ritze and A. Bandilla, Opt. Commun. 29, 126 (1979).
31. S. R. Friberg and S. Machida, to appear in Appl. Phys. Lett.

AMPLITUDE SOLITON SQUEEZING USING AN ASYMMETRIC SAGNAC FIBER LOOP

Dmitriy Krylov and Keren Bergman

Department of Electrical Engineering
Princeton University
B312, E-Quad, Olden Street, Princeton, NJ 08544

We experimentally demonstrate a new scheme for generating amplitude squeezed solitons in an asymmetric fiber Sagnac loop. We measure by direct detection a $5.7 \pm 0.1dB$ (73%) and, with corrections for linear losses $6.2 \pm 0.1dB$ (76%), reduction in the photon number fluctuations below the shot noise level. The same scheme is also shown to generate a significant classical noise reduction and is limited by Raman effects in fiber.

Generation of amplitude-squeezed states using the Kerr nonlinearity in optical fibers has been recently demonstrated in a novel scheme, pioneered by Friberg et al., employing soliton propagation followed by spectral filtering[1]. By launching a soliton-like pulse with energy slightly greater than the fundamental soliton energy ($N > 1$) into a fiber of length equivalent to several soliton periods followed by a spectral filter, it is possible to observe a reduction in the photon number fluctuations by direct detection. In subsequent experiments up to 3.8dB of such reduction was directly detected with femtosecond pulses[2,3]. The filtering action also introduces additional zero-point fluctuations into the system limiting the highest achievable squeezing to approximately 8dB below the shot-noise level for ideal fibers and bandpass filters[4].

In order to eliminate the above limitation, it has been proposed that amplitude-squeezed pulses can be produced by interference between the counter-propagating fields in an asymmetric fiber Sagnac loop[5]. For the case of soliton squeezing, the idea that interfering a high-energy ($N > 1$) soliton pulse with a weaker pulse or a dispersive wave can produce squeezing is also consistent with the soliton perturbation treatment[6] as well as the general quantum theory of soliton propagation[7,8]. The geometry proposed in [5] is well suited for testing this idea in practice and has been recently explored experimentally by Schmitt et al.[9]. Different geometries such as polarization interferometry for achieving amplitude squeezing have also been recently demonstrated[10]. In this paper we experimentally demonstrate the asymmetric fiber Sagnac approach and obtain a record $5.7 \pm 0.1dB$ (73%) directly detected photon number squeezing below the shot noise level. With correction for linear system losses, the actual amplitude

Quantum Communication, Computing, and Measurement 2
Edited by Kumar et al., Kluwer Academic / Plenum Publishers, New York, 2000.

squeezing is $6.2 \pm 0.1dB$ (76%). We also measure a significant reduction in the classical noise inherent in the optical signal.

The experimental setup is shown in Fig. 1. A Spectra-Physics optical parametric oscillator (Opal) is used as a source of 182-fs (FWHM) sech-shaped optical pulses at repetition rate of 82 MHz, and centered at 1550 nm. The corresponding dispersion length ($\beta'' = -19ps^2/km$) in the standard polarization maintaining (PM) fiber, Fujikura SM15-P-8 with an $8-\mu m$ core diameter, used in the experiments is about 54cm, and the soliton period is 86cm. The average power required to produce a fundamental ($N = 1$) soliton as determined by measuring the pulse temporal autocorrelation and optical spectrum following propagation in 10m of the PM fiber, is $20 \pm 1mW$ which is consistent with our calculations. We use an asymmetric Sagnac loop configuration, where the light is split by a free space AR coated beamsplitter and then coupled from both ends into a 3.5m section of the PM optical fiber. The free-space configuration is necessary to preserve the spectrum of the signal in both arms of the interferometer, which is not always possible with fiber splitters for femtosecond pulses. The experiment is not critically sensitive to the splitting ratio, as the same qualitative behavior is consistently observed with a range of splitting ratios. In the experiments reported here, the fiber coupling was varied until the optimal squeezing was observed for the input splitting of 88/12. As a result, the output recombination ratio, determined by the beamsplitter only, is not equal to the initial splitting ratio, but we do not see this additional asymmetry as a limitation to the experiment. Subsequently all the input optical powers specified below incorporate the losses due to the fiber coupling and correspond to the optical powers propagating in the fiber.

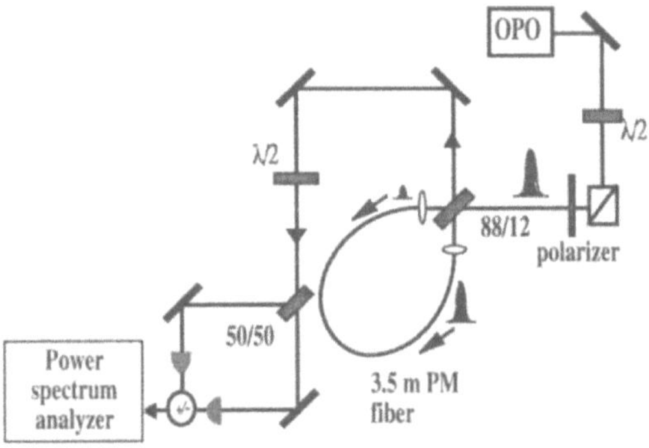

Figure 1. Experimental Setup. (OPO-optical parametric oscillator, PM-polarization maintaining, $\lambda/2$-half wave plate)

With such highly asymmetric splitting, most of the energy propagates in the 88% reflection arm. The noise properties of this pulse are modified in accordance with the quantum nonlinear Schrödinger equation so that the phase fluctuations are coupled to the amplitude fluctuations[5,6,11]. For lower input optical powers, the field in the 12% transmission arm is a dispersive wave which propagates linearly in the fiber loop but begins to exhibit soliton-like behavior for higher input powers. The polarization of the pulses entering and exiting the fiber loop is carefully controlled to assure the optimal interference. The photocurrent fluctuations associated with the pulse, resulting from the interference of the two counter-propagating fields in the loop, are measured by a balanced receiver followed by the HP3588A power spectrum analyzer (PSA) operated in the 'zero-span' regime in a narrow-band interval centered around 5 MHz with a resolution bandwidth of 17 kHz. The PSA measures the noise power as the root mean square fluctuations in the resolvable flat window around 5 MHz which then are normalized to a 1-Hz bandwidth. We use the subtraction mode of the receiver for shot noise calibrations, and the summing mode is used for direct detection of the amplitude fluctuations.

Using the balanced receiver configuration is a convenient way of keeping the maximum power falling on the two reverse-biased photodiodes (Epitaxx ETX-1000T, quantum efficiency 95% at 1550nm) below saturation values. The signal beams are carefully focused to just fit within the 1mm active area diameter of the two photodetectors, since both under-focusing and over-focusing the beam introduces additional noise. In order to avoid the saturation of electronics by the strong 82-MHz frequency component due to the repetition rate of the laser system, we have carefully chosen and tested all the components in our receiver to assure the best performance without saturation. A power amplifier in conjunction with low-pass filtering are used to limit the overall bandwidth of the receiver 35 MHz and protect the PSA from overloading.

Several calibrations were performed within the experimental set-up as well as in free space to accurately establish the relevant noise levels. Firstly, each photodiode was calibrated individually to establish accurately the region where the detector response is linear. Subtracting the photocurrents eliminates the classical fluctuations present in the laser signal with an extinction ratio of about 25dB, and the measured noise levels following this subtraction accurately represent the shot noise. The noise floor of the PSA is at -138 dBm/Hz, which is also the noise magnitude of the dark current. By comparison, the thermal noise due to the shunt resistance of the receiver is approximately -128 dBm/Hz. The dominating noise source is the 45-dB power amplifier used between the receiver and the PSA in conjunction with the 35-MHz low-pass filter with a 50-dB roll-off at 82MHz. The measured noise floor of the whole detection setup, with the photodiodes covered by dark cloth, is -116dBm/Hz. With the optical powers used in the experiment, the shot noise is at least 7dB above this noise floor, so that squeezing can be observed. All the values for the shot noise levels were confirmed to within 0.1dB by measuring the noise power versus the incident optical power for the above experimental setup as well as in free space. Figure 2(a) shows how these noise values scale correctly as the optical power is reduced down to the noise floor of the receiver. The free space calibration was performed to eliminate any potential noise sources associated with the fiber nonlinearities. Comparing the free space calibration values with those obtained for a straight piece of fiber as well as with those obtained within the experimental setup allowed us to accurately determine the shot noise level for the entire range of experimental results. The sum of the photocurrent fluctuations as a function of incident power measured in free space represents the classical noise inherent in the laser signal which can be more than 3dB above the shot noise level.

The results of the squeezing experiments are shown in Fig. 2(b) that is a plot of the classical noise and the noise variations due to amplitude squeezing and anti-squeezing normalized to the shot noise level in units of dB as a function of the incident optical power into the loop. We observe three squeezing resonances (at 30 mW, 52 mW, and 68 mW). The largest resonance occurs at the input power into the loop of 68 mW, which corresponds to an approximate $N = 1.85 \pm 0.05$ in soliton units. The reduction below shot noise is measured to be $5.7 \pm 0.1dB$ (73%). Taking into account 90% overall detection efficiency that includes detector quantum efficiencies and 5% losses that include losses at the fiber outputs and the imperfect mode overlap at the output beamsplitter, this squeezing level corresponds to $6.2 \pm 0.1dB$ (76%) reduction. This result is in very good agreement (within 0.5dB or 10%) with numerical simulations done for an N=1.7 sech pulse propagating through 6.4 dispersion lengths

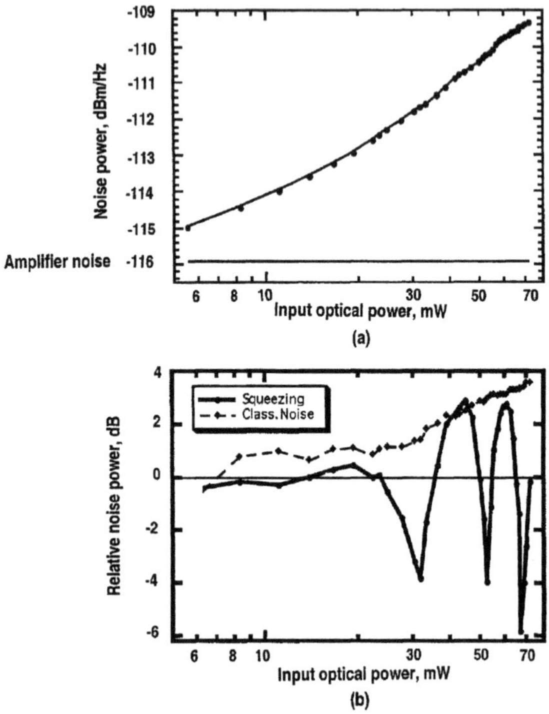

Figure 2. Absolute (a) Shot-noise calibration and (b) Relative noise power fluctuations vs. optical power incident into the fiber loop.

in a loop with a 90/10 splitting ratio[12]. We also note that an important practical advantage if this squeezing scheme is the removal of classical noise inherent in the laser signal in addition to reduction in quantum photon number fluctuations. For the largest squeezing resonance at 68 mW incident power, we observe more than 3dB of classical noise reduction. The total noise reduction including classical and quantum amplitude fluctuations is in excess of 9dB.

The Raman effects in fibers have been predicted to limit the amount of squeezing observed for longer propagation distances[8,12]. To test this, we also performed additional experiments with different fiber lengths (up to 9 meters). Figure 3(a) shows the spectra of the input soliton pulse (N=1.85) and of the same pulse after propagating through 3.5m (b) and 9m (c) of fiber. We note that while after propagating through 3.5m the spectrum is slightly broadened but virtually unchanged, after 9-m fiber, most of the pulse energy has been shifted by approximately 30nm towards the longer wavelengths. Since, the pulse traveling in the 12% transmission arm does not have enough energy to experience a significant Raman shift, one might expect that the interference is no longer optimal and the squeezing is reduced. In fact the maximum amplitude squeezing observed for the long fiber lengths was approximately 2.5dB. In addition to shifting the spectrum, noise from Raman has been shown to limit the maximum observable squeezing with femtosecond pulses to approximately 8dB from a possible 11dB for ideal fibers[12]. The Raman noise imposed limit on squeezing is still a subject of continuing theoretical and experimental investigations.

In conclusion, we have experimentally demonstrated a new scheme to produce non-classical states of light in a highly asymmetric Sagnac loop and measured a record 5.7$\pm$

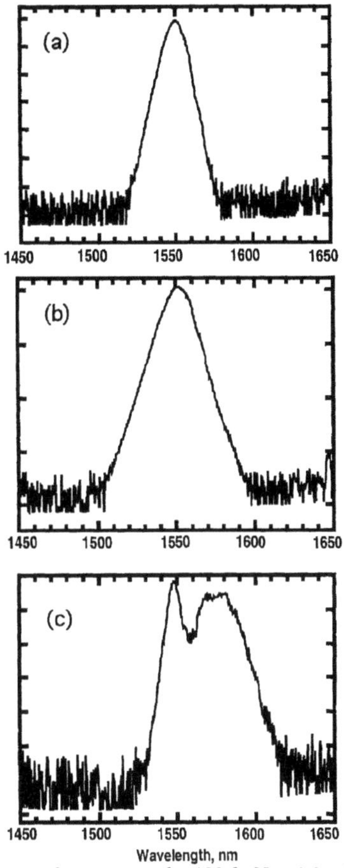

Figure 3. Input (a) and output pulse spectra of a 182-fs $N = 1.8$ soliton after propagating through 3.5m (b) and 9m (c) of standard PM fiber.

$0.1dB$ ($6.2 \pm 0.1dB$ with correction for losses) reduction in photon-number fluctuations of soliton optical pulses. This experiment conclusively shows that the interference between the two counter-propagating fields is the main mechanism that produces the noise reduction. The scheme is also useful in removing classical noise from the signal and appears to be limited by the Raman effect for longer propagation distances.

The authors would like to thank M. J. Werner and Y.Lai for helpful discussions. This work was supported by the U.S. Office of Naval Research (N00014-96-0773).

REFERENCES

1. S. R. Friberg, S. Machida, M. J. Werner, A. Levanon, and T. Mukai, *Phys. Rev. Lett.* **77,** 3775 (1996).
2. S. Spälter, M. Burk, U. Strössner, M. Boehm, A. Sizmann, and G. Leuchs, *Europhysics Lett.* **38,** 335 (1997).
3. S. Spälter, M. Burk, U. Strössner, and G. Leuchs, *Opt. Expr.* **2,** 77 (1998).
4. M. J. Werner, *Phys. Rev. A* **54,** R2567 (1996).
5. M. J. Werner, Symposium on Quantum Optics for Communications, OSA Annual Meeting, (Long Beach, CA 1997).
6. H. A. Haus and Y. Lai, *J. Opt. Soc. Am. B* **9,** 386 (1990).
7. S. J. Carter and P. D. Drummond, *Phys. Rev. Lett.* **67,** 3757 (1991).

8. Y. Lai and S.-S. Yu, *Phys. Rev. A* **51,** 817 (1995).
9. S. Schmitt, F. König, B. Mikulla, S. Spälter, A. Sizmann, and G. Leuchs, International Quantum Electronics Conference, OSA Technical Digest, **7,** 195 (San Francisco, CA 1998).
10. M. Margalit, E. P. Ippen, and H. A. Haus, International Quantum Electronics Conference, OSA Technical Digest, **7,** 170 (San Francisco, CA 1998).
11. S. J. Carter, P. D. Drummond, M. D. Reid, and R. M. Shelby, *Phys. Rev. Lett* **58,** 1841 (1987).
12. M. J. Werner and S. R. Friberg, International Quantum Electronics Conference, OSA Technical Digest, **7,** 130 (San Francisco, CA 1998).

TIME DOMAIN CORRELATIONS AND GATED DETECTION OF QUANTUM SOLITONS

D. Levandovsky, M. Vasilyev, and P. Kumar

Department of Electrical and Computer Engineering
Northwestern University, Evanston, Illinois 60208-3118
Telephone: (847) 467-2261; Fax: (847) 491-4455
E-mail: dmitry@ece.nwu.edu

We study the time-domain quantum-noise properties of solitons in optical fibers. Perturbation theory, including quantum description of the continuum, is used to derive a complete analytical expression for the second-order correlator of the amplitude quadrature in time domain. This correlator is subsequently used to calculate the photon-number noise reduction below the standard quantum limit by means of a time-gating filter. For propagation distances up to three soliton periods, the length at which the best noise reduction occurs, a square band-reject time filter is found to produce approximately 3.2 dB of squeezing.

INTRODUCTION

Recently, sub-Poissonian light was generated by frequency filtering solitons after their propagation through the fiber.[1] Numerical and analytical models of these experiments were developed that are based on a positive P representation,[2] a back propagation approach[3], or a direct method using the second-order correlator of the frequency-domain amplitude quadrature.[4] An experimental study of this correlator was recently reported.[5] In addition, quantum properties of 1-dimensional spatial solitons of the Kerr nonlinearity have been investigated in the context of near- and far-field spatial aperturing[6] and noise amplification through four-wave mixing.[7] In particular, the work in Ref. 6 shows that far-field spatial aperturing is equivalent to frequency filtering of the quantum field of temporal solitons, whereas the near-field aperturing is equivalent to modulating (time-gating) the temporal solitons before their detection by a photon-counting detector. In the latter case, to study the quantum-noise properties of the solitons, one needs to obtain the second-order correlators of the time-domain (or the near-field spatial domain) quadratures. In this paper, we derive analytical expressions for the quantum-noise correlators of the time-domain soliton quadratures from their frequency-domain counterparts, taking into account complete contribution of the continuum.[4, 8] We subsequently apply these correlators to the problem of quantum-noise reduction by means of time gating the soliton field.

Quantum Communication, Computing, and Measurement 2
Edited by Kumar et al., Kluwer Academic / Plenum Publishers, New York, 2000.

CORRELATORS OF THE TIME-DOMAIN QUADRATURE

We start with propagation of the modal annihilation operator $\hat{a}(\tau,\xi)$ inside an optical fiber as described by the quantum nonlinear Schrödinger equation (NLSE), which in dimensionless coordinates is

$$\frac{\partial}{\partial\xi}\hat{a}(\tau,\xi) = i\left[\frac{1}{2}\frac{\partial^2}{\partial\tau^2} + \hat{a}^\dagger(\tau,\xi)\,\hat{a}(\tau,\xi)\right]\hat{a}(\tau,\xi). \tag{1}$$

If the operators in this equation are replaced by c-numbers, Eq. (1) is known to have the classical fundamental-soliton solution

$$a(\tau,\xi) = \underline{f}_n(\tau)\exp(i\xi/2) \equiv \operatorname{sech}(\tau)\exp(i\xi/2), \tag{2}$$

given here in the canonical form corresponding to the average number of photons $\langle\hat{N}\rangle = 2$, where $\hat{N} \equiv \int \hat{a}^\dagger(\tau)\hat{a}(\tau)\mathrm{d}\tau$. To circumvent the problem of solving for the nonlinear evolution of the operator $\hat{a}$, we use the soliton perturbation approach developed in Ref. 8. We write the operator in the following form:

$$\begin{aligned}\hat{a}(\tau,\xi) &= \underline{f}_n(\tau)\exp(i\xi/2) + \Delta\hat{b}(\tau,\xi)\\ &\equiv [\underline{f}_n(\tau) + \Delta\hat{a}(\tau,\xi)]\exp(i\xi/2),\end{aligned} \tag{3}$$

with $\Delta\hat{a}$ subject to usual commutation relations everywhere inside the fiber

$$[\Delta\hat{a}(\tau,\xi),\Delta\hat{a}^\dagger(\tau',\xi)] = \delta(\tau-\tau'), \tag{4}$$

$$[\Delta\hat{a}(\tau,\xi),\Delta\hat{a}(\tau',\xi)] = [\Delta\hat{a}^\dagger(\tau,\xi),\Delta\hat{a}^\dagger(\tau',\xi)] = 0. \tag{5}$$

Substituting Eq. (3) in Eq. (1), and keeping only terms that are linear in $\Delta\hat{b}$, divides the analysis of Eq. (1) into two separate problems: the soliton solution of the classical NLSE, and the solution of the linearized operator equation

$$\frac{\partial\Delta\hat{b}}{\partial\xi} = \frac{i}{2}\frac{\partial^2}{\partial\tau^2}\Delta\hat{b} + 2i\,|a(\tau,\xi)|^2\,\Delta\hat{b} + ia(\tau,\xi)^2\,\Delta\hat{b}^\dagger, \tag{6}$$

where the quantum-mechanical fluctuation operator $\Delta\hat{b}$ represents perturbation of the classical soliton field by the quantum noise. This linearization approach is valid if the photon-number noise is fairly small compared to the mean number of photons in the soliton, which is usually the case in most experiments. Note also that, since we only retain terms up to first order in $\Delta\hat{b}$, the photon-number noise is determined by the fluctuations in the amplitude quadrature (defined below) of the soliton field. Since Eq. (6) is linear, its solution can be written as a normal-mode expansion[4, 8]

$$\Delta\hat{a}(\tau,\xi) = \int\frac{\mathrm{d}\Omega}{2\pi}[\hat{V}_c(\Omega,\xi)\,f_c(\Omega,\tau) + \hat{V}_s(\Omega,\xi)\,f_s(\Omega,\tau)] + \sum_{j=n,p,\tau,\theta}\hat{V}_j(\xi)\,f_j(\tau), \tag{7}$$

wherein the four discrete modes f_n, f_p, f_τ, and f_θ represent perturbations to the soliton shape due to changes in the soliton photon number, momentum (frequency), position (time), and phase, respectively; and f_c and f_s are the symmetric and antisymmetric parts, respectively, of the modes that represent perturbations of the continuum (dispersive radiation) in the fiber. Note that in this formalism all the ξ dependence in Eq. (7) is associated with the operator coefficients $\hat{V}_j$, $\hat{V}_c$, $\hat{V}_s$ (Heisenberg picture), which are Hermitian. The modes f_j, f_c, f_s used in the expansion given by Eq. (7), which are the same as those in Ref. 9, are properly normalized, orthogonal, and form a complete set. Thus, orthogonality of the modes and completeness of this solution allow for the reconstruction of the total quantum field at the output of the optical fiber by solving for the evolution of the operator coefficients in Eq. (7) and applying the principle of superposition. In our model we assume a white coherent-state quantum noise at the input of the fiber ($\xi = 0$), which excites all the modes of the linearized NLSE. Defining two quadrature-like operators in frequency domain as

$$\Delta\hat{a}_c(\omega,\xi) \equiv [\Delta\hat{a}(\omega,\xi) + \Delta\hat{a}^\dagger(\omega,\xi)]/2 \ , \tag{8}$$

$$\Delta\hat{a}_s(\omega,\xi) \equiv [\Delta\hat{a}(\omega,\xi) - \Delta\hat{a}^\dagger(\omega,\xi)]/2 \ , \tag{9}$$

which represent amplitude (cosine) and phase (sine) perturbations of the field at a given frequency ω, we may write the following expressions for the three correlators:

$$G^{cc}(\omega,\omega',\xi) \equiv 4\langle\Delta\hat{a}_c(\omega,\xi)\Delta\hat{a}_c(\omega',\xi)\rangle = G_N^{cc}(\omega,\omega',\xi) + 2\pi\delta(\omega-\omega'), \tag{10}$$

$$G^{ss}(\omega,\omega',\xi) \equiv 4\langle\Delta\hat{a}_s(\omega,\xi)\Delta\hat{a}_s(\omega',\xi)\rangle = G_N^{ss}(\omega,\omega',\xi) - 2\pi\delta(\omega-\omega'), \tag{11}$$

$$G^{cs}(\omega,\omega',\xi) \equiv 4\langle\Delta\hat{a}_c(\omega,\xi)\Delta\hat{a}_s(\omega',\xi)\rangle = G_N^{cs}(\omega,\omega',\xi) - 2\pi\delta(\omega-\omega'), \tag{12}$$

wherein we have used the Fourier transform relation $\Delta\hat{a}(\omega,\xi) = \int \Delta\hat{a}(\tau,\xi)e^{i\omega\tau}\mathrm{d}\tau$, and introduced $G_N^{jk}(\omega,\omega',\xi)$ as the normal-ordered part of the respective correlator. We may also define two time-domain quadrature-like operators as

$$\Delta\hat{a}_r(\tau,\xi) \equiv [\Delta\hat{a}(\tau,\xi) + \Delta\hat{a}^\dagger(\tau,\xi)]/2 \ , \tag{13}$$

$$\Delta\hat{a}_i(\tau,\xi) \equiv [\Delta\hat{a}(\tau,\xi) - \Delta\hat{a}^\dagger(\tau,\xi)]/2 \ , \tag{14}$$

which represent the amplitude (real) and phase (imaginary) perturbations of the field at time τ. We note here that the inverse-Fourier transforms of the operators $\Delta\hat{a}_c(\omega,\xi)$ and $\Delta\hat{a}_s(\omega,\xi)$ are not the same as the two operators in Eqs. (13) and (14), but, instead, are given by

$$\Delta\hat{a}_c(\tau,\xi) = [\Delta\hat{a}(\tau,\xi) + \Delta\hat{a}^\dagger(-\tau,\xi)]/2 \ , \tag{15}$$

$$\Delta\hat{a}_s(\tau,\xi) = [\Delta\hat{a}(\tau,\xi) - \Delta\hat{a}^\dagger(-\tau,\xi)]/2 \ . \tag{16}$$

The three correlators in Eqs. (10)–(12) were derived analytically in Ref. 9 in terms of the normal modes of the linearized NLSE [Eq. (6)]. Their normal-ordered parts are plotted in Fig. 1(left). The three correlators of the time-domain quadratures, Eqs. (13) and (14), can be obtained from the inverse-Fourier transforms of the correlators in Eqs. (10)–(12) for the frequency-domain quadratures as follows:

$$\begin{aligned} G^{rr}(\tau,\tau',\xi) &\equiv 4\langle\Delta\hat{a}_r(\tau,\xi),\Delta\hat{a}_r(\tau',\xi)\rangle \equiv \delta(\tau-\tau') + G_N^{rr}(\tau,\tau',\xi) \ , \\ G_N^{rr}(\tau,\tau',\xi) &= \frac{1}{2}\mathrm{Re}\{G_N^{cc}(\tau,\tau',\xi) + G_N^{cc}(-\tau,\tau',\xi) + G_N^{ss}(\tau,\tau',\xi) - G_N^{ss}(-\tau,\tau',\xi) \\ &\quad + G_N^{cs}(\tau,\tau',\xi) + G_N^{cs}(-\tau,\tau',\xi) + G_N^{cs}(\tau',\tau,\xi) + G_N^{cs}(-\tau',\tau,\xi)\} \ , \end{aligned} \tag{17}$$

$$\begin{aligned} G^{ii}(\tau,\tau',\xi) &\equiv 4\langle\Delta\hat{a}_i(\tau,\xi),\Delta\hat{a}_i(\tau',\xi)\rangle \equiv -\delta(\tau-\tau') + G_N^{ii}(\tau,\tau',\xi) \ , \\ G_N^{ii}(\tau,\tau',\xi) &= \frac{1}{2}\mathrm{Re}\{G_N^{cc}(\tau,\tau',\xi) - G_N^{cc}(-\tau,\tau',\xi) + G_N^{ss}(\tau,\tau',\xi) + G_N^{ss}(-\tau,\tau',\xi) \\ &\quad + G_N^{cs}(\tau,\tau',\xi) - G_N^{cs}(-\tau,\tau',\xi) + G_N^{cs}(\tau',\tau,\xi) - G_N^{cs}(-\tau',\tau,\xi)\} \ , \end{aligned} \tag{18}$$

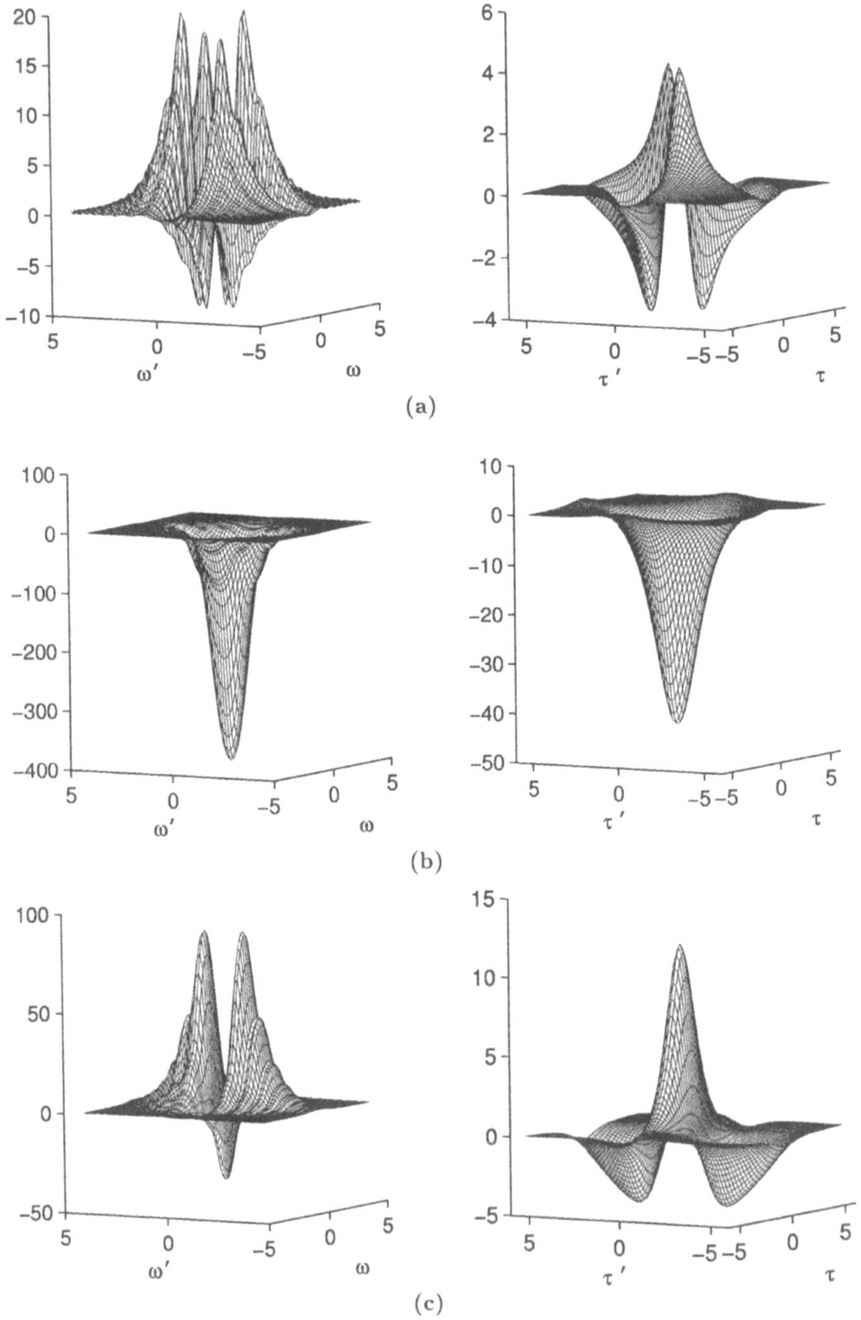

Figure 1. The three correlators of the frequency-domain (left) and time-domain (right) quadratures for a fiber length $z = 3$ soliton periods: (a) G_N^{cc} and G_N^{rr}; (b) G_N^{ss} and G_N^{ii}; and (c) G_N^{cs} and G_N^{ri}.

$$G^{ri}(\tau,\tau',\xi) \equiv 4\langle\Delta\hat{a}_r(\tau,\xi),\Delta\hat{a}_i(\tau',\xi)\rangle \equiv -\delta(\tau-\tau') + G_N^{ri}(\tau,\tau',\xi)\ ,$$
$$G_N^{ri}(\tau,\tau',\xi) = \frac{i}{2}\mathrm{Im}\{G_N^{cc}(\tau,\tau',\xi) + G_N^{cc}(-\tau,\tau',\xi) + G_N^{ss}(\tau,\tau',\xi) - G_N^{ss}(-\tau,\tau',\xi)$$
$$+G_N^{cs}(\tau,\tau',\xi) + G_N^{cs}(-\tau,\tau',\xi) + G_N^{cs}(\tau',\tau,\xi) - G_N^{cs}(-\tau',\tau,\xi)\}\ . \quad (19)$$

The above correlators of the time-domain quadratures, which are plotted in Fig. 1(right), can now be used to solve the problem of quantum-noise reduction by means of time gating the optical field.

QUANTUM-NOISE REDUCTION BY TIME GATING

The time-gating arrangement consists of a length of optical fiber followed by an amplitude modulator. Coherent-state solitons are launched into the fiber and the emerging radiation, after passing through the modulator (time gate), is direct-detected by a photodetector. The action of a modulator, with an amplitude transfer function $H(\tau)$, on the quantum state of light is inevitably accompanied by the mixing-in of the noise of an independent vacuum-state mode $\hat{v}$, arising because of the time-dependent loss introduced. The output fluctuation operator is, then, given by

$$\Delta\hat{a}_{\mathrm{out}}(\tau,\xi) = |H(\tau)|\,\Delta\hat{a}(\tau,\xi) + \sqrt{1-|H(\tau)|^2}\,\hat{v}(\tau)\,, \quad (20)$$

wherein the amplitude transfer function must be restricted in the range $0 < |H(\tau)| < 1$ for it to represent a physically realizable time gate. Therefore, normalizing the output

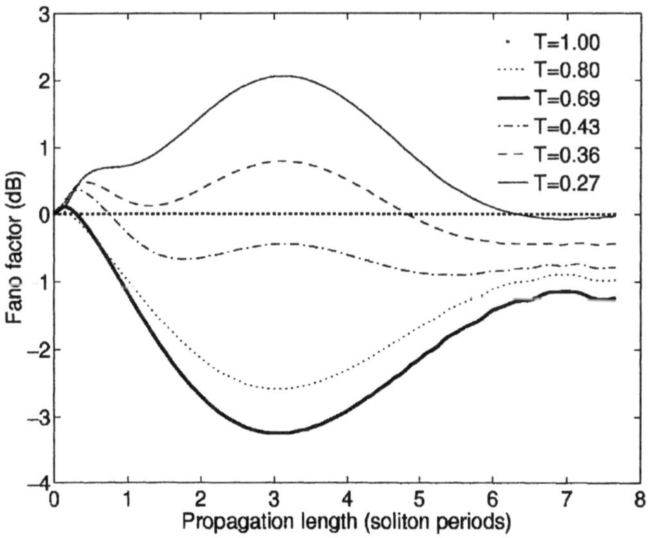

Figure 2. Fano factor for time gates of various widths $\Delta\tau$ plotted as a function of propagation distance in the fiber. The widths $\Delta\tau$ are characterized in terms of the total energy transmission $T \equiv \langle\hat{N}_{\mathrm{out}}\rangle/\langle\hat{N}\rangle$ ($T = 1$ corresponds to the shot noise level).

photon-number variance to the average output photon number, we obtain the desired Fano factor

$$F(\xi) \equiv \langle\Delta\hat{N}_{\mathrm{out}}(\xi)^2\rangle/\langle\hat{N}_{\mathrm{out}}\rangle$$
$$= 1 + \langle\hat{N}_{\mathrm{out}}\rangle^{-1}\iint d\tau d\tau' \underline{f}_n(\tau)\,|H(\tau)|^2\, G_N^{rr}(\tau,\tau',\xi)\,|H(\tau')|^2\,\underline{f}_n(\tau')\,, \quad (21)$$

where $\langle \hat{N}_{\text{out}} \rangle \equiv \int |H(\tau)|^2 \underline{f}_n(\tau)^2 d\tau$ is the average number of output photons and $G_N^{rr}(\tau, \tau', \xi)$ is the normal-ordered part of the second-order correlator of the time-domain amplitude quadrature as given by Eq. (17). This Fano factor is plotted in Fig. 2 as a function of the propagation distance $z = 2\xi/\pi$ in soliton periods for time gates of several different widths $\Delta\tau$, which reject a square region near the center of the soliton. Physical symmetry of the problem suggests that the time during which the output soliton is completely suppressed $[H_{\Delta\tau}(\tau) = 0]$ must be chosen to be symmetric around the $\tau = 0$ point. As seen from the figure, the best Fano factor is obtained for distances that are close to $z \approx 3$ soliton periods, where a noise reduction of ≈ 3.2 dB is achieved. The corresponding optimum square time gate has a total energy transmission $T \equiv \langle N_{\text{out}} \rangle / \langle N \rangle \approx 69\%$. These results are in excellent agreement with those reported in Ref. 6.

CONCLUSIONS

In conclusion, we have derived the shapes of the three second-order correlators of the time-domain quadratures. The resulting correlations in the amplitude quadrature are found to be weaker than the spectral correlations studied in Ref. 4, although their physical origins are closely related. Namely, there exists a negative correlation between the soliton photon number and pulse width. By appropriately choosing the time gate to reject the central portion of the soliton pulse, one is able to take advantage of this correlation and reduce the photon-number noise below the standard quantum limit. This is because a greater photon number results in a narrower pulse-width, which, in turn, causes a greater photon-number loss from the time gate. In our model the correlation between the photon number and pulse width is manifested mostly through non-zero moments $\langle \hat{V}_n \hat{V}_c \rangle$ and $\langle \hat{V}_n \hat{V}_s \rangle$ at the output of the fiber, which have been given explicitly in Ref. 9.

ACKNOWLEDGMENTS

The authors acknowledge useful discussions with A. Mecozzi. This research was supported in part by the U.S. Office of Naval Research.

REFERENCES

1. S. R. Friberg, S. Machida, M. J. Werner, A. Levanon, and T. Mukai, Phys. Rev. Lett. **77**, 3775 (1996); S. Spälter, M. Burk, U. Strössner, M. Böhm, A. Sizmann, and G. Leuchs, Europhys. Lett. **38**, 335 (1997); S. Spälter, M. Burk, U. Strössner, A. Sizmann, and G. Leuchs, Opt. Exp. **2**, 77 (1998).
2. M. J. Werner and S. R. Friberg, Phys. Rev. Lett. **79**, 4143 (1997).
3. A. Mecozzi and P. Kumar, Opt. Lett. **22**, 1232 (1997).
4. D. Levandovsky, M. Vasilyev, and P. Kumar, Opt. Lett. **XX**, XXX (1998).
5. S. Spälter, N. Korolkova, F. König, A. Sizmann, and G. Leuchs, Phys. Rev. Lett. **81**, 786 (1998).
6. A. Mecozzi and P. Kumar, Quantum Semiclass. Opt. **10**, L21 (1998)
7. E. M. Nagasako, R. W. Boyd, and G. S. Agarwal, Phys. Rev. A **55**, 1412 (1997).
8. D. J. Kaup, Phys. Rev. A **42**, 5689 (1990); H. A. Haus and Y. Lai, J. Opt. Soc. Am. B **7**, 386 (1990); H. A. Haus, W. S. Wong, and F. I. Khatri, J. Opt. Soc. Am. B **14**, 304 (1997).
9. D. Levandovsky, M. Vasilyev, and P. Kumar, Opt. Lett. **XX**, XXX (1998); M. Vasilyev, D. Levandovsky, and P. Kumar, submitted to Phys. Rev. A.

OBSERVATION OF SQUEEZING IN FREE-SPACE PHASE-DEPENDENT RESONANCE FLUORESCENCE

Z. H. Lu, S. Bali, and J. E. Thomas

Physics Department
Duke University
Durham, North Carolina 27708-0305

Abstract

We measure squeezing spectra for phase-dependent fluorescence of coherently driven two-level atoms in *free space*. The measured spectra are in excellent agreement with our predictions for a single two-level atom. We find that maximum squeezing occurs for homodyne detection at a phase near ±45° relative to the exciting field, rather than in the in- or out-of-phase quadratures. A new physical picture of phase-dependent noise that incorporates quantum collapses into a Bloch vector model shows why this occurs.

INTRODUCTION

Resonance fluorescence is a central topic in quantum optics that has been carefully explored in theory and experiment.[1, 2, 3] However, the phenomenon of *squeezing* in resonance fluorescence, first predicted in 1981 by Walls and Zoller[4] and in 1982 by Mandel,[5] has eluded observation despite receiving considerable attention. Indeed, it has been suggested recently that "...the observation of phase-sensitive squeezing is the most fundamental open problem with respect to the quantum properties of resonance fluorescence radiation".[6] We report the first measurement of single-atom squeezing spectra in the phase-dependent fluorescence of a beam of driven two-level atoms in free space.

Although squeezing in atomic systems is by now well established,[7] measurements are difficult to perform and precision near-resonant phase-dependent spectra have not been obtained until recently.[8, 9, 10] On the other hand, phase-dependent resonance fluorescence has been investigated theoretically in some detail, especially for the in- and out-of-phase quadratures. Squeezing has been predicted for the in-phase quadrature for on-resonance excitation with small driving fields.[11, 12, 13] It has been pointed out that quantum noise in the projections of the atomic Bloch vector is different for different quadrature axes,[14] leading to reduced noise in some components. Nonclassical phase-dependent correlations between elastically scattered and sideband photons have been predicted.[15] It was suggested that for resonant excitation, the squeezing averages to zero in a many atom system,[16] unless they form a regular array.[17] However, the many

Quantum Communication, Computing, and Measurement 2
Edited by Kumar et al., Kluwer Academic / Plenum Publishers, New York, 2000.

atom phase-dependent fluorescence spectra for scattering *in the forward direction* is predicted to be identical to that of a single atom and exhibits the same squeezing for both on- and off- resonance excitation.[13] This is confirmed by our experiments, where the complete squeezing spectrum measured for the 45° quadrature with a spectrum analyzer is found to be in excellent quantitative agreement with our predictions for the single atom squeezing spectrum.

EXPERIMENT

Our experiments are performed in a qualitatively different regime than has been explored previously. We strongly drive the $^1S_0 \rightarrow ^3P_1$ 556 nm transition of ^{174}Yb, for which the radiative lifetime is 875 ns, long compared to the interaction time. For near resonance excitation with nonzero detuning, we find that maximum squeezing occurs when the local oscillator has a phase near $\phi = \pm 45°$ relative to the driving field. In this case, as described below, squeezing arises from a cross-correlation between in- and out-of-phase fluctuations that peaks for $\sin 2\phi = \pm 1$.

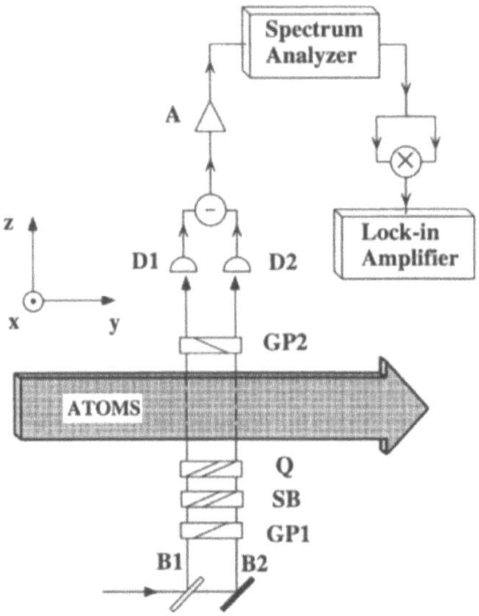

Figure 1. Measurement of phase-dependent resonance fluorescence spectra by subtraction of transmitted power signals from two identically prepared atomic samples.

In our experiments, the radiation field $\mathcal{E}$ of the atoms is mixed with a local oscillator (LO) field $|\mathcal{E}_{LO}|e^{i\phi}$ having a controllable fixed phase ϕ relative to the field that drives the atoms. The interference between the atom field and the LO causes fluctuations in the detected power ΔP that are determined by the quadrature operator

$$\hat{x}_\phi(t) = e^{-i\phi}\,\hat{\mathcal{E}}(t) + e^{i\phi}\,\hat{\mathcal{E}}^\dagger(t), \tag{1}$$

where $\Delta\hat{P}(t) \propto |\mathcal{E}_{LO}|\,\hat{x}_\phi(t)$. Fluctuations in the detected power are monitored with a diode detector. A spectrum analyzer is used to obtain the phase-dependent noise power spectra of the fluctuations in the selected quadrature. With the shot noise contribution removed, the atom contribution to the one-sided noise spectrum is given by

$$S_\phi(\omega) \propto \frac{2}{\pi}\int_0^\infty d\tau \cos\omega\tau\,\langle :\Delta\hat{P}(t)\,\Delta\hat{P}(t+\tau):\rangle, \tag{2}$$

where the double dots denote normal and time ordering, and $\langle \ldots \rangle$ denotes quantum and time t averages.

Phase-dependent resonance fluorescence spectra are measured by using a novel homodyne detection scheme, Fig. 1, that subtracts transmitted power signals from two identically prepared atomic samples.[8, 9, 10] In this method, a 2 cm wide supersonic Yb beam crosses two continuous laser beams, B1, B2 to prepare two samples containing different atoms. Quantum noise adds for the two samples since the quantum fluctuations are independent. Excess noise is suppressed for both the local oscillator and for the nonvanishing quadrature signals,[18] in contrast to homodyne detection with a beam splitter.[19] Large quantum noise signals are obtained by compensating the Doppler shifts of the diverging Yb beam with a linearly varying magnetic field. As in our previous experiments, this method permits excitation of a large sample of atoms with a laser field of well-defined detuning.[8, 9, 10]

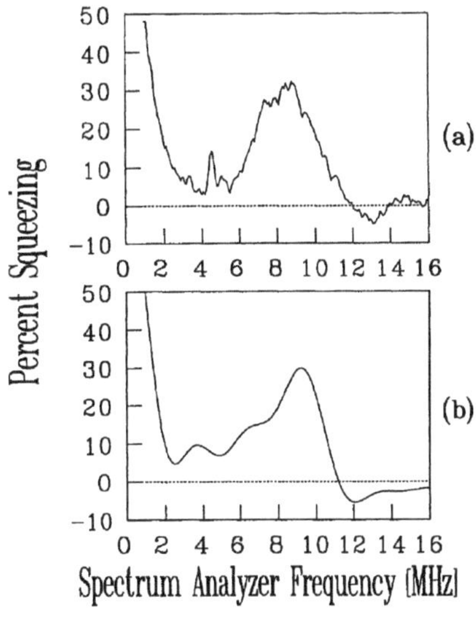

Figure 2. Phase-dependent fluorescence spectrum showing the % squeezing in the $\phi = -45°$ quadrature. (a) Data; (b) Thin sample theory, including only single atom correlations, showing quantitative agreement. Zero corresponds to the level of the shot noise spectral density that has been subtracted. Note that the small peak near 4 MHz in the data arises from the ion laser.

Local oscillator (LO) beams are created in each region that have a well defined phase ϕ relative to the exciting field. Each incident field passes through a polarizer GP1, a Babinet compensator SB, and a quarter wave plate, Q, so that it contains a σ_+ polarization component and a strong σ_- component. A strong magnetic field splits the $J = 1$ excited state magnetic sublevels so that the σ_- component is far away from resonance while the σ_+ component is near resonance and excites the atoms. The σ_+ polarized radiation field $\mathcal{E}$ of the atom and the incident field are mixed on an output Glan polarizer GP2. The net field transmitted through GP2 from each region takes the form $\mathcal{E}_{LO} + \mathcal{E}/\sqrt{2}$, where the effective LO field is $\mathcal{E}_{LO} \equiv |\mathcal{E}_{LO}|e^{i\phi} = \mathcal{E}_+/\sqrt{2} + \mathcal{E}_- e^{i\varphi}/\sqrt{2}$. The ratio of the magnitudes of the $\sigma_\pm$ incident field components $\mathcal{E}_\pm$ is determined by the orientation of GP1. Their relative phase φ is adjusted with SB to give the desired relative phase ϕ between the LO field and the exciting field $\mathcal{E}_+$. This method assures that ϕ is the same for all atoms and permits measurement of noise power spectra for arbitrary phase quadratures.

Fig. 2 (a) shows the measured % squeezing spectrum (linear scale[8, 9, 10]), obtained for $\phi = -45°$ with the laser tuned $-6\pm$ 1 MHz below resonance. The laser fields are focused to a 1/e field radius of 0.13 mm along the atomic beam, and 1 mm in the orthogonal direction, as measured with a 0.025 mm pixel diode array. The exciting field σ_+ power is 2 mW (one region) and the σ_- power is 7.1 mW. Setting $\varphi = 157°$ yields $\phi = -45°$ (actually $180° - 45°$) and an LO power of 1.1 mW. The noise power spectrum exhibits a peak at low frequency and a second peak near the generalized Rabi frequency. At a frequency just above the generalized Rabi frequency, the noise dips below the zero level, indicating squeezing.

THEORY

We have developed a new physical picture of phase-dependent noise, valid for long lived atoms. This picture elucidates how spontaneous emission from the excited state and collapse of the coherent dipole moment contribute to phase-dependent fluorescence spectra. Remarkably, this picture shows that the same physical process, collapse of the coherent dipole, causes the squeezing predicted previously for in-phase, on resonance excitation as well as that observed in our measurements for $\pm 45°$, with off resonance excitation.

For thin samples, the power autocorrelation function can be calculated using operator Bloch vector equations. Spontaneous emission can be neglected in the Bloch operator evolution when the radiative lifetime is long compared to the atom transit time to cross the field 1/e radius.[8, 9, 10] The normal and time ordered correlation function contains only *single atom* contributions[13] that appear as products of functions of t and $t' = t + \tau$. By combining these terms, the correlation function for $t' = t + \tau \geq t$ can be written in the simple form[20]

$$\langle : \Delta\hat{P}(t)\Delta\hat{P}(t') : \rangle = \frac{1}{2} P_e(t)\, M(t'-t) \,+\, \langle \Delta\hat{P}(t) \rangle \, \langle \Delta\hat{P}(t'-t) \rangle_g . \tag{3}$$

Eq. 3 has a simple physical interpretation. The first term is proportional to the excited state probability at time t, $P_e(t)$, and describes spontaneous emission from the excited state. The factor $M(t'-t)$ arises from precession of the quadrature components of the spontaneous dipole about the generalized Rabi vector of the driving field. Essentially, this factor describes the temporal 'shape' of spontaneous photons detected in the LO mode.

Most important for our discussion is the second term in Eq. 3 that describes fluctuations arising from collapse of the coherent dipole moment of an atom. The probability that a coherent dipole radiates into the LO at time t is proportional to the mean power $\langle \Delta\hat{P}(t) \rangle$. Detection of coherent radiation via a fluctuation in the LO mode collapses an atom to the ground state. Starting from the ground state at time t, the same atom produces a correlated power fluctuation with a probability $\propto \langle \Delta\hat{P}(t'-t) \rangle_g$ at time $t' \geq t$. Note that Eq. 3 is valid only when spontaneous emission between coherent emissions at t and t' is negligible, as is the case for long lived atoms.

Squeezing arises from a dispersive feature in the power spectrum that comes from the second (coherent dipole) term in Eq. 3. This term describes both the in-phase, on resonance squeezing predicted for small driving fields,[11, 12, 13] and the squeezing we observe in the $\pm 45°$ quadratures at strong driving fields.

To evaluate the correlation function, we average Eq. 3 over t and multiply by the LO field correlation function for a square pulse of length τ_o as seen in the atom frame. A particularly simple noise spectrum is obtained by evaluating Eq. 2 for large generalized Rabi frequencies Ω' as used in the experiments. To lowest order in $1/\Omega'\tau_o$ we obtain:

$$S_\phi(\omega) \propto \mathrm{L}(\delta\tau_o)\cos^2\phi + 2\,\mathrm{L}(\omega\tau_o)\sin^2\phi + \frac{\Delta}{\Omega'}\frac{\sin(\delta\tau_o) - \delta\tau_o}{(\delta\tau_o)^2}\sin 2\phi\,, \tag{4}$$

where $\mathrm{L}(2x) = (\sin^2 x)/x^2$ and $\delta = \omega - \Omega'$.

Despite its simplicity, Eq. 4 yields the *entire* squeezing spectrum when $\Omega'\tau_o >> 1$. The first term is the in-phase noise spectrum, centered at the generalized Rabi frequency Ω'. The second term is the out-of-phase noise spectrum, centered near spectrum analyzer frequency $\omega = 0$. The third (dispersive) term arises from the cross correlation between in- and out-of-phase fluctuations. For $\Delta\sin 2\phi > 0$, it attains a maximum negative value for a spectrum analyzer frequency $\omega = \Omega' + \pi/\tau_o$ that results in squeezing.

A quantitative prediction of the measured $\phi = -45°$ spectrum is obtained by averaging Eq. 4 over the gaussian intensity distribution of the incident fields in the direction perpendicular to the atomic beam axis. This yields spectra that agree within a few percent with the more complete results of Ref.'s[8, 9, 10] that perform the time average without approximation. In the same notation as those references, the experimental parameters are $\dot{N}\hbar\Omega = 0.25$ mW, as determined by the power change of the transmitted beam, the exciting beam power is 2 mW (one region), and the LO power is 1.1 mW. The transit time (square beam approximation) of the atoms is taken to be $\tau_o = 0.21\sqrt{\pi}\,\mu$s. The complete theoretical spectrum, obtained for $\Delta = -2\pi \times 5.6$ MHz, $\theta_M = \Omega_R\tau_o = 6.5\pi$, is shown in Fig. 2(b). While the detuning falls within the experimental uncertainty, the fitted value of θ_M is somewhat lower than the value 8.2π estimated using the measured parameters. This is expected, as our modelling for on-resonance excitation shows that fits obtained with square beam profiles[8, 9] for $\theta_M = 6.5\pi$ are similar to those obtained with gaussian beams[10] for $\theta_M = 8.2\pi$ as used in the experiment. However, the square beam predictions are in excellent agreement with the data in both shape and magnitude.

We have also observed larger squeezing at higher density. In this case, we find qualitative but not quantitative agreement with thin sample predictions.

CONCLUSIONS

In conclusion, we have observed squeezing in the phase-dependent fluorescence spectrum of coherently driven two-level atoms. The regime of large Rabi frequency and near resonant but nonzero detuning has been investigated. For samples which are not too thick, our experiments measure the squeezing spectrum under conditions where the relative phase between the local oscillator and the exciting field is well-defined and controllable. For strong driving fields and nonzero detuning, we observe maximum squeezing near the $\pm 45°$ quadratures, rather than the $0°$ quadrature that has received most attention. The measured noise power spectra are in excellent agreement with a quantum theory for thin samples of long-lived two-level atoms, that includes only single atom correlations. Theoretical noise spectra take a simple form which shows that both the in-phase and $45°$ quadrature squeezing arise from collapse of the coherent dipole moment. For very thick samples, detailed analysis will require a quantum theory of optical propagation in a dense medium of long-lived atoms.

The authors are indebted to Professor Howard Carmichael, Professor Paul Berman, Professor Leonard Mandel, and Dr. Frank Narducci for stimulating conversations regarding this work. This research has been supported by the National Science Foundation and the Army Research Office.

REFERENCES

1. B. R. Mollow, *Phys. Rev.* 188:1969 (1969); F. Wu, R. E. Grove, and S. Ezekiel, *Phys. Rev. Lett.* 35:1426 (1975).
2. C. Cohen-Tannoudji, *in:* "Frontiers in Laser Spectroscopy," vol. 1, pp.3-104, R. Balian, S. Haroche, and S. Liberman, eds., North Holland, New York (1977).
3. H. J. Kimble, M. Dagenais, and L. Mandel, *Phys. Rev. Lett.* 39:691 (1977); R. Short and L. Mandel, *Phys. Rev. Lett.* 51:384 (1983).
4. D. F. Walls and P. Zoller, *Phys. Rev. Lett.* 47:709 (1981).
5. L. Mandel, *Phys. Rev. Lett.* 49:136 (1982).
6. W. Vogel, *Phys. Rev. Lett.* 67:2450 (1991).
7. For a review, see S. Reynaud, A. Heidmann, E. Giacobino, and C. Fabre, *in:* "Progress in Optics XXX," E. Wolf ed., Elsevier, Amsterdam (1992).
8. H. Z. Zhao, Z. H. Lu, and J. E. Thomas, *Phys. Rev. Lett.* 79:613 (1997).
9. H. Z. Zhao, Z. H. Lu, A. M. Bacon, L. J. Wang, and J. E. Thomas, *Phys. Rev. A* 57:1427 (1998).
10. A. M. Bacon, H. Z. Zhao, L. J. Wang, and J. E. Thomas, *Phys. Rev. Lett.* 75:1296 (1995).
11. J. Collett, D. F. Walls, and P. Zoller, *Opt. Comm.* 52:145 (1984).
12. Z. Y. Ou, C. K. Hong, and L. Mandel, *J. Opt. Soc. Am. B* 4:1574 (1987).
13. A. Heidmann and S. Reynaud, *J. Phys.* 46:1937 (1985).
14. K. Wódkiewicz and J. H. Eberly, *J. Opt. Soc. Am. B* 2:458 (1985).
15. S. Bali, F. A. Narducci, and L. Mandel, *Phys. Rev. A* 47:5056 (1993).
16. R. Loudon, *Opt. Comm.* 49:24 (1984).
17. W. Vogel and D.-G. Welsch, *Phys. Rev. Lett.* 54:1802 (1985).
18. J. H. Shapiro, *IEEE J. Quant. Elec.* QE-21:237 (1985).
19. H. P. Yuen and V. W. S. Chan, *Opt. Lett.* 8:177 (1983).
20. This general result, valid when the spontaneous lifetime is long compared to the interaction time, will be discussed in detail elsewhere.

OBSERVATION OF NOISELESS IMAGE AMPLIFICATION BY AN OPTICAL PARAMETRIC AMPLIFIER

Sang-Kyung Choi, Michael Vasilyev, and Prem Kumar

Department of Electrical and Computer Engineering
Northwestern University, Evanston, Illinois 60208-3118
Telephone: (847) 491-4128; Fax: (847) 491-4455
E-mail: kumarp@nwu.edu

A spatially-broadband optical parametric amplifier is used for image amplification. In the phase-sensitive configuration, we observe a noise figure of $-0.1 \pm 0.3\,\mathrm{dB}$ for a gain of 2.4, which is 2 dB lower than the quantum limit for an ideal phase-insensitive amplifier. Image pre-amplification before loss is shown to improve the detected signal-to-noise ratio.

INTRODUCTION

A coherent light field E can be described by two quadratures, X and Y, as depicted on a phasor diagram in Fig. 1. With a conventional optical amplifier, such as a laser amplifier, gain occurs irrespective of changes in the phase of the input field. For this kind of a phase-insensitive linear amplifier (PIA) with intensity gain g, the output quadrature-noise variance is $2g-1$ times the input-noise variance for either quadrature. Therefore, at the output, the noise is amplified to a greater degree than the mean field, which in the case of a laser amplifier is because of the contribution of noise from spontaneous emission in the amplifier. With a phase-sensitive linear amplifier (PSA), however, the quadrature that is in phase with the pump, say X, gets amplified, and the other quadrature, Y, is attenuated. The deamplified Y quadrature exhibits squeezing, while the X quadrature is amplified without added noise. An example of a PSA is the degenerate optical parametric amplifier (OPA), which is based on the parametric down-conversion process in a $\chi^{(2)}$ medium. The noise performance of the PSA has been shown experimentally to be better than the quantum limit of the PIA, for both CW and pulsed signals.[1, 2]

The $\chi^{(2)}$ parametric process has also been utilized for imaging purposes. Interest was originally focused on upconverting infrared images to the visible domain.[3-5] More recently, work has been done on amplifying images,[6-8] and utilizing pulsed OPAs for time-gated image recovery.[8-11]

Optical imaging is also one of the important potential applications for nonclassical states of light. It has been theoretically shown that spatially-broadband squeezed light

Quantum Communication, Computing, and Measurement 2
Edited by Kumar et al., Kluwer Academic / Plenum Publishers, New York, 2000.

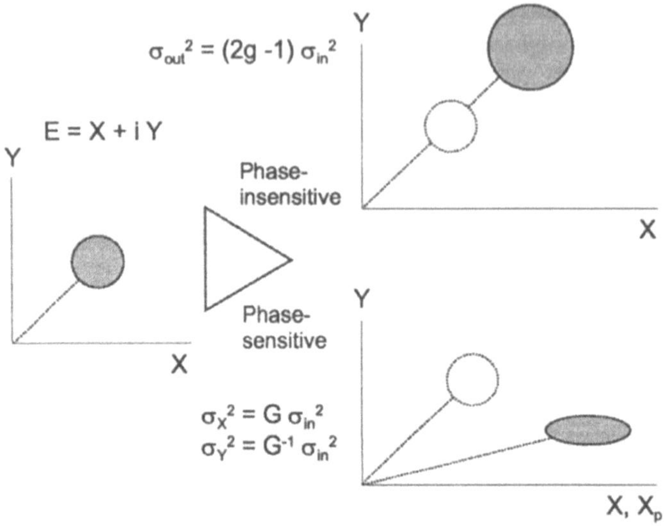

Figure 1. Schematic representation of a PIA and a PSA, for coherent-state input, shown on a phasor diagram with gains of g and G, respectively. The radii of the circles correspond to quadrature noises. Effectively, the PIA amplifies the input noise more than the mean field, irrespective of the signal phase. The PSA amplifies one quadrature, deamplifies the other, and affects the input quadrature noise and the mean field identically.

can be used to image faint objects with sub-shot noise sensitivity.[12] The broad spatial bandwidth of an OPA, which is determined by the requirement of phase-matching in the $\chi^{(2)}$ medium, makes it an ideal candidate for studying the feasibility of imaging with nonclassical states of light. Quantum-noise reduction of more than 6 dB has previously been demonstrated in twin beams generated by a traveling-wave OPA.[13] Recently, quantum-noise correlations between spatial frequencies of the parametrically-amplified signal and idler images have also been observed.[14] Quantum correlations between the parametric photons allow one to achieve amplification without degradation due to noise when the OPA is configured as a PSA. However, an OPA can also be configured as a PIA, in which case it imposes a quantum-noise penalty that approaches 3 dB at high gain.

Here we report, to the best of our knowledge, the first observation of noiseless amplification of an image by means of phase-sensitive parametric amplification.

RELEVANT SIGNAL-TO-NOISE ANALYSIS

The equations describing a traveling-wave OPA are[13]

$$\hat{b}_s = \mu\, \hat{a}_s + \nu\, \hat{a}_i^\dagger, \tag{1}$$

$$\hat{b}_i = \mu\, \hat{a}_i + \nu\, \hat{a}_s^\dagger, \tag{2}$$

where $\hat{a}_s$, $\hat{a}_i$ are the input and $\hat{b}_s$, $\hat{b}_i$ are the output annihilation operators for the signal, idler modes, respectively. Equations (1) and (2) describe not only the typical collinear OPA, but they can be shown to apply also to the non-collinear case wherein the signal and pump waves are incident from different directions. In this case, however, the explicit forms of μ and ν are more complicated compared to the collinear case.[15]

Depending on whether the OPA is configured as a PIA or a PSA, the above equations take the following forms for the signal-bearing mode of interest:

$$\hat{b} = \sqrt{g}\,\hat{a} + \sqrt{g-1}\,\hat{v}^{\dagger}, \qquad \text{(PIA)} \tag{3}$$

$$\hat{b} = \sqrt{g}\,\hat{a} + \sqrt{g-1}\,\hat{a}^{\dagger}, \qquad \text{(PSA)} \tag{4}$$

where $\hat{v}$ is the annihilation operator for a vacuum-state mode, and g is the parametric gain in the phase-insensitive configuration. The mean of the output photon number for an input coherent state $|\alpha\rangle$ is

$$\langle \hat{b}^{\dagger}\hat{b} \rangle = g\,|\alpha|^2 + g - 1, \qquad \text{(PIA)} \tag{5}$$

$$\langle \hat{b}^{\dagger}\hat{b} \rangle = g\,|\alpha|^2 + (g-1)(|\alpha|^2+1) + 2\sqrt{g(g-1)}\,|\alpha|^2\cos(2\theta_\alpha). \qquad \text{(PSA)} \tag{6}$$

Therefore, for a sufficiently large input signal ($|\alpha|^2 \gg 1$), g can be interpreted as the PIA gain of the OPA. In the PSA case, the mean output photon number depends on the phase θ_α of the coherent state $|\alpha\rangle$, and attains the maximum gain (once again, for $|\alpha|^2 \gg 1$) $G = 2g - 1 + 2\sqrt{g(g-1)}$ when $\theta_\alpha = 0$. In addition, the variance of the output photon number is

$$\langle \Delta(\hat{b}^{\dagger}\hat{b})^2 \rangle = g(2g-1)\,|\alpha|^2 + g(g-1), \qquad \text{(PIA)} \tag{7}$$

$$\langle \Delta(\hat{b}^{\dagger}\hat{b})^2 \rangle = G^2\,|\alpha|^2 + 2g(g-1). \qquad \text{(PSA with } \theta_\alpha = 0) \tag{8}$$

If detection is non-ideal, the measured output is described by $\langle \hat{d}^{\dagger}\hat{d} \rangle$ and $\langle \Delta(\hat{d}^{\dagger}\hat{d})^2 \rangle$. Here, $\hat{d}$ is related to $\hat{b}$ and $\hat{v}_\eta$ through

$$\hat{d} = \sqrt{\eta}\,\hat{b} + \sqrt{1-\eta}\,\hat{v}_\eta, \tag{9}$$

where η is the overall detection efficiency, including losses from the amplifier to the photodetector, and $\hat{v}_\eta$ is another vacuum-state mode.

We define the signal-to-noise ratio (SNR) as the ratio of the square of the mean photon number to the photon-number variance. Hence, for a coherent-state input signal field that is sufficiently bright (to render the second terms in Eqs. (7) and (8) negligible), and which is detected with non-ideal efficiency, the SNRs at the input and output of the OPA are related as follows:

$$\mathrm{SNR_{out}} = \frac{\eta\, g}{2\eta\, g + 1 - 2\eta}\,\mathrm{SNR_{in}}, \qquad \text{(PIA)} \tag{10}$$

$$\mathrm{SNR_{out}} = \frac{\eta\, G}{\eta\, G + 1 - \eta}\,\mathrm{SNR_{in}}. \qquad \text{(PSA)} \tag{11}$$

The ratio of $\mathrm{SNR_{in}}$ to $\mathrm{SNR_{out}}$ characterizes the total noise figure (NF) of the OPA, including the effect of optical losses. The intrinsic NF of the OPA is the ratio of $\mathrm{SNR_{in}}$ to $\mathrm{SNR_{out}}$ for $\eta = 1$. Therefore, from Eq. (11), the intrinsic NF of the PSA is 1 (0 dB), implying that the PSA preserves the SNR in the limit of a bright, coherent-state input.

The SNR analysis of this section also applies to spatially multi-mode input signals. This is because the OPA is a linear device with respect to the signal modes. Since Eqs. (1) and (2) also hold in the non-collinear case,[15] the conclusions given by Eqs. (10) and (11) are valid for any spatial frequency of the input signal. Moreover, if the OPA gain is constant over the spatial bandwidth of the signal (as is true for a spatially-broadband OPA), our analysis can be applied to images composed of many spatial features.

EXPERIMENTAL RESULTS

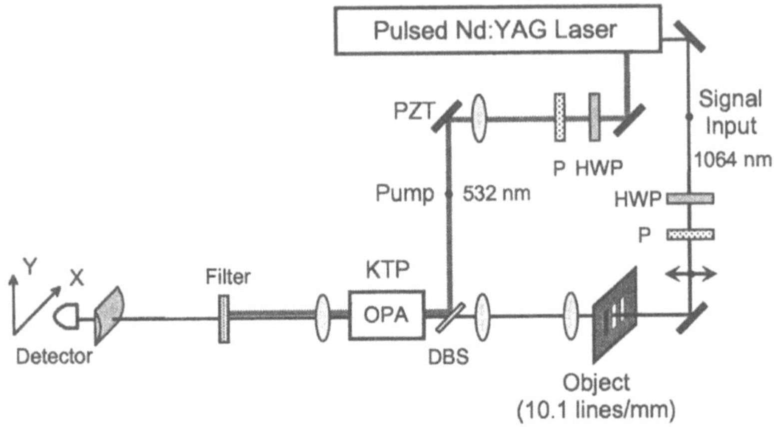

Figure 2. Experimental setup.

Our experimental layout is shown in Fig. 2. The OPA consists of a 1.57-mm-long KTP crystal that is pumped by a p-polarized frequency-doubled beam (532 nm) from a Q-switched and mode-locked Nd:YAG laser. A double-slit object (99 μm slit separation and 50 μm slit width) is illuminated by the signal beam from the fundamental of the Nd:YAG laser (1064 nm) and imaged into the center of the crystal by means of a $\times 1$ telescope that consists of two 10-cm focal-length lenses. The signal field is polarized at 45° with respect to the optic axis of the KTP crystal, which renders the OPA phase-sensitive. The pump beam is weakly focused into the KTP crystal by a lens of sufficiently large focal length (40 cm), so that we can approximate it as an on-axis plane wave (zero spatial frequency). A dichroic beamsplitter directs the pump beam conincidentally with the on-axis component (zero spatial frequency) of the signal beam into the crystal, where the pump amplifies the input image within the spatial bandwidth determined by phase-matching conditions. The amplified image that is created in the crystal is subsequently magnified 25 times by the lens behind the crystal. This is to ensure that the amplified image can be spatially resolved by the InGaAs photodetector of 300 μm diameter. The green pump beam is blocked after the crystal by an RG filter that passes most of the IR signal. Since the spatial features of interest lie along the horizontal direction, we use a cylindrical lens to compress the beam in the vertical direction. This enables the noise of the optical power falling on the photodetector to be above the electronic-noise floor of the photodetection system. The photodetector scans the image along the horizontal direction and measures the mean intensity as well as the 27-MHz noise power at each point, both with and without amplification. While amplifying, the phase of the pump beam relative to the signal is locked to maximize the PSA gain. The uniformity of the parametric gain of our OPA over the photodetector's bandwidth enables estimation of the NF by relating the DC intensity gain to the noise-power gain at 27 MHz.

The experimentally observed spatial profiles of the signal and noise are shown in Fig. 3. The amplified image shows very little spread compared to the bare image, indicating that the spatial spectrum of the image was well within the phase-matching bandwidth of the OPA. For NF measurements, we consider the intensity and noise

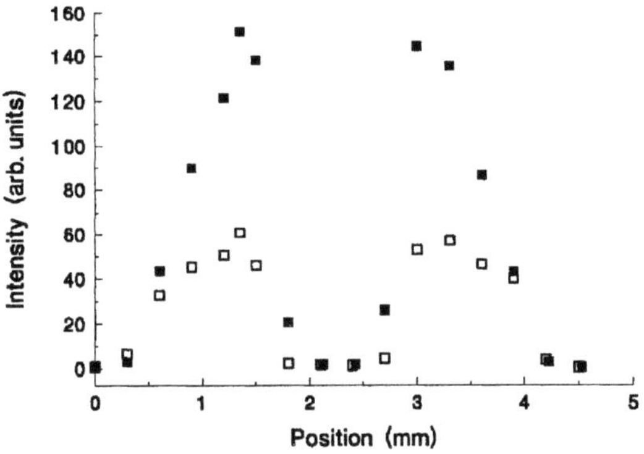

(a)

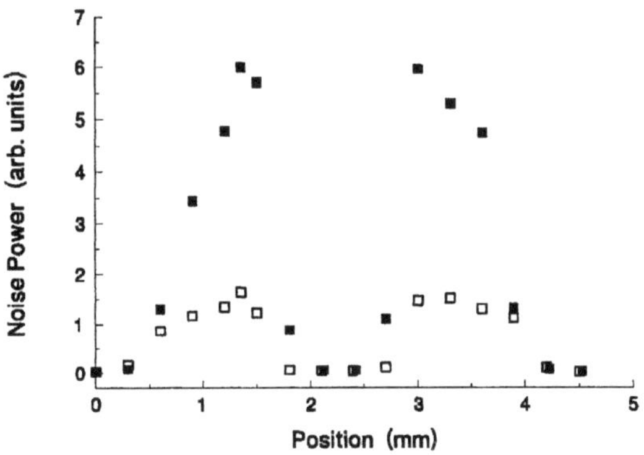

(b)

Figure 3. Spatial profiles of (a) the intensity and (b) the noise power scanned by the photodetector along the horizontal direction. Empty squares—without amplification, solid squares—with amplification. Both vertical scales are linear.

gains at the peaks of the spatial profiles only, in order to avoid the effect of spatial averaging due to the finite size of the photodetector. In the phase-sensitive configuration of the OPA, spatially-broadband gains of 2.4 were obtained at the peaks, with noise power being amplified at a slower rate than the square of the mean intensity, leading to effective improvement in the detected SNR, i.e., the detected amplified signal shows a higher value of SNR than the detected bare signal. This results from the transmission between the OPA and the photodetector in our setup being limited to $\eta = 0.34$. Hence, the OPA functioned essentially as a noiseless pre-amplifier that overcame the effect of optical losses. At the peaks of the spatial profile, the SNR of the detected amplified signal was measured to be 2.2 dB higher than that of the detected bare signal. Therefore, the SNR of the detected amplified signal was smaller than that of the input signal by the amount $2.2 + 10 \log \eta = -2.5$ dB. We use this value and the measured value of η to calculate the NF of the amplifier using Eq. (9), which gives $\mathrm{NF}_\eta = \mathrm{NF}_{\eta=1} + (1-\eta)/\eta G$. We find that the SNR from the input to the output of the phase-sensitive OPA remains unchanged within the experimental margin of error. Thus, at the peaks of the spatial profile, the SNR of the amplified image equals the SNR of the unattenuated image. Moreover, it shows improvement over the SNR degradation of $(2 - 1/g)^{-1}$ for a PIA.

Table 1. NF for gain = 2.4, $\eta = 0.34$.

	$NF_{amp+loss}$	NF_{amp}
PIA Theory	3.8 dB	2.0 dB
PSA Theory	2.6 dB	0 dB
Experiment	2.5±0.2 dB	-0.1±0.3 dB

The experimental result is compared with the theoretical calculations in Table 1, in terms of the NF of the OPA. Theoretical and experimental values of the noise figure of the OPA are shown, for a gain of 2.4, and the detection efficiency of 0.34. The first column shows noise-figure values for the amplifier with subsequent optical losses. The second column shows the intrinsic noise figure of the OPA.

CONCLUSION

We have demonstrated that a spatially-broadband OPA in its phase-sensitive configuration pre-amplifies an image before loss, to enhance the detected SNR of the image. When the detection efficiency is taken into account, the phase-sensitive OPA preserves the SNR. Future work will concentrate on improving the detection efficiency of the optical system, as well as exploring new media with higher nonlinear gains and spatial bandwidths.

Acknowledgments

This research was supported in part by the U.S. Office of Naval Research.

REFERENCES

1. Z. Y. Ou, S. F. Pereira, and H. J. Kimble, Phys. Rev. Lett. **70**, 3239–3242 (1993).
2. J. A. Levenson, I. Abram, T. Rivera, P. Fayolle, J. C. Garreau, and P. Grangier, Phys. Rev. Lett. **70**, 267 (1993); J. A. Levenson, I. Abram, Th. Rivera, and Ph. Grangier, J. Opt. Soc. Am. B **10**, 2233–2238 (1993); K. Bencheikh, O. Lopez, I. Abram, and J. A. Levenson, Appl. Phys. Lett. **66**, 399 (1995).
3. J. E. Midwinter, IEEE J. Quantum Electron. **4**, 716–720 (1968).
4. A. H. Firester, J. Appl. Phys. **40**, 4842–4849 (1969).
5. R. A. Andrews, IEEE J. Quantum Electron. **6**, 68–80 (1970).
6. Y. Fainman, E. Klancnik, and S. H. Lee, Opt. Eng. **25**, 228–234 (1986).
7. P. A. Laferriere, C. J. Wetterer, L. P. Schelonka, and M. A. Kramer, J. Appl. Phys. **65**, 3347–3350 (1989).
8. E. Lantz and F. Devaux, Quantum Semiclass. Opt. **9**, 279–286 (1997).
9. G. W. Faris and M. Banks, Opt. Lett. **19**, 1813–1815 (1994).
10. J. Watson, P. Georges, T. Lépine, B. Alonzi, and A. Brun, Opt. Lett. **20**, 231–233 (1995).
11. S. M. Cameron, D. E. Bliss, and M. W. Kimmel, Proc. SPIE **2679**, 195–203 (1996).
12. M. I. Kolobov and P. Kumar, Opt. Lett. **18**, 849–851 (1993).
13. O. Aytür and P. Kumar, Phys. Rev. Lett. **65**, 1551–1554 (1990).
14. M. L. Marable, S.-K. Choi, and P. Kumar, Opt. Express **2**, 84–92 (1998).
15. A. Gavrielides, P. Peterson, and D. Cardimona, J. Appl. Phys. **62**, 2640–2645 (1987).

TRAVELING-WAVE NOISELESS AMPLIFICATION OF OPTICAL IMAGES

Mikhail I. Kolobov,[1,2] Ivan V. Sokolov,[2] and Luigi A. Lugiato[3]

[1]Fachbereich Physik, Universität-GH Essen,
D-45117 Essen, Deutschland
[2]Physics Institute, St. Petersburg University,
198904 Petrodvorets, St. Petersburg, Russia
[3]INFM, Dipartimento di Fisica, Università degli Studi di Milano,
via Celoria 16, 20133 Milano, Italy

We consider a traveling-wave version of a ring-cavity noiseless amplifier of optical images, proposed recently.[1] This modification, on the one hand, is more realistic for the practical implementations, on the other hand, has several advantages over the cavity-based scheme, such as greater spatial bandwidth. We investigate in detail the quantum fluctuations of the amplifier and formulate criteria for its noiseless performance. We also estimate the spatial resolving power of noiseless amplification.

At present, there is an increased interest in applications of squeezed and other non-classical light to the optical phenomena in spatial domain. One of such applications is noiseless amplification of optical images. The original idea belongs to Kolobov and Lugiato who have demonstrated recently that noiseless image amplification can be realized by a ring-cavity optical parametric oscillator below threshold, therefore, operating as a phase-sensitive amplifier.[1]

In this paper we consider a traveling-wave version of the parametric image amplifier without external cavity. Such a geometry is more realistic for possible practical realizations and has several advantages over its cavity-based counterpart. The most important ones are greater frequency and spatial-frequency amplification bandwidths. Greater frequency bandwidth allows one to amplify pulsed images, while greater spatial-frequency bandwidth provides better spatial resolution. We pay special attention to phase matching inside the nonlinear crystal serving as an active medium for a traveling-wave optical parametric amplifier (OPA). We evaluate conditions for optimum phase matching which provides the best signal-to-noise ratio and spatial resolving power.

Quantum fluctuations in parametric image amplification were studied experimentally in Ref. [2], and in recent experiments by the group of Kumar.[3, 4, 5] In particular, in Ref. [5] noiseless image amplification was for the first time demonstrated in practice. The results of our theoretical analysis can be useful for these and future experiments in domain of noiseless amplification of optical images.

Quantum Communication, Computing, and Measurement 2
Edited by Kumar et al., Kluwer Academic / Plenum Publishers, New York, 2000.

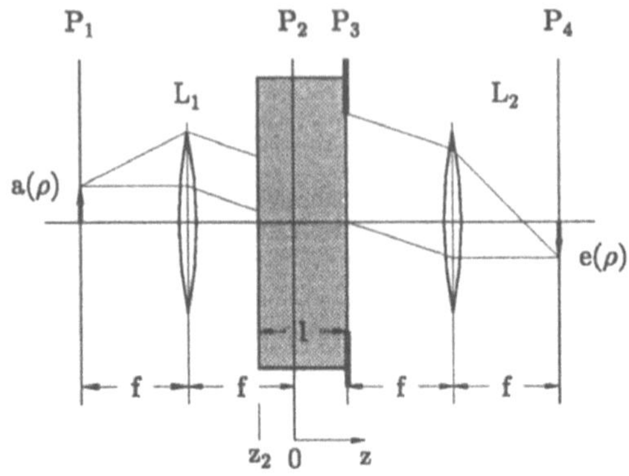

Figure 1. Schematic of a traveling-wave optical image amplifier.

We consider the optical scheme, shown in Fig. 1. An input optical image is imprinted into a faint spatial modulation of the wavefront of a coherent light wave in the object plane P_1. The lens L_1 performs a spatial Fourier transform of the input image into the Fourier plane P_2 laying at the distance z_2 from the input plane of the parametric crystal. This distance will be determined below from the condition of the optimum phase matching. The nonlinear crystal of length l plays the role of a traveling-wave OPA. Parametric interaction takes place between the signal wave of frequency ω, carrying the input image, and the plane monochromatic pump wave of double frequency, $\omega_p = 2\omega$, illuminating the input plane of the crystal. As in the optical scheme of Ref. 1, we place a pupil at the output plane P_3 of nonlinear crystal. This pupil cuts off the noise photons of spontaneous parametric down-conversion, which do not contribute to the signal part of the amplified wavefront in the image plane P_4. The back Fourier transform of the amplified object wave is performed by the lens L_2.

The amplified image is detected by a dense array of small photodetectors, *pixels*, of a finite area S_d during the observation time T_d. The quantity of interest for us is the number of photoelectrons $N_I(\vec{\rho}, t)$ registered by the pixel centered at the point $\vec{\rho}$ in the image plane in the time interval from $t - T_d/2$ to $t + T_d/2$:

$$N_I(\vec{\rho}, t) = \int_{S_d} d\vec{\rho}\,' \int_{T_d} dt'\, i(\vec{\rho}\,', t'). \tag{1}$$

Here $i(\vec{\rho}, t)$ is the surface photocurrent density, i.e., the number of photoelectrons per cm^2 per second. Using the quantum photodetection theory,[6] we evaluate the mean number of photoelectrons $\langle N_I(\vec{\rho}, t)\rangle$, its variance, $\sigma_I^2(\vec{\rho}, t)$, and the signal-to-noise ratio R_I of the amplified image, $R_I = \langle N_I(\vec{\rho}, t)\rangle^2/\sigma_I^2(\vec{\rho}, t)$.

The equivalent quantities $\langle N_O(\vec{\rho}, t)\rangle$, $\sigma_O^2(\vec{\rho}, t)$, and R_O, measured in the object plane, describe the signal and noise of the input image. Our goal is to compare R_I and R_O and to show that under certain conditions they can be equal.

To study quantitatively the noise performance of the amplifier we introduce the noise figure F' as

$$F' = \frac{R_O(\eta = 1)}{R_I}. \tag{2}$$

Since a linear amplifier cannot improve the signal-to-noise ratio in the input image, the noise figure F' is always not smaller than unity. We refer to the case $F' = 1$ as to noiseless amplification. Notice, that in Eq. (2) the input signal-to-noise ratio corresponds to the ideal photodetection in the input plane. As explained in Ref. 1, this correction is necessary to obtain the noise figure which characterizes the noise *added by the amplifier* but not the noise due to imperfections in the pre-amplification apparatus (like non-ideal photodetection array in the object plane).

We describe the evolution of the light field in the scheme in terms of space-time dependent photon creation and annihilation operators. With the help of the photodetection theory, we can express $\langle N_I(\vec{\rho},t)\rangle$ and $\langle \Delta N_I^2(\vec{\rho},t)\rangle$ through the mean irradiance $\langle e^\dagger(\vec{\rho},t)e(\vec{\rho},t)\rangle$ and the intensity correlation function $\langle e^\dagger(\vec{\rho},t)e^\dagger(\vec{\rho}\,',t')e(\vec{\rho}\,',t')e(\vec{\rho},t)\rangle$ of the field in the image plane. To evaluate these observable quantities in terms of the input signal and the gain of our amplifier we need the relation between the field operators $e(\vec{\rho},t)$ in the image plane and $a(\vec{\rho},t)$ in the object plane.

Let us start from the situation when $z_2 = 0$. In this case it can be shown (see Ref. 7) that for the pupil of infinite area the relation between the field amplitudes $e(\vec{\rho},\Omega)$ and $a(\vec{\rho},\Omega)$ is given by

$$e(\vec{\rho},\Omega) = u(\vec{\rho},\Omega)a(\vec{\rho},\Omega) + v(\vec{\rho},\Omega)a^\dagger(-\vec{\rho},-\Omega), \tag{3}$$

with the following coefficients $u(\vec{\rho},\Omega)$ and $v(\vec{\rho},\Omega)$:

$$\begin{aligned} u(\vec{\rho},\Omega) &= \exp\Big\{i\Big[\Omega n/c - (2k-k_p)/2\Big]l\Big\}\left[\cosh\Gamma(\vec{\rho},\Omega) + \frac{i\delta(\vec{\rho},\Omega)}{2\Gamma(\vec{\rho},\Omega)}\sinh\Gamma(\vec{\rho},\Omega)\right], \\ v(\vec{\rho},\Omega) &= \exp\Big\{i\Big[\Omega n/c - (2k-k_p)/2\Big]l\Big\}\frac{g}{\Gamma(\vec{\rho},\Omega)}\sinh\Gamma(\vec{\rho},\Omega). \end{aligned} \tag{4}$$

Here c/n is the velocity of light in the crystal, k and k_p are the wave numbers of the signal and the pump wave. The parameter $\Gamma(\vec{\rho},\Omega)$ is defined as

$$\Gamma(\vec{\rho},\Omega) = \sqrt{g^2 - \delta^2(\vec{\rho},\Omega)/4}, \tag{5}$$

where g is the dimensionless coupling strength of nonlinear interaction, taken as real for simplicity. It is proportional to the nonlinear susceptibility, the length of the crystal, and the amplitude of the pump field. The dimensionless mismatch function $\delta(\vec{\rho},0)$ for $\Omega = 0$ is given by

$$\delta(\vec{\rho},0) = \delta_0 - \rho^2/\rho_0^2, \quad \delta_0 = (2k-k_p)l, \tag{6}$$

where ρ_0 is defined as

$$\rho_0 = f\sqrt{\frac{1}{kl}}. \tag{7}$$

An important case of degenerate phase matching in the crystal corresponds to $\delta_0 = 0$. In this case the maximum parametric amplification takes place around the optical axis of the scheme, $\vec{\rho} = 0$. The spatial scale ρ_0 determines the linear dimension of the effectively amplified area in the object plane. In the case of non-degenerate phase matching with positive δ_0, the dimensionless mismatch (6) is equal to zero for $\rho > 0$. This corresponds to parametric amplification of a region in the object plane, which has the shape of a ring.

We shall assume the temporal evolution of the image to be slow and put $\Omega = 0$ in Eq. (3) for the calculation of the amplified signal. As in Ref. 1, we assume that the input field is in coherent state with complex amplitude $s(\vec{\rho})$, where

$$s(\vec{\rho}) = s(-\vec{\rho}). \tag{8}$$

To gain more physical insight into the squeezing transformation (3), let us consider what happens under this transformation with a classical complex field amplitude α

taken for the moment as constant over the whole object plane. Denoting the output classical field amplitude ϵ, we can write

$$\epsilon = u\alpha + v\alpha^* = e^{i\psi}\left\{|u|e^{-i\phi}\alpha + |v|e^{i\phi}\alpha^*\right\}, \tag{9}$$

where we have introduced the angles ψ and ϕ as,

$$\psi = \frac{1}{2}(\arg u + \arg v), \quad \phi = \frac{1}{2}(\arg v - \arg u). \tag{10}$$

Let us define the quadrature components of the field on the input and the output of the crystal in their *eigen* coordinate systems determined by the angles ϕ and ψ as follows:

$$\alpha = e^{i\phi}(\alpha_1 + i\alpha_2), \tag{11}$$

$$\epsilon = e^{i\psi}(\epsilon_1 + i\epsilon_2). \tag{12}$$

It is easy to see from Eq. (3) that these eigen quadrature components are related by the following transformation:

$$\epsilon_1 = e^r\alpha_1, \qquad \epsilon_2 = e^{-r}\alpha_2, \tag{13}$$

where we have introduced the *squeezing parameter* r:

$$e^{\pm r} = |u| \pm |v|. \tag{14}$$

In our case the coefficients u and v depend on $\vec{\rho}$ [see Eqs. (3), (4)]. Therefore, the phase $\phi(\vec{\rho})$ which determines the amplified quadrature component, is a function of the transverse coordinate $\vec{\rho}$. As a consequence, we obtain that the input classical signal $s(\vec{\rho})$ with constant phase for all $\vec{\rho}$ does not match the condition of maximum amplification everywhere.

There are two possibilities to gain the maximum parametric amplification for all points of the signal wavefront in the object plane. The first one is to introduce the $\vec{\rho}$-dependent phase shift into the incident object wave. In the paraxial approximation this phase shift must be quadratic in ρ, i.e., the wavefront must have certain curvature.

Another possibility is to adjust the optical scheme such as to make the phase $\phi(\vec{\rho})$ independent of $\vec{\rho}$. It follows from Eqs. (4) and (10) that

$$\begin{aligned}\phi(\vec{\rho}) &= -\frac{1}{2}\arg\left\{\cosh\Gamma(\vec{\rho},0) + \frac{i\delta}{2\Gamma(\vec{\rho},0)}\sinh\Gamma(\vec{\rho},0)\right\} \\ &= -\frac{1}{2}\tan^{-1}\left(\frac{\delta}{2\Gamma(\vec{\rho},0)}\frac{\sinh\Gamma(\vec{\rho},0)}{\cosh\Gamma(\vec{\rho},0)}\right).\end{aligned} \tag{15}$$

In the case of interest, when the coupling constant is large, $g \gg 1$, and the dimensionless mismatch is relatively small, $\delta \ll g$, we can approximate $\phi(\vec{\rho})$ as

$$\phi(\vec{\rho}) \approx -\frac{\delta(\vec{\rho},0)}{4g} = \frac{\rho^2/\rho_0^2 - \delta_0}{4g}. \tag{16}$$

Let us shift the crystal together with the lens L_2 and the detection plane P_4 at a certain distance so that the input plane of the crystal is located at $z_2 \neq 0$. As a result, we find that the coefficients u, v in Eq. (3) become

$$\begin{aligned}\tilde{u}(\vec{\rho},\Omega) &= u(\vec{\rho},\Omega)\exp[i\varphi_2(z_2,\vec{\rho})], \\ \tilde{v}(\vec{\rho},\Omega) &= v(\vec{\rho},\Omega)\exp[-i\varphi_2(z_2,\vec{\rho})],\end{aligned} \tag{17}$$

with

$$\varphi_2(z_2,\vec{\rho}) = -\rho^2 \frac{k}{2f^2} z_2. \tag{18}$$

The phase ϕ which determines the amplified and attenuated quadrature components of the signal, now changes to $\tilde{\phi}$ according to

$$\tilde{\phi}(\vec{\rho}) = \phi(\vec{\rho}) - \varphi_2(z_2,\vec{\rho}), \tag{19}$$

Since both $\phi(\vec{\rho})$ and $\varphi_2(z_2,\vec{\rho})$ depend quadratically on $\vec{\rho}$, by appropriate choice of the distance z_2 one can eliminate the ρ dependence in $\tilde{\phi}(\vec{\rho})$. Taking into account Eqs. (16), and (18), we obtain that the optimum choice of z_2 is given by

$$z_2^{(\mathrm{opt})} = -\frac{2f^2}{k\rho_0^2 4g} = -\frac{l}{2g}. \tag{20}$$

If z_2 is chosen according to Eq. (20), the amplified and attenuated quadratures of the signal field in the object plane are defined by Eq. (11) with

$$\tilde{\phi} = -\frac{\delta_0}{4g}. \tag{21}$$

In the case of degenerate phase matching, $\delta_0 = 0$, the input signal with a real classical amplitude $s(\vec{\rho})$ is uniformly amplified for all $\vec{\rho}$, at least in the paraxial approximation.

Let us consider non-degenerate phase matching with $\delta_0 > 0$. Assume that the input image is complex,

$$s(\vec{\rho}) = e^{i\varphi(\vec{\rho})}|s(\vec{\rho})|, \tag{22}$$

with a constant phase $\varphi(\vec{\rho}) = \tilde{\phi}$ given by Eq. (21). As follows from Eq. (11), such a signal has optimally matched quadrature components. Since the phase $\tilde{\phi}$ in Eq. (21) does not depend on transverse coordinate $\vec{\rho}$, this matching can be achieved for the whole image.

In Ref. 7 we calculate analytically the noise figure $F'(\vec{\rho})$. Figure 2 illustrates graphically the results for degenerate phase matching. The solid curve refers to the noise figure with phase correction, while the dotted line is obtained without it. One can see that under phase correction the spatial region of noiseless amplification becomes much larger. Large peaks of excess noise in the peripheral regions in the dotted curve almost disappear in the solid one.

Acknowledgments

This research was carried out in the framework of the activities of the Network "Quantum Structures" (QSTRUCT) of the TMR Program of the European Union, and partially supported by the INTAS Project 93-1914-EXT and by the Russian Foundation for Basic Research Project 98-02-18129.

REFERENCES

1. M. I. Kolobov and L. A. Lugiato, "Noiseless amplification of optical images," *Phys. Rev. A* 52:4930 (1995).
2. D. Guthals and D. Sox, "Quantum limited optical parametric image amplification," in *Proceedings of International Conference on Lasers '89*, eds. D. G. Harris and T. M. Shay (STS, Mclean, Va., 1990), pp. 808-815.

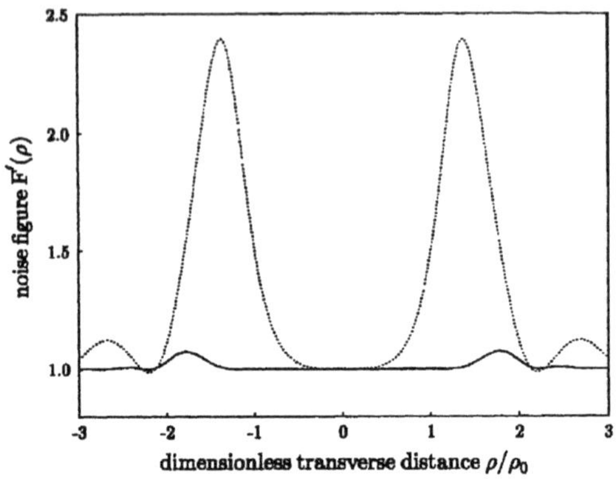

Figure 2. Noise figure $F'(\rho)$ as a function of dimensionless transverse distance ρ/ρ_0 for degenerate phase matching. The maximum gain $G = 100$. The dotted line is obtained without phase correction, the solid line–with correction.

3. S.-K. Choi, M. L. Marable, and P. Kumar, "Observation of quantum noise correlations in parametric image amplification," in *Quantum Electronics and Laser Science*, Vol. 12, 1997 OSA Technical Digest Series, (Optical Society of America, Washington, D.C., 1997), pp. 94-95.
4. M. L. Marable, S.-K. Choi, and P. Kumar, "Measurement of quantum-noise correlations in parametric image amplification," *Optics Express*, Vol. 2, No. 3 (1998).
5. S.-K. Choi, M. Vasilyev, and P. Kumar, "Noiseless image amplification by a phase-sensitive parametric amplifier," in *Conference on Lasers and Electro-Optics*, Vol. 6, 1998 OSA Technical Digest Series (Optical Society of America, Washington DC, 1998), pp. 470-471.
6. D. F. Smirnov and A. S. Troshin, "New phenomena in quantum optics: photon antibunching, sub-Poisson photon statistics, and squeezed states," *Sov. Phys. Usp.* 30:851 (1987).
7. I. V. Sokolov, M. I. Kolobov, and L. A. Lugiato, "Quantum fluctuations in traveling-wave amplification of optical images," (in preparation).

QUANTUM NOISE REDUCTION IN SINGLY RESONANT OPTICAL DEVICES

C. Cabrillo, J. L. Roldán, P. García-Fernández

Instituto de Estructura de la Materia, CSIC
Serrano 123, 28006 Madrid, Spain

Quantum noise in a model of a singly resonant frequency doubler including phase mismatch and a driving in the harmonic mode is analyzed. The squeezing spectra calculated inside the linear approximation are given. For squeezing in the fundamental mode, the optimum working point corresponds to turning points of dispersive bistability induced by cascading of the second order nonlinear response. In the case of the harmonic mode phase mismatch as well as a finite driving in the harmonic allows for surpass the 1/9 limit of the standard SHG scheme. The squeezing improvement when the driving is included is substantial for parameter values corresponding to reported experiments

INTRODUCTION

Second Harmonic Generation has nowadays quite a long tradition as a mean of squeezed light generation.[1] The preferred experimental setup has been the doubly resonant configuration as, at least in principle, permits arbitrarily large squeezing. However, such scheme has been hampered by the technical difficulties arising from keeping the resonance in both modes simultaneously. Thus, in spite of the development of very ingenious stabilizing procedures,[2] for the moment it has been only possible to maintain the double resonance for a few seconds. Certainly, this kind of experimental delicacy can hardly surprise when dealing with the generation of non-classical states of light. In view of such difficulties, some experimental efforts have been recently redirected to singly resonant configurations.[3, 4, 5] Although the maximum noise suppression is then limited to a 90%,[3] the efforts resulted in very stable intense squeezed light sources with degrees of squeezing even surpassing those reported in the doubly resonant counterparts.[5]

At the same time, singly resonant Optical Parametric Oscillation, the most successful method to squeeze the vacuum[6] has been generalized to singly resonant Parametric Amplification, i.e., a laser driving in the harmonic mode has been added, again showing an extraordinary stability at quite high noise suppression values in the fundamental mode.[7] Although the squeezed beams are in this case much less intense than the SHG counterpart, this setup permits also an easy control of the phase of the squeezed quadrature, something which allowed a spectacular demonstration, using quantum tomography, of the different kinds of squeezed states.[8]

Quantum Communication, Computing, and Measurement 2
Edited by Kumar et al., Kluwer Academic / Plenum Publishers, New York, 2000.

In view of this experimental success it seems timely to extend the quantum mechanical model beyond the pure phase matched cases. More specifically, we address here quantum noise reduction in an extension of the conventional singly resonant SHG to include also a coherent input in the harmonic mode as well as phase mismatch between the interacting waves.

QUANTUM MECHANICAL MODEL

The system we want to address consists in a second order nonlinear medium coupling two modes of frequency ω (fundamental) and 2ω (harmonic) respectively and placed inside a ring cavity resonant only with the fundamental mode. We will also assume just one input-output mirror of finite reflectivity. The quantum mechanical evolution equation associated with such a system reads

$$\frac{da}{dt} = -\left[\gamma + i\delta + (\mu + i\Gamma)a^\dagger a\right] a + 2\sqrt{\mu}\, a^\dagger b_{in} + \sqrt{2\gamma_c}\, a_{in} + \sqrt{2\gamma_s}\, w_{in}\,, \tag{1}$$

which are supplemented by the boundary conditions relating the outside modes with the intracavity one

$$a_{out} = \sqrt{2\gamma_c}\, a - a_{in}\,, \tag{2}$$
$$b_{out} = \sqrt{\mu}\, a^2 - b_{in}\,, \tag{3}$$

where a denote the annihilation operators for the fundamental mode while $b_{in,out}$ correspond to that for the harmonic ingoing and outgoing modes. The fluctuations induced by absorption and scattering in the system are represented by the white noise w_{in}. The decay constant γ equals $\gamma_c + \gamma_s$. The input fields are assumed to be in a coherent state. The μ term comes from the two-photon absorption induced by the harmonic generation while the nonlinear dispersion Γ comes from the cascading effect due to finite mismatch. For a plane wave geometry

$$\mu = \nu \left(\mathrm{sinc}\frac{\Delta k L_m}{2}\right)^2 \tag{4}$$
$$\Gamma = \frac{2\nu}{\Delta k L_m}\left[\mathrm{sinc}\frac{\Delta k L_m}{2}\cos\frac{\Delta k L_m}{2} - 1\right]. \tag{5}$$

where ν is the second-order coupling constant, L_m the length of the nonlinear medium and Δk the phase mismatch between the interacting waves.

Defining fluctuation operators as

$$a = \alpha + \delta a\,, \tag{6}$$
$$a_{in,out} = \alpha_{in,out} + \delta a_{in,out}\,, \tag{7}$$
$$b_{in,out} = \beta_{in,out} + \delta b_{in,out}\,, \tag{8}$$

being $\alpha_{in,out}, \beta_{in.out}$ the mean values of the corresponding input and output modes and α a stable fixed point of the classical counterpart of Eq. (1), the corresponding equations for the fluctuations are obtained as

$$\frac{d\,\delta a}{dt} = -\left[\gamma + i\delta + 2(\mu + i\Gamma)|\alpha|^2\right]\delta a + \left[2\sqrt{\mu}\beta_{in} - (\mu + i\Gamma)\alpha^2\right]\delta a^\dagger + 2\sqrt{\mu}\,\alpha^*\delta b_{in} + \sqrt{2\gamma_c}\,\delta a_{in} + \sqrt{2\gamma_s}\, w_{in}\,, \tag{9}$$

and

$$\delta a_{out} = \sqrt{2\gamma_c}\,\delta a - \delta a_{in}\,, \tag{10}$$

$$\delta b_{out} = 2\alpha\sqrt{\mu}\,\delta a - \delta b_{in}\,, \tag{11}$$

The stability of the fixed points is governed by the real part of the eigenvalues of the drift matrix associated to the linearized evolution equation. Very simple algebra yields

$$\lambda_{\pm} = -(\gamma + 2\mu\, n) \pm \sqrt{|(\mu + i\Gamma)\alpha^2 - 2\sqrt{\mu}\,\beta_{in}|^2 - (\delta + 2\Gamma\, n)^2}\,. \tag{12}$$

Provided that the real part of both eigenvalues are negative the fixed point will be stable.

The squeezing spectrum optimized in phase as well as the associated "stretching" spectrum for an outgoing mode are given by

$$S_{-,+}(\omega) = 1 + \langle :\delta a^{\dagger}_{out}(\omega)\delta a_{out}(-\omega):\rangle \mp |\langle :\delta a_{out}(\omega)\delta a_{out}(-\omega):\rangle|\,, \tag{13}$$

corresponding to a quadrature of the mode with a phase

$$\theta_m = \frac{\nu(\omega) + \pi}{2}\,, \tag{14}$$

where $\nu(\omega)$ is the phase of $\langle \delta a_{out}(\omega)\delta a_{out}(-\omega)\rangle$.

Let us define a non-linear and a total decay rate as $\gamma_{nl} \equiv 2\mu n$ and $\gamma_t \equiv \gamma + \gamma_{nl}$ respectively. Scaling Eq. (13) by γ_t and after some algebra the spectra for both modes can be written as

$$S^a_{-,+}(\tilde\omega) = 1 + \frac{\gamma_c}{\gamma_t} : S^{ref}_{-,+}(\tilde\omega) :\,, \tag{15}$$

$$S^b_{-,+}(\tilde\omega) = 1 + \frac{\gamma_{nl}}{\gamma_t} : S^{ref}_{-,+}(\tilde\omega) :\,, \tag{16}$$

where $\tilde\omega = \omega/\gamma_t$ and

$$: S^{ref}_{-,+}(\tilde\omega) : = 4|\tilde B|\frac{2|\tilde B| \mp \sqrt{(1 + \tilde\omega^2 + |\tilde B|^2 - \Delta^2)^2 + 4\Delta^2}}{(1 - \tilde\omega^2 - |\tilde B|^2 + \Delta^2)^2 + 4\tilde\omega^2}\,. \tag{17}$$

with $\Delta = (\delta + 2n\Gamma)/\gamma_t$, $\tilde B = (2\sqrt{\mu}\beta_{in}e^{-i2\theta} - (\mu + i\Gamma)n)/\gamma_t$ and θ being the phase of α. $: S^{ref}_{-}(\tilde\omega) :$ reach perfect squeezing (-1) at the instability (the eigenvalues (12) equaling zero) corresponding to $1 + \Delta^2 = |\tilde B|^2$. The spectra (17) are also Heisenberg limited in the sense that the $: S^{ref}_{+}(\tilde\omega) :$ is the minimum compatible with the Heisenberg principle. It is easy to show that whatever the values of $\tilde B$, $: S^{ref}_{-}(\tilde\omega) :$ is minimum for $\tilde\omega = 0$ and $\Delta = 0$. This last condition implies also a squeezing phase (14) independent of ω.

SQUEEZING AT THE FUNDAMENTAL MODE

Expression (15) makes explicit the role of extra input channels as far as quantum noise suppression is concerned. In view of it, it is clear that the best performance corresponds to $\gamma_{nl} = 0$, that is, either $n = 0$ or $\mu = 0$, as then the losses ratio minimizes to $\eta \equiv \gamma_c/(\gamma_c + \gamma_s)$ (the escape efficiency of the cavity). The case $n = 0$ corresponds to the very well known case of squeezed vacuum generation. For $\mu = 0$ and finite n the system is formally equivalent to a resonant optical Kerr effect system whose quantum noise behavior has been amply studied previously.[9] The instabilities are at $\delta = -2n\Gamma \pm \sqrt{n^2\Gamma^2 - \gamma^2}$, the well known turning points of optical dispersive bistability[9] but with the non-linear dispersion induced by cascading. Indeed, such cascading induced bistability has been experimentally demonstrated in Ref. 10. An optimum path to reach the instability corresponds to increasing n while maintaining $\delta = 2n\Gamma$ and $\omega = 0$. Figure 1 displays the evolution of both the maximum squeezing and the maximum excess noise following such a path in the parameter space for three values of the escape efficiency, namely, 0.9, 0.99 and the ideal 1. The noise is expressed in dB's with respect to the vacuum. A Heisenberg limited excess noise appears in such a case as a specular image of the squeezing. The instability is signaled by the divergence in the excess noise. Above it, the curves shown are unphysical as they correspond to unstable fixed points. The case $\eta = 0.99$ in Fig. 1 shows an excellent behavior with an almost Heisenberg limited excess noise till near the instability. However, it should be taken rather as an ideal illustrative case. The ratio $\gamma_c/(\gamma_c + \gamma_s)$ is difficult to maximize in a resonant mode because, by its own resonant nature, γ_c must be rather low. Thus, in Ref. 10 it was only of 0.52, while in Ref. 3 it was 0.36. Even in Ref. 2, a doubly resonant system specifically designed to squeeze the fundamental mode, the escape efficiency was around 0.9, limiting the maximum squeezing achievable to 90% (in practice, a 52% of noise reduction was reached).

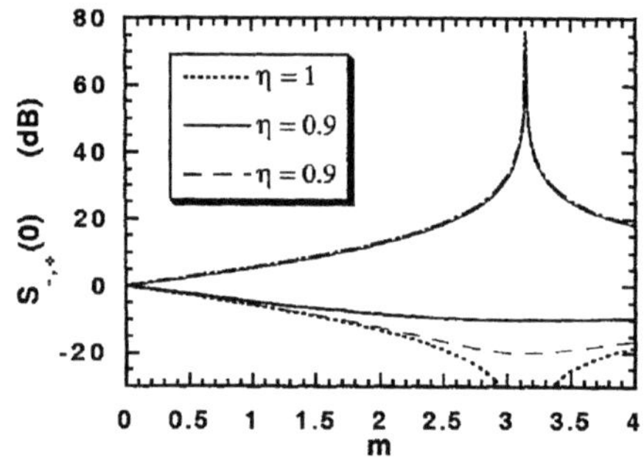

Figure 1. The squeezing and stretching spectra at zero frequency for the fundamental mode following and optimized path as a function of $m = \nu n/\gamma$ for three different values of the escape efficiency.

SQUEEZING AT THE HARMONIC MODE

Now, the situation is the complete opposite: the performance is favored by a finite μ in order to have a non-zero γ_{nl} and a large n to approach the ratio γ_{nl}/γ_t to one. Centering first in the pure SHG case ($\beta_{in} = 0$, the case of the experiments in Ref. 3 and Ref. 5), the instability points are now given by $\delta_\pm = -2n\Gamma \pm \sqrt{n^2(\Gamma^2 - 3\mu^2) - 4\gamma\mu n - \gamma^2}$. Both kinds of non-linearities are in this case necessary as the factor $\Gamma^2 - 3\mu^2$ needs to be positive to allow $\delta_\pm$ to be real. The maximum squeezing available is easily obtained by setting : $S^{ref}_{-,+}(\tilde{\omega})$: to -1 in Eq. (16), that is, $S_M = \gamma/\gamma_t$ dependent on the phase mismatch. Although its lowest value is at null mismatch, the detuning can only be real at a finite mismatch moving the optimum working point (at least for large enough n) far from the matching condition. Now the squeezing is not limited, approaching zero at large n, but the improvement with respect to the standard phase SHG (limited to

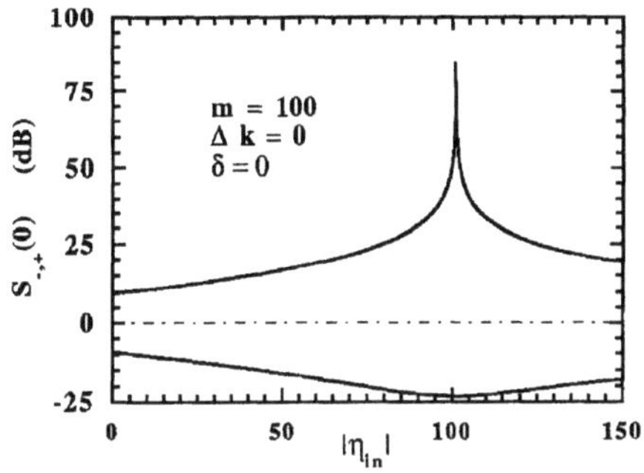

Figure 2. The minimum quantum noise and the associated excess noise for the harmonic mode as a function of $|\eta_{in}| = 2\sqrt{\nu}|\beta_{in}|/\gamma$.

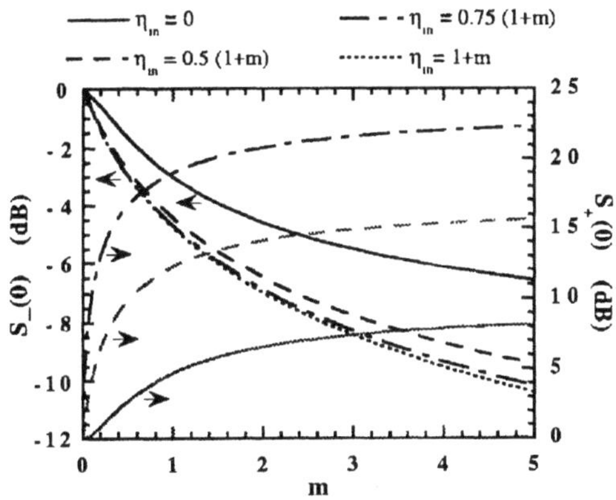

Figure 3. The behavior of minimum quantum noise and the associated excess noise with m for the harmonic mode at different values of the input harmonic amplitude. $|\eta_{in}| = 1 + m$ corresponds to the instability

1/9) is rather low till quite large values of n. Although a finite mismatch allows to reach the ideal limit, the overall optimum working point corresponding to the ideal limit at $\Delta k = 0$ is out of reach. Given the expression for $\tilde{B}$ a driving in the harmonic mode would allow to reach S_M at $\Delta k = 0$. Taking $\Delta = 0$, phase matching ($\Delta k = 0$) implies $\delta = 0$, $\Gamma = 0$ and $\mu = \nu$. We further optimize choosing the phase of β_{in} which maximizes $|\tilde{B}|$. The behavior of the noise with $|\beta_{in}|$ is illustrated in Fig. 2. The excess noise does not depart from the Heisenberg limit until near the instability. The curves shown above the instability are unphysical as they corresponds to calculations around an unstable fixed point. Again this case should be taken as an ideal illustrative case. The physical scale corresponding to the reported experiments with conventional nonlinear crystal can be estimated by comparing the results obtained for $\beta_{in} = \Delta k = 0$. This gives for the experiment of Tsuchida[5] a value of $m = \nu n/\gamma$ of just 2.5. Fig. 3 shows the squeezing and the excess noise behavior for till $m = 5$. A substantial improvement can be achieved by the inclusion of the driving in the harmonic mode even quite far from the instability. The parameter m is related with the energy load inside the cavity. An excess of such should be the cause of degrading the squeezing in the real experiments. It can be expected, therefore, that if the nonlinear material could cope with it with $\beta_{in} = 0$ it will also in this case. This improvement of the squeezing was found previously in Ref. 11 but it was not reckoned as interesting enough because the comparison was made fixing the input power instead of m. In spite of the improvement shown in Fig. 3, it is clear that much larger ratios of ν/γ are needed to exploit the potentialities of the proposed system. In this respect, artificial nonlinear materials such as asymmetric quantum wells look very promising specially schemes like the one proposed in Ref. 12 as they produce resonant enhancement of the nonlinearity while suppressing the absorption via quantum interference.

Acknowledgments

Work supported in part by grants No. TIC95-0563-C05-03, No. PB96-00819, CICYT, Spain, and Comunidad de Madrid 06T/039/96.

REFERENCES

1. S. F. Pereira et al., Phys. Rev. A38, 4931 (1988); A. Sizmann et al, Opt. Commum 80, 138 (1990); P. Kurz et al., Europhys. Lett. 24, 449 (1993); R. Paschotta et al., Phys. Rev. Lett. 72, 3807 (1994); T. C. Ralph et al., Opt. Lett. 20, 1316 (1995); H. Tsuchida, Opt. Lett. 20, 2240 (1995);
2. P. Kurz et al., Europhys. Lett. 24, 449 (1993).
3. R. Paschotta et al., Phys. Rev. Lett. 72, 3807 (1994).
4. T. C. Ralph et al., Opt. Lett. 20, 1316 (1995).
5. H. Tsuchida, Opt. Lett. 20, 2240 (1995).
6. E. S. Polzik, J. Carri, H. J. Kimble, Appl. Phys. B 55, 279 (1992).
7. K. Schneider *et al*, Opt. Lett. 21, 1396 (1996).
8. G. Breitenbach, S. Schiller and J. Mlynek, Nature 387, 471 (1997).
9. S. Reynaud et al, Phys. Rev. A40, 1440 (1989).
10. A. G. White, J. Mlynek and S. Schiller, Europhys. Lett. 35, 425 (1996).
11. S. Schiller *et al*, Appl. Phys. B 60, S77 (1995).
12. H.Schmidt and A. Imamoglu, Opt. Commun.131, 333 (1996).

INFLUENCE OF DISSIPATION EFFECTS ON POLARIZATION - SQUEEZED LIGHT FORMATION AT FREQUENCY DOUBLING

V.V. Volokhovsky,[1] A.S.Chirkin[1]

[1]M.V.Lomonosov Moscow State University
Department of Physics, Moscow, Russia, 119899

The influence of losses and thermal noise on the polarization - squeezed light formation in type II second harmonic generation process has been analysed. New analysis method based on delta-correlation noise properties permitting to calculate correlated and noncorrelated parts of system operators with reservoir ones was employed. Mean values and variances of the Stokes parameter of the light have been calculated.

INTRODUCTION

Recently a set of methods allowing to generate so called polarization - squeezed (PS) light have been proposed. Such a light is characterized by suppressed fluctuation of Stokes parameters by which the radiation polarization state is represented. The possibility to obtain PS light was shown for the first time during an analysis of two orthogonally polarized modes propagation through an anizotropic cubic - nonlinear medium[1]. Lately the PS formation in gyrotropic[2], periodically nonhomogenious media with third order nonlinearity[3] as well as in the process of second harmonic generation (SHG) by mixing[4] have been studied theoretically.

So far only idealized models have been used to describe the polarization fluctuation suppression process. To estimate the real fluctuation suppression in experiment taking into account the real properties of light and media is necessary. The main aim of this work is to analyze the influence of the distributed quantum noise and losses of the medium upon the PS light formation at frequency doubling. Below we study the Stokes parameters statistical characteristics. To determine the noise contribution the linearization procedure was used in combination with method of expansion of system operators into correlated and noncorrelated with heat bath operators contributions. Such approach makes it possible to calculate all necessary parameters to determine polarization squeezing for arbitrary number of bath quanta without employing a time-consuming master equation solution technique.

Quantum Communication, Computing, and Measurement 2
Edited by Kumar et al., Kluwer Academic / Plenum Publishers, New York, 2000.

QUANTUM POLARIZATION STATE DESCRIPTION. STATISTICAL PROPERTIES OF STOKES OPERATORS

We used the radiation polarization state description by means of Stokes operators which in the basis of two orthogonally polarized modes with same frequency and annihilation (creation) operators $\hat{A}_j$ ($\hat{A}_j^+$) have the form

$$S_0 = \hat{A}_1^+\hat{A}_1 + \hat{A}_2^+\hat{A}_2, \qquad S_2 = \hat{A}_1^+\hat{A}_2 + \hat{A}_2^+\hat{A}_1,$$
$$S_1 = \hat{A}_1^+\hat{A}_1 - \hat{A}_2^+\hat{A}_2, \qquad S_3 = i(\hat{A}_2^+\hat{A}_1 - \hat{A}_1^+\hat{A}_2). \tag{1}$$

The operators satisfy commutation relation of SU_2 algebra

$$[S_m, S_n] = 2i\epsilon_{mnl}S_l, \qquad (m,n,l = 1,2,3)$$

which leads to the following uncertainty relation, where V_m is the parameter S_m fluctuation variance, ϵ_{mnl} is antisymmetric Levy - Civita tensor.

$$V_mV_n \geq |\epsilon_{mnl}\langle S_l\rangle|^2.$$

PS light is a light with Stokes parameter fluctuation level less than one for coherent state. Such light is formed due to additional correlation between initially noncorrelated modes. This correlation occurs because of nonlinear interaction of polarization modes.

FLUCTUATION SUPPRESSION MODEL WITH DAMPING EFFECTS

We consider PS light formation in SHG by mixing. In this Process the additional correlation between the modes of fundamental radiation, necessary for PS light formation, occurs due to the fact that into second harmonic only correlated photons are transformed. Thus, the photons of fundamental radiation become predominantly anticorrelated as modes undergo SHG. Let B_j (B_j^+) ($j = 1,2$) are the ordinary and extraordinary mode photon annihilation (creation) operators, B_3 (B_3^+) are those for second harmonic. The interaction hamiltonian for this case has the form

$$H_{int} = \hbar\beta(B_1^+B_2^+B_3 + B_3^+B_1B_2), \tag{2}$$

where β is nonlinear coupling coefficient and $\hbar$ is Planck's constant

To take into account the quantum noise we use the standard model for simulate the thermal noise properties of the medium. We consider it as an infinite boson system with operators b_j (b_j^+) and assume that polarization modes are absorbed independently from each other. That leads to

$$H_{bath} = i\hbar\sum_l\int\eta(\Omega,l)[b^+(\Omega,l)A_l - h..]d\Omega \tag{3}$$

This Hamiltonian shows that the coherent mode can lose energy by creation the reservoir quanta and vice versa.

We consider the case $\langle B_1(t=0)\rangle = \langle B_2(t=0)\rangle$ that corresponds to the linear polarization wave incident on the quadrature - nonlinear crystal at a 45° with respect to the crystal polarization plane. As it was shown[4] this conditions cause maximum polarization squeezing. To simplify future analysis we introduce new basis $\{\hat{A}_j\}$ with accordance of following formula:[5].

$$\hat{A}_1 = -\frac{i}{\sqrt{2}}(\hat{B}_1 + \hat{B}_2), \qquad \hat{A}_2 = -\frac{i}{\sqrt{2}}(\hat{B}_1 - \hat{B}_2), \qquad \hat{A}_3 = i\hat{B}_3$$

In this basis we have an intensive mode $\hat{A}_1$ and vacuum one $\hat{A}_2$.

In accordance with H_{bath} each thermal reservoir is characterized by damping constant γ_j and number of quanta $\bar{N}_j$. We took these parameters to be the same for all of the reservoirs. The following set of Heisenberg - Langevin equations corresponds to the introduced above interaction Hamiltonians

$$\begin{aligned} \frac{d\hat{A}_{1,2}}{dt} &= -\gamma \hat{A}_{1,2} \mp \beta \hat{A}_{1,2}^{+} \hat{A}_3 - \sqrt{2\gamma} b_{1,2}(t), \\ \frac{d\hat{A}_3}{dt} &= -\gamma \hat{A}_3 + \frac{1}{2}\beta(\hat{A}_1^2 - \hat{A}_1^2) - \sqrt{2\gamma} b_3(t). \end{aligned} \tag{4}$$

The fluctuation operators fulfil the relations

$$\begin{aligned} &[b_j(t'), b_k^{+}(t)] = \delta(t'-t)\delta_{jk}, \quad \langle b_j \rangle = \langle b_j^2 \rangle = 0, \\ &\langle b_j^{+}(t') b_k(t) \rangle = \bar{N}_j \delta_{jk} \delta(t'-t), \quad \langle b_j(t') b_k^{+}(t) \rangle = (\bar{N}_j + 1)\delta_{jk}\delta(t'-t), \end{aligned} \tag{5}$$

Since the equations (4) do not have an analytical solution, to determine the statistical characteristics of Stokes parameters we used a perturbation method and procedure of correlators calculation of field and bath operators.[6] We decompose the field operators $\hat{A}_j$ into correlated and noncorrelated parts with bath operators: $\hat{A}_j = \hat{A}_{jnc}(t - \Delta t) + \hat{A}_{jcor}(t)$. Here the time interval Δt should be less than time of considerable change of field operator and more than noise correlation time. Then for $\hat{A}_{jcor}$ we have

$$\hat{A}_{1,2cor} = \lim_{\Delta t \to 0} \int_{t-\Delta t}^{t} (-\gamma \hat{A}_{1,2} \mp \beta \hat{A}_{1,2}^{+} \hat{A}_3 - \sqrt{2\gamma} b_{1,2}(t'))dt' = -\sqrt{2\gamma} \tilde{b}_{1,2}(t)$$

In accordance with perturbation method all of the operators $\hat{A}_j$ were represented as sum of classical number A_j and quantum operator a_j: $\hat{A}_j = A_j + a_j$, $(j = 1,2,3)$. Having used the well-known solution for classical numbers[7]

$$A_1(\tau) = A_0 \cosh^{-1}(\tau), \qquad A_2(\tau) = 0, \qquad A_3(\tau) = \frac{A_0}{\sqrt{2}} \tanh(\tau) \tag{6}$$

where $\tau = \frac{1}{\sqrt{2}} \beta A_0 z = z/L_{nl}$, $z = vt$ where v is phase velocity in the medium,one can obtain the following set of linearized equations

$$\begin{aligned} \frac{da_1}{d\xi} &= -\beta A_1 a_3 - \beta A_3 a_1^{+} - \sqrt{2\gamma}(1-\gamma\xi)^{-2} b_1, \\ \frac{da_2}{d\xi} &= \beta A_3 a_2^{+} - \sqrt{2\gamma}(1-\gamma\xi)^{-2} b_2, \\ \frac{da_1}{d\xi} &= \beta A_1 a_1 - \sqrt{2\gamma}(1-\gamma\xi)^{-2} b_3, \end{aligned} \tag{7}$$

where $\xi = (1 - \exp(-\gamma t))/\gamma$. Then the necessary for future analysis correlators are

$$\begin{aligned} &\langle b_l^{+} a_m \rangle = \langle a_l^{+} b_m \rangle = -\frac{1}{2}\sqrt{2\gamma} e^{2\gamma t} \bar{N}_l \delta_{lm}, \\ &\langle a_l b_m^{+} \rangle = \langle b_l a_m^{+} \rangle = -\frac{1}{2}\sqrt{2\gamma} e^{2\gamma t} (\bar{N}_l + 1)\delta_{lm}, \\ &\langle b_l a_m \rangle = \langle b_l^{+} a_m^{+} \rangle = 0. \end{aligned} \tag{8}$$

In assumption of Gaussian fluctuation statistics the correlators of any higher degree can be expressed as products of just obtained expressions.

SUPPRESSION OF POLARIZATION FLUCTUATIONS

Analytical solution is possible only for parameters S_2 and S_3 fluctuation variances. The rest values could be determined by numerical solution of system (7) with taking into account of the expressions (6) and (8). The analytical expression for normalized variances $\sigma_j^2 = V_j/\langle S_0\rangle$ has the form

$$\sigma_2^2 = \cosh^2(t)(1-\mu t)^2 + 4(1-\mu t)^2 \int_0^t \mu(1-\mu t')^{-2}\cosh^{-2} t' dt',$$
$$\sigma_3^2 = \cosh^{-2}(t)(1-\mu t)^2 + 4(1-\mu t)^2 \int_0^t \mu(1-\mu t')^{-2}\cosh^{2} t' dt', \qquad (9)$$

where $\mu = L_{nl}/L_{dis}$.

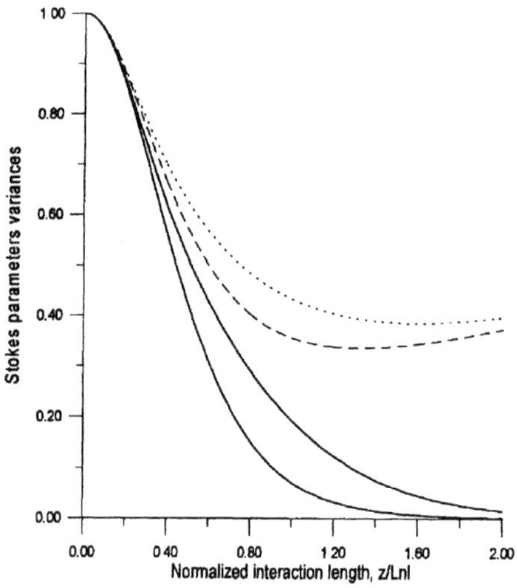

Figure 1. Variances of parameters $S_{0,1}$ and S_3 as a function of normalized interaction length z/L_{nl}. $\sigma_{0,1}^2$ – dashed line, σ_3^2 – dotted line, solid lines correspond to the lossless medium.

As it can be seen from the Fig.1, $\sigma_{0,1,3}^2 \leq 1$ where $\sigma_j^2 = 1$ corresponds to the coherent state. Thus, the PS light is generated. Expressions (9) show that fluctuation variances contain coherent and noncoherent part which decrease and increase while modes propagation through the medium. Increasing of parameters fluctuation is determined by additional noise occurred due to coupling to the reservoirs.

On the Fig. 2 correlation coefficients of quadrature components of fundamental radiation and second harmonic as a function of interaction length are shown. It can be seen that initially noncorrelated modes become correlated due to nonlinear interaction. The correlation is destroyed by thermal noise.

CONCLUSION

1. In the process of second harmonic generation by mixing two initially noncorrelated coherent modes of fundamental radiation become correlated due to nonlinear interaction. This additional correlation causes the PS light generation.

2. The fluctuations of the Stokes parameters contain a coherent part and a noise one which decrease and increase while light propagation through a nonlinear medium, respectively.

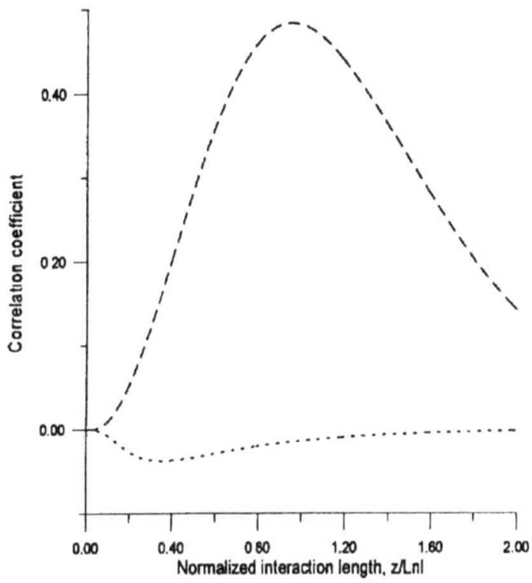

Figure 2. Correlation coefficients of fundamental wave and second harmonic as a function of normalized interaction length z/L_{nl}. $\langle x_1 x_3 \rangle$ – dashed line, $\langle y_1 y_3 \rangle$ – dotted line.

3. The maximum fluctuation suppression in the Stokes parameter is reached on the interaction length. The further interaction length growth leads to the noise level increasing and it exceeds the coherent part fluctuations.

REFERENCES

1. A.S. Chirkin, A.A. Orlov, D.Yu. Parashuk, *Quantum Electron.* 20:999 (1993).
2. A.S. Chirkin, V.V. Volokhovsky, *J. Rus. Laser Research* 16:526 (1995).
3. A.P.Alodjants, S.M. Arakelian, A.S. Chirkin, *Appl. Phys. B* 66:53 (1998).
4. V.N. Beskrovniy, A.S. Chirkin, *Quantum Semiclas. Opt* 10:263 (1998).
5. R-D Li, P.Kumar, *J. Opt. Soc. Am. B* 12:2310 (1995).
6. V.V. Volokhovsky, A.S. Chirkin,*JNOPM* 6:455 (1997).
7. Y.R. Shen, "The Principles of Nonlinear Optics", Wiley, New York, 1984.

ACKNOWLEDGMENTS

This work was supported by Russian Foundation for Basic Research (grant number 96-02-16714a). The authors thank the organizing committee of Forth International Conference on Quantum Communication, Measurement and Computing (QCM'98) for financial support to present the results of the paper at the conference.

DEGENERATE OPTICAL PARAMETRIC AMPLIFICATION WITH FOCUSED GAUSSIAN BEAMS

Kahraman G. Köprülü and Orhan Aytür

Bilkent University
Department of Electrical and Electronics Engineering
TR-06533 Bilkent, Ankara, TURKEY

We present the results of an accurate numerical model for degenerate optical parametric amplification with focused Gaussian beams. This model shows that the optimum phase difference between the pump and the signal beams can be significantly different from that predicted by the plane-wave theory, and that squeezing in excess of 10 dB can be obtained for relatively long interaction lengths and low pump power levels.

INTRODUCTION

The generation of quantum states of light, such as squeezed states, often relies on degenerate optical parametric amplification. The gain of a degenerate optical parametric amplifier (DOPA) depends on the phase difference between the signal and the pump fields; one quadrature of the signal field is amplified while the orthogonal quadrature is deamplified. Quantum fluctuations of the deamplified signal quadrature get squeezed, and this quadrature exhibits a noise level below the quantum limit.[1, 2] Accurate modeling of practical DOPA's is crucial for the optimal design of experiments aimed at generating large levels of squeezing. Various models have brought out the importance of matched local oscillators[2, 3] and diffraction effects due to transverse gain variations.[4, 5] However, the results of these models are valid for a limited range of interaction length values and underestimate the squeezing levels that can be achieved in practice.

In this paper, we present the results of an accurate numerical DOPA model that is valid for any interaction length. We solve the paraxial DOPA equations using finite difference techniques for a pump field that is a Gaussian beam with a Gaussian input signal field. Our results show that squeezing in excess of 10 dB can be obtained for relatively long interaction lengths that are beyond the scope of previous models. We also show that the phase difference yielding maximum squeezing is different from that used by previous models as predicted by the plane-wave theory.

Quantum Communication, Computing, and Measurement 2
Edited by Kumar et al., Kluwer Academic / Plenum Publishers, New York, 2000.

DEGENERATE OPTICAL PARAMETRIC AMPLIFIER

The coupled mode equations that govern the evolution of the pump and signal fields in a phase-matched DOPA can be written as[5]

$$\frac{\partial A_s(\rho,z)}{\partial z} - \frac{1}{2jk_s}\nabla_\perp{}^2 A_s(\rho,z) = -j\frac{\omega d_e}{nc}A_p(\rho,z)A_s^*(\rho,z) \tag{1}$$

$$\frac{\partial A_p(\rho,z)}{\partial z} - \frac{1}{2jk_p}\nabla_\perp{}^2 A_p(\rho,z) = -j\frac{\omega d_e}{nc}A_s^2(\rho,z), \tag{2}$$

where A_s and A_p are the positive frequency components of the signal and the pump electric fields, d_e is the effective nonlinear coefficient of the material, $\nabla_\perp{}^2 = \partial^2/\partial\rho^2 + (1/\rho)\,\partial/\partial\rho$ is the transverse Laplacian in cylindrical coordinates, n is the refractive index, λ is the signal wavelength, and $k = 2\pi n/\lambda$ is the wavenumber. Here we assume that the slowly varying envelope approximation is valid, that there is no Poynting vector walk-off, and the fields have cylindrical symmetry. The transformations

$$z = \xi z_0 \tag{3}$$

$$\rho = rW_0 \tag{4}$$

$$A_{s,p} = u_{s,p}\sqrt{4P_{s,p}/\pi n\epsilon_0 c W_0^2} \tag{5}$$

where W_0 is the beam waist and z_0 is the confocal distance of the pump beam such that $z_0 = k_p W_0^2$, can be used to normalize Equations (1) and (2) as

$$\frac{\partial u_s(r,\xi)}{\partial \xi} - \frac{1}{j}\nabla_\perp{}^2 u_s(r,\xi) = -j\gamma_p\, u_p(r,\xi)\, u_s^*(r,\xi) \tag{6}$$

$$\frac{\partial u_p(r,\xi)}{\partial \xi} - \frac{1}{2j}\nabla_\perp{}^2 u_p(r,\xi) = -j\gamma_s\, u_s^2(r,\xi)\,. \tag{7}$$

Here, the unitless normalized field amplitudes u_p and u_s satisfy

$$\int |u_{s,p}|^2\, r\, dr\, d\varphi = \pi/2 \tag{8}$$

at the input of the nonlinear crystal, and

$$\gamma_p = 8\pi d_e\sqrt{\frac{z_0 P_p}{\lambda^3 n^3 \epsilon_0 c}} \tag{9}$$

$$\gamma_s = 8\pi d_e\sqrt{\frac{z_0 P_s^2/P_p}{\lambda^3 n^3 \epsilon_0 c}} \tag{10}$$

are unitless constants where P_p and P_s are the pump and the signal powers at the input. Given the pump and signal field profiles at the input facet of the crystal and the interaction length in normalized units, the solutions of Equations (6) and (7) describe the evolution of the signal and pump beams along the crystal. Diffraction effects resulting from both propagation and nonuniform transverse gain variations are contained in these solutions.

The DOPA gain is given by

$$G = \frac{\int |u_s(r,\xi_l/2)|^2\, r\, dr\, d\varphi}{\int |u_s(r,-\xi_l/2)|^2\, r\, dr\, d\varphi} \tag{11}$$

where l is the physical and $\xi_l = l/z_0$ is the normalized crystal length, and the origin is at the center of the crystal. Squeezing of the signal field occurs when $G < 1$.

In plane-wave theory, the Laplacian terms on the left-hand sides of Equations (6) and (7) disappear. When the signal power is much smaller than the pump power, we have $\gamma_s \simeq 0$ and the small-signal gain of the DOPA is

$$G = |\cosh\Phi + \exp[j(\Delta\phi - \pi/2)]\sinh\Phi|^2 \tag{12}$$

where ϕ_s and ϕ_p are the phases of the signal and the pump, $\Delta\phi = \phi_p - 2\phi_s$ is the phase difference, and $\Phi = \gamma_p \xi_l$. The small signal gain assumes its maximum and minimum values

$$G = \exp(\pm 2\Phi). \tag{13}$$

at $\Delta\phi = \pi/2$ and $\Delta\phi = -\pi/2$, respectively. Keeping ξ_l constant and $\Delta\phi = -\pi/2$, G is a monotonically decreasing function of γ_p, and hence the pump intensity.

Using the results of the plane-wave theory for confined beams does not yield accurate results. The signal and pump beams experience diffraction due to propagation (linear diffraction) and distortion of the phase fronts due to gain variations along the radial axis (gain induced nonlinear diffraction). These diffraction effects result in nonuniform phase variations in the transverse plane which also depend on ξ. As a result, it becomes impossible to maintain the phase difference required for deamplification throughout the interaction.

In order to model practical DOPA's aimed at generating squeezing, we take the input signal power to be very small ($\gamma_s = 0$), and the pump field as a Gaussian beam given by

$$u_p = \frac{1}{1-2j\xi}\exp\left(\frac{-r^2}{1-2j\xi}\right) \tag{14}$$

with its beam waist located at $\xi = 0$. In this small-signal model, the pump beam does not get influenced by the presence of the nonlinear interaction.

An exact analytical solution $u_s(r,\xi)$ for an arbitrary input signal $u_s(r, -\xi_l/2)$ and γ_p is not available. An approximate analytical solution introduced by Choi *et* al.[5] is valid for $l/z_0 \ll 1$. In this model, the signal is a Gaussian beam with its waist located at $\xi = 0$ and has the same confocal distance as the pump beam when there is no nonlinear interaction. Figure 1 illustrates this beam geometry for $l/z_0 = 1$ where the phase fronts of the signal and the pump beams are shown. (In this example, $\Delta\phi = 0$ at $\xi = 0$.) Note that the phase difference $\Delta\phi$ does not stay constant throughout the interaction, due to the Gouy phase of the signal and pump modes.[6] Although the Gouy phase of both modes have the same $\tan^{-1}(2\xi)$ dependence, they do not cancel each other since $\Delta\phi$ is defined as $\phi_p - 2\phi_s$. This phase slip influences the net gain of the DOPA, resulting in decreased squeezing.

In our model, we solve Equation (6) using finite difference techniques. For that purpose we formed grid points in ξ and r and wrote down finite difference equations using central differencing. We define the signal input as

$$u_s(r, -\xi_l/2) = \frac{\exp(-j\Delta\phi/2)}{\sqrt{2}\,(1-2j\xi)}\exp\left(\frac{-r^2/2}{1-2j\xi}\right)\Bigg|_{\xi=-\xi_l/2} \tag{15}$$

which, in the absence of nonlinear interaction, propagates as a regular Gaussian beam having the same confocal distance as the pump beam. The signal power at the output is calculated by trapezoidal integration of $|u_s|^2$ over the transverse coordinates. Accurate results require a window size that is at least three times larger than the beam waist of

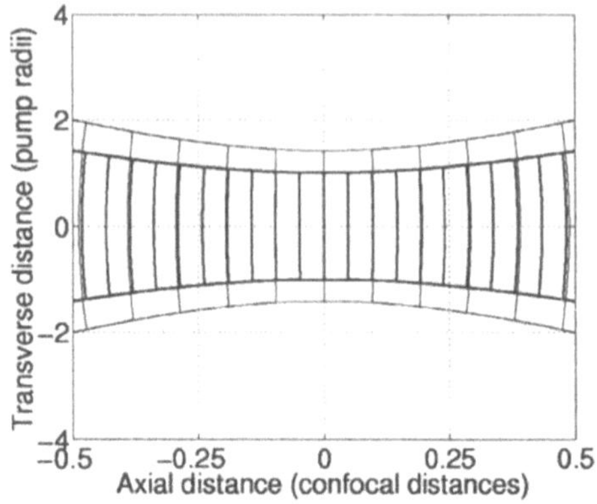

Figure 1. The phase fronts of the pump (thick lines) and the signal (thin lines) fields with no parametric gain. The wavelength is exaggerated to bring out the mismatch.

the pump in the r direction, and an absorbing boundary condition at the edge of this window.

The DOPA gain calculated using Equation (11) is a function of $\Delta\phi$ and γ_p. We find the $\Delta\phi$ and γ_p that minimize the gain (maximize the squeezing) by using an optimization algorithm (the conjugate gradient method), where the cost function is the DOPA gain. Figure 2 shows the maximum squeezing (minimum gain) and the $\Delta\phi$ at which this squeezing level is achieved as functions of l/z_0. Also included in Fig. 2 is the pump power (relative to its minimum value) that is required to achieve maximum squeezing. Note that the pump power and γ_p are related by Equation (9). For small values of l/z_0, the phase difference that maximizes the squeezing level is close to $-\pi/2$, which is the plane-wave theory value. This is expected since diffraction effects are small in this regime. Also note that, the pump power required to obtain maximum squeezing is high.

The squeezing level decreases with increasing l/z_0 up to $l/z_0 = 0.3$, while the optimum $\Delta\phi$ deviates from $-\pi/2$ and the required pump power decreases. However for $l/z_0 > 0.3$, squeezing begins to increase while the required pump power keeps on decreasing. Squeezing levels in excess of 10 dB can be obtained with relatively low pump power for $l/z_0 > 0.9$. In this regime, the $\Delta\phi$ that maximizes squeezing is significantly different from $-\pi/2$.

For a comparison of our model with previous models, we plot the DOPA gain

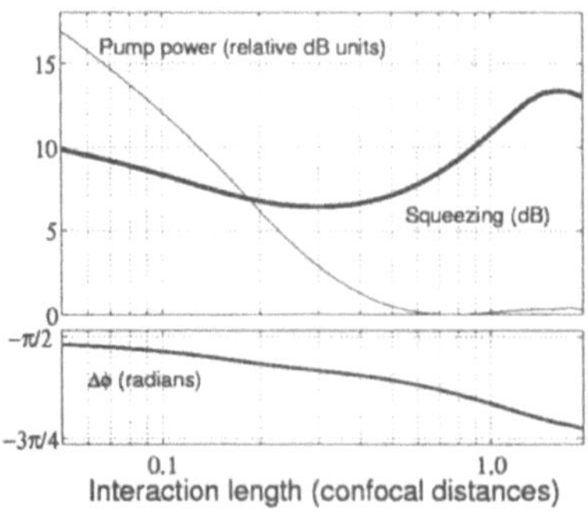

Figure 2. Maximum squeezing, optimum phase difference $\Delta\phi$, and required pump power as functions of normalized distance.

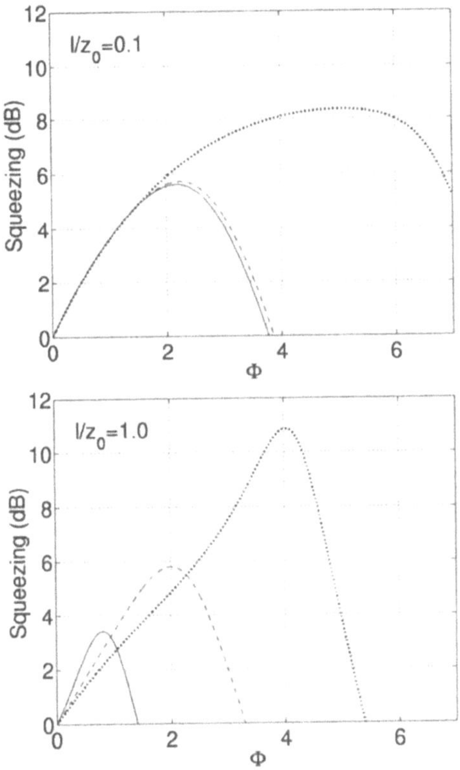

Figure 3. Comparison of analytical results [5] (solid) with numerical results when $\Delta\phi = \pi/2$ (dashed) and when $\Delta\phi$ is set to maximize the squeezing (dotted).

as a function of $\Phi = \gamma_p \xi_l \propto P_p^{1/2} l$ for two different l/z_0 values in Fig. 3. In the $l/z_0 \ll 1$ regime, the analytical results of Ref. 5 (solid line) are reproduced by our model for $\Delta\phi = -\pi/2$ (dashed line), both predicting a maximum squeezing of 5.7 dB at $l/z_0 = 0.1$. However, when $\Delta\phi$ is optimized, our model predicts larger levels of squeezing (dotted line), 8.3 dB in this example.

Analytical results are no longer accurate when the $l/z_0 \ll 1$ condition is relaxed. Fig. 3 shows the predictions of the analytical model (solid line), our model with $\Delta\phi = -\pi/2$ (dashed line), and our model with an optimized $\Delta\phi$, for $l/z_0 = 1$ (dotted line). The importance of optimizing $\Delta\phi$ is apparent in both examples.

Figure 4 illustrates the phase-front distortions due to gain induced diffraction[4] (transverse gain variations), where the phase curvatures of the signal and the pump beam are shown for $l/z_0 = 1$ under optimized conditions. Successful detection of the generated squeezing requires a local oscillator beam whose phase fronts are matched to those of the distorted signal phase fronts.

CONCLUSION

In conclusion, we have presented the results of a numerical model for a Gaussian beam DOPA aimed at squeezed state generation. Our model provides accurate gain (squeezing) calculations with no restriction on the ratio of the crystal length to the confocal distance (l/z_0). We demonstrate that the phase difference that maximizes

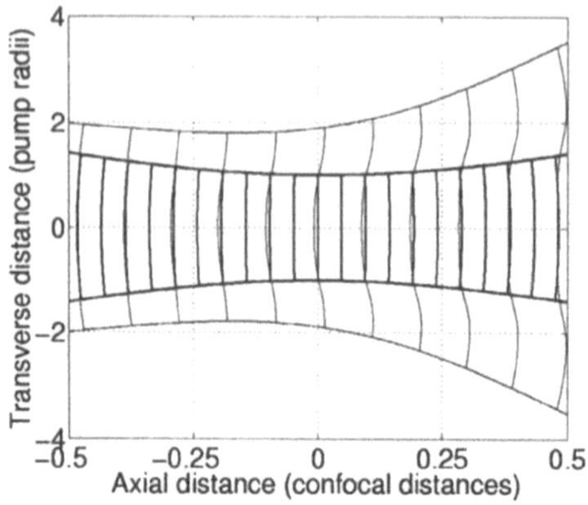

Figure 4. The phase fronts of the pump and the signal fields with small-signal parametric gain.

squeezing may be significantly different from $-\pi/2$, which is the value determined by the plane-wave theory and used in previous Gaussian beam models. Our model predicts squeezing levels in excess of 10 dB for crystal lengths on the order of or larger than the confocal distance z_0, and with relatively low pump power levels. We also calculate the phase fronts of the squeezed mode that a local oscillator beam must match for efficient detection.

REFERENCES

1. L. A. Wu, H. J. Kimble, J. L. Hall, and H. Wu, Generation of squeezed states by parametric down conversion, *Phys. Rev. Lett.* 57:2520 (1986).
2. C. Kim and P. Kumar, Quadrature-squeezed light detection using a self-generated matched local oscillator, *Phys. Rev. Lett.* 73:1065 (1994).
3. A. La Porta and R. E. Slusher, Squeezing limits at high parametric gains, *Phys. Rev. Lett.* 44:2013 (1991).
4. C. Kim, R. D. Li, and P. Kumar, Deamplification response of a travelling-wave phase-sensitive optical parametric amplifier, *Opt. Lett.* 19:132 (1993).
5. S. K. Choi, R. D. Li, C. Kim, and P. Kumar, Travelling-wave optical parametric amplifier: investigation of its phase-sensitive and phase-insensitive gain response, *J. Opt. Soc. Am. B* 14:1564 (1997).
6. B. E. A. Saleh and M. C. Teich, Beam optics, *in*: "Fundamentals of Photonics," John Wiley and Sons, Inc., New York (1991).

EIGEN STATE TRANSMISSION OF BIREFRINGENT OPTICAL FIBERS USING ENTANGLED TWO PHOTONS

Takeshi Ozeki and Atsushi Masubuchi

Department of Electrical and Electronics Engineering
Faculty of Science and Technology, Sophia University, Japan
t-ozeki@gentei.ee.sophia.ac.jp

When two photons are coded into entangled polarization states, one of eigen states of direct product of fiber transfer function matrix is found to be independent of transmission fiber birefringence. The eigen state is the entangled singlet state Ψ^-, one of the Bell states. The entangled polarization state Ψ^- can be transmitted through optical fiber without the degradation in fidelity.

INTRODUCTION

Teleportation, super dense coding, and broadcasting of EPR pairs are expected to take important roles in the quantum network systems.[1-3] One of the EPR pair particle acts like a key or a token for these quantum state transmission. Experimental verification of super dense coding was reported using down-conversion polarization entanglement and the Bell state analyzer,[4] where the increased capacity of a quantum information channel was demonstrated. Also, C. H. Bennett discussed the entanglement-enhanced classical communication on a noisy quantum channel given by the Kraus representation theorem.[5,6] The channel could be an optical fiber with birefringence. The two transmissions would then be represented by two separate photons being sent through the cable. The bit is to be encoded in the polarization degrees of freedom of the photons, each represented by a two-dimensional Hilbert space H_2. The question is whether the optimal encoding for such a channel will be in terms of product states — $|\psi_0\rangle\,|\psi_0\rangle$ and $|\psi_1\rangle\,|\psi_1\rangle$ for 0 and 1, respectively, — or rather $|\psi_0\rangle$ and $|\psi_1\rangle$ on $H_2 \otimes H_2$. In this paper, we extend this discussion to analyze the characteristics of eigen state of a birefringent fiber using polarization entanglement.

SINGLE PHOTON TRANSMISSION

For a single photon transmission, input polarization state E is given as quantum superposition state by

$$\mathrm{E} = c_0 \cdot |0\rangle + c_1 \cdot |1\rangle , \qquad |c_0|^2 + |c_1|^2 = 1, \tag{1}$$

Quantum Communication, Computing, and Measurement 2
Edited by Kumar et al., Kluwer Academic / Plenum Publishers, New York, 2000.

where $|1>$ and $|0>$ are orthogonal polarization bases, and c_1 and c_0 are probability coefficients. Transmission characteristics of loss-less birefringent optical fiber can be described a unitary transfer function matrix with $SU(2)$ symmetry:[7]

$$T(\omega) = \begin{pmatrix} \cos\Theta e^{-j\phi - j\psi} & -\sin\Theta e^{-j\phi + j\psi} \\ \sin\Theta e^{+j\phi - j\psi} & \cos\Theta e^{+j\phi + j\psi} \end{pmatrix}, \tag{2}$$

where Θ is polarization angle. ϕ is called vertical phase difference, which is the phase difference between T_{11} and T_{21}. ψ is denoted lateral phase difference, which means the phase difference between T_{11} and T_{12}. Polarization-independent phase shift, including such as chromatic dispersion, is neglected for simplicity.

By using Pauli spin operators, the transfer function matrix is expanded as follows:[4]

$$\begin{aligned} T(\omega) = &\cos\Theta\cos(\phi+\psi)\cdot\sigma_0 - j\sin\Theta\sin(\phi-\psi)\cdot\sigma_1 \\ &-j\sin\Theta\cos(\phi-\psi)\cdot\sigma_2 - j\cos\Theta\sin(\phi+\psi)\cdot\sigma_3, \end{aligned} \tag{3}$$

where

$$\sigma_0 = \begin{pmatrix} 1 & 0 \\ 0 & 1 \end{pmatrix}, \quad \sigma_1 = \begin{pmatrix} 0 & 1 \\ 1 & 0 \end{pmatrix}, \quad \sigma_2 = \begin{pmatrix} 0 & -j \\ j & 0 \end{pmatrix}, \quad \sigma_3 = \begin{pmatrix} 1 & 0 \\ 0 & -1 \end{pmatrix}. \tag{4}$$

This expansion is the Kraus representation for unitary fiber transmission of quantum state,[5] so that the following are satisfied:

$$T = \sum_{i=0}^{3} A_i\cdot\sigma_i, \qquad \sum_{i=0}^{3} A_i^*\cdot A_i = 1, \tag{5}$$

where

$$\begin{cases} A_0 = \cos\Theta\cos(\phi+\psi), \\ A_1 = +j\sin\Theta\sin(\phi-\psi), \\ A_2 = -j\sin\Theta\cos(\phi-\psi), \\ A_3 = -j\cos\Theta\sin(\phi+\psi). \end{cases} \tag{6}$$

A single photon transmission using eigen states is obtained by the eigen value equation:

$$T(\omega)\cdot \mathrm{E}(\omega) = \Gamma\cdot \mathrm{E}(\omega). \tag{7}$$

The eigen value Γ and eigen states generally depend on Θ, ϕ, and ψ:

$$\Gamma_\pm = \cos\Theta\cos(\phi+\psi) \pm j\sqrt{1-\cos^2\Theta\cos^2(\phi+\psi)}. \tag{8}$$

The eigen state are calculated as

$$\xi_\pm = \rho_\pm\cdot\begin{pmatrix} \alpha \\ j\beta_\pm \end{pmatrix}, \tag{9}$$

where

$$\alpha = \sin\Theta\cdot e^{-j\phi+j\psi}, \tag{10}$$

$$\beta_\pm = -\cos\Theta\cdot\sin(\phi+\psi) \mp \sqrt{1-\cos^2\Theta\cos^2(\phi+\psi)}, \tag{11}$$

$$\rho_\pm = \frac{1}{\sqrt{|\alpha|^2+\beta_\pm^2}}, \tag{12}$$

$$|\alpha|^2 + \beta_+\cdot\beta_- = 0. \tag{13}$$

The eigen states of a fiber transfer function matrix are called principal states by C. D. Poole, which depend on the fiber birefringence generally.[8] The fiber birefringence varies as circumstance parameters, such as temperature, change, so that it is not practical to use these eigen states for quantum state transmission.

ENTANGLED TWIN PHOTON TRANSMISSION

The output-polarization state of two entangled photons is described by direct product $|\psi^{out}{}_1\rangle \otimes |\psi^{out}{}_2\rangle$. In this formulation, transfer function matrix for two entangled photon is given by direct product $T \otimes T$:

$$T \otimes T = \begin{pmatrix} \cos^2\Theta \cdot e^{-j2\phi-j2\psi} & -\cos\Theta\sin\Theta \cdot e^{-j2\phi} & -\cos\Theta\sin\Theta \cdot e^{-j2\phi} & \sin^2\Theta \cdot e^{-j2\phi+j2\psi} \\ \cos\Theta\sin\Theta \cdot e^{-j2\psi} & \cos^2\Theta & -\sin^2\Theta & -\cos\Theta\sin\Theta \cdot e^{+j2\psi} \\ \cos\Theta\sin\Theta \cdot e^{-j2\psi} & -\sin^2\Theta & \cos^2\Theta & -\cos\Theta\sin\Theta \cdot e^{+j2\psi} \\ \sin^2\Theta \cdot e^{+j2\phi-j2\psi} & \cos\Theta\sin\Theta \cdot e^{+j2\phi} & \cos\Theta\sin\Theta \cdot e^{+j2\phi} & \cos^2\Theta \cdot e^{+j2\phi+j2\psi} \end{pmatrix}. \quad (14)$$

The eigen values of this direct product are given by single photon transmission eigen values Γ_+ and Γ_-:

$$\left\{ \begin{array}{ll} \Gamma_1 = \Gamma_+^2, & \Gamma_2 = \Gamma_-^2 \\ \Gamma_3 = \Gamma_4 = \Gamma_+ \cdot \Gamma_- = 1 & \end{array} \right\}. \quad (15)$$

The first pair Γ_1, Γ_2 are dependant on the transmission channel parameters. On the other hands, the second pair Γ_3, Γ_4 are degenerated as unity, which do not depend on the transmission birefringent parameters. The eigen states are obtained by direct products of the eigen vectors for the single photon transmission given by Eq. (9):

$$\xi_+ \otimes \xi_+ = \rho_+{}^2 \begin{pmatrix} \alpha^2 \\ j\alpha \cdot \beta_+ \\ j\alpha \cdot \beta_+ \\ -\beta_+{}^2 \end{pmatrix} \Longleftrightarrow \Gamma_1, \quad (16a)$$

$$\xi_- \otimes \xi_- = \rho_-{}^2 \begin{pmatrix} \alpha^2 \\ j\alpha \cdot \beta_- \\ j\alpha \cdot \beta_- \\ -\beta_-{}^2 \end{pmatrix} \Longleftrightarrow \Gamma_2, \quad (16b)$$

$$\xi_+ \otimes \xi_- = \rho_+\rho_- \begin{pmatrix} \alpha^2 \\ j\alpha \cdot \beta_- \\ j\alpha \cdot \beta_+ \\ -\beta_+{}^2 \end{pmatrix} \Longleftrightarrow \Gamma_3, \quad (16c)$$

$$\xi_- \otimes \xi_+ = \rho_+\rho_- \begin{pmatrix} \alpha^2 \\ j\alpha \cdot \beta_+ \\ j\alpha \cdot \beta_- \\ -\beta_+{}^2 \end{pmatrix} \Longleftrightarrow \Gamma_4. \quad (16d)$$

The eigen states for degenerated eigen value make a subspace in the Hilbert space $H_2 \otimes H_2$. In the subspace, we find an eigen state which is independent of the birefringence of the transmission fiber. This eigen state is the entangled state included in the Bell operator bases:

$$\frac{j \cdot e^{j\phi-j\psi}}{\sqrt{2}}(\xi_+ \otimes \xi_- - \xi_- \otimes \xi_+) = \frac{1}{\sqrt{2}} \cdot \begin{pmatrix} 0 \\ 1 \\ -1 \\ 0 \end{pmatrix} \Longleftrightarrow \Psi^-. \quad (17)$$

The entanglement of the two photons results in realization of eigen state transmission without the degradation in fidelity.[5] The scattering coefficients for the Bell states are shown in Table 1, which are obtained by using Kraus representation of Eq. (3). It is shown that components scattered from Ψ^- to Φ^+, Φ^-, and Ψ^+ vanish by interference.

Table 1. Scattering Coefficients among the Bell States

Input State	Scattered State Φ^+	Φ^-	Ψ^+	Ψ^-
Φ^+	$\cos^2\Theta\cos(\phi+\psi) + \sin^2\Theta\cos 2(\phi-\psi)$	$-j\cos^2\Theta\sin 2(\phi+\psi) - j\sin^2\Theta\sin 2(\phi-\psi)$	$-j\sin 2\Theta\sin 2\psi$	0
Φ^-	$-j\cos^2\Theta\sin 2(\phi+\psi) - j\sin^2\Theta\sin 2(\phi-\psi)$	$\cos^2\Theta\cos 2(\phi+\psi) + \sin^2\Theta\cos 2(\phi-\psi)$	$\sin 2\Theta\cos 2\psi$	0
Ψ^+	$j\sin 2\Theta\sin 2\psi$	$-\sin 2\Theta\cos 2\psi$	$\cos 2\Theta$	0
Ψ^-	0	0	0	1

CONCLUSION

It is found that one pair of degenerated eigen states of direct product of fiber transfer function matrix, are independent of fiber birefringence, when two photons are coded into polarization entanglement. The eigen state is an entangled singlet state Ψ^- included in Bell states. The entangled polarization state Ψ^- can be transmitted through optical fiber without the degradation in fidelity.

REFERENCES

1. C. H. Bennett, "Classical and Quantum Information Transmission and Interaction," Proc. 3rd Int. Conf. Quantum Comm. and Measurement, 25 (1996).
2. C. H. Bennett and S. J. Wiesner, "Communication via One- and Two-Particle Operators on Einstei-Podolsky-Rosen States," Phys. Rev. Letters, vol. 69, 2881 (1992).
3. C. H. Bennett and R. Jozsa, "Teleporting an Unknown Quantum State via Dual Classical and Einstein-Podolsky-Rosen Channels," Phys. Rev. Letters, vol. 70, 1898 (1993).
4. V. Buzek, V. Vedral, M. B. Plenio, P. L. Knight and M. Hillery, "Broadcasting of entanglement via local copying," Phys. Rev. A, vol. 55, 3327 (1997).
5. C. H. Bennett *et al.* "Entanglement-Enhanced Classical Communication on a Noisy Quantum Channel," Proc. 3rd Int. Conf. Quantum Comm. and Measurement, 79 (1996).
6. K. Kraus, "Effect and Operations: Fund. Notation of Quantum Theory," vol. 190, Springer (1983).
7. K. Sano, T. Kudou and T. Ozeki, "Simultaneous Measurement of Polarization Mode Dispersion and Group Delay Dispersion," Proc. ECOC'96, TuP-9 (1996).
8. C. D. Poole and R. E. Wagner, "Phenomenological approach to polarization dispersion in long single mode fibers," Electron. Lett., vol. 22, 1029 (1986).

FOUR-PHOTON PARTITION AT BEAMSPLITTER IN PARAMETRIC DOWN-CONVERSION WITH PULSED PUMPING

Z. Y. Ou,[1] J. -K. Rhee,[2] and L. J. Wang [2]

[1]Department of Physics
Indiana University-Purdue University Indianapolis
402 N. Blackford St.
Indianapolis, IN 46202
[2]NEC Research Institute, Inc.
4 Independence Way
Princeton, NJ 08540

When four photons arrive at a beamsplitter, two from each side, four-photon interference will occur such that the probability for three photons to exit from one side and one photon from the other vanishes while the probability for four photon to exit from one side is hence increased. We will analyze the phenomenon and report an experiment to demonstrate it in parametric down-conversion with pulsed pumping.

INTRODUCTION

Since Greenberger, Horne, and Zeilinger (1989) discovered that quantum states with 3 or more-particles can display more dramatic locality violation, there has been a race to create such multi-particle states. Several proposals have been made based on superposition of independent pairs of photons from parametric down-conversion (Yurke & Stoler, 1992; Zeilinger et al., 1997). However, since independent sources are involved, generating single photon at the exact time becomes a problem. Although it can be solved by ultrafast pulse and spectral filtering, this often leads to dispersion and reduction of the signal. In this paper, we will consider the stimulated process in parametric down-conversion for the generation of a four-photon state and will use the new state to demonstrate a four-photon interference effect, namely, four-photon partition at a 50:50 beamsplitter.

More specifically, we consider the output of a 50:50 beamsplitter with two pairs of photons input ($|2,2\rangle$) into opposite sides of the beamsplitter (Fig.1). We will examine the probability to obtain the states $|4,0\rangle$ or $|0,4\rangle$ and $|3,1\rangle$ or $|1,3\rangle$.

Quantum Communication, Computing, and Measurement 2
Edited by Kumar et al., Kluwer Academic / Plenum Publishers, New York, 2000.

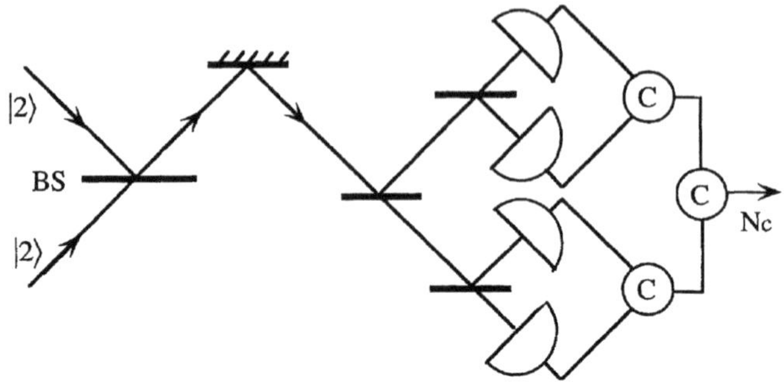

Figure 1. Geometry for four-photon interference with the help of a beamsplitter.

It can be readily shown (Campos et al., 1992) that for an input state of $|2,2\rangle$ at a 50:50 beamsplitter, the output state is

$$|\Psi\rangle = \sqrt{\frac{3}{8}}\,\big(|4,0\rangle + |0,4\rangle\big) + \frac{1}{2}\,|2,2\rangle. \tag{1}$$

On the other hand, if the four incoming photons were classical particles, they follow simple classical Bernoulli probability distribution for partition. The probability for $|4,0\rangle$ would simply be $(1/2)^4 = 1/16$, as compared to $3/8$ in Eq.(1) for the quantum case. Classical probability for $|3,1\rangle$ or $|1,3\rangle$ is a non-zero value of $1/4$. But the $|3,1\rangle$ and $|1,3\rangle$ states are missing from the output state in Eq.(1). While the disappearance of the $|3,1\rangle$ and $|1,3\rangle$ terms can be easily explained by two-photon interference (known as 2×2), the six-fold increase in probability for $|4,0\rangle$ and $|0,4\rangle$ is attributed to quantum interference of the four input photons. For, if it were due to two-photon interference (2×2), the contribution for the probability would be from the cases where two pairs are independent of each other. There exist three possibilities: (1) two photon of each independent pair are from the same side of the BS. Because the two pairs are independent, no interference occurs. This is like the classical case with $1/16$ probability for $|4,0\rangle$; (2) & (3) two photons of the pair come from opposite sides of the BS. Two-photon interference occurs and produces a two-photon quantum state (Hong et al., 1987):

$$|\Psi\rangle = |\psi\rangle \times |\psi\rangle, \quad \text{with} \quad |\psi\rangle = \frac{1}{\sqrt{2}}(|2,0\rangle + |0,2\rangle). \tag{2}$$

This will give rise to a probability of $1/4$ for $|4,0\rangle$. If the three possibilities are equally likely, then the result of two-photon interference would only be $(1/3)(1/16 + 1/4 + 1/4) = 3/16$ for $|4,0\rangle$, a factor of 2 off the prediction from four-photon interference. In the case of parametric down-conversion, which we will discuss in the following, the first possibility is zero. So two-photon interference (2×2) in this case will produce a probability of $1/4$ for $|4,0\rangle$, a four-fold rather than six fold increase when compared to the classical prediction.

Next let us see how we can produce a 4-photon state from nondegenerate parametric down-conversion. It is well known that parametric down-conversion process can be described by the Hamiltonian $\mathcal{H} = \kappa \hat{a}_s^\dagger \hat{a}_i^\dagger + h.c.$. So the state of fields from parametric down-conversion is

$$|\Psi\rangle = \exp(-i\mathcal{H}t/\hbar)\ |vac\rangle \approx [1 - i\mathcal{H}t/\hbar + (-i\mathcal{H}t/\hbar)^2/2]|vac\rangle, \tag{3}$$

where we have dropped the higher order terms in the approximation. We can rewrite the quantum state as

$$|\Psi\rangle = (1 - \eta^2/2)|vac\rangle - i\eta|1_s, 1_i\rangle - \eta^2|2_s, 2_i\rangle, \tag{4}$$

where $\eta \equiv \kappa t/\hbar$. The state is normalized up to the order of η^2. It is obvious that from this state we can realize the required four-photon state. However, the process of parametric down-conversion has a very wide spectrum. The state in Eq.(4) only represents a single mode of down-conversion. In the following, we will consider a multimode treatment.

MULTIMODE THEORY

The four-photon term in Eq.(4) can be thought as a result of stimulated emission (or amplification) from the two-photon term (2nd one). In fact, parametric down-conversion produces thermal fields for each of the down-converted beams (Yurke and Potasek, 1987). Therefore, the four-photon term is due to photon bunching effect or the Hanbury Brown-Twiss effect (Brown and Twiss, 1956). It is well known that in order to observe the bunching effect or equivalently the four-photon contribution in Eq.(4), detectors must have a faster response time than the fluctuations of the fields. Unfortunately, parametric down-conversion has a broad bandwidth so that current technology of photodetector is not advanced enough to resolve it although effort is made toward reducing the bandwidth of the down-conversion (Ou, 1998). One solution to this problem is to use ultrafast pulse to pump the down-conversion process and to use narrow band filter to create a transform-limited (single mode) down-conversion pulses (Rarity, 1995). Such a strategy has been applied in recent experiments on teleportation (Bouwmeester et al., 1997) and quantum state swapping (Pan et al., 1998).

The quantum state for parametric down-conversion pumped by a pulsed field has the following form (Ou, 1997):

$$|\Psi\rangle = (1 - \frac{\eta^2}{2})|vac\rangle - i\eta|\Phi\rangle - \frac{\eta^2}{2}|\Phi\rangle \times |\Phi\rangle, \tag{5}$$

where

$$|\Phi\rangle \equiv \int d\Omega_1 d\Omega_2 F(\Omega_1, \Omega_2)\ \hat{a}_s^\dagger(\Omega_1)\ \hat{a}_i^\dagger(\Omega_2)|vac\rangle \tag{6}$$

is a normalized multimode two-photon state with

$$\int d\Omega_1 d\Omega_2\ \ |F(\Omega_1, \Omega_2)|^2 = 1. \tag{7}$$

This state is not transform-limited but with narrow band filters for photodetection, it will be effectively transform-limited.

We consider the situation shown in Fig.1, where signal and idler beams from down-conversion enter the beamsplitter from opposite sides. We will introduce a delay τ between the two input beams and consider the following four-fold coincidence at one output of the BS:

$$P_4(t_1, t_2, t_3, t_4) = \langle\Psi|\hat{E}^\dagger(t_1)\hat{E}^\dagger(t_2)\hat{E}^\dagger(t_3)\hat{E}^\dagger(t_4)\hat{E}(t_4)\hat{E}(t_3)\hat{E}(t_2)\hat{E}(t_1)|\Psi\rangle, \tag{8}$$

where the quantized electric field has the form of

$$\hat{E}(t) = \frac{1}{\sqrt{2}}[\hat{E}_s(t) + \hat{E}_i(t-\tau)], \tag{9}$$

with a mode expansion,

$$\hat{E}_{s,i}(t) = \int d\Omega\ \hat{a}_{s,i}(\Omega)\ f(\Omega)\ e^{-j\Omega t}. \tag{10}$$

Here τ is the delay between the signal and idler beams and $f(\Omega)$ is the spectral transmission function of the narrow-band filter.

Since the pump pulse is much shorter than the response time of any detectors, the observed four-fold coincidence is a time average of $P_4(t_1, t_2, t_3, t_4)$:

$$P_4(\tau) = \int_{-\infty}^{\infty} dt_1 dt_2 dt_3 dt_4\ P_4(t_1, t_2, t_3, t_4). \tag{11}$$

Because of the limited space here, we will skip the detailed derivation and only present the result of the calculation. Due to nonstationary nature of the fields, we expect that when the delay τ is large, no interference occurs so that it corresponds to the classical situation whereas when $\tau = 0$, maximum interference occurs. For simplicity, we only give the result for the above two extreme cases.

At $\tau = \infty$, we have

$$P_4(\infty) = A + E, \tag{12}$$

while at $\tau = 0$, we have

$$P_4(0) = 4A + 8E, \tag{13}$$

where

$$A = C \int d\Omega_1 d\Omega_2 d\Omega_1' d\Omega_2' |\bar{F}(\Omega_1, \Omega_2)\ \bar{F}(\Omega_1', \Omega_2')|^2, \tag{14}$$

$$E = C \int d\Omega_1 d\Omega_2 d\Omega_1' d\Omega_2' \bar{F}(\Omega_1, \Omega_2)\ \bar{F}(\Omega_1', \Omega_2') \bar{F}^*(\Omega_1, \Omega_2')\ \bar{F}^*(\Omega_1', \Omega_2) \tag{15}$$

with $\bar{F}(\Omega_1, \Omega_2) \equiv F(\Omega_1, \Omega_2) f(\Omega_1) f(\Omega_2)$ and $C \propto \eta^4$. The terms A and E represent the accidental and excess quadruple coincidence. The existence of the excess coincidence E in Eq.(12) is the result of photon bunching (see experimental part later). Notice that $E \leq A$ by Schwartz inequality so that the ratio

$$\frac{P_4(0)}{P_4(\infty)} = 4\Big(1 + \frac{E}{A+E}\Big) \leq 6. \tag{16}$$

In the ideal case when $A = E$, the ratio is exactly six corresponding to the six-fold increase in the probability for $|4, 0\rangle$ from the classical prediction to the quantum prediction as in Eq.(4). So the multimode prediction covers the single mode case. On the other hand, when $E = 0$, we have the ratio equal to 4. This corresponds to the 2×2 situation when only two-photon interference contributes.

EXPERIMENT AND SUMMARY

The outline of the experimental setup is shown in Fig.2. A Ti:sapphire mode-locked laser [Spectra Physics 3600D] produces 860 nm, nearly bandwidth-limited 150 fs pulses at a repetition rate of 80 MHz. The laser pulses are frequency doubled with

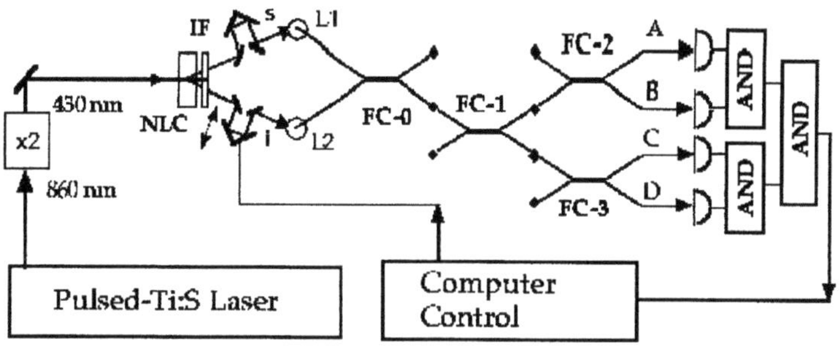

Figure 2. Experimental setup. L1 and L2 are two lens systems that couple the signal and idler beams into the fiber system with high efficiencies.

a 1-mm-long $KNbO_3$ nonlinear crystal, and the resulting 430 nm pulse train is used to pump an identical nonlinear crystal, NLC, serving as a parametric down-converter. We tune the temperature of the down-converter such that the signal and idler beams at 860 nm are emitted at approximately 4 degrees apart. After going through a 0.9 nm bandwidth interference filter (IF), the signal and idler beams are carefully aligned into the input ports of a 50:50 2x2 optical fiber coupler (FC-0). The relative time delay between the signal and idler photon wavepackets is scanned by a computer-controlled translation stage.

As the first stage of the experiment, the two output ports from the fiber coupler FC-0 are fed into two avalanche photodiode single photon detectors [EG&G SPCM-132] and the two-photon coincidence counts are measured as a function of the relative time delay. A coincidence dip of 80% modulation is observed (Hong et al., 1987). One of the two output ports of FC-0 is then connected as shown in Fig.2 and the four outputs from the fiber couplers FC-2 and FC-3 are fed into four single photon detectors for quadruple coincidence measurement. Here, all quadruple coincidence events are accidental as the results of two possibilities. First, two sets of signal-idler photon pairs are emitted from NLC and with a properly combined binary branching at all four fiber couplers, the four photons produce a coincidence count. As a second possibility a pair of signal and idler photons are first generated in NLC and then amplified to produce two pairs of identical signal and idler photons. The relative ratio of the possibilities for the two events manifests in the form of the ratio E/A given in Eqs.(14,15) and can be measured experimentally. For this purpose, we block off the idler beam, and measure the accidental coincidence rate between the detectors A and B. The resulting coincidence rate follows a simple relation $R_{AB} = (1 + E/A) \times R_A R_B / R_L$, where R_{AB}, R_A, R_B, and R_L are the counting rates for coincidence, for detectors A and B, and the laser repetition rate, respectively. The excess coincidence from a non-zero E is a result of photon bunching for thermal field. Ratios of $E/A = 0.7$ and 0.6 are obtained for the signal and idler beams, respectively.

The quadruple coincidences are measured as a function of the relative time delay between the signal and idler beam paths. The result is shown in Fig.3. A least square fitting procedure is employed and gives a ratio of $R_4(0)/R_4(\infty) = 5.09 \pm 0.37$. This is in good agreement with the prediction of $R_4(0)/R_4(\infty) = 5.6$ according to Eq.(16). Furthermore, the observed ratio is in better agreement with a more complicated analysis that takes into account the less than 100% visibility of the two-photon coincidence dip (Hong et al., 1987). The analysis will be published elsewhere.

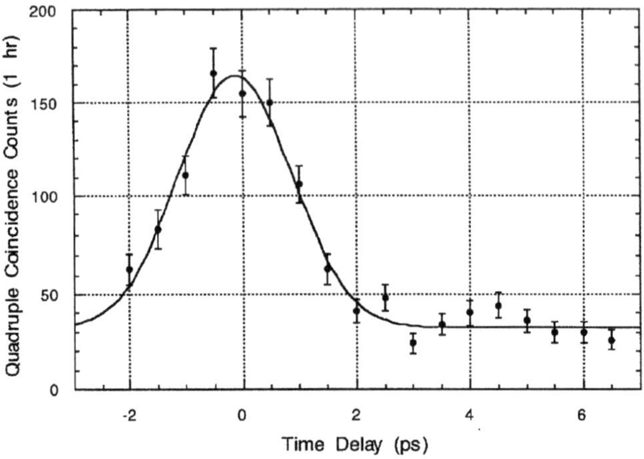

Figure 3. Quadruple coincidence counting vs. relative time delay between the signal and idler photons. The counting time for each data point is one hour. The solid line is the result from a least square fitting procedure.

In conclusion, we analyzed the interesting situation of the partition at a beamsplitter for four photons generated by a parametric down-converter by pulsed pumping. We point out the different partition ratios for identifiable (2×2-photon) and unidentifiable (4-photon) pairs of photons. We demonstrated for the first time such an effect utilizing pulse-pumped parametric down-conversion.

Acknowledgments

ZYO would like to thank the support of the Office of Naval Research. The experiment was conducted at the NEC Research Institute.

REFERENCES

Brown, R. Hanbury and Twiss, R. Q., 1956, Nature **177**:27.

Campos, R. A., Saleh, B. E. A., and Teich, M. C., 1990, *Phys. Rev.* A**40**:1371.

Greenberger, D. M., Horne, M. A. , and Zeilinger, A., 1989, *in*: "Bell's Theorem, Quantum Theory, and Conceptions of the Universe," M. Katafos, ed., Kluwer Academic, Dordrecht, The Netherlands.

Hong, C. K., Ou, Z. Y., and Mandel, L., 1987, *Phys. Rev. Lett.* **59**:2044.

Ou, Z. Y., 1997, *Quantum Semiclass. Opt.* **9**:599.

Ou, Z. Y., 1998, p?? in current book.

Rarity, J. G., 1995, *in*: "Fundamental Problems in Quantum Theory," D. M. Greenberger and A. Zeilinger eds.: Ann. NY Acad. Sci. **755**:624.

Yurke, B. and Potasek, M., 1987, Phys. Rev. **A36**:3464

Yurke, B. and Stoler, D., 1992, *Phys. Rev. Lett.* **68**:1251.

Zeilinger, A., Horne, M. A., Weinfurter, H., and Zukowski, M.,1997, *Phys. Rev. Lett.* **78**:3031.

ENERGY TRANSFER IN PARAMETRIC DOWN-CONVERSION

A. Bandilla [1], G. Drobný [2], I. Jex[3]

[1] AG "Nichtklassische Strahlung", Humboldt-Universität, Institut f. Physik, Invalidenstr. 110, 10115 Berlin, Germany
[2] Institute of Physics, Slovak Academy of Sciences, Dúbravská cesta 9, 842 28 Bratislava, Slovakia
[3] Department of Physics, FNSPE, Czech Technical University Prague Břehová 7, 115 19 Praha 1, Czech Republic

Parametric effects are widely used in physics but they are especially important in nonlinear optics and for the generation of quantum fields without any classical analog. While the second harmonic of a laser beam or the sum-frequency of two strong waves emerge already classically from the vacuum, the reverse process with no fundamental waves but a strong pump (the second harmonic or the sum frequency, respectively) can only start in a quantum mechanical description. In these two contrary situations we can expect besides the differences in the transient regime also remarkable distinctions in the saturation behaviour when at least one wave is strongly depleted. Thus, it is very interesting which amount of energy can in principle be transferred. This optimal energy transfer will be discussed in the following and it will also become clear how the fluctuations occurring thereby can be investigated. Obviously, it is realistic that we limit ourselves to the first optimum and to coherent initial fields.

Spontaneous parametric down-conversion as the decay of pump photons from an intense laser beam into vacuum signal and idler waves takes place during three-wave interaction in a nonlinear $\chi^{(2)}$ crystal. Numerical solutions of the corresponding quantum mechanical equations of motion for weakly excited number-state or coherent pump beams [1,2,3,4] showed already fundamental transfer limits, e.g., during the degenerate process (spontaneous two-photon down-conversion) only less than 2/3 of the pump energy can maximally be pumped into the signal [3] when the pump wave reaches its first intensity minimum. Also in the nondegenerate case, numerical solutions with weak pump states reveal that in average about 7 from 9 photons can be transferred during the first pump minimum into signal and idler [4].

If, in contrast, all three waves are coherently excited at the beginning, then the numerical diagonalization for relatively weak beams shows a phase-sensitive transfer where almost total and almost no energy transfer become possible [5]. This is easily understood by comparison with classical result [5,6]. However, each coherent field has phase and amplitude uncertainties, while the classical solutions are valid for sharp amplitudes and phases. Therefore, it is rather surprising that even the classical no-energy-transfer case can be realized quantum mechanically in a very good approximation [5,10].

Quantum Communication, Computing, and Measurement 2
Edited by Kumar et al., Kluwer Academic / Plenum Publishers, New York, 2000.

On the other hand, the classical solutions of Armstrong et al. [6] can be used as classical trajectories in the truncated Wigner approach [7,8,9,10] to describe also intense beams in parametric down-conversion and gain a more physical understanding of this process. The last point is mentioned here because the exact contour lines of the Wigner or Husimi functions deform during the interaction in a very puzzling way, especially when there is strong depletion.

To begin with, the Hamiltonian of three-wave interaction (in the interaction picture) is

$$\hat{H} = \hbar\kappa(\hat{a}\hat{b}\hat{c}^\dagger + \hat{a}^\dagger\hat{b}^\dagger\hat{c}), \tag{1}$$

where $\hat{a}, \hat{b}$ and $\hat{c}$ are the annihilation operators of the signal (frequency ω_1), idler (ω_2) and pump ($\omega_3 = \omega_1 + \omega_2$), respectively, and κ is the coupling constant containing $\chi^{(2)}$. The initial state we will concentrate mainly on is

$$|\psi(0)\rangle = |\alpha_0 = 0\rangle_a |\beta_0 = 0\rangle_b |\gamma_0\rangle_c \tag{2}$$

where signal and idler start in the vacuum while the pump begins with a coherent state of the strong amplitude γ_0.

To get an estimation of the possible accuracy of the approximate quantum mechanical description via the truncated Wigner approach we consider only one mode and average over the other two. Assuming coherent states in all modes, i. e., $\alpha_0 \neq 0, \beta_0 \neq 0$, the exact Wigner function of the signal is given up to second order in (κt) by

$$\begin{aligned} W_a(\alpha, t) = \frac{2}{\pi} e^{-2|\alpha-\alpha_0|^2} \big[1 - \{i\kappa t 2(\alpha^* - \alpha_0^*)\beta_0^*\gamma_0^* + c.c.\} \\ +(\kappa t)^2 \big(\{-2(\alpha^* - \alpha_0^*)^2 \beta_0^{*2}\gamma_0^2 + c.c.\} + 2|\alpha - \alpha_0|^2 (2|\beta_0|^2|\gamma_0|^2 + |\beta_0|^2 + |\gamma_0|^2) \\ +(|\gamma_0|^2 - |\beta_0|^2)\{\alpha(\alpha^* - \alpha_0^*) + c.c.\} - 2|\gamma_0|^2(|\beta_0|^2 + 1)\big) + O((\kappa t)^3)\big]. \end{aligned} \tag{3}$$

If we use a short-time expansion for the classical amplitudes and insert them as trajectories into the Gaussians of the initial W we find also (3) except that in second order the terms

$$|\alpha - \alpha_0|^2 - 1/2 \tag{4}$$

have to be added. Note that these added terms do not disturb the normalization. Equivalent expressions can be derived for other modes. Likewise, the coherent initial state can be generalized to a big manifold of states by using a P representation.

Before we continue with the extreme initial state (2) we note that a small $\alpha_0 \neq 0$ as a signal and $\beta_0 = 0$ corresponds to phase-insensitive amplification including saturation while an additional coherent idler results in phase-sensitive amplification.

The initial Wigner function corresponding to (2) is

$$W(\alpha, \beta, \gamma) = \left(\frac{2}{\pi}\right)^3 e^{-2|\alpha|^2} e^{-2|\beta|^2} e^{-2|\gamma-\gamma_0|^2}. \tag{5}$$

To get the trajectories we use the exact solutions of the coupled-mode equations [6]

$$u_3^2(\zeta) = u_{3a}^2 + (u_{3b}^2 - u_{3a}^2) sn^2[(u_{3c}^2 - u_{3a}^2)\zeta - X_0, m],$$

$$u_1^2(\zeta) = u_1^2(0) + u_3^2(0) - u_3^2(\zeta) \quad , \quad u_2^2(\zeta) = u_2^2(0) + u_3^2(0) - u_3^2(\zeta). \tag{6}$$

Equation (6) gives the squared real amplitudes of the pump (u_3^2), signal (u_1^2) and idler (u_2^2). The interaction time is $\zeta = \kappa t$ to adjust with (1). The parameter m in the Jacobian elliptic function $sn[x,m]$ [7] is given by $m = (u_{3b}^2 - u_{3a}^2)/(u_{3c}^2 - u_{3a}^2)$ and $u_{3c}^2 \geq u_{3b}^2 \geq u_{3a}^2$ are the ordered roots of a cubic equation [6,5]. X_0 is the initial constant.

Being here interested only in the calculation of different moments of the photon number operator $\hat{n}_i$ $(i = 1,2,3)$, we do not care about the individual phases of signal, idler and pump, that are actually not given by [6]. These single phases could be included without any problems if we numerically integrate the classical equations of motion and follow the complex trajectories in the phase space. Many examples of such calculations were given in [10]. For the photon number moments we do not need the individual phases and can therefore use immediately the solutions (6).

Because the truncated Wigner approach requires integrations over the initial values determined within the limits given by (5), it is very important to calculate the initial constant X_0 in such a way that enables us to simplify expressions (6). For this aim we took in [8] the initial state (2) with $\gamma_0 \gg 1$ (γ_0 real) and accounted for (5) what is equivalent to

$$u_1^2(0), u_2^2(0) \ll u_3^2(0) = \gamma_0^2. \tag{7}$$

Introducing the abbreviations

$$u_1(0) = x_1, \quad \varphi_1(0) = x_2, \quad u_2(0) = y_1, \quad \varphi_2(0) = y_2, \quad u_3(0) = z_1, \quad \varphi_1(0) = z_2, \tag{8}$$

where $\varphi_i(0)$ $(i = 1,2,3)$ are the initial phases of signal, idler and pump, respectively, X_0 was found to be [8]

$$X_0 = \frac{1}{2} \ln \left[\frac{16 z_1^2}{x_1^2 + y_1^2 + 2x_1 y_1 \sin(z_2 - x_2 - y_2)} \right] \tag{9}$$

Limiting on the first minimum of the pump photon number we try now to approximate the elliptic functions by hyperbolic ones giving up the periodicity of the exact solutions. This is possible due to the fact that $m \to 1$. The last property becomes clear by calculating the roots u_{3i}^2, $(i = a,b,c)$ for the initial condition (7) what allows also the classical parametric approximation for certain interaction times [8]. These roots take on a simple structure namely

$$u_{3a}^2 = \frac{x_1^2 y_1^2}{z_1^2} \cos^2(z_2 - x_2 - y_2),$$

$$u_{3b,c}^2 = \frac{1}{2}\Big(2z_1^2 + x_1^2 + y_1^2 - u_{3a}^2 \mp \sqrt{(x_1^2 - y_1^2) + 4x_1^2 y_1^2 \cos^2(z_2 - x_2 - y_2)}\Big). \tag{10}$$

We note that according to (6) u_{3a}^2 determines the energy amount that stays in the pump of a single trajectory. This energy becomes negligible due to (7). For the parameter m we obtain from (10), $m = 1 - \sqrt{(x_1^2 - y_1^2) + 4x_1^2 y_1^2 \cos^2(z_2 - x_2 - y_2)/z_1^2}$. Using the Landen transformation [7], this parameter m comes even closer to unity what allows then to extend the approximation of the elliptic functions by hyperbolic ones to the whole depletion region [7,8]. The squared pump amplitude is eventually very well approximated by

$$u_3^2(x) = z_1^2 + \frac{1}{2}(x_1^2 + y_1^2) -$$

$$\frac{1}{4} e^{-2xz_1/\gamma_0}[x_1^2 + y_1^2 - 2x_1 y_1 \sin(z_2 - y_2 - x_2)] - \frac{z_1 + \frac{1}{2}(x_1^2 + y_1^2)}{\cosh^2\left(\frac{xz_1}{\gamma_0} - X_0\right)}. \tag{11}$$

The intensities $u_1^2(x)$ and $u_2^2(x)$ are given analogously by (6). On the left-hand side of equation (11) the whole dependence on the initial values is suppressed and pump coordinate scaling is introduced by $x = \gamma_0 \kappa t$. Terms of the order $u_i^2(0)/u_3^2(0), (i = 1, 2)$ and $u_1^2(0)u_2^2(0)/u_3^2(0)$ are neglected. Note that for $z_1 x/\gamma_0 = X_0$ each trajectory (11) reveals complete energy transfer to signal and idler.

The photon number moments have now to be calculated according to the general rules via the Wigner function [9]. The pump photon number is obtained, e.g., by

$$\langle \hat{n}_3(x) \rangle = \langle\langle u_3^2(x) \rangle\rangle - \frac{1}{2}, \tag{12}$$

where $\langle\langle u_3^2(x) \rangle\rangle$ means an integration of $u_3^2(x)$ over x_1, y_1, z_1 and x_2, y_2, z_2 with the initial Wigner function (5) in polar coordinates

$$W(x_1, x_2; y_1, y_2; z_1, z_2) = \left(\frac{2}{\pi}\right)^3 e^{-2x_1^2} e^{-2y_1^2} e^{-2(z_1^2 - 2z_1\gamma_0 \cos z_2 + \gamma_0^2)}. \tag{13}$$

To do so amounts to importance sampling [10]. Parts of the integration can be executed analytically, so we have to consider the last term in (11) and insert (9) for X_0. This term is responsible for the depletion. Then we arrive at

$$\frac{1}{\cosh^2\left(\frac{xz_1}{\gamma_0} - X_0\right)} = \frac{4[x_1^2 + y_1^2 + 2x_1y_1 \sin(z_2 - y_2 - x_2)]}{\left([x_1^2 + y_1^2 + 2x_1y_1 \sin(z_2 - y_2 - x_2)]\frac{e^{xz_1/\gamma_0}}{4z_1} + 4z_1 e^{-xz_1/\gamma_0}\right)^2} \tag{14}$$

Due to the initial state (2) no phases in signal and idler are distinguished and we have to integrate (14) from $-\pi$ to π, say over $y_2 = \varphi_2(0)$ [11]. The resulting expression has no dependence on x_2 and z_2. Thus, these integrations can also be done. The integral over z_2 leads, because of (13), to a modified Bessel function $I_0(4z_1\gamma_0)$ that can be replaced with its asymptotic $e^{4z_1\gamma_0}/\sqrt{8\pi z_1 \gamma_0}$ because z_1 is limited to great values around γ_0. In this way only the integrations over x_1, y_1 and z_1 are left to be done numerically without limitations. Consequently, large initial pump photon numbers can be handled.

Doing all these integrations we obtain the result that the maximum pump depletion can reach approximately 78%. The same is found via exact numerical diagonalization of (1) for small ($\gamma_0 = 3, 5$) and medium ($\gamma_0 = 10$) coherent excitations of the pump wave. The case of the spontaneous two-photon down-conversion can be calculated in a similar way. There we get a maximum energy transfer of about 64% from a strong pump wave into the initially empty signal mode.

In addition we can treat higher moments of the number operators of all three modes in a straightforward generalization. For example, the pump photon number fluctuations are characterized by $\langle \hat{n}_3^2 \rangle$ which is to be calculated according to [9]

$$\langle \hat{n}_3^2(x) \rangle = \langle\langle u_3^4(x) \rangle\rangle - \langle\langle u_3^2(x) \rangle\rangle \tag{15}$$

The integrations proceed analogously as for (12). In this way the fluctuations during depletion can be studied. As exact numerical calculations for weak pump fields already show, it turns out that there is a characteristic change of the pump photon number distribution when signal and idler come into the saturation region.

If we start with phase-insensitive amplification then we observe a fast increase of the transfer maximum. This can still be described within condition (7). As an example, we mention that already for $\alpha_0 = 2$ and $\beta_0 = 0$ and a strong γ_0 approximately 10% of the initial pump energy remains minimally in this mode [15]. From this and the above calculations we can conclude that mostly the initial phase fluctuations in signal and idler are responsible for the limited transfer rate from a strong pump to initially weak or empty signal and idler waves.

In the exact quantum mechanical picture the reasons for these remarkable limits in maximum energy transfer during spontaneous down-conversion are very different. The diagonalization shows that the transfer efficiency depends on the overlap of the initial state of the whole system with the eigenstates of the Hamiltonian [3]. In the spontaneous down-conversion only a few eigenstates overlap significantly with the initial state (2). The contributing eigenvalues are therefore relatively small and it takes a long time for this process to gain strength.

The contrary process, e. g., second harmonic generation in the degenerate case, goes much faster. There is a considerable overlap of many eigenstates with the initial state where the fundamental mode is in a strong coherent state while the second harmonic builds up from the vacuum. Hence, also much greater eigenvalues take part in this process and speed it up. The energy transfer to the second harmonic becomes almost complete. According to our calculations with the truncated Wigner method only a fraction proportional to the square root of the initial photon number stays in the fundamental (signal) mode.

REFERENCES

1. Walls, D. F., and Barakat, R., Quantum-mechanical amplification and frequency conversion with a trilinear Hamiltonian, *Phys. Rev. A* 1:446 (1970).
2. Walls, D. F., and Tindle, C. T., Nonlinear quantum effects in optics, *Phys. Rev. A* 8:534 (1972).
3. Drobný, G., Bužek, V., Fundamental limit on energy transfer in k-photon down-conversion, *Phys. Rev. A* 50:3492 (1994).
4. Drobný, G., and Jex, I., Quantum properties of field modes in trilinear optical processes, *Phys. Rev. A* 46:499 (1992).
5. Bandilla, A., Drobný, G., and Jex, I., Nondegenerate parametric interactions and nonclassical effects, *Phys. Rev.A* 53:507 (1996).
6. Armstrong, J. A., Bloembergen, N., Ducuing, J., and Pershan, P. S., Interactions between light waves in a nonlinear dielectric, *Phys. Rev.* 127:1918 (1962).
7. Abramowitz, M., and Stegun, I. A., "Handbook of Mathematical Functions," New York, Dover, (1964).
8. Bandilla, A., Drobný, G., and Jex, I., The classical parametric approximation for three-wave interactions, *Optics Commun.* to be published.
9. Cahill, K. E., and Glauber, R. J., Density operators and quasiprobability distributions, *Phys. Rev.* 177:1882 (1969).
10. Drobný, G., Bandilla, A., and Jex, I., Quantum description of nonlinearly interacting oscillators via classical trajectories, *Phys. Rev.A* 55:78 (1997).
11. Prudnikov, A. P., Brychkov, Yu. A., and Marichev, O. I., "Integrals and Series, vol.1," Gordon and Breach Science Publishers, New York, (1992).
12. Mollow, B. R., and Glauber, R. J., Quantum theory of parametric amplification.I and II, *Phys. Rev.* 160:1076 and 160:1097 (1967), respectively.
13. Kupiszewska, D., and Rzazewski, K., Quantum fluctuations in parametric down-conversion and their classical stochastic description, *Phys. Rev. A* 42:6869 (1990).
14. Kinsler, P., Fernée, M., and Drummond, P. D., Limits to squeezing and phase information in the parametric amplifier, *Phys. Rev. A* 48:3310 (1993).
15. Bandilla, A., Drobný, G., and Jex, I., Quantum and classical description of three-wave interaction, *Acta Physica Slovaca* 47:179 (1997).

A REALISTIC INTERACTION-FREE RESONATOR

Mladen Pavičić[1,2]

[1]Atominstitute of the Austrian Universities, Vienna, Austria
[2]University of Zagreb, Zagreb, Croatia; mpavicic@faust.irb.hr

A realistic interaction-free resonator which makes use of total reflection and frustrated total reflection is presented. Time evolution and overall losses of the interference in the resonator impose restrictions on the experiment which one can partly overcome by detecting two outputs.

INTRODUCTION

Interaction-free measurements are *void detections* of paths within an interference experiment which therefore destroy their indistinguishability. In 1986 I formulated this in the following way.

"Consider a photon experiment shown in figure 1 which results with an interference in the region D provided we do not know whether it arrived to the region by path s_1 or by path s_2. As is well-known, experimental facts are: If we, after a photon passed the beam splitter B and before it could reach the point C, suddenly introduce a detector in the path s_2 in the point C and do *not* detect *anything*, then it follows that the photon must have taken the path s_1—and, really, one can detect it in the region D but it does *not* produce interference there. Quantum mechanically, if we registered the interference in the region D, we could not find an experimental procedure to directly either prove or disprove that the photon uses both paths simultaneously. However, the fact that by detecting *nothing* in point C we destroy the interference implies that the photon *somehow* knows of the other path when it takes the first one." (Pavičić, 1986; pp. 31, 32)

Photon's "knowledge" about the other path one can employ to detect an object (at point C) without transferring even a single quantum of energy to it. The efficiency of such an application with symmetrical Mach-Zehnder interferometer (shown in figure 1) is ideally only 25% for single detections as shown in Elitzur and Vaidman's (1993) first quantitative formulation of void detections in interference experiments. Therefore several more efficient models have been formulated recently, some of which are reviewed in Paul and Pavičić (1998). A possible application to interaction-free interference erasure is given in Pavičić (1996).

Quantum Communication, Computing, and Measurement 2
Edited by Kumar et al., Kluwer Academic / Plenum Publishers, New York, 2000.

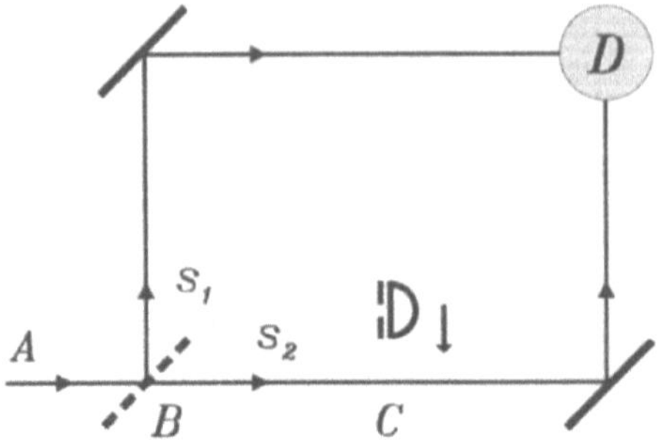

Figure 1. Figure taken from Pavičić (1986). "By detecting *nothing* in the point C we destroy the interference [in the region D]." (Pavičić, 1986; p. 31)

THE EXPERIMENT

Figure 2 shows an outline of the proposed experiment. The resonator used for the experiment consists of Pellin–Broca prisms which are designed so that the entrance and exit faces are at Brewster's angles thus minimizing reflection losses. When there is no object in the resonator, an incoming laser beam is almost totally transmitted into detector D_t and when there is an object, an incoming laser beam is being (ideally) totally reflected into detector D_r. The entrance prism is coupled to the adjacent loop prism by the frustrated total reflection, which is an optical version of quantum mechanical tunneling. Depending on the dimension of the gap between the prisms one can well define reflectivity R within the range from 10^{-5} to 0.99995. The uniqueness of the reflectivity at the gaps and at the same time no reflectivity at the entrance and exit faces of the prisms for each photon is assured by choosing the orientation of the polarization of the incoming laser beam perpendicular to the plane of incidence. As a source of the incoming beam a continuous wave laser (e.g., Nd:YAG) should be used because of its coherence length (up to 300 km) and of its very narrow linewidth (down to 10 kHz in the visible range)

Detailed wave packet calculations carried out by Paul and Pavičić (1997) yield that the efficiency of the suppression of the reflection into D_r is given by

$$\eta = 1 - \frac{I_r}{I_i} = (1 - R_1)(1 - \rho^2 R_2)\,\Phi\,, \tag{1}$$

and the efficiency of the throughput into D_t by:

$$\tau = \frac{I_t}{I_i} = (1 - R_1)(1 - R_2)\,\Phi\,, \tag{2}$$

where

$$\Phi = \frac{\displaystyle\int_0^\infty \frac{\exp[-\mathcal{T}^2(\omega - \omega_{res})^2/2]d\omega}{1 - 2\rho\sqrt{R_1 R_2}\cos[(\omega - \omega_{res})\mathcal{T}/a] + \rho^2 R_1 R_2}}{\displaystyle\int_0^\infty \exp[-\mathcal{T}^2(\omega - \omega_{res})^2]d\omega}\,, \tag{3}$$

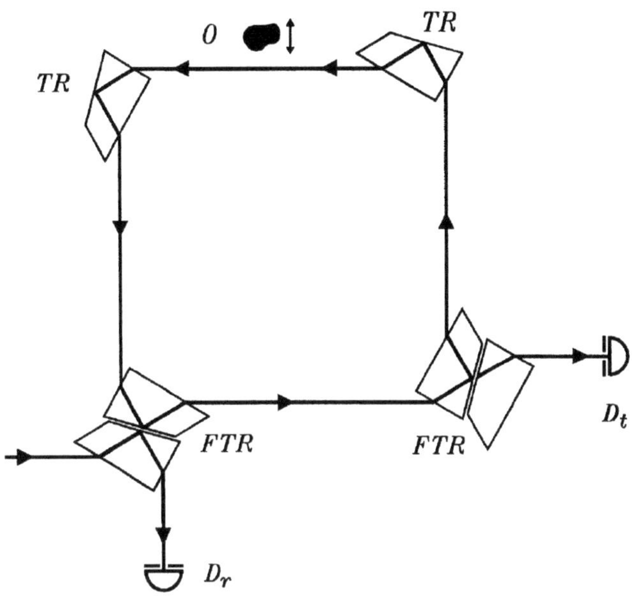

Figure 2. Schematic of the proposed realistic interaction–free device. A single p–polarized photon tunnels (frustrated total reflection) into the resonator made of Pellin–Broca prisms. With a realistic efficiency of over 98% the beam is guided by two total reflections and two frustrated total reflections to exit into D_t when there is no object in the path and is being reflected into D_r when there is.

where $a \equiv \mathcal{T}/T$ is a ratio of the coherence time $\mathcal{T}$ and the round-trip time T; ω_{res} is the selection frequency of the resonator; R is the frustrated reflectivity; $\rho \leq 1$ is a measure of overall losses. For a reliable measurement the coherence length should be long enough to allow sufficiently many round trips. Below we exploit this property to simulate moving of objects in the resonator. Φ turns out to be very susceptible to small changes of ρ so as to yield rather different outputs of τ in opposition to η. Let us have a closer look.

From figure 3 we see that only losses a few percent smaller than the reflectivity at the entrance of the resonator give an acceptable efficiency of the suppression of the reflection into D_r when there is no object in the resonator.

On the other hand the efficiency of the throughput into D_t when there is no object in the resonator is much more susceptible to losses as we can see in the figure 4. This increases the risk of discharging energy stored in the resonator into an inserted object up to 40 % but at the same time enable us to calibrate *grayness* (which can be interpreted as a kind of losses) of objects introduced into the resonator by measuring η. The latter feature reduces the number of required repeating of tests thus balancing the first one.

In figures 5 and 6 we simulate moving of objects inside the resonator by looking at η and τ as functions of low a, the ratio of coherence time and the round-trip time. We can see that even for resonators with the round-trip length long, only objects moving faster than 10^4 m/s can suffer a discharge. One can also see that the efficiencies stabilize for $a > 100$.

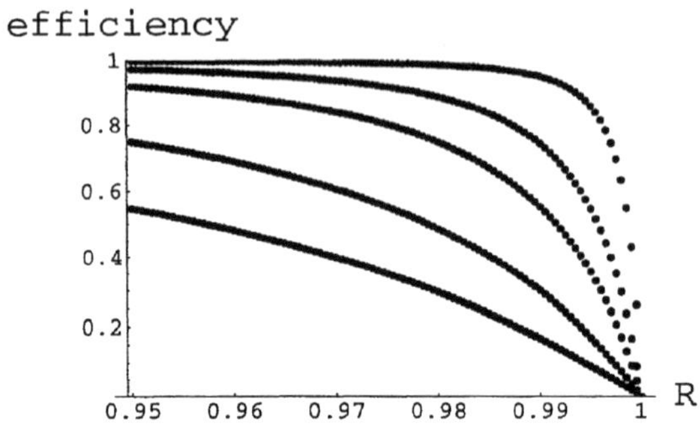

Figure 3. The efficiency of the suppression of the reflection into D_r when there is no object in the resonator, η. R is the frustrated total reflection at the two coupling output prisms. The curves correspond to $\rho = 0.998$ (top), 0.99, 0.98, 0.95, and 0.9 (bottom).

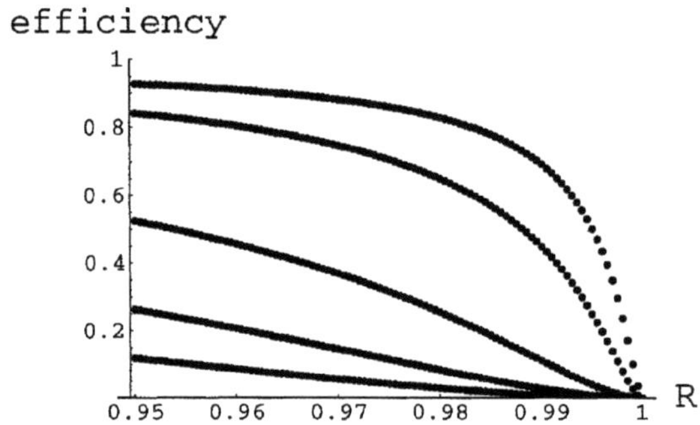

Figure 4. The efficiency of the throughput into D_t when there is no object in the resonator, τ. We again have $\rho = 0.998$ (top), 0.99, 0.98, 0.95, and 0.9 (bottom).

CONCLUSION

We have shown that although with our resonator which makes use of total reflections and frustrated total reflections and Pellin–Broca prisms to minimize reflection losses, thus reducing overall losses to under 2% apparently a realistic interaction-free measurement can hardly be balanced so as to reach an efficiency of over 95%. This nevertheless high efficiency together with a possibility to calibrate grayness of inserted objects by a ratio of reflectivity to transitivity of the resonator makes the device not only very suitable for the foundational experiments but also a good candidate for more general applications. For, although *gray* objects have an increased probability of being hit by a photon, a calibration of the ratio reduces the required repetition of testing objects.

Acknowledgments

I acknowledge supports of the Technical University of Vienna, Austria and the Ministry of Science of Croatia. I would also like to thank Johann Summhammer, Atominstitute of the Austrian Universities, Vienna for many valuable discussions.

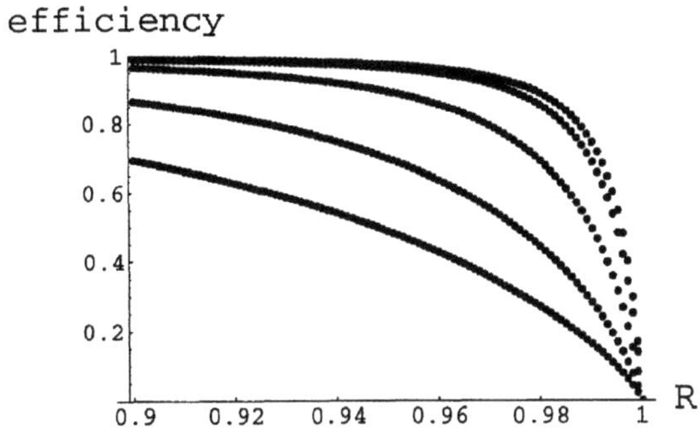

Figure 5. η as a function of a, the ratio of coherence time and the round-tip time for $\rho = 0.99$ and $0.9 \leq R \leq 1$. $a = 500$ (top), 150, 50, 20, and 10 (bottom).

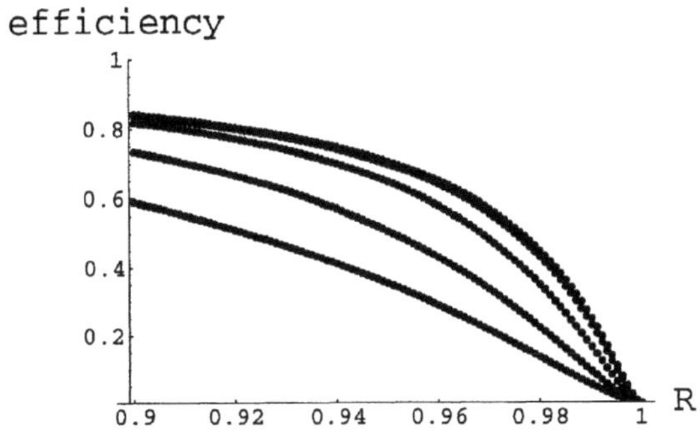

Figure 6. τ as a function of a for $\rho = 0.99$ and $0.9 \leq R \leq 1$.. $a = 500$ (top), 150, 50, 20, and 10 (bottom).

REFERENCES

Elitzur, A.C., and Vaidman, L., 1993, "Quantum mechanical interaction-free measurements," *Found. Phys.* 23:987.

Paul, H., and Pavičić, H., 1996, "Resonance interaction-free measurement," *Int. J. Theor. Phys.* 35:2085.

Paul, H., and Pavičić, H., 1997, "Nonclassical interaction-free detection of objects in a monolithic total-internal-reflection resonator," *J. Opt. Soc. Am. B* 14:1273.

Paul, H., and Pavičić, H., 1998, "Realistic interaction-free detection of objects in a resonator," *Found. Phys.* 28:959.

Pavičić, M., 1986, "Algebraico-Logical Structure of the Interpretations of Quantum Mechanics," Ph. D. Dissertation (in Croatian), University of Belgrade, Zagreb.

Pavičić, M., 1996, "Resonance energy-exchange-free detection and 'welcher Weg' experiment," *Phys. Lett. A* 223:241.

INDEX

Adiabatic approximation, 433
Amplifier, 207, 481
Asymtotic efficiency, 105
Atom laser, 201
Authentication, 389

Bayesian approach, 176
Bell state, 52, 514
Bell's inequality, 163
Berry–Uhlmann's curvature, 105
Binary signal, 20
Birefringent optical fiber, 511
Bloch vector, 475
Bunching, 335

Cassegrain telescope, 369
Channel capacity, 93
Chaos, 207
χ^3 medium, 433
Cloning, 133
Compton scattering, 300
Controlled-NOT gate, 293, 309, 339

Decoherence, 60, 337, 375
Degenerate parametric amplification, 505
Dim coherent state, 387
Dissipation effects, 499
Doppler cooling, 284
Doubly resonant cavity, 413
Downconversion, 381, 413, 515, 521
Dressed state, 331
Dynamical localization, 321

Eavesdropping, 370
Elliptic function, 523
Entangled state, 49, 339, 444, 511
Entropy exchange, 76
Error correction, 337

Fisher information, 100, 109
Floquet solution, 323
Fock state, 390
Fokker-Planck equation, 360
Frequency doubling, 499
Frequency standards, 337

Gate, 291, 333, 339
Gaussian beam, 508
Gaussian state, 75
GHZ state, 52, 515
Gram matrix, 39
Group covariant detection, 20
Group covariant signal, 37
Grover's algorithm, 307

Haar's measure, 144
Hadamard transform, 308
Halting, 241
Hanbury-Brown-Twiss effect, 335
Hydrodynamical moments, 170
Hypersensitivity to perturbation, 257

Image amplifier, 488
Induced transparency, 433
Interaction-free resonator, 527
Ion-trap laser, 274
Ionization spectra, 187

Kerr nonlinearity, 457
Kicked rotator, 321

Lamb shift, 332
Lamb–Dicke regime, 285, 294
Lie Group, 144
Light scattering, 295
Linear pointer, 120
Loop mirror, 453
Louivillian transformation, 182
Lorentz covariant hypersurface, 222

Mach–Zender interferometer, 393
Markov approximation, 195
Master equation, 195, 201, 259, 314
Maximum entropy principle, 176
Maximum likelihood estimation, 176
Measuring apparatus, 117
Modulation based amplifier, 207
Multimode homodyne tomography, 142, 158
Mutual information, 17, 44

NMR, 60, 257, 355
Noise figure, 483, 492
Non-Markovian system, 195
Nonorthogonal states, 11
Number-squeezing, 453

Optical pumping, 302
Oracle, 310

Parametric effects, 521
Parametric oscillation, 493
Pellin–Broca prism, 529
Phase-difference, 215
Photon blockade, 434
Poincaré sphere, 147
Polarization state, 147
Polarization-squeezing, 499
Popper's experiment, 443
POVM, 43, 99, 126
Privacy amplification, 388
Projection synthesis, 127

Quantum
  bus channel, 284
  control, 433
  cryptography, 367, 381, 387, 367, 393, 399
  estimation theory, 99
  feedback, 313, 353
  field theory, 221
  image amplification, 481
  information, 37, 59, 75, 261
    compression, 27
    geometry, 105
    processing, 49
  injection, 375
  machine, general, 252
  network, 263, 348
  noise, 207
  non-demolition measurement, 217
  nonseparability, 380
  parallelism, 250
  phase gate, 264
  processing, 59
  robot, 3
  state reconstruction, 169,
  tomography, 138, 147, 157, 163, 169, 175, 181
    conditioned, 145
  trajectory, 195
Qubit, 60, 316, 381

Raman effects, 455, 463
Ramsey pulse, 339
Relativistic quantum theory, 221
Resonance fluorescence, 271, 475
Retrodiction, 125
Rydberg electrons, 187

Sagnac interferometer, 457, 463
Schroedinger cat, 145, 359, 375
Second harmonic generation, 493
Singlet state, 384
Soliton, 453, 469, 463
Spectral filtering, 453, 463
Square-root detection, 20
Squeezed state, 93, 112
  amplitude, 453, 463, 469
  spin, 424
Squeezing, 453, 475, 487, 499, 508
State
  preparation, 129, 301, 438
  protection, 313
  purification, 78
  reduction, 223
  separation, 131
Stern–Gerlach experiment, 144
Stochastic golden rule, 233
Sudden-ionization approximation, 188

Telecloning, 54
Teleportation, 50, 229, 267
Ternary signal, 21
Thermal state, 99
Three-wave interaction, 522
Trapped
  atom, 265, 295, 301
  ion, 271, 283, 289
Trapping, 303
Travelling-wave amplifier, 488
Two-photon state, 413
Two-photon interaction, 331
Turing machine, 241
Type-I downconversion, 415
Type-II downconversion, 416

Universal compression, 29

Void detection, 527
Von Neumann
  entropy, 93
  measurement, 125, 118

Whispering gallery modes, 266
Wigner function, 123, 314, 363, 378, 522

MIX
Papier aus verantwortungsvollen Quellen
Paper from responsible sources
FSC® C105338

If you have any concerns about our products,
you can contact us on
ProductSafety@springernature.com

In case Publisher is established outside the EU,
the EU authorized representative is:
Springer Nature Customer Service Center GmbH
Europaplatz 3, 69115 Heidelberg, Germany

Printed by Libri Plureos GmbH
in Hamburg, Germany